INTERNATIONAL SERIES ON
MATERIALS SCIENCE AND TECHNOLOGY

EDITOR: G. V. RAYNOR, FRS

VOLUME 15

THE THEORY OF *Transformations in Metals and Alloys*

PART I

THE THEORY OF

Transformations in Metals and Alloys

SECOND EDITION

AN ADVANCED TEXTBOOK IN PHYSICAL METALLURGY

PART I
Equilibrium and General Kinetic Theory

J. W. CHRISTIAN

Professor of Physical Metallurgy, University of Oxford

PERGAMON PRESS

Oxford · New York · Toronto

Sydney · Paris · Braunschweig

Pergamon Press Ltd., Headington Hill Hall, Oxford
Pergamon Press Inc., Maxwell House, Fairview Park, Elmsford, New York 10523
Pergamon of Canada Ltd., 207 Queen's Quay West, Toronto 1
Pergamon Press (Aust.) Pty. Ltd., 19a Boundary Street, Rushcutters Bay, N.S.W. 2011, Australia
Pergamon Press SARL, 24 we des Ecoles, 75240 Paris, Codex 05, france
Pergamon Press GmbH, 3300 Braunschweig, Postfach 2923, Burgplatz 1, West Germany

---

First edition 1965
Second edition 1975

Library of Congress Cataloging in Publication Data

Christian, John Wyrill.

The theory of transformations in metals and alloys.
2nd edition
(International series on materials science and technology; v. 15)
CONTENTS: pt. 1. Equilibrium and general kinetic theory.
Includes indexes.
1. Physical metallurgy. I. Title.
TN690.C54 1975 669'.94 74-22470
ISBN 0-08-018031-0

Printed in Hungary

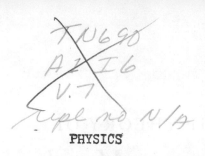

Contents

Preface to the First Edition

SUITABLE thermal or mechanical treatments will produce extensive rearrangements of the atoms in metals and alloys, and corresponding marked variations in physical or chemical properties. In this book, I have tried to describe how such changes in the atomic configuration are effected, and to discuss the associated kinetic and crystallographic features. The most drastic rearrangements accompany phase transformations, and the atomic processes leading to changes in the phase structure of an assembly form the main topic of the book. There are, however, many changes in the solid state which are not phase reactions, but which depend on the same atomic processes. I have also classed these changes as "transformations" since they are distinguished from phase transformations only by the nature of their driving forces. In terms of mechanism, kinetics and crystallography, it should be possible to give a consistent description of all these varied phenomena.

The subject matter so defined includes virtually all major kinetic effects in physical metallurgy and solid state physics, with the notable exception of plastic deformation by slip. A good case could be made for including this also in a unified treatment with the other phenomena, but it so vast a subject that it would have doubled the size of an already large book, and would have been beyond my competence to attempt. When we think about transformations in the solid state, however, the theory of plastic deformation is never far in the background, and the properties of lattice defects are equally vital in both cases. Moreover, some types of phase transformation may also be considered as modes of deformation, so that mechanical twinning, for example, is included in this book because of its close relation to martensitic transformation.

The book does not deal with the static problems of metal physics, notably the prediction of the equilibrium configuration from the known properties of the atoms. Nevertheless, I felt that a self-consistent description of the kinetic theories is not possible without a reasonably detailed description of the (usually crude) models used to describe the equilibrium or metastable state. The book thus contains what I hope are rather complete discussions of such topics as lattice geometry, point defects, dislocations, stacking faults, grain and interphase boundaries, solid solutions, diffusion, etc., and there are short accounts of more controversial topics such as the thermodynamics of the steady state. These introductory chapters form an appreciable fraction of the whole book, and unlike the kinetic sections, they cannot be justified on the grounds that no comparable description exists in other texts. I have two excuses for including them here. One is that the existing accounts, excellent though they are, frequently lack either the emphasis or the detail required for the present discussion of transformation problems. Thus I have found no adequate descriptions (except in original papers and specialized review articles) of such vital topics as the geometry of finite homogeneous deformation, the structure of a crystal surface, the properties of twinning dislocations, or the elastic energy of a constrained transformation. My second excuse is best illustrated by a quotation from Dr. Ziman's preface to his own recent book:† "There is need for treatises covering, in reasonable detail, up to the level of active research, the major branches into which the subject has divided." After attempting such a task, I am very conscious of my own inadequacies, but even with imperfect optics, there is something to be said for the wide-angle lens. In a very tentative sense, I hope that the present volume may find some use as an advanced account not only of phase transformation theory, but of a much wider field of physical metallurgy.

Throughout the book, I have tried to emphasize the theoretical description of the various phenomena, and experimental results are selected more for their ability to illustrate a particular argument than because they are the best data available. As in other branches of solid-state research which are sensitive to defect configurations and interactions, much of the recent active research work is intended to give a detailed description of the observations made on a particular metal or alloy, and this presents some difficulties to the writer of a general text. I have tried to solve these by giving at least one detailed description of each important effect. Thus the theoretical coverage is intended to be reasonably complete, and I may be justly blamed if it is not; the experimental results are only examples from a much wider range. The literature has been thoroughly surveyed up to the end of 1961, and work from some of the more important papers up to September 1962 has also been included.‡ However, my main aim has been to present an account of the established body of theory, so except in one or two special cases (e.g. Cahn's theory of spinodal decomposition), it has not seemed particularly important to include all the latest papers. After a period of very rapid development, it now seems likely that we are entering a stage of consolidation in this subject, so the present time seems well suited to the publication of a more detailed account than is to be found in any of the collections of review articles.

In writing the book, my ideal reader has been the graduate student just beginning research, although I hope that much of it may be useful at an undergraduate level. Some knowledge of elementary crystallography, chemical thermodynamics, and statistical mechanics is assumed, but relevant parts of these subjects are developed almost *ab initio*. The mathematical development should not give much trouble to an undergraduate reading for an honours degree in physics, but the metallurgical student may be a little alarmed by the extensive use of vector analysis and matrix algebra. The matrix method is so powerful in handling the crystallographic problems, and especially the compatibility between changes in the unit cell, distributions of defects, and changes in macroscopic shape, that it would have been inept not to have used it. However, I have given parallel geometrical illustrations of most of the matrix manipulations, so that the physical ideas may be followed even if the details of the calculation appear obscure. Only elementary matrix algebra is used, and the student to whom this is new could learn sufficient about it from any good textbook in a

† *Electrons and Phonons*, Clarendon Press Oxford, 1960.

‡ A few references to important papers published up to mid-1964 are being added in proof.

few hours. The notation is explained rather fully in the text, as are some standard operations such as the procedure for finding the eigenvalues of a matrix.

The only books which have previously covered any appreciable part of the subject matter are Volmer's *Kinetik der Phasenbildung* and Frenkel's *Kinetic Theory of Liquids.* I have greatly enjoyed reading these two classical texts, but they were written before the importance of defects in transformations was fully appreciated, and the main sources for this book are therefore individual papers and review articles. These are listed separately in the references, but I mention here my indebtedness especially to the various volumes of *Progress in Metal Physics* (now *Progress in Materials Science*), edited by B. Chalmers and R. King, and of *Solid State Physics*, edited by F. Seitz and D. Turnbull, and also to *Phase Transformations in Solids* (Wiley, 1951) and *The Mechanism of Phase Transformations in Metals* (The Institute of Metals, 1956). Many authors and publishers have kindly given permission for the reproduction of figures; the sources are individually acknowledged in the text. I am especially grateful to those friends who kindly supplied prints of optical or electron micrographs. Chapter 20 is based largely on unpublished material from Dr. A.G. Crocker's Ph.D. thesis, and I am very grateful to him and to Professor Bilby for allowing me to use it.

I have not shown the text to any of my colleagues, so that I have sole responsibility for any misrepresentations or errors of fact or judgement which may be detected, and I shall be very grateful if these are brought to my notice. On the other hand, I am extremely conscious of the help I have received from very many friends in numerous discussions of specific points, and I should like to mention especially B. A. Bilby, R. Bullough, J. W. Cahn, M. Cohen, A. G. Crocker, A. Hellawell, R. F. Hehemann, M. Hillert, D. Hull, A. Kelly, P. M. Kelly, D. S. Lieberman, J. K. Mackenzie, J. Nutting, W. S. Owen, T. A. Read, C. S. Smith, G. R. Speich, J. A. Venables, C. M. Wayman and M. S. Wechsler. I owe an especial debt to Dr. Z. S. Basinski, whose ideas contributed so much to my first understanding of the problems of martensite crystallography. The book has been written over many years at Oxford, and throughout this time I have had the benefit of Professor Hume-Rothery's help and encouragement. He is so good at writing books himself that he can probably not understand why it has taken me so long, but he will be relieved to discover that even the weariest river winds somewhere safe to sea. The final preparation of the manuscript has been undertaken during a sabbatical leave spent at Case Institute of Technology, and I should like to express my warm gratitude to Professor A. R. Troiano and his staff for their hospitality and many kindnesses.

Cleveland, Ohio
Oxford, England

J. W. CHRISTIAN

Preface to the Second Edition

THE arrangement of chapters and sections is unchanged but the book is now divided into two parts, the first of which covers the general theory whilst the second is concerned with descriptions of specific types of transformation. My original intention was to make only limited revisions and additions but I eventually decided that very extensive rewriting was required in view of the large number of papers published in the last ten years. Part I of the new edition is consequently about 50% longer than the corresponding chapters of the first edition, and is appearing after a lengthy delay, for which I apologize. However, I hope that all the major developments of the interim period have been incorporated, so that the edition may perhaps again serve both as a text book and as a reference source up to the level of active research.

An increased symbol list has necessitated some minor changes of notation (especially Q instead of Z for partition functions) and numerical quantities are now given in SI units. In response to several requests, I have included a few selected problems at the end of part I; these are of very variable standard, but I hope that teachers and students may find some of them useful.

I should like to thank many friends and colleagues for their kind comments and for notes of mistakes or misprints in the first edition. New errors will inevitably have appeared in this volume, and I shall be very grateful if they are also brought to my attention.

Oxford
December 1974

J. W. CHRISTIAN

Note on Symbols used for Physical Quantities

THE number of physical quantities introduced in the text is so large that a separate symbol for each would have required many different type faces. Some duplication has therefore been accepted, but this has been minimized by suitable use of superscripts and subscripts, and it is believed that no confusion can arise. Wherever possible, the recommendations of the appropriate international commissions on nomenclature have been followed, but there are inevitably some variations. The main features of the notation are given in the following list, and a complete index of symbols is given at the end of each volume.

Type of quantity	*Type face*	*Example*
Universal physical constants	Bold face italic	$\boldsymbol{R}, \boldsymbol{k}, \boldsymbol{N}_0$
Scalar quantities	Italic, Roman or Greek	x, T, α
Vectors or higher order tensors	Bold face upright, Roman or Greek	$\mathbf{u}, \mathbf{S}, \boldsymbol{\xi}$
Vector or tensor components	Italic, Roman or Greek	u_i, S_{ij}, ξ_i
Matrix representations of vectors or tensors in particular bases	Sans serif, Roman†	$\mathsf{u} = [\text{A}; \mathbf{u}]$ $\mathsf{S} = (\text{A}\,\mathbf{S}\,\text{A})$
Sets of base vectors	Small capitals	$\text{A} = \mathbf{a}_1, \mathbf{a}_2, \mathbf{a}_3$

The meanings of the main superscript and subscript symbols are given in the following list.

Superscript before main symbol	*Meaning*
v	per unit volume
L	per unit length
B	per unit area of boundary
S	per unit area of substrate
E	per unit length of edge
C	per grain corner

† The bold face symbol **l** is used for both the vector **l** and for its matrix representation. This avoids confusion with the sans serif symbol l which represents the unit matrix. Matrix representations of vectors denoted by Greek letters are avoided if possible; when they occur, the bold face symbol is used.

Superscript after main symbol	*Meaning*
$v, l, s, \alpha, \beta, \gamma$, etc.	relating to vapour, liquid, solid, α, β, γ, etc. phase
o	pure substance
H	homogeneous
B, E, C, S	relating to a grain boundary, edge or corner or to a surface
θ	maximum value
m	matrix or metastable phase
$\alpha\,\beta$	relating to an $\alpha-\beta$ interface
$\alpha\,\beta$ (thermodynamic quantities)	from α to β. Thus $\Delta g^{\alpha\beta} = g^{\beta} - g^{\alpha}$
Subscript after main symbol	
i, j, k	identifying subscript for vector components
ij, kl, etc.	identifying subscript for tensor components, or for transformation matrices
$A, B, C, \ldots$	components of the assembly (elements)
T	at temperature T
□, □□, ●	vacancy, divacancy, interstitial
c	critical value of a quantity
Subscript before main symbol (used only after mathematical symbol Δ)	
a	activation, e.g. $\Delta_a g$ = free energy of activation
m, n	mixing, e.g. $\Delta_m g$ = free energy of mixing
e, f	excess, e.g. $\Delta_e s$ = excess entropy of mixing

The summation convention is used with the subscript notation for vector and tensor components. Contravariant and covariant components are not distinguished by the positions of the sub- or superscript, but components referred to a reciprocal set of base vectors (covariant components) are distinguished by an asterisk, u_i^*. The asterisk is omitted for the components of plane normals, which are always referred to the reciprocal basis. Column and row vectors are distinguished by square and round brackets respectively. The transposed matrix of A is written A′, and the inverse matrix A^{-1}. These features of the tensor and matrix notation are explained more fully in the text.

CHAPTER 1

General Introduction

1. CLASSIFICATION OF TRANSFORMATIONS

An assembly of atoms or molecules which has attained equilibrium under specified external constraints consists of one or more homogeneous and physically distinct regions. The regions of each type may be distinguished by a common set of parameters defining such intrinsic properties as density, composition, etc., and they constitute a phase of the assembly. Two phases are distinguishable if they represent different states of aggregation, different structural arrangements in the solid, or have different compositions. This book deals mainly with changes in the phase structure of metals and alloys when the external constraints are varied.

In a complete discussion of the theory of phase changes, the following two questions must be considered:

(1) Why does a particular phase change occur?
(2) What is the mechanism of the transformation?

To answer the first question we must investigate the properties of different arrangements of a given assembly of atoms or molecules and thus attempt to find the equilibrium configuration. The formal theory of equilibrium, developed by Willard Gibbs, is expressed in terms of macroscopic thermodynamic parameters, and the problem reduces to the evaluation of these quantities from the properties of isolated atoms or molecules. Ideally, this can be tackled from first principles by quantum mechanics, but since the wave equation to be solved contains $\sim 10^{22}$ variables, it is clear that drastic simplifications are required. Although much work has been devoted to finding suitable methods of calculation, the results obtained are not encouraging, and in particular it is not usually possible to predict the relative stabilities of different crystal structures. This is not surprising when it is remembered that the heats of transition in metals which exist in more than one solid form are only of the order of one per cent of the binding energy of the solid, as measured by the latent heat of vaporization. The calculation of the binding energy is itself a very difficult problem; in various approximate methods it is obtained as the sum of several terms, the uncertainty in any of which is usually greater than the difference in energy between alternative structures. It is true that in some favourable cases it has been possible to provide plausible expanations why one structure should be more stable than another, but these explanations are essentially semi-qualitative, being based on physical models rather than detailed calculations.

A less fundamental approach to the question (1) above is to avoid the quantum mechanics by assuming a simple form of atomic interaction. For some solid materials, this procedure is quite justifiable, but it has severe limitations when applied to metals and alloys, since the interactions assumed are very poor approximations to the real binding forces. Moreover, the calculation of the equilibrium properties introduces difficult statistical problems, even with the simplest assumptions. If exact treatments of this part of the calculation are not possible, as is frequently the case, it is not easy to decide whether the properties of an approximate solution should be attributed to the mathematical approximations or to the model itself.

In view of these difficulties, the amount of attention which has been paid to the mathematical development of not very realistic statistical models is perhaps surprising. This type of theory has been most used in work on liquids and on solid solutions, especially in regard to order–disorder (superlattice) changes. It may be justified to some extent by the negative argument that no other theory seems feasible at present. Despite its limitations, it is fair to acknowledge that work of this kind has contributed very substantially to our present understanding of phenomena such as order–disorder changes.

The second fundamental question asks how a transformation occurs. The methods of classical thermodynamics are now of more limited application, since phase transformations are "natural changes". The appropriate theories are essentially kinetic theories, and some model of the atomic processes involved is implicit in any treatment. The models to be used are of two kinds. We first require an acceptable description of the "ideal" structure, using approximations for the interatomic forces to make them amenable to mathematical treatment. This is the problem already mentioned, but fortunately the kinetic properties are not so sensitive to the nature of the binding forces, so that assumptions which would be unwarranted in a study of equilibrium properties may be used with rather more confidence. We have next to consider the deviations from the ideal structure which occur in real solid materials. Vacant lattice sites, interstitial atoms, dislocations, stacking faults, and grain boundaries are comparatively unimportant in a description of the equilibrium state, but their presence may be fundamental to the process of transformation. Properties which depend sensitively on the detailed structure of the assembly are sometimes called structure sensitive.

In this book we shall be concerned mainly with the mechanism of phase transformations, and the reasons for transformation will not be examined in any thorough manner. In order to deal efficiently with the second question, however, the models mentioned above must be developed in some detail, so that we shall also consider some aspects of equilibrium theory. This is necessary, partly to make our treatment reasonably complete and self-contained, and partly because many of the topics which are important in discussing transformations are left out of the usual descriptions of equilibrium properties. It should be obvious from the above discussion that the models to be described refer almost exclusively to the solid state. For the vapour state, the kinetic theory of gases provides an adequate approximation, and there is no readily visualized model which gives a satisfactory description of the liquid state. The first part of this book is thus devoted mainly to an account of the solid state in metals and alloys, and the second part to the ways in which transformations involving any of the states of matter may occur.

Any phase transformation requires a rearrangement of the atomic structure of the assembly. In the solid state, similar rearrangements take place during processes which are not phase reactions, for example during the recrystallization of a deformed metal or its subsequent grain growth. Such reactions are distinguished by their driving forces;† the atoms take up new relative positions under the influence of strain energy, surface energy, or external stress, and not because the free energy of one arrangement is inherently lower than that of the other. The atomic mechanisms involved in all these changes, however, are closely related, and it is advantageous to treat them together. The word "transformation" will thus be used in a general sense to mean any extensive rearrangement of the atomic structure. The definition is intended to exclude mechanical deformation by slip, which (in principle) only alters the atomic arrangement by translation of part of the structure over the remainder. Deformation twinning is included because of the highly ordered nature of the rearrangement and its close relation to one type of phase transformation.

The driving force for any transformation is the difference in free energy (usually Gibbs free energy) of the initial and final states, and is thus determined by thermodynamic parameters appropriate to large regions of the phases concerned. The mode of transformation is very dependent on the effect of small fluctuations from the initial condition, and in particular on the question of whether such a fluctuation raises or lowers the free energy. A metastable assembly is resistant to all possible fluctuations, and any transformation path must pass through states of higher free energy. Conversely, if any infinitesimal fluctuation is able to lower the free energy, the initial condition is unstable, and there is no energetic barrier to transformation along a path represented by this fluctuation. A truly unstable phase only has a transitory existence, but it may also be possible to obtain a phase in such a condition that the only barrier to transformation is that limiting atomic movement or diffusion. Such an assembly may also be described as unstable, although it decomposes at a finite rate determined by the diffusion rate.

In considering the problem of stability, Gibbs distinguished two different kinds of fluctuation, namely those corresponding to fairly drastic atomic rearrangements within very small localized volumes, and those corresponding to very small rearrangements spread over large volumes. An example of the first kind of fluctuation is the formation of a very small droplet of liquid in a supersaturated vapour, whilst an example of the second kind is the development of a periodic variation of composition with a very long wavelength within an initially homogeneous solid solution. Much confusion in discussions of the initiation of phase transformations in supersaturated solid solutions may be attributed to a failure to distinguish clearly between these two kinds of fluctuation.

Most of the transformations with which we shall be concerned are heterogeneous; by this we mean that at an intermediate stage the assembly can be divided into microscopically

† In any natural process the free energy of a closed assembly in thermal contact with its surroundings decreases. The difference in initial and final free energies provides a driving energy or thermodynamic potential for the change; this is often loosely described as a "driving force".

distinct regions of which some have transformed and others have not. Transformation thus begins from identifiable centres in the original phase, a process which is called nucleation. The classical theory of the formation of a nucleus is formulated in terms of the first type of fluctuation, and it may be shown that any assembly is stable to such fluctuations on a sufficiently small scale. The reason for the free energy barrier is usually expressed in the form that the negative free energy change resulting from the formation of a given volume of a more stable phase is opposed by a positive free energy change due to the creation of an interface between the initial phase and the new phase. As the volume of the region transformed decreases, the positive surface term must eventually dominate the negative volume term, so that the whole free energy change becomes positive. Clearly these macroscopic concepts are not really applicable to very small fluctuations, and any separation of the net free energy change into volume and surface contributions is arbitrary. However, the formalism is useful in deriving quantitative expressions for the nucleation rate, provided due caution is exercised in identifying parameters introduced into the nucleation theory with those derived from measurements on bulk phases, and it also illustrates the conditions under which the energy barrier to nucleation might disappear.

For localized fluctuations of any one kind, there is a critical size of nucleus at which the free energy barrier is a maximum, and the magnitude of this maximum increase in free energy determines the rate of nucleation according to the classical theory. If the surface energy decreases, the critical nucleus size and the height of the energy barrier both decrease, and more and more nuclei will be formed in a given volume in a given time. In the limit, the nucleation barrier disappears altogether when the surface energy becomes zero, or at least very small, and the original phase becomes unstable. The transformation will then be homogeneous, taking place in all parts of the assembly simultaneously. The condition for a homogeneous transformation thus appears to be a zero, or near zero, surface energy, and this implies that there must be no abrupt change at the boundary between two phases. In some order–disorder transformations, it is possible that this condition is readily approached, so that homogeneous or near homogeneous transformation is possible. The surface between two phases of similar structure but different composition and lattice parameter may also have a low effective energy if the transition from one phase to the other is spread over a wide region, i.e. if there is a diffuse rather than a sharp interface. However, a diffuse interface is clearly incompatible with a localized fluctuation, so that in attempting to remove the energy barrier from the first kind of fluctuation we have essentially converted it into the second kind of fluctuation. Homogeneous transformations without nucleation are thus generally possible only in certain assemblies which are unstable with respect to the second kind of fluctuation. The most important examples of such homogeneous transformations in non-fluid assemblies are probably those already mentioned, namely the decomposition of supersaturated solutions within certain limits of temperature and composition, and some order–disorder changes. It is generally impossible to avoid nucleation in transitions between solid phases with different structures, since the transition from one arrangement to another cannot then be made continuous through a diffuse boundary.

Heterogeneous transformations are usually divided into two groups, originally distinguished from each other by a different dependence of reaction velocity and amount of transformation on temperature and time. This experimental classification almost certainly corresponds to a real difference in the physical mechanisms of transformation, but experience has shown that the kinetic features cannot always be interpreted unambiguously, and many reactions of intermediate character are now recognized. The two groups are usually known as "nucleation and growth" reactions and "martensitic" reactions respectively. This nomenclature is rather unfortunate, since the formation of stable small regions of product and the subsequent growth of these nuclei may have to be treated as separate stages in both classes of transformation.

In typical transformations in the first group, the new phase grows at the expense of the old by the relatively slow migration of the interphase boundary, and growth results from atom by atom transfers across this boundary. The atoms move independently and at a rate which varies markedly with the temperature. At a given temperature, the reaction proceeds isothermally, and the amount of new phase formed increases with time. Although the volume of a transformed region differs in general from its original volume, its shape is substantially unaltered. Such changes are largely governed by thermal agitation. The unit processes (e.g. the motion of an individual atom within the parent phase or across the interphase boundary) are similar to chemical reactions, and to a first approximation they may be treated formally by the methods of the statistical reaction rate theory.

Nucleation and growth transformations are possible in all metastable phases, and the initial or final condition may be solid, liquid, or gaseous. The second kind of heterogeneous transformation, however, is only possible in the solid state, and utilizes the co-operative movements of many atoms instead of the independent movements of individual atoms. Most atoms have the same nearest neighbours (differently arranged) in the two phases, and the net movements are such that in small enough regions a set of unit cells of the original phase is effectively homogeneously deformed into a corresponding set of cells of the new phase. The transforming regions then change their shapes, and may be recognized, for example, by the disturbances produced on an originally flat polished surface. Discrete regions of the solid usually transform suddenly with a very high velocity which is almost independent of temperature. In most cases, the amount of transformation is characteristic of the temperature, and does not increase with the time. Reactions of this kind are often called diffusionless or shear transformations, but in recent years it has become common practice to refer to them as martensitic transformations. The name is an extension of the nomenclature originally reserved for the hardening process in quenched steels.

In recent years it has become increasingly evident that many transformations do not fit easily into the above classification. Difficulties arise especially when the growth is thermally activated but nevertheless results in the type of shape change and crystallography normally associated with a martensitic transformation. Provided the shape change has been properly measured (Clark and Wayman, 1970), it implies the existence of a correspondence of lattice sites of parent and product structures (see Chapter 2), but in the case of ordering or non-ferrous "bainitic" reactions, it is clear that there must also be some atomic mobility over distances equal to or greater than the nearest neighbour separation. Lieberman (1969, 1970) has proposed that the various kinds of transformation which give rise to a martensitic shape change should be subdivided into ortho-, para-, quasi-, and pseudo-martensites, but these names have not been generally accepted. We feel that it is preferable to restrict the word martensitic to changes which are in principle diffusionless, and to use some more general term for all transformations, including martensite formation, which involve a lattice correspondence or partial correspondence.

The descriptive terms "displacive" and "reconstructive" transformations have a long-established usage, especially in non-metallurgical fields (Buerger, 1951) and may seem more adaptable than their metallurgical equivalents. This classification, however, is rather closely linked to the concept of pairwise atomic interactions, or bonds. An alternative division of a more general kind (Frank, 1963; Christian, 1965) is based on a sustained analogy between the different mechanisms of transformation and the ways in which soldiers and civilians respectively execute some simple task, such as boarding a train. Thus the main categories of transformation are called "military" and "civilian", but rigid classification is not required since soldiers may sometimes be out of step and civilians may sometimes form para-military organizations!

Boundaries in the solid state may be conveniently regarded as either glissile or non-glissile. A glissile boundary can migrate readily under the action of a suitable driving stress, even at very low temperatures, and its movement does not require thermal activation. Examples of the motion of glissile boundaries are provided by the growth of a martensite plate or of a mechanical twin, or by the stress-induced movement of a symmetrical low angle grain boundary. In all cases, the shape of the specimen changes as the boundary is displaced, so that the movement may be regarded as a form of plastic deformation. It follows that a suitable external mechanical stress should be able to produce displacement of any glissile interface.

The remaining types of boundary can move only by passing through transitory states of higher free energy, so that the motion requires the assistance of thermal fluctuations. However mobile such a boundary may be at high temperatures, it must become virtually immobile at sufficiently low temperatures. We subdivide non-glissile boundaries into those in which there is no change of composition across the interface and those dividing regions of different composition. In the first group are any transformations from a metastable single phase to an equilibrium single phase (polymorphic changes), processes such as recrystallization and grain growth which are entirely one-phase, and order–disorder reactions. In all these examples, the rate of growth is determined by atomic processes in the immediate vicinity of the interface, and we may describe such growth as "interface controlled".

Familiar examples of growth in which there is a composition difference across a moving interface are provided by precipitation from supersaturated solid solution and eutectoidal decompositions. The motion of the interface now requires long-range transport of atoms of various species towards or away from the growing regions, so that we have to consider the diffusional processes which lead to the segregation. Two extreme cases can be distinguished in principle. In one of these we have a boundary which can move very slowly, even under the influence of high driving forces. The rate of motion will then be largely independent of the diffusion rate, and we may again describe the growth as interface controlled. The other extreme case is where the boundary is highly mobile when compared with the rate of diffusion, so that it will move as rapidly as the required segregation can be accomplished. The growth rate is then determined almost entirely by the diffusion conditions, and is said to be "diffusion controlled". Since on the average an atom will have to make many hundreds or thousands of atomic jumps in the parent phase, and only one or two jumps in crossing the interphase boundary, it is rather more likely that growth will be diffusion controlled when the composition difference is appreciable. In many diffusion controlled reactions, a linear dimension of the growing product region is proportional to the square root of the time of growth, whereas in an interface controlled reaction, it is linearly proportional to the time. Thus an interface controlled boundary of a very small product region may become diffusion controlled as the region grows larger.

In an interface controlled transformation, regions in the immediate vicinity of the interface, but on opposite sides of it, must have different free energies per atom, this being the driving potential for the growth. It follows that the temperature–composition conditions at the interface cannot correspond to equilibrium between the bulk phases, and the interface velocity will be some function of the deviation from equilibrium, as measured, for example, by the supercooling or supersaturation at the interface. In contrast to this, the deviation from equilibrium at the interface is negligible when the growth is controlled by the diffusion of matter (or energy). The overall growth rate is then a function of the difference in composition (or temperature) between regions of the primary phase near the interface and metastable regions remote from the interface. Obviously there is a range of interface mobilities over which the growth rate cannot correspond to either extreme, and this range must be encountered if there is a transition from interface control to diffusion control as a product region grows. When comparable driving forces are needed to maintain the interface mobility and to remove or supply solute (or heat), the growth velocity will be dependent on both the local deviation from equilibrium at the interface and the gradients of composition (or temperature) in the parent phase. It is usually assumed that this range is quite small, so that the motion of most non-glissile interfaces is essentially controlled either by atomic kinetics at the interface or by diffusion.

Linear growth does not necessarily imply interface control Dendritic growth is mainly diffusion controlled, but dendrites lengthen at constant rate. Certain solid state reactions (discontinuous precipitation, eutectoidal decomposition), in which a duplex product grows into a single phase parent, also merit special consideration. The product and parent have the same mean composition, but the product consists of alternate lamellae differing in composition. These reactions have linear growth rates, but have usually been described as diffusion controlled, the operative diffusion path being either through the parent phase, or along the boundary between parent and product. In contrast to this, some theories imply that two parameters are needed to specify the growth conditions in such transformations, the interface velocity not being controlled completely by either the diffusion coefficient or the interface mobility.

In the description of growth processes, we have neglected the heat which is released or absorbed during the growth. In most transformations, the rate of growth is not limited by the rate at which this heat is supplied or removed, but this is not true when we extend the classification to reactions involving phases which are not solid. In particular, the rate of growth of a solid crystal from the melt, or from a liquid solution, is often controlled by the energy flow conditions, and this is also true of the reverse process of melting. Solidification under conditions in which the heat can be readily removed, as in a thin-walled capillary tube, is interface controlled. Growth from the vapour phase, whether or not it involves long range transport, is almost always interface controlled. Energy flow control is always equivalent to a form of diffusion control, and is covered by including variations of temperature as well as composition in the above statements.

The relations between the various forms of transformation to be discussed in this book are shown schematically in Table I, which summarizes the growth classification we have just described. Some further subdivisions are made for convenience, and the significance of these will become apparent later. It should be obvious from the above discussion that the category in which a given type of transformation is placed will be dependent to some extent on the experimental conditions under which it is observed, so that the divisions indicated are by no means rigid.

Despite its limitations, we believe the scheme shown in Table I still gives the most useful physical picture of the interrelations among the various transformation mechanisms. This scheme is approximately the same as that adopted by the committee on phase transformations for the survey *Perspectives in Materials Research* (Cohen *et al.*, 1965), but Guy (1972) has suggested a slightly different classification in which the initial division into homogeneous and heterogeneous transformations is discarded by adopting a more general definition of a phase (Guggenheim, 1967). An inhomogeneous assembly is then treated as an infinity of infinitesimal phases, and a surface is regarded as a separate phase. This formalism may have certain advantages, but we believe that the Gibbs concept of a phase is clearer, and that the distinction between homogeneous and heterogeneous transformations is useful because the nucleation process may be very important in heterogeneous transformations.

Our classification focuses attention on the growth process, but either the growth rate or the nucleation rate may be effective in determining the overall kinetics of a heterogeneous transformation. The net transformation rate will depend mainly on the slower of the two stages, becoming virtually zero if either the nucleation rate or the growth rate is sufficiently slow. We should thus consider whether or not there are differences in the physical mechanisms leading to the formation of nuclei in particular transformations; if there are such differences, a classification of transformations based on the nucleation stage will be equally valid. We shall find, in fact, that there are various ways in which nuclei may be formed, but these are not readily distinguished by experimental criteria. The growth classification is convenient largely because most reactions in the solid state are greatly influenced by the growth mechanism.

In some transformations, there is no doubt that thermal fluctuations resulting in atomic rearrangements are responsible for the appearance of stable nuclei of a new phase. Such fluctuations may occur randomly throughout the volume of the assembly, or at preferred sites where impurities or structural defects act as catalysts for the phase change. Many solid assemblies contain defects which are not in thermal equilibrium with the structure but are "frozen in". Non-equilibrium defects of this kind may be able to form suitable nuclei for some changes without the aid of thermal agitation.

The classical theory of random (bimolecular) fluctuations provides an adequate descrip-

TABLE I. CLASSIFICATION OF TRANSFORMATIONS ACCORDING TO GROWTH PROCESSES

- Heterogeneous transformations
  - Athermal growth (glissile interfaces)
    - Coherent interface
      - Coherent martensite
      - Mechanical twinning
    - Semi-coherent interaface
      - Martensite
      - Low-angle boundary
  - Thermally activated growth
    - No long-range transport (interface controlled)
      - Polymorphic transitions, growth from vapour
      - Recrystallization grain growth
      - Order–Disorder changes
    - Long-range transport
      - Continuous reactions
        - Interface controlled
          - Crystal growth from vapour
          - Precipitation dissolution
        - Diffusion controlled
          - Precipitation dissolution
      - Discontinuous reactions (diffusion plus interface controlled ?)
        - Eutectoidal reactions
        - Discontinuous precipitation
  - Growth controlled by heat transport
    - Solidification melting

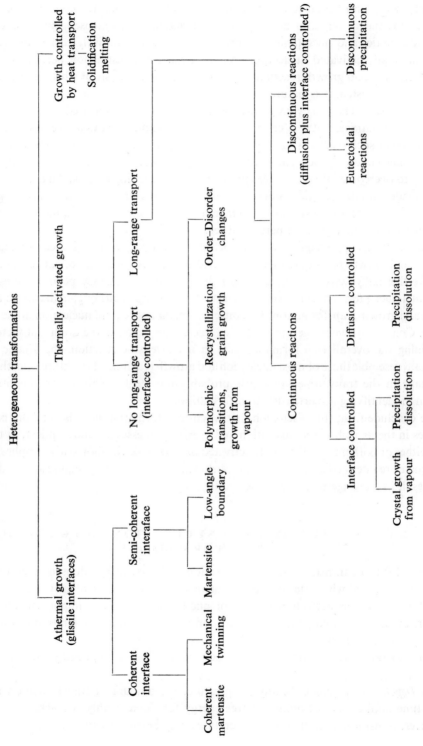

tion of nucleation phenomena when both phases are fluid (e.g. the condensation of a vapour). This theory has also been applied to reactions in condensed phases, and despite many difficulties and uncertainties, it probably gives a qualitatively correct description of most thermally activated nucleation processes. Since the distinction between martensitic and nucleation and growth reactions has been made on the basis of whether or not growth is thermally assisted, it is natural to enquire whether the same distinction applies to the nucleation stage. There is evidence that thermal nucleation is not needed for some martensitic transformations, but it is certain that thermal agitation plays a role in the nucleation stage of other such reactions. It does not follow from this that the classical theory of homogeneous nucleation applies, since thermal vibrations (or zero point energy) may only be helping to overcome initial obstacles to the growth of existing defects. An analogy is sometimes made with deformation processes, where thermal vibrations enable existing glide dislocations to overcome obstacles to their motion, although they are quite unable to form new dislocations in regions of perfect lattice.

These remarks apply equally to any nucleation process, including those of many nucleation and growth reactions. In suitable conditions, it is conceivable that reactions which require thermally activated migration for the growth process may begin from essentially athermal nuclei formed from existing defects. Such changes, although still classed as nucleation and growth transformations for convenience, do not require nucleation in the classical sense. Conversely, when thermally assisted nucleation becomes the dominant factor in determining the overall transformation rate in a martensitic reaction, the kinetic features naturally resemble those of most nucleation and growth reactions. For this reason, the change of shape in the transformed region is probably the most definitive experimental observation in identifying a martensitic transformation.

We conclude from this discussion that when transformation does begin from particular centres in the assembly, the nucleation and growth processes are independent, in the sense that either or both may be thermally activated or athermal. We should also emphasize again that there are intermediate categories of transformation, which may require atomic diffusion, but result in a change of shape in the product regions.

2. CHARACTERISTICS OF NUCLEATION AND GROWTH TRANSFORMATIONS

The velocity of transformation in nucleation and growth reactions may be dependent both on the rate at which stable nuclei form and on their subsequent growth rates. In some transformations, the activation energy for nucleation is the only important rate limiting factor, but in condensed phases the activation energy for atomic migration or diffusion is usually equally important.

The general characteristics of these reactions may be summarized as follows.

(1) *Dependence on time*. At any temperature, the amount of transformation increases with time until a state of minimum free energy for the assembly is reached. In practice, however, transformation at some temperatures may be so slow that it cannot be detected

in any observable period of time. The isothermal transformation laws are considered in more detail in Section 4 below.

(2) *Dependence on temperature.* If given sufficient time, the transformation will, in principle, continue until complete. The amount of transformation does not, therefore, depend on temperature, except in the trivial sense that the equilibrium state is itself a function of temperature. The velocity of transformation varies enormously, and for any transformation a temperature can be found below which the change proceeds at a negligible rate. For homogeneous changes, the reaction velocity increases approximately exponentially with the temperature over the whole range in which transformation may occur. The theory of rate processes of this kind is treated in Chapter 3. For nucleation and growth reactions, however, the rate of reaction becomes zero at the thermodynamical transformation temperature, since the energy required for nucleation is then infinite. Consider a transformation which occurs on cooling. As the temperature is lowered, the free energy of formation of a critically sized nucleus decreases much more rapidly than does the available thermal energy. The probability of nucleation may thus increase rapidly with decreasing temperature, at least until very low temperatures are reached. In condensed phases, however, the formation energy of the critical nucleus constitutes only part of the required activation energy, and there may be additional terms representing the energy needed for an atom to cross the interphase boundary or for interdiffusion of the species, where a composition change is involved. Since the activation energy for this type of process is nearly temperature independent, the rate of growth of both sub-critical and super-critical nuclei decreases as the temperature is lowered. The decreasing formation energy and decreasing growth rate produce a characteristic dependence of reaction velocity on temperature, in which the rate increases to a maximum at a temperature below the equilibrium transformation temperature, and then decreases again.

This type of variation often makes it possible to obtain an alloy in a thermodynamically metastable state. This can be done if the specimen can be cooled through the region of rapid transformation so quickly that there is effectively no reaction during the cooling; the high temperature structure will then be preserved indefinitely at a sufficiently low temperature. "Quenching" to retain the high temperature structure is a standard method used in the experimental determination of metallurgical equilibrium diagrams. In general, all reactions which involve changes of composition in different regions of the specimen, i.e. long-range diffusion, will be slow, and can be prevented by rapid cooling. If no martensitic transformation intervenes, the quenching method is successful in such cases. Some reactions, for example the solidification of a liquid alloy, cannot be inhibited by drastic cooling, but equilibrium, which requires diffusion, can be prevented, and the non-equilibrium structure formed during the quench is readily recognized.

(3) *Irreversibility of the transformation.* Since the individual atoms move independently, there is no correlation between the initial and final positions of the atoms after retransforming to the original phase. For example, if we convert a crystalline phase α into a new structure β by heating through a transformation range, and then reconvert to α by cooling, the assembly is thermodynamically in its original state (neglecting grain boundary and surface energies). Usually, however, the crystals of α will be unrelated in size, shape, or orientation

to the original α crystals. The transformation is thus irreversible, not only thermodynamically, but also in the special sense that the assembly never returns to its initial configuration. In practice this may not always be obvious because nucleation may be heterogeneous and take place at certain preferred sites.

(4) *Effect of plastic deformation*. Reactions in the solid state are often accelerated by cold-working the material prior to transformation. There are several possible explanations for this effect. Nucleation may be easier in plastically deformed regions of the crystal lattice, since the effective driving force is increased and the free energy barrier decreased. Another factor is the activation energy for atomic diffusion, which may similarly be lowered in heavily deformed regions. The process of deformation leads to a large temporary increase in the number of vacant lattice sites, and this will also temporarily increase the diffusion rate, independently of the activation energy. Plastic deformation at temperatures where internal stresses can be removed rapidly by recovery or recrystallization does not affect the transformation rate.

(5) *Composition, atomic volume, and shape of the new phase*. The composition and atomic volume of the reaction products need not be related in any way to those of the original phase. Only in pure metals, polymorphic changes, order–disorder reactions, and transformations not involving thermodynamic phase changes will there be no change in composition. In nucleation and growth transformations, the shape of the particles of the new phase when transformation is incomplete, or when the equilibrium configuration is polyphase, varies considerably. If surface energies are important, the shape will be spherical for isotropic surface energy (e.g. liquid droplets), and roughly equi-axed polyhedral for anisotropic energies. If there is a volume change, however, the strain energy in a condensed assembly will often be more important than the surface energies. The crystals of the new phase will then be in the form of plates or needles, orientated with respect to the original lattice so that the best atomic fit is obtained across the interface. The resulting microstructure is called a Widmanstätten structure.

(6) *Orientation relations*. In some nucleation and growth transformations in the solid state there is no relation between the orientations of the two lattices. When Widmanstätten structures are produced, or in the early stages of coherent growth of precipitates from solid solution, it is, however, usually found that one lattice has a fixed orientation relative to the other. Orientation relations are also commonly found when two phases are formed together, as in the solidification of a eutectic alloy or a eutectoidal reaction in the solid state.

3. CHARACTERISTICS OF MARTENSITIC TRANSFORMATIONS

Martensitic reactions are possible only in the solid state. They do not involve diffusion, and the composition of the product is necessarily the same as that of the original phase. In alloys which are ordered, it has been shown that the phase formed by a martensitic reaction is also ordered. Negligible mixing of the atoms thus takes place during the transformation, and it follows that the thermodynamically stable configuration of the assembly often cannot be produced by such a change.

In a martensitic reaction, a co-operative movement of many thousands of atoms occurs with a velocity approaching that of sound waves in the crystal. Thermally supplied activation energy could not account for such a process, and the concept of activation energy is not generally so useful in martensitic reactions, except in the nucleation stage. The reaction starts spontaneously at some temperature, and the parent structure is then effectively mechanically unstable. In a similar way, mechanical twinning may be regarded as a special kind of martensitic reaction in which the driving force is an externally applied stress, rather than an internal free energy difference.

The main characteristics of martensitic reactions are summarized below:

(1) *Dependence on time.* The amount of transformation is virtually independent of time. At a constant temperature, a fraction of the original phase transforms very rapidly, after which there is usually no further change. This is a primary characteristic of martensitic transformations, but in some reactions there is also a small amount of isothermal transformation to the martensite phase, and in a few cases, the change is almost completely isothermal. These differences will be discussed in detail later, but as already indicated, the isothermal characteristics are the result of thermally assisted nucleation processes.

(2) *Dependence on temperature.* The amount of transformation is characteristic of temperature, providing other variables such as grain size are held constant. The velocity of transformation is probably independent of temperature and is usually very rapid. Transformation on cooling begins spontaneously at a fixed temperature (M_s), and as the temperature is changed, more and more material transforms, until the temperature (M_f) is reached, at which the change is complete. In some assemblies it is doubtful whether complete transformation ever occurs spontaneously.

At any temperature a number of single crystals of the new phase form rapidly within an original grain. On cooling to another temperature these crystals usually do not grow but new crystals are formed. In favourable circumstances, however, a single crystal of the original phase can be continuously converted to a single crystal of the martensite phase.

(3) *Reversibility of the transformation.* Martensitic reactions are very reversible in the sense that an initial atomic configuration can be repeatedly obtained. A single crystal of the original phase may, for example, transform on cooling into several crystals of the new phase. The reverse change on heating will usually result in a single crystal of the same size, shape, and orientation as the original crystal. The reversibility is associated with a temperature hysteresis and the reverse reaction begins at a temperature above M_s. Moreover, in repeated transformations, the plates (single crystals) which form on cooling have the same size and shape, and appear in the same regions of the original crystal. This behaviour probably applies in principle to all martensitic reactions. Apparent exceptions, where reversibility has not been observed, can always be attributed to interfering secondary effects: in iron–carbon alloys, for example, the martensitic phase is thermodynamically unstable and starts decomposing into stable phases (tempering) before the reverse transformation can begin.

(4) *Effect of applied stress.* Plastic deformation is much more important in martensitic reactions than in nucleation and growth changes. Application of plastic stresses at any temperature in the transformation range usually increases the amount of transformation,

and the reaction can often be completed by this means. In some transformations, elastic stresses have a similar effect. When single crystals are used, the direction of the applied stress is important, and some reactions may be inhibited as well as aided by a suitably orientated stress.

Deformation above M_s may also result in the formation of the product phase, even though the temperature is too high for spontaneous reaction. The highest temperature at which martensite may be formed under stress is called M_d. In general, the reverse reaction can be aided in the same way, and a suitable stress will induce transformation below the temperature at which it begins spontaneously.

If the original phase is cold-worked in a temperature range where it is stable, e.g. at temperatures sufficiently above M_s, the resultant deformation often inhibits transformation. Providing the temperature of deformation is not high enough to permit self-annealing, the M_s temperature is depressed, and the amount of transformation found at any temperature is reduced.

(5) *Composition, atomic volume, and shape of the new phase.* In a martensitic reaction, each crystal transforms to new crystals of the same chemical composition. Volume changes are often, though not invariably, small, and in some cases are zero to within the limits of experimental error. The martensite crystals are usually flat plates, which thin towards the extremities and so have a lenticular cross-section. There are exceptions in certain simple transformations where parallel-sided bands are formed. The plates or bands are orientated with respect to the original lattice; the plane of the lattice on which they are formed is called the habit plane. It has also been possible in some transformations, as mentioned above, to change a single crystal of the original phase into a martensite single crystal by the migration of an interface from one side of the crystal to the other. This interface lies along the habit plane.

(6) *Orientation relations.* In martensitic changes, there is always a definite relation between the orientation of the original structure and that of the new phase. As with the habit planes it is usually possible to find all crystallographically equivalent variants of the relation under suitable conditions. A single martensite plate may be a single crystal or may contain two twin orientations. In the latter case, the orientations of the twins relative to the matrix are not necessarily equivalent.

(7) *Stabilization.* We have already referred to the effects produced by cold-working the original structure. The reaction is also inhibited in another way. If the specimen is cooled to a temperature in the transformation range, held there for a period of time, and then cooled again, transformation does not begin again immediately. At all subsequent temperatures the amount of transformation is less than that produced by direct cooling to the temperature concerned. This phenomenon is referred to as stabilization. The degree of stabilization increases with the time for which the specimen is held at the temperature. Slight variations in amount of transformation with cooling velocity are also presumably to be attributed to stabilization. There is no general agreement whether or not stabilization is produced by halting the cooling above the M_s temperature.

4. ISOTHERMAL TRANSFORMATION CURVES

In this section we give an introductory description of the theory of the relation between the fraction of the assembly transformed and the time at constant temperature. This formal theory is largely independent of the particular models used in detailed descriptions of the mechanism of transformation, and can therefore be given here before these models are discussed. The concept of nucleation rate will be given an operational definition, which may later be compared with the theoretical quantity introduced in Chapter 10.

In a homogeneous reaction, the probability of any small region transforming in a given time interval will be the same in all parts of the untransformed volume. The volume transforming in a short time interval will thus be proportional to the volume remaining untransformed at the beginning of this interval, and this leads to a first-order rate process. Suppose the total volume is V and the volume which has transformed from α to β at any time is V^{β}. Then:

$$\frac{dV^{\beta}}{dt} = k(V - V^{\beta}) \quad \text{or} \quad V^{\beta}/V = 1 - \exp(-kt). \tag{4.1}$$

The constant k is called the rate constant. The rate of transformation decreases continuously with time (see Fig. 1.1).

For nucleation and growth reactions, the situation is more complex. Consider first the size of an individual transformed region. The region is formed at a time $t = \tau$ (τ may

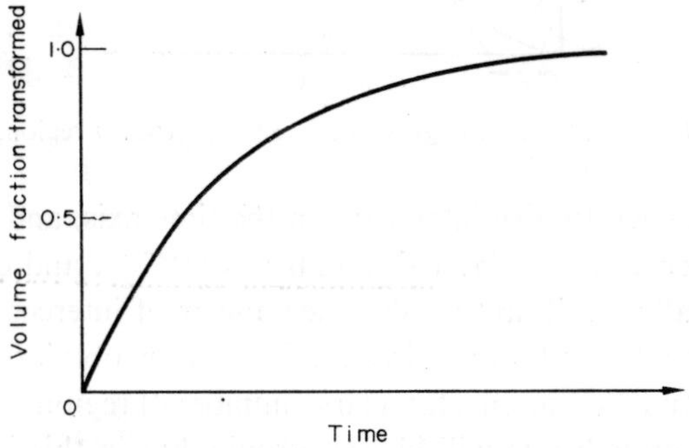

FIG. 1.1. Reaction curve for a homogeneous transformation.

be called the induction period), and thereafter its size increases continuously. If the transformation product has the same composition as the original phase, it is found experimentally that in nearly all reactions any dimension of the transformed region is a linear function of time. This is shown schematically in Fig. 1.2. The reduction of growth rate when t becomes large is due to the mutual impingement of regions transforming from separate nuclei, which must ultimately interfere with each other's growth. The intercept on the time axis, which gives the induction period τ for nucleation of the region, naturally cannot be observed experimentally, and is inferred by extrapolating back the linear portion of the curve. In most transformations, this procedure probably represents the actual way in which the regions form, but in some cases alternative extrapolations, such as that shown to zero time, may be more correct. Such a curve implies that stable nuclei of the new phase are already present in the assembly at the beginning of the transformation, but that the initial growth rate is an increasing function of the time.

In developing a formal theory of transformation kinetics, the distinction between these alternative physical processes leading to the formation of nuclei is irrelevant. We adopt an operational definition of the nucleation rate per unit volume, ^vI, which is related to the reciprocal of a mean value of the period τ. Suppose that at time $t = \tau$ the untransformed volume is V^α, and that between times $t = \tau$, $\tau + \delta\tau$ a number of new regions are nucleated (i.e. reach some arbitrary minimum size). In principle, this number can be determined experimentally by plotting curves of the type shown in Fig. 1.2 for a large number of

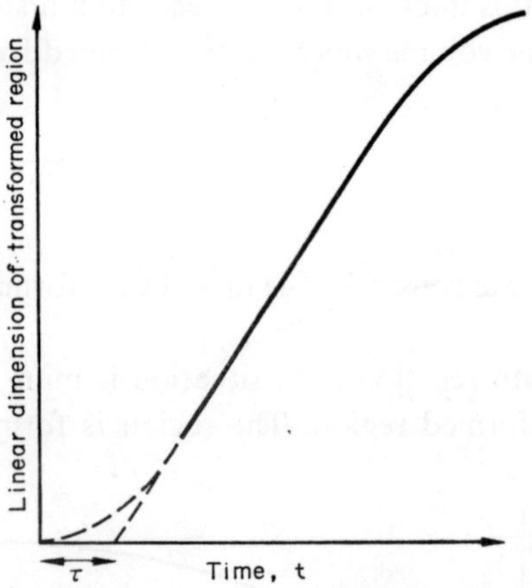

FIG. 1.2. Schematic growth curve for a product region.

regions, extrapolating back to give intercepts on the time axis, and finding the number of such intercepts between τ and $\tau + \delta\tau$. This number is $^vIV^\alpha\,\delta\tau$, and defines ^vI at time τ; in order to give statistically significant results, the number of intercepts per unit time ($^vIV^\alpha$) must be large. The growth rate for any direction is similarly obtained by plotting the length for that direction as a function of time for a large number of regions, and finding the average slope. In general, the growth rate will be anisotropic, but in this introductory treatment, we assume an isotropic growth rate Υ so that the transformed regions are spherical. This is, in fact, a good approximation in many actual changes. The volume of a β region originating at time $t = \tau$ is then

$$\left.\begin{array}{ll} v_\tau = (4\pi/3)\,\Upsilon^3(t-\tau)^3 & (t > \tau), \\ v_\tau = 0 & (t < \tau). \end{array}\right\} \tag{4.2}$$

When the mean composition of the matrix must be changed during the transformation, the size of a transformed region in any direction is often found to be proportional to $(t-\tau)^{1/2}$, this parabolic growth law being dependent on diffusion rates. An equivalent operational definition of the nucleation rate can obviously be given, but in the present section only linear growth rates are considered.

$$V_\tau = \frac{4\pi}{3}\left(k_\rho\sqrt{t}\right)^3$$

In the whole assembly, the number of new β regions nucleated in the time interval between τ and $\tau+\mathrm{d}\tau$ is ${}^vIV^\alpha\,\mathrm{d}\tau$. During the initial stages of transformation, when $V^\beta \ll V^\alpha$, the nuclei are widely spaced, and the interference of neighbouring nuclei is negligible. Under these conditions, the transformed volume at time t resulting from regions nucleated between times τ and $\tau+\mathrm{d}\tau$ is $\mathrm{d}V^\beta = v_\tau\,{}^vIV^\alpha\,\mathrm{d}\tau$. Since V^α is effectively constant and equal to V, the total volume transformed at time t is thus

$$V^\beta = (4\pi V/3) \int_{\tau=0}^{t} {}^vI\Upsilon^3(t-\tau)^3\,\mathrm{d}\tau.$$

This equation can be integrated only by making some assumption about the variation of ${}^vI$ with time. The simplest assumption is that ${}^vI$ is constant, and this leads to the result

$$\zeta = V^\beta/V = (\pi/3)\,{}^vI\Upsilon^3t^4, \tag{4.3}$$

where we have now introduced the symbol ζ for the volume fraction transformed at time t. The rate of transformation according to this equation rises rapidly in the initial stages.

In a more exact treatment, we must consider the mutual interference of regions growing from seperate nuclei. When two such regions impinge on each other, there are three possible consequences. The regions may unite to form a single region, as often happens with liquid droplets forming from the vapour, or they may separate, each continuing to grow (until a late stage in the transformation) as though there were no impingement. Separation can obviously occur only if the primary phase is fluid, and preferably gaseous. The third possibility is that the two regions develop a common interface, over which growth ceases, although it continues normally elsewhere. This must happen in all solid transformations, and it is the case which we shall consider. The problem is primarily geometrical, and was first treated by Johnson and Mehl (1939) and Avrami (1939, 1940, 1941).

During the time $\mathrm{d}\tau$, when ${}^vIV^\alpha\,\mathrm{d}\tau$ new transformed regions are nucleated, we may also consider that ${}^vIV^\beta\,\mathrm{d}\tau$ regions would have nucleated in the transformed portion of the assembly, had not transformation previously occurred there. Avrami described these as phantom nuclei, and went on to define an "extended" volume of transformed material, V_e^β, by the relation

$$\left.\begin{aligned} \mathrm{d}V_e^\beta &= v_\tau\,{}^vI(V^\alpha+V^\beta)\,\mathrm{d}\tau, \\ \text{i.e.} \qquad V_e^\beta &= (4\pi V/3) \int_0^t \Upsilon^3\,{}^vI(t-\tau)^3\,\mathrm{d}\tau. \end{aligned}\right\} \tag{4.4}$$

V_e^β differs from the actual volume of transformed material in two ways. Firstly, we have counted phantom regions, nucleated in already transformed material. Secondly, we have treated all regions as though they continued growing, irrespective of other regions. The extended volume can thus be visualized as a series of volume elements having the same limiting surface as the actual transformed volume but all growing "through" each other. Some elements of the transformed volume are counted twice, others three times, and so on, in order to obtain the extended volume. It follows that the extended volume may be larger than the real volume of the whole assembly V.

The significance of V_e^β is that it is simply related to the kinetic laws of growth, which may thus be separated from the geometrical problem of impingement. We have now to find a relation between V_e^β and V^β. Consider any small random region, of which a fraction $(1-V^\beta/V)$ remains untransformed at time t. During a further time dt, the extended volume of β in the region will increase by dV_e^β, and the true volume by dV^β. Of the new elements of volume which make up dV_e^β, a fraction $(1-V^\beta/V)$ on the average will lie in previously untransformed material, and thus contribute to dV^β, whilst the remainder of dV_e^β will be in already transformed material. This result clearly follows only if dV_e^β can be treated as a completely random volume element, and it is for this reason that phantom nuclei have to be included in the definition of V_e^β.

The above arguments are based on the assumption that nucleation is random, in the sense that if we divide the assembly into small equal volume elements, the probability of forming a nucleus in unit time is the same for all these elements. The treatment does not preclude the possibility that nuclei form preferentially at certain sites in the β phase, but as developed here, it may be applied to experimental observations only if the minimum resolvable β volume contains several such sites. All equal volume elements of size greater than this observational limit then have equal nucleation probabilities. In practice, nucleation may occur preferentially along macroscopic surfaces (grain boundaries) of the assembly, so that volume elements in different regions may have quite different nucleation probabilities. The formal theory can be extended to cover such cases, but discussion is postponed to Chapter 12.

We now write the relation between V^β and V_e^β in the form

$$dV^\beta = (1-V^\beta/V)\, dV_e^\beta$$

or

$$V_e^\beta = -V \ln(1-V^\beta/V). \tag{4.5}$$

Substituting into (4.4)

$$-\ln(1-\zeta) = (4\pi/3)\Upsilon^3 \int_0^t {}^vI(t-\tau)^3\, d\tau. \tag{4.6}$$

This equation may be integrated only by making specific assumptions about the variation of ${}^vI$ with time. In particular, if ${}^vI$ is constant

$$\zeta = 1-\exp(-\pi\Upsilon^3\, {}^vIt^4/3). \tag{4.7}$$

Note that eqn. (4.3) is given by the first term in the expansion of the right-hand side of (4.7), so that the two expressions become identical as $t \to 0$, in accordance with the assumptions made in deriving (4.3).

In general, ${}^vI$ may not be constant, but may either increase or decrease with time. The physical processes leading to this will be described fully later, but we emphasize here that eqn. (4.6) and not (4.7) must be used for ζ in the general case. This is made clear by an alternative assumption, used by Avrami. He supposes that nucleation occurs only at certain preferred sites in the assembly, which are gradually exhausted. If there are ${}^vN_0$ sites per unit volume of the α phase initially, and ${}^vN$ remaining after time t, the number disappearing in a further small time interval dt is $d^vN = -{}^vN\nu_1\, dt$, where the frequency ν_1 gives the rate at which an individual site becomes a nucleus. Thus ${}^{v}N = {}^{v}N_0 \exp(-\nu_1 t)$, and the nucleation rate per unit volume is

$$ {}^{v}I = -\mathrm{d}{}^{v}N/\mathrm{d}t = {}^{v}N_0 \nu_1 \exp(-\nu_1 t). \tag{4.8} $$

Substituting $({}^{v}I)_{t=\tau}$ into eqn. (4.6) and integrating by parts gives

$$ \zeta = 1 - \exp\left[(8\pi\, {}^{v}N_0\, \Upsilon^3/\nu_1^3)\left\{\exp(-\nu_1 t) - 1 + \nu_1 t - \frac{\nu_1^2 t^2}{2} + \frac{\nu_1^3 t^3}{6}\right\}\right]. \tag{4.9} $$

There are two limiting forms of this equation, corresponding to very small or very large values of $\nu_1 t$. Small values of $\nu_1 t$ imply that ${}^{v}I$ (eqn. (4.8)) is effectively constant, and the limiting value of (4.9), obtained by expanding $\exp(-\nu_1 t)$, is identical with eqn. (4.7). Large values of $\nu_1 t$, in contrast, mean that ${}^{v}N$ quickly becomes zero, all nucleation centres being exhausted at an early stage in the reaction. The limiting value of eqn. (4.9) is then

$$ \zeta = 1 - \exp\{-(4\pi\, {}^{v}N_0/3)\Upsilon^3 t^3\}. \tag{4.10} $$

Avrami proposed that for a three-dimensional nucleation and growth process, we should use the general relation

$$ \zeta = 1 - \exp(-kt^n), \tag{4.11} $$

where $3 \leqslant n \leqslant 4$. This should cover all cases in which ${}^{v}I$ is some decreasing function of time, up to the limit when ${}^{v}I$ is constant.

The above treatment, whilst including the effects of impingement, neglects the effect of the free surface. Thus if transformation occurs in a thin sheet of solid material, it may happen that the average dimension of a transformed region is much greater than the thickness of the sheet. Growth in this direction must soon cease, and thereafter the growth is essentially two-dimensional. Instead of (4.2), the expression for the volume of an individual transformed region becomes

$$ v'_\tau = \pi\, \delta \Upsilon^2 (t-\tau)^2, \tag{4.12} $$

where δ is the sheet thickness. Similarly, for a wire of diameter δ, the growth would be one-dimensional, the volume of each transformed region being

$$ v''_\tau = (\pi/4)\, \delta^2 \Upsilon (t-\tau). \tag{4.13} $$

The use of v' or v'' in the definition of the extended volume (eqn. (4.4)) will introduce corresponding modifications in the expressions for ζ. In the Avrami theory of nucleation at preferred sites, it is readily seen that the general expression for the volume transformed (eqn. (4.11)) remains valid if $2 \leqslant n \leqslant 3$ (two-dimensional growth) and $1 \leqslant n \leqslant 2$ (one-dimensional growth).

The isothermal transformation curves obtained by substituting $n = 1$ in eqn. (4.11) is equivalent to that for a first-order homogeneous reaction. All the other possibilities, however, give sigmoidal curves for ζ against t, in which the fractional volume transformed increases slowly at first, then much more rapidly, and, finally, slowly again. Most experimental transformation curves are sigmoidal in shape; a typical example is shown in Fig. 1.3.

Avrami pointed out that if we plot ζ against ln t, all the curves with the same value of n will have the same shape, and will differ only in the value of k, which is equivalent to a change of scale. He therefore proposed that the "shape" of a reaction curve should be defined by a $\zeta - \ln t$ plot. A more useful plot is $\log\log[1/(1-\zeta)]$ against $\log t$, the slope of which gives n.

In the general form of eqn. (4.11), the above theory applies to many real transformations. However, it cannot be safely applied in practice unless the assumptions about ^{v}I and Υ

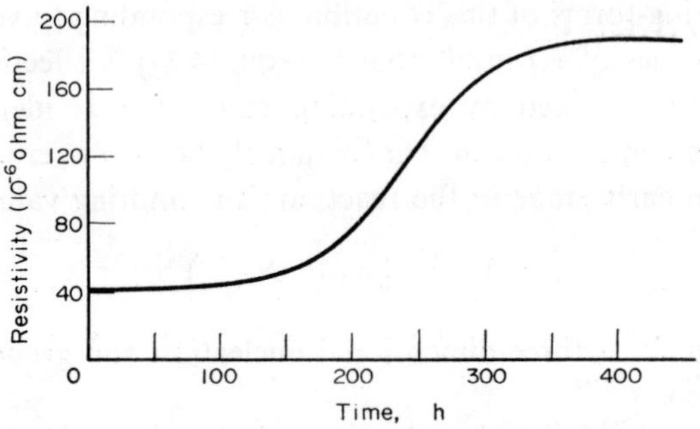

FIG. 1.3. Kinetics of transformation $\beta \rightarrow \alpha$ manganese at 25°C, as shown by electrical resistivity measurements (after Potter *et al.*, 1949).

can be verified. The Avrami theory assumes that the nucleation frequency is either constant, or else is a maximum at the beginning of transformation and decreases (slowly or rapidly) during the course of transformation. In contrast to this, there may be an operational rate of nucleation which increases with time. The Avrami equation is also often applied to transformations in which the growth rate is diffusion controlled, but in most cases there is no adequate theoretical sanction for this. These questions are fully discussed in Chapter 12, which takes up the formal theory of this section in greater detail.

REFERENCES

AVRAMI, M. (1939) *J. Chem. Phys.* **7**, 1103; (1940) *Ibid.* **8**, 212; (1941) *Ibid.* **9**, 177.

BUERGER, M. J. (1951) *Phase Transformations in Solids*, p. 183, Wiley, New York.

CHRISTIAN, J. W. (1965) *Physical Properties of Martensite and Bainite*, Iron and Steel Institute Spec. Rep. **93**, 1.

CLARK, H. M. and WAYMAN, C. M. (1970) *Phase Transformations*, p. 59, American Society for Metals, Metals Park, Ohio.

COHEN, M., CAHN, J. W., CHRISTIAN, J. W., FLINN, P. A., HILLERT, M., KAUFMAN, L., and READ, T. A. (1965) *Perspectives in Materials Research*, p. 309, Office of Naval Research, Washington.

FRANK, F. C. (1963) *Relation between Structure and Strength in Metals and Alloys*, p. 248, HMSO, London.

GUGGENHEIM, E. A. (1967) *Thermodynamics*, 5th ed., North-Holland, Amsterdam.

GUY, A. G. (1972) *Met. Trans.* **3**. 2535.

JOHNSON, W. A. and MEHL, R. F. (1939) *Trans. Am. Inst. Min. (Metall.) Engrs.* **135**, 416.

LIEBERMAN, D. S. (1969) *The Mechanism of Phase Transformations in Crystalline Solids*, p. 167, Institute of Metals, London.

LIEBERMAN, D. S. (1970) *Phase Transformations*, p. 1, American Society for Metals, Metals Park, Ohio.

POTTER, E. V., LUKENS, H. C., and HUBER, R. W. (1949) *Trans. Am. Inst. Min. (Metall.) Engrs.* **185**, 399.

CHAPTER 2

Formal Geometry of Crystal Lattices

5. DESCRIPTION OF THE IDEAL CRYSTAL

Solid metals are crystalline, and some appreciation of crystallography is essential to a study of metallic transformations. The scientific concept of a crystal has evolved gradually from the original classification by external shape to modern views on the internal atomic arrangement. The recognition that the distinguishing feature of crystalline solids lies in their regular internal arrangement led to a description which Zachariasen (1945) has termed the macroscopic concept of a crystal. The macroscopic crystal is defined in terms of physical properties which have precise meaning (or, at least, are measurable) only for regions containing appreciable numbers of atoms. Such properties are invariant with respect to a translation within an infinite crystal, but (except for scalar properties) not with respect to a rotation. A crystal is thus a homogeneous, anisotropic solid; a noncrystalline, or amorphous, solid is both homogeneous and isotropic.

The development of X-ray methods enabled the structure of a crystal to be investigated on a finer scale. It was then found that crystals are not truly homogeneous, but the arrangement of atoms is periodic in three dimensions. This is the familiar modern picture, which we shall take as our starting point, but we emphasize here that it is only an abstraction from the much less ordered situation in a real crystal. In recent years, attention has been directed especially to the imperfections in real crystal structures. These imperfections represent comparatively small deviations from the mathematical concept of an ideal crystal, but they nevertheless control many of the most important physical properties. In this chapter we are concerned only with the ideal crystal; the nature of the approximations involved in this description, and the extent to which a real crystal may be considered to be an ideal crystal containing imperfections, will be considered in detail later.

The ideal crystal may be regarded as the repetition in three dimensions of some unit of structure, within which the position of each atom is specified exactly by a set of spatial coordinates. Let us choose an origin within the crystal. This will be one of an infinite set of points, each possessing an identical configuration of surrounding atoms. The positions of these points may be represented by the vectors

$$\mathbf{u} = u_i \mathbf{a}_i \qquad (i = 1, 2, 3), \tag{5.1}$$

where u_i have only integral values, and the summation convention is used. The translation between any two lattice points is a lattice vector, and the three non-coplanar lattice vectors

$\mathbf{a}_i$ outline a parallelepiped known as the unit cell. Any other set of lattice vectors, formed from linear combinations of the set $\mathbf{a}_i$, may also be used as basic vectors, so there is an infinite number of possible unit cells. The volume of the unit cell is given by the scalar triple product of the vectors which outline it; when this volume is a minimum the vectors are primitive vectors, and define a unit cell of the Bravais lattice. Each such primitive cell contains only one lattice point; cells of larger volume contain two or more lattice points. The primitive unit cell may still be chosen in an infinity of ways, since we have placed no restriction on its shape. The most useful unit cell is usually that in which the vectors $\mathbf{a}_i$ are as nearly as possible of equal scalar magnitude.

The quantities u_i of eqn. (5.1) give the position of the end point of $\mathbf{u}$ in an oblique Cartesian coordinate system, in which distances along the axes are measured in multiples of the lengths of the basic vectors $\mathbf{a}_i$. Such a coordinate system forms a natural framework for representation of the crystal lattice, and is generally preferable to the alternative method of using coordinate axes parallel to the vectors $\mathbf{a}_i$, but having unit measure lengths. In the latter system, the coordinates of a lattice point are $u_{(i)}|\mathbf{a}_{(i)}|$ (the brackets indicating suspension of the summation rule), and these are sometimes called the physical components of the vector $\mathbf{u}$. However, it is often convenient to use a coordinate system with orthogonal axes of equal base lengths, and this is called an orthonormal system. It is defined by the three unit vectors $\mathbf{i}_i$, which satisfy the relations

$$\mathbf{i}_i \cdot \mathbf{i}_j = \delta_{ij}, \tag{5.2}$$

where δ_{ij}, called the Kronecker delta, is equal to unity when $i = j$, and to zero when $i \neq j$. In terms of the orthonormal system, a vector $\mathbf{x}$ may be written as

$$\mathbf{x} = x_i \mathbf{i}_i. \tag{5.3}$$

In tensor analysis or matrix algebra,† the vector $\mathbf{u}$ is regarded as the array of numbers u_i, which form its components. We may, for example, write $\mathbf{u}$ as a single row matrix $(u_1\ u_2\ u_3)$ or as a single column matrix $\begin{pmatrix} u_1 \\ u_2 \\ u_3 \end{pmatrix}$, and these arrays would conventionally be given the same symbol as $\mathbf{u}$. This sometimes leads to confusion, and it is desirable to have a way of distinguishing between the vector $\mathbf{u}$, which is a physical entity, and its matrix representation, which depends on a particular coordinate system. We shall do this by using a notation in which **bold face type** is used for symbols representing physical quantities (vectors and tensors), and $\mathsf{sans\ serif}$ type is used for their matrix representations. In addition, we specify that column matrices may be written $\begin{pmatrix} u_1 \\ u_2 \\ u_3 \end{pmatrix} = [u_1\, u_2\, u_3]$ for convenience, whilst row matrices

† The formal theory of this chapter may be expressed in either tensor or matrix notation. We find it more convenient to use the latter, but many of the equations will be given in both forms to facilitate reference to other books and papers. A brief description of the more important features of tensor notation is given on pp. 33–6.

are written $(u_1 u_2 u_3)$. We thus have

$$\mathsf{u} = [u_1 u_2 u_3] \qquad \mathsf{x} = [x_1 x_2 x_3], \tag{5.4}$$

where u and x are representations of the vectors **u** and **x**. Wherever possible we shall keep to the practice that a letter such as **u** (and later **v**) will represent a vector which is most conveniently expressed in a lattice coordinate system, while a letter such as **x** (and later **y**) will represent a vector which is usually referred to an orthonormal system.

Any matrix of n rows and m columns may be formed into a new matrix of m rows and n columns (the *transposed* matrix) by interchanging the rows and columns. Clearly, the single row and single column matrix representations of **u** are the transposes of each other, so that eqn. (5.4) also implies the notation

$$\mathsf{u}' = (u_1 u_2 u_3) \qquad \mathsf{x}' = (x_1 x_2 x_3), \tag{5.5}$$

where u' is the symbol for the transposed matrix formed from u.

Later in this chapter we need to refer the vector **u** to other sets of coordinate axes. We define a second set of axes by the base vectors $\mathbf{b}_i$, and the vector **u** then has a different matrix representation. We may distinguish between the two representations when required by describing the set of vectors $\mathbf{a}_i$ as the basis A, and the set of vectors $\mathbf{b}_i$ as the basis B. The column matrix representations are then written as $^A\mathsf{u} = [^Au_1\, ^Au_2\, ^Au_3]$ and $^B\mathsf{u} = [^Bu_1\, ^Bu_2\, ^Bu_3]$. Symbols of this type are sufficient for most purposes, but occasionally extra clarity is achieved by use of an extended notation, such as that used by Bowles and Mackenzie (1954). In the extended notation, the representations of **u** as column matrices are

$$^A\mathsf{u} = [\mathrm{A}; \mathbf{u}] \qquad ^B\mathsf{u} = [\mathrm{B}; \mathbf{u}], \tag{5.6}$$

and the corresponding row matrices are

$$^A\mathsf{u}' = (\mathbf{u}; \mathrm{A}) \qquad ^B\mathsf{u}' = (\mathbf{u}; \mathrm{B}) \tag{5.7}$$

in which both the round brackets and the reversal of the order of the vector symbol **u** and the base symbol A or B signify the tranposition of the column matrices $^A\mathsf{u}$, $^B\mathsf{u}$. The same notation is applied to vectors referred to an orthonormal basis, for which we shall usually use the symbol I. Hence two different vectors **u** and **x** have representations

$$^I\mathsf{u} = [\mathrm{I}; \mathbf{u}] \qquad ^I\mathsf{x} = [\mathrm{I}; \mathbf{x}]$$

in such a basis.

The advantages of the extended notation will become apparent later, but it is often sufficient to use the sans serif symbols. When no confusion about the bases is possible, the identifying superscripts will be omitted, as in eqn. (5.4).

A clear distinction must be made between the lattice points of the unit cell, and the positions of the atoms within the cell. The simplest types of crystal structure are obtained by placing an atom at each point of the lattice, and this category includes two of the three common metallic structures. More generally, the primitive Bravais lattice only gives the interval over which the unit of pattern, or motif unit, is repeated. This unit may be a single atom or a more complex atom group; in the latter case, the structure is said to have a basis. The repeating properties of the lattice then require that if an atom of kind A is situated in a given position with respect to the origin, a similar atom is similarly situated with respect to each of the lattice points.† If there are r atoms of this kind within the unit cell, their positions with respect to the origin of the cell may be specified by the vectors $\boldsymbol{\xi}_{A,1}$, $\boldsymbol{\xi}_{A,2}, \ldots, \boldsymbol{\xi}_{A,r}$, and hence the positions of all the atoms of type A are given by

$$\mathbf{u}_{A,n} = \mathbf{u}+\boldsymbol{\xi}_{A,n} \qquad (n = 1, 2, \ldots, r).$$

Often we do not need to specify the type of atom, so that the subscript A may be omitted and the equation may be written

$$\mathbf{u}_n = \mathbf{u}+\boldsymbol{\xi}_n = (u_i+\xi_{n,i})\mathbf{a}_i \qquad (n = 1, 2, \ldots, r). \tag{5.8}$$

Whilst the components u_i are all integers, the components $\xi_{n,i}$ are all less than unity. For a realistic choice of motif unit, the restriction $|\xi_{n,i}| \leqslant \frac{1}{2}$ will usually be valid. When the atomic arrangement is centrosymmetric, it is possible to choose an origin so that for each atom in a position $\boldsymbol{\xi}$, there is a similar atom in a position $-\boldsymbol{\xi}$.

A structure which contains r atoms in the unit cell may be discussed in two different ways. We may think of a single lattice framework, at each point of which is situated a motif unit of r atoms, or we may consider the whole structure to be composed of r interpenetrating simple Bravais lattices. In many inorganic and organic crystals, the motif units have some physical significance, since they are the molecules of the compound. This is not true for most metallic structures, and it is sometimes possible to choose motif units in various different ways, all having equal validity. The alternative description may then be useful, and we shall write of single, double, etc., lattice structures, meaning structures in which the primitive unit cell contains one, two, etc., atoms. All single and double lattice structures are centro-symmetric, and the structures of most metals fall into one of these two groups.

The ideal crystal is classified by considering the symmetry properties of the atomic arrangement. There are 230 space groups, or combinations of symmetry elements, but most of these are obtained from relations between the vectors $\boldsymbol{\xi}_{A,n}$, $\boldsymbol{\xi}_{B,n}$, etc. The symmetry properties of the lattice are much more restricted, and there are only fourteen Bravais lattices, obtained from relations between the vectors $\mathbf{a}_i$, Instead of the primitive unit cell, it is often convenient to use a larger unit cell which illustrates the symmetry of the lattice positions. For example, if the unit cell of the Bravais lattice has rhombohedral shape, and the angles between the axes are either 109°30′ or 60°, it is readily shown that the lattice positions have cubic symmetry. The conventional unit cells are cubic, and contain two and four points of the Bravais lattices respectively; the lattices are called body-centred and face-centred cubic. If we place an atom on each point of these lattices, we obtain the two common cubic metallic structures.

† This statement has to be modified in the case of a substitutional solid solution in which the possible atomic positions are occupied more or less randomly. The structure is then periodic only if the differences between the atoms of different species are ignored, so that for these atoms, $\boldsymbol{\xi}_A = \boldsymbol{\xi}_B =$, etc.

The set of integers u_i defines a translation from the origin to some lattice point. An alternative set of integers, $U_i = Cu_i$ gives a parallel translation to a lattice point which is C times more distant from the origin. If we assume that the set u_i contains no common factor, we may use the equation

$$\mathbf{u} = C_1 u_i \mathbf{a}_i = U_i \mathbf{a}_i \tag{5.9}$$

in which C_1 takes all possible integral values from $-\infty$ to $+\infty$ to represent all lattice points in a straight line passing through the origin. In a similar way, the equation

$$\mathbf{u} = \mathbf{u}_k + U_i \mathbf{a}_i \tag{5.10}$$

represents all lattice points in a parallel straight line passing through a lattice point with position vector $\mathbf{u}_k$. For any given set of rational values u_i, eqn. (5.10) represents all the lattice points. The lattice structure may thus be regarded as rows of points on parallel straight lines. In conventional crystallographic notation, the quantities u_i are called the direction indices of the line, and are enclosed in square brackets $[u_1 u_2 u_3]$. The direction indices of a line are thus given by its representation as a column matrix u. When the symmetry of the lattice requires that certain directions are equivalent, the whole set of such directions may be represented by the symbol $\langle u_1 u_2 u_3 \rangle$. When u_i have non-integral values, the direction they specify does not lie along a row of lattice points, and is called irrational.

Since the lattice points are arranged along straight lines, it is also possible to regard them as situated on planes. Consider first the plane defined by the basic vectors $\mathbf{a}_1$ and $\mathbf{a}_2$. The normal to this plane is parallel to the vector $\mathbf{a}_1 \wedge \mathbf{a}_2$, and may be denoted by a vector $\mathbf{a}_3^* = (\mathbf{a}_1 \wedge \mathbf{a}_2)/v_a$, where v_a is a scalar constant. The area of the face of the unit cell formed by the vectors $\mathbf{a}_1$ and $\mathbf{a}_2$ is numerically equal to the length of the vector $\mathbf{a}_1 \wedge \mathbf{a}_2$, and since the volume of the unit cell is given by the scalar triple product $(\mathbf{a}_1\mathbf{a}_2\mathbf{a}_3) = \mathbf{a}_3 \cdot \mathbf{a}_1 \wedge \mathbf{a}_2$, we have the distance between adjacent lattice planes in the $\mathbf{a}_3^*$ direction is equal to $\mathbf{a}_3 \cdot \mathbf{a}_1 \wedge \mathbf{a}_2 / |\mathbf{a}_1 \wedge \mathbf{a}_2|$. For reasons which will soon be evident, it is convenient to take the constant v_a equal to the volume of the unit cell, so that $|\mathbf{a}_3^*|$ is equal to the reciprocal of the interplanar spacing. We have then $\mathbf{a}_3 \cdot \mathbf{a}_3^* = 1$. Proceeding in the same way for the other two faces of the unit cell, we obtain a set of vectors $\mathbf{a}_i^*$ perpendicular to the faces of the cell and satisfying the relations:

$$\mathbf{a}_i \cdot \mathbf{a}_j^* = \delta_{ij}. \tag{5.11}$$

The vectors $\mathbf{a}_i^*$ are said to be reciprocal to the vectors $\mathbf{a}_i$. We can now exactly reverse the above reasoning by writing $\mathbf{a}_1 = \mathbf{a}_2^* \wedge \mathbf{a}_3^* / v_a^*$, and since $\mathbf{a}_1 \cdot \mathbf{a}_1^* = 1$, it follows that $v_a^* = (\mathbf{a}_1^* \mathbf{a}_2^* \mathbf{a}_3^*)$, the volume of the parallelepiped formed by the reciprocal set $\mathbf{a}_i^*$. The relations between the two sets of vectors are thus symmetrical, and each is reciprocal to the other. The two volumes are also reciprocal and

$$v_a v_a^* = 1. \tag{5.12}$$

The set of reciprocal vectors, $\mathbf{a}_i^*$, may be used to define a new coordinate system, which we describe as basis A^*, and we then have

$$\mathbf{u} = u_i^* \mathbf{a}_i^*,$$

so that **u** is represented by†

$$\mathsf{u}^* = [u_1^* \, u_2^* \, u_3^*] = [\mathrm{A}^*; \mathbf{u}].$$

Suppose we form the scalar product of **u** and one of the reciprocal vectors $\mathbf{a}_i^*$. This gives

$$\mathbf{u} \cdot \mathbf{a}_i^* = u_j \mathbf{a}_j \cdot \mathbf{a}_i^* = u_i \tag{5.13}$$

and, similarly,

$$\mathbf{u} \cdot \mathbf{a}_i = u_i^*. \tag{5.14}$$

These relations are shown in two dimensional form in Fig. 2.1. In the axis system A, the components u_i measured in units of $|\mathbf{a}_i|$ give the displacements parallel to the axes which

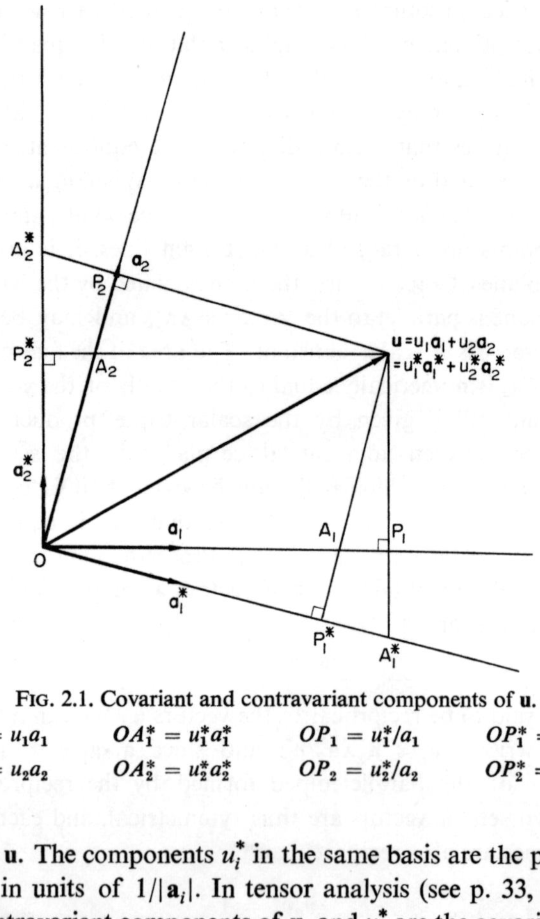

FIG. 2.1. Covariant and contravariant components of **u**.

$OA_1 = u_1 a_1$ $\quad OA_1^* = u_1^* a_1^*$ $\quad OP_1 = u_1^*/a_1$ $\quad OP_1^* = u_1/a_1^*$

$OA_2 = u_2 a_2$ $\quad OA_2^* = u_2^* a_2^*$ $\quad OP_2 = u_2^*/a_2$ $\quad OP_2^* = u_2/a_2^*$

add together to give **u**. The components u_i^* in the same basis are the projections of **u** along the axes, measured in units of $1/|\mathbf{a}_i|$. In tensor analysis (see p. 33, the quantities u_i are referred to as the contravariant components of **u**, and u_i^* are the covariant components of **u**. In the basis A*, the interpretations of u_i and u_i^* are reversed, as shown in Fig. 2.1.

† In matrix algebra, the notation U* commonly means the matrix which is the complex conjugate of the matrix U. In this book, all matrices are real, so no confusion arises.

We can now write useful expressions for the scalar and vector product of two vectors **u** and **v**. For the scalar product

$$\mathbf{u}\cdot\mathbf{v} = (u_i\mathbf{a}_i)\cdot(v_j^*\mathbf{a}_j^*) = u_i v_i^* = u_i^* v_i. \tag{5.15}$$

In matrix form this equation is

$$\mathbf{u}\cdot\mathbf{v} = \mathsf{u}'\,\mathsf{v}^* = \mathsf{v}'\,\mathsf{u}^*$$

and, in particular, the length of the vector **u** is given by

$$|\mathbf{u}|^2 = \mathsf{u}'\,\mathsf{u}^*. \tag{5.16}$$

Equation (5.16) involves both sets of components u_i and u_i^*, but it is obvious that the length of the vector can be expressed in terms of the components u_i alone. In order to do this, we must find the relation between the bases A and A*, and this is discussed in the next section.

For the vector product of **u** and **v**, we have

$$\mathbf{u}\wedge\mathbf{v} = (u_j\mathbf{a}_j)\wedge(v_k\mathbf{a}_k) = v_a\varepsilon_{ijk}u_j v_k\mathbf{a}_i^*, \tag{5.17}$$

where $\varepsilon_{ijk} = 0$ unless i, j, k are all different, and $\varepsilon_{ijk} = \pm 1$ according to whether i, j, k are even or odd permutations of 1, 2, 3. If we write $\mathbf{w} = (\mathbf{u}\wedge\mathbf{v})$, eqn. (5.17) gives the components of **w** in the basis A* as

$$w_i^* = v_a\varepsilon_{ijk}u_j v_k. \tag{5.18}$$

Clearly the components of **w** in the basis A are given by

$$w_i = v_a^*\varepsilon_{ijk}u_j^* v_k^*.$$

The orientation of any plane is completely specified by the relative lengths of the intercepts it makes on the axes $\mathbf{a}_i$. Consider a plane which intersects the axes at points distant h_1^{-1}, h_2^{-1}, h_3^{-1} from the origin. This plane contains the two vectors $(\mathbf{a}_1/h_1-\mathbf{a}_2/h_2)$ and $(\mathbf{a}_2/h_2-\mathbf{a}_3/h_3)$, and its normal is then parallel to their vector product, i.e. to a vector

$$C\mathbf{h} = \frac{(\mathbf{a}_1\mathbf{a}_2\mathbf{a}_3)}{h_1h_2h_3}(h_1\mathbf{a}_1^*+h_2\mathbf{a}_2^*+h_3\mathbf{a}_3^*).$$

Leaving out the scalar multiplying factor, we may use the normal vector

$$\mathbf{h} = h_i\mathbf{a}_i^* \tag{5.19}$$

to represent the plane. The numbers h_i are the components of the vector **h** in the basis A*, and for consistency of notation we should have written them h_i^*. The matrix representation of **h** is then

$$\mathsf{h}^* = [h_1^*\ h_2^*\ h_3^*] = [\mathrm{A}^*;\mathbf{h}].$$

Since we shall always refer vectors which represent plane normals to the reciprocal basis A*, we may omit the asterisks and use h_i for these components. Note that the magnitude of **h** has, at present, been chosen arbitrarily.

The triad of numbers h_i represent the orientations of all parallel planes which intersect the axes at distances C_2/h_i. The scalar equation of these planes is

$$h_1u_1+h_2u_2+h_3u_3-C_2 = 0, \tag{5.20}$$

where each u_i is now regarded as a continuous variable. We are interested in those planes which contain sets of equivalent lattice points. A plane containing three lattice points includes two lattice directions, and hence an infinite set of points. The coordinates of any three points are $[U_1\,U_2\,U_3]$, $[V_1\,V_2\,V_3]$, and $[W_1\,W_2\,W_3]$. From eqn. (5.20) and the three equations obtained by substituting these values into it, we obtain the determinal equation

$$\begin{vmatrix} u_1 & u_2 & u_3 & -1 \\ U_1 & U_2 & U_3 & -1 \\ V_1 & V_2 & V_3 & -1 \\ W_1 & W_2 & W_3 & -1 \end{vmatrix} = 0, \tag{5.21}$$

and since the quantities U_i, V_i, W_i are all integers, it follows from (5.20) and (5.21) that h_1, h_2, h_3 and C_2 are also integral. Planes in which h_i are not integral do not contain lattice points, and are termed irrational. Suppose first that h_i contains no common factor. For each integral value of C_2 between $-\infty$ and $+\infty$, there is an infinite number of points $u_i = U_i$ satisfying eqn. (5.20). As C_2 varies, (5.20) thus represents an infinite set of equally spaced parallel planes, containing all the lattice points. One of these planes ($C_2 = 0$) passes through the origin, and the next plane of the set has $C_2 = 1$. As already shown, the vector **h** is perpendicular to the set of planes, and the interplanar spacing is thus given by the projection of the vector $\mathbf{a}_1/h_1$ on the direction of **h**, i.e. by the relation

$$d_h = \frac{\mathbf{h}\cdot\mathbf{a}_1/h_1}{|\mathbf{h}|} = \frac{1}{|\mathbf{h}|}\,. \tag{5.22}$$

The vectors **h** are thus not only perpendicular to the planes having indices h_i, but are of length equal to the reciprocal of the interplanar spacing. The orientation of a plane is usually specified in crystallography by placing the indices h_i in round brackets, $(h_1\,h_2\,h_3)$, and this is equivalent to writing the vector as a row matrix $\mathrm{h}' = (\mathbf{h};\,\mathrm{A}^*)$. When there are several equivalent planes, they are indicated by the symbol $\{h_1\,h_2\,h_3\}$.

We now consider the set of equations obtained by replacing the h_i in eqn. (5.19) by quantities $H_i = C_3h_i$ (C_3 integral), it being assumed as before that the set h_i contains no common factor. As C_3 varies from $-\infty$ to $+\infty$, we again obtain an infinite set of planes parallel to the first set but spaced C_3 times more closely. Equation (5.20) can now be satisfied with quantities $u_i = U_i$ only if C_2 is an integral multiple of C_3. When this happens, the plane is one of the original set h_i. We thus see that the vector:

$$\mathbf{h} = C_3h_i\mathbf{a}_i^* = H_i\mathbf{a}_i^* \tag{5.23}$$

represents a set of parallel planes of spacing, $1/|\mathbf{h}|$, but that only every C_3th plane of the set passes through equivalent lattice points.

Equation (5.23) shows that the vectors **h** define a lattice, and this is called the reciprocal lattice. Each point of the reciprocal lattice corresponds to an infinite set of parallel crystal planes; similarly, each set of reciprocal lattice planes is associated with a point of the real lattice. The linear vector space which is defined by the vectors $\mathbf{a}_i^*$ is called reciprocal space. The concepts of reciprocal space, and of the reciprocal lattice, have proved very useful in crystal geometry and in the theory of X-ray diffraction.

6. LINEAR TRANSFORMATIONS OF THE COORDINATE SYSTEM

The position of any lattice point is given by its coordinates u_i with respect to some chosen origin and set of base vectors A. As previously emphasized, the choice of basis is arbitrary, and it is often desirable to use a new basis B defined by the vectors $\mathbf{b}_i$. We then have to develop the appropriate transformation formulae connecting ${}^A u_i$ and ${}^B u_i$ for any direction, and ${}^A h_i$ and ${}^B h_i$ for any plane.

The new basic vector $\mathbf{b}_1$ will differ from $\mathbf{a}_1$ both in magnitude and direction. Since any four vectors are linearly dependent, however, we can write $\mathbf{a}_1$ as a linear function of the new vectors $\mathbf{b}_i$:

$$\mathbf{a}_1 = J_{11}\mathbf{b}_1 + J_{21}\mathbf{b}_2 + J_{31}\mathbf{b}_3 .$$

This relation signifies that the direction indices of the vector $\mathbf{a}_1$ referred to the basis B are $[J_{11} J_{21} J_{31}]$.† The transformation from one basis to the other may thus be represented completely by three equations of the above form, or in the usual summation convention

$$\mathbf{a}_i = J_{ji}\mathbf{b}_j . \qquad (6.1)$$

Note that although written in subscript form, this is a vector equation and not a relation between vector components. When we deal with base vectors, the subscript identifies a particular vector of the set A or B, rather than a particular component. The vector $\mathbf{a}_i$ may be represented in its own basis system A by the column matrix $\mathbf{a}_i$; obviously the kth component of $\mathbf{a}_i$ in this representation is $(a_i)_k = \delta_{ik}$.

If we now write the set of vectors $\mathbf{a}_i$ which constitute the basis A as a column matrix of *vectors*, $\mathsf{A} = [\mathbf{a}_1\mathbf{a}_2\mathbf{a}_3]$, and the set $\mathbf{b}_i$ as a similar matrix B, the three equations may be combined in the matrix equation

$$\mathsf{A} = \mathsf{J}' \,\mathsf{B}. \qquad (6.2)$$

Here J is the 3×3 array with elements J_{ij}, and the transposed matrix J' is the corresponding array with elements J_{ji}. A is obtained by multiplying J and B in accordance with the laws of matrix algebra.

The components of each individual vector of the set $\mathbf{a}_i$ are given by the representations

$$[\text{A};\, \mathbf{a}_i] = [\delta_{1i}\delta_{2i}\delta_{3i}] \quad \text{and} \quad [\text{B};\, \mathbf{a}_i] = [J_{1i}J_{2i}J_{3i}].$$

† We write the direction indices of $\mathbf{a}_1$ in this form, rather than as $[J_{11}\ J_{12}\ J_{13}]$ in order that the matrix array $\mathsf{J} = J_{ij}$ introduced below shall give directly the transformation of vector *components*.

By comparing coefficients, it is readily seen that these quantities are related by the matrix equation

$$[\text{B}; \mathbf{a}_i] = \mathsf{J}[\text{A}; \mathbf{a}_i]. \tag{6.3}$$

We shall now show that this is a particular example of the general equation relating the components ${}^{A}u_i$. ${}^{B}u_i$ of any vector **u**. Writing $\mathbf{u} = {}^{A}u_i\,\mathbf{a}_i = {}^{B}u_i\,\mathbf{b}_i$, and substituting for $\mathbf{a}_i$ from (6.1),

$${}^{B}u_i\,\mathbf{b}_i = {}^{A}u_j\,J_{ji}\mathbf{b}_j = {}^{A}u_j\,J_{ij}\mathbf{b}_i$$

since both *i* and *j* are dummy indices (to be summed) on the right-hand side. The expression is now an identity, and the coefficients of corresponding vectors $\mathbf{b}_i$ on both sides may be equated to give

$${}^{B}u_i = J_{ij}\,{}^{A}u_j \tag{6.4}$$

or in matrix form

$${}^{B}\mathsf{u} = \mathsf{J}\,{}^{A}\mathsf{u}. \tag{6.5}$$

This equation reduces to (6.3) when we put $\mathbf{u} = \mathbf{a}_i$. The matrix J is a representation of the transformation from the basis A to the basis B; its columns are the direction indices of $\mathbf{a}_i$ referred to B. Although this notation is often adequate, it is sometimes necessary to show explicitly that J represents the operation changing the reference basis, and we then use the extended notation

$$\mathsf{J} = (\text{B}\,\mathbf{J}\,\text{A})$$

so that (6.5) may be written in full as

$$[\text{B}; \mathbf{u}] = (\text{B}\,\mathbf{J}\,\text{A})\,[\text{A}; \mathbf{u}]. \tag{6.6}$$

The bold face symbol **J** is to be regarded as an operator which transforms the representation ${}^{A}\mathsf{u}$ into ${}^{B}\mathsf{u}$; equivalently, we may say that **J** is a function of A and B which gives the transforming matrix J.† The fact that eqn. (6.6) represents a change of axes is emphasized by the use of different base symbols on either side of **J**; note also that the base symbols A occur in juxtaposition on the right-hand side.

Provided that the vectors $\mathbf{a}_i$, $\mathbf{b}_i$ are non-coplanar sets (which is a necessary condition for them to define unit cells), the determinant $|\mathsf{J}|$ of the matrix array J does not vanish. The three simultaneous eqn. (6.5) may then be solved for ${}^{A}\mathsf{u}$ in terms of ${}^{B}\mathsf{u}$, giving the reverse transformation

$${}^{A}\mathsf{u} = \mathsf{J}^{-1}\,{}^{B}\mathsf{u}, \tag{6.7}$$

where the *reciprocal* matrix J^{-1} has elements $J^{-1}{}_{ij} = J^{ji}/|\mathsf{J}|$ and J^{ji} means the cofactor of the element J_{ij} of the matrix J. In the fuller matrix notation, the reverse transformation is

$$[\text{A}; \mathbf{u}] = (\text{A}\,\mathbf{J}\,\text{B})\,[\text{B}; \mathbf{u}]$$

so that our notation implies

$$(\text{A}\,\mathbf{J}\,\text{B}) = \mathsf{J}^{-1} = (\text{B}\,\mathbf{J}\,\text{A})^{-1}. \tag{6.8}$$

† An alternative notation sometimes used (e.g. Bullough and Bilby, 1956) is to write (B **J** A) as (B/A). This has the slight disadavantage of not allowing the use of the single symbol J when convenient.

Similarly, the transformation between sets of vectors has an inverse

$$\mathsf{B} = \mathsf{J}'^{-1}\mathsf{A}.$$

If we wish to write eqn. (6.5) in terms of row vectors, we must transpose both sides of the equation. From the rule for transposing matrix products, this gives

$${}^{B}\mathsf{u}' = {}^{A}\mathsf{u}'\,\mathsf{J}',$$

but if we write this in extended notation, it is convenient to have a symbol for the transpose of $(\mathrm{B}\,\mathbf{J}\,\mathrm{A})$ which will preserve the juxtaposition of like base symbols. We thus introduce the notation

$$(\mathrm{B}\,\mathbf{J}\,\mathrm{A})' = (\mathrm{A}\,\mathbf{J}'\,\mathrm{B}) = \mathsf{J}'$$

and the transpose of (6.5) is then

$$(\mathbf{u};\mathrm{B}) = (\mathbf{u};\mathrm{A})\,(\mathrm{A}\,\mathbf{J}'\,\mathrm{B}). \tag{6.9}$$

The new base vectors $\mathbf{b}_i$ are associated with a new reciprocal set of vectors $\mathbf{b}_i^*$ where

$$\mathbf{b}_i\cdot\mathbf{b}_j^* = \delta_{ij}.$$

The square of the length of $\mathbf{u}$ may be written from (5.16) as

$$|\mathbf{u}|^2 = (\mathbf{u};\mathrm{B}^*)\,[\mathrm{B};\mathbf{u}] = (\mathbf{u};\mathrm{B}^*)\,(\mathrm{B}\,\mathbf{J}\,\mathrm{A})\,[\mathrm{A};\mathbf{u}] = (\mathbf{u};\mathrm{A}^*)\,[\mathrm{A};\mathbf{u}]$$

so that

$$(\mathbf{u};\mathrm{A}^*) = (\mathbf{u};\mathrm{B}^*)\,(\mathrm{B}\,\mathbf{J}\,\mathrm{A}). \tag{6.10}$$

However,

$$(\mathbf{u};\mathrm{A}^*) = (\mathbf{u};\mathrm{B}^*)\,(\mathrm{B}^*\,\mathbf{J}'\,\mathrm{A}^*),$$

so that our notation implies

$$(\mathrm{B}^*\,\mathbf{J}'\,\mathrm{A}^*) = (\mathrm{B}\,\mathbf{J}\,\mathrm{A}) \quad \text{and} \quad (\mathrm{A}\,\mathbf{J}'\,\mathrm{B}) = (\mathrm{A}^*\,\mathbf{J}\,\mathrm{B}^*). \tag{6.11}$$

The law for the transformation of vector components referred to the reciprocal bases is thus identical with that for transforming base vectors (eqns. (6.1) and (6.2)); if J is the representation of the change $\mathrm{A} \to \mathrm{B}$, J' is the corresponding representation of the change $\mathrm{B}^* \to \mathrm{A}^*$.

From eqn. (6.10), the reverse transformation may be written

$$[\mathrm{B}^*;\mathbf{u}] = (\mathrm{A}\,\mathbf{J}'\,\mathrm{B})^{-1}\,[\mathrm{A}^*;\mathbf{u}] = (\mathrm{B}\,\mathbf{J}'\,\mathrm{A})\,[\mathrm{A}^*;u]. \tag{6.12}$$

In particular, the quantities ${}^{A}h_i$ must transform like this, since they are the components of $\mathbf{h}$ referred to the basis A^* (see p. 27). The transformation law is, however, usually more conveniently expressed in the form (6.10), since it is generally preferable to write the vector $\mathbf{h}$ representing a plane normal as a row matrix h'. This gives

$$(\mathbf{h};\mathrm{B}^*) = (\mathbf{h};\mathrm{A}^*)\,(\mathrm{A}\,\mathbf{J}\,\mathrm{B}) \tag{6.13}$$

or in component form

$$^{B}h_j = J^{-1}{}_{ij}\,{}^{A}h_i.$$

We also have

$$\mathbf{h} = {}^{B}h_i\,\mathbf{b}_i^* = {}^{A}h_j\,\mathbf{a}_j^* = J_{ij}\,{}^{B}h_i\,\mathbf{a}_j^*,$$

so that

$$\mathbf{b}_i^* = J_{ij}\mathbf{a}_j^*$$

or

$$\mathsf{B}^* = \mathsf{J}\,\mathsf{A}^*, \tag{6.14}$$

which gives the relation between reciprocal vectors, corresponding to (6.2) for lattice vectors.

Finally, it is useful to examine the transformation from a base A to its own reciprocal base A*. Let the components of a vector **u** referred to the two bases be related by

$$\mathsf{u}^* = \mathsf{G}\,\mathsf{u}, \tag{6.15}$$

where $\mathsf{G} = (\text{A}^*\,\mathbf{G}\,\text{A})$ is the matrix representation of the transformation from A to A*. From eqn. (5.15) the scalar product of two vectors may be written

$$\mathbf{u}\cdot\mathbf{v} = u_j^* v_j = G_{ji} u_i v_j,$$

but this product may also be expanded as

$$\mathbf{u}\cdot\mathbf{v} = u_i\mathbf{a}_i\cdot v_j\mathbf{a}_j = u_i v_j \mathbf{a}_i\cdot\mathbf{a}_j$$

so that the elements of the matrix G are

$$G_{ij} = G_{ji} = \mathbf{a}_i\cdot\mathbf{a}_j. \tag{6.16}$$

The function symbol **G** is called the metric,† and is represented by the symmetrical square matrix G. Its importance arises from its fundamental connection with the distance between two points, since the square of the length of a vector is

$$|\mathbf{u}|^2 = \mathsf{u}'\,\mathsf{u}^* = \mathsf{u}'\,\mathsf{G}\,\mathsf{u} = G_{ij}u_i u_j. \tag{6.17}$$

The reverse transformation is

$$\mathsf{u} = \mathsf{G}^{-1}\,\mathsf{u}^*, \tag{6.18}$$

where G^{-1} has components $G^{-1}{}_{ij} = G^{-1}{}_{ji} = \mathbf{a}_i^*\cdot\mathbf{a}_j^*$. The length of **u** may then also be ex pressed as

$$|\mathbf{u}|^2 = \mathsf{u}^{*\prime}\,\mathsf{G}^{-1}\,\mathsf{u}^* = G^{-1}{}_{ij}\,u_i^* u^*.$$

Similarly, eqn. (5.22) gives for the interplanar spacing d of the planes **h**

$$(d)^{-2} = |\mathbf{h}|^2 = G^{-1}{}_{ij}\,h_i h_j. \tag{6.19}$$

† Since the relation between A and A* is defined by A alone, **G** may also be regarded as a second order tensor (see next section) and the matrix $\mathsf{G} = G_{ij}$ then gives the components of this tensor in the system A. **G** is usually called the metric tensor. The function symbol **J**, connecting arbitrary axis systems, is not a tensor.

It follows from a result to be proved later (Section 7, p. 46) that the determinant of the matrix G is equal to the ratio of the volume of the cell defined by $\mathbf{a}_i$ to that defined by $\mathbf{a}_i^*$. Combining this result with eqn. (5.12) shows that

$$|\mathsf{G}|^{1/2} = v_a \quad \text{and similarly} \quad |\mathsf{G}^{-1}|^{1/2} = v_a^*, \tag{6.20}$$

where, as before, v_a, v_a^* are the volumes of the respective unit cells.

Since G is symmetric, $\mathsf{G}' = \mathsf{G}$, or in full

$$(\text{A}\,\mathbf{G}'\,\text{A}^*) = (\text{A}^*\,\mathbf{G}\,\text{A}).$$

We now write the scalar product of $\mathbf{u}$ and $\mathbf{v}$ in the basis A and transform to the basis B using eqns. (6.17) and (6.6)

$$\mathbf{u}\cdot\mathbf{v} = (\mathbf{u};\text{A})(\text{A}^*\,\mathbf{G}\,\text{A})[\text{A};\mathbf{v}] = (\mathbf{u};\text{B})(\text{B}\,\mathbf{J}'\,\text{A})(\text{A}^*\,\mathbf{G}\,\text{A})(\text{A}\,\mathbf{J}\,\text{B})[\text{B};\mathbf{v}]$$

so that

$$(\text{B}^*\,\mathbf{G}\,\text{B}) = (\text{B}\,\mathbf{J}'\,\text{A})(\text{A}^*\,\mathbf{G}\,\text{A})(\text{A}\,\mathbf{J}\,\text{B}), \tag{6.21}$$

which gives the relation between the representations of $\mathbf{G}$ in two different bases. We could also have derived (6.21) by making use of the identity $(\text{B}^*\,\mathbf{G}\,\text{B}) = (\text{B}^*\,\mathbf{J}\,\text{A}^*)(\text{A}^*\,\mathbf{G}\,\text{A})(\text{A}\,\mathbf{J}\,\text{B})$ and substituting for $(\text{B}^*\,\mathbf{J}\,\text{A}^*)$ from (6.11).

Consider now the orthonormal basis I. The basic vectors $\mathbf{i}_i$ then satisfy eqn. (5.2) and $\mathsf{G}^{-1} = \mathsf{G} = \mathsf{I}$, the unit matrix having elements $I_{ij} = \delta_{ij}$. The metric of an orthonormal system is thus unity. This corresponds to the fact that the bases I and I* are identical in such a system.

Examination of the above equations shows that the laws of transformation are of two kinds. Quantities which transform like the vector components u_i or the reciprocal base vectors $\mathbf{a}_i^*$ are called *contravariant* in tensor analysis; those which transform like the base vectors $\mathbf{a}_i$ or the plane indices (reciprocal vector components) h_i are called *covariant*. In addition, there are scalar quantities which are *invariant* with respect to an axis transformation. Thus u_i form the components of a contravariant tensor of the first order (a vector), or more simply, may be described as the contravariant components of the real vector $\mathbf{u}$; u_i^* are the covariant components of $\mathbf{u}$. Now let ${}^Au_i$, ${}^Bu_i$ represent continuous variables along the axes $\mathbf{a}_i$, $\mathbf{b}_i$, so that they define coordinate systems. The linear relation between the coordinates, given by (6.4), may be written

$${}^Bu_i = (\partial^B u_i/\partial^A u_j)\,{}^Au_j, \tag{6.22}$$

where $J_{ij} = \partial^B u_i/\partial^A u_j$, etc. This is the law for the transformation of the contravariant components of $\mathbf{u}$. The corresponding law for the covariant components u_i^* is

$${}^Bu_i^* = (\partial^A u_j/\partial^B u_i)\,{}^Au_j^*. \tag{6.23}$$

In tensor notation, contravariant quantities are distinguished by writing the identifying suffix as a superscript, and there is then no need for a separate notation for the bases A and A*. Thus eqn. (5.1) would be written

$$\mathbf{u} = u^i\mathbf{a}_i, \tag{6.24}$$

where the summation convention applies as before. Note that the superscript i is not a power index.

Covariant quantities are written with the suffix as subscript, so that the covariant components u_i^* are now written simply u_i. From eqn. (5.13) it follows that the covariant components of $\mathbf{u}$ in base A are equal to the orthogonal projections of $\mathbf{u}$ on the axes $\mathbf{a}_i$ (see Fig. 2.1). From (5.12) we also have

$$\mathbf{u} = u_i \mathbf{a}^i, \tag{6.25}$$

where $\mathbf{a}^i$ are the reciprocal vectors, formerly written $\mathbf{a}_i^*$. The covariant components of $\mathbf{u}$ in the base A are thus the contravariant components in the base A*, and vice versa. In general, we may alter any equation of the form (6.24) by lowering the dummy index in one place and raising it in another.

In terms of the new notation, the transformation laws (6.22) and (6.23) become

$$^{B}u^i = (\partial ^{B}u^i/\partial ^{A}u^j)\, ^{A}u^j \tag{6.26}$$

and

$$^{B}u_i = (\partial ^{A}u^j/\partial ^{B}u^i)\, ^{A}u_j = (\partial ^{B}u_i/\partial ^{A}u_j)\, ^{A}u_j. \tag{6.27}$$

In the above discussion, we have been careful to write about the covariant and contravariant components of one vector $\mathbf{u}$, since we wish to emphasize the idea of the vector as the physical entity. In tensor algebra, it is customary to treat the components as separate covariant and contravariant vectors, and this leads to economy of description. The vector $\mathbf{u}$ is sometimes called the real vector.

Anticipating the results of the next section a little, we may also form covariant, contravariant and mixed tensors of the second order, with components T_{ij}, T^{ij}, and T^i_j respectively. These tensors are arrays of nine quantities, each depending on two directions of the coordinate axes, which represent a linear relation between two vectors (e.g. the stress tensor relates the linear force on a surface element and the vector normal to the element). When the vectors are referred to another coordinate system, the quantities in the tensor transform according to laws which are contravariant for both suffices, covariant for both suffices, or mixed, thus:

$$\left.\begin{aligned}
&\text{(Contravariant)} && ^{B}T^{ij} = (\partial ^{B}u^i/\partial ^{A}u^k)(\partial ^{B}u^j/\partial ^{A}u^l)\, ^{A}T^{kl},\\
&\text{(Covariant)} && ^{B}T_{ij} = (\partial ^{A}u^k/\partial ^{B}u^i)(\partial ^{A}u^l/\partial ^{B}u^j)\, ^{A}T_{kl}\\
& && \qquad = (\partial ^{B}u_i/\partial ^{A}u_k)(\partial ^{B}u_j/\partial ^{A}u_l)\, ^{A}T_{kl},\\
&\text{(Mixed)} && ^{B}T^i_j = (\partial ^{B}u^i/\partial ^{A}u^k)(\partial ^{A}u^l/\partial ^{B}u^j)\, ^{A}T^k_l\\
& && \qquad = (\partial ^{B}u^i/\partial ^{A}u^k)(\partial ^{B}u_j/\partial ^{A}u_l)\, ^{A}T^k_l.
\end{aligned}\right\} \tag{6.28}$$

In the remainder of this book we shall not use tensor notation, preferring the matrix representation when it is necessary to distinguish covariance from contravariance. When components are required, they will therefore be written with all suffices as subscripts, except for the standard notation for the cofactor of a matrix element, used previously. Tensor notation is especially powerful in problems where curvilinear coordinates are required, so that the linear relations (6.2) and (6.4) have to be replaced by the more general

$$^{B}u^i = f^i(^{A}u^1, {}^{A}u^2, {}^{A}u^3)$$

so that

$$\mathrm{d}^B u^i = (\partial^B u^i / \partial^A u^j)\, \mathrm{d}^A u^j. \tag{6.29}$$

The general properties of contravariance or covariance are then still defined by eqns. (6.26)–(6.28), but is should be noted that the coordinates u^i themselves no longer transform according to the contravariant law. The transformation of coordinates is only contravariant when the relation (6.29) is linear of type (6.4); this is called an affine transformation (see next section) and the coordinate systems are all Cartesian. In all other cases, it is clearly to some extent arbitrary whether we write the coordinates as u^i or u_i.

It is perhaps of interest to show that the components of the matrix **G** may be regarded as the representation of a tensor. If the coordinates of two neighbouring points in the reference system A are $^A u^i$ and $^A u^i + \mathrm{d}^A u^i$, we know that the separation of the points, dw, is given by

$$(\mathrm{d}w)^2 = {}^A G_{ij}\, \mathrm{d}^A u^i\, \mathrm{d}^A u^j.$$

In the system B, the corresponding equation is given from (6.29) as

$$(\mathrm{d}w)^2 = {}^B G_{ij}\, \mathrm{d}^B u^j\, \mathrm{d}^B u^j = {}^B G_{ij} (\partial^B u^i / \partial^A u^k) (\partial^B u^j / \partial^A u^l)\, \mathrm{d}^A u^k\, \mathrm{d}^A u^l,$$

and since the separation is an invariant,

$$^A G_{ij} = (\partial^B u^i / \partial^A u^k) (\partial^B u^j / \partial^A u^l)\, {}^B G_{ij}. \tag{6.30}$$

Comparison with (6.28) shows that the components of **G** transform according to the law for covariant tensors of the second order, so that we are justified in regarding **G** as a tensor. In effect, **G** relates two vectors such that the contravariant components of the second are the covariant components of the first. In the same way, we can show that the components of the matrix $\mathbf{G}^{-1}$ transform according to the law for contravariant tensors (6.28), and would be written G^{ij} in tensor notation.

Note that if both sets of axes are orthogonal and Cartesian, the covariant and contravariant components coincide, and the two transformation laws are identical. This is why contravariance and covariance are not distinguished in elementary vector analysis. We shall frequently use oblique Cartesian coordinates, but curvilinear coordinates are not needed for most problems in crystal geometry. As already emphasized, we shall now employ matrix rather than tensor notation.

For reference, we give in Table II a summary comparison of the main features of the different notations, and of the standard equations of crystal geometry expressed in these notations. Most of the results in the table have been obtained in the text above, but there is one additional point to note.

Equations (6.17) and (6.19) for the length of a vector and the spacing of a set of planes respectively are valid for any choice of the base vectors $\mathbf{a}_i$, including vectors which do not define a primitive unit cell. However, we may need to know the distance between adjacent lattice points (identity distance) in the direction **u**, or between adjacent lattice planes normal to **h**. It we use components u_i, h_i which are integers with no common factor, these dis-

TABLE II. SUMMARY OF LATTICE GEOMETRY

	Matrix notation	Tensor notation
Base vectors	$\mathsf{A} = \mathbf{a}_i$	$\mathbf{a}_i$
Reciprocal base vectors	$\mathsf{A}^* = \mathbf{a}_i^*$	$\mathbf{a}^i$
Contravariant components of **u**	$\mathsf{u} = [\mathsf{A}; \mathbf{u}]$	u^i
Covariant components of **u**	$\mathsf{u}^* = [\mathsf{A}^*; \mathbf{u}]$	u_i
Covariant components of plane normal, **h**	h' (strictly $\mathsf{h}^{*\prime}$) $= (\mathbf{h}; \mathsf{A}^*)$	h_i
Metric $\mathbf{a}_i \cdot \mathbf{a}_j$	$\mathsf{G} = (\mathsf{A}^* \,\mathbf{G}\, \mathsf{A})$	G_{ij}
Reciprocal metric, $\mathbf{a}^i \cdot \mathbf{a}^j$	$\mathsf{G}^{-1} = (\mathsf{A}\, \mathbf{G}^{-1}\, \mathsf{A}^*)$	G^{ij}
Volume of cell, v_a	$\lvert \mathsf{G} \rvert^{1/2}$	
Volume of reciprocal cell, v_a^*	$\lvert \mathsf{G}^{-1} \rvert^{1/2}$	
Scalar product, $\mathbf{u} \cdot \mathbf{v}$	$\mathsf{u}'\mathsf{v}^* = \mathsf{u}'\mathsf{G}\mathsf{v}$ $= \mathsf{u}^{*\prime}\mathsf{G}^{-1}\mathsf{v}^*$	$G_{ij}u^iv^j = G^{ij}u_iv_j$
Contravariant components of $\mathbf{w} = \mathbf{u} \wedge \mathbf{v}$		$w^i = \lvert \mathbf{G}^{-1} \rvert^{1/2} \varepsilon^{ijk} u_j v_k$ †
Covariant components of $\mathbf{w} = \mathbf{u} \wedge \mathbf{v}$		$w_i = \lvert \mathbf{G} \rvert^{1/2} \varepsilon_{ijk} u^j v^k$ †
Repeat distance along **u** (u^i relatively prime)	$I'(\mathsf{u}'\mathsf{G}\mathsf{u})^{1/2}$	$I'(G_{ij}u^iu^j)^{1/2}$
Interplanar spacing, d of planes with h_i relatively prime	$I(\mathsf{h}'\mathsf{G}^{-1}\mathsf{h})^{-1/2}$	$I(G^{ij}h_ih_j)^{-1/2}$
Cosine of angle between **u** and **v**	$\dfrac{\mathsf{u}'\mathsf{G}\mathsf{v}}{(\mathsf{u}'\mathsf{G}\mathsf{u})^{1/2}(\mathsf{v}'\mathsf{G}\mathsf{v})^{1/2}}$	$\dfrac{G_{ij}u^iv^j}{(G_{ij}u^iu^j)^{1/2}(G_{ij}v^iv^j)^{1/2}}$
Cosine of angle between planes **h** and **k**	$\dfrac{\mathsf{h}'\mathsf{G}^{-1}\mathsf{k}}{(\mathsf{h}'\mathsf{G}^{-1}\mathsf{h})^{1/2}(\mathsf{k}'\mathsf{G}^{-1}\mathsf{k})^{1/2}}$	$\dfrac{G^{ij}h_ik_j}{(G^{ij}h_ih_j)^{1/2}(G^{ij}k_ik_j)^{1/2}}$
Cosine of angle between direction **u** and plane **h**	$\dfrac{\mathsf{h}'\mathsf{u}}{(\mathsf{h}'\mathsf{G}^{-1}\mathsf{h})^{1/2}(\mathsf{u}'\mathsf{G}\mathsf{u})^{1/2}}$	$\dfrac{u^ih^j}{(G_{ij}u^iu^j)^{1/2}(G^{ij}h_ih_j)^{1/2}}$
Zone axis of **h** and **k**	**u** parallel $\mathbf{h} \wedge \mathbf{k}$	$u^i \propto$ to $\varepsilon^{ijk}h_jh_k$
Plane containing **u** and **v**	**h** parallel $\mathbf{u} \wedge \mathbf{v}$	$h_i \propto$ to $\varepsilon_{ijk}{}^j v^k$

† The factors $\lvert \mathbf{G} \rvert^{1/2}$ and $\lvert \mathbf{G}^{-1} \rvert^{1/2}$ are often included in the definition of ε_{ijk} and ε^{ijk} respectively.

tances are given by the same expressions, (6.17) and (6.19), provided the basis A is primitive. As we have already stated, however, it is often convenient to use centred (non-primitive) cells, so that the vectors $u_i\mathbf{a}_i$ with u_i taking all possible integral values do not represent *all* the lattice points. For example, in the body-centred cubic structure with the conventional cubic basis, the lattice points are given by the vectors $\frac{1}{2}u_i\mathbf{a}_i$ with the restriction that the quantities u_i must be all odd or all even integers. This complication is allowed for in Table II by introducing the "cell factors" I, I' into the equations for identity distance and planar spacing. The cell factors for the important structures are given separately in Table III.

TABLE III. CELL FACTORS FOR CENTRED LATTICES

Type of cell	Plane indices	I	Direction indices	I'
Primitive	All h_i	1	All u_i	1
Body-centred	Σh_i even	1	u_i all odd	$\frac{1}{2}$
	Σh_i odd	$\frac{1}{2}$	u_i mixed odd and even	1
Base-centred	h_1+h_2 even	1	u_1+u_2 even	$\frac{1}{2}$
	h_1+h_2 odd	$\frac{1}{2}$	u_1+u_2 odd	1
Face-centred	h_i all odd	1	Σu_i even	$\frac{1}{2}$
	h_i mixed odd and even	$\frac{1}{2}$	Σu_i odd	1

We conclude this section by referring to the well-known Miller–Bravais four-axis system for indexing directions and planes in hexagonal structures. The four-axis convention is described in most elementary textbooks on crystallography; it has the advantage that symmetry-related planes and directions are obtained by a simple permutation of indices, but it is clearly inconvenient when a hexagonal lattice has to be related to a lattice of different symmetry. However, the relations between the direct and reciprocal lattices when a four-axis system is used, together with associated problems of crystal geometry, are not obvious, and we shall briefly describe some aspects of these relations in order to complete our description of lattice geometry.

The hexagonal lattice may be regarded as a series of planar hexagonal nets of edge a stacked vertically above each other with a separation c. The most useful bases of conventional type are defined by the vector sets $\mathbf{a}_1$, $\mathbf{a}_2$, $\mathbf{c}$ and $\mathbf{a}_1$, $\mathbf{a}_1+2\mathbf{a}_2$, $\mathbf{c}$, which describe respectively a primitive cell and a C-centred orthorhombic cell. Various rhombohedral bases may also be used, but for hexagonal lattices their disadvantages outweigh their advantages. The non-conventional four-axis basis consists of the vectors $\mathbf{a}_1$, $\mathbf{a}_2$, $\mathbf{a}_3$, $\mathbf{c}$ and a vector direction $\mathbf{u}$ is represented by

$$\mathbf{u} = u_i\mathbf{a}_i \qquad (i = 1, \ldots, 4), \tag{6.31}$$

where it is a requirement that

$$u_1+u_2+u_3 = 0. \tag{6.32}$$

Rational lattice directions are represented by relatively prime integral values of u_i, and the coordinates of all the lattice points along such a direction through the origin are then given by $I'nu_i$, where n is any integer and I' is equal to $\frac{1}{3}$ when u_1-u_2 is divisible by 3 and otherwise is unity. (I' is analogous to the cell factors of Table II.)

If lattice planes are now defined in terms of their reciprocal intercepts h_i on the axes of the direct basis, we find correspondingly that

$$h_1+h_2+h_3 = 0. \tag{6.33}$$

However, some complications arise when we represent plane normals as directions of the reciprocal lattice. With a four-axis system it is not possible to define a reciprocal basis by means of eqns. (5.11), since no vector can be simultaneously perpendicular to three non-coplanar vectors. Although a suitable four-axis basis for the reciprocal lattice can readily be found, the fact that this basis is not strictly reciprocal to the direct basis has caused much confusion in the literature.

Clearly the reciprocal lattice itself is completely defined by the direct lattice; it may be introduced through eqns. (5.11) applied to a conventional direct basis, or by using the geometrical interpretation discussed at the end of Section 5, which is independent of any choice of basis. The reciprocal lattice is also hexagonal and consists of planar nets of edge $(2/3)^{\frac{1}{2}}a^{-1}$ stacked at a separation of c^{-1}. The hexagonal nets of the direct and reciprocal lattices have parallel normals, but are rotated through 30° relative to each other. The appropriate four-axis reciprocal basis is defined by the vectors

$$\mathbf{a}_i^+ = (\tfrac{2}{3}a^{-2})\mathbf{a}_i \qquad (i = 1, \ldots, 3), \qquad \mathbf{a}_4^+ = c^{-2}\mathbf{a}_4, \tag{6.34}$$

where, following Nicholas and Segall (1970), we have used the notation $\mathbf{a}_i^+$ rather than $\mathbf{a}_i^*$ because eqns. (5.11) are not valid. The normal to the plane h_i is now a vector $\mathbf{h}$ of the reciprocal lattice, where

$$\mathbf{h} = h_i\mathbf{a}_i^+ \qquad (i = 1, \ldots, 4). \tag{6.35}$$

This equation is strictly analogous to (5.19). It also follows that if the direction u_i is contained in the plane h_i, then

$$h_iu_i = 0 \qquad (i = 1, \ldots, 4), \tag{6.36}$$

which is the obvious analogue of the three-axis condition.

It should be noted that the four-axis basis of the direct lattice defines a hexagonal prism, of volume $3^{\frac{3}{2}}\, a^2c/2$, which contains three lattice points; this is why the cell factor I' has to be introduced when repeat distances are calculated. However, the vectors $\mathbf{a}_i^+$ define a hexagonal prism of the reciprocal lattice which contains only one reciprocal lattice point, and the reciprocal spacings of lattice planes are therefore given by $|\mathbf{h}|$, where h_i are relatively prime integers, without the need to introduce a cell factor I.

Since eqns. (5.11) are invalid, the existence of the reciprocal basis (6.34) for which eqns. (6.35) and (6.36) are satisfied, appears almost fortuitous. That this is not so was shown by Frank (1965) and Nicholas and Segall (1970). Frank's approach is to consider a four-dimensional lattice with an orthogonal basis $\boldsymbol{\alpha}_i$ with $|\boldsymbol{\alpha}_i| = \alpha$ for $i = 1, \ldots, 3$ and $|\boldsymbol{\alpha}_4| = \gamma$. Equations (5.11), with a range of 1,...,4, for i, then enable an orthogonal reciprocal basis $\boldsymbol{\alpha}_i^*$ to be defined. The real three-dimensional lattice is obtained either by projecting the four-dimensional lattice along some direction or by sectioning it in a particular "hyperplane". Frank showed that a projection along [1110] gives the hexagonal lattice with parameters $a = (\frac{2}{3})^{1/2}\ \alpha$ and $c = \gamma$. Similarly, a section of the four-dimensional reciprocal lattice in (1110) gives the reciprocal of the real hexagonal lattice. Moreover the four-axis bases of the direct and reciprocal lattices are simply obtained by projecting the orthogonal bases $\boldsymbol{\alpha}_i$ and $\boldsymbol{\alpha}_i^*$. The significance of the above procedure is readily appreciated by lowering the dimensionality by one; it then becomes equivalent to projecting a simple cubic lattice along [111] or sectioning its reciprocal lattice in (111) to give the direct and reciprocal nets, and the three-axis bases in these nets are obtained from the projections of the cubic direct and reciprocal axes.

Nicholas and Segall (1970) have given a more complete description of the general case of a redundent base vector, i.e. the use in an n-dimensional space of a basis $\mathbf{a}_i$ with $i = 1, \ldots n+1$. They show that (6.36) is universally valid (whatever i) provided that h_i are defined by the reciprocal intercepts of a hyperplane on $\mathbf{a}_i$, and also that it is always possible to define a reciprocal basis $\mathbf{a}_i^+$ for which (6.35) is valid. Moreover, this choice is not unique; in the hexagonal lattice, for example, any other basis $\mathbf{a}_i^{\ddagger} = \mathbf{a}_i^+ + \mathbf{d}^+$ $(i = 1, \ldots 3)$, $\mathbf{a}_4^{\ddagger} = \mathbf{a}_4^+$, where $\mathbf{d}^+$ is any vector, will not disturb the validity of (6.35). The real justification for the choice of axes which is customarily made (and hence for the conditions (6.32) and (6.33)) is in order to ensure that cyclically permuted indices represent crystallographically equivalent directions and planes.

The advantage of linking the four-axis systems to the orthogonal vectors of a four-dimensional lattice is that crystallographic formulae can be derived rather simply for the four-dimensional lattice and then transformed to the four-axis representation of a three-

TABLE IIIA. SUMMARY OF LATTICE GEOMETRY WITH MILLER–BRAVAIS INDICES

Base vectors	$\mathbf{a}_i \quad (i = 1,\ldots,4)$
Reciprocal lattice basis	$\mathbf{a}_i^+ \quad (i = 1,\ldots,4)$
Direction vector, **u**	$u_i\mathbf{a}_i$
Plane normal, **h**	$h_i\mathbf{a}_i^+$
Metric of four-space, **G**	$G_{ij}^{-1} = \mathrm{diag}\,\{\tfrac{3}{2}a^2, \tfrac{3}{2}a^2, \tfrac{3}{2}a^2, c^2\}$
Reciprocal metric, $\mathbf{G}^{-1}$	$G_{ij}\,^{1} = \mathrm{diag}\,\{\tfrac{2}{3}a^{-2}, \tfrac{2}{3}a^{-2}, \tfrac{2}{3}a^{-2}, c^{-2}\}$
Volume of hexagonal prism (3 lattice points) defined by $\mathbf{a}_i$	$3^{3/2}\, a^2c/2$
Volume of primitive hexagonal prism defined by $\mathbf{a}_i^+$	$(\tfrac{2}{3})^{\frac{1}{2}}a^{-2}c^{-1}$
Scalar product, $\mathbf{u}\cdot\mathbf{v}$	$G_{ij}u_iv_j$
Repeat distance along **u** (u_i relatively prime)	$I'(G_{ij}u_iu_j)^{1/2}$
"Cell factor"	$I' = \tfrac{1}{3}$ if $u_1 - u_2 = 3n$ $I' = 1$ if $u_1 - u_2 \neq 3n$
Interplanar spacing (h_i relatively prime)	$(G_{ij}^{-1}\, h_ih_j)^{-1/2}$
Cosine of angles between **u** and **v**, **h** and **k**, **u** and **h**	See Table II – the equations are identical with $i, j = 1, \ldots, 4$
Zone axis, **u**, of **h** and **k**	$u_1 \propto h_4(k_2-k_3)-k_4(h_2-h_3)$ $u_2 \propto h_4(k_3-k_1)-k_4(h_3-h_1)$ $u_3 \propto h_4(k_1-k_2)-k_4(h_1-h_2)$ $u_4 \propto -3(h_1k_2-k_1h_2)$
Normal, **h**, to plane containing **u** and **v**	$h_1 \propto u_4(v_2-v_3)-v_4(u_2-u_3)$ $h_2 \propto u_4(v_3-v_1)-v_4(u_3-u_1)$ $h_3 \propto u_4(v_1-v_2)-v_4(u_1-u_2)$ $h_4 \propto -3(u_1v_2-v_1u_2)$

dimensional lattice. Particular equations (not usually derived in this way) are given in various papers (e.g. Otte and Crocker, 1965, 1966; Nicholas, 1966; Neumann, 1966; Okamoto and Thomas, 1968); most of these papers contain mistakes as pointed out by Nicholas (1970). The most important formulae are summarized in Table IIIA, which may be looked upon as a supplement to Table II. By introducing the metric of the four-dimensional space, formulae containing scalar products may be written in vector form, exactly as in Table II. Vector products, however, cannot be expressed quite so simply because the two bases are not properly reciprocal, and we have therefore written out the components of two typical vector products in full. The full tensor notation with contravariance and covariance distinguished by superscript and subscript indices has not been used in Table IIIA, but may readily be derived as in Table II.

Finally, it should perhaps be emphasized that the whole of this decriptions is concerned with the geometry of the hexagonal *lattice*. The hexagonal close-packed (h.c.p.) structure is derived from that lattice by placing two atoms around each lattice point, e.g. in sites 0, 0, 0, 0 and $\frac{1}{3}$, 0, $-\frac{1}{3}$, $\frac{1}{2}$ and their equivalents (see footnote on p. 119).

7. AFFINE TRANSFORMATIONS: HOMOGENEOUS DEFORMATION

Equations (6.5) or (6.7) are commonly interpreted in two ways. In the first of these, used in the last section, the quantities ${}^{A}u_i$, ${}^{B}u_i$ are the components of the same vector **u** referred to two different sets of base vectors. An alternative interpretation is to suppose that we have a fixed reference system $\mathbf{a}_i$, and the equations then represent a physical transformation which changes a vector ${}^{A}\mathbf{u}$ into another vector ${}^{B}\mathbf{u}$, where ${}^{B}u_i$ are the components of ${}^{B}\mathbf{u}$ in the fixed reference system. This second interpretation may be used, for example, to specify the relations between two different crystal lattices, which are in fixed orientations with respect to each other. The possibility of interpreting (6.5) in these two ways is a result of the linear nature of both axis transformations and homogeneous deformations. Nevertheless, some confusion may arise if the equation is freely interpreted in either sense, as is sometimes done, and one advantage of the extended matrix notation is that it completely avoids ambiguity of this kind.

Consider the general linear transformation

$$\mathbf{v} = \mathbf{S}\mathbf{u}, \tag{7.1}$$

where **S** is a physical entity (a tensor) which converts the vector **u** into a new vector **v**. In matrix notation, this equation is written

$$\mathsf{v} = \mathsf{S}\,\mathsf{u} \quad \text{or} \quad [\text{A}; \mathbf{v}] = (\text{A}\,\mathbf{S}\,\text{A})\,[\text{A}; \mathbf{u}]. \tag{7.2}$$

The extended notation emphasizes that all physical quantities, **u**, **v**, and **S** are referred to the basis A, and a clear distinction is obtained between square matrix quantities of the type (A **S** A), which are the representations of a tensor in some particular coordinate system, and those of type (B **S** A), which are the representations of a function operator connecting two coordinate systems.

During a transformation of type (7.1), points which were originally collinear remain collinear, and lines which were originally coplanar remain coplanar. Such a transformation is called affine; it represents a homogeneous deformation of space, or of the crystal lattice.

Equation (7.2) is the matrix representation of a homogeneous deformation referred to the system A. In the basis B, there will be a corresponding representation

$$[\mathrm{B}; \mathbf{v}] = (\mathrm{B}\,\mathbf{S}\,\mathrm{B})\,[\mathrm{B}; \mathbf{u}]. \tag{7.3}$$

Suppose the relation between A and B is given by

$$[\mathrm{B}; \mathbf{u}] = (\mathrm{B}\,\mathbf{J}\,\mathrm{A})\,[\mathrm{A}; \mathbf{u}]$$

so that (7.2) may be written

$$(\mathrm{A}\,\mathbf{J}\,\mathrm{B})\,[\mathrm{B}; \mathbf{v}] = (\mathrm{A}\,\mathbf{S}\,\mathrm{A})\,(\mathrm{A}\,\mathbf{J}\,\mathrm{B})\,[\mathrm{B}; \mathbf{u}]$$

and multiplying both sides by $(\mathrm{B}\,\mathbf{J}\,\mathrm{A})$

$$[\mathrm{B}; \mathbf{v}] = (\mathrm{B}\,\mathbf{J}\,\mathrm{A})\,(\mathrm{A}\,\mathbf{S}\,\mathrm{A})\,(\mathrm{A}\,\mathbf{J}\,\mathrm{B})\,[\mathrm{B}; \mathbf{u}].$$

Comparing this equation with (7.3) we see that

$$(\mathrm{B}\,\mathbf{S}\,\mathrm{B}) = (\mathrm{B}\,\mathbf{J}\,\mathrm{A})\,(\mathrm{A}\,\mathbf{S}\,\mathrm{A})\,(\mathrm{A}\,\mathbf{J}\,\mathrm{B}), \tag{7.4}$$

and this is called the similarity transform of $(\mathrm{A}\,\mathbf{S}\,\mathrm{A})$ into $(\mathrm{B}\,\mathbf{S}\,\mathrm{B})$, both of these matrices being representations of the tensor **S**. In shortened form, the equation is

$$^{B}\mathsf{S} = \mathsf{J}\,^{A}\mathsf{S}\,\mathsf{J}^{-1}.$$

The usefulness of the juxtaposition of the bases in the extended form of the equation should be noted.†

Now consider **u** to be any vector in the plane having normal **h** so that $\mathbf{h}\cdot\mathbf{u} = \mathsf{h}'\,\mathsf{u} = 0$. After deformation, the vector **u** is changed into a vector **v**, and the plane has a new normal **k** where $\mathbf{k}\cdot\mathbf{v} = 0$. Writing the matrix representations of these two equations in the basis A, we have

$$\mathsf{h}'\,\mathsf{u} = \mathsf{k}'\,\mathsf{S}\,\mathsf{u} = 0 \quad \text{or} \quad \mathsf{k}' = \mathsf{h}'\,\mathsf{S}^{-1}. \tag{7.5}$$

This gives the effect of the tensor **S** on the components of vectors normal to planes, and may be written in full as

$$(\mathbf{k}; \mathrm{A}^*) = (\mathbf{h}; \mathrm{A}^*)\,(\mathrm{A}\,\mathbf{S}\,\mathrm{A})^{-1}.$$

Let us now consider the properties of the most general form of homogeneous deformation. The following statements will be taken as self-evident, though formal proofs occur incidentally later in this section. If we imagine a sphere inscribed in the material before deformation, this would be distorted into a triaxial ellipsoid, called the strain ellipsoid. The principal axes of the strain ellipsoid would be mutually perpendicular before deformation, and would thus have suffered no relative change in orientation, although in general

† It will be noted that the components of the matrix S are those of a mixed tensor, and the tensor form of (7.2) is $v^i = S^i_j u^j$.

each would have been rotated from its original position, and changed in length. It follows also that we could have inscribed an ellipsoid in the material before deformation, such that after deformation it became a sphere; this is called the reciprocal strain ellipsoid. Lines in the directions of the axes of the reciprocal strain ellipsoid before deformation lie in the directions of the axes of the strain ellipsoid after deformation; the axes of the reciprocal strain ellipsoid are called the directions of principal strain.

In the most general deformation, all vectors change their length, but there is at least one vector (and generally three) which is unrotated. We may prove this as follows. Suppose for some vector **u** the transformation leaves **v** parallel to **u**. Then the only effect of **S** is to multiply the components of **u** by a constant scalar factor, say λ_i. Then

$$\mathsf{S}\,\mathsf{u} = \lambda_i\,\mathsf{u} \quad \text{or} \quad (\mathsf{S}-\lambda_i\,\mathsf{I}) = \mathsf{u}\,0, \tag{7.6}$$

where I is the unit matrix. This equation has non-trivial solutions ($U_i \neq 0$ for all i) only when

$$|\mathsf{S}-\lambda_i\,\mathsf{I}| = 0. \tag{7.7}$$

Equation (7.7) is a cubic in λ_i, and is called the characteristic equation of the matrix S. If S is real, there are three roots, of which one must necessarily be real; if the matrix S is symmetric, all three must be real. There is thus always one possible solution of (7.6), giving a vector which is unchanged in direction (if λ_i is negative, it is reversed in sign, but this does not correspond to a physically achievable deformation). If all three roots are real, there are three such directions. The values of the roots, which are called the eigenvalues of the matrix, are given by the following equations, obtained by expanding (7.7):

$$\left.\begin{aligned} \lambda_1+\lambda_2+\lambda_3 &= S_{11}+S_{22}+S_{33} = S_{ii}, \\ \lambda_1\lambda_2+\lambda_2\lambda_3+\lambda_3\lambda_1 &= S_{11}S_{22}+S_{22}S_{33}+S_{33}S_{11}-S_{12}S_{21}-S_{23}S_{32}-S_{31}S_{13} \\ \lambda_1\lambda_2\lambda_3 &= |\mathsf{S}|. \end{aligned}\right\} \tag{7.8}$$

It is readily proved that two matrices related by a similarity transformation have the same eigenvalues. Since the eigenvalues of a diagonal matrix are simply its non-zero components, it follows that provided the matrix has three distinct, real eigenvalues, it can always be reduced to diagonal form by a similarity transformation. Physically, this corresponds to an axis transformation to a new set of coordinates lying along the unrotated directions. It is obvious that the matrix representation of **S** in this system will be a diagonal matrix.

Having found the eigenvalues, we can determine an axis transformation which will diagonalize S as follows. Choose the first root, λ_1, and write (7.6) in the form:

$$(\mathsf{S}-\lambda_1\,\mathsf{I})\,\bar{\mathsf{a}}_1 = 0 \quad \text{or} \quad \mathsf{E}\,\bar{\mathsf{a}}_1 = 0, \tag{7.9}$$

where $\bar{\mathsf{a}}_1$ is the matrix representation of one of the unrotated vectors, which we now call $\bar{\mathbf{a}}_1$. Since (7.9) represents three linear homogeneous equations, we cannot determine the components $(\bar{a}_1)_i$ of $\bar{\mathbf{a}}_1$ uniquely, but we may find their ratio. If eqns. (7.9) are written in full, we see that a possible solution is

$$(\bar{a}_1)_i = CE^{ki}, \tag{7.10}$$

where C is an undetermined constant, and k may be 1, 2, or 3. Some of the cofactors E^{ki} may be zero, but there are always sufficient non-zero ones to give a solution for the vector components $(\bar{\mathbf{a}}_1)_i$. We repeat this process with the other two roots, and obtain the components of two other unrotated vectors, $\bar{\mathbf{a}}_2$ and $\bar{\mathbf{a}}_3$. The vectors $\bar{\mathbf{a}}_i$ are called the eigenvectors of **S**, and may be used as a new basis $\bar{\text{A}}$. If the transformation from the new basis to the old is represented by $\mathsf{J} = (\text{A}\,\mathbf{J}\,\bar{\text{A}})$, the matrix J has columns J_{ji} consisting of the components $(\bar{a}_i)_j$ of the eigenvectors $\bar{\mathbf{a}}_i$ (see p. 30). The deformation is represented in the new basis by the matrix $\bar{\mathsf{S}} = (\bar{\text{A}}\,\mathbf{S}\,\bar{\text{A}})$, and from (7.4)

$$\bar{\mathsf{S}} = (\bar{\text{A}}\,\mathbf{J}\,\text{A})(\text{A}\,\mathbf{S}\,\text{A})(\text{A}\,\mathbf{J}\,\bar{\text{A}}) = \mathsf{J}^{-1}\,\mathsf{S}\,\mathsf{J}.$$

From (7.9)

$$(S\,J)_{ik} = S_{ij}J_{jk} = (S_{ij}-\lambda_k\delta_{ij})J_{jk}+\lambda_k\delta_{ij}J_{jk} = \lambda_k J_{ik},$$

so that

$$\mathsf{J}^{-1}\mathsf{S}J = \lambda_k\delta_{jk}. \tag{7.11}$$

The matrix $\bar{\mathsf{S}}$ is thus a diagonal representation of the deformation, as concluded above. The quantities λ_i give the ratios of the lengths of the vectors $\bar{\mathbf{a}}_i$ after the deformation to their lengths before deformation.

We have seen that the components of J are not determined absolutely, since each eigenvector contains an arbitrary constant. The diagonal matrix $\bar{\mathsf{S}}$ represents the strain **S** in all coordinate systems with axes parallel to $\bar{\mathbf{a}}_i$, and the magnitudes of the measure lengths may be chosen arbitrarily. There is also an arbitrary choice of the order of the columns of J, and correspondingly of the elements of $\bar{\mathsf{S}}$, since we may label any root of (7.7) as λ_1. This arises because we are free to label our coordinate axes in the basis $\bar{\text{A}}$ in any way we please, giving six different axis transformations from A to $\bar{\text{A}}$. We shall return to this question later, when discussing the idea of correspondence between directions and planes in different lattices.

When there are three real roots of (7.7), the unrotated directions may be used to specify three unrotated planes. Alternatively, these may be obtained by considering the condition that the vectors **h** and **k** which represent a plane normal before and after deformation are parallel. If h and k are the representations of **h** and **k** as column matrices in the bases A^* (as before) we have

$$\mathsf{k}' = \mathsf{h}'\,\mathsf{S}^{-1} = (1/\lambda_i)\,\mathsf{h}', \tag{7.12}$$

which corresponds to (7.6) and has non-trivial solutions only when (7.7) is satisfied. The vectors obtained by substituting the roots λ_i into (7.12) give the directions of the unrotated plane normals, and are reciprocal to the eigenvectors $\bar{\mathbf{a}}_i$. The quantities λ_i give the ratios of the initial to the final spacings of the unrotated planes.

Note that if any $\lambda_i = 1$, the corresponding eigenvector is an invariant line, i.e. a direction which is both unrotated and undistorted, and there is correspondingly a plane with an invariant normal. If there are two invariant lines, they define an invariant plane, and the basis $\bar{\text{A}}$ is no longer unique. The normal to an invariant plane is necessarily unrotated, but need not itself be invariant.

A special case of homogeneous deformation arises when there are three orthogonal directions which are unrotated by the deformation. The axes of the strain ellipsoid and the reciprocal strain ellipsoid then coincide, and the deformation is said to be a pure strain. The formal definition of a pure strain is that it is a deformation such that the three orthogonal directions which remain orthogonal retain their directions and senses.

Let us now consider a deformation $\mathbf{v} = \mathbf{Pu}$ which is represented in an orthonormal basis I by the equation

$$[\text{I}; \mathbf{v}] = [\text{I}\, \mathbf{P}\, \text{I}]\, [\text{I}; \mathbf{u}]. \tag{7.13}$$

Consider a second orthonormal basis K, related to I by

$$[\text{K}; \mathbf{u}] = (\text{K}\, \mathbf{L}\, \text{I})\, [\text{I}; \mathbf{u}]. \tag{7.14}$$

The matrix $(\text{K}\,\mathbf{L}\,\text{I}) \equiv \mathsf{L}$ represents an axis transformation which is merely equivalent to rotating the basis I into a new position.† The components L_{ij} are thus the cosines of the angles between $\mathbf{k}_i$ and $\mathbf{i}_j$. By writing $\mathbf{i}_j = L_{ij}\mathbf{k}_i$, we obtain

$$\mathbf{i}_j \cdot \mathbf{i}_k = L_{ij} L_{lk} \mathbf{k}_i \cdot \mathbf{k}_l = L_{ij} L_{lk} \delta_{il} = L_{ij} L_{ik}$$

and similarly $\mathbf{k}_j \cdot \mathbf{k}_k = L_{ji} L_{ki}$. Since both these scalar products are equal to δ_{jk}, we obtain the well known relations between the direction cosines of the axes

$$L_{ij} L_{ik} = L_{ji} L_{ki} = \delta_{jk}\,,$$

or in matrix form

$$\mathsf{L}\,\mathsf{L}' = \mathsf{I}, \quad \mathsf{L}' = \mathsf{L}^{-1}, \quad |\mathsf{L}| = \pm 1. \tag{7.15}$$

Such a matrix is called orthogonal. When $|\mathsf{L}| = +1$, the bases I and K both correspond to right-handed (or left-handed) sets of base vectors, and the transformation of axes represented by L is a proper rotation. When $|\mathsf{L}| = -1$, a right-handed set of base vectors is converted into a left-handed set, and vice versa; this is equivalent to a rotation plus a reflection in some plane, and is called an improper rotation. We consider only proper rotations.

If the strain (7.13) is now referred to the basis K, its representation is given by (7.4)

$$(\text{K}\,\mathbf{P}\,\text{K}) = (\text{K}\,\mathbf{L}\,\text{I})(\text{I}\,\mathbf{P}\,\text{I})(\text{I}\,\mathbf{L}\,\text{K}) = (\text{K}\,\mathbf{L}\,\text{I})(\text{I}\,\mathbf{P}\,\text{I})(\text{I}\,\mathbf{L}'\,\text{K}).$$

Hence taking the transpose of both sides

$$(\text{K}\,\mathbf{P}'\,\text{K}) = (\text{K}\,\mathbf{L}\,\text{I})(\text{I}\,\mathbf{P}'\,\text{I})(\text{I}\,\mathbf{L}'\,\text{K}),$$

and it follows that if $(\text{I}\,\mathbf{P}\,\text{I})$ is a symmetric matrix, so also is $(\text{K}\,\mathbf{P}\,\text{K})$. Symmetric matrices thus remain symmetric as a result of an orthogonal transformation.

If the deformation is a pure strain, we may choose the vectors $\mathbf{k}_i$ to lie along the principal axes of this strain (i.e. $\text{K} = \bar{\text{I}}$). The matrix $(\text{K}\,\mathbf{P}\,\text{K}) = \bar{\mathsf{P}}$ is then diagonal, and $(\text{I}\,\mathbf{P}\,\text{I})$ must therefore be symmetrical. A pure strain is thus characterized by a symmetric representation of the tensor $\mathbf{P}$ in any orthonormal basis.

† We use the symbol **L** for axis transformations which are pure rotations, and **J** for more general axis transformations.

A pure deformation is equivalent to simple extensions or contractions along the three principal axes of strain. If the material consists of a rectangular parallelepiped with edges along the principal axes, it will thus remain a rectangular parallelepiped after deformation. The ratio of the new volume to the original volume is then $\lambda_1\lambda_2\lambda_3 = |\mathsf{P}|$, where P is the matrix representation of the strain in any orthonormal basis, and λ_i are the eigenvalues of P.

In simple cases, the pure strain matrix P may be reduced to diagonal form by inspection. Thus if the basis I may be transformed into the basis $\bar{\mathrm{I}}$ by a single rotation about one of the vectors $\mathbf{i}_i$ the components L_{ij} $(i \neq j)$ of the rotation matrix are zero. In the general case, the problem is equivalent to finding the principal axes of a quadric surface. We use the procedure of p. 42, and since P is symmetric, the roots of the equation

$$(\mathsf{P} - \lambda_i \mathsf{I})\,\mathsf{x} = 0, \tag{7.16}$$

are necessarily all real. This equation is identical with an equation known as the discriminating cubic of the geometrical surface with scalar equation

$$\mathsf{x}'\,\mathsf{P}\,\mathsf{x} = P_{ij}x_ix_j = \text{const.} \tag{7.17}$$

As before, the columns of the matrix J which diagonalizes P are multiplied by undetermined constants. However, we wish the basis $\bar{\mathrm{I}}$ to be orthonormal, so we must normalize the eigenvectors by choosing the constants so that eqns. (7.15) are satisfied. In this way, J becomes an orthogonal matrix L which is unique, except for the order of its columns. Three of the six ways in which the columns can be arranged correspond to improper rotations; of the remaining three, the most obvious choice is to label the principal axis which makes the smallest angle with $\mathbf{i}_1$ as $\bar{\mathbf{i}}_1$, etc.

A pure strain is one of the two component deformations into which any homogeneous deformation may be analysed. The other type is a pure rotation, characterized by the condition that all vectors remain the same length. Obviously, a pure rotation is given by a tensor relation $\mathbf{v} = \mathbf{Ru}$, in which the components of the tensor $\mathbf{R}$ form an orthogonal matrix in an orthonormal basis. This follows since if v and u are the vector representations in the orthonormal basis

$$|\mathbf{v}|^2 = \mathsf{v}'\,\mathsf{v} = \mathsf{u}'\,\mathsf{R}'\,\mathsf{R}\,\mathsf{u} = \mathsf{u}'\,\mathsf{u} = |\mathbf{u}|^2,$$

provided R is orthogonal.

Any homogeneous deformation may be regarded as the result of a pure strain combined with a pure rotation. Thus we may write $\mathbf{v} = \mathbf{Su}$ as

$$\mathsf{v} = \mathsf{S}\,\mathsf{u} = \mathsf{P}_1\,\mathsf{R}_1\,\mathsf{u} = \mathsf{R}\,\mathsf{P}\,\mathsf{u},$$

where P_1, P represent pure strains, and R_1, R pure rotations.

Note that $\mathsf{P}_1 \neq \mathsf{P}$ and $\mathsf{R}_2 \neq \mathsf{R}$; there are two ways of resolving the deformation, depending on whether the rotation or the pure strain is considered to occur (mathematically) first. For the present, we find it convenient to use the second resolution, in which:

$$\mathsf{y} = \mathsf{R}\,\mathsf{P}\,\mathsf{x}, \tag{7.18}$$

means that the vector **x** is first given a pure strain **P** and then a pure rotation **R** to form a new vector **y**.

We have already seen that the ratio of the transformed volume to the original volume during the pure strain is given by $|\mathbf{P}|$. Since the rotation cannot change the volume, this quantity is also equal to the volume ratio for the whole deformation **S**. Moreover, since $|\mathbf{R}| = 1$, $|\mathbf{S}| = |\mathbf{P}|$, so that for any affine transformation, the volume ratio is given by $|\mathbf{S}|$. Finally, we note from eqn. (7.4) that for any axis transformation, $|{}^{B}\mathbf{S}| = |{}^{A}\mathbf{S}|$. The volume ratio is thus given by the determinant of *any* matrix representation of **S**.

The geometrical relations involved in the general deformation may be appreciated by reference to Fig. 2.2, which is, however, two-dimensional. As a result of the deformation **S**,

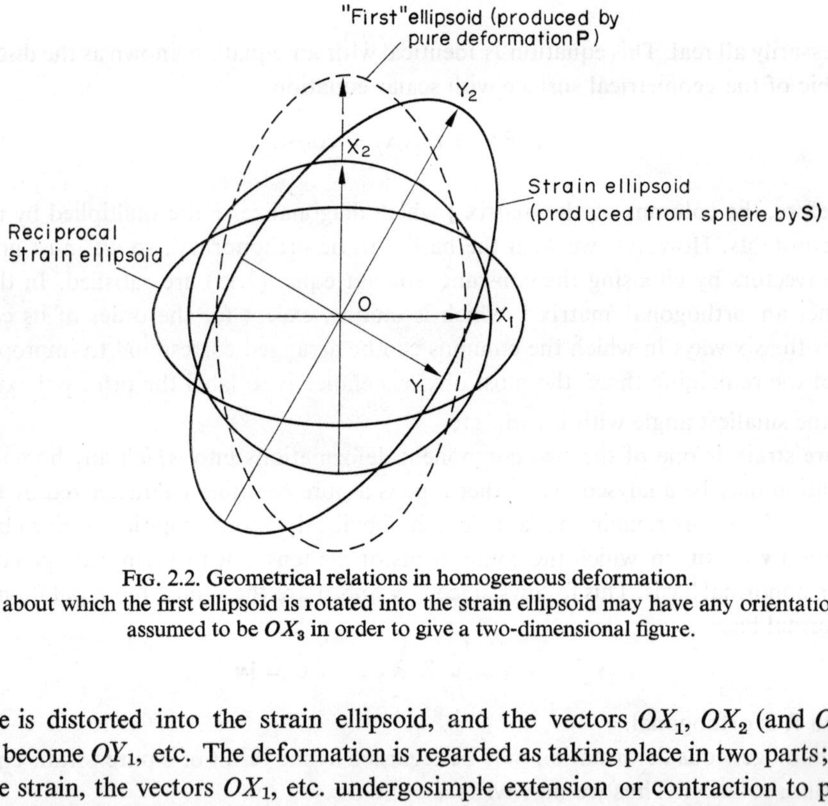

FIG. 2.2. Geometrical relations in homogeneous deformation.
The axis about which the first ellipsoid is rotated into the strain ellipsoid may have any orientation, but is assumed to be OX_3 in order to give a two-dimensional figure.

a sphere is distorted into the strain ellipsoid, and the vectors OX_1, OX_2 (and OX_3 not shown) become OY_1, etc. The deformation is regarded as taking place in two parts; during the pure strain, the vectors OX_1, etc. undergosimple extension or contraction to produce an ellipsoid which is shown dotted. There is no standard name for this figure, but we shall refer to it as the *first* ellipsoid. Following this, there is a rotation (not necessarily in the plane of OX_1 and OX_2) to the positions OY_1, etc.

The reciprocal strain ellipsoid will obviously be changed into a sphere by the pure strain **P**, since **R** produces no change in shape. The axes of this ellipsoid will thus be coincident with the principal axes of strain, as previously assumed. Given the matrix representation of the general deformation, in any set of orthonormal axes, we may resolve into **P** and **R** as follows. Any vector **x** is converted into $\mathbf{y} = \mathbf{S}\,\mathbf{x}$. Now suppose **x** represents a radius vector of the reciprocal strain ellipsoid. Then after deformation, the components y satisfy the equation

$$y_1^2+y_2^2+y_3^2 = \text{const.}$$

or in matrix form

$$\mathsf{y}'\,\mathsf{y} = \text{const.}$$

Substituting for y in terms of the original components x, we obtain the scalar equation of the reciprocal strain ellipsoid

$$\mathsf{x}'\,\mathsf{S}'\,\mathsf{S}\,\mathsf{x} = S_{ki}S_{kj}x_ix_j = \text{const.} \tag{7.19}$$

By expanding S into its components R and P

$$\mathsf{S}'\,\mathsf{S} = \mathsf{P}'\,\mathsf{R}'\,\mathsf{R}\,\mathsf{P} = \mathsf{P}'\,\mathsf{P} = \mathsf{P}^2 \tag{7.20}$$

and this set of equations is sufficient to determine P. The symmetric matrix S′ S has eigenvalues λ_i^2 which are the squares of the eigenvalues of P. The reciprocal strain ellipsoid has the property that the ratio of the length of any deformed vector to its original length is proportional to the inverse radius vector of the ellipsoid drawn in the original direction.

From the eigenvalues of S′ S, we can construct the matrix $\bar{\mathsf{P}}$ which is the diagonal representation of **P** in an orthonormal system along the principal axes, and we can also find the orthogonal transformation $\mathsf{L} = (\bar{\mathrm{I}}\,\mathbf{L}\,\mathrm{I})$ which transforms the basis I into the principal basis $\bar{\mathrm{I}}$. The representations of the components of S in the original basis are then given by

$$\left.\begin{aligned} \mathsf{P} &= \mathsf{L}^{-1}\bar{\mathsf{P}}\,\mathsf{L} \\ \text{and}\qquad \mathsf{R} &= \mathsf{S}\,\mathsf{P}^{-1}. \end{aligned}\right\} \tag{7.21}$$

When referred to principal axes, the whole deformation takes the form

$$\bar{\mathsf{y}} = \bar{\mathsf{S}}\,\bar{\mathsf{x}} = \mathsf{L}\,\mathsf{R}\,\mathsf{L}^{-1}\bar{\mathsf{P}}\,\bar{\mathsf{x}}. \tag{7.22}$$

We can also show that the above procedure gives the principal axes without explicit reference to the reciprocal strain ellipsoid. Suppose we have two vectors, $\mathbf{x}_1$ and $\mathbf{x}_2$, which are converted into two orthogonal vectors, $\mathbf{y}_1$, and $\mathbf{y}_2$. Then in the basis I

$$\mathbf{y}_1\cdot\mathbf{y}_2 = \mathsf{y}_1'\,\mathsf{y}_2 = \mathsf{x}_1'\,\mathsf{S}'\,\mathsf{S}\,\mathsf{x}_2 = 0.$$

Now if the two vectors were perpendicular before the deformation, $\mathsf{x}_1'\,\mathsf{x}_2 = 0$, so that

$$\mathsf{x}_1'\,\mathsf{S}'\,\mathsf{S}\,\mathsf{x}_2 = \lambda_i^2\,\mathsf{x}_1'\,\mathsf{x}_2.$$

where λ_i is a scalar. The vectors $\mathbf{x}_1$ and $\mathbf{x}_2$ are then both solutions of the equation

$$(\mathsf{S}'\,\mathsf{S}-\lambda_i^2\,\mathsf{I})\,\mathsf{x} = 0 \tag{7.23}$$

and for non-trivial solutions

$$|\mathsf{S}'\,\mathsf{S}-\lambda_i^2\,\mathsf{I}| = 0. \tag{7.24}$$

This gives three orthogonal vectors which define the principal axes, and the procedure is equivalent to diagonalizing the matrix S′ S, as described above. Moreover, if the three roots are $\lambda_1^2, \lambda_2^2, \lambda_3^2$, we have

$$|\mathbf{y}_1|^2 = y_1' y_1 = \lambda_1^2 x_1' x_1,$$

so that the principal deformations are λ_i.

The above equations have to be modified when the deformation is expressed in a general basis A. The scalar product of $\mathbf{y}_1$ and $\mathbf{y}_2$ is now written

$$\mathbf{y}_1 \cdot \mathbf{y}_2 = (\mathbf{y}_1;\, \mathrm{A})(\mathrm{A}^* \mathbf{G}\, \mathrm{A})[\mathrm{A};\, \mathbf{y}_2] = (\mathbf{x}_1;\, \mathrm{A})(\mathrm{A}\, \mathbf{S}\, \mathrm{A})'(\mathrm{A}^* \mathbf{G}\, \mathrm{A})(\mathrm{A}\, \mathbf{S}\, \mathrm{A})[\mathrm{A};\, \mathbf{x}_2]$$

so that the general equation corresponding to (7.23) is

$$\{(\mathrm{A}\, \mathbf{S}\, \mathrm{A})'(\mathrm{A}^* \mathbf{G}\, \mathrm{A})(\mathrm{A}\, \mathbf{S}\, \mathrm{A}) - \lambda_i^2(\mathrm{A}^* \mathbf{G}\, \mathrm{A})\}[\mathrm{A};\, \mathbf{x}] = 0, \tag{7.25}$$

and there is an obvious corresponding equation for finding the characteristic roots.

The strain ellipsoid has a scalar equation which may be found by making $|\mathbf{x}|^2$ constant. Thus in the basis I

$$y'(S^{-1})' S^{-1} y = \text{const.} \tag{7.26}$$

This surface has the geometrical interpretation that the ratio of the length of a deformed vector to its original length is proportional to the radius vector of the ellipsoid drawn in the final direction of the vector.

Finally, we note that the surface

$$x'(S - I)x = \text{const}, \tag{7.27}$$

is called the elongation quadric, or in linear elasticity theory (Section 10), the strain quadric. The surface may be either an ellipsoid or a hyperboloid; it has axes in the same directions as those given by eqn. (7.17), and in general these do not coincide with the principal axes of strain. The elongation quadric has the geometrical interpretation that the *extension* of any line, resolved in the original direction of that line, is inversely proportional to the square of the radius vector to the surface, drawn in the original direction of the line.

8. TWIN CRYSTALS

Solid metals are usually composed of a compact mass of separate crystals or grains, joined along arbitrary internal surfaces, and randomly orientated with respect to each other. The orientation relation between any two grains having the same crystal structure may be expressed by a tensor relation representing a pure rotation, and the transformation may always be achieved by a proper rotation, with $|R| = +1$. In crystals of fairly high symmetry, the relation may also be expressed as an improper rotation, if so desired. In certain crystals which possess no centre of symmetry and few planes of symmetry, an improper rotation may produce an atomic arrangement which is not obtainable by a proper rotation. Such arrangements are called optical isomorphs, and show optical activity, i.e. the ability to rotate the plane of polarized light. They do not occur in metals.

The relation between two randomly orientated crystals thus requires three degrees of freedom for its specification, since there are three independent quantities in a rotation matrix. If these crystals meet along a grain boundary surface, two further parameters are needed to specify the orientation of this surface at any point. The general grain boundary thus has five degrees of freedom. The concept of grain boundaries really belongs to the subject of crystal imperfections, and is considered in Chapter VIII. We may usefully consider here, however, the transformation between two orientations which are related in a well specified manner to the symmetry of the structure, so that the crystals are said to be twins of each other. Two crystals in twinned orientation may still be joined along any surface, but there is always some plane which will give a boundary of very low energy, and this composition plane then has no degrees of freedom.

Two crystals are twins of each other when they may be brought into coincidence either by a rotation of 180° about some axis (the twinning axis) or by reflection across some plane (the twinning plane).† The possible orientation relations may be further classified by the relation of the axis of symmetry to the composition plane of low energy along which the twin crystals are usually joined. The rotation axis may be normal to this composition plane (normal twins) or parallel to the composition plane (parallel twins). If the atomic structure is centrosymmetric, it follows that a normal rotation twin is equivalent to a reflection in the composition plane, which is then the twinning plane. For non-centrosymmetric structures, the operations of normal rotation and reflection will produce different twins having the same composition plane. Similarly, a parallel rotation twin is equivalent to a twin produced by a reflection in a plane normal to the rotation axis, and hence to the composition plane, for centrosymmetric structures, but these two operations produce different twins in non-centrosymmetric structures. There are thus four possible types of orientation relation between two twins, reducing to two equivalent pairs for centrosymmetric structures (and, of course, lattices).

Since most metallic structures are centrosymmetric, articles on twinning sometimes refer only to two types of twin orientation. These are then designated as "reflection" twins (equivalent to normal rotation twins) and "rotation" twins (equivalent to reflection in a plane normal to the composition plane). We have presented these results in axiomatic form, but they follow naturally from the condition that the two lattices fit together exactly along the composition plane. We shall emphasize this aspect of twinning in a more complete treatment in Chapter XX.

Obviously, it is not possible to have reflection twins in which the twinning plane is a plane of symmetry in the crystal structure, since the twinning operation then merely reproduces the original orientation. In the same way, the twinning axis in a rotation twin can never be an even axis of symmetry. When two crystals are joined in twin orientation, the twinning plane and/or the twinning axis become pseudo-symmetry elements of the composite structure.

Most metallic crystals form twins which may be regarded both as reflection and normal rotation twins, though there are other types in metals of low symmetry. We could specify the twin relations by the rotation matrix R required to bring the crystals into coincidence

† A more general description of a mechanical twin (Crocker, 1962) is any region of the parent which has undergone a homogeneous shear to give a re-orientated region with the same crystal structure. The above orientation relations are then not necessarily valid; this is discussed in Chapter XX.

with each other, but it is often more useful to employ another type of deformation tensor. In many metals, and some non-metallic crystals, twins may be formed by a physical deformation of the original structure, and this is known as glide twinning. The process is macroscopically equivalent to a homogeneous shear of the original structure, and we therefore use this kind of deformation to describe the twinning law.

Figure 2.3 shows a section through a three-dimensional lattice which has undergone glide twinning. The open circles represent the lattice points in their original positions; the

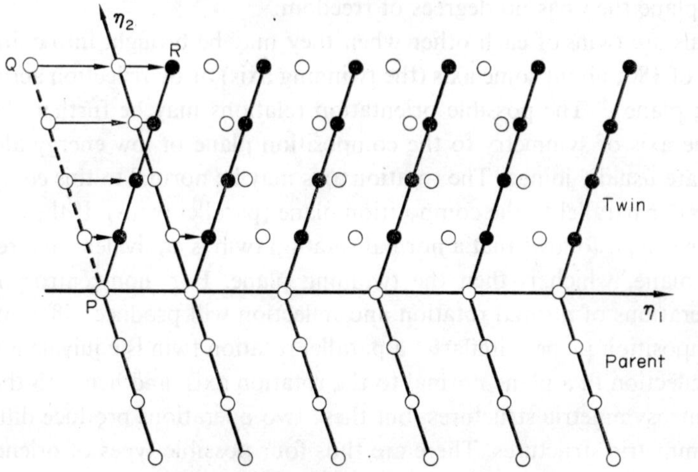

FIG. 2.3. Simple illustration of glide twinning.
PQ, *PR* are corresponding directions in parent and twin.

filled circles are the final positions to which they move. It will be seen that, in the twinned region, each lattice point moves in the same direction through a distance proportional to its distance from the composition plane, which is parallel to the direction of movement. In the simple example shown in the figure, the twinned structure is a reflection of the original structure in the composition plane, and the twinned lattice is obtained by a homogeneous simple shear of the original lattice. The composition, or twinning, plane is conventionally denoted K_1 and the direction of shear η_1. The plane containing η_1 and perpendicular to K_1 (i.e. the plane of the diagram) is called the plane of shear.† Note that each lattice point moves only a fraction of the lattice repeat distance.

In the last paragraph, we emphasized that the two *lattices* in glide twinning are related by a homogeneous shear. The two *structures* are not so related unless the primitive unit cell contains only one atom, i.e. the structure may be obtained by placing an atom on each point of the lattice. More generally, some of the atoms must move in different directions from the lattice points; inhomogeneous movements of this kind are sometimes called atomic "shuffles". The *macroscopic* effect of the deformation is unaffected by the shuffles.

† The K_1 plane is sometimes referred to as the shearing plane. This usage is better avoided, because of the possibility of confusion between "shearing plane" and "plane of shear".

We can thus describe the twin orientation by a matrix representing a simple shear, providing we confine our attention to the lattice and ignore the vectors $\xi_{A,n}$ of eqn. (5.8).

Figure 2.4 shows the section of an original sphere which becomes an ellipsoid after the deformation. The section is in the plane of shear; vectors perpendicular to this plane are unaffected by the deformation, so the problem is essentially two-dimensional. If we use an orthonormal basis, with axes x_i parallel to η_1, perpendicular to the plane of shear and perpendicular to K_1 respectively, we may specify the twin relation as

$$\mathsf{y} = \mathsf{S}\,\mathsf{x},$$

where

$$\mathsf{S} \equiv \begin{pmatrix} 1 & 0 & s \\ 0 & 1 & 0 \\ 0 & 0 & 1 \end{pmatrix} \tag{8.1}$$

and s, the amount of shear, is the distance moved by a lattice point at unit distance from the plane K_1. We note that $|\mathsf{S}| = 1$, so there is no volume change in the transformation, as is physically obvious.

From Fig. 2.4., we see that the original sphere and the strain ellipsoid meet in two circ-

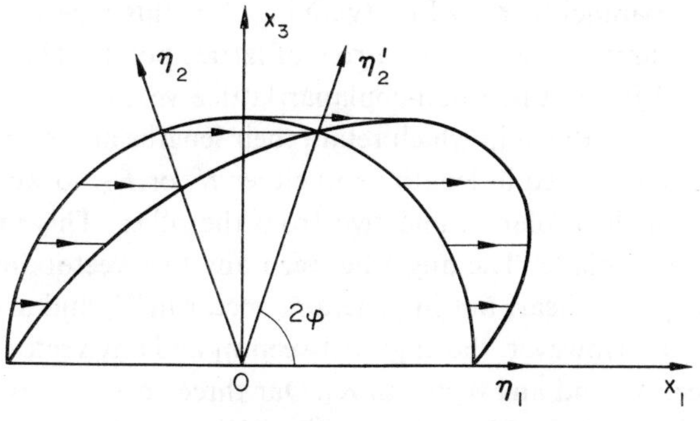

FIG. 2.4. Geometry of glide twinning.
s is the displacement at unit distance along OX_3.

les, which thus represent planes which are undistorted by the transformation. One of these is the composition plane K_1 and this is also unrotated; the second plane is rotated through an angle $\pi - 4\varphi = 2 \text{ arc tan}(s/2)$ where 2φ is the angle between the undistorted planes in their final positions. The second undistorted plane is denoted K_2, and its intersection with the plane of shear is in the direction η_2. The relation between s and φ is

$$s = 2 \cot 2\varphi \tag{8.2}$$

and the equation of the K_2 plane in its initial and final positions is

$$x_1/x_3 = \mp s/2. \tag{8.3}$$

The matrix S is not symmetrical, and the deformation is thus not a pure shear. Using the method of pp. 42–45 or more simply from the geometry of Fig. 2.4, we find that the principal axes of strain are obtained by a right-handed rotation of x_i through an angle φ about x_2. Thus maximum extension takes place in a direction at $\pi/2-\varphi = \arctan[\frac{1}{2}\{s+(s^2+4)^{1/2}\}]$ to η_1, and the change in length is in the ratio $[1+\frac{1}{2}s\{s+(s^2+4)^{1/2}\}]^{1/2} : 1$. Maximum contraction takes place in a direction at an angle $\varphi = \arctan[\frac{1}{2}\{s-(s^2+4)^{1/2}\}]$ to η_1, and the change in length is in the ratio $[1+\frac{1}{2}s\{s-(s^2+4)^{1/2}\}]^{1/2} : 1$.

The simple shear is equivalent to a pure shear, specified by the above extension and contraction, followed by a right-handed rotation of $\pi/2-2\varphi$ about x_2. During the pure shear, the K_1 and K_2 planes rotate through equal angles $\pi/2-2\varphi$ in opposite directions. It is thus clear that if we superimpose a left-handed rotation of this amount on the pure shear, the plane hitherto denoted K_2 will be unrotated in the total deformation. The role of the K_1 and K_2 planes, and also of the η_1 and η_2 directions is thus interchanged; twins so related are called reciprocal or conjugate twins. Two crystals in twin orientation may be transformed into reciprocal twin orientation by a relative rotation of $\pi-4\varphi = 2 \arctan (s/2)$ about the normal to the plane of shear.

The above specification of glide twinning does not yet include the most important condition, namely that the twinning deformation should produce an equivalent lattice to the original lattice. If the twin is a mirror image in the composition plane, it is geometrically obvious that lattice points which are reflections of each other after twinning must have been separated by a vector parallel to η_2 before twinning. For this type of twinning, therefore, η_2 must be a rational direction parallel to a row of lattice points. The lattice structure will obviously be preserved if any three non-coplanar lattice vectors in the parent crystal are transformed into vectors in the twin which retain their lengths and mutual inclinations. All vectors which remain unchanged in length lie in either K_1 or K_2, so we must select one lattice vector from one of these planes, and two from the other. This means that either K_1 or K_2 must be a rational plane. The angle between any two vectors in either K_1 or K_2 is unchanged by the twinning shear, but in general a vector in K_1 and a vector in K_2 change their relative inclination. However, the angle between η_1 and any vector in K_2 is unchanged, as is the angle between η_2 and any vector in K_1. Our three vectors may thus be η_2 and any two lattice vectors in K_1, or η_1 and any two lattice vectors in K_2. In the first case, we have η_2 and K_1 rational, and this is called a twin of the first kind; a twin of the second kind has η_1 and K_2 rational.†

The two possibilities are illustrated in Figs. 2.5 and 2.6. Twins of the first kind are a simple reflection of the parent crystal in the K_1 plane, and the twin *lattice* may equivalently be obtained by a rotation of 180° about the normal to K_1. As mentioned above, these two descriptions of the *structure* are also equivalent if this is centro-symmetric. Twins of the second kind are related by a rotation of 180° about η_1, and for centrosymmetric structures, this is equivalent to a mirror reflection in the plane perpendicular to η_1, as shown in the figure.

† If the more general definition of mechanical twinning mentioned on p. 50 is adopted, twins with all four elements irrational are possible. In practice, such twinning modes probably result only from 'double twinning', that is, from the combination of two twinning operations of the type discussed in the text. More rigorous proofs of the statements in the text are given in Chapter XX.

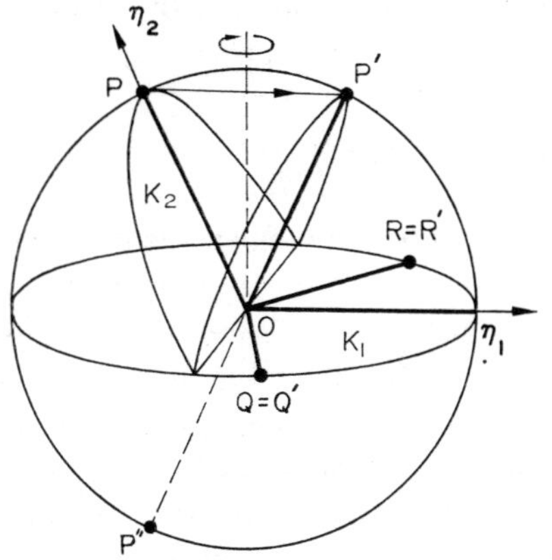

FIG. 2.5. Type I twinning.

P, Q, R represent lattice vectors before twinning. *P′, Q′, R′* represent the corresponding vectors after twinning (after Cahn, 1953).

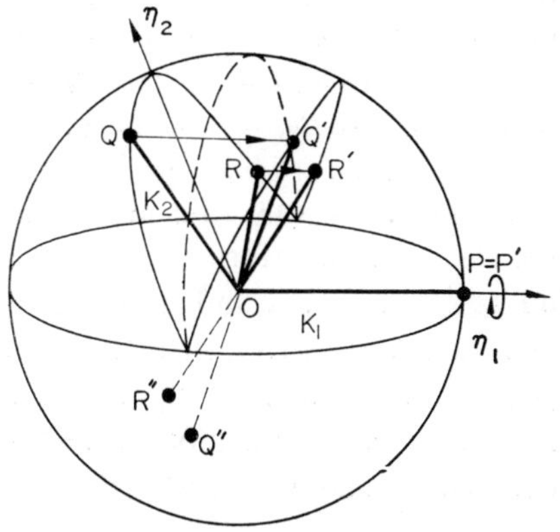

FIG. 2.6. Type II twinning.

P, Q, R represent lattice vectors before twinning. *P′, Q′, R′* represent the corresponding vectors after twinning (after Cahn, 1953).

If the basic vectors used to define the twinning elements are all primitive vectors of the lattice, the homogeneous deformation will produce the twinned lattice, as assumed above. However, it is possible to produce twinned structures even when all these vectors are not primitive. In this case, some of the atomic movements must be inhomogeneous, even in a structure without a basis, since only a fraction of the lattice points move to their twinned positions as the result of the shear. In effect, the macroscopic shear now converts a superlattice of the structure into its twinned orientation.

For any crystal structure, the twinning elements and the amount of shear are completely specified by either η_2 and K_1 or by η_1 and K_2. All four elements may be rational, and we then have compound twins. Most twins are compound in metals of cubic, trigonal and tetragonal symmetry, but both type I and type II twins have been found in orthorhombic α-uranium. In metals of the highest symmetry, K_1 and K_2 and η_1 and η_2 are frequently crystallographically equivalent, so that the twin and its reciprocal represent equivalent twinning modes.

If the deformation is represented in a basis A of the parent crystal by S, we may now use the transformation formulae previously derived to find the new indices of any plane or lattice direction relative to A. If these new indices are $\mathsf{k} = [\mathrm{A}^*; \mathbf{k}]$ and $\mathsf{v} = [\mathrm{A}; \mathbf{v}]$ where **k** and **v** are the new plane normal and lattice vector, we have

$$\mathsf{k}' = \mathsf{h}'\mathsf{S}^{-1} \quad \text{and} \quad \mathsf{v} = \mathsf{S}\,\mathsf{u}.$$

In general, both directions and planes in the twin have irrational indices when referred to a basis in the parent. We then find it useful to introduce the idea of *correspondence* between the two lattices. Refer the vector **v** to a new basis which forms a unit cell in the twin. Since twin and parent have the same structure, it will be natural to choose a unit cell of the same size and shape as that outlined by the vectors $\mathbf{a}_i$; the new basis B will then differ from A only by a rotation (a proper rotation if both sets of base vectors are right-handed). We now have

$$^{B}\mathsf{v} = [\mathrm{B};\ \mathbf{v}] = (\mathrm{B}\ \mathbf{L}\ \mathrm{A})[\mathrm{A};\ \mathbf{v}]$$

and

$$^{B}\mathsf{v} = \mathsf{L}\ {}^{A}\mathsf{v} = \mathsf{L}\,\mathsf{S}\,\mathsf{u}^{A} = \mathsf{C}^{A}\,\mathsf{u}. \tag{8.4}$$

The matrix $\mathsf{C} = (\mathrm{B}\ \mathbf{C}\ \mathrm{A})$ is called the correspondence between the two lattices; it combines the effects of the deformation and the change of basis. Clearly, since $\mathbf{v}$ is rational in B and $\mathbf{u}$ is rational in A, $^{B}\mathsf{v}$ and $^{A}\mathsf{u}$ both have rational components, and the components of C are all rational. Moreover, since the bases A and B both refer to a unit cell of the same volume, $|\mathsf{C}| = \pm 1$. From the result on p. 43, it follows that the columns of $(\mathrm{B}\ \mathbf{C}\ \mathrm{A})$ are the components referred to base B of the vectors which are formed from the base vectors $\mathbf{a}_i$ of A by the transformation $\mathbf{S}$. If we use the transformed base vectors of A as the base vectors of B (symbolically, $\mathrm{B} = \mathrm{CA}$), the correspondence matrix is $(\mathrm{CA}\ \mathbf{C}\ \mathrm{A}) = \mathsf{I}$. We can thus always establish a unitary correspondence between direction indices in the parent and in a suitable basis in the twin. In type I twinning, for example, we could choose the basis A to have $\mathbf{a}_1$ and $\mathbf{a}_2$ in K_1 and $\mathbf{a}_3 = \mathbf{h} - \frac{1}{2}\mathbf{s}$ where $\mathbf{h}$ is the vector normal to K_1 and $\mathbf{s}$ is a vector in K_1 chosen so that $\mathbf{a}_3$ is parallel to η_2. After twinning, the vectors $\mathbf{a}_1$ and $\mathbf{a}_2$ are unchanged and $\mathbf{a}_3$ becomes $\mathbf{h} + \frac{1}{2}\mathbf{s}$. If these three vectors are used as the basis B, then $\mathsf{C} = \mathsf{I}$. However, it is sometimes more convenient to choose bases so that the correspondence matrix is not the unit matrix. When this is done, the transformation of direction indices is given by eqn. (8.4), and the correspondence between plane indies is specified by

$$\left.\begin{aligned}(\mathbf{k};\ \mathrm{B}^*) &= (\mathbf{k};\ \mathrm{A}^*)(\mathrm{A}^*\ \mathbf{L}\ \mathrm{B}^*)\\ &= (\mathbf{h};\ \mathrm{A}^*)(\mathrm{A}\ \mathbf{S}^{-1}\ \mathrm{A})(\mathrm{A}\ \mathbf{L}\ \mathrm{B})\\ &= (\mathbf{h};\ \mathrm{E}^*)(\mathrm{A}\ \mathbf{C}\ \mathrm{B})\end{aligned}\right\} \tag{8.5}$$

using the result of (6.11). This equation may be written more briefly as

$$^{B}\mathsf{k}' = {}^{A}\mathsf{h}'\,\mathsf{C}^{-1}.$$

9. RELATIONS BETWEEN DIFFERENT LATTICES

In deformation twinning, the magnitude of the shear is fixed, once the twinning elements have been specified. An arbitrary shear would produce a new lattice with a different symmetry, but having the same volume. More generally, any space lattice may be converted into any other lattice by a homogeneous deformation $\mathbf{S}$. The deformation determines not only the symmetry and parameters of the new lattice, but also its orientation relative to the original lattice. The orientation of the interphase boundary (i.e. the surface of separation) must be specified separately.

When we have two crystal structures in contact with each other in a fixed relative orientation, it is often convenient to choose a deformation tensor $\mathbf{S}$ to describe the relation between the lattices. We have already emphasized that whilst any two unit cells of the lattices may be connected by a suitable homogeneous deformation, the positions of the atoms cannot necessarily all be described in this way. If the primitive unit cells of the two structures contain different numbers of atoms, the conversion of one structure into another involves a net loss or gain of lattice points. In such cases, it seems sensible to choose a tensor **S** which relates unit cells containing the same number of atoms, and this is essential in transformations where there is a correspondence (see below).

In the case of twinning, we emphasized that shuffles may be produced either because the structure contains more than one atom per unit cell, or because the simple shear **S** does not relate all the lattice points. The same conclusion applies to more general lattice deformations; the smallest unit cells related by **S** need not be primitive cells of either structure, and shuffling is then required to complete the transformation even if both structures have only one atom per lattice point. A distinction is sometimes made between these latter type "lattice shuffles", which are determined by choice of **S**, and the more general "structure shuffles", which arise from the atomic position vectors $\boldsymbol{\xi}$ of (5.8).

For complete generality, the relation between any two lattices should be written in the form

$$\mathbf{v} = \mathbf{t} + \mathbf{S}\mathbf{u}, \tag{9.1}$$

where the tensor **S** specifies the relative sizes and orientations of the unit cells, and the constant vector **t** represents a translation of the lattice points of one crystal relative to those of the other. In general, such translations are not detectable by ordinary crystallographic methods, and are of interest only when the actual atomic positions in two lattices separated by an interface are being considered (Section 36). A special case is when $\mathbf{S} = \mathbf{I}$; the relation then describes a surface defect, known as a stacking fault, in a single lattice (Section 16). For the remainder of this section, we assume $\mathbf{t} = 0$ and consider only the orientation and structural relations specified by **S**.

In any structure, there is an infinite number of operations which will bring the lattice into self-coincidence. Correspondingly, there is an infinite number of deformations **S** which will convert a specified unit cell into another specified unit cell with given orientation relation. It is clear that as long as we wish merely to give a formal statement of the relative positions of the two sets of lattice points, any deformation **S** which gives the desired relation is valid. In some phase transformations, however, the atoms in a region of a product crystal have moved from their original positions in the parent crystal in such a way that the lattice of the parent has been effectively deformed into the lattice of the product. There is then a particular tensor **S** which not only specifies the relations of the lattices, but also the way in which one lattice may change into another. This tensor has usually to be selected by some external physical assumption, e.g. that each point of the original lattice moves to the nearest point of the final lattice. The simplest example of the physical significance of the choice of **S** occurs in twinning. A particular twinning law could be represented by a rotation about a suitable axis by any odd multiple of π, or alternatively by a simple shear deformation. For mechanical twinning, the latter statement of the law is more meaningful, since one lattice is physically sheared into the other.

A deformation which is physically significant implies a one to one correspondence between vectors in the two lattices. Each vector in one lattice may be associated unambiguously with a "corresponding" vector of the other lattice into which it is converted by the transformation. We summarize these relationships by means of a correspondence matrix, as already used for twin crystals.

Suppose we use bases A and B in the two lattices which we call α and β respectively; the unit cells defined by A and B need not contain the same numbers of atoms. The relation between vectors in the two lattices is given by

$$[\mathrm{A}; \mathbf{v}] = (\mathrm{A}\,\mathbf{S}\,\mathrm{A})\,[\mathrm{A}; \mathbf{u}],$$

and the relation between the two bases is

$$[\mathrm{B}; \mathbf{u}] = (\mathrm{B}\,\mathbf{J}\,\mathrm{A})\,[\mathrm{A}; \mathbf{u}].$$

Combining these two equations,

$$[\mathrm{B}; \mathbf{v}] = (\mathrm{B}\,\mathbf{J}\,\mathrm{A})(\mathrm{A}\,\mathbf{S}\,\mathrm{A})\,[\mathrm{A}; \mathbf{u}] = (\mathrm{B}\,\mathbf{C}\,\mathrm{A})\,[\mathrm{A}; \mathbf{u}], \tag{9.2}$$

or briefly

$$^{B}\mathsf{v} = \mathsf{C}\,^{A}\mathsf{u}.$$

As on p. 54, the columns of the correspondence matrix $\mathsf{C} = (\mathrm{B}\,\mathbf{C}\,\mathrm{A})$ are the components referred to basis B of the vectors which are formed from the base vectors $\mathbf{a}_i$ by the deformation **S**. If these transformed vectors are used to define the new unit cell (B = CA), the correspondence matrix is I. However, this will not usually happen if B and A are derived from conventional unit cells in the two structures.

On p. 46 we showed that the determinant of $(\mathrm{A}\,\mathbf{S}\,\mathrm{A})$ gives the ratio of the volume of the β structure to that of the α structure. If the bases A and B contain the same number of atoms, the determinant of $(\mathrm{B}\,\mathbf{J}\,\mathrm{A})$ will give the ratio of the volume per atom in the α structure to that in the β structure, and hence the determinant of $(\mathrm{B}\,\mathbf{C}\,\mathrm{A})$ will be unity.† Correspondingly, if there are different numbers of atoms in the two bases (as may often happen if $\mathbf{a}_i$ and $\mathbf{b}_i$ define primitive or conventional unit cells), $|\mathsf{C}|$ will equal the ratio of the number of atoms in the unit cell defined by A to the number in the unit cell defined by B. The elements of $(\mathrm{B}\,\mathbf{C}\,\mathrm{A})$ will all be rational; that is, they are small integers or fractions.

The reverse transformation is clearly specified by

$$[\mathrm{A}; \mathbf{u}] = (\mathrm{A}\,\mathbf{C}\,\mathrm{B})\,(\mathrm{B}; \mathbf{v}),$$

where $(\mathrm{A}\,\mathbf{C}\,\mathrm{B})$ is the reciprocal matrix $(\mathrm{B}\,\mathbf{C}\,\mathrm{A})^{-1}$.

The correspondence of directions also implies a one to one correspondence of lattice planes, since from eqns. (7.5) and (6.10)

$$(\mathbf{k}; \mathrm{A}^*) = (\mathbf{h}; \mathrm{A}^*)\,(\mathrm{A}\,\mathbf{S}\,\mathrm{A})^{-1},$$

and

$$(\mathbf{k}; \mathrm{B}^*) = (\mathbf{h}; \mathrm{A}^*)\,(\mathrm{A}\,\mathbf{S}\,\mathrm{A})^{-1}(\mathrm{A}\,\mathbf{J}\,\mathrm{B}) = (\mathbf{h}; \mathrm{A}^*)\,(\mathrm{A}\,\mathbf{C}\,\mathrm{B}), \tag{9.3}$$

or briefly

$$\mathsf{k}' = \mathsf{h}'\,\mathsf{C}^{-1}.$$

† The determinant $|\mathsf{C}|$ is ± 1 depending on whether or not both bases are defined by equal-handed sets of vectors.

In simple shear, the lattice points move parallel to the K_1 plane through distances which are proportional to their separation from this plane. This is an example of plane strain in which all displacements are coplanar. A more general type of homogeneous plane strain is shown in Fig. 2.7. Once again all lattice points move in the same direction through distances which are proportional to their separation from a fixed plane, but the direction of movement is no longer parallel to this plane. There are again two undistorted planes; one is also unrotated, and corresponds to the K_1 plane in mechanical twinning. The whole

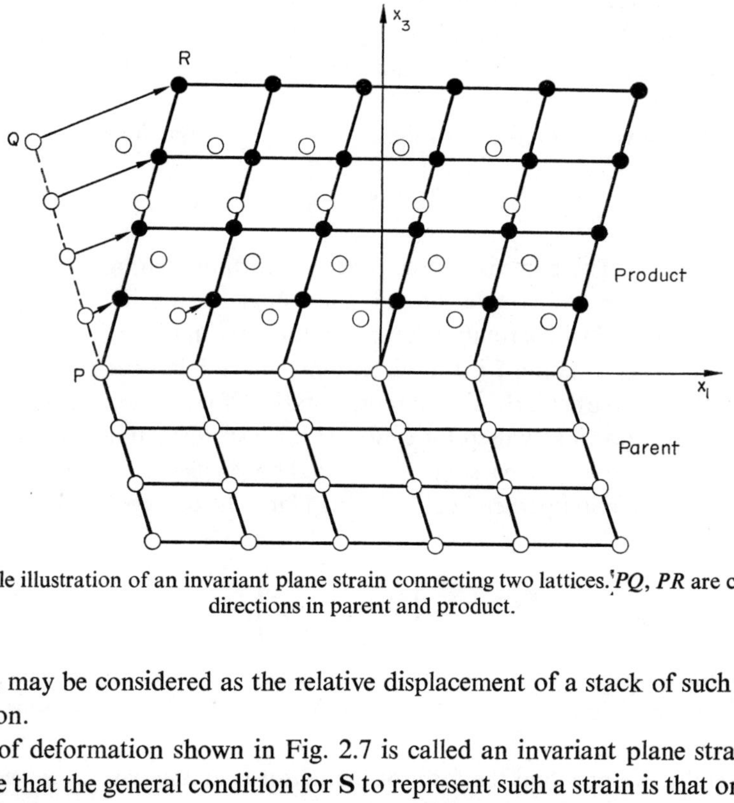

FIG. 2.7. Simple illustration of an invariant plane strain connecting two lattices. *PQ*, *PR* are corresponding directions in parent and product.

deformation may be considered as the relative displacement of a stack of such planes in a fixed direction.

The type of deformation shown in Fig. 2.7 is called an invariant plane strain. We can readily prove that the general condition for **S** to represent such a strain is that one principal strain be zero, and the other two have opposite signs. Thus in an orthonormal basis in which the deformation is $\mathsf{y} = \mathsf{S}\,\mathsf{x} = \mathsf{R}\,\mathsf{P}\,\mathsf{x}$, we refer the representations to the principal axes of strain (basis $\bar{\imath}$) and obtain (see eqn. (7.22))

$$\bar{\mathsf{y}} = \mathsf{L}\,\mathsf{R}\,\mathsf{L}\,\bar{\mathsf{P}}\,\bar{\mathsf{x}},$$

where $\bar{\mathsf{P}} = (\lambda_i \delta_{ij})$ is a diagonal matrix. A vector is unchanged in length if

$$\bar{\mathsf{y}}'\,\bar{\mathsf{y}} = \bar{\mathsf{x}}'\,\bar{\mathsf{P}}'\,\bar{\mathsf{P}}\,\bar{\mathsf{x}} = \bar{\mathsf{x}}'\,\bar{\mathsf{x}},$$

since L and R are both orthogonal. In scalar form, this equation is

$$(\lambda_i^2 - 1)\bar{x}_i^2 = 0, \tag{9.4}$$

and gives the locus of the surface containing lines of unchanging length. If there are two such lines which are coplanar, they define an undistorted plane. By considering the intersection of (9.4) with the plane $\bar{x}_k = 0$, we obtain the equations

$$\frac{\lambda_i^2-1}{\lambda_j^2-1} = -\frac{\bar{x}_j^2}{\bar{x}_i^2}. \tag{9.5}$$

For a real solution, each λ_i^2-1 must differ in sign from each of the other two, because $\bar{x}_i^2$ are necessarily positive. This is only possible if one of the λ_i^2-1 is zero, and the other two have opposite signs. Note that in contrast to this, the only condition for a plane strain is that one λ_i should equal 1. The invariant plane strain is thus not the most general kind of plane strain.

Equation (9.5) now represents a straight line which is the intersection of the plane surface

$$(\lambda_1^2-1)\bar{x}_1^2+(\lambda_3^2-1)\bar{x}_3^2 = 0, \quad \bar{x}_2 = \text{const}, \tag{9.6}$$

with the plane $\bar{x}_2 = 0$. This plane is a plane of zero distortion. During the pure deformation **P**, the undistorted plane will rotate through some angle, so an invariant plane strain is obtained by combining **P** with a rotation **R** which returns the undistorted plane to its original position. The total deformation **S** is thus determined, once **P** has been specified.

The tensor **S** has a particularly simple representation if we choose axes x_i in Fig. 2.7 to correspond to those previously used for simple shear in deformation twinning. The whole deformation is then equivalent to a simple shear on the x_1x_2 plane, combined with a uniaxial expansion or contradiction perpendicular to this plane. In this system the deformation is thus given by

$$\mathsf{S} = \begin{pmatrix} 1 & 0 & s \\ 0 & 1 & 0 \\ 0 & 0 & 1+\Delta \end{pmatrix}, \tag{9.7}$$

where s and Δ specify the shear and the expansion, and $1+\Delta$ is the volume ratio of the transformation.

More generally, we now find the form of the tensor **S** in any basis A. A vector **u** in the undistorted plane must satisfy the relation

$$\mathsf{S}\,\mathsf{u} = \mathsf{u}. \tag{9.8}$$

The normal to the undistorted plane is $\boldsymbol{\nu}$ where $\boldsymbol{\nu}'\,\mathsf{u} = 0$, and after deformation the normal is given by

$$\boldsymbol{\nu}'\,\mathsf{S}^{-1} = \{1/(1+\Delta)\}\boldsymbol{\nu}', \tag{9.9}$$

where Δ is a scalar.

Quite generally, we may write $\mathsf{S} = \mathsf{I}+\mathsf{S}^\circ$, and since $\mathsf{S}^\circ\,\mathsf{u} = 0$ it follows that S° must have the form $\mathsf{e}\,\boldsymbol{\nu}'$, where e is any constant vector. The above relations are thus satisfied if

$$\mathsf{S} = \mathsf{I}+\mathsf{e}\,\boldsymbol{\nu}' \tag{9.10}$$

and this is the general representation of the invariant plane strain on the plane with normal $\mathbf{v}$. In suffix form

$$S_{ij} = \delta_{ij} + e_i \nu_j. \tag{9.11}$$

We also see from (9.9) that the scalar product $\mathsf{v}'\,\mathsf{e} = \Delta$. By expansions of $|\mathsf{S}|$, the volume ratio of the new phase to the old is found to be $1+\Delta$; Δ is called the cubical dilatation, or simply the dilatation.

The reciprocal matrix to S is found similarly to be

$$\mathsf{S}^{-1} = \mathsf{I} - \frac{\mathsf{e}\,\mathsf{v}'}{1+\Delta}. \tag{9.12}$$

Any point with coordinates given by the position vector **u** moves to a position **Su**, and its displacement is thus

$$\mathsf{S}\,\mathsf{u} - \mathsf{u} = \mathsf{e}\,\mathsf{v}'\,\mathsf{u}, \tag{9.13}$$

which is always in the direction **e**. The amount of displacement is proportional to the perpendicular distance from the plane $\mathsf{v}'\,\mathsf{u} = 0$.

Similarly, any normal **h** becomes a vector **k** where $\mathsf{k}' = \mathsf{h}'\,\mathsf{S}^{-1}$, and the displacement of the end point of this vector is thus

$$\mathsf{h}'\left(\mathsf{I} - \frac{\mathsf{e}\,\mathsf{v}'}{1+\Delta}\right) - \mathsf{h}' = -\frac{\mathsf{h}'\,\mathsf{e}\,\mathsf{v}'}{1+\Delta}. \tag{9.14}$$

This is always in the direction $\mathbf{v}$, and hence each plane normal rotates in a plane containing $\mathbf{v}$.

Crystals of a new phase frequently form inside an existing solid phase in the form of flat plates. In nucleation and growth reactions, this shape is usually adopted to minimize the elastic strain energy due to the volume change in the transformation, and the boundary surface need not have special significance in any representation of the orientation relations. In martensitic transformations, however, we have emphasized that some choice of **S** has physical significance in defining the atom movements. By analogy with the situation in mechanical twinning, we might expect that when the correct choice is made, S has the form (9.10), and the plane $\mathbf{v}$ specifies the interfacial boundary or habit plane of the martensite crystals. However, although there are an infinite number of ways in which two lattices may be related by a tensor S, it is seldom possible to find a representation of type (9.10). This is because the lattice parameters of the two structures are determined mainly by short range interactions, and the condition (9.5) will be satisfied for two different structures only coincidentally. In fact, similar planes of almost identical atomic arrangement do occur in certain transformations between the closely related f.c.c. and h.c.p. structures and the interface plane is then rational in both lattices. Apart from these isolated examples, however, the interpretation of the martensite habit plane cannot be as simple as the interpretation of the K_1 plane in mechanical twinning.

As previously described, a general homogeneous deformation is characterized by at least one and possibly three unrotated planes. Jaswon and Wheeler (1948) pointed out that if the martensite lattice is produced from the parent lattice by a homogeneous deformation, the associated disturbance of the untransformed parent phase in the region round a martensite crystal will be very large, unless the boundary between the phases is unrotated. They therefore suggested that the habit plane is an unrotated plane, and should be found by diagonalizing S in accordance with the procedure on p. 42. However, these unrotated planes are distorted, and the rotation of rows of atoms within them would equally give rise to an extremely large strain energy.

In considering martensitic transformation, we are thus faced with the difficulty that finite homogeneous deformation of a macroscopic region of the parent crystal is only feasible if the boundary between the deformed and undeformed regions at any stage at least approximates to an invariant plane of the deformation. This requirement can be reconciled with the impossibility of relating the lattices by a homogeneous deformation of this kind only by assuming that a martensite plate is not produced physically by a *homogeneous* deformation of the parent lattice. It is now generally accepted that the atom movements during transformation are such that the deformation of the parent lattice into the martensite lattice is homogeneous only over a localized region. An adjacent region of the plate is formed by a different physical deformation, which also, of course, generates the martensite structure. Each of these deformations may be factorized into two components, of which one is common to both regions, and the other is opposite in the two regions, in the sense that the combined effect of these components is to produce no net change in shape or volume. The whole transformation thus consists of a homogeneous deformation, together with deformations which are locally homogeneous, but which have zero macroscopic effect.

The geometry and crystallography of martensite transformations is considered in detail in Part II, Chapter 22; the subject is introduced here only in order to show the importance of analysing lattice deformations **S** into component deformations. From the above description, we see that the component of the lattice deformation common to all regions of a plate must approximate closely to an invariant plane strain. The component with zero macroscopic effect must be such that there is no dilatation (i.e. the determinant of its matrix representation must be unity), since any volume change has to be identical in all regions, and would thus accumulate to produce a macroscopic effect. It is generally assumed that this component is also a plane strain, that is, in this case, a simple shear. The description of this component depends on the order in which the separate deformations are considered to be applied, and this emphasizes the purely mathematical character of the factorization. The terms "first" and "second" strain, which are often used for convenience, have no physical significance, and this is true of all factorizations of this type.

Two invariant plane strains applied successively will not give a resultant deformation with an invariant plane unless the plane $\mathbf{v}$ or the direction $\mathbf{e}$ (eqn. (9.10)) is common to both components. This may be seen by writing the components D, T in subscript form

$$D_{ik} = \delta_{ik} + (e_1)_i (\nu_1)_k, \qquad T_{kj} = \delta_{kj} + (e_2)_k (\nu_2)_j.$$

The product $S_{ij} = D_{ik}T_{kj}$ can only be written in the form $\delta_{ij} + (e_3)_i (\nu_3)_j$ if either $\mathbf{e}_1 = \mathbf{e}_2 = \mathbf{e}_3$ or $\mathbf{v}_1 = \mathbf{v}_2 = \mathbf{v}_3$. There must, however, always be an invariant line in such a transformation, since one line is common to the two invariant planes. Any deformation tensor which is to be factorized into two invariant plane strains must thus be an invariant line strain. For any deformation, there is at least one unrotated line, but this need not be unchanged in length, and an invariant line strain is not the most general form of homogeneous deformation.

The condition for two lattices to be related by an invariant line strain is simply that one principal strain either be zero, or have opposite sign from the other two. This follows immediately from eqn. (9.5) since we then have for any point on an unextended line

$$(\overline{x_1})^2 = -\frac{(\lambda_2^2-1)(\overline{x_2})^2-(\lambda_3^2-1)(\overline{x_3})^2}{(\lambda_1^2-1)}, \tag{9.15}$$

and if the two principal strains λ_2-1, λ_3-1 have the same sign, λ_1-1 must have the opposite sign. Provided this condition is satisfied, eqn. (9.15) represents a cone of directions of unchanging length. By addition of a suitable rotation, any line in the cone may be returned to its original position, so that it is an invariant line of the whole deformation. The rotation only affects the orientations of the two lattices, so that the condition above is sufficient to ensure that two lattices can be related by an invariant line strain.

In discussing invariant plane and line strains, we have imposed no restrictions on the nature of these planes or lines, since (in contrast to the situation in twinning) we make no general assumptions about the relations of the lattice symmetries or constants. It follows that the invariant planes may be irrational.

10. INFINITESIMAL DEFORMATIONS

The deformations considered previously in this chapter have been homogeneous and finite. In this section we briefly consider the relation of the results to the theory of linear elasticity, which deals with infinitesimal inhomogeneous displacements of the atoms. Let us first consider an affine transformation. Throughout this section, we shall use an orthogonal basis i in which the representation of a general position vector is x. Since we shall not be concerned with the crystal lattice, we use the symbol $\mathbf{x}$ rather than $\mathbf{u}$ for the vector itself, and the deformation

$$\mathbf{y} = \mathbf{S}\mathbf{x}$$

is represented by the matrix equation

$$\mathsf{y} = \mathsf{S}\,\mathsf{x},$$

in the basis i. As a result of the deformation, a point with coordinates x_i moves to a position with coordinates $y_i = S_{ij}x_j$. We define the displacement vector $\mathbf{w}$ of this point as the vector joining its initial and final positions. The displacement vector thus has components $w_i = y_i - x_i$, and is given by the equation

$$\left.\begin{aligned} \mathbf{w} &= \mathbf{S}\mathbf{x}-\mathbf{x}, \\ \text{i.e.} \quad \mathsf{w} &= (\mathsf{S}-\mathsf{I})\,\mathsf{x} \\ \text{or} \quad w_i &= (S_{ij}-\delta_{ij})x_j. \end{aligned}\right\} \tag{10.1}$$

Now consider a further small deformation with matrix representation $\mathsf{z} = \mathsf{T}\,\mathsf{y}$. The total deformation is thus

$$\mathsf{z} = \mathsf{T}\,\mathsf{S}\,\mathsf{x}$$

or in suffix form,

$$z_i = T_{ik}S_{kj}x_j = [(T_{ik}-\delta_{ik})(S_{kj}-\delta_{kj})+(S_{ij}+T_{ij})-\delta_{ij}]x_j. \tag{10.2}$$

Now suppose the deformations **S** and **T** are both infinitesimal, in the sense that the elements of the displacement vectors $(\mathsf{S}-\mathsf{I})\,\mathsf{x}$, $(\mathsf{T}-\mathsf{I})\,\mathsf{y}$ are so small that the product of any term in the matrix $(\mathsf{S}-\mathsf{I})$ with any other term, or with any term in the matrix $(\mathsf{T}-\mathsf{I})$, may be neglected. This means that the first term in the square brackets of eqn. (10.2) is zero. The total displacement is thus given by

$$w_i = z_i - x_i = (S_{ij}+T_{ij}-2\delta_{ij})x_j$$

or, in matrix form,

$$\mathsf{w} = (\mathsf{S}+\mathsf{T}-2\,\mathsf{I})\,\mathsf{x}. \tag{10.3}$$

The components of the displacement vector for two successive strains may thus be obtained by adding the components of the vectors for the separate strains. This result is obviously only valid for small displacements, and is known as the principle of superposition. When it is applicable, the resultant displacement is independent of the order of the strains. Conversely, we may factorize any infinitesimal deformation into two separate and independent components.

In principle, the infinitesimal affine deformation applied to any lattice will produce a new lattice. However, when the displacements are small enough for the principle of superposition to be applied, each lattice point in the deformed structure can be associated clearly and unambiguously with its original position. It is then more useful to regard the deformed structure as a slightly imperfect (strained) version of the original lattice, rather than as a new lattice. This point of view is, moreover, essential when later in this section we consider inhomogeneous displacements. In their deformed positions, the lattice points then no longer constitute a space lattice in our former mathematical use of the term. They may still be regarded as forming a slightly imperfect lattice of the original type.

Consider the equation

$$\mathsf{w} = (\mathsf{S}-\mathsf{I})\,\mathsf{x} = \mathsf{Q}\,\mathsf{x}, \tag{10.4}$$

where the components of Q are all infinitesimal. In general, Q will not be symmetric, since S need not represent a pure deformation. However, we may always write

$$\mathsf{Q} = \mathsf{e}+\boldsymbol{\omega},$$

where

$$\left.\begin{aligned} e_{ij} &= \tfrac{1}{2}(Q_{ij}+Q_{ji}),\\ \omega_{ij} &= \tfrac{1}{2}(Q_{ij}-Q_{ji}). \end{aligned}\right\} \tag{10.5}$$

The components e_{ij} form a symmetric matrix e which is a representation of the strain tensor, whilst the components ω_{ij} form an antisymmetric matrix $\boldsymbol{\omega}$. We shall now show that for infinitesimal deformations, an antisymmetric matrix is the representation of a rigid-body rotation.

The condition that the length of a vector should remain unchanged as the result of a deformation R was shown on p. 45 to be

$$\mathsf{x}'\,\mathsf{R}'\,\mathsf{R}\,\mathsf{x} = \mathsf{x}'\,\mathsf{x},$$

i.e.

$$R_{ki}R_{kj}x_i x_j = \delta_{ij}x_i x_j .$$

Expanding $R_{ki}R_{kj}$, this gives

$$\delta_{ij} = (R_{ki}-\delta_{ki})(R_{kj}-\delta_{kj})+R_{ji}+R_{ij}-\delta_{ij}. \tag{10.6}$$

Our previous assumption shows that the first term on the right may be neglected, so that

$$R_{ij}+R_{ji} = 2\delta_{ij}. \tag{10.7}$$

If the matrix R satisfies this condition, the displacement vector is given by

$$\mathsf{w} = (\mathsf{R}-\mathsf{I})\,\mathsf{x},$$

where the matrix $\mathsf{R}-\mathsf{I}$ is antisymmetric. An antisymmetric displacement matrix thus represents an equal rotation of all vectors about the origin of coordinates, i.e. a rigid-body rotation.

From the Principle of Superposition, the general displacement vector may now be analysed into two components. The change $\mathsf{w}_1 = \boldsymbol{\omega}\,\mathsf{x}$ represents a rotation, whilst the change $\mathsf{w}_2 = \mathsf{e}\,\mathsf{x}$ represents a pure deformation. Since only three independent quantities are needed to specify the antisymmetric tensor $\boldsymbol{\omega}$, the elements ω_{ij} $(i \neq j)$ can also be regarded as the components of an axial vector

$$\boldsymbol{\omega} = [\mathrm{I};\boldsymbol{\omega}] = [\omega_{32}\omega_{13}\omega_{21}]. \tag{10.8}$$

The linear transformation $\mathbf{w}_1 = \boldsymbol{\omega}\mathbf{x}$, where $\boldsymbol{\omega}$ is the representation of the antisymmetric tensor, may also be written

$$\mathbf{w}_1 = \boldsymbol{\omega}\wedge\mathbf{x}, \tag{10.9}$$

where $\boldsymbol{\omega}$ is the vector of (10.8), as may readily be seen by comparing the coefficients of $\mathbf{w}_1$. It is unimportant whether we use the tensor or vector methods of representing the small rotation; note that the components of the vector give the component rotations about the three axes. For the infinitesimal rotations considered in linear elasticity theory, the component rotations may be considered as vectors, and added to give the net rotation.

The geometrical interpretation of the components of the strain tensor is readily obtained. Neglecting products of these components, we find that the diagonal elements represent the extensions (i.e. changes in length per unit length) of lines originally parallel to the coordinate axes. The components e_{ij} $(i \neq j)$ give half the cosines of the angles between vectors originally parallel to x_i and x_j; since e_{ij} is small, $2e_{ij}$ thus gives the relative rotation (the change in the mutual orientation) of such vectors. The quantities $2e_{ij}$ are commonly called the shear strains.

During the deformation, a vector of original length $|\mathbf{x}|$ changes into a vector of length $|\mathbf{y}| = |\mathbf{x}|+\delta|\mathbf{x}|$. Since $\delta|\mathbf{x}|$ is small, $|\mathbf{y}|^2-|\mathbf{x}|^2 = 2|\mathbf{x}|\,\delta|\mathbf{x}|$, and

$$2|\mathbf{x}|\,\delta|\mathbf{x}| = \mathsf{x}'\,\mathsf{S}'\,\mathsf{S}\,\mathsf{x}-\mathsf{x}'\,\mathsf{x} = (S_{ij}+S_{ji}-2\delta_{ij})x_i x_j,$$

where cross products of the terms of S have again been neglected. Since $S_{ij}+S_{ji} = 2(e_{ij}+\delta_{ij})$, we have

$$|\mathbf{x}|\,\delta|\mathbf{x}| = e_{ij}x_ix_j.$$

Consider the surface

$$e_{ij}x_ix_j = \text{const.} \tag{10.10}$$

By comparison with the previous equation, we see that

$$\frac{\delta|\mathbf{x}|}{|\mathbf{x}|} = \frac{\text{const}}{|\mathbf{x}|^2},$$

where the coordinates of the end point of **x** satisfy eqn. (10.10). The surface (10.10) thus has the property that the extension of any vector is proportional to the inverse square of the radius vector to the surface in the corresponding direction. The direction of the displacement is normal to the tangent plane to the surface at the point **x**. We have already met this equation in the theory of finite deformations, and we noted there that $\delta|\mathbf{x}|$ should strictly be replaced by the resolved elongation of **x**. When the displacements are infinitesimal, the elongation quadric is known as the strain quadric, and its axes are the principal axes of the strain.

So far, we have assumed that the deformations, though infinitesimal, are homogeneous. The theory may be extended to inhomogeneous infinitesimal deformations in the following way. A point with coordinates x_i will move during deformation to a new position y_i, and the displacement may again be specified by

$$\mathbf{w} = \mathbf{y}-\mathbf{x}.$$

Consider a neighbouring point with coordinates $x_i+\zeta_i$ (Fig. 2.8), so that before deformation the two points were related by a vector $\boldsymbol{\zeta} = [\text{I}; \zeta]$. This point will move to a position

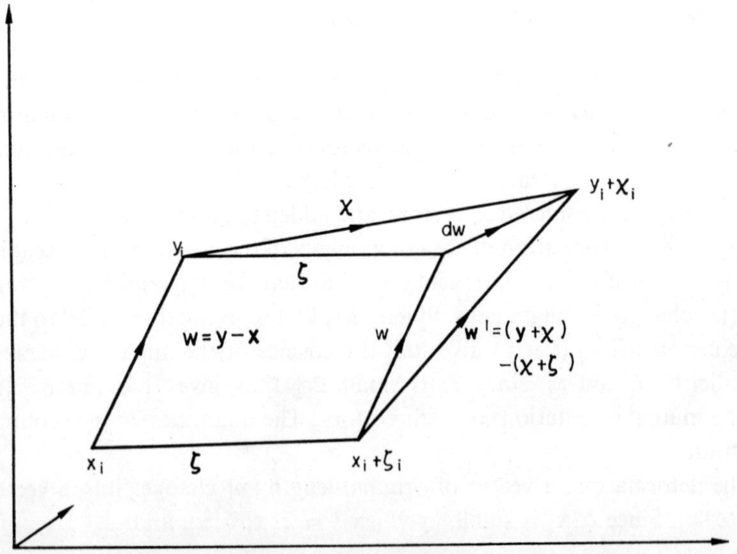

FIG. 2.8. Displacements in infinitesimal deformation.

with coordinates $y_i+\chi_i$, so that its displacement vector is $\mathbf{w}' = (\mathbf{y}+\boldsymbol{\chi})-(\mathbf{x}+\boldsymbol{\zeta})$. We may write the components of this displacement as a Taylor series

$$(y_i+\chi_i)-(x_i+\zeta_i) = (y_i-x_i)+\frac{\partial(y_i-x_i)}{\partial x_j}\zeta_j+0(\zeta_j^2), \quad \text{etc.}$$

The differentials in this expression are to be evaluated at the point $\mathbf{x}$. If the components of $\boldsymbol{\zeta}$ are small enough for the squares to be neglected, this gives

$$\chi_i-\zeta_i = \frac{\partial w_i}{\partial x_j}\zeta_j,$$

and the change in the separation of the two points is

$$d\mathbf{w} = \mathbf{w}'-\mathbf{w} = \boldsymbol{\chi}-\boldsymbol{\zeta} = \mathbf{Q}\boldsymbol{\zeta}, \tag{10.11}$$

where the matrix $\mathbf{Q}$ has components $\partial w_i/\partial x_j$. Comparison of eqn. (10.11) with (10.4) shows that they are of the same form; the vector $d\mathbf{w}$ represents the displacement of points near $\mathbf{x}$ relative to an origin which moves with $\mathbf{x}$. The transformation is affine in the small region round $\mathbf{x}$, and $\mathbf{Q}$ may again be separated into a strain tensor $\mathbf{e}$ and a rotation tensor $\boldsymbol{\omega}$, where now

$$\left.\begin{aligned} e_{ij} &= \tfrac{1}{2}(\partial w_i/\partial x_j+\partial w_j/\partial x_i), \\ \omega_{ij} &= \tfrac{1}{2}(\partial w_i/\partial x_j-\partial w_j/\partial x_i). \end{aligned}\right\} \tag{10.12}$$

In the region around $\mathbf{x}$, $\mathbf{e}$ and $\boldsymbol{\omega}$ may be interpreted in the same way as we previously interpreted them for the whole crystal. An inhomogeneous deformation may thus still be considered affine in a small region, and is characterized by the pure strain and rotation of each region. In contrast to the previous results, the components of $\mathbf{e}$ and $\boldsymbol{\omega}$ are not constants, but are functions of the coordinates x_i.

If we use the vector representation of the rotation $\boldsymbol{\omega}$, we have the components of this vector as

$$\omega_i = \tfrac{1}{2}(\partial w_k/\partial x_j-\partial w_j/\partial x_k) \qquad (i \neq j \neq k)$$

or

$$2\boldsymbol{\omega} = \text{curl }\mathbf{w}. \tag{10.13}$$

The rotation is then represented by a vector which is a function of the coordinates x_i, but is independent of the choice of coordinate system.

Utilizing the result on p. 46, it follows from (10.11) that the ratio of the new volume of a small region to its old volume is given by

$$\frac{v+\Delta v}{v} = |\mathbf{I}+\mathbf{Q}| = 1+\frac{\partial w_i}{\partial x_i},$$

since all the remaining terms vanish for an infinitesimal deformation. The quantity

$$\text{div }\mathbf{w} = \frac{\partial w_i}{\partial x_i} = e_{ii} = \frac{\Delta v}{v} = \Delta, \tag{10.14}$$

6

is called the elastic dilatation, and may be given the same symbol Δ, already used for finite homogeneous deformations. For infinitesimal deformations, Δ is clearly a scalar property of the vector **w**, and is independent of the coordinate system. If e is *any* representation of the strain tensor **e**, then

$$\Delta = \text{trace}\,(\mathrm{e}) = e_{ii}. \tag{10.15}$$

The components of e are not able to vary in an arbitrary manner, since they determine the displacements w of the points of the material. Provided the volume considered is singly connected (i.e. any closed curve can be shrunk continuously down to a point without crossing the boundaries), w must be everywhere single-valued and continuous. This restriction results in six equations which must be satisfied by the second differential coefficients of the tensor components e_{ij}; they are known as the equations of compatibility. We shall not derive them here, but they may all be expressed in the shortened form

$$\frac{\partial^2 e_{ij}}{\partial x_k\,\partial x_l}+\frac{\partial^2 e_{kl}}{\partial x_i\,\partial x_j}-\frac{\partial^2 e_{ik}}{\partial x_j\,\partial x_l}-\frac{\partial^2 e_{jl}}{\partial x_i\,\partial x_k}=0. \tag{10.16}$$

11. STRESS–STRAIN RELATIONS: THEORY OF ELASTICITY

The equations of elastic equilibrium do not, of course, form part of the geometry of crystal lattices, but it is convenient to consider them here, since the theory of linear elasticity is so dependent on the notion of infinitesimal strain introduced in the last section. We first clarify the concept of stress. Any volume element v of a body is subject in general to forces of two kinds. There are first forces which act on all the particles of the volume element, and these are called body forces. They are caused by the presence of some external field of force, for example the gravitational field. In general, this force will not be uniform, but its effect on the whole of the volume element may be summed to give a resultant body force $\int_v \mathbf{g}\,dv$, where **g** is the body force vector, and a resultant moment about any origin of $\int_v (\mathbf{g}\wedge\mathbf{x})\,dv$, where **x** is the position vector with respect to the origin.

The second type of force acting on the volume v arises from internal forces between particles of the material. Thus, consider a small planar element δO of the (interior) surface O separating v from the rest of the body. In general, the two parts of the body on either side of δO will be in a strained condition, and will exert equal and opposite forces on each other. These forces are used to specify the state of stress at a point within the element δO. Let the outward normal to δO be **n**, a unit vector. In this section, we use **n** rather than **h** as the symbol for a vector normal to a plane, since we are interested in both rational and irrational planes. For the most part, we are not concerned with the crystallographic nature of the structure, so we take **n** as a unit normal. The forces exerted by the material on one side of the element δO across δO will then reduce to a single force acting at a point within the element, and a moment about some axis. As $\delta O \to 0$, the direction of the resultant force approaches a fixed value, though the magnitude of both force and couple tend, of course, to zero. However, the force divided by the area remains finite, and approaches a limit as $\delta O \to 0$, whilst the moment divided by δO tends tozero. The limiting vector, having the dimensions force per unit area, is called the stress vector **p**, or simply the stress acting across δO. In a fluid, the only surface force is a uniform hydrostatic pressure, and **p** is necessarily along the direction $-\mathbf{n}$. In a solid, however, **p** may make any angle with the direction **n**. The quantity $\mathbf{p}\,\delta O$ when δO is small, is called the traction across δO. Forces of this second kind are called surface forces.

The resultant of the surface forces on the volume v is given by the surface integral $\int_O \mathbf{p}\,dO$. This follows since the internal surface forces across all elements δv within v cancel out. In the same way, there will be a resultant moment of $\int_O (\mathbf{p}\wedge\mathbf{x})\,dO$ on the origin, as a result of the surface forces.

The complete specification of the surface forces at a point requires a knowledge of the tractions across all planar elements at the point. We shall now show that this specification may be obtained by a set of quantities X_{ij} which form a symmetric matrix, and which are a representation of a second order tensor called the stress tensor. Consider a rectangular parallelepiped with faces perpendicular to the coordinate axes. The surface forces exerted by the material outside the parallelepiped on the material inside it are specified by a stress vector **p** at each point of the surface. At any point on the surface normal to the axis x_i which has a positive normal (i.e. outward normal along $+x_i$), we write the stress vector as

$$\mathbf{p}_i = X_{ij}\mathbf{i}_j, \tag{11.1}$$

where $\mathbf{i}_i$ define the orthonormal system of coordinates x_i. The quantity $X_{(i)(i)}$ thus gives the stress acting normal to the face at the point considered, and the other two components are shear stresses.

Note that if **p** is directed outwards from the face, $X_{(i)(i)}$ is positive, and hence the sign convention is such that tensile stresses exerted by the surrounding material on the volume element are positive. Equally, if we had taken a point on the surface normal to $\mathbf{i}_i$ and having negative surface normal, we should have written

$$\mathbf{p}_i = -X_{ij}\mathbf{i}_j,$$

so that tensile forces would again be specified by positive $X_{(i)(i)}$.

Now consider a small tetrahedron at the point P (Fig. 2.9), formed from three planes normal to the coordinate axes, and a fourth plane, ABC, which has unit normal **n** and area δO. The areas of the other three faces of the tetrahedron are $\delta O_i = n_i\,\delta O$, where n_i are the components of **n**. The outward normals to all these faces are negative, if the components of **n** are all positive. In equilibrium, there will be no resultant force on the whole tetrahedron, so that

$$\int_v \mathbf{g}\,dv + \int_O \mathbf{p}\,dO = 0. \tag{11.2}$$

Suppose now that the volume of the tetrahedron is allowed to shrink continuously to zero. The forces proportional to the volume then vanish more rapidly than those proportional to the surface, and in the limit we need consider only the latter. The traction across

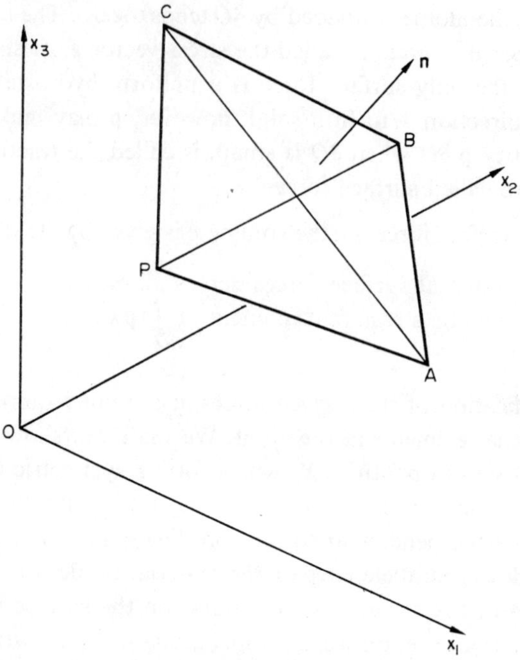

FIG. 2.9. To illustrate the meaning of the stress components at *P*.

each plane surface of the tetrahedron approaches the limit of the product of its area and a constant stress. Thus across the surfaces normal to the axes, the sum of the forces acting is

$$\mathbf{p}_i\,\delta O_i = -n_i X_{ij}\mathbf{i}_j\,\delta O.$$

If we suppose that the stress vector across *ABC* has the value **p** in the limit, the equilibrium condition is

$$\mathbf{p}\,\delta O + \mathbf{p}_i\,\delta O_i = 0$$

or

$$p_j = X_{ij}n_i. \tag{11.3}$$

In matrix form, we may write this equation as

$$(\mathbf{p};\, \mathrm{I}) = (\mathbf{n};\, \mathrm{I})\,(\mathrm{I}\,\mathbf{X}\,\mathrm{I})$$

or

$$\mathrm{p}' = \mathrm{n}'\,\mathrm{X}. \tag{11.4}$$

The stress vector across any plane at a point *P* is thus specified by the positive normal to the plane, and the array of quantities which represent the stress tensor at this point.

We now proceed to calculate the restrictive conditions which the components of **X** must satisfy if the body is to be in internal equilibrium. The condition of equilibrium within any volume v in the material is expressed by eqn. (11.2). If we take the component of force in the direction $\mathbf{i}_i$,

$$\int_v g_i\,dv + \int_O X_{ji}n_j\,dO = 0.$$

The second part of this expression may be transformed into a volume integral by means of Gauss's theorem to give

$$\int_v (g_i + \partial X_{ji}/\partial x_j)\, dv = 0. \tag{11.5}$$

Since the region v is arbitrary, the expression in the integral must vanish identically, and we therefore have a set of equations

$$\partial X_{ji}/\partial x_j = -g_i. \tag{11.6}$$

which are known as the equations of equilibrium.

In many important physical problems the particles of the body are not held in static equilibrium but are in states of motion. Returning to Fig. 2.9 we see that if there is a resultant force on the tetrahedron, its centre of mass will have an acceleration in the direction of this resultant. Taking the component of acceleration in the direction $\mathbf{i}_i$, eqn. (11.5) is replaced by

$$\int_v (g_i + \partial X_{ji}/\partial x_j)\, dv = \varrho v\, \partial^2 x_i/\partial t^2,$$

where ϱ is the density of the material at the point P. This gives a set of equations

$$\frac{\partial X_{ji}}{\partial x_j} + g_i = \varrho \frac{\partial^2 x_i}{\partial t^2}, \tag{11.7}$$

which are known as the equations of motion.

When the body is in equilibrium, the resultant moment of the forces acting on v must vanish at any point. Choose an arbitrary origin, and let $\mathbf{x}$ be the position vector of some point within the volume v. The moment about the origin is zero if

$$\int_v (\mathbf{g} \wedge \mathbf{x})\, dv + \int_O (\mathbf{p} \wedge \mathbf{x})\, dO = 0, \tag{11.8}$$

and taking the component in the direction $\mathbf{i}_1$,

$$\int_v (x_3 g_2 - x_2 g_3)\, dv + \int_O (x_3 X_{i2} n_i - x_2 X_{i3} n_i)\, dO = 0. \tag{11.9}$$

By using Gauss's theorem and eqn. (11.6), the second part of this expression may be written as the volume integral:

$$\int_v [-(g_2 x_3 - g_3 x_2) + X_{32} - X_{23}]\, dv,$$

and substituting this into (11.9), finally,

$$X_{23} = X_{32}. \tag{11.10}$$

By similarly writing down the other resolved components of (11.8), we can show that the quantities X_{ij} form a symmetric matrix with $X_{ij} = X_{ji}$. Equation (11.4) can thus be written in the equivalent form

$$\mathsf{p} = \mathsf{X}\,\mathsf{n}. \tag{11.11}$$

Using a suitable orthogonal transformation, $\mathbf{X}$ may be referred to a new set of axes in which it has diagonal form. The values of the diagonal elements, i.e. the eigenvalues of X, are called the principal stresses.

In the same way as we defined the strain quadric, we may define a surface by the relation

$$X_{ij}x_i x_j = \text{const}, \tag{11.12}$$

and this is called the stress quadric. It has the property that if a vector is drawn from the origin to a point **x** of the stress quadric in the direction of the normal **n** to any plane, the stress vector acting across that plane is in the direction of the normal to the tangent surface at x, and the normal component of the stress vector is proportional to the inverse square of the radius vector **x**.

For any body in equilibrium under the action of external surface forces and body forces, the six components of the stress tensor must satisfy the three eqns. (11.6) at all points in the interior. In addition, eqn. (11.11) must be satisfied over the external surface, **p** now being the externally applied stresses. This gives three boundary conditions. These equations are insufficient to determine the state of stress; the information which is still required is the connection between the state of stress and the state of strain. This relation is obtained in the generalized form of Hooke's law, which states that the stress components are linear functions of the strain components, and vice versa. For our purpose, this is best regarded as an empirical law, based on experiment; it is approximately valid for small deformations.

Since the stress components X and the strain components e are both representations of second rank tensors, the general linear relation is of the form

$$X_{ij} = c_{ijkl}e_{kl}, \qquad e_{ij} = s_{ijkl}X_{kl}, \tag{11.13}$$

where the quantities c_{ijkl}, s_{ijkl} are representations of fourth rank tensors. Fortunately, however, the 81 components of the representations c and s are readily reduced to more manageable numbers. In the first place, both X and e are symmetric matrices with only 6 independent components. This reduces the number of independent quantities in c and s to 36. It is usual to adopt a simplified notation, due to Voigt, and write the independent components of X and e as

$$\mathsf{X} = \begin{pmatrix} X_1 & X_4 & X_5 \\ X_4 & X_2 & X_6 \\ X_5 & X_6 & X_3 \end{pmatrix} \quad \text{and} \quad \mathsf{e} = \begin{pmatrix} e_1 & \frac{1}{2}e_4 & \frac{1}{2}e_5 \\ \frac{1}{2}e_4 & e_2 & \frac{1}{2}e_6 \\ \frac{1}{2}e_5 & \frac{1}{2}e_6 & e_3 \end{pmatrix}. \tag{11.14}$$

The factors of one half are introduced into the off-diagonal elements of the strain tensor to conform to long-established standard notation. Before the formal theory of tensors was developed, the shear strain components e_4, e_5 and e_6 were defined so as to give directly the changes in mutual inclination of vectors along the coordinate axes. These quantities are commonly called the shear strains, or simply the shears; to avoid confusion, it is better to designate them the engineering shear strains, and to distinguish them from the tensor shear strains $e_{ij} (i \neq j)$. Note that the array of engineering strains

$$\begin{pmatrix} e_1 & e_4 & e_5 \\ e_4 & e_2 & e_6 \\ e_5 & e_6 & e_3 \end{pmatrix}$$

does not constitute a tensor, and it is thus rather unfortunate that the older terminology is so well established. Although it is now common practice to use the tensor representation, the older component strains have been retained in defining the relations between stress and strain, which may now be written

$$\left. \begin{aligned} X_i &= c_{ij} e_j \\ e_i &= s_{ij} X_j \end{aligned} \quad (i, j = 1, 2, \ldots, 6), \right\} \tag{11.15}$$

where, in contrast to previous equations, the range of i and j is from 1 to 6. The quantities c_{ij} and s_{ij} have been variously called elastic constants, moduli, or coefficients, with little agreement amongst different authors about which names are appropriate to the two sets. All these names are avoided in the modern (and descriptive) American terminology, in which the quantities c_{ij} are called stiffness constants, and the quantities s_{ij} are called compliances. In general, it is easy to examine a body under a uniaxial stress, but it is not possible to produce a uniaxial strain. The compliances may thus be determined directly by experiment; the stiffness constants only indirectly. With the above notation, the tensors **c** and **s** may be represented by 6×6 square matrices c and s.

Now suppose the deformation at any point is changed by an infinitesimal amount, so that the strains vary from e_i to $e_i + \mathrm{d}e_i$. During this change, work in done by the external forces, and the potential energy in the deformed region increases. The work per unit volume may be obtained by summing the products of each stress component and the corresponding change in strain, as may be seen by considering first a unit cube of material with only one stress component, X_i, acting. If the state of strain changes by an infinitesimal amount, the only work done by X_i is $X_i \, \mathrm{d}e_i$, where e_i is the corresponding strain component. The total work done on the unit cube in increasing the stress from 0 to X_i whilst the corresponding strain component increases from 0 to e_i is thus $\frac{1}{2} X_{(i)} e_{(i)}$. The principle of superposition now allows us to treat the work done by a system of forces acting on a unit cube as the sum of the amounts attributable to each stress component acting separately, so that for an incremental strain, the increase in strain energy per unit volume may be written

$$\mathrm{d}W_s / v = X_i \, \mathrm{d}e_i \qquad (i = 1, \ldots, 6). \tag{11.16}$$

The strain energy W_s must be a single valued function of the strains, so that

$$\mathrm{d}W_s = \frac{\partial W_s}{\partial e_i} \, \mathrm{d}e_i$$

and the condition for $\mathrm{d}W_s$ to be a perfect differential is

$$\frac{\partial^2 W_s}{\partial e_i \, \partial e_j} = \frac{\partial^2 W_s}{\partial e_j \, \partial e_i},$$

i.e. from (11.16)

$$\partial X_i / \partial e_j = \partial X_j / \partial e_i .$$

From eqn. (11.15), we see that this implies

$$c_{ij} = c_{ji} \tag{11.17}$$

and the stiffness constants thus form a symmetric matrix. Similarly, the compliances s_{ij} form a symmetric matrix **s**. The form of the strain energy function may be written

$$W_s/v = \tfrac{1}{2}c_{ij}e_ie_j + \text{const.} \tag{11.18}$$

The constant is zero if the zero energy is the undeformed state. If W_s is known, the components of the stress tensor may be obtained from the relation

$$X_i = \frac{1}{v}\frac{\partial W_s}{\partial e_i} = \frac{1}{2}(c_{ij}+c_{ji})e_j = c_{ij}e_j. \tag{11.19}$$

Leaving out the constant, the quantity W_s/v gives the strain energy per unit volume stored in the material. We note in passing that we have not specified the conditions for the change dW_s. The deformation may be carried out either adiabatically or isothermally; the function W_s will be different in the two cases, and the corresponding stiffness and compliance constants are also different.

The symmetry of the components of **c** and **s** reduces the number of independent quantities in each to 21, and this number of parameters is needed to specify the elastic properties of crystals of the lowest symmetry. Further reductions in the number of independent terms are due either to the symmetry properties of the crystal structure, or to an assumed law of force. A possible assumption is that the forces between atoms are all central, i.e. they act along the lines joining two atoms, and are a function only of the separation of these atoms. If all forces are of this kind, and in addition the crystal structure is such that each atom is at a centre of symmetry, the following equations may be derived:

$$\left.\begin{array}{lll} c_{44} = c_{23}, & c_{55} = c_{31}, & c_{66} = c_{12}, \\ c_{56} = c_{14}, & c_{64} = c_{25}, & c_{45} = c_{36}. \end{array}\right\} \tag{11.20}$$

For cubic crystals, as shown below, the equations all reduce to $c_{44} = c_{12}$. The equations are known as the Cauchy relations; they are not valid for metals.

The existence of symmetry elements in a crystal structure leads to a reduction in the number of independent elastic stiffness constants. Thus an n-fold axis of symmetry means that the crystal is brought into self-coincidence by a rotation of $2\pi/n$ about the axis, and such a rotation must therefore leave the elastic properties unchanged. If, therefore, we refer **X** and **e** to the new axes obtained by such a rotation, we obtain relations between the c_{ij} (or the s_{ij}). We shall not work through the examples, but merely quote the results.

If the x_3 axis is a twofold axis of symmetry (the x_1x_2 plane is a plane of symmetry), then

$$c_{14} = c_{15} = c_{24} = c_{25} = c_{34} = c_{35} = c_{64} = c_{65} = 0. \tag{11.21}$$

This symmetry element thus reduces the number of independent stiffness constants to 13. If the x_3 axis is a threefold axis of symmetry

$$\left.\begin{array}{l} c_{16} = c_{26} = c_{34} = c_{35} = c_{36} = c_{45} = 0, \\ c_{11} = c_{22}, \quad c_{13} = c_{23}, \quad c_{14} = -c_{24} = c_{56}, \\ c_{15} = -c_{25} = -c_{46}, \quad c_{44} = c_{55}, \quad c_{66} = \tfrac{1}{2}(c_{11}-c_{12}), \end{array}\right\} \tag{11.22}$$

and there are seven independent constants. For a fourfold axis along x_3, the following relations are additional to those for a twofold axis:

$$c_{36} = c_{45} = 0, \quad c_{11} = c_{22}, \quad c_{13} = c_{23}, \quad c_{16} = -c_{26}, \quad c_{44} = c_{55}, \tag{11.23}$$

which again gives seven independent constants. A sixfold axis about x_3 is obtained by combining the two- and threefold axis relations, and leaves five stiffness constants.

With the aid of these equations, the form of the matrix c for any symmetry can be determined. Thus a cubic crystal has fourfold axes x_1, x_2 and x_3. If the coordinate axes coincide with the edges of the cubic unit cell, the matrix has the form

$$\mathsf{c} = \begin{pmatrix} c_{11} & c_{12} & c_{12} & 0 & 0 & 0 \\ c_{12} & c_{11} & c_{12} & 0 & 0 & 0 \\ c_{12} & c_{12} & c_{11} & 0 & 0 & 0 \\ 0 & 0 & 0 & c_{44} & 0 & 0 \\ 0 & 0 & 0 & 0 & c_{44} & 0 \\ 0 & 0 & 0 & 0 & 0 & c_{44} \end{pmatrix} \tag{11.24}$$

and there are thus only three independent components. The behaviour of cubic crystals is most readily considered in relation to the combinations c_{44}, $(c_{11}-c_{12})/2$ and $(c_{11}+2c_{12})/3$. These measure respectively the resistance to deformation when a shearing stress is applied across a {100} plane in a ⟨010⟩ direction, across a {110} plane in a ⟨110⟩ direction, and when a uniform hydrostatic pressure is applied to the crystal. The cubic elastic properties are thus represented by two shear moduli and a bulk modulus. The two shear moduli become equal in an elastically isotropic material, and the ratio $2c_{44}/(c_{11}-c_{12})$ may thus be used as an elastic anisotropy factor for cubic crystals. Values of this factor are shown in Table IV, and provide an estimate of the validity of calculations using isotropic elastic theory for the metals concerned.

In an elastically isotropic body, the stiffness constants and compliances must be invariant with respect to any rotation of the axes. This leads to the further relation $c_{44} = (c_{11}-c_{12})/2$, and there are only two independent quantities to be specified. These are usually given the symbols

$$\mu = c_{44}, \qquad \lambda = c_{12}.$$

μ is called the shear modulus; λ has no special name. If we now write out the components of stress for the isotropic medium, we find they can be expressed in the form

$$X_i = c_{ij}e_j = \lambda(e_1+e_2+e_3)+2\mu e_i = \lambda\Delta+2\mu e_j \qquad (i = 1, 2, 3)$$

and

$$X_i = c_{(i)(i)}e_i = \mu e_i \qquad (i = 4, 5, 6). \tag{11.25}$$

Using eqn. (11.14) to transform back to the components of the stress and strain tensors,

$$X_{ij} = \lambda\delta_{ij}\Delta+2\mu e_{ij}, \tag{11.26}$$

and, in particular,

$$X_{ii} = X_{11}+X_{22}+X_{33} = (3\lambda+2\mu)\Delta.$$

TABLE IV. VALUES OF ANISOTROPY FACTOR, $2c_{44}/(c_{11}-c_{12})$ FOR CUBIC METALS
(Elastic stiffnessses taken from Huntington, 1958)

Metal	Structure	Anisotropy factor
Lithium	b.c.c.	9·39
Sodium	b.c.c.	8·14
Potassium	b.c.c.	6·34
Copper	f.c.c.	3·20
Silver	f.c.c.	3·01
Gold	f.c.c.	2·90
Aluminium	f.c.c.	1·29
Silicon	Diamond cubic	1·56
Germanium	Diamond cubic	1·66
Lead	f.c.c.	3·89
Nickel	f.c.c.	2·51
Thorium	f.c.c.	3·62
Molybdenum	b.c.c.	0·77
Tungsten	b.c.c.	1·00
β-Brass	b.c.c.	8·40

The components of the strain tensor, e_{ij}, may similarly be obtained in terms of the stress components, using (11.13) which gives

$$e_{ij} = -\frac{\lambda}{2\mu(3\lambda+2\mu)}\delta_{ij}X_{kk}+\frac{1}{2\mu}X_{ij}. \tag{11.27}$$

If e is a diagonal matrix, X is also diagonal, so that the principal axes of stress and strain must coincide for an isotropic body.

Real crystals are not elastically isotropic, although some cubic crystals have small anisotropy factors. However, very many important problems are concerned not with the properties of isolated crystals, but with polycrystalline aggregates. If the distribution of orientations is effectively random, and the grain size small enough, these assemblies behave as isotropic bodies. Experimental results on elastic properties are then usually expressed in the form of various moduli of elasticity. If the body is deformed by a uniaxial stress, then all the X_{ij} except X_{11} are zero. The corresponding strains are

$$e_{11} = \frac{\lambda+\mu}{\mu(3\lambda+2\mu)}X_{11}, \quad e_{22} = e_{33} = -\frac{\lambda}{2\mu(3\lambda+2\mu)}X_{11}, \quad e_{ij} = 0 \quad (i \neq j).$$

The ratio $Y = X_{11}/e_{11} = \mu(3\lambda+2\mu)/(\lambda+\mu)$ is called Young's modulus of elasticity. The ratio of the lateral contraction to the longitudinal expansion in such an experiment, $\nu = -e_{22}/e_{11} = \lambda/2(\lambda+\mu)$ is called Poisson's ratio.

A body subject to pure shear has $X_{23} = X_{32}$, and all the other components of stress are zero. This gives

$$e_{23} = e_{32} = X_{23}/2\mu.$$

The quantity μ thus represents the ratio of the shearing stress to the change in angle between the x_2 and x_3 axes; this latter quantity is the (engineering) shear strain, and μ is the shear modulus. It is also readily seen that when a uniform hydrostatic pressure is applied, so that $X_{11} = X_{22} = X_{33}$, the ratio of the compressive stress to the cubical compression, $-\Delta$, is given by $K = \lambda + 2\mu/3$, where K is called the bulk modulus.

The theory of elasticity has to be applied to two principal classes of problem. These are the calculation of the distribution of stress and the displacements in the interior of an elastic body when the body forces are known, and either (a) the displacements over the surface of the body or (b) the forces acting on the surface, are specified. The equations to be solved are the stress equations of equilibrium (in more general problems, the equations of motion), together with the auxiliary conditions represented by the stress–strain relations, and where required (in the second type of problem), the compatibility conditions. For isotropic media, we may substitute eqns. (11.26) into the equations of equilibrium, to give

$$(\lambda+\mu)\frac{\partial \Delta}{\partial x_i} + \mu \nabla^2 w_i = -g_i \tag{11.28}$$

or, using (10.14),

$$(\lambda+\mu) \text{ grad div } \mathbf{w} + \mu \nabla^2 \mathbf{w} = -\mathbf{g}. \tag{11.29}$$

Equations (11.28) and (11.29) are very useful in many elementary problems of the first type. Sometimes, however, it is convenient to use the identity $\nabla^2 = \text{grad div} - \text{curl curl}$ to transform (11.29) into

$$\left.\begin{aligned} (\lambda+2\mu) \text{ grad div } \mathbf{w} - \mu \text{ curl curl } \mathbf{w} &= -\mathbf{g} \\ \text{or} \qquad (\lambda+2\mu) \text{ grad } \Delta - \mu \text{ curl } \boldsymbol{\omega} &= -\mathbf{g}. \end{aligned}\right\} \tag{11.30}$$

The first problem is completely solved if solutions of one of the equivalent forms (11.28)–(11.30) are obtained, subject to the boundary conditions in which $\mathbf{w}$ is specified over the limiting surface of the solid considered.

To solve the second type of problem, the differential equations have to be expressed in terms of stresses rather than displacements. Since not every solution of the stress equations of equilibrium represents a possible state of strain, it is necessary to incorporate the compatibility equations. After some manipulation, eqns. (10.16) and (11.27) give the following set of six differential equations:

$$\nabla^2 X_{ij} + \frac{1}{1+\nu}\frac{\partial^2}{\partial x_i \partial x_j}(X_{ii}) = -\frac{\nu}{1+\nu}\delta_{ij} \text{ div } \mathbf{g} - \left(\frac{\partial g_i}{\partial x_j} + \frac{\partial g_j}{\partial x_i}\right). \tag{11.31}$$

These are known as the Beltrami–Michell compatibility equations. The second main class of elastic problem then involves the solution of these equations, subject to the boundary conditions in which the forces acting over the surface of the body are specified. In this book, we shall not be concerned with problems of this type, nor with the mixed class when the boundary conditions are specified partially as forces and partially as displacements.

REFERENCES

BOWLES, J. S. and MACKENZIE, J. K. (1954) *Acta metall.* **2,** 129, 138.
BULLOUGH, R. and BILBY, B. A. (1956) *Proc. Phys. Soc. B,* **69,** 1276.
CAHN, R. W. (1953) *Acta metall.* **1,** 49.
CROCKER, A. G. (1962) *Phil. Mag.* **7,** 1901.
FRANK, F. C. (1965) *Acta crystallogr.* **18,** 862.
HUNTINGTON, H. B. (1958) *Solid State Phys.* **7,** 214.
JASWON, M. A. and WHEELER, J. A. (1948) *Acta crystallogr.* **1,** 216.
NEUMANN, P. (1966) *Phys. stat. sol.* **17,** K71.
NICHOLAS, J. F. (1966) *Acta crystallogr.* **21,** 880; (1970); *Phys. stat. sol.* (a) **1,** 563.
NICHOLAS, J. F. and SEGALL, R. L. (1970) *Acta crystallogr.* **A26,** 522.
NYE, J. F. (1957) *Physical Properties of Crystals*, Clarendon Press, Oxford.
OKAMATO, P. R. and THOMAS, G. (1968) *Phys. stat. sol.* **25,** 81.
OTTE, H. M. and CROCKER, A. G. (1965) *Phys. stat. sol.* **9,** 44; (1966) *Ibid.* **16,** K25.
SOKOLNIKOFF, I. S. (1946) *Mathematical Theory of Elasticity*, McGraw-Hill, New York; (1951) *Tensor Analysis*, Wiley, New York.
ZACHARIASEN, W. H. (1945) *Theory of X-Ray Diffraction in Crystals*, Wiley, New York.

CHAPTER 3

The Theory of Reaction Rates

12. CHEMICAL KINETICS AND ACTIVATION ENERGY

A typical problem in the theory of transformation is the motion of an interface separating two regions of different composition. Provided we interpret the term "chemical reaction" in a rather more liberal sense than is usual, we may regard the growth of the more stable region as the combination of a transport process (atomic diffusion) with a chemical reaction (formation of a new phase). Moreover, the transport process may itself be regarded as a series of chemical reactions, and this concept has proved very useful. Under appropriate circumstances, we may use theories which have been developed to describe the kinetics of chemical reactions and extend them to analogous phenomena such as diffusion.

Most descriptions of chemical kinetics are based on a statistical theory developed by Eyring and his co-workers (Glasstone *et al.*, 1941). This theory depends on the assignation of thermodynamic properties to non-equilibrium states, and is now usually known as the "absolute reaction rate theory" or as the "transition state theory". Despite its success in chemical problems, the claim implicit in the first name seems a little too ambitious, since calculation of reaction rates from first principles is seldom possible in practice. When applied to metallic assemblies, the method has yielded useful results, but we must emphasize at once that there are many solid state phenomena for which it is clearly not valid. Reaction rate theory should therefore be applied only after careful consideration of the probable physical mechanism of the process considered. When this mechanism appears to be consistent with the postulates of the theory, as discussed below, the chemical analogy may be useful, but in other cases it may be misleading.

The basic assumption of Eyring's theory is that the whole reaction consists of the repetition of a fundamental unit step, or series of such steps. Each unit step arises from the interaction of a *small* number of atoms or molecules to form a new configuration; before this is achieved, the group passes through intermediate situations which have higher energies than either the initial or final states. There is a critical configuration which must be attained if the final state is to be reached, and the increase in free energy required to form this critical configuration determines the time taken to make the unit step. There is thus an activation energy for the process which must be supplied by a local fluctuation in the thermal energy of the assembly.

These conditions are obviously satisfied by most ordinary chemical reactions, both in the gaseous phase and in liquid solution. There are many additional difficulties in the solid state, but the general description of Chapter 1 suggests that nucleation and growth transformations can usually be split up into unit steps governed by thermally supplied activation energies, at least as a first approximation. On the other hand, changes in which the movements of large numbers of atoms are closely co-ordinated are unlikely to be controlled by thermal fluctuations, so that this kind of theory is inappropriate to martensitic growth. Early attempts were occasionally made to apply the theory to phenomena to which it is unsuited and were incorrect mainly because of insufficient attention to the physical mechanisms. Thus a phenomenological model of plastic deformation in which individual units of flow are treated as single thermally activated processes is not generally valid, although there are many dislocation interactions contributing to plastic deformation in which the concept of activation energy is useful.

Once this danger has been recognized it is not difficult to select processes which are genuinely thermally activated, and to which the chemical analogy may be supposed to apply. It is very much more difficult to decide whether the rather restrictive assumptions of Eyring's theory are valid in any particular thermally activated process. Here we encounter difficulties of several kinds. The original form of the reaction rate theory applied to gases and to liquid solutions which behave ideally. Attempts to extend the theory to non-ideal assemblies are generally made by introducing activity coefficients, as in equilibrium thermodynamics, but the equations then contain quantities of rather uncertain physical significance. This problem arises in most applications of the theory to the solid state, and the logical difficulties involved are frequently ignored. Fortunately the difficulty can often be avoided by making use of a treatment due to Wert and Zener (1949) and Zener (1952), which permits a direct correlation of theory and experiment.

Another difficulty which arises with solids, and which was first emphasized by Crussard (1948), is that the atoms or molecules cannot be legitimately treated as a set of independent quantum systems, especially in problems in which thermal energy plays an important role. The simplest form of the reaction rate theory effectively considers the thermal energy of motion of a solid to be the energy of a number of independent atomic or molecular oscillators, whereas in fact the motions are coupled so that much of the energy may be contained in relatively long wavelength vibrations. Thus in considering the probability of a small number of atoms or molecules surmounting an energy barrier, we should treat the energy increase as a temporary local concentration of phonons, caused by interference of wave packets. Some progress has been made in developing a theory of this kind, but most existing treatments of molecular processes in solids use the ordinary Eyring theory. This approximation should not be too seriously wrong, except at very low temperatures; it is roughly equivalent to using an Einstein model of a solid instead of a Debye model, or a more sophisticated model. Vineyard (1957) has clarified the many-body aspects of thermally activated rate processes in solids and has shown that in a harmonic approximation the formalism of the Eyring theory is still acceptable. This paper is very important in the development of a satisfactory conceptual theory, but discussion of it is deferred to the end of this chapter so that it may more readily be related to the work of Wert and Zener.

These other assumptions, which are listed in the next section, provide the final and greatest difficulty in the application of the theory. The problem concerns the validity of thermodynamic concepts applied to the critical transition state at the top of the free energy barrier in the unit step and of the assumption that reactants in this transition state are in equilibrium with the reactants in the initial state. This requires that the lifetime of the transition state should be long in comparison with the time of thermal relaxation in the lattice, and it is doubtful if this is so. Clearly theories of the process which are based on the dynamic properties of the molecules participating in the unit step would be preferable to the pseudo-thermodynamic theories with which we usually have to be satisfied. Some progress in developing such theories has been made by Rice (1958) for the case of diffusion, and is discussed on p. 92. We may remark parenthetically that although we have kept the discussion in this section general, almost all advances in the theory of rate processes in solids have been made with reference to the theory of diffusion in the first instance.

In the remainder of this section we shall summarize some well-known results of chemical kinetics as a prelude to consideration of the rate theories in the next section. The unit step in a chemical reaction may be represented by the equation

$$aA+bB+\ \dots\ = pP+qQ+\ \dots, \tag{12.1}$$

where the reacting systems (atoms, molecules, etc.) are represented by A, B, ..., the products by P, Q, ..., and the numbers a, b, ... p, q, ..., are integers. The rate of reaction is specified by the velocity constant or specific rate constant k_f which is so defined that the number of product systems formed per unit volume in unit time is equal to k_f multiplied by some function of the concentrations of the reacting systems. This may seem rather a vague definition; it arises because the form of the dependence of the rate of reaction on the concentrations of A, B, etc., in the general case can only be determined by experiment. However, the function is usually a simple one, and it is then possible to assign an *order* to the reaction.

The order is defined as the power of the concentration terms which determine the rate law. Suppose, for example, that we have a single species of reacting molecule, A. If the rate of reaction is proportional to the concentration c_A (e.g a radioactive decomposition) it is first order; if it is proportional to c_A^2, it is second order. Similarly, a gaseous reaction between two molecules A and B, is of the second order if it is proportional to $c_A c_B$. It follows from this that although the rate constant k_f is always numerically equal to the rate of reaction at unit concentration of each of the reactants, the physical dimensions of k_f depend on the order of the reaction.

In complex reactions involving several unit steps, it is not always possible to give an unambiguous meaning to the order of the reaction. It is then preferable to specify the order with respect to a particular reactant, this being the power of the concentration term of that reactant which enters into the rate law. In the example above, the reaction between A and B is of first order with respect to each reactant.

The molecularity of a reaction is defined as the number of reacting molecules (or atoms) which take a direct part in the change. Before reaction can occur, these molecules must be raised to higher energy states, and a particular configuration which is critical for the change is said to form the activated complex. The molecularity of the reaction can thus be stated only when the mechanism is known, and only molecules which form part of the activated complex are considered to take a direct part in the reaction. If there are several unit steps, the molecularity can only be specified for each of these separately.

The meaning of direct in the above definition may be clarified by the following example in which we suppose the unit step to consist of the decomposition of a single molecule A into two molecules P and Q. The decomposition can occur only if the A molecule acquires a higher energy than it possesses in the ground state, and in this condition which we write $A^{\ddagger}$ the molecule constitutes the activated complex. Now suppose the reaction occurs in the gaseous phase. Interchanges of energy among the molecules then result only from collisions, and the whole unit step may be written

$$A+A = A^{\ddagger}+A,$$
$$A^{\ddagger} = P+Q.$$

The molecularity of this reaction is one because one of the molecules in the collision plays only an indirect role in the reaction.

If the whole reaction consists of one unit step, we might expect the molecularity to equal the order of the reaction. This is generally true, but there is one important exception which we shall mention below. When there are several atomic processes, there is usually one which is much slower than the others, and the overall reaction rate is then determined almost entirely by the rate applicable to this unit step. In this case, the overall order of the reaction will be equal to the molecularity of this rate determining step. In complex reactions there may be two or more slow steps, and the whole reaction does not then have a well-defined order, though each unit step still has a definite molecularity. The one exception to these simple rules is the monomolecular gas reaction discussed above. The kinetic theory of gases shows that this is first order at high pressures and second order at low pressures.

The distinction between reaction order which refers to the overall change, and molecularity which refers to the mechanism should be remembered, but will not concern us further. In metallic transformations we shall encounter only first-order unimolecular reactions and second-order bimolecular reactions. In the equations of this chapter we shall use an undetermined number of concentration terms in the expressions for the reaction rate in order to facilitate comparison with standard chemical textbooks. The actual number of terms must be determined for the unit step in any process considered.

A chemical reaction such as that represented by eqn. (12.1) does not usually proceed to completion, but is reversible. The assembly then tends to an equilibrium state in which the concentrations of products and reactants remain constant. This equilibrium is the result of a dynamic balance between the forward and reverse reactions, which continue on an atomic scale, but with equal velocities. There must then be a connection between the specific rate constants for the two reactions and the thermodynamic equilibrium constant which describes the macroscopic state of the assembly.

The rates of the forward and back reactions may be written $k_f f_f(c_A, c_B, \ldots)$ and $k_b f_b(c_P, c_Q, \ldots)$, where the form of the functions f is not specified. Equating the two rates gives

$$k_f/k_b = f_f(c_A, c_B, \ldots)/f_b(c_P, c_Q, \ldots) = f(c_A, c_B, \ldots, c_P, c_Q, \ldots).$$

Under specified constraints, k_f and k_b are both constants, and it follows therefore that there must be some function of the concentrations of reactants and products which is constant at equilibrium. However, the equilibrium constant K specifies the equilibrium state in terms of the activities and hence of the concentrations of the products and reactants. It follows that K must equal $f(c_A, c_B, \ldots, c_P, c_Q, \ldots)$, or else be some power of this function. In practice k_f/k_b almost always equals K, though, in a few cases, this quotient has the value $K^{1/2}$ or K^2.

We have assumed that there is only one reaction path leading from the initial to the final state. If there are several alternative paths, the functions f_f, f_b will contain as many separate terms as there are paths. At equilibrium, the sum of the rates from reactants to products by all paths must equal the sum of the reverse rates, i.e.

$$k_f f_f + k_f' f_f' + k_f'' f_f'' + \ldots = k_b f_b + k_b' f_b' + k_b'' f_b'' + \ldots \quad (12.2)$$

This equation satisfies the condition for stationary values of the thermodynamic parameters, and if the macroscopic thermodynamic laws are regarded as postulates, they do not impose any further restrictions. However, these laws may be related to the kinetic behaviour of the systems of the assembly, and statistical considerations show that (12.2) is insufficient to ensure equilibrium; in addition it is necessary that the forward rate along any one reaction path shall equal the reverse rate along that path. This condition, expressed in the equations

$$k_f/k_b = k_f'/k_b' = k_f''/k_b'' = K, \quad (12.3)$$

is known as the principle of detailed balancing. It is one aspect of a more fundamental statistical law called the principle of microscopical reversibility, which we shall discuss in the next chapter. The fundamental reason for the principle of detailed balancing may be illustrated by the following argument. Let there be two reaction paths from the initial to the final condition, one of which depends on the presence of a surface catalyst, and assume that eqns. (12.3) are not satisfied at equilibrium. We may now consider a virtual change in which the catalyst is removed thereby closing one reaction path and displacing the equilibrium. This will result in a spontaneous change of the assembly to a new condition. If the catalyst is now re-introduced, there will be a further spontaneous change, and by performing such cycles repeatedly we would be able to draw energy continuously from the assembly, in contradiction to the second law of thermodynamics.

Consider now the unit step in any reaction, which we may represent symbolically as

$$\left.\begin{aligned} A+B+\ldots &= M^{\ddagger} \\ M^{\ddagger} &= P+Q+\ldots \end{aligned}\right\} \quad (12.4)$$

Whatever the nature of this step, it is found experimentally that for a very large number of reactions, the rate of reaction may be expressed in the form

$$\left.\begin{aligned} \ln k_f &= \ln C_4 - E/\boldsymbol{R}T, \\ \text{i.e.} \quad k_f &= C_4 \exp(-E/\boldsymbol{R}T) = C_4 \exp(-\varepsilon/\boldsymbol{k}T), \end{aligned}\right\} \quad (12.5)$$

where C_4 is a constant and E, ε are called the experimental activation energies per mole and per molecule respectively. This is the Arrhenius equation. Provided the temperature range considered is not large, C_4 and E are effectively constants.

Consider now the energy diagram of Fig. 3.1 which shows the variation of internal or potential energy during the course of the reaction. The initial and final states are stable and hence correspond to minima in the energy field; at some intermediate stage the energy

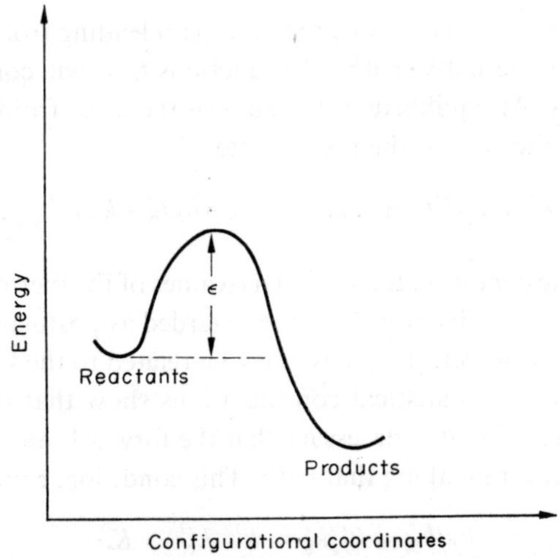

FIG. 3.1. Schematic energy diagram.

rises to a maximum. The abscissae of the figure represent configurational coordinates along the reaction path. In the simplest example the reaction might consist of the approach of two molecules along a linear path until at a critical separation they form the complex $M^{\ddagger}$. The first part of the reaction path would then be a linear spatial coordinate, with the intermolecular distance decreasing; similarly, the second part of the reaction path might be another linear spatial coordinate with increasing intermolecular distance. In more complex cases, the behaviour of the molecules cannot be visualized so readily, but there will always be some continuous series of configurations in terms of which we may interpret the abscissae of the figure.

From Fig. 3.1, we see that before the reactants can travel along the reaction path and thus form the products, they must acquire temporarily an amount of energy equal to the height of the energy barrier. If the probability of the systems having this energy were proportional to $\exp(-\varepsilon/kT)$, the Arrhenius equation would follow naturally. In fact this is not true, but it gives an approximate interpretation of the exponential factor of eqn. (12.5).

In order that the product $C_4 \exp(-E/\boldsymbol{R}T)$ shall have the right dimensions, C_4 must have the dimensions of a frequency. The early theory of gaseous reactions assumed that C_4 was the frequency of collision of the reacting molecules, collisions being necessary for transfers of energy to be possible. In the same way, for a bound particle in the solid state, C_4 would be the frequency of vibration in the direction leading to the reaction path. In effect, C_4 measures the number of times per second that the reacting systems can attempt to traverse the reaction path, and the exponential term gives the probability of their having sufficient energy to be successful at any one attempt. In the next section we shall obtain more correct interpretations of both these two factors.

13. THE REACTION RATE THEORY

We begin this section with a description of the statistical theory of reaction rates developed by Eyring *et al.*, and take up later the special difficulties mentioned on p. 78. The basic assumptions of Eyring's theory are:

(1) The reaction is characterized by some initial configuration which proceeds to the final configuration by a continuous change of coordinates. This change constitutes the unit step of the process. The initial and final configurations are stable states situated at relative minima in the energy field, so that a system in one of these states tends to return to its original position when given a small displacement.

(2) There is an energy barrier between the initial and final states along any reaction path, and the most-favoured reaction path will generally be that for which this energy barrier is least. The energy maximum along the most-favoured reaction path will be at a lower energy level than adjacent points lying on other reaction paths, and hence this maximum will be a saddle point in the energy field. It is usually tacitly assumed that there is only one such saddle point along the chosen reaction path; this statement seems reasonable, although no firm proof can be given. The energy field concerned was originally considered to be the potential energy field, but later developments have shown that we must consider free energies (see below).

(3) There is a critical configuration intermediate between the initial and final states and situated at the energy maximum (strictly the free energy maximum) of the reaction path. This configuration is called the activated complex or activated state, and once it is attained there is a high probability of the reaction proceeding to completion.

(4) The reacting systems in the initial state are regarded as being in equilibrium with the activated complexes $M^{\ddagger}$, even though these do not form equilibrium states.

(5) The activated complex is assumed to possess all the properties of a normal equilibrium configuration, except that it has no degree of freedom corresponding to vibrational motion along the reaction path. Such vibrational motion leads in fact to decomposition of the complex to form the final stable configuration. In place of this vibrational degree of freedom, we may thus suppose the complex to have an additional degree of freedom corresponding to translational motion along the reaction path.[†] The velocity of the reaction is determined by the rate at which products form from $M^{\ddagger}$, i.e. by the rate at which the activated complexes travel along the reaction path on the top of the energy barrier.

In order to calculate absolute reaction rates from first principles it is necessary to evaluate the energy along the reaction path. This is a very difficult problem, which has been

---

[†] This must not be confused with spatial translational degree of freedom of the complex. In the rather strange terminology of this theory, translational motion of the complex along the reaction path is synonymous with disintegration of the complex.

approximately solved only for the very simplest reactions. We emphasize again that either experiment or other considerations must decide whether or not the above assumptions are valid for any particular reaction process. As already stated, wrong results have occasionally been obtained by application of the theory to phenomena which do not satisfy the first two assumptions, but in most cases it is relatively easy to decide whether or not they are approximately satisfied. When (1) and (2) are valid, it has generally been assumed that the remaining conditions are not unduly restrictive, the main alternative being that the activated complexes decompose as rapidly as the reacting systems accumulate the necessary thermal energy, so that there is no equilibrium.

Let us consider the average "velocity" of an activated complex in the $+x$ direction, which we regard as the direction of the reaction path, not as a fixed spatial coordinate. A complex of effective mass m and velocity $\dot{x}$ has kinetic energy $\frac{1}{2}m\dot{x}^2$. If we assume an equilibrium distribution of velocities, the probability of the complex having a velocity between $\dot{x}$ and $\dot{x}+d\dot{x}$ is proportional to $\exp(-m\dot{x}^2/2kT)\,d\dot{x}$, and the average velocity of a complex in the $+x$ direction will be given by

$$\bar{\dot{x}} = \frac{\int_0^\infty \exp(-m\dot{x}^2/2kT)\dot{x}\,d\dot{x}}{\int_{-\infty}^\infty \exp(-m\dot{x}^2/2kT)\,d\dot{x}}.$$

The integration limits of the dividend are 0 to ∞ since we are not considering velocities in the $-x$ direction. Evaluation of the integrals gives

$$\bar{\dot{x}} = (kT/2\pi m)^{1/2}.$$

The activated complexes may be regarded as being stable within an (unspecified) length δx lying along the top of the reaction path. The average time taken for a complex to cross the barrier is thus given by $\tau = \delta x/\bar{\dot{x}}$, and if $c^{\ddagger}$ is the number of complexes per unit volume which are within the length δx of the reaction path, the number of complexes decomposing per unit volume per unit time is $c^{\ddagger}/\tau$. This is the rate of the unit step reaction, and is equal to the velocity constant multiplied by the concentrations of the reacting systems, i.e.

$$k_f c_A c_B \ldots = (c^{\ddagger}/\delta x)(kT/2\pi m)^{1/2}.$$

For both ideal and non-ideal assemblies,

$$c^{\ddagger}/c_A c_B \ldots = Q^{\ddagger\|}/Q'_A Q'_B \ldots$$

in which the quantities on the right-hand side are the partition functions for the activated complex and for the reactants, all referred to the same energy zero. If we refer each partition function to its own zero energy level, we obtain

$$c^{\ddagger}/c_A c_B \ldots = Q^{\ddagger|}\exp(-\varepsilon_0/kT)/Q_A Q_B \ldots,$$

where ε_0 is the difference in the zero energy levels of the activated complex and the reacting systems, and is therefore the activation energy per system for the reaction at 0 K. We have written the partition function for the activated complex in the form $Q^{\ddagger|}$ purely for convenience, since we wish to reserve the symbol $Q^{\ddagger}$ for another related quantity.

We now have

$$k_f = \frac{Q^{\ddagger|}}{Q_A Q_B \ldots}\left(\frac{kT}{2\pi m}\right)^{1/2} \frac{\exp(-\varepsilon_0/kT)}{\delta x}. \tag{13.1}$$

$Q^{\ddagger|}$ is the complete partition function for the activated complex, and may be factorized into the partition function for translational motion along the reaction path, and the partition function for the remaining degrees of freedom. We thus have

$$Q^{\ddagger|} = Q^{\ddagger} Q_x^{\ddagger},$$

where if we use the standard expression for a translational partition function

$$Q_x^{\ddagger} = (2\pi mkT/h)^{1/2}\,\delta x. \tag{13.2}$$

Combining eqns. (13.1) and (13.2),

$$k_f = (kT/h)(Q^{\ddagger}/Q_A Q_B \ldots) \exp(-\varepsilon_0/kT). \tag{13.3}$$

We note that the (kT/h) factor in this equation arises from the combination of the mean velocity in the "direction" of decomposition with the translational partition function in this direction. Since δx does not appear in the result, we may choose it, if we wish, so as to make $Q_x^{\ddagger}$ equal to unity. $Q^{\ddagger}$ and $Q^{\ddagger|}$ are then numerically equal, but they refer of course to different degrees of freedom. In calculation of $Q^{\ddagger}$, motion of the complex along the reaction path must be excluded. The form of the derivation shows that kT/h is a universal factor entering into all such rate equations. An alternative way of obtaining eqn. (13.3) illustrates this point. Instead of regarding the activated complex as possessing no vibrational degree of freedom along the reaction path, we may suppose that its vibration in this direction is so weak that each vibration is transformed into a translation and leads to decomposition. The rate of reaction is then equal to the frequency of vibration along the reaction path, multiplied by the concentration of activated complexes. Hence

$$\left.\begin{aligned} k_f c_A c_B \ldots &= c^{\ddagger}\nu \\ \text{or}\qquad k_f &= (Q^{\ddagger|||}/Q_A Q_B \ldots)\nu \exp(-\varepsilon_0/kT). \end{aligned}\right\} \tag{13.4}$$

We now factorize $Q^{\ddagger|||}$ into $Q^{\ddagger}$ and the partition function corresponding to the vibrational freedom along the reaction path. Since the activated state is thoroughly excited, we may take the classical value of $kT/h\nu$ for this, and combining with eqn. (13.4) we again obtain eqn. (13.3).

Eqn. (13.3) may be regarded as the fundamental equation of the reaction rate theory, and applies to both ideal and non-ideal assemblies. There is, however, a factor which we have neglected. Quantum mechanics show that there is a finite probability that a system with sufficient energy to surmount the energy barrier and which is travelling along the forward reaction path will nevertheless be reflected back to the initial state. For completeness, eqn. (13.3) should therefore be multiplied by a quantum mechanical transmission probability. In normal reactions, this factor is close to unity and we shall ignore it. In the same way, there is a finite probability of a set of systems reaching the final state even although they do not possess the requisite activation energy. This "tunnel effect" is quite negligible except for nuclear phenomena and for extremely low temperatures.

For an assembly which behaves ideally, we may introduce an equilibrium constant K_c, where

$$K_c = (c^{+}/c_A c_B c_C \ldots) = (Q^{+\prime}/Q_A Q_B \ldots) \exp(-\varepsilon_0/\boldsymbol{k}T).$$

We may also define a similar "equilibrium constant" K_c^{+} by the equation

$$K_c^{+} = (Q^{+}/Q_A Q_B \ldots) \exp(-\varepsilon_0/\boldsymbol{k}T), \tag{13.5}$$

where K_c^{+} differs only slightly from K_c, and may indeed be made numerically equal to K_c by suitable choice of δx. This fact is frequently allowed to obscure the difficulty of giving a physical interpretation to K_c^{+}. Since the number of degrees of freedom in the divisor and dividend of eqn. (13.5) are different, K_c^{+} cannot be a true equilibrium constant and the definition is made purely for convenience. The constant refers to the hypothetical equilibrium between the reactants and the activated complexes which have, as it were, been deprived of a degree of freedom.

We can now write eqn. (13.3) in the form

$$k_f = (\boldsymbol{k}T/\boldsymbol{h})K_c^{+} \qquad \text{(ideal assemblies only).} \tag{13.6}$$

In fact, since we shall be more interested in applying the rate theory to solutions which are not dilute, it is often more convenient to use the equilibrium constant K^{+} expressed in terms of atom fractions rather than concentrations. Since the activated complex is to be regarded as a single molecule, it follows that

$$K^{+} = c^{i-1} K_c^{+},$$

where i is the molecularity of the unit step, and c is the total concentration. This gives

$$k_f = (\boldsymbol{k}T/\boldsymbol{h}) c^{1-i} K^{+} \qquad \text{(ideal assemblies only).} \tag{13.7}$$

For reactions in non-ideal solutions, the equilibrium constant is normally given in terms of activities rather than atomic fractions. Thus for the equilibrium between reactants and activated complexes,

$$k_f = (\boldsymbol{k}T/\boldsymbol{h})(\gamma_A \gamma_B \ldots/\gamma^{+}) c^{1-i} K^{+}. \tag{13.8}$$

where the γ are activity coefficients. This equation, though quoted in standard textbooks and papers, raises the problem of the meaning to be attributed to γ^{+}, and this difficulty may be linked with that of defining K^{+}. A better formalism which provides a more direct link with experimental observations was introduced by Wert and Zener (1949) in connection with the theory of diffusion; we consider this below.

By means of the thermodynamic equation

$$\boldsymbol{R}T \ln K^{+} = -\Delta G^{+}$$

we can express the reaction rate in the form

$$k_f = (\boldsymbol{k}T/\boldsymbol{h})(\gamma_A\gamma_B \ldots /c^{i-1}\gamma^{\ddagger}) \exp(-\Delta G^{\ddagger}/\boldsymbol{R}T)$$
$$= (\boldsymbol{k}T/\boldsymbol{h})(\gamma_A\gamma_B \ldots /c^{i-1}\gamma^{\ddagger}) \exp(\Delta S^{\ddagger}/\boldsymbol{R}) \exp(-\Delta H^{\ddagger}/\boldsymbol{R}T). \quad (13.9)$$

The quantities $\Delta G^{\ddagger}$, $\Delta S^{\ddagger}$, $\Delta H^{\ddagger}$ are usually called the molar free energy, molar entropy and molar heat of activation respectively. They are then stated to be the differences in free energies, entropies, and enthalpies of the activated complexes and the reactants when all are in their standard states.† It follows from the discussion above, however, that their physical meaning is rather less certain than this statement suggests. The standard state of the activated complexes is not well defined, and must refer to complexes with one degree of freedom less than the reactants. Nevertheless, this equation is perhaps the most useful form of expressing the result of the rate theory, and the physical significance is certainly celarer than in the expression involving partition functions. We note especially that the important rate determining factor is the free energy of activation.

In some particular examples it is possible to obtain an expression with a more precisely defined physical meaning. If we return to eqn. (13.3), we may factorize the divisor on the right-hand side by taking out the partition function which corresponds to vibrational motion of the reactants along the reaction path. The quotient of the two kinds of partition function remaining may then be put equal to $\exp(-\Delta_a G/\boldsymbol{R}T)$, where $\Delta_a G$ differs from $\Delta G^{\ddagger}$ defined above. We then obtain

$$k_f = (\boldsymbol{k}T/\boldsymbol{h})(1/Q_v) \exp(-\Delta_a G/\boldsymbol{R}T), \quad (13.10)$$

where Q_v is the vibrational partition function of the reactants. $\Delta_a G$ has a physical interpretation. It is the isothermal work needed slowly to transfer the configuration of molecules which form the reactants from their initial positions to their critical positions, subject to the restriction that they are allowed at all times to vibrate only in the plane perpendicular to the reaction path. Moreover, the effect of the activity coefficients is then included in the definition of $\Delta_a G$, so that eqn. (13.10) is valid for both ideal and non-ideal solutions.

Equation (13.10) was obtained for the particular case of diffusion by Wert and Zener, as already mentioned. Its utility obviously depends on whether a simple expression is available for Q_v, and in chemical reactions this will not often be possible. The form (13.9) has thus been used extensively, although (13.10) seems logically preferable, especially for reactions in non-ideal solutions. Equations (13.9) differ from those usually given in chemical textbooks only by the inclusion of the factor c^{1-i}. This is merely a result of our using equilibrium constants relating to atomic fractions rather than concentrations, i.e. of a choice of a different standard state (the pure components).

In many reactions, the entropy of activation does not vary greatly, and hence the rate of reaction appears to be governed by the heat of activation. This differs only slightly from the experimental activation energy. The entropy of activation may be positive or negative; a high positive entropy implies a high probability of formation of an activated complex,

† Since there is equilibrium between reactants and activated complexes, the actual free energy difference is zero, of course.

and vice versa. On p. 82 we interpreted C_4 as giving the frequency with which the unit step can be attempted. In many gaseous reactions, it is found that C_4 is approximately equal to the collision frequency, but in some reactions the two differ by many powers of ten. This is because the collision theory fails to take into account the distribution of energy between the different degrees of freedom. A specific example is provided by two polyatomic molecules which combine together to form a single molecule. During the reaction, three translational and three rotational degrees of freedom disappear and are replaced by six new vibrational degrees of freedom. There is thus a relatively small probability of transition, even when the total energy is available, unless the coordinated motions happen to be suitably related. This is merely another way of saying that $\Delta S^{\ddagger}$ is negative. Conversely, for a unit step in which a relatively highly organized form of energy becomes more randomly distributed, $\Delta S^{\ddagger}$ is positive and there is a high probability of transition.

We now have to correlate the experimental activation energy with the results of reaction rate theory. E is determined experimentally from the equation

$$\varepsilon/\boldsymbol{k} = E/\boldsymbol{R} = -\mathrm{d}(\ln k)/\mathrm{d}(1/T) = T^2\,\mathrm{d}(\ln k)/\mathrm{d}T.$$

In some textbooks, (13.8) is used, so that

$$\frac{E}{\boldsymbol{R}} = T+T^2\frac{\partial}{\partial T}\left\{\ln\frac{K^{\ddagger}}{c^{i-1}}\right\}+T^2\frac{\partial}{\partial T}\left\{\ln\frac{\gamma_A\gamma_B\ldots}{\gamma^{\ddagger}}\right\}.$$

Now

$$\frac{\partial}{\partial T}\left\{\ln\frac{K^{\ddagger}}{c^{i-1}}\right\} = \frac{\partial}{\partial T}(\ln K_c^{\ddagger}) = \frac{\Delta U^{\ddagger}}{\boldsymbol{R}T^2},$$

where $\Delta U^{\ddagger}$ is the internal energy of activation per mole for the unit step reaction. If we neglect the variation of the activity coefficients with temperature, we have

$$E = \boldsymbol{R}T+\Delta U^{\ddagger} = \boldsymbol{R}T+\Delta H^{\ddagger}-p\,\Delta v^{\ddagger}, \tag{13.11}$$

where $\Delta v^{\ddagger}$ is the change in volume in forming the activated complex and may be neglected for condensed assemblies. Since the $\boldsymbol{R}T$ term is negligible in comparison with E the experimental activation energy is very nearly equal to the theoretical "activation enthalpy" in condensed assemblies.

This treatment is obviously unsatisfactory, and the Zener method (13.10) leads to better results. In cases where it may be applied, e.g. in solids, it is usually permissible to write Q_ν as $\boldsymbol{k}T/\boldsymbol{h}\nu$, where ν is a characteristic vibration frequency for the reaction concerned. Equation (13.10) thus becomes

$$k_f = \nu\exp(-\Delta_a G/\boldsymbol{R}T) = \nu\exp(\Delta_a S/\boldsymbol{R})\exp(-\Delta_a H/\boldsymbol{R}T) \tag{13.12}$$

and differentiation gives directly

$$E = \Delta_a H, \tag{13.13}$$

so that the physical quantity in the rate equation given by Zener can not only be defined more clearly, but also has a clear identification with the experimental measurement.

We now discuss the difficulty mentioned on p. 78, namely that in the solid state, the reactants necessarily interact with all the other atoms present in forming the activated complexes. The reactant is frequently a single atom, as in most diffusion mechanisms for example, and the approximation involved is thus that of treating a many body problem as a single body problem. Vineyard (1957) developed a more general treatment which is based only on the rate theory assumptions that well-defined initial and saddle-point configurations exist in quasi-equilibrium, and on the additional assumption that the motions of atoms in both configurations may be treated as small, simple, harmonic oscillations. We abandon the rather artificial chemical description in terms of concentrations of reacting species and consider instead a crystal of N atoms with $3N$ degrees of freedom represented by spatial coordinates $x_1, \ldots, x_{3N}$. Let the potential energy of the crystal be $U(y_1, \ldots, y_{3N})$, where $y_1 = m_1^{1/2} x_1$, etc., and m_i are the masses associated with each x_i. Then in a $3N$-dimensional configurational space y_i, U has a minimum at some initial and final configurations (A and P) and there is also a critical configuration M^{+} for the transition $A \rightarrow P$. A hypersurface S of dimensionality $3N-1$ may be defined so that it passes through M^{+} and is perpendicular to the contours of U elsewhere. A transition from A to P occurs if a system crosses S with finite velocity.

By arguments essentially similar to those developed above for the Eyring theory, Vineyard showed that the rate constant for the transition $A \rightarrow P$ is given by

$$k_f = \left(\frac{kT}{2\pi}\right)^{1/2} \frac{\int_S \exp(-U/kT)\,\mathrm{d}S}{\int_V \exp(-U/kT)\,\mathrm{d}V}, \tag{13.14}$$

where the upper integration is over the hypersurface S and the lower integral is over the region of configuration space bounded by S. This equation may be compared with (13.1); it contains the ratio of two configurational partition function integrals, which must be calculated so as to include all atoms and degrees of freedom.† The theory of small vibrations is now used to expand U as a Taylor series about its values $U(A)$ and $U(M^{+})$ respectively. By means of an axis transformation, the normal coordinates $q_1, \ldots, q_{3N}$ and frequencies $\nu_1, \ldots, \nu_{3N}$ for the initial state A are introduced, and the value of U in the volume integral becomes

$$U \simeq U(A) + \tfrac{1}{2} \sum_{i=1}^{3N} (2\pi \nu_i q_i)^2. \tag{13.15}$$

Hence the partition function for the initial state may be written

$$\int_V \exp(-U/kT)\,\mathrm{d}V = \prod_{i=1}^{3N} (kT/2\pi \nu_i^2) \exp\{-U(A)/kT\}. \tag{13.16}$$

Similarly, the normal coordinates q_i' and frequencies ν_i' for vibrations about M^{+} but confined to the hypersurface S give the approximation for the activated complex

$$U \simeq U(M^{+}) + \tfrac{1}{2} \sum_{i=1}^{3N-1} (2\pi \nu_i' q_i')^2 \tag{13.16}$$

† Note that the partition function integrals in (13.14) have dimensions of area and volume in the hyperspace y_i; i.e. the ratio of the two integrals has dimension $(\text{mass})^{-1/2}$ $(\text{length})^{-1}$.

and

$$\int_S \exp(-U/kT)\,dS = \prod_{i=1}^{3N-1} (kT/2\pi\nu_i')^2 \exp(-U(M^{\neq})/kT). \tag{13.17}$$

Thus, finally, from (13.14)

$$k_f = \nu^* \exp-[\{U(M^{\neq})-U(A)\}/kT], \tag{13.18}$$

where the effective frequency is

$$\nu^* = \left(\prod_{i=1}^{3N} \nu_i\right) \Big/ \left(\prod_{i=1}^{3N-1} \nu_i'\right). \tag{13.19}$$

The frequency ν^* is thus the ratio of the product of the $3N$ normal frequencies of the assembly in the initial configuration to the $3N-1$ normal frequencies of the assembly constrained in the saddle point configuration. In general the frequencies ν_i' will differ from the corresponding frequencies ν_i, and it will not be justifiable to identify ν^* with, for example, the best Einstein or Debye frequency obtained from specific heat data.

Vineyard's expression may also be written in the form of eqn. (13.12) by a device similar to that used by Wert and Zener. Consider a hypersurface S_0 which is similar to S and passes through the point A of configuration space so that its normal at A is along the line of force leading to $M^{\neq}$. By defining a new partition function for vibrations restricted to the hypersurface S_0, Vineyard showed that (13.12) is valid with

$$\nu^* = \nu \exp(\Delta S/k), \tag{13.20}$$

$$\Delta S = k \ln\left\{\left(\prod_{i=1}^{3N-1} \nu_i^0\right) \Big/ \left(\prod_{i=1}^{3N-1} \nu_i'\right)\right\}, \tag{13.21}$$

where ν_i^0 are the normal frequencies for the assembly constrained to lie in S_0. The frequency ν is thus given by

$$\nu = \left(\prod_{i=1}^{3N} \nu_i\right) \Big/ \left(\prod_{i=1}^{3N-1} \nu_i^0\right) \tag{13.22}$$

and corresponds approximately to its previous description as the frequency of vibration in the initial state along the reaction path (normal to S_0). The interpretation of $\Delta_a G$, $\Delta_a H$, $\Delta_a S$ is also similar to that given above, the assembly being constrained not to vibrate except in a hypersurface which is everywhere normal to the path from A to P in the $3N$ dimensional configuration space.

When reaction rate theory is applied to diffusion it is important to consider the effects of using different isotopes as tracer elements. The activation energies may then be assumed to be almost equal, but the frequency factors differ because of the effects of the isotopic mass on the normal modes. The simplest assumption is that the migrating mechanism involves only one atom, and that there is little or no coupling between this atom and the rest of the crystal. If the rates of migration are then calculated for two isotopes, all frequencies cancel with the exception of those localized frequencies which are associated with the migrating atoms. Thus for isotopes A and α, $\nu_{iA}/\nu_{i\alpha} = \nu_{iA}'/\nu_{i\alpha}'$ and

$$k_A/k_\alpha = \nu_A/\nu_\alpha = (m_\alpha/m_A)^{1/2}, \tag{13.23}$$

where ν_A and ν_α are corresponding frequencies of vibration for the two isotopes, and the result follows since atomic frequencies are inversely proportional to the square roots of the masses.

The assumption that the migrating atom is effectively decoupled from the rest of the crystal is rather extreme. Normally there will be a change of volume in the activated state, and the displacements of the surrounding atoms then produce changes in the vibrational modes. Vineyard treated this complication by defining an effective mass for the migrating atom; a further development due to Le Claire (1966) shows that the ratio of rate constants can be expressed in the form

$$k_A/k_\alpha = \nu'_{1A}/\nu'_{1\alpha}, \tag{13.24}$$

where ν'_{1A} is the frequency of the unstable mode normal to the hypersurface S_0. Introducing

$$\Delta K = \{(\nu'_{1A}/\nu'_{1\alpha})-1\}/\{(m_\alpha/m_A)^{1/2}-1\}, \tag{13.25}$$

Le Claire showed that ΔK is that part of the kinetic energy in the mode of frequency ν'_1 which is possessed by the migrating atom, divided by the total kinetic energy in this mode. Equation (13.27) was first derived by Mullen (1961), who defined ΔK as the fraction of the translational kinetic energy which is possessed by the migrating atom as it passes over the saddle point; Le Claire's derivation shows that only the energy in the mode ν'_1 is to be taken into account in defining this fraction.

Quantum effects are neglected in all the above derivations but may become important in certain circumstances, e.g. in thermally activated processes below the Debye temperature θ_D. If the classical partition functions are replaced by quantum partition functions, eqn. (13.19) is modified to

$$\nu^* = \frac{\prod_{i=2}^{3N}\left[\exp-(\tfrac{1}{2}h\nu'_i/kT)/\{1-\exp-(\tfrac{1}{2}h\nu'_i/kT)\}\right]}{\prod_{i=1}^{3N}\left[\exp-(\tfrac{1}{2}h\nu_i/kT)/\{1-\exp-(\tfrac{1}{2}h\nu_i/kT)\}\right]}\,\frac{kT}{h}. \tag{13.26}$$

Taking the first term of the expansion

$$e^{-z}/(1-e^{-z}) = \frac{1}{2}\operatorname{cosech} z = \left(\frac{1}{2z}\right)\left(1-\frac{z^2}{6}+\dots\right)$$

we obtain (13.19), whilst if the first two terms are retained,

$$\nu^* = \frac{\prod_{i=1}^{3N}\nu_i}{\prod_{i=2}^{3N}\nu'_i}\left\{1-\frac{1}{24}\left(\frac{h}{kT}\right)^2\left(\sum_2^{3N}(\nu'_i)^2-\sum_1^{3N}(\nu_i)^2\right)\right\}. \tag{13.27}$$

Le Claire shows that this equation leads to a revised form of (13.23):

$$\begin{aligned}\frac{k_A}{k_\alpha}-1 &= \left[\left(\frac{m_\alpha}{m_A}\right)^{1/2}-1\right]\left\{1-\frac{1}{12}\left[\frac{m_\alpha}{m_A}+\left(\frac{m_\alpha}{m_A}\right)^{1/2}\right]\left(\frac{h\nu_\alpha}{kT}\right)^2\left[\left(\frac{\nu'_\alpha}{\nu_\alpha}\right)^2-\frac{3}{2}\right]\right\}\\ &\simeq \left[\left(\frac{m_\alpha}{m_A}\right)^{1/2}-1\right]\left\{1-\frac{1}{6}\left(\frac{h\nu}{kT}\right)^2\left[\left(\frac{\nu'}{\nu}\right)^2-\frac{3}{2}\right]\right\}\end{aligned} \tag{13.28}$$

when $m_A \simeq m_\alpha$ so that $\nu_A \simeq \nu_\alpha \simeq \nu$. For numerical estimates, $h\nu/k$ is replaced by the Debye temperature θ_D, and Le Claire concludes that for typical diffusion measurements the maximum correction, represented by the term in braces in (13.28), is likely to be only a few per cent except perhaps in alkali metals.

Vineyard's theory gives the best description currently available for activated processes in the solid state. We may also, however, briefly mention the dynamical approach to rate theory, which has been developed for the particular case of diffusion by Rice and his collaborators (Rice, 1958; Rice and Nachtrieb, 1959). In an attempt to avoid the difficulty mentioned on p. 79, this theory considers explicitly the probability that a diffusing atom will have sufficient amplitude of vibration to move in a given direction, and that the neighbouring atoms will move out of the way. The parameters which define this model are shown by Rice to be related to the normal coordinates of the crystal, a normal mode vibrational analysis which includes the effect of lattice defects being supposed to have been made. The details of this theory are rather complex, and will not be discussed here since the validity of the approach is not generally accepted. The model does not seem to be amenable to an exact calculation, except under rather restrictive conditions, for example that the atoms are harmonically bound. With such assumptions, the expression for the rate constant may be put into a form equivalent to that given by the chemical rate theory, although the physical interpretations of the terms are slightly different.

We have already emphasized that a dynamical theory may be superior in principle to a thermodynamic theory, since the assumption of an activated state in equilibrium with the reactants is a marked over-simplification. The activated state has so short a lifetime (at most a few vibrations) that the assumptions of the chemical theory are at best unverifiable. On the other hand, the dynamical theory as developed at present is forced to make other assumptions, which are also rather unrealistic.

REFERENCES

CRUSSARD, C. (1948) *Compt. Rend. Semaine d'Études de la Physique des Métaux*, 39; (1952) *L'État solide* (Report 9th Solvay Conference Brussels), 346.

GLASSTONE, S., LAIDLER, K. J., and EYRING, H. (1941) *The Theory of Rate Processes*, McGraw-Hill, New York.

LE CLAIRE, A. D. (1966) *Phil. Mag.* **14,** 1271.

MULLEN, J. G. (1961) *Phys. Rev.* **121,** 1649

RICE, S. A. (1958) *Phys. Rev.* **112,** 804.

RICE, S. A. and NACHTRIEB, N. H. (1959) *J. Chem. Phys.* **31,** 139

VINEYARD, G. H. (1957) *J. Phys. Chem. Solids* **3,** 121.

WERT, C. A. and ZENER, C. (1949) *Phys. Rev.* **76,** 1169.

ZENER, C. (1952) *Imperfections in Nearly Perfect Crystals*, *p.* 305, Wiley, New York.

CHAPTER 4

The Thermodynamics of Irreversible Processes

14. MICROSCOPICAL REVERSIBILITY: THE ONSAGER RECIPROCAL RELATIONS

Classical thermodynamics is concerned primarily with the interdependence of certain well-defined macroscopic concepts (temperature, pressure, entropy, energy, composition, etc.) possessed by a closed assembly. The usual thermodynamic equations are valid only for assemblies at equilibrium and for reversible transitions between such equilibrated assemblies. When thermodynamic considerations are applied to irreversible (i.e. "natural") processes, the equations become inequalities, and are much less useful. For example, the principle of the increase in entropy during an adiabatic irreversible process provides information only about the direction of the change.

The theory of transformations is a description of a particular class of irreversible processes. The details of such a description, as emphasized in Chapter 1, depend on assumptions about atomic interactions and the nature of the elementary steps of the reaction. This type of theory will be our main concern, and may appropriately be described as kinetic. Kinetic theory is not, however, confined to the non-equilibrium state; thermodynamic equilibrium is characterized by stationary values of the macroscopic parameters, but each of these is effectively the average value of some continuously varying microscopic concept. The connection between the microscopic properties of the systems of the assembly and the macroscopic (measurable) properties is made by statistical mechanics. The second law of thermodynamics then follows from a single basic postulate, and may be regarded as the necessary consequence of the averaging process applied to an assembly which has reached dynamic equilibrium.

Following this approach, it is natural to enquire whether there is any statistical principle which when applied to the kinetic laws of an irreversible process also imposes restrictions on the macroscopic parameters of the assembly. In fact, such restrictions have been proposed for certain kinds of change which can be described as steady state phenomena, and the attempt to produce a formal theory of this development has become known as the thermodynamics of irreversible processes. Although this theory has been discussed mainly in the last 30 years, the one useful and reasonably well-accepted result was obtained by Onsager in in 1931. It is as well to emphasize here that much work in this field is highly controversial, and almost all the "results" or "principles" enunciated by some workers are thought invalid by others. We shall be mainly concerned with the Onsager relations, which are widely used, but even here we shall see that the usual way of applying the result rests on doubtful foundations.

A steady state implies that at each part of the assembly, which is now "open" rather than "closed", the thermodynamic quantities are time independent, even though energy dissipation is occurring. An example of such a process is the conduction of heat down a bar, one end of which is maintained at a uniform high temperature. The theory has, in fact, been applied mainly to transport phenomena in which there is a flow of heat, matter, electricity, etc. In the study of transformations, the theory is most important in the treatment of diffusion in the solid state, but attempts have also been made to use the theory directly in more complex phenomena, such as the migration of an interface. The thermodynamics of irreversible processes has also been applied to chemical reactions, though only in the limiting case of very close approach to equilibrium.

The macroscopic laws describing a steady-state process are often known or assumed to require a linear relation between the rate at which the process occurs and the "thermodynamic force" which causes the process. The most familiar example is Ohm's law, which states there is a linear relation between the flow of electrical current and the potential gradient or electromotive force. In general, the flow rates may be called currents (diffusion currents, heat currents, etc.) and are linearly related to the forces or affinities. The forces are the gradients of some potential function (temperature, chemical potential, etc.) which may be regarded as responsible for the currents.

The terms forces and currents used in this thermodynamic sense are rather misleading. For example, although both these quantities will usually be vectors, as in transport phenomena, they may in principle be tensors of any order. An example of scalar forces and currents occurs in chemical reactions; the forces are proportional to the differences in chemical potential, and the currents to the corresponding rates of reaction.

The assumption that a current $\mathbf{I}$ and its conjugate force $\mathbf{Z}$ are linearly related may be expressed

$$\mathbf{I} = M\mathbf{Z} \text{ (vector quantities)} \quad \text{or} \quad I = MZ \text{ (scalar quantities).} \tag{14.1}$$

In this equation we have assumed that when the force and current are both vectors they are parallel to each other. The scalar quantity M is given different names in different steady-state phenomena, but is always of the nature of a conductance, since its reciprocal expresses the resistance to flow.

Equation (14.1) will be satisfied in isotropic media, e.g. in randomly orientated polycrystalline solids, but in single crystals it is necessary to introduce a more general relation

$$I_i = M_{ij}Z_j \quad (i, j = 1, 2, 3) \tag{14.2}$$

or, in matrix form,

$$\mathsf{I} = \mathsf{M}\,\mathsf{Z},$$

where M_{ij} form the components of a second-order tensor, so that we now have linear relations between the components of the currents and the components of the conjugate forces.

The situation of interest in the thermodynamic theory arises either when we have a three-dimensional transport process (as in eqn. (14.2)) or when we have two or more currents and forces of different kinds occurring together. There may then be mutual coupling or interference between the separate effects. For example, if two metallic junctions are maintained at different temperatures, there will be a flow of electricity as well as of heat, and conversely if there is an electrical potential difference maintained between them, heat is absorbed or liberated at the junctions.

In general, we may suppose that we have different kinds of vector flow $\mathbf{I}_1, \mathbf{I}_2, \ldots, \mathbf{I}_n$ and different kinds of vector force $\mathbf{Z}_1, \mathbf{Z}_2, \ldots, \mathbf{Z}_n$. If the currents and their conjugate forces are parallel, as in eqn. (14.1), we may write for the interaction of all these quantities

$$\mathbf{I}_i = M_{ij}\mathbf{Z}_j \qquad (i, j = 1, 2, \ldots, n), \tag{14.3a}$$

where the diagonal elements of the matrix M are the conjugate conductances, and the non-diagonal elements are the interference or coupling conductances. Similarly if the flows are scalars we have

$$I_i = M_{ij}Z_j \qquad (i, j = 1, 2, \ldots, n). \tag{14.3b}$$

Note that we assume all forces to be either scalars or vectors; there is, of course, non inter-action between currents and forces of different tensor order.

If there is a tensor relation between each current and the conjugate force, we expect the *component* of each current to be a linear function of all the components of each force. We may write this equation in the form

$$I_i = M_{ij}Z_j \qquad (i, j = 1, 2, \ldots, 3n) \tag{14.3c}$$

if we interpret each subscript, i or j, as giving a double specification—the type of flow and the axis to which the vector component refers. This notation is similar to that used for the Hooke's law eqns. (11.15), and is useful because it preserves the relation between currents and forces in square matrix form. The diagonal elements of the matrix are now the conductances relating the component of any flow along one of the axes with the component of its conjugate force along the same axis; the non-diagonal elements relate the component of any flow along one axis with the components of its conjugate force along the other axes, or with the components of the non-conjugate forces along any axis. Clearly, we have reduced the vector problem to a scalar representation, in which component flows and forces are conjugate to each other.

Equations (14.3a, b, or c) are called the phenomenological equations or the thermodynamic equations of motion. We emphasize here that thermodynamics can say nothing of their validity; this must be determined either by experiment or by arguments based on kinetic theory.

The important result on which the thermodynamics of irreversible processes largely depends was first obtained by Onsager (1931a, b). This states that provided a proper choice has been made of conjugate currents and forces, the matrix M must be symmetrical and

$$M_{ij} = M_{ji}. \tag{14.4}$$

We thus see that the theory only yields useful results when there is mutual interference of effects. The proper choice of currents and forces is considered below; this aspect of the theory can lead to real difficulties.

We shall not attempt to give a full derivation of Onsager's relations, as this would mean a lengthy diversion into statistical theory. The basis of the equations lies in the principle of microscopical reversibility, which states that under equilibrium conditions, any microscopic (i.e. molecular) process and its reverse take place on the average at the same rate. Expressed mathematically, this means that the mechanical equations of motion of individual particles are symmetric with respect to the time, so that the transformation $t \to -t$ leaves them unchanged. This result is founded on quantum statistics and is of very general validity; the only exceptions are effects in non-conservative force fields (electromagnetic induction, Coriolis forces), in which the direction of the field has to be reversed to obtain an invariant transformation from t to $-t$. We have already noted that the chemical principle of detailed balancing is a special case of the above principle.

Onsager obtained the reciprocal relations by considering fluctuations in an assembly at equilibrium, and relating these to macroscopic flows. Suppose we measure these fluctuations by a set of variables† e_i which represent the local deviations in pressure, temperature, density, etc., from the mean or equilibrium values of these quantities. Thus the mean value of e_i over a large region, or for a small region over an appreciable period of time, is zero. Now if at time t the local deviation of type i in any region is e_i, we may use the symbol $\overline{e_i(t)\, e^j(t+\tau)}$ to represent the average probability of a fluctuation e_i being followed by a fluctuation e_j at a time τ later. According to the principle of microscopical reversibility, this sequence of two fluctuations must occur just as frequently as its reverse, so that

$$\overline{e_i(t)\, e_j(t+\tau)} = \overline{e_j(t)\, e_i(t+\tau)}.$$

We now consider how the averages are to be evaluated. If we first take a fixed fluctuation at time t, then at time $t+\tau$ there will in general be an unpredictable series of deviations. We may, however, take the average value of $e_j(t+\tau)$ by considering a large number of situations in which the fluctuation at t is fixed, and finding the mean of the values of e_j after a time interval τ. We denote this average by $\left[e_j(t+\tau)\right]_t$, the square brackets indicating that the situation at time t is fixed whilst the average is evaluated. We can then obtain the average value of $e_i(t)\, e_j(t+\tau)$ by allowing the fixed fluctuation at t to vary and averaging over these variations. We thus have

$$\overline{e_i(t)\left[e_j(t+\tau)\right]_t} = \overline{e_j(t)\left[e_i(t+\tau)\right]_t}.$$

Now subtracting from each side the average product of $e_i(t)$ and $e_j(t)$, which may be written $\overline{e_i(t)\, e_j(t)} = \overline{e_i(t)\left[e_j(t)\right]_t} = \overline{e_j(t)\left[e_i(t)\right]_t}$,

$$\overline{e_i(t)\left[e_j(t+\tau)-e_j(t)\right]_t} = \overline{e_j(t)\left[e_i(t+\tau)-e_i(t)\right]_t}. \tag{14.5}$$

† The symbol e has this meaning only in this chapter.

Onsager now makes an important assumption, namely that the average rates of change of the quantities e_i with respect to time are equal to the corresponding macroscopic flows. The fluctuations are thus supposed to be small transient flows of heat, matter, etc. This type of assumption is implicit in the treatment of such fluctuation phenomena as Brownian motion, and may be shown to be highly probable by kinetic arguments. We may thus write

$$\mathrm{d}e_i/\mathrm{d}t = (1/\tau)\,\overline{[e_i(t+\tau)-e_i(t)]} = I_i = M_{ik}Z_k$$

and $$\mathrm{d}e_j/\mathrm{d}t = I_j = M_{jk}Z_k. \tag{14.6}$$

Substituting into (14.5),

$$\overline{e_i(t)\,[M_{jk}Z_k]_t} = \overline{e_j(t)\,[M_{ik}Z_k]_t}. \tag{14.7}$$

Now let us consider the deviation in the entropy associated with a fluctuation specified by $e_1, e_2, \ldots$ Since S is a maximum in the equilibrium state ($\Delta S = 0$), we may express the entropy fluctuation ΔS as a quadratic function of the state parameters e_j, so that

$$\Delta S = -\tfrac{1}{2}g_{ij}e_ie_j, \tag{14.8}$$

where g_{ij} is a positive definite form. Using Boltzmann statistics, it follows that the probability of a particular fluctuation in which the state parameters (regarded as continuous) have values lying between e_1 and $e_1+\mathrm{d}e_1$, e_2 and $e_2+\mathrm{d}e_2$, etc., is given by

$$P\,\mathrm{d}e_1, \mathrm{d}e_2, \ldots \mathrm{d}e_n = \frac{\exp(\Delta S/k)\,\mathrm{d}e_1\,\mathrm{d}e_2 \ldots \mathrm{d}e_n}{\int\limits_{e_1}\int\limits_{e_2}\ldots\int\limits_{e_n}\exp(\Delta S/k)\,\mathrm{d}e_1\,\mathrm{d}e_2\ldots\mathrm{d}e_n}$$

the denominator on the right-hand side merely being a normalizing factor to make the total integral of P equal to unity. With this equation, it is easily proved that the mean value of the product $e_i(\partial\Delta S/\partial e_j)$ is given by

$$\overline{e_i(\partial\Delta S/\partial e_j)} = -k\delta_{ij}.$$

Let us make the identification

$$Z_j = (\partial\Delta S/\partial e_j) = -g_{ij}e_i \tag{14.9}$$

so that

$$\overline{e_iZ_j} = -k\delta_{ij}. \tag{14.10}$$

Now substituting into (14.7),

$$-k\delta_{ik}M_{jk} = -k\delta_{jk}M_{ik} \quad \text{or} \quad M_{ji} = M_{ij},$$

which is Onsager's relation. The above identification of Z will be considered further in the next section; in effect, this is the meaning of a proper choice of conjugate forces.

We note that eqn. (14.4), although founded on a microscopical principle, expresses relations between macroscopic quantities. This situation is analogous to that encountered in classical thermodynamics since the second law is also a macroscopic law based on a microscopic principle (that of equal *a priori* probability of all microstates). We also note the formal similarity between the result obtained by Onsager for steady-state processes, and the symmetry of the scheme of elastic compliances or stiffnesses, eqn. (11.17). The symmetry of the latter, as does that of other equilibrium properties, depends on the assumption of an energy function; in a sense, the principle of microscopical reversibility is a variational form of this assumption for non-static processes.

15. ENTROPY PRODUCTION IN NATURAL PROCESSES

Thermodynamic quantities are strictly defined only for assemblies which are in equilibrium. We begin this section with a discussion of the circumstances under which it is meaningful to assign thermodynamic parameters to non-equilibrium assemblies. Consider an irreversible transformation in an assembly which is initially and finally at equilibrium. The change in entropy is then defined precisely in terms of other functions which describe the equilibrium states, and the second law tells us that entropy is created during the process. Now if it is possible to know the local pressure, volume, temperature, chemical composition, etc., at each stage in the transformation to a sufficient degree of accuracy, the entropy is also defined, at least to within narrow limits. The local values of the macroscopic parameters are obtained in principle by isolating small regions of the assembly (which still contain large numbers of molecules) and allowing each to come to equilibrium. The procedure is valid only when the deviations from equilibrium are everywhere small, and it is only under these conditions that we are justified in thinking of the assembly as possessing macroscopic properties.

In most processes, the difficulty of defining local parameters arises mainly from the lack of equilibrium between the different forms of energy which are present. The temperature is determined by the mean value of the kinetic energy, and has meaning only when there are sufficient collisions or other molecular interactions for the excited forms of energy to attain local equilibrium very rapidly. This means that the free path for energy transfer must be much smaller than the distances over which the temperature varies, and is another way of stating that the situation must not be too far removed from equilibrium.

The second law shows that entropy, unlike energy, is not conserved during natural changes. During a reversible change in which an amount of heat dq enters the assembly from the surroundings, the entropy of the assembly increases by dS, where

$$T\,dS = dq = -T\,d_eS$$

and d_eS is the change in entropy of the environment. More generally, for any reversible change involving flows of energy, matter, etc., we have

$$dS + d_eS = 0,$$

and for an irreversible change

$$dS + d_eS > 0.$$

We may define d_iS as the irreversible entropy which is *created* during the process, where

$$d_iS = dS + d_eS. \tag{15.1}$$

Note that dS and d_eS may have either sign, but d_iS must be positive or zero.

There is a particularly simple expression for d_iS if we consider irreversible processes at constant temperature and pressure. Then we have

$$dG = dU + p\,dV - T\,dS = -T\,d_eS - T\,dS,$$

so that

$$d_iS = -dG/T,$$

where G is the Gibbs function for the assembly. The rate at which entropy is produced during the process is then

$$d_iS/dt = -(1/T)\,dG/dt. \tag{15.2}$$

This quantity may be called the entropy source strength.

More generally, the above discussion shows that we may write the increment of entropy during any process which is never too far removed from equilibrium as

$$T\,dS = dU + p\,dV - g_i\,dN_i, \tag{15.3}$$

where g_i, N_i represent the chemical potential per molecule and the number of molecules of the ith substance within the assembly at any time during the change. If we use this equation for dS and subtract from it the contribution to the change in entropy of the assembly produced by the transfer of work, heat, matter, etc., from the surroundings, we obtain the irreversible entropy produced, d_iS.

We may now use Onsager's identification of the macroscopic flows with fluctuation phenomena, eqn. (14.6), to relate the rate of entropy production to the forces and fluxes of the last section. The quantity d_iS/dt of this section is to be identified with $d\Delta S/dt$ for a fluctuation. Using eqns. (14.8) and (14.9), we find that we can express the entropy source strength in the form

$$d_iS/dt = I_k Z_k, \tag{15.4}$$

where I_k, Z_k are conjugate currents and thermodynamic forces, and, as explained above, may be either scalars or parallel vector components.

As an example, we consider a single chemical reaction at constant temperature and pressure. From either (15.2) or (15.3)

$$d_iS/dt = -(1/T)\,g_i(dN_i/dt).$$

Suppose the reaction is written

$$a_iA_i = 0,$$

where a_i represents any of the quantities $a, b, p, q, \ldots$, and A_i any of the components $A, B, \ldots, P, Q, \ldots$, in eqn. (12.1). If there were initially $N_{i,0}$ molecules of species i in the assembly, we may use the quantity $\xi = (N_i - N_{i,0})/a_i$ as a measure of the amount of the reaction, and this quantity is moreover the same for all species at a given stage in the reaction. If we adopt the convention of calling a_i positive for the products of the reaction, then ξ, called the degree of advancement, is also positive. Substituting $d\xi = dN_i/a_i$

$$d_iS/dt = -(1/T)(d\xi/dt)a_i g_i. \tag{15.5}$$

The quantity $-a_i g_i$ may now be called the affinity of the reaction; it is a measure of the free energy difference between reactants and products, and is thus positive. The quantity $d\xi/dt$ is a measure of the velocity of the reaction. The right-hand side may thus be written as the product of a single thermodynamic force and its associated current, and provides an example of eqn. (15.4) when there is no coupling.

Now suppose we have a set of chemical components amongst which r different reactions are possible. We write the reactions

$$\begin{aligned} a_{1i}A_i &= 0, \\ a_{2i}A_i &= 0, \\ &\ldots, \\ a_{ni}A_i &= 0. \end{aligned}$$

The degree of advancement of each reaction may be represented by $\xi_j = (N_i - N_{i,0})/a_{ji}$ and the rate of this reaction is thus $d\xi_j/dt = (1/a_{ji})\, dN_i/dt$. This gives

$$d_iS/dt = -(1/T)(d\xi_j/dt)a_{ji}g_i. \tag{15.6}$$

Each quantity $-a_{ji}g_i$ may now be called the affinity for the reaction j, and the entropy source strength is thus equal to the sum of the products of the affinities and the corresponding flows, as in eqn. (15.4). Note that the sum of the products of the affinities and reaction rates must be positive, but there is no need for any individual product to be positive. When the reactions are coupled, some of them may be driven "backwards".

We have quoted the example of chemical reactions to show how eqn. (15.4) is often used in phenomenological theories to identify the forces and fluxes of eqns. (14.3). We have now to emphasize that the expression for the entropy source strength, although it appears to be very convenient in phenomenological theories, is in fact a most unsatisfactory way of selecting forces and fluxes. It is implied in many books, e.g. de Groot (1951), that the Onsager relations eqn. (14.4) are valid for all forces and fluxes which simultaneously satisfy eqns. (14.3) and (15.4). Coleman and Truesdell (1960) have shown clearly that this cannot be true, the difficulty being that the fact that (15.4) follows from (14.6) to (14.9) does not ensure the validity of (14.6) to (14.9) given (15.4).

Suppose we have a set of forces and fluxes which satisfy eqns. (14.3), (15.4), and (14.4). Now let W be any non-zero antisymmetrical matrix with the same number of rows and columns as M, so that $W_{ij} = -W_{ji}$. We can now define new fluxes and forces by the linear transformation

$$\begin{aligned} I_i' &= I_i + W_{ij}Z_j, \\ Z_i' &= Z_i. \end{aligned} \tag{15.7}$$

Then it follows that

$$I_i'Z_i' = I_iZ_i + W_{ij}Z_iZ_j,$$

and since $W_{ij} = -W_{ji}$, the last term is zero and

$$I_i'Z_i' = I_iZ_i = \mathrm{d}_iS/\mathrm{d}t. \tag{15.8}$$

We also have from (15.7)

$$I_i' = (M_{ij} + W_{ij})Z_j = M_{ij}'Z_{ij}',$$

which is equivalent to (14.3) but has new mobilities

$$M_{ij}' = M_{ij} + W_{ij} \tag{15.9}$$

which do not form a symmetric matrix. Thus by beginning with forces and fluxes which satisfy (14.3) and (15.4) and the Onsager relations (14.4), we can construct by a simple linear transformation a new set of forces and fluxes which satisfy (14.3) and (15.4), but which do not satisfy the Onsager relations. Moreover, this argument is reversible in that if we have forces and fluxes for which the Onsager relation is not valid, we can construct new forces and fluxes for which it is valid by splitting the matrix of coefficients M_{ij}' into symmetrical and antisymmetrical parts, and redefining the fluxes by a transformation which is the inverse of (15.7). It might be thought that this is a satisfactory way of obtaining the correct forces and fluxes which are both physically significant and satisfy the Onsager relations: unfortunately this choice is not unique, since any transformation of form

$$\begin{aligned} Z_i^* &= A_{ik}Z_k \\ I_i^* &= A_{ki}^{-1}I_k, \end{aligned} \tag{15.10}$$

will yield another set of forces and fluxes which satisfy eqns. (14.3), (15.4), and (14.4) in the same way as the first set.

The general conclusion is thus that there is no real meaning to the Onsager result (14.4) if it is based on a choice of forces and fluxes which satisfy eqn. (15.4) and which are assumed to be linearly related. This does not mean that the Onsager relations themselves are not useful; the logical difficulties disappear if it can be shown that the forces and fluxes satisfy eqns. (14.8) and (14.9). Unfortunately, in most applications of the theory this has not been done, so that phenomenological descriptions which use the Onsager relations are frequently rather unstisfactory at present; they depend on physical intuition in selecting appropriate forces and fluxes rather than on a logical treatment. This is certainly the position in the theory of diffusion, where Coleman and Truesdell point out that there is no convincing evidence that the relations are valid. However, Howard and Lidiard (1966) have pointed out that some in the theories of diffusion include the Onsager relations as a consequence of the assumptions.

We also emphasize again that the thermodynamic theory does not tell us anything about the validity of the linear relations (14.3). Thus even if suitable forces and fluxes have been found, the linear relation between them is a matter for experiment or for kinetic theory. In particular, there is not generally a linear relation between the affinity of a chemical reaction and the time derivative of the degree of advancement, unless the reaction is very close to equilibrium. This may be see, from the theory of the last chapter. The forward rate of the reaction is proportional to the factor $\exp(-\Delta_a g/kT)$, where $\Delta_a g$ is the free energy of activation per molecule. The back reaction will have the same activated complex, and will thus have a reaction rate proportional to $\exp\{(-\Delta_a g+g_i a_i)/kT\}$, since the activation energy is increased by the difference in the standard free energies of reactants and products. The net rate of reaction will thus be proportional to

$$\exp(-\Delta_a g/kT)\,[1-\exp(g_i a_i/kT)] \qquad (15.11)$$

and will thus be linear in $-g_i a_i$ only if the reaction is so close to equilibrium that $-g_i a_i \ll kT$.

It may be shown quite generally that the validity of the assumption of linear relations and the validity of the proof of Onsager's theorem both depend on the existence of small and definable deviations from equilibrium. In chemical reactions, the free energy per molecule released during the process must be less than kT. Transport phenomena are usually observed in the linear region; chemical reactions seldom are.

It follows from our discussion above that there is a general class of linear transformations of forces and currents which leaves the reciprocal relations valid. In particular, it is possible to find a linear transformation which will reduce the matrix M to diagonal form. This means that by suitably combining the forces and currents, we obtain phenomenological relations without cross terms. Such combinations have no particular physical significance.

We note that for any single physical property of a three-dimensional crystal, e.g. properties of the type (14.3c), the Onsager relations must be combined with the symmetry properties of the crystal to obtain the scheme of coefficients. Thus we find that for vector–vector properties, the second order tensor reduces to a scalar if the symmetry is cubic. Properties like the electrical conductivity and the coefficient of diffusion are thus isotropic in cubic materials, although three coefficients are needed to specify properties like the elastic stiffnesses which relate two second order tensors. A detailed survey of the effect of crystal anisotropy on physical properties is given by Nye (1957).

The description which we have given of the thermodynamics of irreversible processes has been concerned almost exclusively with the Onsager relations. In equilibrium thermodynamics, the equilibrium state is characterized by stationary values of certain extensive thermodynamic parameters, the appropriate condition (maximum entropy, minimum free energy, etc.) depending on the external constraints. The discovery of the Onsager relations suggests the possibility of finding similar conditions which will define the steady state, and such a generally applicable principle would undoubtedly be of great assistance in formulating kinetic theories of steady state or quasi-steady-state processes. In fact, a variational principle has been suggested for this purpose by de Groot (1951), following a theorem due to Prigogine, and has been much discussed by these and other workers. This principle is that the rate of entropy production, d_iS/dt, has a minimum value in the stationary state, consistent with the auxiliary imposed conditions which can be regarded as fixing certain of the forces Z_i at constant levels.

The principle of minimum entropy production is one of the most controversial developments of the formal theory of this chapter, and we shall not discuss it further here. In recent years, some workers have used the principle in theories of metallurgical growth processes. However, it is evident that the theory does not hold in all circumstances, as may be shown by choosing particular examples where the answer is known, and no general rule for defining the limits of applicability seems to have been discovered. It thus seems unjustified to apply the theory to the rather complex conditions of a growth process, where it may or may not be valid.

REFERENCES

COLEMAN, B. D. and TRUESDELL, C. (1960) *J. Chem. Phys.* **33,** 28.
DE GROOT, S R. (1951) *Thermodynamics of Irreversible Processes,* North-Holland, Amsterdam.
HOWARD, R. E. and LIDIARD, A. B. (1964) *Reports* on *Progress Phys.* **27,** 161.
NYE, J. F. (1957) *Physical Properties of Crystals,* Clarendon Press, Oxford.
ONSAGER, L. (1931a) *Phys. Rev.* **37,** 405; (1931b) *Ibid.* **38,** 2265.

CHAPTER 5

The Structure of Real Metals

16. METALLIC STRUCTURES: POLYMORPHISM

For most purposes, the kinetic theory of gases provides a sufficiently accurate description of the metallic vapour state. In this chapter, and in Chapters 6–8, we shall discuss approximate models of the structure of ideal solids. We shall also briefly describe our present ideas about liquids, but unfortunately a satisfactory simple model of the liquid state has not yet been developed.

The description of the ideal crystal given in Chapter 2 is evidently incorrect. Even at 0 K the atoms are not rigidly fixed, but vibrate about mean positions because of the residual zero-point energy. The root mean square of the amplitude of thermal vibrations increases with temperature, and may become an appreciable fraction of the interatomic distance. These thermal vibrations are of fundamental importance in almost all kinetic phenomena; however, they may be eliminated from our discussion of the structure of a real crystal merely by taking a suitable time average of the atomic positions. When this is done, we find that the actual atomic arrangement in any real crystal differs only slightly from the ideal mathematical description of Chapter 2, and it is then convenient to describe the real crystal in terms of an ideal reference crystal possessing a number of "defects".

The concept of a defect implies that a localized region of the actual crystal can be associated unambiguously with a corresponding region of the ideal crystal. Although the two sets of atoms will not have identical mean positions, the differences can be described by small (elastic) displacements, and it is always possible to establish a unique one to one correspondence between the atoms of the two crystals. Regions of crystal where this local correspondence is possible are called "good" crystal (Frank, 1951a); the remaining regions of "bad" crystal may be considered to be localized defects in the structure.

Defects may be classified into point, line, or planar types, according to whether the severely distorted region extends many interatomic distances in zero, one, or two dimensions. Point defects are surrounded entirely by good crystal, and if a real crystal contains only point defects, it will be possible to establish a unique correspondence between all its atoms and those of the reference crystal. This is no longer true when line defects (dislocations) are considered, and we have been careful in general to claim only a local correspondence between regions of good crystal and of the reference crystal. In this chapter, we shall consider the properties of point defects, but discussion of dislocations is deferred to Chapter 7.

We assume that the vast majority of the atoms present in a real crystal lie in regions of good crystal, so that we are justified in treating this real crystal as an ideal crystal with a distribution of defects. The justification for this assumption is that it accords with experimental evidence; for example, we may cite the way in which the ideal structure may be inferred from X-ray diffraction experiments on real crystals.

We shall be concerned mainly with the three common metallic structures, namely body-centred cubic, face-centred cubic, and hexagonal close-packed.† The two cubic structures have primitive unit cells of rhombohedral shape, in which the angles between the axes are 109°30′ for the b.c.c. structure, and 60° for the f.c.c. structure. These structures are single lattice structures, each atom being a point of the Bravais lattice. The conventional unit cells, chosen to show the symmetry of the atomic positions, contain two and four atoms respectively. In the h.c.p. structure the unit cell of the Bravais lattice is also the conventional unit cell; it has two equal vectors at 120° to each other, and a third perpendicular vector. This primitive unit cell contains two atoms, so that this is a double lattice structure. Some metals crystallize in other double lattice structures, and these will be of interest in our discussion of mechanical twinning (Part II, Chapter 20).

Many pure metals undergo polymorphic transitions from one equilibrium structure to another, and the frequency of such transitions in elements requires some comment. For two possible structures, α and β, the difference in the corresponding Gibbs functions per atom at temperature T is $\Delta g_T^{\alpha\beta} = g_T^{\beta} - g_T^{\alpha}$, where

$$\Delta g_T^{\alpha\beta} = \Delta h_0^{\alpha\beta} - \int_0^T \int_0^T \Delta C_p^{\alpha\beta}\, \mathrm{d}(\ln T)\, \mathrm{d}T. \qquad (16.1)$$

Here $\Delta h_0^{\alpha\beta}$ is the difference in the enthalpies per atom in the two structures, $h^{\beta} - h^{\alpha}$, measured at 0 K, and $\Delta C_p^{\alpha\beta}$ is the difference, $C_p^{\beta} - C_p^{\alpha}$, in the specific heats at constant pressure for the two structures. The main contribution to the specific heats arises from the lattice terms; if the Debye approximation is used, it follows that a positive $\Delta h_0^{\alpha\beta}$ implies a lower Debye temperature for the β phase, and hence a positive ΔC_p. Thus $\Delta g^{\alpha\beta}$, which is positive at low temperatures, decreases as the temperature rises and may become negative. If this happens below the melting point, there is an equilibrium transition from α to β.

Figure 5.1 shows the expected form of the curve of $\Delta g^{\alpha\beta}$ against T according to (16.1). The slope of the curve changes from zero at 0 K to a maximum negative value $\int_0^{T'} \Delta C_p^{\alpha\beta}\, \mathrm{d}T$, where T' is the temperature at which $C_p^{\beta} = C_p^{\alpha}$. If the lattice terms are the only important contributions, $\Delta C_p^{\alpha\beta}$ is effectively zero at all temperatures $T > T'$, so that the curve becomes a straight line of fixed slope.

For the transition elements, and for ferro- or antiferromagnetics, more detailed discussion is required, since electronic and magnetic terms can then contribute appreciably to ΔC_p. Thus the double transition in iron

$$\text{b.c.c.}\ (\alpha) \rightleftharpoons \text{f.c.c.}\ (\gamma) \rightleftharpoons \text{b.c.c.}\ (\delta),$$

† We shall henceforth use the abbreviations b.c.c., f.c.c., and h.c.p. for the common metallic structures. For details of these structures and of the other metallic structures which may be mentioned, see W. Hume-Rothery and G. V. Raynor, *The Structure of Metals and Alloys*, and C. S. Barrett and T. B. Massalski, *The Structure of Metals*, etc.

was explained by Seitz (1940) in terms of the lower Debye temperature of the f.c.c. phase and the greater electronic specific heat of the b.c.c. phase. It is evident from eqn. (16.1) that such a double transition implies that ΔC_p is positive at low temperatures and negative at high temperatures. According to Seitz's theory, the positive contribution comes from the lattice vibrations, and the negative contribution from the electronic terms.

A slightly different and more complete discussion was given by Zener (1955). He pointed out that the $\Delta g^{\alpha\gamma}$ curve is linear over the approximate temperature range 100–500°C, and

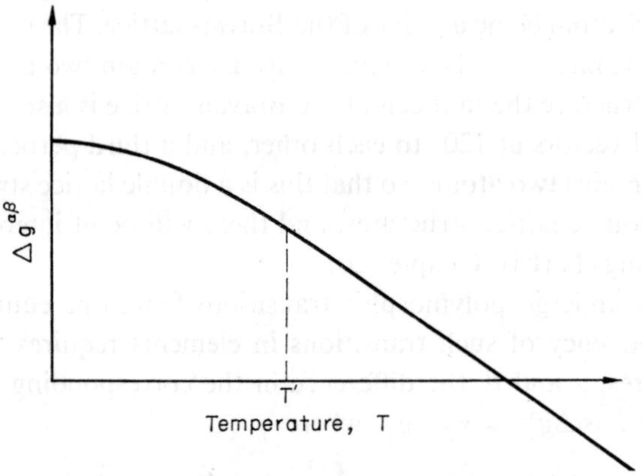

FIG. 5.1. Free energy vs. temperature relation according to. (16.1) when lattice terms are dominant.

that it then gradually acquires an increasing positive curvature which reaches a maximum in the region of the Curie temperature (760°C). Since (eqn. (16.1)) the positive curvature implies a negative $\Delta C_p^{\alpha\gamma}$, the results show that there must be an increase in C_p^{α} above 500°C, reaching a maximum near the Curie temperature. This increase may be identified quite naturally with the anomalous specific heat associated with the magnetic disordering of the ferromagnetic α phase. Since the total heat effect at the Curie temperature is greater than the latent heat of the $\alpha \rightarrow \gamma$ transition, it is obvious that the magnetic disordering must appreciably affect the equilibrium state. Zener estimates that the total entropy change due to magnetic disordering is $\boldsymbol{R} \ln 3 = 9{\cdot}2$ J mole^{-1} K^{-1} (2·2 cal mole^{-1} °C^{-1}), and this is actually slightly greater than the observed change in the entropy difference between the phases in the temperature range from 500°C to the melting point.

Figure 5.2 shows the experimental curve for $\Delta g^{\alpha\gamma}$ in pure iron, together with the suggested analysis into non-magnetic and magnetic components. Note that although the magnetic disordering does not prevent the $\alpha \rightarrow \gamma$ transition, it raises the temperature appreciably. It follows from the curve that if α-iron had a lower Curie temperature, there would be no equilibrium γ field in pure iron.

This description of the iron transition enables the effects of alloying elements on the $\alpha \rightleftharpoons \gamma \rightleftharpoons \delta$ equilibrium and on the martensitic start temperature M_s to be understood. Iron equilibrium diagrams are usually discussed in terms of whether the alloying element lowers the free energy of the α phase relative to that of the γ phase (leading to equilibrium diagrams with "γ loops"), or raises the relative free energy of the α phase (leading to open γ fields). However, Zener pointed out that this conventional interpretation is sometimes inconsistent with observed effects on the M_s temperature. For example, molybdenum gives a γ-loop type diagram, but also lowers the M_s temperature, the latter effect suggesting a lower relative free energy for the γ phase. Difficulties of this type can be reconciled when it is realized that there are two independent effects contributing to $\Delta g^{\alpha\gamma}$, and these may be differently affected by the alloying elements.

Zener's suggestion was further developed by Weiss and Tauer (1956), who proposed methods for separating the thermodynamic functions such as enthalpy, entropy, free energy,

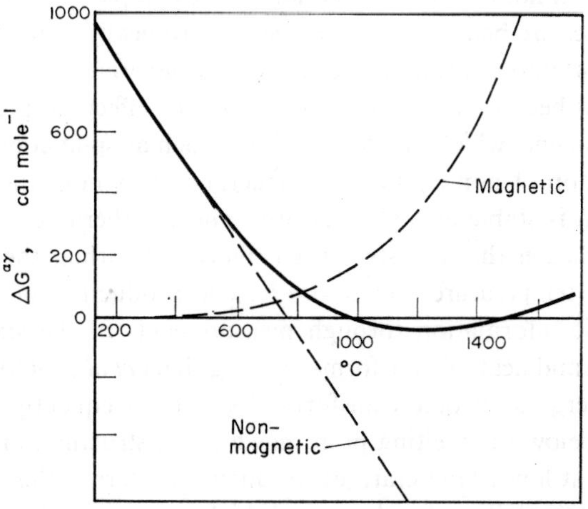

FIG. 5.2. Suggested division of the free energy vs. temperature relation for α and γ iron into non-magnetic and magnetic components (after Zener, 1955).

etc., into lattice, magnetic, and electronic terms, the separation being based on assumed additivity of the corresponding specific heat contributions. Their analysis shows that the total magnetic entropy at the melting point approaches the theoretical value $\mathbf{R}\ \ln(2s+1)$, where s is the unpaired spin per atom, and the magnetic enthalpy is of order $\mathbf{k}T_c$, where T_c is the Curie temperature. They conclude, as did Zener, that the b.c.c. α phase is stabilized only because of the magnetic terms, in the absence of which γ would be the stable low temperature form. However, Zener's detailed assumptions are criticized as inadequate; for example, the straight lines obtained for the non-magnetic free energy difference in Figs. 5.1 and 5.2 depend on the assumption of equal coefficients of linear expansion as well as equal electronic specific heats for the two phases. The effect of alloying elements is rather complex, since changes in the Bohr magneton numbers and in the Debye temperatures must be included in the magnetic and non-magnetic terms respectively.

In a later analysis of the magnetic properties of iron and its alloys, Kaufmann *et al.* (1963) and Weiss (1963) suggested that atoms in the γ phase may exist in two distinct electronic states, separated by a small energy gap. The ground state is supposed to be ferromagnetic with a small magnetic moment, whilst the higher energy state is antiferromagnetic with a higher moment. In pure γ-iron, the proportion of atoms in the ground state is never large enough to allow long-range coupling, and hence macroscopic ferromagnetism, but the addition of nickel (Weiss, 1963) or chromium (Miodownik, 1970) stabilizes the ferromagnetic f.c.c. structure. Anomalous magnetic, electrical, and thermodynamic properties may be rationalized in this way, but the model is obviously based on a rather artificial assumption.

Weiss and Tauer (1958) and Kaufman (1959a, b) also applied the procedure of separating the thermodynamic functions into component terms to a number of non-ferrous metals and alloys. The results for manganese show that the b.c.c. structure which is stable as the δ phase just below the melting point, would also be the equilibrium structure at low temperatures, as in iron, were it not for the intervention of the complex α- and, β-manganese structures. These structures are believed to be related electronically to the b.c.c. structure. More recently, experimental information on the relative free energies of different structural form of various metals has been obtained from studies of the effects of pressure on phase stability, and from experiments which utilize techniques such as splat cooling or vapour deposition for the production of non-equilibrium structures. It is now known, for example, that a h.c.p. form of iron is stable at high pressures, and the thermodynamic analysis (Blackburn *et al.*, 1965) indicates that this structure is more stable than the f.c.c. form at atmospheric pressure and temperatures below 400 K. The production of metastable structures gives thermodynamic information through measurements of the specific volume change (Chopra *et al.*, 1967) and heat of transformation (e.g. Jena *et al.*, 1968).

When metals undergo allotropic transformations, it is frequently found that the phase stable immediately below the melting point has a b.c.c. structure, and more close-packed structures are stable at lower temperatures. In qualitative terms, this is readily understood, since the b.c.c. structure will tend to have both higher energy and higher entropy than the close-packed structures. For example, in a central force, hard sphere model for the binding forces, the b.c.c. structure has zero resistance to shear on a $\{110\}$ plane in a $\langle 1\bar{1}0\rangle$ direction, and this should lead to a large amplitude of thermal vibration in this mode in real crystals. This led Zener (1947) to suggest that in fact many metals which appear to be b.c.c. at all temperatures might become close-packed at sufficiently low temperatures, where entropy effects are unimportant. Subsequent search (Barrett, 1947, 1956) led to the discovery of the martensitic transformations to h.c.p. or f.c.c. structures in lithium below $\sim$77 K and in sodium below $\sim$36 K. The apparent anomaly of high melting-point materials like chromium which have b.c.c. structures at all temperatures, and of iron which has a b.c.c. structure stable at lower temperatures than the f.c.c. phase, are now explained by the low energy and entropy of this structure in ferromagnetic and antiferromagnetic materials below the Curie or Néel temperature. Measurements of the specific heat of h.c.p. iron–ruthenium alloys (Stepakoff and Kaufman, 1968) when extrapolated to pure iron indicate a Debye temperature of $\sim$375 K, which is appreciably lower than the estimate ($\sim$432 K) for α-iron; this result thus confirms that in this case the thermal entropy of the close-packed phase is higher than that of the b.c.c. phase.

It is worth noting here that although we use the common term close-packed structure for f.c.c. and h.c.p. phases, we do not intend to imply that the volume per atom is necessarily smaller in these structures than in a b.c.c. structure of the same metal. Rudman (1965) has emphasized that for many metals, including the alkalis but excluding iron and manganese, the b.c.c. structure has the *smaller* atomic volume. In most cases, the change in volume at an equilibrium transformation temperature is $\lesssim 1\%$; a notable exception is tin, where the low temperature (α) phase is a diamond-cubic structure with a volume per atom more than 20% larger than the tetragonal high temperature (β) phase. The measurements of Chopra *et al.* also indicate that the metastable f.c.c. phase in several metals has a much larger specific volume than the equilibrium structure. It is a common misconception that the b.c.c. phase has a more "open" structure and that this leads to a high thermal entropy.

In the theory of phase transformations we are often able to make direct use of thermodynamic data on the free energies of the phases concerned. Even when such data are available, however, it is not always possible to relate it to the atomic scale phenomena we wish to describe, and we have then to use either theoretical or empirical models for the interatomic forces and energies. In addition to the common metallic structures, models of this kind are sometimes used with an assumed simple cubic structure, although this does not occur in practice in metals. An important theoretical concept is the binding energy of an atom in various positions in the interior or on the surface of a crystal. The simplest possible model uses the assumption that this energy can be expressed as the sum of a series of independent interaction energies with the atoms which are nearest to the atom considered. For example, suppose we have an atom in the interior of a simple cubic crystal. Such an atom has six nearest neighbours each at distance a in $\langle 100 \rangle$ directions, twelve next nearest neighbours each at distance $a\sqrt{2}$ in $\langle 110 \rangle$ directions, and eight third nearest neighbours each at distance $a\sqrt{3}$ in $\langle 111 \rangle$ directions. If we assign characteristic interaction energies $-2\Xi_1$, $-2\Xi_2$, $-2\Xi_3$ to atom pairs at these distances, we may write the energy of the atom as $-(6\Xi_1+12\Xi_2+8\Xi_3)$, and this is the potential energy of the structure per atom. The potential energies are usually expressed relative to the free atoms (i.e. the vapour phase). They are then all negative, so that Ξ_1, Ξ_2 and Ξ_3 are positive quantities as defined here, with $\Xi_1 > \Xi_2 > \Xi_3$. We are assuming that we have a pure metal, so that Ξ_1, Ξ_2, Ξ_3 do not vary with the atoms forming the pair; solid solutions are considered in the next chapter. We also assume that the interaction energy between two atoms is not affected by their environment, and in particular by lattice defects or by the presence of a surface.

This simple model of a crystal thus consists of an assembly of atoms held together by short range central forces. There are solids for which this is a good approximation, for example, the solid rare gases in which the molecules are bound together by van der Waals (polarization) forces, but it is obviously a very poor representation of the metallic solid state. The failure of the Cauchy relations (see p. 72) for almost all metals shows at once that interatomic forces are not even approximately central, but recent work on the electron theory of non-transition metals shows, nevertheless, that the major part of the change in energy with atomic configuration at *constant atomic volume* may be equivalent to a central force interaction. General accounts of the electron theory of metals are given by Callaway (1964), Ziman (1964), Lomer (1969), Lomer and Gardner (1969), and Altmann (1970), whilst more advanced accounts are collected in the volume presented to Sir Nevill Mott (Ziman, 1969b). Theory and experiment both suggest that the most appropriate picture of metallic cohesion, at least for the simpler metals, is that of a lattice of positive ions held together by a "gas" of negative electrons. The complex electron–electron interactions of the quantum-mechanical many-body problem are then replaced by the approximation of one electron in a self-consistent field (band theory), and experimental justification for the validity of this assumption has been provided by measurements of the shape of the Fermi surface and related properties (Cracknell, 1969, 1971). The energy of the crystal, which is made up of the potential energies of all the particles plus the kinetic energies of the electrons, may then be expressed approximately as the sum of several nearly independent terms which are:

(a) The energy of the lowest state of a valency electron in the field of the positive ions.

(b) The kinetic or Fermi energy of the valency electrons, resulting from the Pauli principle which prevents all electrons from occupying the lowest state.

(c) A short-range repulsive energy arising from the overlap of closed shell positive ion cores. This energy includes the purely Coulomb interaction of the charge distribution, and an exchange energy, the latter being the more important.

(d) An attractive potential resulting from van der Waals forces between the ion cores.

(e) The remaining electrostatic potential energy of the lattice and electrons, excluding the interaction between a positive ion and its "own" electron, already included in (a).

(f) Terms arising from positional correlation and exchange effects of the valence electrons.

Many methods are now used for the calculation of one-electron (Bloch) functions, and an extensive survey is given by Ziman (1971). Despite much progress, the difficulty of making first principle calculations remains formidable, and in particular a self-consistent type of solution for the effective potential is required. Moreover, the electron structures of some metals, especially the transition metals, are so complex in comparison with those of the univalent alkali and noble metals, that it is by no means certain that the one electron band description is appropriate. In these metals, the ions have incomplete outer shells, and it is generally accepted that additional attractive forces may arise from exchange spin interactions, giving ferromagnetic or antiferromagnetic structures.

The terms (a) and (b) above do not correspond in any way to central forces, although the gradient of term (a) may be a maximum along the line of centres of the two atoms. The energy terms (c) and (d) correspond to repulsive and attractive forces acting along the line of centres, but are not independent of the number and positions of the surrounding atoms. However, we may expect the central force approximation to be most valid when one or both of these terms are important, as in the case of the univalent noble metals (Fuchs, 1935). These metals (copper, silver, gold) have "full" structures which may be regarded approximately as the close packing of rather hard atomic spheres. The term (e) is usually small for a structure at equilibrium under zero stress, but Fuchs (1935) has shown that it may become appreciable when the lattice is distorted. The electrostatic energy can thus be significant in determining the elastic constants, and must be included when mechanical stability of the lattice is considered.

The Fermi energy (term (b)) is proportional to the atomic volume to the power $(-\frac{2}{3})$ if the electrons behave as free electrons. The energy of the lowest state is also dependent mainly on the atomic volume, so that a model of the metallic state with a central force interaction between neighbouring atoms plus an energy term which is a function only of the atomic volume is a better approximation than the simpler central force model. In real metals the forces between atoms depend on the number and distribution of the other neighbours, and the Fermi energy, for example, is a function both of the atomic volume and of the shape of the Fermi surface. Nevertheless, many effective calculations have been made by the method of pseudopotentials (Harrison, 1966, 1972; Heine, 1970), which depends on the result that for non-transition metals the energy vs. wave-vector relations in the regions between spherically symmetrical atomic cores conform to the nearly free electron form (relative small pertubations to planar Bloch functions) even though the electrons close to the atom centres are very tightly bound. For many purposes, the complex real potential may be replaced by a much simpler pseudopotential which reproduces the interactions in the important interstitial regions and which may be regarded as the superposition of spherically symmetrical fields centred on an array of "pseudoatoms". The essential feature of the pseudopotential $\Xi_p(\mathbf{r})$ is that its Fourier transform $\Xi_p(\mathbf{q})$, where $\mathbf{q}$ is a reciprocal lattice vector, must give the correct scattering of an electron wave $\mathbf{k}$ to $\mathbf{k}+\mathbf{q}$. This theory has been reviewed by Heine (1970) and Cohen and Heine (1970), and its application to the calculation of the energies of the structures of different specific metals is discussed at length by Heine and Weaire (1970).

Pseudopotential calculations may be carried out either in reciprocal space ($\mathbf{k}$ space) or in real space. When the latter representation is used, the calculations show that for simpler metals, the part of the energy which depends on atomic positions (as distinct from that which depends on the average density of matter) is equivalent to a central force interaction. These results thus provide some justification for models based on pairwise interactions, and the development of high-speed computing techniques has led to appreciable advances in such model calculations of structural energies and of defect configurations. Suitable potentials are presently available only for a few metals, and the central force approximation is in any case of very doubtful validity for the transition metals. Nevertheless, the use of the pairwise approximation with various empirical potentials may still be useful inasmuch as it is believed that certain qualitative features of the predictions (e.g. the "shape" of a dislocation core, or the possible existence of a metastable fault configuration) are determined mainly by the crystal structure. The value of the empirical pairwise method is that the stability of the structure may be ensured so that these qualitative properties may be studied. However, it is very important to emphasize that the method is valid only for the study of atomic rearrangements at constant density. In kinetic problems we may often be interested only in order of magnitude estimates of energies, and the effects to be discussed are frequently not very sensitive to the model used. We shall therefore use the central force approximation rather freely, and we shall consider the relative magnitudes of the potentials Ξ_1, Ξ_2, Ξ_3, etc.

The mutual potential energy of two atoms in a molecule, in a liquid, or in a solid held together by central forces must always be of the general form shown in Fig. 5.3. This corresponds to a repulsive force which dominates at small separations of the atoms, and an attractive force which is more important at larger separations. These two forces are balanced at some equilibrium separation r_0, which would be the actual distance between the atoms in a di-atomic molecule. In a solid, r_0 will not necessarily be equal to the nearest neighbour separation r_1, because of the influence of the more distant neighbours, but it will be near

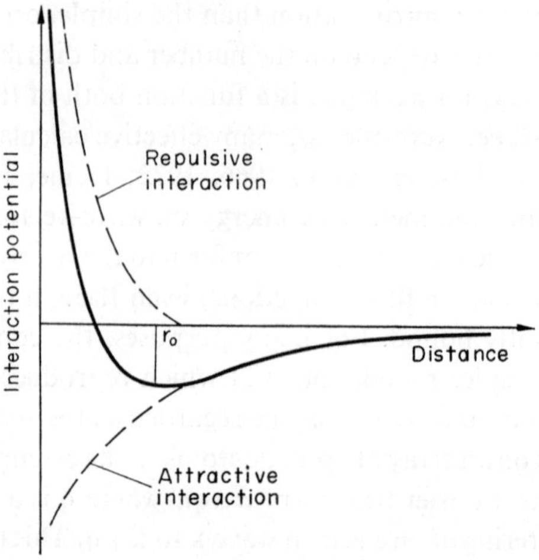

FIG. 5.3. Schematic form for the interaction potential between two atoms which is assumed in central force models.

to r_1 if the forces are all central. If non-central forces are present in addition, r_1 may be appreciably different from r_0, and r_0 may not exist at all. As we have already noted, the central forces in copper are wholly repulsive if van der Waals forces are neglected, so that the pairwise potential probably decreases sharply towards zero with increasing r, without ever becoming negative. We write the potential as $-2\Xi(r)$ for an atom pair at separation r, so that the potential energy per atom in a solid, W, may be written as a lattice sum, $\Sigma - \Xi(r)$, taken over all possible pairs which one atom forms with the remaining atoms. When the forces are short-range in character, only the near neighbours of an atom need be included in this sum, as assumed above.

The form of $-\Xi(r)$ when this is the only potential present must be such that the lattice is mechanically stable, and the energy must rise for any small perturbation of the atoms from their equilibrium positions. When there are several alternative structures which are mechanically stable, the force law will show that one of these structures has a lower internal energy than the others. Ideally, this structure should be that actually observed at very low temperatures, whilst at finite temperatures, the thermodynamically stable structure will be that of lowest free energy.

The conditions that the potential $-\Xi(r)$ shall give a structure which is mechanically stable are first that it has a minimum at finite r, and secondly that for large r the magnitude of Ξ tends to zero more rapidly than r^{-3}. These two conditions together guarantee the stability of any structure with respect to infinitesimal homogeneous expansions or contractions of the lattice, the second of them ensuring that the total lattice energy is finite. Consider a crystal of N atoms in positions $\mathbf{y}_i$ and let the corresponding defect free crystal have atom positions $\mathbf{x}_i$. Then if the pairwise ineraction energy is $-2\Xi(r)$, the lattice energy per atom W is given by the sum

$$NW = \sum_{i,j}(-\Xi(|\mathbf{y}_i - \mathbf{y}_j|)) + U_D(e_{kl}), \tag{16.2}$$

where non-central forces are included only through the "deformation energy" U_D, which is zero for the perfect lattice. For small deviations from the perfect lattice, the linear approximation

$$U_D = \tfrac{1}{2}\sigma_{kl}e_{kl}, \tag{16.3}$$

where

$$\sigma_{kl} = \frac{1}{2v_a}\sum_{i,j}\left[\frac{(x_i)_k(x_j)_l}{|\mathbf{x}_i-\mathbf{x}_j|}\;\frac{\partial\Xi(|\mathbf{x}_i-\mathbf{x}_j|)}{\partial(|\mathbf{x}_i-\mathbf{x}_j|)}\right] \tag{16.4}$$

may be used (Born and Huang, 1954). The components σ_{kl} are those of an effective stress which must be supposed to be applied externally in order to stabilize the structure. Suppose, for example, that the potential $\Xi(r)$ has been adjusted to match measured elastic constants for the (assumed centrosymmetric) structure defined by $\mathbf{x}_i$. Then unless these constants satisfy the Cauchy relations (11.20), the minimum of energy will not correspond to the atomic positions $\mathbf{x}_i$ but to some other set of atomic positions. The addition of the external stress system returns the atoms to positions $\mathbf{x}_i$, and it follows that the six components σ_{kl} may be expressed in terms of the failure of the six Cauchy relations. In particular, for cubic symmetry eqn. (16.3) reduces to

$$U_D = p\,\Delta V, \tag{16.5}$$

where $p = \frac{1}{3}\sigma_{kk}$ is the effective hydrostatic pressure and ΔV is the total lattice dilatation. To ensure the stability of the perfect lattice, p must equal $\frac{1}{2}(c_{44}-c_{12})$; it is important to note that this effective pressure can be quite large. It is also necessary to consider mechanical stability against infinitesimal shear deformations. For cubic crystals, Born (1940) showed that the required conditions are that all the (non-zero) elastic stiffnesses are positive, and that $c_{11}-c_{12} > 0$. Equivalent expressions for h.c.p. crystals were given in a later paper (Born, 1942). A positive value of c_{44} is needed to ensure resistance to shear along $\{100\}$ planes in $\langle 010\rangle$ directions, whilst a positive value of $c_{11}-c_{12}$ stabilizes the lattice against $\{110\}\langle 1\bar{1}0\rangle$ shear. The four conditions taken together ensure that the strain energy function for the most general infinitesimal deformation is always positive definite.

The most used approximation for $\Xi(r)$ in the early work was

$$-\Xi(r) = A(r_0/r)^n - B(r_0/r)^m \qquad (n > m), \tag{16.6}$$

where the first term represents the repulsive potential, and the second term the attractive potential, r_0 being the equilibrium separation discussed above. According to Born's calculations, and to those of Misra (1940), this law of force always gives a mechanically stable f.c.c. structure, but the simple cubic structure is always mechanically unstable with respect to a $\{100\}\langle 010\rangle$ shear. A b.c.c. structure is mechanically unstable with respect to a $\{110\}$ $\langle 1\bar{1}0\rangle$ shear, except for very slowly varying interactions ($m < n < 5$).

It is instructive to examine the reasons for mechanical instability. The elastic constants are derived from the potential energy function by using the relations (see pp. 72)

$$vc_{ij} = (\partial^2 W/\partial e_i\,\partial e_j),$$

where v is the atomic volume. This results in expressions in which the elastic constants are given as the sums of series of terms of the form $-(r/v)(\partial\Xi/\partial r)$ and $(r^2/v^2)(\partial^2\Xi/\partial r^2)$ evaluated for $r = r_1, r_2, \ldots$, etc., corresponding to nearest neighbour, next nearest neighbour, etc., separations. In first order elasticity theory (neglect of terms in e^2), there is no contribution from nearest neighbour interactions to c_{44} in a simple cubic structure, or to $c_{11}-c_{12}$ in a b.c.c. structure, so that the shear modulus is zero in first approximation. This is a simple consequence of the fact that in these two cases the nearest neighbour bonds are all either in the invariant plane of the shear or normal to it, giving zero changes in nearest neighbour distances in first approximation. There is no such set of planes in the f.c.c. structure. When interactions with second nearest neighbours are considered, the dominant term is usually that in $r_2(\partial^2\Xi/\partial r^2)_{r=r_2}$, and as can be seen from Fig. 5.3, this will be negative unless the second nearest neighbour distance lies to the left of the inflection point of the $-\Xi(r)$ curve. This is the basis of the Born and Misra calculations, which include interactions with all neighbours, but are based on first-order elasticity theory. The possible stability of the b.c.c. structure then depends on second nearest neighbour interactions.

An alternative procedure is to focus attention on near neighbour interactions but to use second-order elasticity theory. It can then be shown that

$$v(c_{11}-c_{12}) = -(8r_1/3)\,(\partial\Xi/\partial r)_{r_1} + r_2^2(\partial^2\Xi/\partial r^2)_{r_2} - r_2(\partial\Xi/\partial r)_{r_2} + \ldots \qquad (16.7)$$

(Fuchs, 1935, 1936; Isenberg, 1951). A contribution from nearest neighbours now appears, although if $r_1 \approx r_0$ this will be very small since $(\partial\Xi/\partial r)$ is zero when $r = r_0$. The essential reason for the instability of the b.c.c. structure is shown by the absence of a term in $(\partial^2\Xi/\partial r^2)_{r_1}$ even in second-order theory.†

The mechanical instability of the b.c.c. structure in the central force approximation cannot apply to real metals which have this structure. We should expect such instability, whenever we have rapidly varying repulsive forces resulting from closed shell overlap. The quantum mechanical calculation of Fuchs (1935) show that the electrostatic energy (term (e)) gives a positive contribution to $c_{11}-c_{12}$, but in the case of the noble metals, this is outweighed by the large repulsive force (term (c)). When the central force interaction is repulsive, the first term in eqn. (16.7) becomes dominant and gives a negative contribution to $c_{11}-c_{12}$; a hypothetical b.c.c. form of copper would thus be mechanically unstable, and would shear immediately to the f.c.c. form.

For the alkali metals, closed shell interaction is much smaller because of the large separations of the ions, and the electrostatic term stabilizes the b.c.c. structure. Nevertheless, an incipient tendency towards instability may be detected in several real structures which have abnormally low values of $c_{11}-c_{12}$. Absolute values of shear constants vary, of course, from one metal to another, so a convenient measure of the tendency towards instability is the anisotropy factor $2c_{44}/(c_{11}-2c_{12})$ (Zener, 1948). This is equal to unity for an isotropic elastic body (p. 73), and is large when there is a low resistance to $\{110\}\langle 1\bar{1}0\rangle$ shear. Some values are given in Table IV, p. 74; the f.c.c. metals indicate the range expected when there is no tendency towards instability. Clearly the b.c.c. transition metals, which do not have

† Equation (16.7), also given by Mott and Jones (1936) for nearest neighbour interactions, is not without ambiguity. Difficulties arise with shears in second order elasticity theory, since either the deformation is not dilatationless, or else the strains are not symmetrical with respect to the x- and x- axes.

closed shell ions, are in a different category from the alkali metals, and from β-brass. As we have noted, the stability of the b.c.c. structure in the transition metals is probably due to energy terms arising from the interactions of electrons in the incomplete outer shells.

A low value of $c_{11}-c_{12}$ implies a high entropy contribution from the corresponding vibrational modes of the lattice, and hence a free energy which varies rapidly with temperature. This is the basis of Zener's prediction, subsequently verified for lithium and sodium, that the equilibrium structures of the alkali metals at low temperatures should be close-packed (Zener, 1947, 1948). It is also noteworthy that β-brass, the cubic structure showing greatest elastic anisotropy, is only stable at high temperatures. According to Jones (1952), the mechanical stability of this structure may be due to the long-range (Fermi) energy term. When this phase is quenched, the order–disorder transition cannot be suppressed, and Jones suggested that this may mean the Fermi energy cannot stabilize the structure against "localized shears" sufficient to cause ordering.

Returning to the central force model, we know that for crystals of the solidified rare gases the attractive potential results from van der Waals forces, and it can be shown that $m = 6$. The most appropriate repulsive potentials seem to correspond to $n = 10$–12; when 12 is chosen, the interactions are sometimes known as Lennard-Jones forces (Lennard-Jones, 1937). We then find that Ξ_3 is negligible, and for most crystal structures $\Xi_2 \sim 0{\cdot}1\Xi_1$. Second nearest neighbour interactions may thus also be ignored, except for some special purposes, and we have the nearest neighbour model of a crystal structure with a single interatomic potential $\Xi_1 = \Xi$. As we shall see in the next chapter, this model is especially convenient for solid solutions, since it allows the variation of thermodynamic quantities with composition to be expressed by a single parameter.

Instead of assuming the form of the force law, we may retain eqn. (16.6) in general form and evaluate the parameters by comparison with experimentally determined quantities for a particular metal. Furth (1944) showed that A and B may be deduced from the known lattice spacing and sublimation energy, and the temperature and pressure variation of the sublimation energy, compressibility, and coefficient of thermal expansion may then be used to obtain best values of m and n. In this way, Furth deduced empirical equations of state for many solids. He found, for example, that the b.c.c. transition metals chromium, molybdenum, and tungsten may be represented by a potential of the form of (16.6) with $m = 5$ and $n = 7$, although as we have seen this is inconsistent with the criterion for mechanical stability of this structure.

Many calculations have also been made with empirical expressions of the type

$$-\Xi(r) = D[\exp\{-2\alpha(r-r_0)\}-2\exp\{-\alpha(r-r_0)\}] \tag{16.8}$$

in which the repulsive and attractive potentials are exponential in form, rather than powers of r. This is the Morse potential, first used in discussing the hydrogen molecule. Girifalco and Weizer (1959) have shown that it makes both f.c.c. and b.c.c. structures mechanically stable, and that the parameters may be evaluated from experimental quantities in a manner similar to that used by Furth. They claim that the theoretical equation of state agrees well with experimental quantities, as also do the elastic stiffnesses, although of course the theoretical predictions include the Cauchy relation $c_{12} = c_{44}$ which is not valid experimentally.

The Morse function has been used by various authors for calculations on point defects, and its applicability to the case of copper was critically investigated by Lomer (1959). Lomer's results are shown in Fig. 5.4, and emphasize the relatively long tail of the Morse function, so that interactions with all neighbours up to the fourth nearest have to be included. The greater emphasis on long-range interactions presumably accounts for the ability of the Morse function to predict stable b.c.c. structures, but it can lead to serious errors if these do not correspond to the genuine long-range forces in a metal (which are, of course,

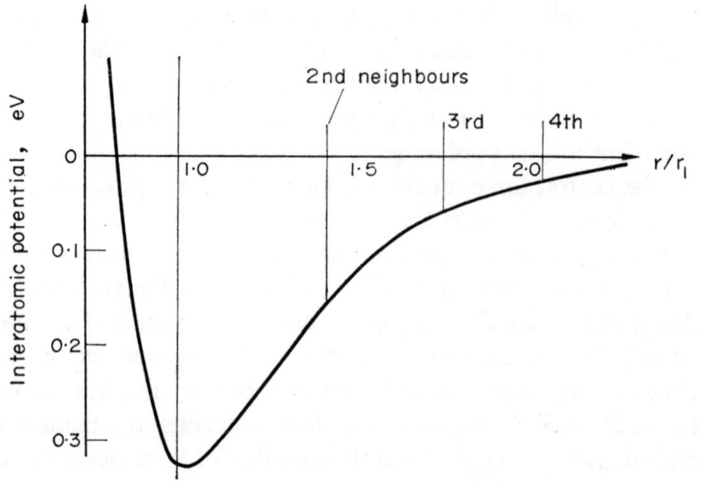

FIG. 5.4. Morse interaction potential for copper (after Lomer, 1959). The three constants of eqn. (16.8) were fixed empirically from the binding energy, lattice constant, and bulk compressibility of the metal.

non-central). It thus seems doubtful whether the use of a Morse potential represents any real improvement on the simple nearest neighbour model.

Realistic potentials $-\Xi(r)$ have been calculated for a few metals either directly (Duesbery and Taylor, 1969) or from the theories of pseudopotentials and model potentials (Heine 1970). Such potentials do not have simple analytical shapes for small r, but at large r, they are oscillatory with forms such as

$$-2\Xi(r) = Ar^{-3}\cos(2k_F r), \tag{16.9}$$

where k_F is the wave vector at the Fermi surface. This is the leading term in an asymptotic expansion, but Basinski *et al.* (1970) show that it does not become dominant in sodium until $r \gtrsim 6a$, where a is the lattice parameter. For $2a < r < 6a$ the potential derived by Duesbery and Taylor has the form

$$-2\Xi(r) \simeq B(2k_F r)^{-5}\sin(2k_F r + \alpha). \tag{16.10}$$

This potential is plotted in Fig. 5.5.

When realistic potentials are not available, calculations are made with simple analytical forms, which are sometimes matched to observable properties of particular metals (elastic

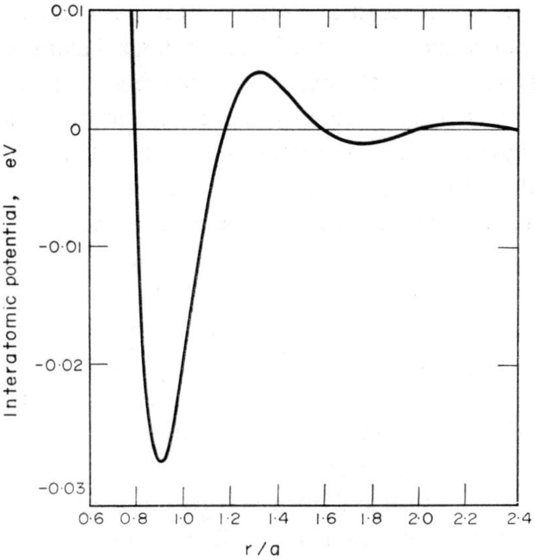

FIG. 5.5. The effective ion–ion interaction energy for sodium as a function of ion–ion separation, in units of the lattice parameter $a = 0{\cdot}4234$ nm (4·234 Å) after Duesbery and Taylor (1969).

stiffnesses, phonon dispersion curves, etc.) as outlined above. Morse potentials have continued to be used, but for b.c.c. metals an empirical potential developed by Johnson (1964) for α-iron

$$-2\Xi(r) = a_1(a_2-r)^3+a_3r+a_4, \tag{16.11}$$

where a_i are constants, have proved more useful. In particular, Vítek (1968) and Vítek *et al.* (1970) have used the general form of the Johnson potential

$$-2\Xi(r) = \sum_{k=0}^{n} a_k r^k \tag{16.12}$$

to investigate the properties of stacking faults and dislocations in b.c.c. structures.

The linear approximation (16.2) is probably not valid even for simple metals, but the part of the lattice energy which is not given by the first sum in (16.2) still depends mainly on the volume and is thus nearly constant when the volume is conserved. It follows that even though the non-central interactions may contribute a major part of the lattice energy, some properties may still be calculated satisfactorily by the pairwise model. Actually the numerical results for the cohesive energy are also encouraging in the case of sodium, for which the potential derived by Duesbery and Taylor (1969) ensures stability of both b.c.c. and h.c.p. forms with almost equal lattice energies (Basinski *et al.*, 1970).

We now consider the common metallic structures in terms of the central force approximations. An important concept is the coordination number of an atom, z, which may be defined in two different ways. The most usual definition takes z as the number of nearest neighbours possessed by any atom, so that $z = 12$ in a f.c.c. structure, and $z = 8$ in a b.c.c. structure. For most metals with the h.c.p. structure, we should in principle have $z = 6$, except when the axial ratio has its ideal value of $(\frac{8}{3})^{1/2}$, which gives $z = 12$. In practice, the differences between first and second nearest neighbour distances for these metals are so small that they are disregarded for the purpose of defining z, which is generally considered to be 12 for all h.c.p. metals. An extension of this argument suggests that a better value for the co-ordination number of a b.c.c. structure would be 14, since the difference between nearest neighbour and second nearest neighbour distances is also quite small here. This cannot be justified on the rigorous definition of z as the number of nearest neighbours, but is obtained if an alternative definition due to Frank and Kaspar (1958) is adopted.

Suppose an atom is joined by a set of lines to all other atoms in the structure, and the planes normal to these lines and bisecting them are drawn. The innermost polyhedron defined by these planes may be regarded as the *domain* of the atom considered; it is a region of space within which all points are nearer to the centre of this atom than to any other atom. The domain, as here defined, is identical with the atomic polyhedron used in the Wigner–Seitz–Slater method of solving the wave equation; when applied to a random set of points it is also known as a "Dirichlet region" or as a "Voronoi polygon , and the concept then has applications to the theory of liquids (p. 158). Each face of the domain is equidistant between the chosen atom and some other atom which is now considered to be a *neighbour;* the number of such neighbours was defined by Frank and Kaspar as the coordination number z. This automatically gives values of 12 for z in both f.c.c. and h.c.p. structures, even with non-ideal axial ratios, and of 14 for the b.c.c. structure. The definition is especially useful in dealing with complex alloy structures, which can often be understood in terms of sphere packing (Kaspar, 1956; Frank and Kasper, 1958, 1959; Pearson, 1973; Sinha, 1972). These structures have high coordination numbers according to the definition, whereas rigid application of the nearest neighbour concept would give $z = 1$ or 2 in many structures. The whole set of neighbours surrounding an atom may be referred to as the coordination shell, and in using a central force approximation for the binding energy we should expect to include at least all atoms in the first coordination shell.

Although the Frank–Kaspar definition of z overcomes some disadvantages of the nearest neighbour definition, it sometimes introduces opposite difficulties by giving values which seem physically to be too large. Thus Laves (1967) pointed out that the definition would give $z = 16$ for the diamond cubic structure, and to avoid this difficulty Frank (1967) suggested that a distinction between "indirect" and "direct" neighbours first made by Meijering (1953) should be incorporated into the definition of z. The straight line joining an atom to a direct neighbour intersects the face of the Dirichlet cell which has been constructed as the plane normal to this line; the straight line joining the atom to an indirect neighbour intersects some other face of the Dirichlet cell. This means that two atoms are neighbours if some line exists along which any point is closer to one of the two atoms than to any other atom of the structure, but they are direct neighbours only if this line is a straight line.

Indirect neighbours normally arise only in rather open structures, and the restriction to direct neighbours does not affect z for the f.c.c. and b.c.c. structures, but reduces the diamond structure value of z to 4. However, even this definition can be unsatisfactory; for example, a slight tetragonal distortion of the f.c.c. structure increases the number of direct neighbours to 14 ($c/a < 1$) or to 16 ($c/a > 1$), but the additional direct neighbours are at much larger distances than the original direct neighbours. Thus it seems difficult to define z geometrically, i.e. without reference to some arbitrary cut-off distance, in a way that will describe the physical situation in all cases.

As is well known, the f.c.c. structure and the h.c.p. structure with $c/a = (\frac{8}{3})^{1/2}$ are closely related; they represent alternative ways in which equi-sized spherical atoms may be stacked together as closely as possible. Close-packed planes in the two structures have identical atomic arrangements, each atom being at the corner of six equilateral triangles and surrounded by six other atoms in hexagonal array. A second layer may be stacked on the first in either of two positions, which are denoted by B and C, the original layer being A. Repetition of the sequence ... *ABCABCABC* ... gives the f.c.c. structure, and of the sequence ... *ABABAB* ... gives the h.c.p. structure. There are four sets of equivalent close-packed $\{111\}$ planes in the more symmetrical cubic structure, and the hexagonal crystal has a single set of close-packed $\{0001\}$ planes.

In both close-packed structures, an atom has twelve nearest neighbours, of which six lie in a close-packed plane through the atom and three in each of the neighbouring parallel close-packed planes. In the cubic crystal, these atoms are at distances $a_0/\sqrt{2}$ along $\langle 110 \rangle$ directions, where a_0 is the cell edge of the cubic unit cell, whilst in the h.c.p. structure they are at distances a along the six $\langle 11\bar{2}0 \rangle$ directions and at distances $a(4+3\gamma^2)^{1/2}/2\sqrt{3}$ along six of the twelve $\langle 20\bar{2}3 \rangle$ directions, where a is the usual parameter of the hexagonal cell, and $\gamma = c/a$ is the axial ratio. The twelve neighbours of an atom in the h.c.p. structure are at identical distances only when γ has its ideal value of 1·633. The nearest neighbour model is thus unable to distinguish between the lattice energy of the two ideal structures, and this is also true when second nearest neighbour interactions are included. This is because the second nearest neighbours of an atom are all situated in adjacent close-packed planes, and the difference between the structures only appears when at least three planes are considered. The six second nearest neighbours in the f.c.c. structure are at distances a_0 along $\langle 100 \rangle$ directions, whilst in the h.c.p. structure they are at distances $a(15+3\gamma^2)^{1/2}/2\sqrt{3}$ along six of the twelve $\langle 40\bar{4}3 \rangle$ directions.† These separations are again identical when γ has its ideal value and $a = a_0/\sqrt{2}$. A difference in energy of the structures appears when third nearest neighbour interactions are introduced. In the f.c.c. structure, each atom has twenty-four such neighbours at distances $(\frac{3}{2})^{1/2} a_0$ along $\langle 211 \rangle$ directions, whilst in the h.c.p. structure, an atom has two neighbours at a distance $c = \gamma a$ in the $\langle 0001 \rangle$ directions, six neighbours at a distance $\sqrt{(3)}a$ in $\langle 10\bar{1}0 \rangle$ directions, and twelve neighbours at a distance $a(28+3\gamma^2)^{1/2}/2\sqrt{3}$ in $\langle 24\bar{6}3 \rangle$ directions. The latter two distances are equal, and are equivalent to the cubic third nearest neighbour distance, when $\gamma = (\frac{8}{3})^{1/2}$.

In some real metals, the structure is cubic, in others it is hexagonal, and in metals like cobalt both structures exist with very small energy differences. The main difference in the

† The rule for selecting the directions for nearest neighbours and second nearest neighbours in the h.c.p. structure is that in the four index notation $\langle uv\overline{u+v}w \rangle$, u and v may not be interchanged. Thus if there is a nearest neighbour in the direction $[20\bar{2}3]$, there will be no nearest neighbour in the direction $[02\bar{2}3]$, but there will be a nearest neighbour in the direction $[0\bar{2}23]$. If the nearest neighbours are in the directions $[20\bar{2}3]$ and corresponding directions, the second nearest neighbours are in direction $[04\bar{4}3]$ and corresponding directions.

energies of the two structures does not arise from third nearest neighbour interactions or other long-range forces, however, but from the breakdown of the central force model. It seems likely that the Fermi energy may be the determining factor, although there will also be a difference in the energy of the lowest electron state. This arises, for example, in the Wigner–Seitz–Slater method of solving the wave equation since although the two structures will have identical atomic polyhedra, the boundary conditions are different.

In the same way, a central force model predicts that for the h.c.p. structure, only the ideal axial ratio $\gamma = (\frac{8}{3})^{1/2} = 1{\cdot}63299$ will be stable. Observed values of γ vary from 1·57 to 1·89, and this cannot be explained by the simple model. Nabarro and Varley (1951) have shown that a deviation from the close-packed value for γ may be produced by adding to the central force energies an energy term depending only on the volume. However, their results show that this model leads to $\gamma > (\frac{8}{3})^{1/2}$, whereas in practice it may be either greater or smaller.

Since the nearest neighbour relations are independent of the stacking sequence used in forming the structures, it is clear that any other stacking sequence, regular or irregular, will give the same nearest neighbour relations provided any two layers are arranged in relative positions *AB* or *AC*. Thus if the regular sequence of the f.c.c. or h.c.p. structure is disturbed in some way, we shall not appreciably change the energy according to our simple model. In many materials, it is true that stacking irregularities of this kind, usually called stacking faults, do not increase the energy very appreciably, and such faults form an important class of defects in real structures.

A sequence of layers ... *ABCABCBACBAC* ..., represents a f.c.c. crystal and its twin. The composition plane is an atomic {111} plane of low energy, and the two crystals are mirror images in this plane. We now consider a situation in which a low energy composition plane separates two parts of a crystal having the same orientation, but which do not form a continuous lattice. Such a plane of separation is called a stacking fault, and the situation is also described as translation twinning (Frank, 1951a). Stacking faults may be detected by X-ray methods if the faults occur at intervals of $\sim$100 atom planes or less; they also give rise to characteristic interference fringes when thin films are examined by transmission electron-microscopy. They are planar defects separating two regions of good crystal which differ by a rigid translation smaller than a lattice vector in the direction of translation. Planar defects separating two regions of good crystal which differ in orientation (grain boundaries) are defects of quite a different kind.

Frank further classified stacking faults into intrinsic and extrinsic types. In intrinsic faults, the atomic pattern of each half of the crystal extends right up to the composition plane, irrespective of whether or not this is an atomic plane; in extrinsic twins, the composition plane is an atomic plane which does not belong to the lattice structure on either side of it. Consider a perfect f.c.c. crystal. We may introduce a stacking fault either by removing a layer of atoms, or by inserting a layer of atoms between two existing layers. This gives the following stacking sequences:

... *ABCABCACABCABC* ... (remove a *B* layer),
... *ABCABCACBCABCA* ... (insert a *C* layer).

In the first case, we have an intrinsic fault with the composition plane between an *A* and a *C* layer of atoms, and in the second case, we have an extrinsic fault with a *C* layer as the composition plane. There is no definite way of estimating which fault has lowest energy.

Instead of using the *ABC* notation, the stacking sequences may be specified more simply by using stacking operators (Frank, 1951b). We represent any atomic sequence of type *A–B*, *B–C*, *C–A* by the symbol $\triangle$, which is to be regarded as an operator which acts on the first layer to give the second layer. The three reverse sequences are similarly represented by the reverse operator $\triangledown$, which gives *A–C*, *C–B*, *B–A* stacking. Any regular sequence of planes may now be written $n\triangle m\triangledown$, where n, m give the number of times each operator is repeated. Thus the f.c.c. crystal is represented simply by an indefinite number of repetitions of either $\triangle$ or $\triangledown$ (... *ABC* ..., or ... *ACB* ...) and the h.c.p. structure by an indefinite number of repetitions of $\triangle\triangledown$ (... *ABAB* ..., or ... *BCBC* ..., or ... *ACAC* ...). In non-metallic crystals, more complex repeating units are found; e.g. in one form of carborundum, the repeating unit is $3\triangle 3\triangledown$.

In this notation, a f.c.c. crystal and its twin might be represented by the symbol $p\triangle q\triangledown$, where p, q are simply the number of layers in each crystal. The f.c.c. stacking faults will be represented by operators which are all (say) $\triangle$ on either side of the fault, but in which there are some $\triangledown$ operations in the neighbourhood of the fault. Thus the first stacking fault above is represented by $p\triangle\triangledown q\triangle$, and the second one by $p\triangle\triangledown\triangledown q\triangle$. We may abbreviate this notation still further by referring only to the number of extra $\triangledown$ (or $\triangle$) operations introduced into the perfect crystal, the stacking sequence of which is understood. Thus the intrinsic fault would be a ($1\triangledown$) and the extrinsic fault a ($2\triangledown$) fault. Note that a ($3\triangledown$) fault might be more appropriately described as a three layer twin sandwiched between the two crystals of $\triangle$ orientation. Similarly, the ($1\triangledown$) and ($2\triangledown$) faults may be regarded formally as monolayer and two layer twins respectively.

Let us now consider the h.c.p. structure, which is ... $\triangle\triangledown\triangle\triangledown\triangle\triangledown$ ... If we again specify faults by the extra $\triangledown$ (or $\triangle$) introduced, we have the following types.

($1\triangledown$) (e.g. ... *ABABACACAC* ...),
($2\triangledown$) (e.g. ... *ABABACBCBC* ...),
($3\triangledown$) (e.g. ... *ABABACBABA* ...).

The first two of these are intrinsic faults with atomic and non-atomic composition planes respectively; the third is an extrinsic fault. It is readily seen that the faults ($1\triangle$), ($2\triangle$), ($3\triangle$) are equivalent to the above set; indeed the distinction between $\triangle$ and $\triangledown$ is reversed if the basic vectors in the close-packed planes are rotated through 60° about the c axis.

Stacking faults may be introduced into both f.c.c. and h.c.p. crystals by mechanical deformation, provided the fault energy is sufficiently low, and in some metals they are also produced by martensitic transformation. It also seems likely that atomic processes such as growth from the vapour or electrodeposition can lead to appreciable numbers of stacking faults in appropriate circumstances. Anomalous X-ray diffraction effects are observed from crystals containing large numbers of stacking faults, and these may be used to identify the type and frequency of faulting (Warren, 1959). In discussing X-ray diffraction effects, it has been usual to classify faults into growth and deformation faults. Growth faults are defined as single interruptions of the stacking law. Thus for f.c.c. structures, each layer is different from the two preceding layers except at a fault, when it is stacked above the layer next but one beneath it. In h.c.p. crystals, each layer is above the one next but one below, except at a fault. From this definition, it is readily seen that f.c.c. growth faults are really twin boundaries, but that h.c.p. growth faults are the boundaries of translation twins of type (1 ▽). When f.c.c. twinning occurs on a very fine scale, the X-ray diffraction effects are similar to those of faulting, and the distinction between twins and faults disappears when the twins are only a few atoms thick; this situation is sometimes described as "twin faulting".

Deformation faults are defined by considering two parts of an initially perfect lattice to be displaced with reference to each other in a direction parallel to the close-packed planes. This produces f.c.c. faults of type (1 ▽) and h.c.p. faults of type (2 ▽). Evidently, deformation faults in these structures may be regarded as clusters of two growth faults on successive atomic planes. As the name implies, deformation faults are produced by mechanical deformation of a structure, whilst h.c.p. (1 ▽) faults are produced by martensitic transformation or atomic growth processes. A growth fault introduced at every plane of a f.c.c. structure gives a h.c.p. structure and vice versa, so there is a continuous transition between the two structures if an increasing density of random growth faults is introduced. A deformation fault at every plane of a f.c.c. structure gives the f.c.c. twin crystal, whilst such a fault at every other plane gives the h.c.p. structure, and vice versa. We shall find in Section 33 that extrinsic faults might also be introduced by mechanical deformation. Comprehensive reviews of experimental work on stacking faults are given by Christian and Swann (1965) and Saada (1966).

Stacking faults are planar defects of low energy, which must either cross a crystal completely or must end in regions of bad crystal. Stacking faults which are limited in this way are bounded by "imperfect dislocations", and discussion is deferred until Chapter VII. However, we may usefully mention here the possibility that stacking faults are equilibrium defects in the structure, in the sense that the structure containing faults has a lower free energy than the perfect structure. This was originally proposed by Edwards and Lipson (1943) as an explanation of the observed structure of hexagonal cobalt, which may contain a stacking fault on the average every ten atomic planes. In fact, the configurational entropy introduced by the faults is far too small to compensate for the internal energy increase, so that there is no possibility that any appreciable concentration of faults can be in thermal equilibrium with the lattice (Christian, 1951). This is true also of line defects, but not of point defects, a finite concentration of which is always present in the state of thermodynamic equilibrium.

In the b.c.c. structure, each atom has eight nearest neighbours at distances $(\sqrt{3}/2)a_0$ in $\langle 111 \rangle$ directions and six next nearest neighbours at distances a_0 in $\langle 100 \rangle$ directions. From the geometry of the structure it appears most probable that any stacking faults will form either on {112} planes, where a fault could be considered as a monolayer twin, or on {110} planes, where faults might be expected on the basis of a hard sphere model. However, it appears improbable at present that there are any large, single-layer stacking faults which are mechanically stable in this structure under normal conditions. The basis for this assertion is a calculation by Vítek (1968) of the relative energies and stabilities of stacking faults in f.c.c. and b.c.c. structures, using interatomic force laws of the types discussed above.

The problem of finding the equilibrium configuration for a particular type of crystal defect usually involves an initial non-equilibrium configuration which is then relaxed by a static or a dynamical method until equilibrium is attained. A finite crystal block is used with imposed boundary conditions which are determined by the nature of the defect, and it is usually necessary to imagine the block to be continuous with an anisotropic elastic continuum. In principle, rather large external forces must also be applied to the crystal; as we have seen, these reduce to a pressure for cubic crystals, but must, for example, give the correct axial ratio to a hexagonal crystal. Provided these forces do no work during the atomic rearrangements, they may be ignored. In the static relaxation (or variational) method, the force acting on each atom, in a specified order, is computed successively, the atom is given a displacement to reduce this force to zero and the procedure is iterated until the net force on each atom is effectively zero. Tests must be made to ensure that the final configuration is independent of the starting point. The dynamical procedure is an iterative integration of the classical equations of motion, so that the positions and velocities of the atoms are calculated after successive intervals of time until a stable configuration is attained. In all procedures, the lattice sum (16.2) is not evaluated over all possible neighbours, but is truncated at some finite value of $r = |\mathbf{y}_i - \mathbf{y}_j|$. The number of neighbours which must be included for meaningful results varies considerably with the crystal structure and the form of the potential. A general review of these defect calculations has been given by Beeler (1970); see also the conference report edited by Gehlen *et al.* (1972).

In his work on faults, Vítek used the static relaxation method to calculate the relative energies and stabilities of monolayer stacking faults in f.c.c. and b.c.c. structures. We have implicitly assumed above that the relative displacement **f** of the two parts of a lattice separated by a fault is a rational fraction of a lattice vector, but in order to test stability it is necessary to consider all possible faults produced by allowing **f** to vary continuously. Vítek assumed that **f** lies in the plane of the fault **n**, and thus defined a generalized shear fault by the relations

$$\mathbf{f} = x\mathbf{u} + y\mathbf{v}, \quad \mathbf{f} \cdot \mathbf{n} = \mathbf{u} \cdot \mathbf{n} = \mathbf{v} \cdot \mathbf{n} = 0, \tag{16.13}$$

where $0 \leqslant x, y \leqslant 1$ and **u** and **v** are the two shortest lattice vectors which characterize the plane **n**. (Vítek called this a generalized intrinsic fault, but this definition is slightly more restrictive than Frank's usage of intrinsic.) He then used a lattice sum to evaluate the fault energy $\sigma^f(x, y)$, which has absolute minima at integral values of x and y, and also may have minima relative to all immediately adjacent configurations at some other values of x and y. A structure cannot form large stacking faults unless such relative minima exist, since all other configurations are mechanically unstable.

The $1 \bigtriangledown$ f.c.c. stacking fault in f.c.c. metals was found by Vítek to be a metastable configuration of the above type, but no metastable faults were found on either the {110} or {112 planes of the b.c.c. structure. Figure 5.6, for example, shows the form of the energy surface for **f** vectors in the {112} plane; there is no minimum in the energy at the configuration $\mathbf{f} = (a/6) \langle \bar{1}\bar{1}1 \rangle$ which corresponds to a twin monolayer. The result is believed to be a property of the crystal structure and to be independent of the form of the potential; stable faults can only be produced by artificially truncating the potential immediately beyond the second nearest neighbour separation. In particular, it has been found that stable faults are not predicted when the much more realistic atomic potential for sodium is used (Basinski *et al*, 1971). A separate calculation (Vítek, 1970a) shows, however, that a three-layer fault on {112} is a stable configuration, and may, indeed, by regarded as the nucleus of a twin.

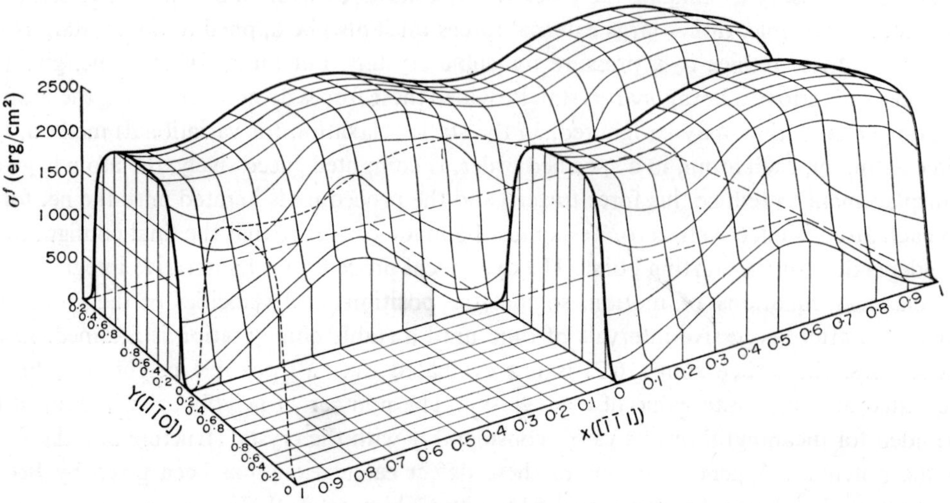

FIG. 5.6. The σ^f surface for the [112] plane in a b.c.c. lattice calculated using Johnson's potential J_0, (after Vítek, 1968).

Various claims have been made that the presence of faults in b.c.c. metals has been demonstrated experimentally. Evidence based on X-ray diffraction or transmission electron-microscopy is not convincing (in the latter case, the faults were probably plate-shaped clusters of impurity atoms), but observations made by field-ion-microscopy (e.g. Smith and Gallot, 1969) are not dismissed so readily. It seems possible that the large hydrostatic tension in the field ion tip is able to stabilize a monolayer fault which could not exist in the stress-free condition; the effect is similar to that of truncating the potential mentioned above (Vítek, 1970b).

The form of the σ^f surfaces is one example of a qualitative feature which may be investigated mith a rather arbitrary potential. Quantitative results are much harder to obtain; however carefully the parameters of a Morse or a Johnson potential are matched to the measured properties of some real metal, the calculations of fault energy would not be expected to have much numerical significance. Calculations of the absolute value of the fault energy have been attempted for some f.c.c. metals for which good potentials are available (for summary, see Heine and Weaire, 1970). A basic difficulty in many of these calculations is that when applied to monovalent metals, they always give a lower energy for the h.c.p. structure than for the f.c.c. structure; this implies that the particular theory is not applicable to the noble metals.

17. POINT DEFECTS IN CRYSTALS

The important point defects in metal crystals are believed to be the vacancy, the divacancy, and the interstitial atom. A vacancy is normally considered to be a region of atomic dimensions characterized by the absence of an atom from a site which should be occupied. Clearly this will affect the mean positions of the other atoms in the crystal, and the vacancy may be considered to be a negative centre of pressure with an elastic strain field. The displacements produced will be very small, except for the atoms immediately adjacent to the vacant site. For these atoms, however, the changes in position may be so large that it is no longer possible to specify unambiguously from which site an atom is missing. Thus a vacancy should properly be considered as an atomic region of bad crystal, within which there is one fewer atom than in a corresponding volume of good crystal. In the same way, a divacancy is an atomic region containing two fewer atoms than a corresponding volume of good crystal; it may be formed by the amalgamation of two single vacancies, but its internal structure need not resemble that of the single vacancy.

An interstitial atom is normally considered to be an atom located on a site which should be unoccupied. The more formal concept is clearly that of a region of bad crystal containing one more atom than the same volume of good crystal. These reservations do not mean that we shall not be concerned with the structure of point defects, consideration of which is essential in some problems, but only that we wish to separate the general properties of point defects, which are common to all materials, from those properties which depend specifically on structure, and vary from one metal to another.

As we shall discuss below, detailed calculations of the atomic configurations and energies of point defects have been attempted only for the noble metals with "full" structures. Since such metals may be regarded as formed by the stacking of hard atomic spheres in contact, the atomic spacings around a defect approximate to the equilibrium spacings. Thus there is little inward relaxation around a vacancy, but there may be appreciable outward movement of the atoms around an interstitial to make room for the extra atom. Larger relaxations may be expected around a vacancy in an open alkali metal, leading to the concept of the extra volume of the vacant site being shared amongst a small number (say 14) of neighbouring atoms. There is then a small volume of bad crystal, the local structure of which is akin to that of a glass; Nachtrieb and Handler (1954) have given the name of "relaxion" to this type of defect. A related concept is that of local melting, in which there is a region of bad crystal containing the same number of atoms as the corresponding volume of good crystal, but randomly arranged. Locally melted regions may be important in radiation damage, but we shall not consider them further since it seems improbable that such configurations could have significant lifetimes in normal circumstances.

The atomic displacements around an interstitial atom may be roughly spherically symmetrical, as usually assumed, but it is possible that a lowering of energy is obtained by adopting the crowdion configuration suggested by Paneth (1950). A crowdion is a kind of linear spreading of an interstitial atom, so that n atoms in a row extend over $(n-1)$ normal sites. Paneth suggested that crowdion defects lie along b.c.c. $\langle 111 \rangle$ directions with $n \sim 8$; in a later application to f.c.c. metals, Lomer and Cottrell (1955) envisaged $\langle 110 \rangle$ crowdions with $n \sim 5$. In discussing interstitials, a distinction is sometimes made between physical defects, or self-interstitials, in which the extra atom is identical with all the other atoms, and chemical defects in which the extra atom is of a different chemical species. A chemical defect of this type is really a very dilute interstitial solution, and the corresponding disturbance around a substitutional solute atom could also be considered as a point defect. In this section, we shall confine the discussion to physical defects.

The important characteristics of a point defect are its free energy of formation, and its free energy of motion. The former governs the number of point defects present at equilibrium, and the latter the rate at which a point defect can move inside the structure. An individual defect will usually have a small energy of formation, and since the presence of a few defects gives a large increase in the configurational entropy of a crystal, real crystals at finite temperatures will always contain a small number of defects in equilibrium with the lattice.

Let us first consider isolated vacancies, which were once believed to originate at the free surface of a crystal, or at grain boundaries, when atoms leave their original sites to build up new layers. The unoccupied sites thus created were supposed to diffuse into the interior. It is now recognized that equivalent processes may occur within a crystal, since dislocation lines may sometimes act as sources or sinks for vacancies. In a crystal containing N atoms and n vacancies, the number of independent arrangements of the assembly is $(N+n)!/N!\,n!$ and the extra configurational entropy resulting from the presence of the vacancies is thus

$$\Delta S = k\{\ln(N+n)! - \ln N! - \ln n!\} \simeq k\{(N+n)\ln(N+n) - N\ln N - n\ln n\}, \qquad (17.1)$$

using Stirling's theorem. If the enthalpy of formation of an individual defect is $\Delta h_{\square}$, and the number of defects is small, so that $\Delta h_{\square}$ is independent of n, the heat of formation of the n defects may be written $n\,\Delta h_{\square}$.

We may also associate an entropy of formation, $\Delta s_{\square}$, with each individual vacancy, so that the free energy of formation (excluding the configurational entropy above) is $\Delta g_{\square} = \Delta h_{\square} - T\,\Delta s_{\square}$. The free energy of formation and the entropy of formation are related by the thermodynamic equation

$$\Delta s_{\square} = -(\partial\,\Delta g_{\square}/\partial T)_p. \qquad (17.2)$$

The entropy of formation is essentially due to the change in the lattice vibrational spectrum around the vacancy configuration. There is some confusion in the literature about a second possible effect arising from the thermal expansion, which would seem to give a contribution to eqn. (17.2) from volume dependent atomic forces (Mott and Gurney, 1948). This is formally correct, but in most theoretical models the energy actually calculated is the internal energy at absolute zero, or the Helmholtz free energy at finite temperatures. Although an energy of this type is a function of volume at absolute zero, the variation of volume with temperature can make no contribution to $\Delta s_{\square}$, since the equation corresponding to (17.2) is

$$\Delta s_{\square} = -(\partial\,\Delta f_{\square}/\partial T)_v.$$

Thus the variation of internal energy with volume does not give rise to an entropy term, even though there may be an apparent variation of enthalpy with temperature on such a model. This point was first emphasized by Vineyard and Dienes (1954).

In the immediate neighbourhood of a vacancy, the residual lattice strain may be regarded as a local lowering of the elastic stiffnesses, giving lower frequencies of vibration and a positive contribution to $\Delta s_\square$. There may also be contributions to $\Delta s_\square$ arising from displacements of atoms in more remote regions of the crystal, and one way of estimating these is to regard $\Delta g_\square$ as an elastic energy caused by the field of a point singularity (centre of dilatation). We discuss this type of theory for the analogous case of a solute atom in Section 25; the elastic field is nearly a pure shear, and the energy (see eqn. (25.16)) may be written as $(C/v)\mu$, where v is the atomic volume and μ the shear modulus. The entropy term is thus proportional to $-(\partial\mu/\partial T)$, and is always positive.

Zener (1951) has used this approach to argue that the entropy of motion of a point defect (see below) is necessarily positive, since it is reasonable to assume that the free energy of activation may be regarded as the work done in elastically straining the lattice. Applied to the problem of the formation of point defects, the elastic approach would equally predict that these should all be positive, since centres of pressure and centres of dilatation have long-range strain fields of the same form. However, the elastic model is a very poor approximation, and the main contribution to Δg comes from local changes around the defect which can only be calculated by quantum mechanical methods. No general conclusion about the sign of Δs is thus possible (Huntington *et al.*, 1955), although as we have seen $\Delta s_\square$ will be positive.

The total change in free energy caused by introducing n vacancies a crystal is thus

$$\Delta G = n\,\Delta g_\square - \boldsymbol{k}T\{(N+n)\ln(N+n) - N\ln N - n\ln n\}$$

and the condition for this to be a minimum, $\partial\,\Delta G/\partial n = 0$, gives

$$\left.\begin{aligned} n/(N+n) &= \exp(-\Delta g_\square/\boldsymbol{k}T) \\ \text{or}\qquad x_\square = n/N &\approx \exp(-\Delta g_\square/\boldsymbol{k}T) \\ &\approx \exp(\Delta s_\square/\boldsymbol{k})\exp(-\Delta h_\square/\boldsymbol{k}T), \end{aligned}\right\} \tag{17.3}$$

where $x_\square$ is the atomic concentration of vacancies which is in equilibrium at any temperature T.

An identical procedure gives the expression for the atomic concentration of interstitials. Thus if there are N atoms on normal sites, and n interstitial atoms, the number of arrangements of the assembly is given by $(z_\bullet N!)/(z_\bullet N-n)!\,n!$, where $z_\bullet$ is a geometrical factor which specifies the number of possible interstitial sites per normal site. The equilibrium concentration of interstitials is then

$$x_\bullet = z_\bullet \exp(-\Delta g_\bullet/\boldsymbol{k}T) = z_\bullet \exp(\Delta s_\bullet/\boldsymbol{k})\exp(-\Delta h_\bullet/\boldsymbol{k}T), \tag{17.4}$$

where $\Delta g_\bullet$, $\Delta h_\bullet$ and $\Delta s_\bullet$ are now the free energy, enthalpy and entropy of formation (excluding the configurational entropy) of an interstitial.

Finally, we consider the possibility of having finite equilibrium concentrations of vacancy pairs. If the free energy of formation of a divacancy is twice that of a single vacancy, the equilibrium concentration of divacancies will be of the order of the square of that of single vacancies, and hence will be negligible in comparison. However, we may expect a short-range force between two isolated vacancies, giving a binding energy which is an appreciable fraction of the free energy of formation of a single vacancy. We thus have an equivalent expression for the concentration of divacancies, in which the quantities $\Delta g_{\square\square}$, $\Delta h_{\square\square}$ and $\Delta s_{\square\square}$ are all probably intermediate between one and two times the corresponding quantities for single vacancies. The equation for $x_{\square\square}$ will also contain an extra entropy term, since the divacancy will probably not have the spherical symmetry of a single vacancy, but may be "orientated" in different lattice directions. This term will result in a small numerical factor in the equation for the concentration, analogous to the factor $z_{\bullet}$ in eqn. (17.4) above.

The equilibrium concentration of defects, as given by eqns. (17.3) and (17.4), may be obtained from theoretical estimates of the free energies of formation, or may be measured experimentally. There are several calculations of the energy of formation of a single vacancy in copper, one of the first and most generally accepted of which is that of Huntington (1942). After making certain corrections (see Lomer, 1959), this gives $\Delta h_{\square} = 1{\cdot}5 \pm 0{\cdot}5$ eV, in reasonable agreement with the results obtained by Fumi (1955) and Seeger and Bross (1956). Although different assumptions were made in these three papers, the authors are all agreed that there is no appreciable relaxation of atomic positions around a vacancy. According to Tewordt (1958), however, the nearest neighbours move inwards by about 2% of the interatomic distance, and the next nearest neighbours by about 0·2%. Tewordt's value for the formation energy, using two alternative semi-empirical repulsive potentials, is $\Delta h_{\square} \approx 1{\cdot}0$ eV. The experimental result is $\sim 1{\cdot}1$ eV, and the small discrepancy is attributed by Tewordt to an overestimate of the relaxations in his calculation.

Many less-accurate estimates of formation energies and related properties of point defects have been based on pairwise interaction models with empirical potentials matched to measured properties, as discussed in Section 16. One difficulty with this procedure is that certain results, e.g. the relaxed positions of the near neighbours of a vacancy, are sensitive to the detailed shape of the potential and may thus be varied within rather wide limits without disturbing the experimental fit. Another difficulty is the large contribution to the energy made by the non-central (volume-dependent) terms in the case of point defects. When a point defect is formed from a perfect cubic lattice, the change in volume is usually of the order one atomic volume, and the effective pressure, $p = \frac{1}{2}(c_{12} - c_{44})$, is typically of order 10^{10} N m^{-2}. Thus the $p\,\Delta v$ term in (16.4) is of order 10^{-19} J ($\sim$ 1 eV), and is thus a large fraction of the total formation energy.

For this and other reasons, absolute values of calculated formation energies can have very little significance except in those rare cases where a realistic potential has been obtained. However, the calculations may often be valuable in giving useful qualitative information about the relative properties of various different defects, especially in cases (such as small vacancy clusters) where these properties cannot readily be disentangled from experimental data. Calculations of this type were reviewed by Beeler (1970) and were discussed at the 1971 Battelle Conference (Gehlen, *et al.*, 1972).

Girifalco and Weizer (1959–60) used empirical Morse potentials to calculate the energies of formation and the atomic relaxations for several f.c.c. and b.c.c. metals, and their results suggested that the displacements around a vacancy in sodium may be large enough for Nachtrieb's relaxion concept to be valid. Various later calculations show rather small relaxations in atomic positions around a f.c.c. vacancy; typically, the nearest neighbours relax inwards along close-packed directions by $\sim 3\%$ and the second neighbours relax outwards along $\langle 100 \rangle$ directions by about the same amount. Calculations have been made by the dynamic method using Born–Mayer repulsive potentials (Gibson *et al.*, 1960) and by the static method using Morse-type potentials (e.g. Doyama and Cotterill, 1965) and Johnson potentials (Johnson, 1966a). Johnson considered a f.c.c. structure with interaction parameters originally selected for α-iron, so that the calculation should be appropriate for a metal like γ-iron or nickel. He found the formation volume of a vacancy to be $\sim 0{\cdot}85$ atomic volumes, and the formation energy, $\Delta h_{\square}$ to be $\sim 1{\cdot}49$ eV.

The stable divacancy in a f.c.c. metal has atoms missing from two adjacent sites, but calculations of the binding energy of a vacancy pair are much more difficult than are those of the formation energy of a single vacancy. An early estimate (Seeger and Bross, 1956) gave a binding energy of 0·3–0·4 eV for copper, and according to Johnson's calculations nickel has a binding energy of 0·25 eV and a binding volume of 0·02 atomic volumes. With larger clusters, several configurations may be metastable, and the most stable structure is by no means obvious. Four types of compact trivacancy have been considered, for example, which are denoted by Cotterill and Doyama (1965) following de Jong and Koehler (1963) as 60°, 90°, 120°, and 180° types, where the angle specified is that between the lines joining the two outer vacant sites to the central site. However, another possibility is a tetrahedron of vacant sites with an atom in the centre (i.e. three vacancies shared equally among four sites) and earlier calculations (Damask *et al.*, 1959; Vineyard, 1961) indicated that the nearest neighbour 60° configuration would relax into this more stable structure. However, Johnson (1966b) found the 60° configuration in nickel to have a binding energy of 0·76 eV, whilst the tetrahedral vacancy had a binding energy of only 0·59 eV. Similarly, the stable tetravacancy in Johnson's calculations has the tetrahedral configurations in which all vacant sites are nearest neighbours; this is in contrast to the results of Cotterill and Doyama (1965) in which the lowest energy configuration calculated from a Morse-function potential is a "diamond" configuration of four vacancies in a rhombus on a $\{111\}$ plane accompanied by appreciable inward relaxation of two atoms.

Johnson's results are especially simple since the binding energy of any cluster may be obtained to good approximation by simply counting the decrease in the number of "broken bonds" in the cluster relative to the separated vacancies. Longer-range interactions, relaxations in atomic positions, and the effects of the stress field all contribute very small amounts to the binding energies. This result is essentially built in to the model because of the short-range nature of the potential; it is difficult to assess its general validity. It leads to a formation energy $\Delta h_{n\square}$ for a cluster of n vacancies of the form

$$\Delta h_{n\square} \simeq \Delta h_{\square}(n - b/6),$$

where b is the number of vacancy–vacancy bonds in the cluster and $b\,\Delta h_{\square}/6$ is thus the binding energy. The factor in brackets is well approximated as $n^{2/3}$ for spherical holes, and an effective surface energy can be derived; Johnson obtained $\sigma \sim 1{\cdot}38$ J m^{-2} (1380 erg cm^{-2}) in comparison with the experimental value of $\sim 1{\cdot}6$ J m^{-2} for nickel. In a later paper, Johnson (1967a) also made computations of the formation energy of various types of vacancy aggregate. Aggregation of vacancies into stacking fault tetrahedra and dislocation loops is further discussed in Section 33.

The potential used by Johnson in the above calculations was originally developed to represent the atomic interactions in b.c.c. α-iron, and the properties of vacancies and interstitials in that metal were computed by Johnson (1964). The stable vacancy has an atom missing from a normal site, but the most stable divacancy consists of a second nearest neighbour pair of vacancies rather than a nearest neighbour pair. This structure reflects the relatively much greater importance of second neighbour interactions in b.c.c. structures; in the case of iron, the binding energy was estimated as $\sim 0{\cdot}2$ eV.

Beeler and Johnson (1967) considered stable vacancy clusters of higher order. The stable trivacancy is an isosceles triangle arrangement with two nearest neighbour separations and one second neighbour separation of the three vacant sites. This configuration, unlike the f.c.c. trivacancy, can migrate without dissociation. The smallest immobile cluster is the tetrahedron formed from four vacancies with four nearest neighbour and two second nearest neighbour edges. For any cluster size, a configuration is possible which has a greater binding energy than that of any smaller cluster, and the gain in binding energy per additional vacancy increases for 2–6 vacancies, approaches an upper limit of $\sim 0{\cdot}8$ eV for 6–10 vacancies, and remains at this value for larger clusters. The binding energy of any configuration can be approximated (within 10%) by counting vacancy–vacancy bonds, as for f.c.c., but higher-order neighbour interactions must be included.

Huntington (1953) considered two alternative configurations for an interstitial atom in a f.c.c. metal, in one of which the extra atom is in an octahedral site at the centre of the cubic unit cell, or equivalently of a cell edge (Fig. 5.7 (a)), whilst in the other, two atoms are symmetrically situated on either side of one normal site (Fig. 5.7 (d)). The energies of the two configurations are probably nearly identical, and were estimated by Huntington as $4{\cdot}0 \pm 1{\cdot}0$ eV.

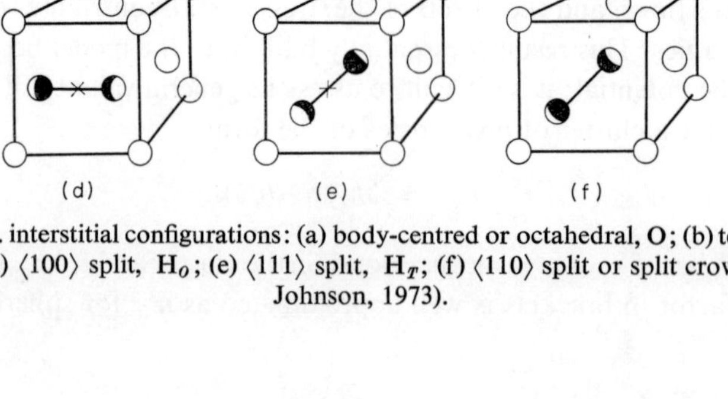

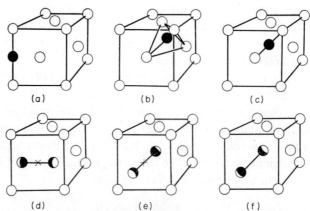

FIG. 5.7. Six f.c.c. interstitial configurations: (a) body-centred or octahedral, O; (b) tetrahedral, T; (c) crowdion, C; (d) ⟨100⟩ split, H_O; (e) ⟨111⟩ split, H_T; (f) ⟨110⟩ split or split crowdion, H_O (after Johnson, 1973).

A third configuration is the $\langle 110 \rangle$ crowdion suggested by Lomer and Cottrell, and according to Lomer (1959) this has a slightly lower energy than either of Huntington's configurations. However, Tewordt found the formation energy of an interstitial in copper to be $\sim 2{\cdot}5$ eV, and the energy of the crowdion configuration to be $\sim 0{\cdot}6$ eV higher than this, whilst Gibson *et al.* (1960) found that with a repulsive potential representing a copper-like metal, the split interstitial or $\langle 100 \rangle$ dumbell of Fig. 5.7(d) is stable. Seeger *et al.* (1962) and Johnson (1966a) considered other possible configurations, most of which are shown in Fig. 5.7. It is found that the smallest possible separation of two atoms in any configuration is $\sim 0{\cdot}85 r_1$, i.e. $\sim 0{\cdot}6a$, and the configurations are then defined by constructing a sphere of radius $0{\cdot}3a$ around each normal site. Each such sphere will contain the centre of one atom, and there will also be one additional atom with its centre not in any sphere; this is defined to be the interstitial. In the special cases where two atoms have centres at opposite diameters of a sphere, thus sharing the lattice site, the configuration is desribed as a split interstitial. In the notation used in these papers, we then have the three basic configurations O, T, and C shown in (a)–(c) of Fig. 5.7 and the corresponding split configurations H_O, H_T, and H_C shown in (d)–(f). Two further configurations which are not shown in the figure were designated OC and OT because the centre of the additional atom is between the O and C or O and T positions, but rather closer to O in each case.

In Johnson's calculations, the $\langle 100 \rangle$ split interstitial H_O was again found to be the most stable form, with a formation energy (in a nickel-like metal) of $\sim 4{\cdot}1$ eV. Three of the configurations (O, OC and H_C) are unstable; of the various possible metastable configurations the $\langle 111 \rangle$ split interstitial (H_T) and the crowdion (C) are most interesting because they lead to independent migration paths. The energy of H_T was computed to be $\sim 0{\cdot}3$ eV above H_O and it is metastable by $0{\cdot}16$ eV; the energy of C is $\sim 0{\cdot}63$ eV above H_O but it is only just metastable ($\sim 0{\cdot}02$ eV).

A comparison of various calculations on f.c.c. interstitials has been given by Johnson (1967b). Repulsive potentials favour roomy configurations and bonding interactions favour configurations with atomic separations near the minimum in the $\Xi(r)$ curve. The $\langle 100 \rangle$ split interstitial satisfies both conditions and is the lowest energy state in all calculations, but the energy of the body-centred configuration O, changes from only slightly greater than H_O for purely repulsive interactions to the highest energy of any configuration in Fig. 5.7 for curves with a deep potential well.

Using an empirical potential fitted to iron, Johnson (1964) found that the smallest separation of two atoms in the b.c.c. structure is also $\sim 0{\cdot}85 r_1$ ($\sim 0{\cdot}75a$). The six basic configurations which correspond to those of Fig. 5.7 are shown in Fig. 5.8. Johnson found that the stable interstitial is split along $\langle 110 \rangle$, whilst the $\langle 111 \rangle$ crowdion is only just metastable and has an energy which is larger by $\sim 0{\cdot}3$ eV.

These theoretical estimates of formation energies imply that for the noble metals only vacancies and possibly divacancies will ever be present in appreciable concentrations at equilibrium. In principle, the relative concentrations of point defects at different temperatures, and hence the formation energy, may be deduced from experimental measurements of physical properties such as electrical resistivity. Usually, however, the intrinsic temperature variation of the chosen property is so large that it is very difficult to separate the contribution due to varying concentrations of point defects. To circumvent this difficulty, many experiments have been made with drastically quenched specimens, the physical property changes due to non-equilibrium concentrations of point defects retained by the quench being then measured at a fixed temperature as a function of the temperature from which the specimens were quenched. Quenching experiments suffer from the disadvantage that the actual concentrations of defects are not known, unless a good calculation can be made of the property change (e.g. the electrical resistivity change) due to a single defect. They also have to be interpreted with great care because of changes which may take place during the quench. Thus the majority of the defects present at the high temperature before the quench are probably single vacancies but these may aggregate to from divacancies or larg-

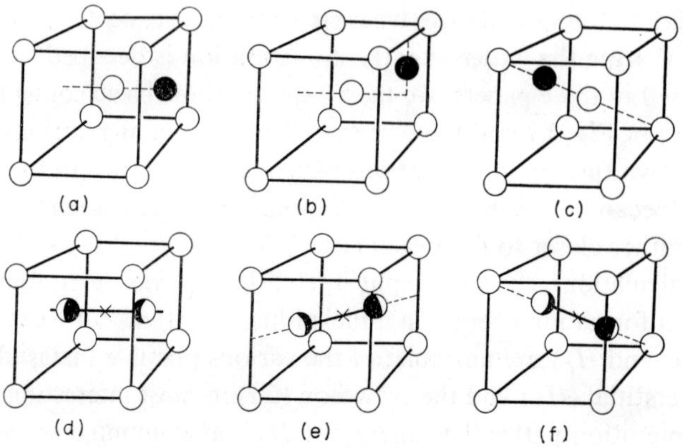

FIG. 5.8. Six b.c.c. interstitial configurations: (a) face-centred or octahedral; (b) tetrahedral; (c) crowdion; (d) ⟨100⟩ split; (e) ⟨110⟩ split; (f) ⟨111⟩ split or split crowdion (after Johnson, 1973).

er groupings during the quench. The problem is further discussed below in connection with the annealing of quenched or irradiated specimens.

Direct measurement of point defect concentrations is possible if the macroscopic density and the X-ray density (size of the unit cell) can be measured simultaneously and accurately. The principle of this method is simply that the X-rays measure the average dilatation of the interatomic spacing due to thermal expansion and to the introduction of point defects, whilst the macroscopic density measurement includes this dilatation and also the dimensional changes due to the creation or destruction of atomic sites within the crystal. Thus the effect of thermal expansion, and of atomic relaxation around the defects, are both eliminated, since both measurements are affected to the same extent (Eshelby, 1954).

Accurate measurements of length and X-ray lattice spacing at temperatures near the melting point have been made for aluminium, copper, silver, gold, lead, cadmium, magnesium, sodium, and lithium (Simmons and Balluffi, 1960 a, b, 1962, 1963; Bianchi *et al.*, 1966; Feder and Charbnau, 1966; Sullivan and Weymouth, 1964; Feder and Nowick, 1967; Feder, 1970; Janot *et al.*, 1970). For cubic metals, the relative expansions of cell edge $\Delta a/a$ and of macroscopic length $\Delta L/L$, are related by

$$\Delta N/N = 3\{(\Delta L/L) - (\Delta a/a)\}, \qquad (17.5)$$

where $\Delta N/N$ is the fractional concentration of additional lattice sites. For non-cubic metals it is necessary to make measurements on crystallographically non-equivalent directions in order to determine $\Delta N/N$ (Nowick and Feder, 1972). Clearly $\Delta N/N = (x_{\square}+2x_{\square\square}+3x_{\square\square\square}+\ldots)-(x_{\bullet}+2x_{\bullet\bullet}+3x_{\bullet\bullet\bullet}+\ldots)$ and the right hand side of eqn. (17.5) will be positive if vacancies are the dominant defects. This is experimentally verified for all metals; the highest concentration of vacant sites is found for aluminium, where $\Delta N/N \simeq 9{\cdot}4\times 10^{-4}$ at the melting point.

Interpretation of the experiments would be straightforward if single vacancies were the only defects present under equilibrium conditions; with aluminium, for example, this assumption leads to $\Delta s_{\square}/k = 2{\cdot}4$ and a heat of formation $\Delta h_{\square} \sim 0{\cdot}76$ eV. The present theory of close-packed metals implies that only vacancies and divacancies need be considered, but reasonable values for the binding energy of a divacancy suggest that up to 15% of the vacant sites near the melting point may represent divacancies. There is thus some uncertainty in detailed interpretation, which arises partly from the limited accuracy which eqn. (17.5) imposes on the derived $\Delta N/N$ values, and partly from the difficulty of finding an unambiguous procedure for separating $\Delta N/N$ into $x_{\square}$ and $2x_{\square\square}$. At least four energetic parameters are needed to specify the equilibrium concentration of vacant sites at any one temperature, and in the more recent analyses it is recognized that the temperature dependencies of these parameters cannot necessarily be ignored.

For many years the lattice expansion method was the only practicable procedure for the determination of equilibrium vacancy concentrations directly from "at temperature" measurements, but it has always been recognized in principle that measurements of the temperature variation of the diffusion coefficient may be analysed in terms of the sums of the formation energies and the migration energies for single vacancies and divacancies (see Chapter 9). Prior to the discovery of the positron annihilation technique (to be described below), two different viewpoints developed, namely (1) that best-fit values of energetic parameters relating to vacancies should be selected only from measurements of equilibrium (or rather steady-state) values of physical properties, or (2) that these properties should be combined with measurements on quenched metals. Thus Seeger and Mehrer (1968, 1970) concluded from an analysis of self-diffusion and equilibrium measurements that $\Delta h_{\square} = 0{\cdot}87$ eV for gold at room temperature, whilst Koehler (1970) included quenching experiments and obtained $\Delta h_{\square} = 0{\cdot}94\pm 0{\cdot}2$ eV.

Modern techniques now allow measurements of the self-diffusion coefficient D to be made over wide temperature ranges, and such data provide valuable information on vacancy behaviour in pure metals, although care is needed to avoid complications caused by dislocation short circuits at low temperatures. Since the migration energy of a divacancy is smaller than that of a monovacancy (see below), any appreciable divacancy binding energy will mean that divacancies make a measurable contribution to D, especially at high temperatures. If an Arrhenius plot of $\ln D$ vs. $1/T$ shows detectable curvature, this may be attributed either to divacancies or to temperature variations of the various enthalpies and entropies. Temperature-dependent energies may be expected on the basis of physical arguments; according to Zener's model (p. 127), for example, they would result from a non-linear dependence of elastic stiffnesses on temperature (Flynn, 1968). The temperature variations of each energy and its corresponding entropy are not, of course, independent, but the number of parameters is, nevertheless, rather large. By making some approximations, Seeger and Mehrer arrive at an expression for the diffusion coefficient which contains only five adjustable parameters, and these are fitted to experimental data by a least-squares procedure. Combination of these results with equilibrium measurements of $\Delta N/N$ then enable the combined parameters to be separated into components. The analysis has been applied to gold, nickel, aluminium, and copper (for a survey see Seeger and Mehrer, 1970).

A completely new method of determining vacancy concentrations from reversible changes of a physical property with temperature has been developed from the discovery by Mackenzie *et al.* (1967) that thermalized positrons in several low melting-point metals have lifetimes which increase much more rapidly with temperature than would be expected from thermal expansion effects. The positrons are injected into the metals as high-energy particles, but their energies are reduced to thermal energies in a time much shorter than the total lifetimes, which are of the order of 10^{-10} s. Annihilation of the positrons with electrons eventually occurs, with the emission of a pair of γ-rays, and Mackenzie *et al.* attributed their observations to the attraction of positrons to vacancies where the probability of annihilation is reduced and the lifetime increased. The mean lifetime then depends on the vacancy concentration. Related measurements which show a similar anomalous temperature variation are the angular correlation of the photons emitted during the 2γ annihilation, and the Doppler broadening of the photon lines.

The effects of vacancies on positron annihilation are rather complex. A positive effect is not found for all metals, and where an effect exists it saturates at high temperatures. This is attributed to the possibility that the positron may be either free or trapped in a bound state localized at one of the vacancies; in the latter case, its lifetime is increased. According to this model, the average lifetime τ varies from the free lifetime τ_f at low vacancy concentrations (low temperatures) to the trapped lifetime τ_t at high vacancy concentrations (high temperatures) in a way which can be predicted, apart from some uncertainty about the correct temperature variation of the trapping rate for unit vacancy concentration. This is the observed behaviour, and the saturation effect is advantageous since the greatest sensitivity to vacancy concentration occurs in the temperature range where equilibrium divacancy concentrations are negligible. Thus the method measures $x_\square$ directly, and it is sufficiently sensitive to measure concentrations of 10^{-6} or lower.

The positron annihilation method has now been applied to many metals, and the results have been surveyed by Seeger (1972, 1973). Various measurements on aluminium give $\Delta h_\square = 0{\cdot}65$ eV with a claimed uncertainty in some cases as low as 0·02 eV. The agreement of the different measurements by both lifetime and angular correlation techniques is impressive, and the results are in excellent agreement with the values of $\Delta h_\square = 0{\cdot}65$ eV, $\Delta s_\square/k = 0{\cdot}8$ deduced by Seeger and Mehrer (1970) by the analysis of diffusion and lattice expansion data. Results obtained from various quenching experiments are much higher, ~0·76 eV, and the possible reasons for this discrepancy are discussed later in this section.

We now turn to consider the mobility of a point defect in the lattice. The atomic configuration which represents the equilibrium state of the defect may be centred about any one of an infinite number of equivalent positions, or lattice points. Other configurations may also exist as transitory states, and as the defect passes from one equilibrium configuration to an adjacent equilibrium configuration, it must pass through such transitory states of higher energy. If the assumptions of the reaction rate theory are valid, the frequency of such movements will be determined by the intermediate configuration of highest free energy, the choice of intermediate configurations (the "path" of the defect) being determined by the condition that this highest free energy shall be smaller than that corresponding to any alternative series of intermediate configurations. The critical configuration thus corresponds to a saddle point in the free energy field, and is the activated complex of Chapter 3. In simple terms, we describe the defect as moving through an interatomic distance, the activation energy for this motion being the highest free energy at some intermediate state. Writing the difference in free energies, enthalpies and entropies of the saddle point configuration as $\Delta_a g$, $\Delta_a h$ and $\Delta_a s$ respectively, we have

$$\begin{aligned} \Gamma &= z\nu \exp(-\Delta_a g/\boldsymbol{k}T) \\ &= z\nu \exp(\Delta_a s/\boldsymbol{k}) \exp(-\Delta_a h/\boldsymbol{k}T), \end{aligned} \tag{17.6}$$

where Γ is the number of jumps per second made by the defect, and ν is the atomic frequency. The factor z allows for the possibility of the defect moving in different directions; for example, if the defect is a vacancy, z will be the number of nearest neighbour atom sites around any one site. The equation obtained by putting $z = 1$ gives the frequency with which the point defect will move in any particular direction. The ways in which a particular defect can move, and hence the appropriate value of z, may be restricted by the configuration of the defect rather than by its neighbours. A crowdion, for example, can probably move easily only along its own line, so that z will probably be two.

If the saddle-point configuration is known, the activation enthalpy may be calculated approximately for full metals, using similar methods to those already mentioned for the calculation of formation energies. In fact, the calculation of $\Delta_a h_\square$ for single vacancies is more difficult than the corresponding calculation of $\Delta h_\square$ because the "path" of the vacancy is not known. For the single vacancy in the f.c.c. structure, Johnson (1966a) found the saddle point to be a halfway displacement of the moving atom towards the vacancy, but divacancy migration is more complex. According to the calculations thought to be appropriate for nickel, the path does not have much symmetry and does not lie in the $\{111\}$ plane defined by the two vacancies and the moving atom. Johnson (1966b) estimated $\Delta_a h_\square$ to be $\sim 0{\cdot}90$ eV for nickel, but the numerical value probably has little significance. An early attempt to calculate this energy was made by Bartlett and Dienes (1953) using a Morse potential to represent the interatomic forces, but Lomer (1959) concluded that the model was very inaccurate.

The single vacancy migration process for b.c.c. structures is also probably a straight-line motion of a nearest neighbour atom into the vacant site, but in contrast to the f.c.c. case, the saddle point is not at the mid-point of this path (Johnson, 1964). The calculations indicate that there are two equivalent saddle-point configurations each side of a local energy minimum at the mid-point, which thus represents a metastable vacancy configuration. However in α-iron where $\Delta_a h_\square$ is calculated as $\sim 0{\cdot}68$ eV, the decrease in energy from the saddle point to the mid-point is estimated as only $\sim 0{\cdot}02$ eV, and is thus scarcely significant.

According to Johnson, divacancy migration in b.c.c. structures is a double-step process in which the configuration first changes to a metastable fourth nearest neighbour pair† and then back to a second nearest neighbour pair. The energy displacement curve for each step resembles that for a single vacancy, and so may be regarded as a perturbed single vacancy migration; the computed activation energy for the more difficult first step was $\Delta_a h_\square = 0{\cdot}66$ eV. An alternative migration process produces a metastable first nearest neighbour pair as the intermediate configuration; this has a higher activation energy (0·78 eV) but changes the orientation of the divacancy in contrast to the alternative process. The metastable nearest neighbour configuration might be regarded as a vacancy trap, since $\sim 0{\cdot}7$ eV are required to convert it into the stable divacancy configuration, in contrast to $\sim 0{\cdot}5$ eV for the transition from fourth neighbour to second neighbour configurations.

Beeler and Johnson (1967) considered in more detail the problem of vacancy pair formation and separation. They found that even when the interaction energy of two separated vacancies is as small as 0·005 eV, the activation energy for a transition to a more tightly bound configuration is lower than that for a change to a more stable configuration by as much as 0·1 eV. This interaction energy covers vacancies separated by up to tenth nearest neighbour distances (2·6*a*, where *a* is the lattice constant), within which range vacancy motion should be correlated, and the rate at which stable divacancies form should be greatly accelerated. This applies, however, only to low temperatures; at the temperatures at which vacancies are mobile in α-iron ($\sim$ 500 K) correlation effects should be important only up to $\sim$ sixth nearest neighbour separations.

The entropies of activation $\Delta_a s_\square$ and $\Delta_a s_{\square\square}$ are very difficult to estimate. These quantities are usually positive but, contrary to earlier arguments (Zener, 1951), they may have either sign since they are determined by local atomic forces (Huntington *et al.*, 1955).

Huntington (1953) considered that the O and H_O configurations of Fig. 5.7 could represent the equilibrium and activated states for the motion of an interstitial atom in an f.c.c. metal, but his calculation did not indicate which had the lower energy, although they suggested an upper limit for the energy difference of $\sim 0{\cdot}25$ eV. It should be noted that if the interstitial atom does move in this way, it is only the configuration which travels, since each atom in turn pushes the next atom off its equilibrium site. This is also true for crowdion motion, which may be expected to be easy along the line of the defect, and difficult in other directions.

According to the later calculations made by Johnson and others, the O configuration of Fig. 7. 5 represents an energy maximum, 0·49 eV above H_O, but it does not represent the saddle point for migration of H_O. A much smaller activation energy is obtained if one atom of the split pair migrates along a ⟨111⟩ direction towards the metastable configuration OT. The computed activation energy for this process is $\sim 0{\cdot}15$ eV, and the migration is completed by the change $OT - H_T$. During the migration, each individual atom never moves far from its original position, but successive different atoms become the interstitial, and the centre of mass moves a nearest neighbour distance. This is also true for other possible migration processes (see "interstitialcy diffusion", p. 376). If the configuration H_T exists,

† Incorrectly designated a third neighbour pair in the original paper.

it may migrate through H_T—T—H_T with a computed activation energy of $\sim$ 0·13 eV The activation energy for straight-line crowdion motion is very low ($\sim$ 0·02 eV).

Similar considerations were applied by Johnson (1964) to interstitial motion in b.c.c. structures. The migration process suggested for the stable interstitial had an estimated migration energy (for α-iron) $\sim$ 0·33 eV, whilst the nearly stable $\langle 111 \rangle$ crowdion has an activation energy for migration estimated at only $\sim$ 0·04 eV.

Experimental information on the mobilities of point defects is more reliable than are the theoretical estimates, although difficulties in interpretation arise, and the appropriate experiments have only been made for a few metals. The method most used is to introduce a non-equilibrium concentration of point defects at a low temperature, and then to raise the temperature until the defects become mobile. The excess defects then migrate through the lattice until they disappear at "sinks", or encounter other defects with which they combine. The movement of the defects is studied by observing the "recovery" of some physical property which depends on the excess defect concentration. The initial excess concentrations are introduced in three main ways, namely by quenching, by irradiation with neutrons or with charged particles, and by deformation.

If the excess values of some physical property are attributable to point defects, and are recovered in a well defined temperature interval, it will usually be possible to assign an experimental activation energy to the recovery process by measuring its rate as a function of temperature. The simplest interpretation of such experiments is to assume that a defect becomes mobile at the temperatures concerned, and the experimental activation energy may then be identified with the activation enthalpy $\Delta_a h$ for motion of this defect (see p. 88). Suppose the defect migrates through the lattice until it reaches a sink at which it can disappear completely, and that the observed change in physical properties is due solely to the removal of defects. The concentration of defects present at any time may then be correlated with the remaining excess value of the physical property at any stage.

Let ΔP be the measured excess value of the property at any stage of an isothermal anneal. Then the number of defects which are removed in unit time is proportional to $\mathrm{d}\Delta P/\mathrm{d}t$, and hence the mean time which an individual defect requires to find a sink is $\Delta P/(\mathrm{d}\Delta P/\mathrm{d}t)$. Since the frequency of a defect jump is given by (17.6), the mean number of jumps made by a defect before it disappears is

$$\{\Delta P/(\mathrm{d}\Delta P/\mathrm{d}t)\} z\nu \exp(\Delta_a s/\boldsymbol{k}) \exp(-\Delta_a h/\boldsymbol{k}T). \tag{17.7}$$

The frequency ν may be taken as 10^{13}, and all other quantities are known from experiment, except for $\Delta_a s$, which is only likely to contribute a small numerical factor or fraction to the number of jumps.

Estimates of the number of jumps made in this way for various recovery processes vary from $\sim$ 1 to 10^{10}, and illustrate the complexity of some of these processes. One simple case to consider is when the defects disappear at a number of randomly distributed sinks, the concentration of which does not change throughout the recovery. If the motion of each defect is a "random walk", and the defects and sinks are initially randomly distributed, the average number of jumps made by a defect before disappearing will simply be the reciprocal of the atomic concentration of sinks. Since this result is independent of the concentration of defects, the mean number of jumps as calculated by (17.7) will be constant throughout the recovery, so that $\Delta P/(\mathrm{d}P/\mathrm{d}t)$ will be approximately constant. In accordance with the discussion in Section 12, we see that this defines a first order recovery process, in which the excess value of the physical property follows a simple exponential decay curve at constant temperature. The results are self-consistent if the measured number of jumps can be correlated with known concentrations of probable sinks.

The random-walk assumption for the motion of an individual defect will be valid if the defect has a symmetrical configuration, provided there is no appreciable long-range interaction between the defect and the sink. An attractive force between the sink and the defect will impose a mean drift flow on the otherwise random motion of the defect, and thus lower the number of jumps before annihilation; the reverse is true for long-range repulsion. It seems likely that any elastic interaction is very small for all conceivable sinks, becoming appreciable only when the defect is very close (say within about 10 atomic distances) to the sink, so that the drift flow may be neglected. The further assumption above that there is uniform random distribution of sinks is not, however, very realistic, and the initial stages of recovery for a fixed set of sinks will generally not correspond to first-order kinetics. The value obtained for the number of jumps will increase in the initial stages, but will become constant. We should note here that first order kinetics do not necessarily imply constancy of the measured $\Delta P/(\mathrm{d}\Delta P/\mathrm{d}t)$, since the assumption that ΔP is proportional to the number of defects remaining may not be valid. When appreciable numbers of defects disappear at internal sinks, for example, the change in configuration of the sink may make an appreciable contribution to ΔP.

Migration of point defects to fixed sinks is one limiting type of recovery process; another simple limiting form is the mutual annihilation of opposite defects. Consider a crystal containing both interstitials and vacancies. If these are produced in closely associated pairs (as might be true in radiation-damaged crystals), recovery will take place when such pairs recombine by migration of the interstitial through a very small number of interatomic distances. Mutual annihilation is also a possible recovery process when there is a random distribution of the two kinds of defect, not necessarily in equal concentrations. The situation is formally similar to the disappearance of a defect at a number of fixed sinks, but the concentration of sinks now decreases with that of the defects, so that the more mobile defect has to sample more and more sites in order to find a sink as the recovery proceeds. The rate of recovery is thus proportional to $x_{\bullet}x_{\square}$, whereas for fixed sinks it is simply proportional to the concentration of the defect being removed. In the special case that the two concentrations are initially equal, they will remain equal at all times, and we have a second order kinetic process. In practice the distributions are either non-random or the concentrations of opposite defects are not equal, so simple second-order kinetics are not observed.

There are many possibilities intermediate between the fixed sinks and mutual annihilation models we have just described. For example, vacancies which encounter one another during annealing may amalgamate to form divacancies, and the divacancies may then migrate to fixed sinks. Assuming the initial distribution of vacancies to be random, the rate of divacancy formation will be proportional to the square of the vacancy concentration, and hence will show second-order kinetics. If the binding energy of a vacancy pair is sufficiently large, and its mobility sufficiently great, the chances of the divacancy dissociating again before it reaches a sink may be small. In this case, the overall kinetics might be determined largely by the rate of pair formation.

In principle, the simplest experiments on point defect migration should be those using quenched specimens, since the defect state before quenching is known to consist of a random distribution of single vacancies with a small number of divacancies. Fixed sinks for these vacancies could be the free surface, grain boundaries, and internal line defects (dislocations). Since the number of jumps before annihilation rarely exceeds 10^9, the first two are likely to be important only for specimens with grain sizes less than 10^4–10^5 atom distances, i.e. $\sim 10^{-3}$ cm. Experimental evidence on the efficiency of dislocations as sources and sinks for vacancies (see pp. 248–50) is rather confusing, and will be discussed further in Chapter 7. The results of some quenching experiments, especially those on platinum (Bradshaw and Pearson, 1956, 1957) seem consistent with the assumption that the main annealing process is the migration of individual vacancies to fixed dislocation sinks. However, careful experiments with gold and other metals have led to rather different conclusions.

Bauerle and Koehler (1957) found that gold wires quenched from below 750°C recovered their excess resistivity at an exponential rate at constant temperature, and with an activation energy of $\sim 0{\cdot}82$ eV. Wires quenched from above this temperature gave a fast decay curve which was much more complex, and the activation energy was only 0·63 eV. These results were interpreted by Koehler *et al.* (1957) as indicating that the recovery process after quenching from low temperatures was the migration of single vacancies to dislocations, whereas divacancies were formed during the quench from higher temperatures, and were then removed at the same fixed sites. A different interpretation of the results was given by Kimura *et al.* (1959), who assumed that a substantial number of divacancies and larger clusters were formed during the quench from higher temperatures. Vacancy clusters can be spherical voids or flat discs; in the latter case, the lattice relaxes around the cluster to form a "prismatic dislocation loop" at its periphery (see p. 298). This flat cluster can continue to grow by addition of further vacancies to its edge, and Kimura *et al.* supposed that the important sinks after a high-temperature quench are dislocation loops of this kind, to which the remaining divacancies migrate. Since the sink density is effectively determined by the total length of dislocation line, this assumption corresponds to an increasing density of sinks, in contrast to the fixed density provided by a grown-in dislocation network.

Striking confirmation of the formation of vacancy clusters which lead to dislocation loops was obtained in the early thin film electron-microscopy of quenched materials (Hirsch *et al.*, 1958; Silcox and Hirsch, 1959; Smallman and Westmacott, 1959). The microscope reveals the dislocations formed from flat vacancy clusters, and also shows evidence in some cases of absorption of vacancies by existing dislocations; we discuss this more fully in Chapter 7. The presence of small spherical clusters, which have also been postulated (see Kuhlmann-Wilsdorf *et al.*, 1962) cannot be shown directly. The behaviour of vacancies in quenched materials is of great importance in low-temperature phase transformations, and will arise again in later chapters.

The interpretation of quenching experiments and of annealing experiments on quenched metals was initially thought to be reasonably straightforward, but is now known to be complex. When the vacancy supersaturation is relatively small, and the metal is very pure, it is probable that only single vacancies, divacancies, and dislocation and sub-boundary sinks need be considered in determining vacancy loss during the quench or a subsequent anneal. Nevertheless, the kinetics are usually quite complex, and the usual methods of determining activation energies for defect migration are not easy to interpret; moreover, the purities required are very high in order that the concentration of solute atoms (which may associate with vacancies) shall be significantly smaller than the concentration of vacancies. High vacancy supersaturations avoid this particular problem, but many higher order vacancy clusters are now expected, and knowledge of the properties of such clusters is very scanty. Large clusters, such as dislocation loops, formed from collapsed discs, or stacking fault tetrahedra (see Chapter 7) may be studied by electron-microscopy, but the properties of small clusters are virtually unknown. Impurities have been found to influence the nucleation of large clusters, but the effects vary considerably with different impurities, presumably because of varying vacancy–impurity interactions, and the experimental evidence is sometimes conflicting (for a survey, see Chik, 1970).

It has been evident since the work of Bauerle and Koehler that single vacancies aggregate or migrate to sinks during a quench, and that the fractional vacancy loss will be greater as the temperature of the quench is increased because of the greater mobility at high temperatures. An attempt to correct for this effect was first made by Mori *et al.* (1962); they deliberately varied the quench rate and by plotting the measured residual electrical resistivity of their gold-wire specimens against reciprocal quench rate for each fixed quenching temperature, they were able to obtain extrapolated values corresponding to infinite quench rates. The method has now been used by several authors; the collected data (Balluffi *et al.*, 1970) fall on a single, but slightly non-linear curve of $\ln \Delta P_\infty$ vs. $1/T$.

An alternative procedure for correcting for vacancies lost during the quench (Takamura, 1961; Takamura *et al.*, 1963) is to plot against specimen size rather than quenching rate. According to a later theoretical analysis by Flynn *et al.* (1965), the fractional vacancy loss depends only on $D_q T_q \tau_q$, where D_q is the diffusion coefficient of vacancies at the quench temperature T_q and τ_q is the time for the temperature to reach 0 K, assuming a linear cooling rate. Flynn *et al.* claimed that experiments in which τ_q varied then allow $\Delta_a h_\square$ to be deduced in addition to $\Delta h_\square$, and their own experiments together with previous data were consistent with $\Delta h_\square = 0{\cdot}98$ eV and $\Delta_a h_\square = 0{\cdot}83$ eV. The fractional vacancy loss, plotted as a function of $\ln(D_q T_q \tau_q)$ suggested that diffusion to either dislocations or to subgrain boundaries is the important process during the quench. Kino and Koehler (1967) considered possible vacancy loss to long dislocations and to clusters nucleated on impurities; in both cases the theoretical loss depends only on $D_q T_q \tau_q$, but their experimental results do not agree with this conclusion. These authors consider that the discrepancy may arise because at any temperature the vacancy sinks are able to operate only when the supersaturation exceeds some finite value; in the case of dislocations, for example, it must be large enough to overcome the line tension and thus bow out the dislocations.

Seeger and Mehrer (1968) have criticized the procedure of extrapolating against quenching rate since this could enhance another source of error, namely the production of plastic strains in rapidly quenched specimens. Such strains may, in principle, cause errors of either sign because (a) additional point defects above the equilibrium concentration are produced by the straining, and (b) additional line defects or other sinks are created, and thus increase the rate of removal of vacancies. Jackson (1965) showed that effect (b) generally predominates, and Balluffi *et al.* (1970) conclude that the extrapolation procedure is then valid. A possible advantage claimed for the Takamura procedure of extrapolating against specimen size is that the effects of both quenching rate and quenching strains are simultaneously eliminated.

In principle, a sufficiently rapid quench of a pure specimen with a low dislocation density and a low vacancy density should be able to retain virtually all of the vacancies. Kino and Koehler (1967) were able to quench very fine gold from ~ 600°C, so as to obtain a residual resistivity independent of quench rate, although the residual resistivities of specimens quenched from 700°C and 800°C varied with quench rate in the usual way. These results, combined with diffusion and $\Delta N/N$ data, are the basis of Koehler's estimate for $\Delta h_{\square}$ quoted above. However, in the discussion of the paper by Seeger and Mehrer (1970), Seeger pointed out that a low-temperature tangent to the Arrhennius plot of quenching data given by Balluffi *et al.* (1970) indicates a value of $\Delta h_{\square}$ of ~ 0·91 eV at room temperature, which is closer to the value deduced from the Seeger and Mehrer analysis.

We now consider some of the problems in determining migration energies and divacancy binding energies from annealing studies on quenched metals; once again, the much studied case of gold will serve as the main example. The difficulties in measuring activation energies and interpreting kinetics even with the simplest assumptions (monovacancy–divacancy–sink systems) have been discussed by many authors, e.g. Balluffi and Siegel (1965), Doyama (1965), Koehler (1970), Chik (1970), and Burton and Lazarus (1970). The simplest estimates are those in which the initial vacancy concentration is obtained by quenching from a low temperature, and is thus relatively small.

In the early experiments of Bauerle and Koehler, the annealing of gold specimens quenched from 700°C followed approximately first-order kinetics, and from the change of slope of the log ΔP vs. time curves produced by a change of temperature, they deduced $\Delta_a h_{\square} = 0{\cdot}82$ eV. According to Balluffi and Siegel (1965) the same data replotted on a linear ΔP vs. t basis, without any assumption about kinetics, give an apparent activation energy of ~ 0·7 eV. This same value has been found in specimens quenched from ~ 700°C by Mori *et al.* (1962), Ytterhus and Balluffi (1965), Siegel (1966), and several other workers, and has generally been attributed to divacancy motion since diffusion data and measurements of equilibrium concentrations require $\Delta_a h_{\square}$ to be appreciably larger. More recently, Wang *et al.* (1968) found the same energy in experiments on annealing at higher temperatures than had hitherto been possible, and they point out that an activation energy of 0·7 eV has been found to dominate the annealing over a wide range of temperatures and in conditions in which different sinks and sink densities are operative. The conclusions, however, do not entirely agree with these of Kino and Koehler (1967), who observed that pure gold, fast-quenched from 700°C, gave second-order annealing kinetics and an energy (attributed to $\Delta_a h_{\square}$) of 0·87 eV, which is in good agreement with Seeger and Mehrer's deductions from diffusion data. According to Koehler (1970), Sharma has succeeded in quenching the same specimens to give either second-order kinetics with the 0·87 eV energy or first-order kinetics with an activation energy of $\sim 0{\cdot}7$ eV; the smaller energy is observed if the cooling rate is less than 2×10^4°C s^{-1}. Sharma *et al.* (1967) also found the apparent activation energy to decrease with decreasing purity of the gold, in contrast to the results of Siegel (1966).

Koehler attributes the change of activation energy with quench rate to a change in the rate-controlling process, and specifically to a change in the sink density which is high for fast-quenched specimens. With a high sink-density, the rate of annealing is controlled by the rate at which divacancies form by the motion of single vacancies; with a lower density, the rate of divacancy motion controls the annealing rate. In support of the value for $\Delta_a h_{\square}$, Koehler points out that the same energy has been observed in the annealing of quenched, deformed, and electron-irradiated gold.

The results of all these experiments seem to establish $\Delta_a h_{\square}$ and $\Delta_a h_{\square\square}$ as respectively $\sim 0{\cdot}9$ and $\sim 0{\cdot}7$ eV, but the interpretation has nevertheless recently been placed in considerable doubt because of the work of Johnson (1968) and Burton and Lazarus (1970). Johnson used a computer to simulate complex annealing processes with immobile single vacancies and mobile clusters. He showed that it is very difficult to extract meaningful information about the assumed processes by any of the analytical techniques normally used. Burton and Lazarus used a simple model based on single vacancies, divacancies, and fixed sinks, and applied computer generated data to consider especially the slope change method. They showed that there are two well-defined activation energies, one obtained from the instantaneous values of the slope before and after the temperature change, and one by extrapolation of the slope from longer times. These two energies may differ because the behaviour of the slope, $d\Delta P/dt$, following a temperature change is not necessarily monotonic. The instantaneous slopes give the activation energy for the defect-annihilation step of the whole process, whilst the apparent activation energy, obtained by extrapolating the curves to the time of the change, is characteristic of the rate-limiting step. In order to distinguish these two possibilities, it is necessary to make continuous measurements of the physical property ΔP, which is used to monitor the anneal. Burton and Lazarus report experiments of this kind which on analysis proved to be consistent with a monovacancy–divacancy-increasing sink-density model, with an instantaneous activation energy (identified as $\Delta_a h_{\square\square}$) of only 0·54 eV. The apparent activation energy in their experiments varied from $\sim 0{\cdot}7$ to $\sim 0{\cdot}95$ eV, and is attributed to an unknown rate-limiting step.

It is now very apparent that one of the difficulties in interpreting quenching experiments is lack of knowledge of which mobile defects and sinks are contributing most effectively to the annealing. The simple assumption of only two mobile species (monovacancies and divacancies) is often made for convenience, but the observations of stacking fault tetrahedra in gold specimens (e.g. Siegel, 1966) show that clustering on a larger scale also takes place, and Koehler estimates the sink density in gold slowly quenched from 700 °C to correspond to the density of tetrahedra observed by Siegel. According to Chik (1970), mobile trivacancies probably play a significant role in the annealing of gold at higher vacancy concentrations, and this may also be the case for other metals. Very little is known about clusters intermediate in size between the divacancy and the smallest tetrahedra or prismatic loops, and there is also much uncertainty about the role of impurities. The observations of Siegel suggest that fault tetrahedra are readily formed only in the presence of impurities, but the evidence on whether or not they are always heterogeneously nucleated is conflicting (see the discussion by Chik, 1970).

Perhaps the most controversial vacancy parameter has been the binding energy of a divacancy, the preferred value of which has been a complex function of both the research group and time. The high estimates for the binding energy in gold ($\sim$ 0·3–0·6 eV) imply that at high temperatures about 10% of the self-diffusion rate is due to divacancy motion, whilst the low estimates ($\sim$ 0·1–0·15 eV) imply a much smaller contribution. The higher values were deduced partly from the experiments discussed above which seemed to show that excess vacancy annealing involved divacancy motion as the predominating process over rather wide temperature ranges. In recent work, a value of $\sim$ 0·6 eV was selected by Wang *et al.* (1968) and accepted by Sharma *et al.* (1967), in contrast to the low value of $\sim$ 0·15 eV deduced by Kino and Koehler. It follows from the above discussion, however, that the basis for this selection is again in considerable doubt, and Seeger and Mehrer (1968, 1970) argue strongly that the self-diffusion data for gold is inconsistent with a high value of the binding energy. The same result is obtained for copper, but the curvature of the Arrhenius plot for silver is such that a high value of $2\,\Delta h_{\square} - \Delta h_{\square\square}$ is much more probable.

Vacancy studies in the b.c.c. metals are difficult because of the high formation energy and low migration energy and because of the pronounced effects of interstitial impurities; thus, most of the work has been in the interpretation of annealing stages in irradiated or cold-worked metals. However, Schultz (1965) introduced a technique of quenching into liquid helium which is claimed to preserve a very high purity, and he was apparently successful in retaining quenched vacancies in tungsten but not in tantalum. He deduced a vacancy concentration at the melting point of $\sim 10^{-6}$, with $\Delta h_{\square} \sim 3{\cdot}3$ eV; later measurements by the same method (Gripshover *et al.*, 1970), gave $\Delta h_{\square} \sim 3{\cdot}6$ eV and hence $\Delta_a h_{\square} \sim 2{\cdot}5$ eV, assuming the activation enthalpy for diffusion to be 6·1 eV. Meakin *et al.* (1964) quenched molybdenum specimens and subsequently observed contrast effects in the electron-microscope which they identified as collapsed vacancy discs. In view of later work (e.g. Evans, 1970) it now seems very probable that contamination by some impurity was responsible for these observations, and Evans's experiments indicate that it is impossible to retain vacancies in pure molybdenum at the highest quenching rates which have been attained. This result also calls into question the interpretation of the liquid-helium quenching results for tungsten, since there is no obvious reason why these two metals should differ appreciably in vacancy properties. If contamination occurs during a quench, the heat of solution of an impurity may be measured, rather than $\Delta h_{\square}$, as shown by Evans and Eyre (1969) for nitrogen in molybdenum. Some observations of vacancies in high melting point b.c.c. metals, including estimates of vacancy concentration and of $\Delta h_{\square}$, have also been made by field-ion microscopy (Müller, 1970; Galligan, 1970). Such experiments are very interesting since individual defects are observed, but there is the usual difficulty that the field-ion tip may not be representative of bulk properties. A review of point defect work on b.c.c. metals up to $\sim$ 1968 is given by Nihoul (1970), and a similar review for h.c.p. metals by Schumacher 1970).

The high formation energy of self-interstitial defects precludes their study by equilibrium measurements, and appreciable concentrations of such defects are therefore formed only in irradiated or cold-worked materials. There is a long-standing controversy over the interpretation of annealing experiments in such materials, which largely centres around the question of whether there is only one mobile interstitial defect (the single interstitial) or whether there is an additional metastable configuration, probably a crowdion, which is also mobile. We can include only a very brief discussion of this complex subject, but it is clearly of considerable importance in the description of metallurgical processes at low temperatures in irradiated or deformed materials.

Irradiation with fast particles at low temperatures (close to 4·2 K) is now known to produce Frenkel defects, i.e. interstitials and vacancies in equal numbers, but the local arrangements differ with the type of particle. On warming to higher temperatures, the mobility of the defects leads to progressive recovery of the excess values of some physical property, such as electrical resistivity, and such recovery usually occurs in various successive temperature ranges. Following a notation introduced by van Bueren (1961), these stages in f.c.c. metals are usually labelled I–V in ascending order of temperature, and in cases where the annealing peaks have a fine structure then substages are usually labelled $I_A, \ldots, I_E$, etc. In the low-temperature stage I, recovery is initially due to the annihilation of close interstitial vacancy pairs, but in the later substages there is evidence for the long-range migration of some interstitial defect. According to one interpretation it is the stable interstitial which migrates, and there is no free migration of any interstitial defect at higher temperatures, whilst according to a rival view it is the metastable interstitial modification which is mobile in stage I. The interpretation of stage II is uncertain in detail, but it is generally agreed that both impurity interactions and intrinsic processes are important, and that release of interstitials from some traps and subsequent formation of deeper traps, combined with rearrangements of configurations and possibly divacancy annealing, are the dominant effects. The remaining controversy centres largely around the interpretation of stage III. The workers who believe normal single interstitials migrate during stage I_E attribute stage II, at least in part, to the long-range migration of vacancies; the other school attributes much of the stage III annealing to the release of interstitials from deep traps and/or the migration of single interstitials in the stable form. The controversy has existed for a least a decade; a summary of the respective viewpoints is given by Schilling *et al.* (1970), van den Beukel (1970), Corbett (1970), Koehler (1970), and Seeger (1970). Some of the difficulties arise from the seeming lack of consistency in the detailed behaviour of similar individual metals, and it is by no means obvious for a particular metal (e.g. platinum) how the observed annealing stages should be labelled so as to correspond to those for another metal (e.g. copper).

18. CRYSTAL SURFACES: THE EQUILIBRIUM STRUCTURE OF A STEP

The preceding sections have dealt with the atomic structure in the interior of a single crystal of a metal, and the boundary structure has been ignored. Internal boundaries in the solid are best considered as regions of bad crystal, and we postpone discussion of their properties to Chapter 8. In the next two sections, we shall consider in some detail the structure of an external (free) surface in contact with its own vapour, and we shall also refer briefly to surfaces in contact with the melt.

An element of crystal surface has an orientation defined in relation to the crystal axes by its Miller indices h_i. A flat surface thus has two degrees of freedom, and its structure and energy will be functions of two independent variables. In general, we do not expect the surface to be flat on an atomic scale even at 0 K. Consider, for example, the (170) face of a simple cubic crystal. Figure 5.9(a) shows a section through this surface in which

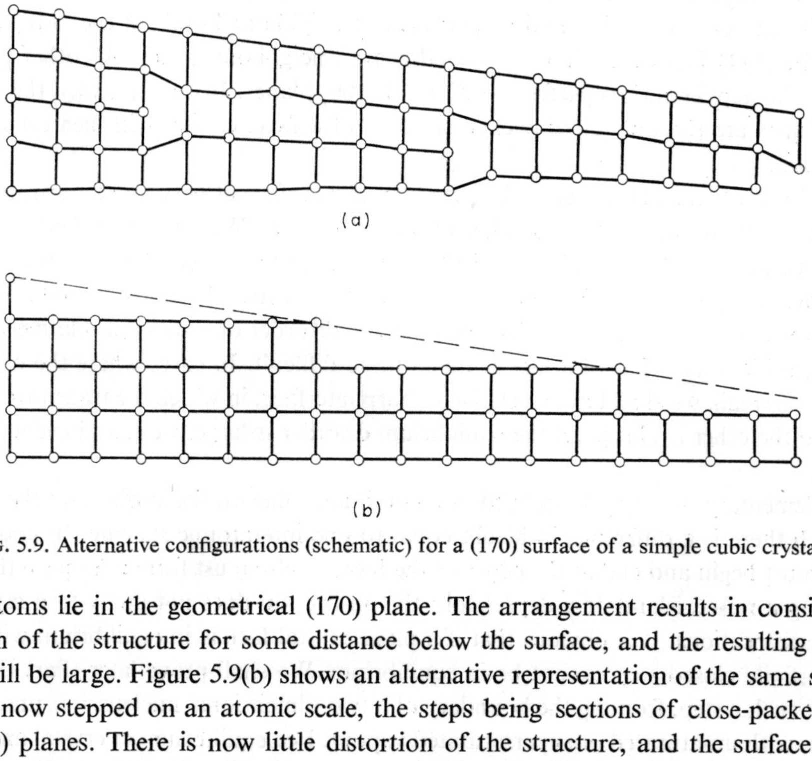

FIG. 5.9. Alternative configurations (schematic) for a (170) surface of a simple cubic crystal.

all the atoms lie in the geometrical (170) plane. The arrangement results in considerable distortion of the structure for some distance below the surface, and the resulting surface energy will be large. Figure 5.9(b) shows an alternative representation of the same surface, which is now stepped on an atomic scale, the steps being sections of close-packed (100) and (010) planes. There is now little distortion of the structure, and the surface energy will be correspondingly lower. Surface energy effects will thus always ensure that a surface which is described macroscopically by relatively high indices h_i is composed of an atomic hill and valley structure, the sides of which are parallel to close-packed planes. For the present, we shall assume that this conclusion is self-evident; a general proof will be given in Section 20, where we discuss singular and non-singular surfaces. Moreover, observations with the field-ion microscope have given direct experimental evidence of the stepped structure of non-close-packed surfaces. It follows from the model that the surface enthalpy per unit area of the (170) plane will not be very different from that of the (100) plane, and surface energy effects are therefore unimportant in deciding the external shape of a macroscopic crystal. The small dependence of surface energy on orientation is a property only of an external surface; the energy of a grain boundary may depend markedly on the parameters describing the boundary.

The surfaces of a perfect crystal can now be divided into close-packed and stepped planes. For the present purpose, it is convenient to re-define a close-packed plane as one in which all surface atoms are equidistant from a geometrical plane parallel to the surface (Burton and Cabrera, 1949). The division into surfaces of the two kinds can be made approximately by counting atomic bonds and estimating the energies of the alternative arrangements on the assumption of a central force model. In the simple cubic structure, the assumption that only nearest neighbour interaction is important leads to the conclusion that only {100} faces are close-packed, whilst if second nearest neighbour interaction is included the planes {110} and {111} may also form close-packed surfaces. The {110} face has nearest neighbour bonds in the ⟨001⟩ direction and second nearest neighbour bonds in the $\langle 1\bar{1}0\rangle$ direction, whilst the {111} faces contain only second-nearest neighbour bonds. All other planes are stepped, the steps being formed from sections of {100} planes. In the same way, the {111} and {100} planes are the only close-packed surfaces of a f.c.c. crystal with nearest neighbour interaction.

The above remarks all concern the structure of a surface at very low temperatures. At a finite temperature, we have two further problems to solve. We must first find the equilibrium structure of a step on a close-packed surface, and then the equilibrium structure of the close-packed region itself. At a finite temperature, we expect both a step and a flat crystal face to exhibit a certain disorder. The equilibrium disorder of a step is a relatively easy statistical problem, but that of a face is much more difficult. In cases where the equilibrium disorder is small, we shall have a relatively sharp interface, in which the transition from one phase to the other is abrupt; if the equilibrium disorder is large, we shall have a diffuse interface.

An elementary step may be defined as a continuous line on the surface, on the two sides of which there is a difference in height equal to one interatomic spacing. In good crystal, a step must begin and end at the edges of the face, or else must form a loop on the surface enclosing a raised island or a depression; this restriction does not apply to a crystal containing dislocations. It is obvious that if a plane crystal face is in equilibrium with the vapour, a finite closed step cannot be in equilibrium. We shall prove later (Part II, Chapter 13) that such a step, forming the boundary of a two-dimensional nucleus, is in unstable equilibrium with unsaturated or supersaturated vapour. For equilibrium of crystal and vapour, we shall therefore have steps only of the type shown in Fig. 5.9(b), extending from one side of the crystal to the other. These steps will be straight, except on an atomic scale, i.e. they will have a constant mean direction.

For simplicity, we use the model of a simple cubic structure with nearest neighbour interaction $-2\Xi_1$, and we consider a (001) face of the crystal with a step lying in the close-packed [100] direction. At 0 K this step will be as straight as possible, but as the temperature is raised a number of kinks will appear as a result of thermal fluctuations. This is illustrated in Fig. 5.10, which shows a number of abrupt changes in direction (+ and −), some of which are associated with single atoms (A) or holes (B) adsorbed on the step. The effect was first predicted by Frenkel (1945), and the changes in direction are sometimes known as Frenkel kinks. They are very important in crystal growth, since it is obvious that they constitute the places from which atoms may most readily be removed during evaporation, and

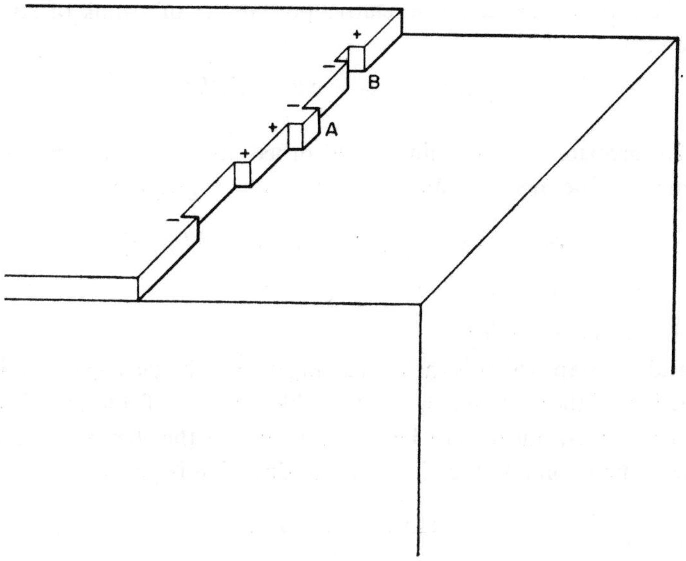

FIG. 5.10. Positive and negative kinks in a surface step.

to which they can most readily be added during growth. The energy gained in adding an atom to a kink (also known as an "exchange" or "Kossel" site) is equal to the evaporation energy per atom.

Suppose the probability of having a kink in one direction at a given point in the step is α_+, and of an oppositely directed kink is α_-, distances being measured along the [100] direction. Then if α_0 is the probability of having no kink, we must have, for a [100] step

$$\alpha_+ = \alpha_-, \qquad \alpha_0+\alpha_++\alpha_- = 1.$$

In these equations, we are assuming that only kinks of one interatomic distance in height occur; there is a finite probability of multiple kinks which we shall neglect for the present.

Burton and Cabrera calculated the probabilities α_+ and α_- by considering various atomic processes and applying the principle of detailed balancing. Referring to Fig. 5.11, we

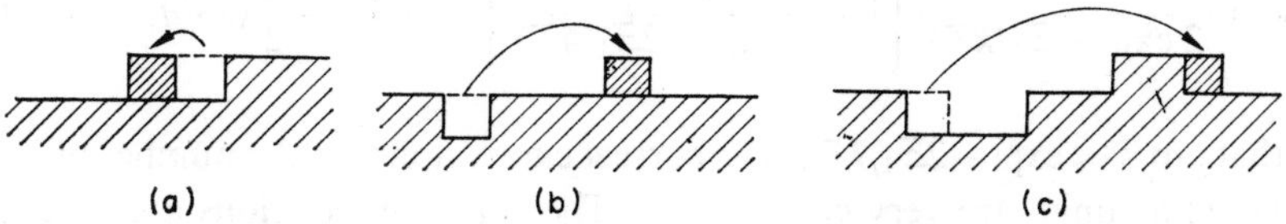

FIG. 5.11. Atom processes on a crystal surface (after Burton *et al.*, 1950–1).

see that the energy required to form an adsorbed atom on the step (Fig. 5.11 (a)) will be $2\Xi_1$ in terms of our simple model, and the energy required to form an adsorbed atom plus a hole will be $4\Xi_1$ (Fig. 5.11 (b)); there is no energy change associated with Fig. 5.11 (c)). The numbers of kinks formed in these processes are 2, 4 and 0 respectively, and the energy of formation of a kink is thus Ξ_1. The relative probability of a kink of given kind at any place in the step is

$$\alpha_+/\alpha_0 = \alpha_-/\alpha_0 = \exp(-\Xi_1/kT), \tag{18.1}$$

whilst the relative probability of forming an adsorbed atom on the step is $\exp(-2\Xi_1/kT)$, and is much smaller. The mean distance between kinks is given by

$$\begin{aligned} x_0 &= a/(\alpha_+ + \alpha_-) = \tfrac{1}{2}a \exp\{(\Xi_1/kT)+2\} \\ &\approx \tfrac{1}{2}a \exp(\Xi_1/kT), \end{aligned} \tag{18.2}$$

where a is the interatomic spacing.

We now consider a step which is inclined at an angle to the close-packed direction [100]. The mean direction of the step over an appreciable number of interatomic distances need not have rational indices, but it must be constant so that the step is straight. The angle θ between the mean direction and the close-packed direction is given by

$$\tan\theta = \alpha_+ - \alpha_-. \tag{18.3}$$

A step of this kind will have more kinks than a [100] step, and as θ increases from zero, α_+ increases and α_- decreases. Calculation using the principle of detailed balancing (Burton *et al.*, 1950–1) leads to the result that the product of the relative probabilities α_+/α_0 and α_-/α_0 is unchanged, i.e.

$$\alpha_+\alpha_- = \alpha_0^2 \exp(-2\Xi_1/kT) \approx \exp(-2\Xi_1/kT). \tag{18.4}$$

The derivation of this result is valid when both the possibility of multiple kinks and the effects of second nearest neighbour interaction Ξ_2 is considered.

From eqns. (18.3) and (18.4), the mean kink density, i.e. the probability of having a kink of either sign at any place in the step, is seen to be

$$\alpha_+ + \alpha_- = [4\exp(-2\Xi_1/kT) + \tan^2\theta]^{1/2}, \tag{18.5}$$

if α_0 is taken as unity. The mean distance between kinks, measured along the principal direction, is now x_θ, where

$$x_\theta = \frac{a}{2\exp(-\Xi_1/kT)}\left\{1 - \frac{\tan^2\theta}{8\exp(-2\Xi_1/kT)}\right\} \approx x_0\left[1 - \frac{1}{2}(x_0/a)^2\tan^2\theta\right], \tag{18.6}$$

provided that $\tan\theta \ll \exp(-2\Xi_1/kT)$. The correct expressions for the number of kinks when α_0 is not taken as unity are very complicated. The simple expressions are strictly valid at low temperatures, where α_+ and α_- are small.

For a [100] step, we know that at 0 K, $\alpha_+ = \alpha_- = 0$, and at low temperatures

$$\alpha_+ + \alpha_- = 2\exp(-\Xi_1/kT). \tag{18.7}$$

This equation was first given by Frenkel (1945), who deduced it by calculating the configurational entropy of the kinked step. Similarly, for a step at an angle θ to the [100] direction we have $\alpha_- = 0$, $\alpha_+ = \tan\theta$ at 0 K. The step is as straight as possible, and contains only the number of kinks geometrically necessary to maintain the direction θ. At low temperatures, eqn. (18.5) is valid, and for small θ can be approximated to give

$$\alpha_+ + \alpha_- = 2\exp(-\Xi_1/kT)\,[1+\tan^2\theta/\{8\exp(-2\Xi_1/kT)\}]. \tag{18.8}$$

According to this equation, $\alpha_+ + \alpha_-$ continually increases with increasing θ, whereas in fact the step lies along a close-packed direction again when $\theta = 90°$. On general grounds we should expect the kink density to pass through a maximum and reach a second relative minimum for the [110] direction at $\theta = 45°$; in order to allow for this, it is necessary to include the effects of second nearest neighbour interactions.

The order of magnitude of the above quantities can now be calculated. The evaporation energy Δh^{sv} is the energy required to remove an atom from a kink in a step, and a value for this energy may be estimated simply by counting the number of nearest and next neighbour bonds which disappear during the process. For the simple cubic structure, we thus have $\Delta h^{sv} = 6\Xi_1 + 12\Xi_2$, and if we neglect Ξ_2, the energy to form a kink is about one-sixth of the evaporation energy per atom. Burton and Cabrera (1949) assume a typical value of $\Xi_1 \approx 0{\cdot}1$ eV on this basis, and at 600 K, eqn. (18.2) shows that the mean distance between kinks will be only 4–5 atoms, even for the close-packed directions.

Although the above equations were derived for the simple cubic crystal (Kossel crystal), they will clearly be valid in general form for steps on the close-packed surfaces of other crys-

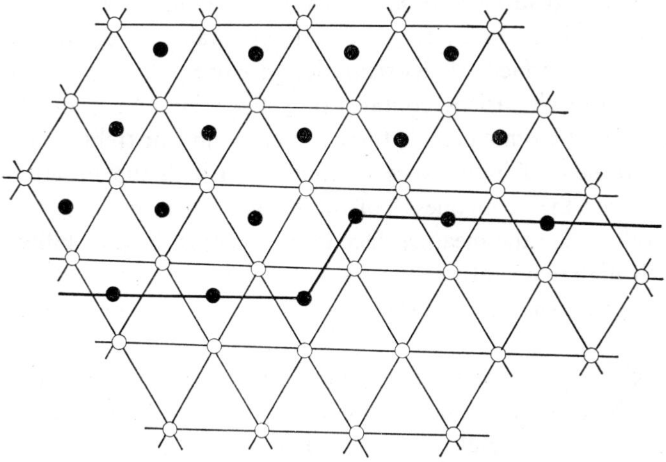

FIG. 5.12. A kink in a step on a close-packed surface. Open and filled circles represent atoms on two successive close-packed planes.

tal structures. For crystals which have an atomic arrangement corresponding to the close packing of spheres, the energy required to form a kink in a close-packed step (e.g. the $[1\bar{1}0]$ step of a (111) face of a f.c.c. structure) will be ~ one-twelfth of the evaporation energy per atom, since each atom has twelve nearest neighbours. This can be seen by reference to Fig. 5.12 which shows such a kink. By considering atomic processes in the same way, it is clear that the energy required to form a kink is again Ξ_1, where $-2\Xi_1$ is the energy of a nearest neighbour pair, but the evaporation energy per atom is now $12\Xi_1$. Calculation of the evaporation energy for a particular metal is rather difficult, but is fortunately not necessary. The main point of this section is to estimate the degree of disorder in a step, and we are mainly interested in temperatures sufficiently high for the crystal to be grown from the vapour. The kinks will be spaced at about four atom distances if $\Xi_1/\boldsymbol{k}T \sim 2$, and it is easy to show that this condition is satisfied at the minimum temperatures for crystal growth. Thus Frank (1952a) estimates that the lowest temperature for growth is that at which the vapour pressure is 10^{-10} atm. Using the thermodynamic expression $\mathrm{d} \ln p/\mathrm{d}(1/T) = \Delta h^{sv}/\boldsymbol{k}$, this gives $\Delta h^{sv}/\boldsymbol{k}T \sim 23$, and $\Xi_1/\boldsymbol{k}T \sim 2$, since $\Delta h^{sv} = 12\Xi_1$ for close-packed structures.

It follows that for all crystal structures, the steps in principal directions have a very high density of kinks at the lowest temperatures at which growth from the vapour is possible, and the proportion of kinks becomes greater as the temperature is raised. Steps which do not lie along close-packed directions have even larger numbers of kinks.

19. THE STRUCTURE OF CLOSE-PACKED SURFACES

The results obtained for the disordering of a step do not apply to the disordering of a surface. The energy of formation of a step is proportional to its length and is usually very large. Except possibly very close to the melting point, there is a negligible possibility of a step in a surface being formed by thermal fluctuations in the same way as a kink is formed in a step. A close-packed surface does not therefore contain an equal number of opposite steps, although a step in a close-packed direction contains an equal number of opposite kinks. In the same way, surfaces of all orientations are in equilibrium with the same external vapour pressure, but each surface contains only the minimum (geometrically necessary) number of steps of unit atomic height; there are no double or treble steps.

The physical problem of importance in connection with the structure of a low index surface is to calculate the "roughness" of the surface, corresponding to the excitation of atoms to levels other than the mean surface level. There is the possibility to be considered that at sufficiently high temperatures there are many atomic levels with incomplete populations of atoms, so that the mean surface level can no longer be defined, and we have a diffuse interface.

The statistics of kinks in steps are simple because the problem is one-dimensional. At each place in the step there are a number of possibilities, i.e. no kinks, or positive or negative kinks of unit or multiple heights, and each of these possibilities occurs independently at each atomic site along the step. The probability of a given configuration for the whole step is thus the product of the probabilities of the individual states at the various atomic sites. Since a surface is two-dimensional, however, the individual probabilities of finding atoms at levels other than a reference level are not independent of each other, and if we describe any closed path on the surface, we must arrive back at the original level. As there are innumerable closed paths, the probability of an atomic jump at any one point is a function of the atomic height of all other points on the surface. This results in a co-operational problem, analogous to ferromagnetism, or to order–disorder changes in alloys.

The statistics of co-operative phenomena are very difficult, but exact mathematical solutions have been obtained for some well defined two-dimensional models. We shall consider these solutions in more detail in connection with order–disorder changes in Section 26; for the present, we shall merely quote the results of Burton *et al.* (1950–1). If, as an approximation, the surface atoms are assumed to be confined to two layers only, the problem becomes equivalent to that presented by the two-dimensional Ising model of a ferromagnetic lattice, and exact solutions have been given by Onsager (1944). Actually, of course, each atom is capable of placing itself on many levels, and in order to investigate the effect of the extra levels, Burton and Cabrera applied a generalized form of the approximate treatment introduced by Bethe (1935).

Burton and Cabrera calculate the surface roughness as a function of temperature by minimizing the free energy of the surface region. The surface layers are assumed to consist of some atoms in normal sites and of some unoccupied sites, thus increasing both the energy and the configurational entropy. The state of the surface is thus a function only of temperature, and transitions of atoms between the solid and some other phase are not considered. Though strictly applicable to a surface maintained *in vacuo*, this calculation should give a good representation of the structure of a surface in contact with is own vapour. This structure should be little influenced by the pressure of the vapour, i.e. it will depend on the temperature but not on whether the solid and vapour are in equilibrium with each other. Clearly, however, the structure of a solid surface in contact with another condensed phase will depend on the properties of both phases.

The surface roughness is defined by Burton and Cabrera as the number of free bonds per atom or molecule parallel to the surface. This has the slight disadvantage that the maximum roughness varies with the approximation used. In the two layer model, for example, there are as many atoms in the second layer as holes in the first layer, so that the maximum roughness on a {100} face of a simple cubic crystal is 1·0, for three layers, the maximum roughness is 1·8, and so on. The detailed results in any approximation show that the roughness is very small at low temperatures, but rises abruptly in a critical temperature region, after which it rises rapidly with increasing temperature. The critical temperature is analogous to the Curie point in a ferromagnetic transition, or to the disordering temperature of a superlattice. The Onsager method gives accurate expressions for this critical temperature in the two level approximation for a symmetrical surface in which each atom has z nearest neighbours and the interaction energy Ξ_1 is equal for all these neighbours. For a square lattice of this type, e.g. the {100} faces of both simple cubic and f.c.c. crystals, the critical temperature T_c is given by

$$kT_c/2\Xi_1 = (2 \ln \cot(\pi/8))^{-1} \sim 0{\cdot}57, \qquad (19.1)$$

whilst for a triangular close-packed plane (e.g. the {111} face of a f.c.c. crystal), the corresponding equation is

$$kT_c/2\Xi_1 = (\ln 3)^{-1} \sim 0{\cdot}91. \qquad (19.2)$$

In both cases, the interaction between nearest neighbour atoms in the surface is the same as in the interior of the crystal. The transition temperatures are of the same orders, or higher than, the melting point of the crystal. For the solid rare gases, for example, the melting point T^{sl} is given by $kT^{sl}/2\Xi_1 \sim 0{\cdot}7$.

As pointed out on p. 146, some close-packed planes involve neighbours with both first and second nearest neighbour interactions. The {110} faces of both simple cubic and b.c.c. structures are rectangular lattices with energies of interaction $-2\Xi_1$ in one direction and $-2\Xi_2$ in the other. The formula for the critical temperature can then be written as

$$kT_c/\{2\Xi_1 \ln \coth(\Xi_1/2kT_c)\} = \Xi_2/\Xi_1 \tag{19.3}$$

and the critical temperature depends on Ξ_2/Ξ_1. A reasonable value for this ratio is 0·1, and the transition temperature given by (19.3) is then about half that given by (19.1). For these faces, therefore, the critical temperature is about half that of the melting point.

Finally, we might regard planes containing only second nearest neighbour bonds, e.g. the {111} planes of a simple cubic structure, as being close-packed. In such cases, the critical temperature is $\sim \Xi_2/\Xi_1$ times the melting point, and is hence much lower. At most normal temperatures, faces parallel to such planes cannot be regarded as atomically flat.

The Onsager method applied to the two-level approximation thus indicates that the critical temperature of disordering depends on the types of bonding in the surface. Burton *et al.* (1949) call the critical temperature the surface melting temperature, and introduce the idea of an atomically rough surface as a "melted" surface layer. For all crystals there are probably some faces which remain "unmelted" right up to the melting point when in contact with their own vapour; these are planes involving only nearest neighbour interactions of surface atoms. Other surfaces become very disordered at temperatures about half those of the melting point, and this may be regarded as the melting of the second nearest neighbour bonds.

The effect of introducing extra levels will be to reduce the transition temperatures, and the aim of using Bethe's method is to obtain an estimate of the reduction. For the two-level problem, Bethe's method gives

$$kT_c/2\Xi_1 = (2 \ln 2)^{-1} \sim 0{\cdot}72 \tag{19.4}$$

for a {100} face of a simple cubic crystal. This result is considerably larger than the correct result (eqn. (19.1)) and also leads to conclusions that are physically inadmissible. The three-level problem for the {100} plane gives the result

$$kT_c/2\Xi_1 \sim 0{\cdot}63, \tag{19.5}$$

which is smaller than (19.4) but still larger than (19.1). The use of more levels has little further influence on T_c, and the correct transition temperature (subject always to the severe limitations of the central force model) would probably be obtained if the three-level problem could be solved exactly. It is not certain whether the decrease in transition temperature on going from two to three levels is as great as eqns. (19.4) and (19.5) suggest, since the behaviour of the two-level approximation is so anomalous in the Bethe approximation. However, the general results of the Onsager treatment are evidently still valid, despite some doubt about the correct value of the critical temperature.

Rather similar calculations have been used by Jackson (1958) to estimate the roughness of the interface between liquid and solid phases. He used the still more approximate treatment of Bragg and Williams, which will also be described in the next chapter. We shall defer discussion of the solid–liquid interface to Section 21, but we may clearly expect the roughness of such an interface to be at least partially determined by the value of $\Delta h^{sl}/kT$, where Δh^{sl} is the latent heat of melting per atom. In the present section, we have found that for a close-packed surface the solid–vapour interface is reasonably sharp at all temperatures. A different conclusion will be reached for the solid–liquid interface.

20. THE EQUILIBRIUM SHAPE OF A CRYSTAL: WULFF'S THEOREM

In this section we shall discuss a special problem which is of interest in the theory of nucleation, and which also leads to a more rigid classification of crystal surfaces into close-packed and stepped surfaces than that given previously. In a finite isolated crystal, the free energy of a surface will depend on its orientation relative to the crystal lattice. The total surface free energy will thus be given by an integral of form $\int \sigma(\mathbf{n})\, dO$ where $\sigma(\mathbf{n})$ is the free energy per unit area of an element of area dO having unit normal $\mathbf{n}$. In thermodynamic equilibrium the crystal will adopt that shape which minimizes the surface free energy, and hence the total free energy. We require a method for calculating this shape, assuming we know the variation of $\sigma(\mathbf{n})$ with $\mathbf{n}$. A construction for this purpose was first given by Wulff (1901), and is now known as Wulff's theorem.

From an origin O we erect normals to all possible crystal faces, making the length of each normal proportional to the surface free energy of the corresponding face. This means that we construct a polar free energy diagram, plotting $\sigma(\mathbf{n})$ radially as a function of $\mathbf{n}$. Wulff's theorem then states that the equilibrium shape, giving the minimum surface energy for a given volume, is obtained by taking the inner envelope of the normals. To find this envelope, we construct a plane at each point of the polar diagram normal to the radius vector at that point. The minimum volume around the origin enclosed by these planes is then geometrically similar to the equilibrium shape. The process is illustrated in Fig. 5.13 for a hypothetical two-dimensional structure, the line energy of which shows prominent cusped minima in the $\langle 10 \rangle$ directions, and slightly less prominent minima in the $\langle 11 \rangle$ directions; the shape giving the minimum energy is seen to be compounded of these two directions. Obviously if the free energy per unit length of the $\langle 11 \rangle$ line had been greater than $\sqrt{2}$ times that of the $\langle 10 \rangle$ line, only the latter would have appeared in the equilibrium shape, which would have been square. Fig. 5.13 can be regarded alternatively as a section through the polar free energy plot and corresponding equilibrium shape of a three-dimensional crystal.

The example of Fig 5.13 shows a polar free energy diagram with prominent cusps, and the corresponding equilibrium shape is polygonal (polyhedral in three dimensions). Most "proofs" of Wulff's theorem assume in fact that the shape is polyhedral, and are therefore logically unsatisfactory. The possibility of the equilibrium crystal being bounded by smoothly curved surfaces seems to have been appreciated independently by Burton *et al.* (1950–1) and Herring (1951). The former workers gave a proof of an extended two-dimensional

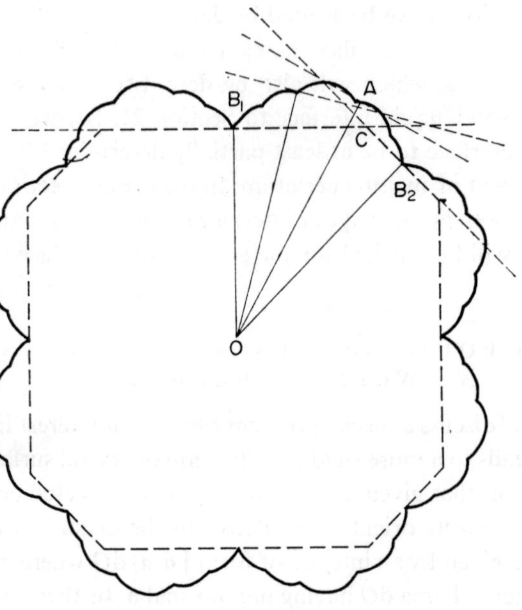

FIG. 5.13. Polar surface free energy plot, and its use in the Wulff construction (after Herring, 1951).

version of Wulff's theorem, whilst according to Herring, a rigorous proof of the general three-dimensional theorem may be obtained by use of an inequality due to Brunn and Minkowski. We shall not attempt to prove the result, but its approximate validity for polyhedral shapes may be seen in the following way. If the area of a plane face is O_1 and the length of the normal to this face is h_1 in the polar diagram, the surface energy of the face is proportional to O_1h_1. The volume in the polar diagram of the pyramid on base O_1, and having its apex at the origin, is $O_1h_1/3$, so that the surface energy of the face is proportional to this pyramidal volume in polar space. Minimum surface energy is thus obtained by finding the minimum volume around the origin of polar space, using the inner envelope construction.

An interesting application of the Wulff diagram, due to Herring, is its use to discuss the question of the relative free energy of an atomically flat surface and a surface having a hill and valley structure. This is obviously related to the distinction between close-packed and stepped surfaces discussed in the previous sections. In Fig. 5.13 we show part of the Wulff diagram in which OA is the direction normal to a surface of a macroscopic crystal, the shape of which need not be related to the equilibrium shape. Suppose the surface free energies used in constructing the diagram are those appropriate to atomically flat surfaces, and consider any other three directions, OB_1, OB_2, OB_3, which have positive projections on OA. Only OB_1 and OB_2 are shown in the figure; the three directions should not be coplanar. The planes normal to these directions at the points where they intersect the Wulff diagram will meet at some point C; this is shown in the figure, but need not lie in the plane OAB_1. We now consider the energy of a hill and valley structure formed from atomically flat planes normal to OB_1, OB_2 and OB_3, and so arranged that the macroscopic surface remains normal to OA. Let the unit vector in the direction OA be $\mathbf{n}$, and in each direction OB_i be $\mathbf{n}_i$. Then we have

$$O\mathbf{n} = O_1\mathbf{n}_1 + O_2\mathbf{n}_2 + O_3\mathbf{n}_3 = O_i\mathbf{n}_i, \tag{20.1}$$

where O_1, O_2, O_3 are the areas of the atomic faces which make up an area O of the macroscopic face. The surface energy per unit area of the macroscopic surface is thus

$$\sigma = O_i\sigma_i/O. \tag{20.2}$$

If we now introduce the unit vectors $\mathbf{m}_i$ which are reciprocal to $\mathbf{n}_i$, defined by the relations

$$\mathbf{m}_i \cdot \mathbf{n}_j = \delta_{ij}, \tag{20.3}$$

we can write eqn. (20.2) in the form

$$\sigma = \sigma_i \mathbf{m}_i \cdot \mathbf{n}. \tag{20.4}$$

Consider the vector $\mathbf{c}$ from the origin to the point C. Since $\widehat{OB_1C} = 90°$, the projection of $\mathbf{c}$ on to OB_1 is of magnitude σ_1, i.e. $\mathbf{c}\cdot\mathbf{n}_1 = \sigma_1$, and in general

$$\mathbf{c}\cdot\mathbf{n}_i = \sigma_i.$$

Using the result of eqn. (5.14), we thus have

$$\mathbf{c} = \sigma_i \mathbf{m}_i,$$

so that

$$\sigma = \mathbf{c}\cdot\mathbf{n}, \tag{20.5}$$

which is the length of the line OM where M is the perpendicular from C on to OA. The condition for OM to be less than OA is that C lies inside the plane through A which is normal to OA. If this condition is satisfied, the surface free energy σ as given by eqn. (20.2) is smaller than that of the atomically flat surface normal to OA, and the macroscopic surface can lower its energy by forming a ridged hill and valley structure. In general, there will always be a set of planes normal to directions OB_i for which C lies inside the plane at A, unless that plane (or an infinitesimal part of it) appears in the equilibrium shape. Any surface which does not appear in the equilibrium shape can therefore lower its energy by forming a stepped structure, composed of sections of planes which do appear in the equilibrium shape, and which give a minimum value to OM. Generally, the polar free energy plot for flat surfaces will show pronounced cusps at orientations corresponding to close-packed planes, so that we expect the equilibrium shape to consist of close-packed faces. Surfaces of other orientations will then be composed of sections of these close-packed surfaces, as concluded intuitively in Section 18.

Frank (1958) has suggested the use of the term "singular" for those planes which give pronounced cusps in the polar free energy diagram, so that they appear in the equilibrium shape. He also points out that the above type of argument is perhaps more readily visualized if the reciprocal to the polar free energy diagram of surface energy is plotted. At finite temperatures, it is possible that there are no singular surfaces, the equilibrium shape being composed of smoothly rounded portions. We now examine the conditions under which this may be possible.

In Fig. 5.14 let OA be any radius vector of the Wulff diagram, and let OA' be a neighbouring vector. The planes normal to OA, OA' at A, A' will intersect along a line through $P(A, A')$. Suppose OA'' is the radius vector to another neighbouring point A'' (not shown), where OA'' is not coplanar with OA and OA'. The plane at A'' normal to OA'' will intersect the other two planes at some point Q through the line at $P(A, A')$. Since $\widehat{OAQ}$, $\widehat{OA'Q}$, and $\widehat{OA''Q}$ are all 90°, it follows that the sphere with OQ as diameter will pass through A, A', and A''. If the polar free energy plot has a continuous slope everywhere in the vicinity

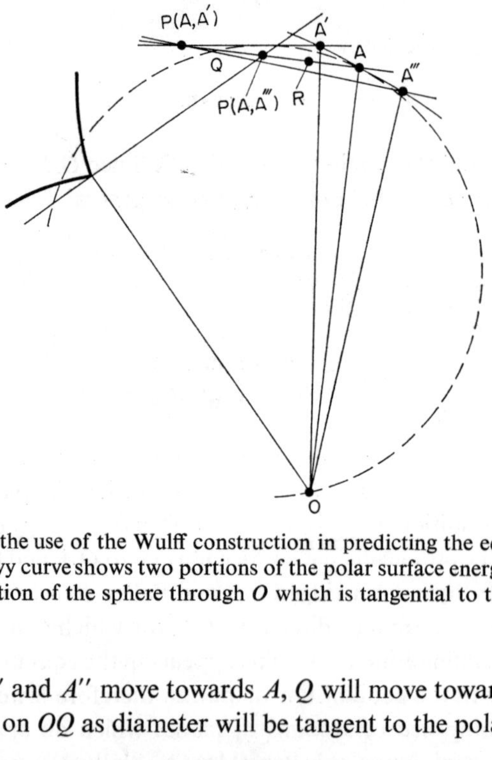

FIG. 5. 14. To illustrate the use of the Wulff construction in predicting the equilibrium shape (after Herring, 1951). The heavy curve shows two portions of the polar surface energy plot, and the broken circle is a section of the sphere through O which is tangential to this plot at A.

of A, and we now let A' and A'' move towards A, Q will move towards a limiting position. In this limit, the sphere on OQ as diameter will be tangent to the polar free energy diagram at A.

Suppose now that we construct the equilibrium shape. At A we draw the tangent sphere to OA which passes through Q. If *any* part of the polar free energy plot lies inside this sphere, a plane normal to the radius vector will cut OQ between O and Q (see Fig. 5.14). In this case, the point Q does not lie on the surface of the equilibrium shape. If the sphere through O and Q lies entirely inside the free energy diagram, then at Q there will be an infinitesimal area of surface having the direction of the unit normal OA. No other point of the plane AQ can be on the periphery of the equilibrium shape, provided (as assumed above) the slope of the free energy plot is continuous near A. For if we consider any point R a finite distance from Q and lying on the plane AQ, it will always lie outside the plane normal to some vector OA''' where A''' is near A.

Clearly the point Q will not approach a limiting position if there is a cusp at A, and we then expect a finite region of the surface normal to OA to appear in the equilibrium shape. In three dimensions, we may have either a point cusp or a line cusp. Detailed consideration shows that a point cusp will lead to a finite region of flat surface, whereas a line cusp will give a cylindrical surface as part of the equilibrium shape. Thus we expect singular surfaces provided the polar free energy diagram includes point cusps in three dimensions.

We summarize these results as follows. If OA is any radius vector of the Wulff diagram, and we construct the sphere passing through the origin and tangential to the polar free energy plot at A, the direction OA will be among the unit normals to the equilibrium surface if this sphere lies entirely within the free energy plot, and not otherwise. If the sphere is within the plot, the surface normal to OA will curve smoothly if the free energy gradient is smooth at A, will curve in one direction only if there is a line cusp at A, and will be flat if there is a point cusp at A. We emphasize that the equilibrium shape is not necessarily tangent to the radius vector, and in general will not be when the surface is smoothly curving. The relation between the point Q, occurring on the equilibrium periphery, and the point A, is that A is the "pedal" of Q. When the equilibrium shape is a smooth curve, the pedal of this curve then gives the polar free energy diagram. This result was obtained first by Burton *et al.*, who gave a general proof of Wulff's theorem for two dimensions, and showed that $h(\theta)/f(\theta)$ is constant, where $h(\theta)$ is the pedal to the equilibrium shape, and $f(\theta)$ is the polar free energy curve.

Wulff's theorem was originally stated in the form: *In the crystal at equilibrium, the distance of any face from the origin is proportional to the surface free energy of that face.* This statement is only valid for polyhedral equilibrium shapes, and for particular points where the radius vector is perpendicular to the face concerned. For a metal crystal at low temperatures, all surfaces with rational indices should produce pointed cusps in the free energy diagram, whilst surfaces with two rational indices should produce line cusps. As the temperature is raised, the non-close-packed planes will become disordered, and the sharp-pointed cusps will gradually disappear. At temperatures near to the melting point, only the close-packed planes will produce cusps.

A polyhedral shape will be expected if the cusps present are so pronounced that no tangent sphere lying inside all the cusps can be drawn. We expect this at low temperatures. If, on the other hand, the cusps are so slight that none of the tangent spheres is intersected by the free energy plot, the equilibrium shape will contain small flat regions from the cusps, joined by smoothly rounded regions with no sharp edges or corners. In the intermediate case, the shape will contain flat and curved surfaces, joined by sharp edges. The general effect of raising the temperature is thus to decrease the flat portions of the equilibrium shape, first the corners and then the edges disappearing.

In Part II, Chapter 13, we shall be concerned with the problem of the two-dimensional nucleation of a sheet of atoms on a close-packed crystal surface. We use a two-dimensional version of Wulff's theorem and examine the edge energy of the sheet. Burton and Cabrera

(1949) calculated the equilibrium nucleus shape for the (001) face of a Kossel crystal. At low temperatures the shape is square, having sides along the [100] and [010] directions of the face, and if the configurational entropy due to kinks is neglected, this shape is unchanged at all temperatures. Inclusion of the entropy terms leads to a decrease in the prominence of the cusps, and some of the tangent surfaces lie entirely inside the energy plot. The corners thus become progressively more rounded as the temperature is raised until, near the melting point, the shape is nearly circular.

21. LIQUID METALS: THE SOLID–LIQUID INTERFACE

In solids and gases, the arrangements of atoms are respectively almost completely regular and almost completely random. It is much more difficult to obtain a picture of the intermediate liquid condition. Although the liquid and gaseous states are continuous at temperatures and pressure above those of the critical point, the structure of a liquid under normal conditions appears to be more readily related to the solid than to the vapour. This conclusion is based largely on the results of X-ray diffraction experiments, but it is also illustrated by the relative values of the changes in macroscopic properties produced by melting and by vaporization. Thus the change of volume on melting is small (usually 3—5% expansion for metals), and the latent heat of fusion is only a few per cent of the latent heat of vaporization. There is, however, a substantial increase in entropy on melting, and this is attributable to the destruction of the long-range order of position in the solid.

X-ray diffraction photographs from liquids consist of a few diffuse haloes, which correspond to statistical tendencies for the scattering centres to be separated by particular distances. From an analysis of the variation of intensity with diffraction angle, a radial distribution function $\varrho(r)$ may be evaluated, this function giving the number of atoms per unit volume at a distance r from a reference atom. In a solid, $\varrho(r)$ is a periodic function of r even for large r (strictly, it is a discontinuous function for an ideal solid at low temperatures), but in a liquid it is a strongly damped periodic function which rapidly approaches the value appropriate to the average liquid density. Thus only two genuine maxima can usually be found in the $\varrho(r)$ curve for a liquid.

The diffraction studies indicate that the distance between neighbouring atoms, and the average number of neighbours possessed by any atom in the liquid correspond closely to these quantities in the solid, especially if the solid is close-packed. In terms of the distribution function, the number of neighbours with centres between r and $r+\mathrm{d}r$ from any atom is given by $4\pi r^2\varrho(r)\,\mathrm{d}r$, so that the first peak defines a co-ordination number which represents the number of neighbours close to the average separation of nearest neighbours. This average separation is usually slightly greater than the interatomic distance in the solid, and the co-ordination number, which should strictly be defined in a manner similar to that used by Frank and Kaspar (see p. 118), is usually between 8 and 11. The liquid co-ordination number tends to be greater if the solid is close-packed, but the differences between liquids are less marked than are those between the corresponding solids. Furukawa (1959) has pointed out that the radial distribution functions for all liquids are remarkably similar when calculated in the same way. All liquids seem to have a tendency towards high coordination of near neighbours (close-packing), and some metals with very open structures (germanium, gallium, bismuth, and antimony) actually contract on melting.

Similar information to that given by X-rays may also be obtained by means of neutron diffraction techniques, which have certain advantages in the study of liquids. Neither method, however, gives direct information about the details of the local groupings of atoms and the way in which these form and break up again. This absence of detailed information is one obstacle to the development of satisfactory models of the liquid state.

In addition to describing the structure, theories of liquids attempt to reproduce the measurable statistical properties of the liquid state, and to predict the solid $\rightleftharpoons$ liquid transition. Although this transition is thermodynamically a phase change of the first order, many of these theories treat the solid and liquid states as quasi-continuous, and the solid $\rightleftharpoons$ liquid transition is then analogous in many ways to a second-order phase change. The models used in this type of theory and the statistical problems involved are quite similar to those encountered in the theory of ordering processes in alloys. A familiar difficulty is that of distinguishing the properties of the rather unrealistic model from properties which may be a result only of mathematical approximations used to solve the appropriate equations.

We may divide theories of liquids into three general classes. The first of these includes attempts to derive equations of state and to deduce the existence of transitions from assumed general laws of molecular interactions without making any specific assumptions about the structure on an atomic scale. The predictions of such theories are all expressed in statistical terms, and may, for example, deal with the structure of the liquid and the transition from the solid by specifying the behaviour of the atomic distribution function. Theories of this kind are logically satisfying in that they make the minimum number of assumptions, and they have been considerably developed by Kirkwood and Monroe (1941), Kirkwood (1951), Born and Green (1946), Green (1947), Mayer (1947), and others. However, the mathematical analyses are usually very complex, and it is not certain whether or not the approximations used to obtain solutions introduce significant errors. The theories suffer from the defect of all very general theories, namely that they are not well adapted to the description of specific physical phenomena, such as the nucleation of the solid from the liquid. A further possible limitation is the extent of the dependence of the behaviour of real liquids on the actual binding forces. The binding forces in liquid metals are probably qualitatively similar to those in solid metals, so that objections similar to those expressed on p. 109 might also apply. It is interesting to note here that rigorous mathematical solutions show transitions in assemblies of the simplest type imaginable. Thus Alder and Wainwright (1957) have shown by machine calculations that a gas of non-attracting rigid spheres undergoes an abrupt transition to an ordered state as the volume is decreased steadily. So-called "lattice gas" theories of melting are reviewed by Runnels (1972), and transitions of this kind in two dimensions have also been studied experimentally by Turnbull and Cormia (1960).

The other two classes of theories of liquids use specific models for the atomic or molecular configurations. Most of the models are based on the experimental evidence of the similarity of the liquid and solid states, which (as mentioned above) are regarded as quasi-continuous. We may group these together as lattice theories, since in all cases the concept of a crystalline lattice which partially retains its identity in the liquid is used.† More recently, there has been considerable interest in geometrical theories of liquids which emphasize the differences as well as the similarities between liquids and solids, and in which the concept of the lattice is discarded at the outset.

As we have noted, the co-ordination of near neighbours in most liquids is only slightly less than twelve, and this suggests the possibility of the existence of local units in which twelve atoms surround a central atom in the arrangement corresponding to cubic or hexagonal close-packing. Frank (1952b) has pointed out, however, that there is an alternative close-packed group of thirteen atoms in which the twelve surrounding atoms are at the face centres of a regular dodecahedron (i.e. at the vertices of a regular icosahedron). In a hard sphere model, the twelve surrounding atoms are in contact only with the central atom, and not with each other. Alternatively, if the atoms are regarded as slightly deformable spheres, the interatomic distance in the icosahedral arrangement will decrease, giving it a lower energy than the other close-packed arrangements. Thus most groups might tend to assume this form.

The icosahedral arrangement is not found in simple solid-state structures because it has fivefold symmetry, and hence a space lattice cannot be formed entirely by primary coordination of this type. Another way of stating this is that a set of regular dodecahedra cannot be stacked together so as to fill all space. Of course, combination of this form of stacking with other forms can lead to permissible space-lattice structures, and many of the more complex intermetallic compounds may be described in terms of sphere-packing in this way (Kaspar, 1956; Frank and Kaspar, 1958, 1959). Frank's suggestion seems to give a clue to liquid structure, the differences between solid and liquid being due not so much to defects in the solid lattice eventually destroying the long-range order (as in lattice theories), as to the existence in the liquid of a large number of groupings of the type that cannot give a lattice structure by primary coordination.

This type of approach was further developed by Bernal (1959, 1960, 1964, 1965). His theory is based on the assumption that a liquid is an essentially irregular assembly of atoms or molecules which contains no holes large enough to accommodate another molecule. In order to achieve a dense irregularly packed array of spheres, many of the local groupings must have symmetries which are not consistent with the establishment of long-range order. This array will have, of course, a volume rather greater than that of the fully ordered close-packed crystal structures Bernal suggested, though without proof, that if the structure is homogeneous there is a range of forbidden densities intermediate between that corresponding to crystal close packing (CCP) and that corresponding to random close packing (RCP). He also introduced the idea of random loose packing (RLP) of spheres as the packing which gives the lowest density consistent with a random arrangement of hard spheres in contact. The relative densities of these three arrangements are 0·74, 0·637, and 0·60 respectively. Physical experiments in which steel ball-bearings are packed into evacuated balloons or into dimpled spherical or cylindrical containers have been carried out by Bernal and his coworkers, and also by Scott (1962) and Scott and Kilgour (1969). Such experiments enable

† This concept is sometimes called cybotaxis (Stewart, 1930), but the word is not in general use.

conclusions to be drawn about the relative frequencies of different arrangements, and they confirm that random packing of spheres is possible for any density within the RCP and RLP limits, but not outside these limits.

Bernal (1964, 1965) also constructed ball and spoke models of the atomic arrangements, using either spokes of unequal lengths in order to simulate the results of X-ray diffraction experiments on real liquids, or else spokes of equal length in order to simulate the structure of an "ideal fluid". In Bernal's geometrical model of an ideal liquid, the atoms or molecules are supposed to lie at the apices of a set of empty polyhedra (holes) with edges forming links between neighbouring atoms. All the edges are as nearly as possible equal in length, although deviations of about 15% from this condition are required if the assembly is to fill space without any polyhedron being sufficiently large to contain an extra spherical atom. Considerations of statistical geometry (Collins, 1972) then show that each polyhedron must be a deltahedron, i.e. it must have triangular faces, and Bernal proved there are five basic or canonical forms. Two of these forms are the tetrahedron and the octahedron; the other three have respectively 8, 9, and 10 vertices, and in each of them surface coordinations of five predominate and thus prevent long-range order. It is believed to be this prevention of long-range order which leads to the fluidity of liquids. Although the idealized model clearly can not represent the structure of real liquids, it may nevertheless be a better approximation than many of the pseudolattice models in which some long-range order is maintained.

Computer calculations of liquid structure in which empirical interatomic forces are used in place of hard sphere models give statistical information about parameters such as the mean number of neighbours $\bar{z}$ possessed by a given atom. It is readily shown that for all statistically homogeneous planar arrays of atoms, the average value of the number of neighbours, as defined on p. 118, is $\bar{z} = 6$ in two dimensions, and it follows that the number of direct neighbours is less than 6. However, in three dimensions, homogeneous aggregates can be constructed with mean values of z which are arbitrarily high (Smith, 1964, 1965), although a large body of evidence indicates that the actual values of $\bar{z}$ for liquids are close to 14. The theoretical value for slight perturbations of CCP is also 14, and may be compared with the value for the gas or random distribution which is (Meijering, 1953)

$$\bar{z} = 2+(48\pi^2/35) \simeq 15{\cdot}54.$$

Rahman (1966) and Finney (1970) have made computer calculations with (different) pair potentials intended to represent interatomic forces in argon. Rahman obtained $\bar{z} = 15{\cdot}67$ for the random array and $\bar{z} = 14{\cdot}45$ and $14{\cdot}26$ for liquid and solid argon. Finney's more detailed results again show a small but significant difference in the values of $\bar{z}$ for solid and liquid states.

The model of a liquid as an atomic arrangement which is close-packed but without long-range order receives some support from the experimental sphere model of Turnbull and Cormia mentioned above. They studied a two-dimensional array of randomly moving spheres, and gradually reduced the free volume (or rather area) by adding more spheres. When the density of particles is low, each sphere moves many times its own diameter before making a collision. As the free volume is reduced, a condition is reached in which the mean free path between collisions is less than one sphere diameter. No long-range order is apparent, however, and new spheres readily find places anywhere within the model; this state is thus approximately liquid-like. Further reduction in the free volume leads to the formation of domains of long-range order, and each individual sphere is then confined to a single position for most of the time.

The restriction of an atom or molecule to an effective volume in the vicinity of a lattice site of the solid is characteristic of the lattice models of the liquid state, which may also be referred to as free volume theories. In a lattice theory, the liquid may be represented as a solid containing a large concentration of lattice defects. For example, the introduction of vacancies into an ideal crystal results at first in a decrease in free energy (Section 17), and there is an equilibrium vacancy concentration which may be quite large at temperatures near to the melting point. If further vacancies are added, the free energy will rise so long as the assumed independence of the vacancy energy on the existing vacancies is valid. If the number of vacancies becomes an appreciable fraction of the number of sites, however, formation of new vacancies will involve only slight increases in the internal energy. Under these circumstances, it is conceivable that the free energy of a pseudocrystal containing a very large number of vacancies could be lower than that of a nearly perfect crystal containing its equilibrium concentration of vacancies. The pseudocrystal with large disorder is a possible lattice model of the liquid state.

One of the earliest and best-known theories of this type is that of Lennard-Jones and Devonshire (1939 a, b), who described the melting process as a kind of order–disorder reaction. They used a model in which the atoms may occupy an extra set of interlattice sites, in addition to the normal sites. In the solid state, the atoms are confined to the normal sites, whilst in the liquid state they are randomly distributed among the two sets of sites. This model is obviously very artificial, but various modifications have been suggested to bring it nearer to the physical picture furnished by the experimental evidence.

An alternative model is to regard a liquid as a solid with a very small grain size. The randomly orientated individual crystals of a polycrystalline aggregate are then supposed to have sizes of the order of atomic dimensions and a possible advantage over the hole theories is that long range order is preserved only over distances of the order of the grain diameter. Mott (1952) has suggested that if the vacancy theory is correct, the latent heat of melting will represent the energy involved in creating the large number of vacancies characteristic of the liquid state, whilst in the very small grain theory, the latent heat would represent the extra energies of the grain boundaries. As we shall describe in Chapter 8, the energies of the grain boundaries are proportional to the value of the shear modulus μ. Mott concludes that the grain boundary model is more appropriate, since the ratio of the latent heat of melting to μ is more nearly constant. The argument is rather unconvincing, since the validity of using the energy to form isolated vacancies and of the ordinary grain boundary formula is far from established. Turnbull (1956) has pointed out that the magnitudes of the known liquid–solid interfacial energies are not inconsistent with the naive supposition that the liquid consists of randomly orientated crystal domains two to three molecules in diameter. If this is correct, the surface energy should be rather less than the maximum grain boundary energy of the solid. In fact, some measurements indicate that the surface energy is less than half the maximum grain boundary energy.

As already indicated, one very unsatisfactory feature of vacancy and interstitial models of liquids is their failure to destroy the long-range order of the lattice. It is claimed that dislocation models are superior in this respect and it has been estimated that a dislocation grid of spacing about four atom diameters would be required to produce the observed enthalpy of melting (Rudman, 1967). Atomistic calculations in which the melting process is simulated in a computer (Jensen *et al.* 1973) have been interpreted as showing that melting begins by spontaneous nucleation of large numbers of dislocations, and this leads naturally to the dislocation model of the liquid. However, although some authors have used the concept of dislocations in liquids, or in non-crystalline polymers, it is very difficult to see how the characteristics of a dislocation line can still remain in the absence of reference regions of good crystal. Dislocation lines as physical entities cease to exist whenever the length of line per unit volume becomes too large, and purely formal descriptions of structures (or interfaces; see Chapter 8) as lattices with very high dislocation densities then have little to commend them.

In general, all lattice theories of liquids are unsatisfactory, and they frequently conflict with experimental data (see, for example, Hildebrand, 1958). It is sometimes argued that lattice theories would predict little or no supercooling when a liquid begins to freeze, whereas the supercooling is usually appreciable. A possible explanation of this is that the local configurations in the liquid are of the type suggested by Frank and Bernal, complete rearrangement being necessary before these can form into solid nuclei. The opposite observation, namely that a solid cannot be superheated, might be quoted in favour of a supposed near-continuity of the solid and liquid states, but it is believed that this is merely the result of the surface acting as a nucleation catalyst.

The similarity of the solid and liquid states envisaged in the lattice theories does raise the question of a possible continuity of state. Above the critical point, the liquid and gaseous phases merge into a single fluid phase, and it is not impossible that there is a similar continuity of the solid and liquid phases. Simon (1952) has emphasized that this problem can most readily be examined experimentally with substances like solid helium, for which pressures and temperatures which are many times those of the critical point may be obtained. From thermodynamic measurements of the trend in the entropy differences between solid and liquid as the pressure and temperature are raised, it may be concluded that the existence of such a continuity of state is most improbable. This is in agreement with the conclusions of the structural models mentioned above, which suppose that there are essential differences in atomic configurations in the two states of matter.

We now turn to consider the nature of the solid–liquid interface. In any multiphase assembly, there must be transition regions in which the atomic configuration changes from that characteristic of one phase to that characteristic of another. These regions are frequently only a very few interatomic distances thick, and the concept of an interphase boundary which may be regarded as a geometrical surface is then quite appropriate. Not all boundaries are sharp, however, since in some circumstances the change in free energy due to the boundary may be lower if the transition takes place gradually over many interatomic distances. One example of a diffuse interface is the Bloch wall between two antiphase ferromagnetic domains; it seems likely that the liquid–solid interface provides another.

In terms of the discussion above, the distinction between liquid and solid phases depends mainly on the existence or non-existence of long-range order of position, and this is only defined adequately in a finite volume. Thus there may be intrinsic difficulties in the concept of a sharp liquid–solid interface, since we may only be able to state definitely that the structure is that of the liquid or solid for atoms well removed from the transition region. Since the X-ray evidence shows that the long-range order disappears in a very few interatomic distances, however, this conceptual difficulty represents quite a small effective width of the transition region, and even this might be avoided in principle by taking long time averages of the density as a function of position. A reasonably sharp interface would then be indicated by a discontinuity in the average density. Frank (1952b) has emphasized that there need not be a density difference, so that the uncritical assumption of a sharp boundary which has sometimes been made is certainly incorrect.

Indirect evidence that the solid–liquid interface is diffuse is provided by a thermodynamic argument due to Hilliard and Cahn (1958). They pointed out that if the grain boundary energy $\sigma^{\alpha\alpha}$ is greater than twice the solid–liquid boundary energy σ^{sl}, a grain boundary would become unstable at temperatures near the melting point with respect to a composite sandwich containing an intermediate layer of supercooled liquid. If the liquid–solid interface is sharp, the thickness of the intermediate layer can tend to zero, and it follows that the maximum value of $\sigma^{\alpha\alpha}$ will be $2\sigma^{sl}$ at all temperatures. Experimental measurements for the noble metals do not support this conclusion, $\sigma^{\alpha\alpha}/\sigma^{sl}$ being apparently appreciably greater than two, even after allowing for possible effects of anisotropy. This argument thus suggests that the solid–liquid interface is in fact diffuse, but it does not necessarily imply that a liquid sandwich is actually formed at high temperatures. The experimental measurement of σ^{sl} (from experiments on the supercooling of liquid droplets) depends on the assumption of a sharp interface, and the true values will be larger if the interface is in fact diffuse.

A more satisfactory approach to the problem of the nature of the interface would be to calculate the atomic structure of a close-packed surface in contact with its melt. This problem has been considered by Jackson (1958), using methods which are broadly similar to the more elaborate calculations of Burton and Cabrera for surfaces in contact with the vapour. In Jackson's formulation, the Bragg–Williams theory (p. 215) is used in place of the more complex statistical treatments, and the latent heat of fusion replaces the energy of volatilization. The significance of the exact calculations is rather doubtful, but the main conclusion is generally accepted, namely that all metallic surfaces in contact with a melt are heavily disordered. This seems to be the description on an atomic scale of the diffuse interface envisaged by Hilliard and Cahn.

REFERENCES

ALDER, B. J. and WAINWRIGHT, T. E. (1957) *J. Chem. Phys.* **27**, 1208.

ALTMANN, S. L. (1970) *Band Theory of Metals — The Elements*, Pergamon Press, Oxford.

BALLUFFI, R. W. and SIEGEL, R. W. (1965) *Lattice Defects in Quenched Metals*, p. 693, Academic Press, New York.

BALLUFFI, R. W., LIE, K. H., SEIDMAN, D. N. and SIEGEL, R. W. (1970) *Vacancies and Interstitials in Metals*, p. 125, North-Holland, Amsterdam.

Barrett, C. S. (1947) *Phys. Rev.* **72,** 245; (1956) *Acta crystallogr.* **9,** 671.

Barrett, C. S. and Massalski, T. B. (1966) *The Structure of Metals,* McGraw-Hill, New York.

Bartlett, J. H. and Dienes, G. J. (1953) *Phys. Rev.* **89,** 848.

Basinski, Z. S., Duesbery, M. S., and Taylor R. (1970) *Phil. Mag.* **21,** 1201; (1971) *Can. J. Phys.* **49,** 2160.

Bauerle, J. E. and Koehler, J. S. (1957) *Phys. Rev.* **107,** 1493.

Beeler, J. R. (1970) *Adv. Mater. Res.* **4,** 295.

Beeler, J. R. and Johnson, R. A. (1967) *Phys. Rev.* **156,** 677.

Bernal, J. D. (1959) *Nature, Lond.* **183,** 141; (1960) *Ibid.* **185,** 68; (1964) *Proc. R. Soc.* A, **280,** 299; (1965) *Structure and Properties of Liquids,* p. 25, Elsevier, London.

Bethe, H. (1935) *Proc. R. Soc.* A, **150,** 552.

Bianchi, G., Mallejac, D., Janot, C., and Champier, G. (1966) *Compt. Rend.* **263,** 1404.

Blackburn, L. D., Kaufman, L., and Cohen, M. (1965) *Acta metall.* **13,** 533.

Born, M. (1940) *Proc. Camb. Phil. Soc.* **36,** 160; (1942) *Ibid.* **38,** 82.

Born, M. and Green H. S. (1946) *Proc. R. Soc.* A, **188,** 10.

Born, M. and Huang, K. (1954) *Dynamical Theory of Crystal Lattices,* Clarendon Press, Oxford.

Bradshaw, F. J. and Pearson, S. (1956) *Phil. Mag.* **1,** 812; (1957) *Ibid.* **2,** 379, 540.

Burton, J. J. and Lazarus, D. (1970) *Phys. Rev.* B, **2,** 787.

Burton, W. K. and Cabrera, N. (1949) *Disc. Faraday Soc.* **5,** 33.

Burton, W. K., Cabrera, N., and Frank, F. C. (1949) *Nature, Lond.* **163,** 398; (1950–1) *Phil. Trans. R. Soc.* A, **243,** 51.

Cahn, J. W. and Hilliard, J. E. (1958) *J. Chem. Phys.* **28,** 258; (1959) *Ibid.* **31,** 688.

Callaway, J. (1964) *Energy Band Theory,* Academic Press, New York.

Chik, K. P. (1970) *Vacancies and Interstitials in Metals,* p. 183, North-Holland, Amsterdam.

Chopra, K. L., Randlett, M. R., and Duff, R. H. (1967) *Phil. Mag.* **16,** 621.

Christian, J. W. (1951) *Proc. R. Soc.* A, **206,** 51.

Christian, J. W. and Swann, P. R. (1965) *The Decomposition of Austenite by Diffusional Processes,* p. 371, Wiley, New York.

Cohen, M. L. and Heine, V. (1970) *Adv. Phys.* **24,** 37.

Collins, R. (1972) *Phase Transitions and Critical Phenomena,* vol. 2, p. 271, Academic Press, London and New York.

Corbett, J. W. (1970) *Vacancies and Interstitials in Metals,* p. 975, North-Holland, Amsterdam.

Cotterill, R. M. J. and Doyama, M. (1965) *Lattice Defects in Quenched Metals,* p. 653, Academic Press, New York.

Cottrell, A. H. and Bilby, B. A. (1951) *Phil. Mag.* **42,** 809.

Cracknell, A. P. (1969) *Adv. Phys.* **18,** 681; (1971) *Ibid.* **20,** 747.

Damask, A. C., Dienes, G. J., and Weizer, V. G. (1959) *Phys. Rev.* **113,** 781.

De Jong, M. and Koehler, J. S. (1963) *Phys. Rev.* **129,** 40.

Doyama, M. (1965) *Lattice Defects in Quenched Metals,* p. 163, Academic Press, New York.

Doyama, M. and Cotterill, R. J. M. (1965) *Phys. Rev.* **137,** A994.

Duesbery, M. S. and Taylor, R. (1969) *Phys. Lett.* A, **30,** 496.

Edwards, O. S. and Lipson, H. (1943) *J. Inst. Met.* **69,** 177.

Eshelby, J. D. (1954) *J. Appl. Phys.* **25,** 255.

Evans, J. H. (1970) *Phil. Mag.* **22,** 1261.

Evans, J. L. and Eyre, B. L. (1969) *Acta Metall* **17,** 1109.

Feder, R. (1970) *Phys. Rev.* B, **2,** 828.

Feder, R. and Charbnau, H. P. (1966) *Phys. Rev.* **144,** 404.

Feder, R. and Nowick, A. S. (1967) *Phil. Mag.* **15,** 805.

Finney, J. L. (1970) *Proc. R. Soc.* A, **319,** 479, 495.

Flynn, C. P. (1968) *Phys. Rev.* **171,** 682.

Flynn, C. P. Bass, J., and Lazarus, D. (1965) *Phil Mag.* **11,** 521.

Frank, F. C. (1951a) *Phil. Mag.* **42,** 809; (1951b) *Ibid.* **42,** 1014; (1952a) *Adv. Phys.* **1,** 91; (1952b) *Proc. R. Soc.* A, **215,** 43; (1958) *Growth and Perfection of Crystals,* p. 1, Wiley, New York; (1967) *Phase Stability in Metals and Alloys,* p. 521, McGraw-Hill, New York.

Frank, F. C. and Kaspar, J. S. (1958) *Acta Crystallogr.* **11,** 184; (1959) *Ibid.* **12,** 483.

Frenkel, J. H. (1945) *J. Phys. USSR* **9,** 392.

Fuchs, K. (1935) *Proc. R. Soc.* A, **151,** 585; (1936) *Ibid.* **153,** 633; **157,** 444.

FUMI, F. G. (1955) *Phil. Mag.* **46,** 1007.
FURTH, R. (1944) *Proc. R. Soc.* A, **183,** 87.
FURUKAWA, K. (1959) *Nature, Lond.* **184,** 1209.
GALLIGAN, J. M. (1970) *Vacancies and Interstitials in Metals*, p. 575, North-Holland, Amsterdam.
GEHLEN, P. C., BEELER, J. R., and JAFFEE, R. I. (1972), *Interatomic Potentials and Simulation of Lattice Defects*, Plenum Press, New York.
GIBSON, J. B., GOLAND, A. N., MILGRAM, M., and VINEYARD, G. H. (1960) *Phys. Rev.* **120,** 1229.
GIRIFALCO, L. A. and WEIZER, V. G. (1959) *Phys. Rev.* **114,** 687; (1959–60) *J. Phys. Chem. Solids* **12,** 265.
GREEN, H. S. (1947) *Proc. R. Soc.* A, **189,** 103.
GRIPSHOVER, R. J., KHOSHNEVISAN, M., ZETTS, J. S., and BASS, J. (1970) *Phil. Mag.* **22,** 757.
HARRISON, W. A. (1966) *Pseudopotentials in the Theory of Metals*, Benjamin, New York; (1972) *Interatomic Potentials and Simulation of Lattice Defects*, p. 69, Plenum Press, New York
HEINE, V. (1970) *Solid State Phys.* **24,** 1.
HEINE, V. and WEAIRE, D. (1970) *Solid State Phys.* **24,** 249.
HERRING, C. (1951) *Phys. Rev.* **82,** 87.
HILDEBRAND, J. H. (1958) *Growth and Perfection of Crystals*, p. 310, Wiley, New York.
HILLIARD, J. E. and CAHN, J. W. (1958) *Acta Metall.* **6,** 772.
HIRSCH, P. B., SILCOX, J., SMALLMAN, R. E., and WESTMACOTT, K. H. (1958) *Phil. Mag.* **3,** 897.
HUME-ROTHERY, W. and RAYNOR, G. V. (1954) *The Structure of Metals and Alloys*, Institute of Metals, London.
HUNTINGTON, H. B. (1942) *Phys. Rev.* **61,** 315; (1953) *Ibid.* **91,** 1092.
HUNTINGTON, H. B., SHIRN, G. A., and WAJDA, E. S. (1955) *Phys. Rev.* **99,** 1085.
ISENBERG, I. (1951) *Phys. Rev.* **83,** 637.
JACKSON, J. J. (1965) *Lattice Defects in Quenched Metals*, p. 467, Academic, Press, New York.
JACKSON, K. A. (1958) *Liquid Metals and Solidification*, p. 175, American Society for Metals, Cleveland, Ohio.
JANOT, C., MALLEJAC, D. and GEORGE, B. (1970) *Phys. Rev.* B, **2,** 3088.
JENA, A. K., GIESSEN, B. C., BEVER, M. B., and GRANT, N. J. (1968) *Acta Metall.* **16,** 1047.
JENSEN, E. J., KRISTENSEN, W. D., and COTTERILL, R. M. J. (1973) *Phil. Mag.* **27,** 623.
JOHNSON, R. A. (1964) *Phys. Rev.* **134,** A1329; (1966a) *Phys. Rev.* **145,** 425; (1966b) *Ibid.* **152,** 629; (1967a), *J. Phys. Chem. Solids* **28,** 275; (1967b) *Acta Metall.* **15,** 513; (1968) *Phys. Rev.* **174,** 691; (1973) *J. Phys.* F, **3,** 295.
JONES, H. (1952) *Phil. Mag.* **43,** 105.
KASPAR, J. S. (1956) *Theory of Alloy Phases*, p. 269, American Society for Metals, Cleveland, Ohio.
KAUFMAN, L. (1959a) *Acta Metall.* **7,** 575; (1959b) *Bull. Amer. Phys. Soc.* **4,** 181.
KAUFMAN, L., CLOUGHERTY, E. V., and WEISS, R. J. (1963) *Acta Metall.* **11,** 323.
KIMURA, H., MADDIN, R., and KUHLMANN-WILSDORF, D. (1959) *Acta Metall.* **7,** 145, 154.
KINO, T. and KOEHLER, J. S. (1967) *Phys. Rev.* **162,** 632.
KIRKWOOD, J. G. (1951) *Phase Transformations in Solids*, p. 67, Wiley, New York.
KIRKWOOD, J. S. and MONROE, E. (1941) *J. Chem. Phys.* **9,** 514.
KOEHLER, J. S. (1970) *Vacancies and Interstitials in Metals*, p. 169; North-Holland, Amsterdam.
KOEHLER, J. S., SEITZ, F., and BAUERLE, J. E. (1957) *Phys. Rev.* **107,** 1499.
KUHLMANN-WILSDORF, D., MADDIN, R., and WILSDORF, H. G. F. (1962) *Strengthening Mechanisms in Solids*, American Society for Metals, Cleveland, Ohio.
LAVES, F. (1967) *Phase Stability in Metals and Alloys*, pp. 85, 521. McGraw-Hill, New York.
LENNARD-JONES, J. E. (1937) *Physica* **10,** 941.
LENNARD-JONES, J. E. and DEVONSHIRE, A. F. (1939a) *Proc. R. Soc.* A, **169,** 317; (1939b) *Ibid.* **170**. 464.
LOMER, W. M. (1959) *Prog. Metal Phys.* **8,** 255; (1969) *Prog. Mater. Sci.* **14,** 97.
LOMER, W. M. and COTTRELL, A. H. (1955) *Phil. Mag.* **46,** 711.
LOMER, W. M. and GARDNER, W. E. (1969) *Prog. in Mater. Sci.* **14,** 145.
MACKENZIE, I. K., KHOO, T. L., MCDONALD, A. B., and MCKEE, B. T. A. (1967) *Phys. Rev.* **19,** 946.
MAYER, J. E. (1947) *J. Chem. Phys.* **15,** 187.
MEAKIN, J. D., LAWLEY, A., and KOSS, R. C. (1964) *Appl. Phys. Lett.* **5,** 133.
MEIJERING, J. L. (1953) *Philips Res. Rep.* **8,** 270.
MIODOWNIK, A. P. (1970) *Acta Metall.* **18,** 541.
MISRA, R. D. (1940) *Proc. Camb. Phil. Soc.* **36,** 173.
MORI, T., MESHII, M., and KAUFFMAN, J. W. (1962) *J. Appl. Phys.* **33,** 2776.

MOTT, N. F. (1952) *Proc. R. Soc.* A, **215,** 1.
MOTT, N. F. and GURNEY, R. W. (1948) *Electronic Processes in Jonic Crystals.* p. 29, Clarendon Press, Oxford
MOTT, N. F. and JONES, H. (1936) *The Theory of the Properties of Metals and Alloys,* p. 149, Clarendon Press, Oxford.
MÜLLER, E. W. (1970) *Vacancies and Interstitials in Metals,* p. 557, North-Holland, Amsterdam.
NABARRO, F. R. N. and VARLEY, J. H. O. (1951) *Rev. Met.* **48,** 681.
NACHTRIEB, N. H. and HANDLER, G. S. (1954) *Acta Metall.* **2,** 797.
NIHOUL, J. (1970) *Vacancies and Interstitials in Metals,* p. 839, North-Holland, Amsterdam.
NOWICK, A. S. and FEDER, R. (1972) *Phys. Rev.* B, **5,** 1238.
ONSAGER, L. (1944) *Phys. Rev.* **65,** 117.
PANETH, H. R. (1950) *Phys. Rev.* **80,** 708.
PEARSON, W. B. (1973) *The Chemistry and Physics of Metals and Alloys,* Wiley, New York.
RAHMAN, A. (1966) *J. Chem. Phys.* **45,** 2585.
RUDMAN, P. S. (1965) *Trans. Am. Inst. Min. (Metall.) Engrs.* **233,** 864; (1967) *Phase Stability in Metals and Alloys,* p. 539, McGraw-Hill. New York.
RUNNELS, L. K. (1972) *Phase Transitions and Critical Phenomena,* vol. 2, p. 305, Academic Press, London and New York.
SAADA, G. (1966) *Theory of Crystal Defects,* p. 167, Academia, Prague.
SCHILLING, W. BURGER, G., ISEBECK, K., and WENZL, H. (1970) *Lattice Defects in Quenchal Metals.* p. 255, North-Holland, Amsterdam.
SCHULTZ, H. (1965) *Lattice Defects in Quenched Metals,* p. 761, Academic Press, New York.
SCHUMACHER, D. (1970) *Vacancies and Interstitials in Metals,* p. 889, North-Holland, Amsterdam.
SCOTT, G. D. (1962) *Nature, Lond.* **194,** 956.
SCOTT, G. D. and KILGOUR, D. M. (1969) *J. Phys.* D, **2,** 863.
SEEGER, A. (1970) *Vacancies and Interstitials in Metals,* p. 999, North-Holland, Amsterdam; (1972) *Comments on Solid State Physics* **4,** 121; (1973) *J. Phys.* F, **3,** 248.
SEEGER, A. and BROSS, H. (1956) *Z. Physik* **145,** 161.
SEEGER, A and MEHRER, H. (1968) *Phys. Stat. Sol.* b**48,** 481; (1970) *Vacancies and Interstitials in Metals,* p. 1, North-Holland, Amsterdam.
SEEGER, A., MANN, E. and VON JAN, R. (1962) *J. Phys. Chem. Solids* **12,** 326.
SEITZ, F. (1940) *Modern Theory of Solids,* McGraw-Hill, New York.
SHARMA, R. K., LEE, C., and KOEHLER, J. S. (1967) *Phys. Rev. Lett.* **19,** 1379.
SIEGEL, R. W. (1966) *Phil. Mag.* **13, 337,** 359.
SILCOX, J. and HIRSCH, P. B. (1959) *Phil. Mag.* **4,** 72.
SIMMONS, R. O. and BALLUFFI, R. W. (1960a) *Phys. Rev.* **117,** 52; (1960b) *Ibid.* **119,** 600; (1962) *Ibid.* **125,** 862; (1963) *Ibid.* **129,** 1533.
SIMON, F. E. (1952) *Low Temperature Physics (Four Lectures).* Pergamon Press, Oxford.
SINHA, A. K., (1972) *Prog. in Mater. Sci.* **15,** 79.
SMALLMAN, R. E. and WESTMACOTT, K. H. (1959) *J. Appl. Phys.* **30,** 603.
SMITH, D. A. and GALLOT, J. (1969) *Metal Sci. J.* **3,** 80.
SMITH, F. W. (1964) *Can. J. Phys.* **42,** 304; (1965) *Ibid.* **43,** 2052.
STEPAKOFF, G. and KAUFMAN, L. (1968) *Acta Metall.* **16,** 13.
STEWART, G. W. (1930) *Phys. Rev.* **35,** 726; (1941) *Rev. Mod. Phys.* **2,** 116.
SULLIVAN, G. A. and WEYMOUTH, J. W. (1964) *Phys. Rev.* **136,** A1141.
TAKAMURA, J. (1961) *Acta Metall.* **9,** 547.
TAKAMURA, J., FURUKAWA, K., MIURA, S., and SHINGU, P. H. (1963) *J. Phys. Sci. Japan.* Suppl. III, **18,** 7.
TEWORDT, L. (1958) *Phys. Rev.* **109,** 67.
TURNBULL, D. (1956) *Solid. State Phys.* **3,** 225.
TURNBULL, D. and CORMIA, R. L. (1960) *J. Appl. Phys.* **31,** 674
VAN DEN BEUKEL, A. (1970) *Vacancies and Interstitials in Metals,* p. 427, North-Holland, Amsterdam.
VAN BUEREN, H. G. (1961) *Imperfections in Crystals,* North-Holland, Amsterdam.
VINEYARD, G. H. (1961) *Disc. Faraday Soc.* **31,** 7.
VINEYARD, G. H. and DIENES, G. J. (1954) *Phys. Rev.* **93,** 265.
VITEK, V. (1968) *Phil. Mag.* **18,** 773; (1970a) *Scripta Metall.* **4,** 725; (1970b) *Phil. Mag.* **21,** 174.
VITEK, V., PERRIN, R. C., and BOWEN, D. K. (1970) *Ibid* **21,** 1049.
WANG, C. G., SEIDMAN, D. N., and BALLUFFI, R. W. (1968) *Phys. Rev.* **169,** 553.

Warren, B. E. (1959) *Prog. in Metal Phys.* **8,** 147.
Weiss, R. J. (1963) *Proc. Phys. Soc.* **82,** 281.
Weiss, R. J. and Tauer, K. J. (1956) *Phys. Rev.* **102,** 1490; (1958) *J. Phys Chem. Solids* **4,** 135.
Wulff, G. (1901) *Z. Kristallogr.* **34,** 449.
Ytterhus, A. and Balluffi, R. W. (1965) *Phil. Mag.* **11,** 707.
Zener, C. (1947) *Phys. Rev.* **71,** 846; (1948) *Elasticity and Anelasticity of Metals*, University of Chicago Press; (1951) *J. Appl. Phys.* **22,** 372; (1955) *Trans. Am. Inst. Min. (Metall.) Engrs.* **203,** 619.
Ziman, J. M. (1964) *Adv. in Phys.* **13,** 89; (1969a) *Principles of the Theory of Solids*, Cambridge University Press, London; (1969b) *The Physics of Solids*, Cambridge University Press, London; (1971) *Adv. in Phys.* **26,** 1.

CHAPTER 6

Solid Solutions

22. PAIR PROBABILITY FUNCTIONS: THERMODYNAMIC PROPERTIES

A solid phase containing two or more kinds of atom, the relative proportions of which may be varied within limits, is described as a solid solution. Terminal solid solutions are based on the structures of the component metals; intermediate solid solutions may have structures which are different from any of those of the constituents. Most solid solutions are of the substitutional type, in which the different atoms are distributed over one or more sets of common sites, and may interchange positions on the sites. In interstitial solutions, the solute atoms occupy sites in the spaces between the positions of the atoms of the solvent metal; this can only happen when the solute atoms are much smaller than the atoms of the solvent.

We must also distinguish between ordered and disordered solid solutions. In the fully ordered state each set of atoms occupies one set of positions, so that the atomic arrangement is similar to that of a compound. This is only possible at compositions where the ratios of the numbers of atoms of different kinds are small integral numbers, but the atomic arrangement may still be predominantly ordered in this way for alloys of arbitrary composition. In disordered solid solutions, the atoms are distributed among the sites they occupy in a nearly random manner. This classification is only approximate, and we shall formulate these concepts more precisely.

The definition of the unit cell, and the concept of the translational periodicity of the lattice, lose their strict validity when applied to a disordered solid solution. The mean positions of the atoms, considered as mathematical points, will no longer be specified exactly by (5.8), since there will be local distortions depending on the details of the local configurations. Moreover, a knowledge of the type of atom at one end of a given interatomic vector no longer implies knowledge of the atom at the other end, as it does for a pure component or a fully ordered structure. In a solid solution, precise statements of this nature have to be replaced by statements in terms of the probability of the atom being of a certain type.

For many purposes, the strict non-periodicity of the structure is not important, since most physical properties are averages over reasonably large numbers of atoms. Thus the positions of X-ray diffraction maxima depend only on the average unit cell dimensions, and their intensities depend only on the mean concentrations of atoms of different kinds and the mean interatomic distances. An approximate description in which the structure is regarded as having a unit cell of fixed size, with atomic positions occupied by identical scattering centres of averaged atomic scattering factor, thus suffices for a description of the main features of the X-ray diffraction pattern. A lattice vector of this structure may actually connect two unlike atoms, but is regarded as connecting two average atoms.

If two metals have the same crystal structure, they may form a single solid solution, and the lattice parameter then varies continuously with composition from the value characteristic of one pure metal to that of the other. In some assemblies, the edge length of the unit cell, or the volume per atom, is approximately linear with the atomic fraction of solute; this is described as Vegard's law. The work of Hume-Rothery and his collaborators (see Hume-Rothery and Raynor, 1954) has shown that a necessary (but not sufficient) condition for the formation of extensive terminal solutions is that the sizes of the atoms in the pure components shall not differ by more than $\sim 15\%$. This is called the size factor rule.

The localized distortion around a solute atom cannot be observed easily, but it may be studied by X-ray diffraction techniques in which the diffuse background scattering is measured. This type of X-ray measurement also enables us to deduce the probability of a given interatomic vector connecting atoms of given types. In theoretical treatments of the properties of solid solutions, it is usual to treat all atoms as situated on the sites obtained from the mean lattice, and this approximation is adequate in view of the severe limitations of the theories in other respects. The way in which the sites are occupied may then be described by a set of probability factors.

Consider a solid solution containing atomic fractions $x_A, x_B, x_C, \ldots,$ of components $A, B, C, \ldots,$ all of which may occupy any of the atomic sites. Then the probability that a particular site is occupied by a particular atom is just equal to the atomic fraction of that kind of atom, whatever the nature of the atomic forces. The probability that a particular vector joins two atoms of specified types, however, is dependent on the details of the distribution, and hence on the interatomic forces. Consider a vector $\mathbf{r}$ linking two atomic positions. There will be many such vectors in a crystal; suppose that in a group $N(\mathbf{r})$ of them, there are $N_{AB}(\mathbf{r})$ which connect together an A atom and a B atom. Then we can define the probability

$$P_{AB}(\mathbf{r}) = \lim_{N(\mathbf{r}) \to \infty} \frac{N_{AB}(\mathbf{r})}{N(\mathbf{r})}. \tag{22.1}$$

This is sometimes called a pair density function, by analogy with similar functions used in the descriptions of non-crystalline materials.

Obviously we may define a pair probability function for each possible kind of occupancy of the two sites, $A-A$, $A-C$, $B-B$, $B-C$, etc. However, the probabilities so defined are not completely independent, since their values must be consistent with the overall composition of the solid solution. Consider, for example, $N(\mathbf{r})$ vectors $\mathbf{r}$ in a binary assembly. They will join together $2N(\mathbf{r})$ atoms, of which $2x_A N(\mathbf{r})$ must be of type A. The vectors linking $A-A$ atom pairs will contribute $2N_{AA}(\mathbf{r})$ A atoms to this total, and those linking

$A-B$ pairs will contribute $N_{AB}(\mathbf{r})$ A atoms. Hence

$$\left.\begin{aligned} 2N_{AA}(\mathbf{r})+N_{AB}(\mathbf{r}) &= 2x_A N(\mathbf{r}) \\ \text{or} \qquad P_{AB}(\mathbf{r}) &= 2\{x_A - P_{AA}(\mathbf{r})\}. \end{aligned}\right\} \tag{22.2}$$

Similarly,

$$P_{AB}(\mathbf{r}) = 2\{x_B - P_{BB}(\mathbf{r})\}.$$

Only one of the three probabilities is independent in this case; we shall take it to be $P_{AB}(\mathbf{r})$. There are equivalent relations reducing the number of independent probabilities for ternary and higher assemblies, but we shall only consider binary assemblies in detail. In such an assembly, $x_A+x_B = 1$, and it is useful to employ a single composition variable $x = x_B$, $(1-x) = x_A$. The composition of the alloy then changes from pure A to pure B as x increases from 0 to 1. Occasionally, however, it is preferable to retain the x_A, x_B notation in order to show whether complex relations are symmetrical with regard to the components. We shall also need the notation $N_A = N(1-x), N_B = Nx$ for the numbers of A and B atoms in a crystal of N atoms.

Suppose we have an assembly in which we have values of $P(\mathbf{r})$ for all possible vectors $\mathbf{r}$ and all possible combinations of atoms. The question then arises whether or not this is a a complete statistical description of the occupancy of the sites. In general, it is not, and we should also write down the probabilities of groups of three, four, five, or more sites being occupied by atom groups of specified types. Such probability factors are introduced in some theoretical treatments, but we shall not have occasion to consider them in detail. The pair probability functions may be deduced from X-ray measurements, as already mentioned, but no experimental method is known for determining triplet or higher probability functions.

A pair probability function must possess the point group symmetry of the Bravais lattice; that is, $P(\mathbf{r})$ and $P(\mathbf{r}')$ must be equal if $\mathbf{r}$ and $\mathbf{r}'$ are vectors which can be transformed into each other by a rotation or reflection symmetry operation. The behaviour of $P(\mathbf{r})$ as $\mathbf{r}$ becomes large falls into one of two categories. In a solid solution which is described as disordered, $P(\mathbf{r})$ in the limit of large $\mathbf{r}$ approaches the value it would have for a completely random distribution of atoms over the available sites. This is the situation we shall discuss in most of this chapter. In a structure described as possessing long-range order, the limit of $P(\mathbf{r})$ as $\mathbf{r} \to \infty$ is different from the random value; this is discussed further in Section 26.

In determining the form of an alloy phase diagram, we are interested in the variation of the free energy of a phase with composition. This means that at fixed temperature and pressure, we need to know the heat and entropy of mixing, and various derived functions. The thermodynamic quantities are related to the parameters which specify the distribution of the atoms over the available sites, but any attempt at precise calculation of the energies of different distributions is formidably difficult. Thus, when analytic expressions are required it is necessary to use a simple model. Much of the theoretical work on solid solutions is restricted to binary solutions formed from two metals of the same crystal structure and nearly equal atomic diameters. Under these conditions, the drastic approximations which are used to obtain the required energy expressions are reasonably valid.

We shall assume throughout that the entropy of a solid solution may be separated into configurational and thermal parts. The thermal or vibrational entropy is obtained from the heat motion of the atoms; the configurational entropy results from the randomness of location of the atoms on their sites, and may be defined as the extra entropy possessed by a hypothetical solid solution with all atoms occupying point sites over the corresponding entropy of the pure components. This separation of the entropy is equivalent to a factorization of the partition function.

In the simplest model (the "ideal solution"), the thermal partition function is determined only by the number of atoms of each kind present, and the configurational partition function corresponds to a completely random distribution of the atoms. The formation of an alloy from its constituents then involves no change in thermal entropy, and the entropy of mixing is equal to the configurational entropy of a random arrangement. The assumption of zero thermal entropy of mixing, though unjustified, is often made in more complex treatments, which attempt to calculate the configurational partition function when the arrangement is not random. The model used in most calculations of this type is that of central force nearest neighbour interactions, and the approach is known as the quasi-chemical theory. There are also statistical models in which volume changes and changes in thermal entropy are not neglected, but other simplifying assumptions have to be made in order to obtain useful results. A completely different model focuses attention on the effects of differences in the sizes of the atoms, and calculates the energy term by treating each solute atom as a centre of dilatation in an elastic medium.

These various models are discussed in some detail in the remaining sections of this chapter. None of them is very satisfactory, and at present it seems that experimental measurements of thermodynamic functions are much more accurate than theoretical predictions. In the remainder of this section we examine the important thermodynamic properties of solutions and their relation to the equilibrium diagram.

The thermodynamic properties of a single component are completely determined by the external constraints but have to be measured from an arbitrary zero; that is, a standard reference state must be specified. This is also true of a solution, and it is often convenient to use the pure components under the same temperature and pressure as the reference state. Thus suppose that $g_A^{\alpha 0}$, $g_B^{\alpha 0}$ are the free energies per atom of the pure components in the α phase, each measured from an arbitrary zero.† (The α phase may, of course, be metastable in either or both components.) The free energy per atom of a homogeneous α solution of composition x may then be expressed relative to the same zeros as

$$g^{\alpha} = (1-x)g_A^{\alpha 0} + x g_B^{\alpha 0} + \Delta_m g^{\alpha}, \tag{22.3}$$

where $\Delta_m g^{\alpha}$ is called the free energy of mixing or the free energy of formation of the α solution, and is measured from the standard state of the unmixed components.

Suppose Fig. 6.1 represents the curve of g^{α} against x at fixed temperature and pressure. An alloy of composition x_4 has a free energy per atom represented by point P if it exists

† The zeros may be chosen so that at some particular temperature and pressure $g_A^{\alpha 0} = g_B^{\alpha}$, but this is not then true for other temperatures and pressures.

as a homogeneous α solid solution, and by point Q if it is a mixture of pure A and pure B (each in the α form). If the assembly contains two α solutions of different composition, the g^{α} curve gives the free energy per atom in each, and hence the mean free energy per atom. Thus the mean free energy per atom of an alloy of composition x_4 is represented by point R if it exists as a two-phase mixture of α solutions of compositions x_5 and x_6. Clearly, the relative stability of the single solid solution and the phase mixture depends on whether R is above or below P. It is readily seen that if the g^{α} curve is concave upwards for all x (i.e. if $\partial^2 g^{\alpha}/\partial x^2 > 0$ for all x), the single solid solution is always more stable than a phase mixture

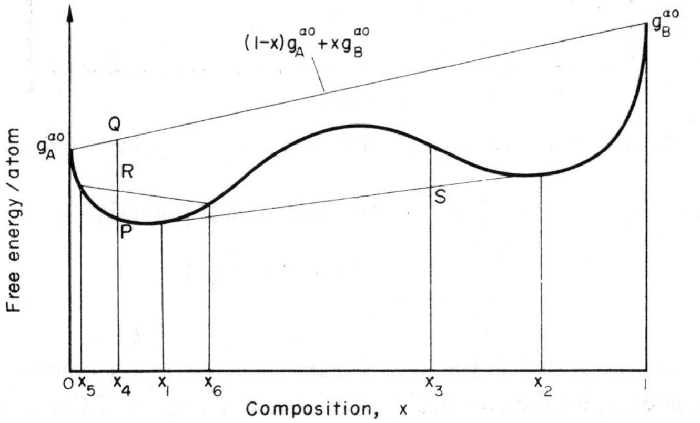

FIG. 6.1. Schematic free energy vs. composition curve for a solid solution.

of two α solutions. However, if the g^{α} curve contains a region of negative $\partial^2 g^{\alpha}/\partial x^2$, as shown in Fig. 6.1, the equilibrium condition will be a phase mixture for certain compositions. For the alloy of composition x_3, the lowest free energy is represented by the point S, and is obtained from a mixture of α phases of compositions x_1 and x_2. These compositions are those corresponding to the points of contact of the common tangent to the two parts of the g^{α} curve. All other alloys in which $x_1 < x < x_2$ also exist in equilibrium as a mixture of α phases of compositions x_1 and x_2, the relative proportions of the phases being fixed by the overall composition.

From this discussion, we see that the limits of solubility x_1 and x_2 are defined by the condition

$$(\partial g^{\alpha}/\partial x)_{x_1} = (\partial g^{\alpha}/\partial x)_{x_2} = \{g^{\alpha}(x_2)-g^{\alpha}(x_1)\}/(x_2-x_1). \tag{22.4}$$

Since the g^{α} curve is essentially equivalent to the $\Delta_m g^{\alpha}$ curve tilted about the straight line $(1-x)g_A^{\alpha 0}+xg_B^{\alpha 0}$, this condition may also be written

$$(\partial \Delta_m g^{\alpha}/\partial x)_{x_1} = (\partial \Delta_m g^{\alpha}/\partial x)_{x_2}. \tag{22.5}$$

So far as the equilibrium properties are concerned, it is immaterial whether we consider the variation of g^{α} or $\Delta_m g^{\alpha}$ with x.

The chemical potential (partial free energy per atom) of A in the α-phase is defined by the equation

$$g_A^{\alpha} = \partial G/\partial N_A = g_A^{\alpha 0}+\partial(N\Delta_m g^{\alpha})/\partial N_A. \tag{22.6}$$

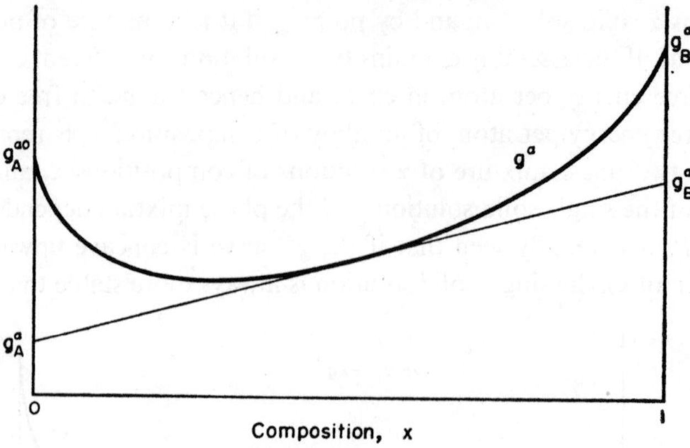

FIG. 6.2. Relation of chemical potentials to free energy vs. composition curve.

The mean free energy per atom is related to the chemical potentials by the equation

$$\left.\begin{aligned} g^\alpha &= (1-x)g_A^\alpha + xg_B^\alpha \\ \partial g^\alpha/\partial x &= g_B^\alpha - g_A^\alpha . \end{aligned}\right\} \quad (22.7)$$

and

These relations are illustrated in Fig. 6.2. The tangent to g^α at the composition x intercepts the $x = 0$ ordinate to give the chemical potential g_A^α and the $x = 1$ line to give the chemical potential g_B^α. Clearly, for the common tangent of Fig. 6.1,

$$g_A^\alpha(x_1) = g_A^\alpha(x_2), \quad g_B^\alpha(x_1) = g_B^\alpha(x_2). \quad (22.8)$$

This is the condition for phase equilibrium in terms of the appropriate intensive thermodynamic quantities (chemical potentials), corresponding to the condition of a minimum value of the extensive quantity G.

We now consider the possibility of two different phases, α and β, existing in the assembly. These phases may represent different states of matter, or may be solid solutions of diffe-

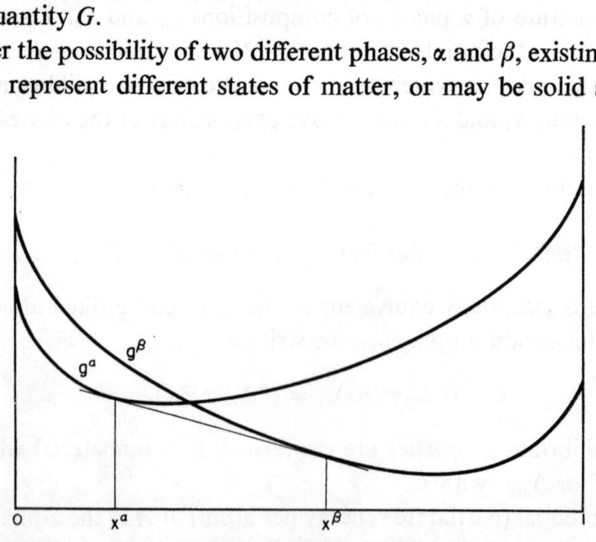

FIG. 6.3. Free energy vs. composition curves for phases α and β.

rent crystal structure. The free energy of the β phase will be

$$g^\beta = (1-x)g_A^{\beta 0}+xg_B^{\beta 0}+\Delta_m g^\beta, \tag{22.9}$$

where $g_A^{\beta 0}$, $g_B^{\beta 0}$ are measured from the same zeros as $g_A^{\alpha 0}$, $g_B^{\alpha 0}$ respectively. Thus $g_A^{\beta 0}-g_A^{\alpha 0}$ is the free energy per atom required to change pure A from the α phase to the β phase. A possible configuration of the free energy curves g^α and g^β is shown in Fig. 6.3. It follows from the previous discussion that the equilibrium state of the assembly is homogeneous α phase for $x < x^\alpha$, homogeneous β phase for $x > x^\beta$ and a mixture of α and β phases for $x^\alpha < x < x^\beta$. The solubilities or solvus lines x^α, x^β are determined by the relation

$$\left(\frac{\partial g^\alpha}{\partial x}\right)_{x^\alpha} = \left(\frac{\partial g^\beta}{\partial x}\right)_{x^\beta} = \frac{g^\beta(x^\beta)-g^\alpha(x^\alpha)}{x^\beta - x^\alpha}. \tag{22.10}$$

It should be noted that eqn. (22.10) is not equivalent to $(\partial\Delta_m g^\alpha/\partial x) = (\partial\Delta_m g^\beta/\partial x)$ because of the different reference states used for $\Delta_m g^\alpha$ and $\Delta_m g^\beta$. It is sometimes convenient to rewrite eqns. (22.3) and (22.9) as

$$\left.\begin{aligned} g^\alpha &= (1-x)g_A^{\alpha 0}+xg_B^{\beta 0}+\Delta_n g^\alpha, \\ g^\beta &= (1-x)g_A^{\alpha 0}+xg_B^{\beta 0}+\Delta_n g^\beta, \end{aligned}\right\} \tag{22.11}$$

where $\Delta_n g^\alpha$, $\Delta_n g^\beta$ are both referred to pure A and pure B in their equilibrium states. Equation (22.10) is then equivalent to

$$(\partial\Delta_n g^\alpha/\partial x)_{x^\alpha} = (\partial\Delta_n g^\beta/\partial x)_{x^\beta}. \tag{22.12}$$

Free-energy–composition curves of the type shown in Fig. 6.3 are typical of those expected in a simple eutectic assembly in which the components have different crystal structures α and β; those shown in Fig. 6.1 are typical of those from a eutectic assembly in which both metals have the same crystal structure. At temperatures below the eutectic temperature, the liquid free-energy curve lies entirely above the common tangent. Above the eutectic temperature, this curve intersects the common tangent, as shown in Fig. 6.4. Alterna-

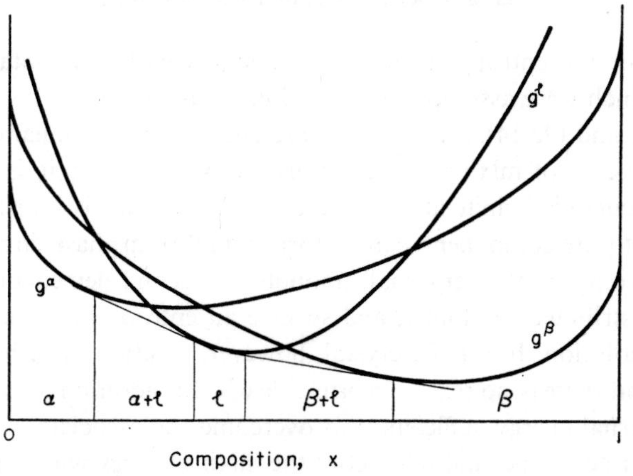

FIG. 6.4. Free energy vs. composition curves for a eutectic reaction.

tively, if the liquid free-energy curve first crosses the g^{α} curve at $x = 0$, or the g^{β} curve at $x = 1$, a peritectic reaction will result. The same principles apply when further (intermediate) phases are considered; these may be solid solutions having relatively flat free-energy curves, or compounds of fixed composition, at which the free energy has a very sharp minimum. The relation of free-energy–composition curves to the equilibrium diagram is discussed in many textbooks; see, for example, Cottrell (1955) for an elementary account. It is now possible to store the basic thermodynamic information for a given system in a computer and to use a program which will enable any required information on phase equilibria or derived thermodynamic functions to be obtained. The necessary free-energy functions are generally obtained from a mixture of experimental measurements, empirical expansions, and information derived from measured phase diagrams. Of particular value for work on phase transformations is the ability in this way to obtain reliable estimates of metastable phase relations in otherwise inaccessible regions of the phase diagram; the usual theory of pearlite formation, for example, requires knowledge of the metastable austenite–ferrite and austenite–cementite equilibria below the eutectoid temperature. This kind of work was pioneered by Kaufman (1967, 1969) and has been surveyed by Kaufman and Bernstein (1970) and Hillert (1970).

It is often convenient to refer the thermodynamic properties of a solution to those of an ideal solution. For our present purpose, the simplest definition of an ideal solution is that the heat of mixing is zero, and the entropy of mixing is equal to the configurational entropy of a random arrangement. Consider a simple structure in which N sites are occupied by N_A A atoms and N_B B atoms. The number of distinguishable ways in which the atoms may be arranged on the sites is $N!/(N_A)!(N_B)!$, and the configurational entropy of mixing is thus

$$\begin{aligned}\Delta_m S &= \boldsymbol{k}\{\ln N! - \ln N(1-x)! - \ln(Nx)!\}\\ &= -N\boldsymbol{k}\{(1-x)\ln(1-x) + x\ln x\}\end{aligned} \tag{22.13}$$

and the free energy of mixing per atom is

$$\Delta_m g = \boldsymbol{k}T\{(1-x)\ln(1-x) + x\ln x\}. \tag{22.14}$$

The expression for the entropy of mixing is identical with that obtained previously for lattice defects, which were assumed to be randomly distributed.

In eqns (22.13) and (22.14), x and $(1-x)$ are necessarily fractional quantities, less than unity. The free energy of mixing, $\Delta_m g$, is thus always negative; it is plotted in Fig. 6.5. The slope of the curve is infinite at $x = 0$ and $x = 1$, and this is sometimes used to justify the statement that pure components cannot form equilibrium phases in an assembly containing two or more components. Very small amounts of a second element dissolved in the lattice will always be distributed randomly, and so give an entropy of mixing corresponding to that of an ideal solution. In a finite crystal, however, solution must be discontinuous, i.e. atom by atom, and there is no reason in principle why the addition of only one atom should not raise the internal energy sufficiently to overcome the associated rise in entropy. This argument is rather academic, and in practice the mixing energy will almost certainly always lead to the presence of a small number of solute atoms in solution at equilibrium, just as

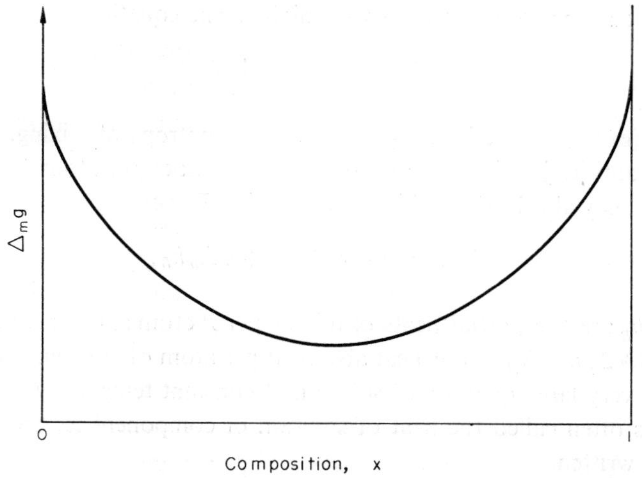

FIG. 6.5. Free energy of mixing vs. composition curve for an ideal solution.

there is always a small number of lattice vacancies. For practical purposes, the solubility may be so small as to be negligible, and we are justified in expecting some solubilities to be too small to be measured by the most sensitive techniques available.

From eqns. (22.6) and (22.4) we may now write expressions for the chemical potentials per atom of an ideal solution. Leaving out the phase identifying superscript, α or β, since this is common to all quantities, we have

$$\left.\begin{aligned} g_A - g_A^0 &= \boldsymbol{k}T \ln(1-x), \\ g_B - g_B^0 &= \boldsymbol{k}T \ln x. \end{aligned}\right\} \tag{22.15}$$

Guggenheim, and other workers, use a function called the absolute activity, λ, which is related to the chemical potential by the equation $g_A = \boldsymbol{k}T \ln \lambda_A$. In terms of these absolute activities, eqn. (22.15) becomes

$$\lambda_A/\lambda_A^0 = 1-x, \quad \lambda_B/\lambda_B^0 = x. \tag{22.16}$$

This equation is useful because it forms the basis of one convenient way of defining the thermodynamic properties of a real solution by means of activity coefficients. For a non-ideal solution, we write

$$\lambda_A/\lambda_A^0 = \gamma_A(1-x), \quad \lambda_B/\lambda_B^0 = \gamma_B x. \tag{22.17}$$

The activity coefficients γ_A and γ_B are functions of composition, and may be determined experimentally for a real solution.

An alternative way of measuring the deviations from "ideal" behaviour is in terms of the "excess" thermodynamic functions. For example, the free energy of mixing may be written

$$\Delta_m g = \Delta_e g + \boldsymbol{k}T\{(1-x)\ln(1-x) + x \ln x\}, \tag{22.18}$$

where $\Delta_e g$ is called the excess free energy of mixing. The equation

$$\Delta_e g = \Delta_m h - T\,\Delta_e s \tag{22.19}$$

then defines the (excess) heat of mixing, and the excess entropy of mixing.

The excess quantities just introduced are the mean values per atom. The corresponding partial functions may also be defined in the usual way. For example

$$\Delta_m h = (1-x)\,\Delta_m h_A + x\,\Delta_m h_B, \tag{22.20}$$

where $\Delta_m h_A$, $\Delta_m h_B$ are the partial heats of mixing per A atom and per B atom respectively. Thus $\Delta_m h_A = \partial(N\Delta_m h)/\partial N_A$ is the heat absorbed per atom of A when a small quantity of A is added to a very large quantity of solution at constant temperature and pressure; for this reason, it is often called the heat of solution of component A. Similarly, the excess entropy may be written

$$\Delta_e s = (1-x)\,\Delta_e s_A + x\,\Delta_e s_B, \tag{22.21}$$

where $\Delta_e s_A$, $\Delta_e s_B$ are the excess partial entropies of mixing, or entropies of solution.

In a very dilute solution, the partial excess quantities of the solvent must tend to zero. Thus, there is no heat of solution when A in the α phase is dissolved in an α solution which is already nearly entirely A, and the entropy of mixing will be represented completely by the configurational term in (22.18). It follows that for a dilute solution

$$\Delta_e g = x(\Delta_m h_B - T\,\Delta_e s_B) \qquad (x \ll 1). \tag{22.22}$$

The B atoms are sufficiently separated to be considered as non-interacting, and each has associated with it an intrinsic energy, which is the limiting value of $\Delta_m h_B$ in very dilute solution, and an intrinsic entropy, which is the corresponding limiting value of $\Delta_e s_B$. Both of these terms are due essentially to the localized disturbance round each solute atom, and the excess entropy of mixing is entirely due to thermal entropy. The situation in a very dilute solution is exactly analogous to that of a metal containing lattice defects, and the term $\Delta_m h_B$ corresponds to the intrinsic energy of a defect (e.g. $\Delta h_{\square}$), and $\Delta_e s_B$ corresponds to the intrinsic thermal entropy of a defect (e.g. $\Delta s_{\square}$).

For very dilute solutions in which $x_1 \ll 1$, $(1-x_2) \ll 1$, the equilibrium condition (eqns. (22.4) or (22.5)) becomes

$$(\partial\Delta_m g/\partial x)_{x_1} = (\partial\Delta_m g/\partial x)_{x_2} = 0,$$

so that the variation of the solubility of B in A when this solubility is very small is obtained from (22.18) as

$$x_1 = \exp(\Delta_e s_B/k)\exp(-\Delta_m h_B/kT) \tag{22.23}$$

with a similar expression for x_2.

A similar treatment may be given for the solubility limits in a simple eutectic assembly in which the component metals have different crystal structures, and are very sparingly soluble in each other in the solid. We have now to use the free energies of mixing defined relative to A in the α phase and B in the β phase (eqn. (22.11)), and the corresponding excess functions which are defined by

$$\Delta_n g = \Delta_f g + kT\{(1-x)\ln(1-x) + x\ln x\}. \tag{22.24}$$

The quantities $\Delta_n g^\alpha$, $\Delta_f g^\alpha$ differ from $\Delta_m g^\alpha$ and $\Delta_e g^\alpha$ respectively by $x(g_B^{\alpha 0} - g_B^{\beta 0}) = x\Delta g_B^{\beta\alpha}$, where $\Delta g_B^{\beta\alpha}$ is the free energy per atom required to change pure B from the β phase to the α phase. Similarly, $\Delta_n g^\beta$ and $\Delta_f g^\beta$ differ from $\Delta_m g^\beta$ and $\Delta_e g^\beta$ respectively by $(1-x)\Delta g_A^{\alpha\beta}$.

Writing

$$\Delta_f g^\alpha = x(\Delta_n h_B - T\Delta_f s_B)$$

and using eqn. (22.12), it is readily seen that in very dilute solution the solubility becomes

$$x^\alpha = \exp(\Delta_f s_B/k)\exp(-\Delta_n h_B/kT) \qquad (x^\alpha \ll 1), \tag{22.25}$$

which is of the same form as (22.23), and differs only in that

$$\Delta_n h_B = \Delta_m h_B + \Delta h_B^{\beta\alpha}, \quad \Delta_f s_B = \Delta_e s_B + \Delta s_B^{\beta\alpha}. \tag{22.26}$$

It is also possible to obtain an equation of the form (22.23) for the solubility limit when a primary solid solution is in equilibrium with an intermetallic compound; the same procedure is followed, but the free energy of mixing and the corresponding excess functions are defined with reference to the pure metal and the compound as standard state.

Equations (22.23) or (22.25) are only valid for very dilute solutions, when each solute atom has an intrinsic energy and entropy. For moderately dilute solutions ($x \gtrsim 0{\cdot}01$) Freedman and Nowick (1958) have developed a more general expression by treating eqn. (22.22) as the first term of a Taylor expansion for both $\Delta_m h$ and $\Delta_e s$. For the case of terminal α and β solutions, they obtain the result

$$\frac{1}{(1-2x^\alpha)}\ln\left\{\frac{x^\alpha}{1-x^\beta}\right\} = \frac{\Delta_f s_B}{k} - \frac{\Delta_n h_B}{kT}, \tag{22.27}$$

and the equivalent result for two solid solutions of the same structure is clearly obtained by writing $x^\alpha = x_1$, $x^\beta = x_2$, $\Delta_n h_B = \Delta_m h_B$ and $\Delta_f s_B = \Delta_e s_B$. In using this expression to investigate experimental solubility data, the left-hand side is equated to $\ln x_{\text{corr}}$ thus reducing it to the form of (22.23). A plot of $\ln x_{\text{corr}}$ against $1/T$ then enables $\Delta_n h_B$ and $\Delta_f s_B$ to be deduced from the slope and the intercept on the $1/T = 0$ axis of the straight line which should be obtained. In practice, the factor $1-x^\beta$ may be equated to unity without affecting the results, but the $1-2x^\alpha$ correction cannot be ignored even in quite dilute solutions, and much better straight lines are obtained from $\ln x_{\text{corr}}$ than from $\ln x$.

An important effect in theories of nucleation (Chapter 10) and growth (Chapter 11) arises from the presence of interfaces in a two-phase assembly. The solubility limits calculated above are strictly valid only for planar interfaces; when interfaces are curved there are additional energy terms which can not be neglected. Consider an assembly in which β particles of surface area O and surface free energy per unit area σ are in equilibrium with the α matrix. The energy of the interfaces displaces the equilibrium condition (22.8) or (22.12) and may be considered to contribute an additional thermodynamic potential so that both the solubility limit x^α and the equilibrium composition x^β of the β phase vary with the size of the particles. We now denote these quantities as x_r^α and x_r^β and the equilibrium compositions for infinite planar interfaces by x_∞^α and x_∞^β. Most published treatments of this size effect (Gibbs–Thomson effect) are oversimplified and assume that x^β remains constant.

In a virtual change in which dn atoms are transferred from the α phase to the β phase, there is now an additional energy term $\sigma\,\mathrm{d}O$ due to the increase in surface area of the β particles. As shown in Fig. 6.6, this may be represented as a displacement of g^β to $g_r^\beta = g^\beta + \sigma(\mathrm{d}O/\mathrm{d}n)$ and the new equilibrium compositions are given by the points of contact of the tangent common to the curves of g^α and g_r^β against x. Let the effective chemical potentials per atom for curved interfaces, defined by the construction of Fig. 6.2 applied

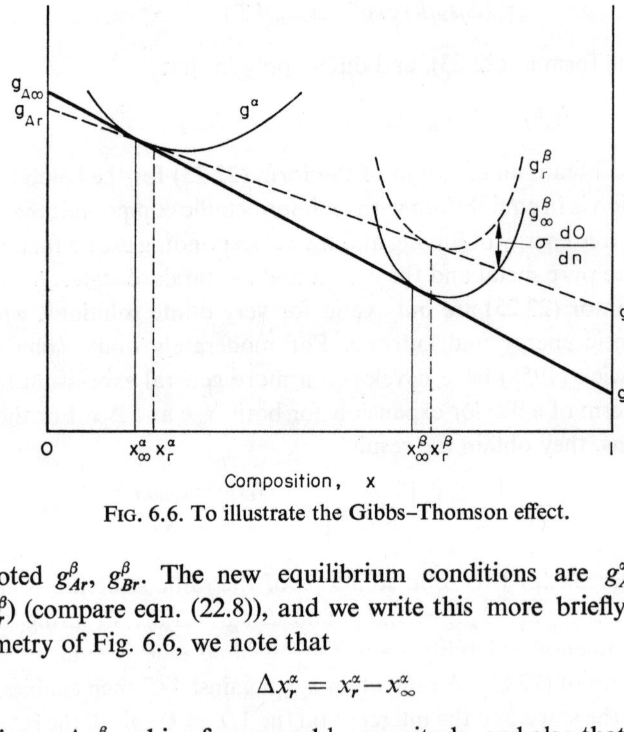

FIG. 6.6. To illustrate the Gibbs–Thomson effect.

to g_r^β, be denoted g_{Ar}^β, g_{Br}^β. The new equilibrium conditions are $g_A^\alpha(x_r^\alpha) = g_{Ar}^\beta(x_r^\beta)$ and $g_B^\alpha(x_r^\alpha) = g_{Br}^\beta(x_r^\beta)$ (compare eqn. (22.8)), and we write this more briefly as $g_{Ar}^\alpha = g_{Ar}^\beta$, etc. From the geometry of Fig. 6.6, we note that

$$\Delta x_r^\alpha = x_r^\alpha - x_\infty^\alpha \tag{22.28}$$

has the same sign as Δx_r^β and is of comparable magnitude, and also that to a good approximation

$$g_{Br}^\alpha - g_{B\infty}^\alpha = \{(1-x_\infty^\alpha)/(x_\infty^\beta - x_\infty)\}\,\sigma(\mathrm{d}O/\mathrm{d}n) \tag{22.29}$$

In terms of the activity coefficients (eqn. (22.17)), we also have

$$g_{Br}^\alpha - g_{B\infty}^\alpha = kT\ln(\gamma_{Br}^\alpha x_r^\alpha/\gamma_{B\infty} x_\infty^\alpha) \tag{22.30}$$

and for dilute solutions or solutions which obey Henry's law γ_B^α is a constant. Equations (22.29) and (22.30) thus give

$$\frac{\Delta x_r^\alpha}{x_\infty^\alpha} \simeq \ln\frac{x_r^\alpha}{x_\infty^\alpha} = \frac{\sigma}{kT}\left(\frac{\mathrm{d}O}{\mathrm{d}n}\right)\left(\frac{1-x_\infty^\alpha}{x_\infty^\beta - x_\infty^\alpha}\right). \tag{22.31}$$

In the often used approximation of spherical particles

$$dO/dn = 2v^{\beta}/r \tag{22.32}$$

and

$$\Delta x_r^{\alpha} = (2\Gamma/r)x_{\infty}^{\alpha}, \tag{22.33}$$

where

$$\Gamma = (\sigma v^{\beta}/kT)(1-x_{\infty}^{\alpha})/(x_{\infty}^{\beta}-x_{\infty}^{\alpha}). \tag{22.34}$$

For sufficiently small values of x_{∞}^{α} (dilute solutions)

$$\Gamma = (\sigma v^{\beta}/kT)/(x_{\infty}^{\beta}-x_{\infty}^{\alpha}), \tag{22.35}$$

and this form which has been used by Trivedi (1970) in theories of diffusion-controlled growth (see Chapter 11). A still more extreme assumption is that $x^{\beta} = 1$ in which case

$$\Gamma = \sigma v^{\beta}/kT. \tag{22.36}$$

Equations (22.33) and (22.36) together constitute a frequently quoted (but obviously inaccurate) statement of the Gibbs–Thomson theory. Finally, in the general case where Henry's law does not apply, Γ in eqn. (22.33) is given by

$$\Gamma = (\sigma v^{\beta}/kT)(1-x_{\infty}^{\alpha})/(x_{\infty}^{\beta}-x_{\infty}^{\alpha})\{1+(d \ln \gamma_B^{\alpha}/ d \ln x)\}. \tag{22.37}$$

Many results obtained with the special assumption of eqn. (22.36) remain valid under more general conditions if Γ from (22.37) is used in place of $(\sigma v^{\beta}/kT)$ in the appropriate equations.

If the particles are cylinders (needles) rather than spheres $dO/dn = v^{\beta}/r$ and eqn. (22.33) is modified to

$$\Delta x_r^{\alpha} = (\Gamma/r)x_{\infty}^{\alpha}. \tag{22.38}$$

In the general case of equilibrium at a curved interface

$$dO/dn = v^{\beta}(r_1^{-1}+r_2^{-1}), \tag{22.39}$$

where r_1 and r_2 are the principal radii of curvature. Similarly, for polyhedral particles of equilibrium shape, the surface energy term becomes $d(O_i h_i)/dn$ where O_i is a surface in the equilibrium shape at distance h_i from the centre of the particle so that according to the theory of Section 20, σ_i/h_i is constant. This gives a more general form of (22.33) as

$$\Delta x_r^{\alpha} = (2\sigma_i/h_i)(\Gamma/\sigma)x_{\infty}^{\alpha}. \tag{22.40}$$

Note that since σ is positive x_r^{α} is always greater than x_{∞}^{α}. We may usefully digress here to point out that there is an equivalent relationship between the vapour pressure in equilibrium with a liquid droplet of radius r and that in equilibrium with a flat liquid surface (Chapter 10, eqn. (46.14)). The curvature of a surface may in fact be regarded as introducing a pressure differential

$$p = \sigma(r_1^{-1}+r_2^{-1}) \tag{22.41}$$

since for a fluid medium the surface tension is numerically equal to the surface free energy σ (Section 35). This concept leads to the same expression for the change in solubility with particle size since the free energy charge per atom is now pV^{β}.

Equations (22.31), (22.33), (22.37), and (22.40) and the equivalent vapour pressure eqn. (46.14) are variously known as the Gibbs–Thomson, Thomson–Helmholtz, or Thomson–Freundlich equation (Gibbs 1878; Thomson 1888; Freundlich 1926) or sometimes simply as the effects of "capillarity", which is the term used by Gibbs.† The equations are often written in terms of concentrations c_∞^α, c_r^α, etc., instead of atomic fractions x_∞^α, x_r^α, etc, but this is strictly valid only when there is negligible volume change on transformation. The above treatment is implicitly based on this assumption since no allowance is made for strain energies due to transformation (see below).

We conclude this section by examining the thermodynamics of inhomogeneous phases and the criteria for distinguishing between a metastable phase and an unstable phase. A solid solution of fluctuating composition may be assigned a free energy which contains three different terms. The first contribution is the sum of the free energies which the individual atoms would have if they were present in homogeneous solutions of the same compositions as those of their local environments. Each term in the sum is thus represented by a point on the free-energy curve of the homogeneous solution. In the particular case that there are regions of only two different compositions, the change in free energy per atom in comparison with the homogeneous solution is given graphically by the construction shown in Fig. 6.1, and the sum over N atoms is just a generalization of this procedure. The second term arises because the free energy of any small volume element of given composition will be changed if it is surrounded by material of different composition. When the transition is sharp, the extra energy appears macroscopically as the surface free energy we have discussed above; Cahn and Hilliard (1958) showed how to express the analogous energy which arises from a continually fluctuating composition. The third term was mentioned briefly above; it is the elastic coherency strain energy which is due to a variation of lattice parameter with composition. If all regions of the inhomogeneous phase are coherent with each other, the assembly will be in a self-stressed state, the elastic energy of which must be considered part of the overall free energy.

We consider first the chemical or volume free energy. For *small* variations about the mean composition, the generalized construction of Fig. 6.1 will always leave R above P in those composition ranges where $(\partial^2 g/\partial x^2) > 0$, and this includes some compositions between x_1 and x_2. However, there is a net decrease in free energy, even for very small fluctuations, if $(\partial^2 g/\partial x^2) < 0$. The curve

$$(\partial^2 g/\partial x^2) = (\partial^2 \Delta_m g/\partial x^2) = 0 \tag{22.42}$$

was called by Gibbs the limit of metastability, since inside this curve a homogeneous solution should decompose spontaneously at a rate limited only by the rate of atomic migration. The curve is now usually referred to as the spinodal, or more strictly as the chemical spinodal. As we shall describe in later chapters, the decomposition of a supersaturated solid solution inside the spinodal does not require nucleation in the classical sense and occurs by a process known as "uphill" or "negative" diffusion. Note that there is no spinodal in free-energy diagrams such as those shown in Figs. 6.2–6.5.

† An interesting survey of the early history of this subject is given by Swan and Urquhart (1927).

Cahn and Hilliard used a multivariable Taylor expansion to express the free energy of an inhomogeneous solution in terms of g^α and of the various spatial derivatives of composition x (or of concentration c). The first term in this expansion is the sum already considered and for centrosymmetric crystals the energy per atom in the next most important term may be written as $Kv(\nabla x)^2 = Kv^3(\nabla c)^2$, where v is the atomic volume and K is a second rank tensor and is thus isotropic for cubic crystals. This "gradient energy" is a continuum analogue of the surface free energy of a discontinuous interface; the Cahn and Hilliard expansion is, of course, invalid if the gradient ∇c becomes large. There has been some dispute over the validity of the symmetry argument for the absence of a term in $|\nabla c|$ (see Tiller *et al.* 1970; Cahn and Hilliard, 1971), but we believe the Cahn and Hilliard treatment is correct.

An exact theory of the coherency energy is difficult when allowance is made for elastic anisotropy (see Cahn 1969), and we mention here only the isotropic result. The elastic energy depends on $\varepsilon = \mathrm{d}\ln a/\mathrm{d}x$, where a is the lattice parameter, and on an elastic modulus $Y' = Y/(1-\nu)$, where Y is Young's modulus and ν is Poisson's ratio. Cahn (1961) showed that if the actual composition is x and the mean composition is x_0 the coherency energy per atom may be written as $\varepsilon^2 Y' v(x-x_0)^2$. This changes both the solubility limits and the limits of metastability. In place of eqn. (22.4), the coherent miscibility gap is defined by x_{1c} and x_{2c}, where

$$\left(\frac{\partial g^\alpha}{\partial x}\right)_{x_{1c}} = \frac{g(x_{2c})-g(x_{1c})}{x_{2c}-x_{1c}} + v\varepsilon^2 Y'(x_{2c}-x_{1c}) \tag{22.43}$$

with a similar equation for x_{2c} and

$$x_{1c} > x_1, \quad x_{2c} < x_2. \tag{22.44}$$

Similarly, the coherent spinodal which represents the true limit of metastability is defined by

$$(\partial^2 g/\partial x^2) + 2\varepsilon^2 Y' v = 0, \tag{22.45}$$

and lies entirely inside the chemical spinodal. The increase in solubility $x_{1c}-x_1$ at fixed temperature, or equivalently the depression of the temperature of the coherent curves below the chemical curve, depends sensitively on the magnitude of ε; the supercooling from the chemical to the coherent spinodal was estimated by Cahn as 40°C for aluminium–zinc alloys and 200°C for gold–nickel alloys.

Note that although the coherency strain energy is a function of the mean composition x_0 of the alloy, the positions of the coherent phase boundary and spinodal are not dependent on x_0. However, directional effects appear when anisotropic elastic theory is used; for example, the solid solution may be unstable to fluctuations in composition in some directions in a crystal but metastable for other directions (see Part II, Chapter 18).

Equation (22.44) shows that the coherent spinodal only exists when there is a negative region of $(\partial^2 g/\partial x^2)$, i.e. when there is also a chemical spinodal. However, precipitates of one phase in another which involve chemical free energy curves of the type shown in Fig. 6.3 may sometimes be obtained in either coherent or incoherent forms, and there is then in principle both a coherent solvus and an equilibrium solvus. Intermediate or metastable phases which are coherent with the matrix are also often encountered in low-temperature precipitation reactions. Such phases frequently have a lower symmetry than the equilibrium phase and it may occasionally be illuminating to regard them as coherent versions of the equilibrium precipitate.

23. THE NEAREST NEIGHBOUR MODEL: REGULAR SOLUTIONS

The forces between the atoms in a solid solution, as in a pure metal, are mainly short-range in character, and the most used model is that in which only nearest neighbour interactions are considered significant. The energy of the crystal is then the sum of the pair interaction energies of the $A-A$, $B-B$, and $A-B$ contacts. This is often referred to as the Ising model, although Ising actually treated only the magnetic analogue of this situation; the first application to alloys was made by Bethe.

The limitations of the nearest neighbour model in providing a satisfactory description of the metallic state have already been emphasized in the previous chapter. When applied to solid solutions, an additional difficulty is at once apparent, since the theory predicts that all properties of the solid solution are symmetrical about the 50 atomic% composition, and this is rarely true in practice. This limitation may be avoided by modifying the theory, but at least two parameters are then required to specify the energy of the assembly. A major attraction of the simple central force model (to the mathematician, at least!) is that all properties of the solid solution at fixed temperature, and of the equilibrium diagram, can be described in terms of a single arbitrary parameter.

Let us consider a binary alloy containing A and B atoms. In place of the characteristic nearest neighbour interaction energy, -2Ξ, we now have the energies $-2\Xi_{AA}$, $-2\Xi_{BB}$, and $-2\Xi_{AB}$ representing the binding energies of two A atoms, two B atoms and an A atom and a B atom respectively. The model assumes that Ξ_{AA}, Ξ_{BB}, and Ξ_{AB} are all independent of the surrounding configuration, and this is probably its most serious limitation.

The three potential energy terms are independent of each other, but in the nearest neighbour model, the properties of the solid solution depend only on the combination

$$\Xi = \Xi_{AA}+\Xi_{BB}-2\Xi_{AB}, \tag{23.1}$$

The nature of the quantity Ξ may be seen by a hypothetical process in which we interchange an A atom and a B atom on any two sites in the crystal. If the number of $A-A$ contacts increases or decreases by X, it follows that whatever the arrangement around the two sites, the number of $B-B$ contacts also changes by X, and the number of $A-B$ contacts by $-2X$. The change in energy is thus $-X(\Xi_{AA}+\Xi_{BB}-2\Xi_{AB})$.

In an ideal solid solution, the internal energy of the crystal is independent of the atomic arrangement. This condition is satisfied if $\Xi_{AA} = \Xi_{BB} = \Xi_{AB}$, and (less restrictively) if $\Xi = 0$. The condition for the formation of an ideal solid solution with this model is thus that the force between unlike atoms is equal to the average of the forces between two like atoms. When longer-range forces are considered, the corresponding conditions are obvious; in any interchange of A and B atoms within the interior of the crystal, the change in energy must vanish. One way of including longer-range forces is to use a pairwise interaction model with a series of interchange potentials Ξ_1, Ξ_2, etc., defined by

$$\Xi_i = \Xi_{AA,i} + \Xi_{BB,i} - 2\Xi_{AB,i} \quad (23.1a)$$

where each potential is defined at a separation corresponding to the *ith* nearest neighbour distance. Note that although, according to this model, there are three interaction potentials for each distance of separation, it is impossible to imagine an experiment which would enable the individual potential to be measured separately, and only the interchange potentials Ξ_i have operational significance. In any interchange of an A atom and a B atom, the change in energy is now given by the sum $-X_i\Xi_i$, where X_i is the increase in the number of *i*th neighbour A–A or B–B bonds, and $-2X_i$ is the corresponding increase in A–B bonds.

For many solid solutions of close-packed structures, the limitations of the nearest neighbour model lie more in the assumption of central forces and the independence of the Ξ terms on the environment than in the neglect of second and third nearest neighbour interactions. Thus Guggenheim (1952) showed that if the interaction energy varies as r^{-6}, the effect of second nearest neighbours on measurable thermodynamic quantities is negligible. The use of such interactions makes equations unwieldy and only in special circumstances does it lead to any essential improvement in the physical description. We shall therefore confine ourselves to nearest neighbour forces in this section, although there are then obvious difficulties in applying the theory to b.c.c. structures which are not mechanically stable under nearest neighbour forces. It will be necessary to consider higher neighbour interactions in connection with order–disorder changes (Section 26), and some comments will also be made there on their influence in systems exhibiting phase segregation.

Non-ideal solutions, which have $\Xi \neq 0$, may be classified qualitatively by the sign of Ξ. If $\Xi_{AA}+\Xi_{BB}<2\Xi_{AB}$, the attractive forces between like atoms are weaker than those between unlike atoms, and there will be a tendency for each atom to surround itself with as many atoms of the opposite kind as possible. This ordering tendency may produce a superlattice at low temperatures; at high temperatures, it is opposed by the thermal energy, which always tends to produce a random arrangement of high entropy. For Ξ positive, the opposite result is valid, and at low temperatures the solid solution tends to segregate into A-rich and B-rich regions.

When the solution is not ideal, theoretical expressions can most readily be obtained on the assumption that it is "regular". This term was first introduced by Hildebrand to describe a class of solutions having physical properties which vary with composition in a regular manner; the definition of the regular solution has, however, varied considerably amongst different workers. The earliest approach was to define a regular solution as a solution in which the configurational entropy of mixing is still given by eqn. (22.13), even though the heat of mixing is not zero. The atomic arrangement is thus considered to be effectively random, although interchanges of atoms lead to changes in the internal energy. This assumption is obviously roughly justified if the magnitude of Ξ is small. In this sense, a regular solution is one which deviates only slightly from ideal conditions.

Guggenheim (1952) uses a less restricted definition of regular solution, and regards the above definition as a crude or "zeroth order" approximation. This approach has some advantage in emphasizing the relation between the simple theory and the higher approximations of the quasi-chemical theory.

For a simple structure in which each site has a common coordination number z, there will be $\frac{1}{2}Nz$ nearest neighbour bonds in a crystal of N atoms, provided N is sufficiently large for surface effects to be negligible. We write the total number of nearest neighbour $A-B$ contacts as zN_{AB} (i.e. zN_{AB} is the value of $N_{AB}(\mathbf{r})$ for the whole crystal when $\mathbf{r}$ is a nearest neighbour vector).† Similarly, zN_{AA} and zN_{BB} are the total numbers of $A-A$ and $B-B$ contacts in the crystal. By considering the neighbours of the A atoms and the B atoms separately, we find

$$\left.\begin{aligned} &\text{Number of } A-A \text{ contacts} = zN_{AA} = \tfrac{1}{2}z(N_A-N_{AB}),\\ &\text{Number of } B-B \text{ contacts} = zN_{BB} = \tfrac{1}{2}z(N_B-N_{AB}),\\ &\text{Number of } A-B \text{ contacts} = zN_{AB}. \end{aligned}\right\} \tag{23.2}$$

For a random distribution, $P_{AA}(\mathbf{r}) = (1-x)^2$, $P_{BB}(\mathbf{r}) = x^2$, and $P_{AB}(\mathbf{r}) = 2x(1-x)$ for all possible interatomic vectors $\mathbf{r}$. In particular, the numbers of nearest neighbour pairs of the types $A-A$, $B-B$, $A-B$ will be $\frac{1}{2}Nz(1-x)^2$, $\frac{1}{2}Nzx^2$, and $Nzx(1-x)$. The potential energy of the crystal is thus

$$\left.\begin{aligned} U &= -Nz\{(1-x)^2\Xi_{AA}+x^2\Xi_{BB}+2x(1-x)\Xi_{AB}\}\\ &= -Nz\{(1-x)\Xi_{AA}+x\Xi_{BB}-x(1-x)\Xi\}. \end{aligned}\right\} \tag{23.3}$$

The value of U for the components is $-Nz(1-x)\Xi_{AA}-Nzx\Xi_{BB}$, so that we have

$$\Delta_m H = \Delta_m U = Nzx(1-x)\Xi \tag{23.4}$$

as the heat or energy of mixing.‡ This curve has a simple U-shape, with a maximum or minimum at $x = \frac{1}{2}$, according to the sign of Ξ. The configurational entropy of mixing is given by eqn. (22.13), since the atomic arrangement is assumed to be random. If we assume for simplicity that the thermal entropy of mixing is zero, we obtain for the free energy of mixing per atom

$$\Delta_m g = zx(1-x)\Xi+kT\{(1-x)\ln(1-x)+x\ln x\}. \tag{23.5}$$

The form of the curve of $\Delta_m g$ against composition thus depends on the sign of Ξ. The relation is shown in Fig. 6.7 for Ξ negative (tendency to form superlattices), Ξ zero (ideal solutions) and Ξ positive (tendency for phase segregation). The latter case corresponds to Fig. 6.1, and the compositions of the two phases in equilibrium are given by eqn. (22.5). In the present model, eqn. (23.5) is symmetrical about $x = \frac{1}{2}$, and the common tangent to the two minima in the curve has zero slope, so that

$$(\partial\Delta_m g/\partial x)_{x_1} = (\partial\Delta_m g/\partial x)_{x_2} = 0. \tag{23.6}$$

† It is more convenient to write the number of $A-B$ contacts az zN_{AB} than as N_{AB}; this means that $P_{AB}(\mathbf{r}) = 2N_{AB}/N$ for nearest neighbour contacts, and simplifies the algebra of the quasi-chemical theory.

‡ In condensed phases, we need not distinguish between ΔU and ΔH, or between ΔF and ΔG.

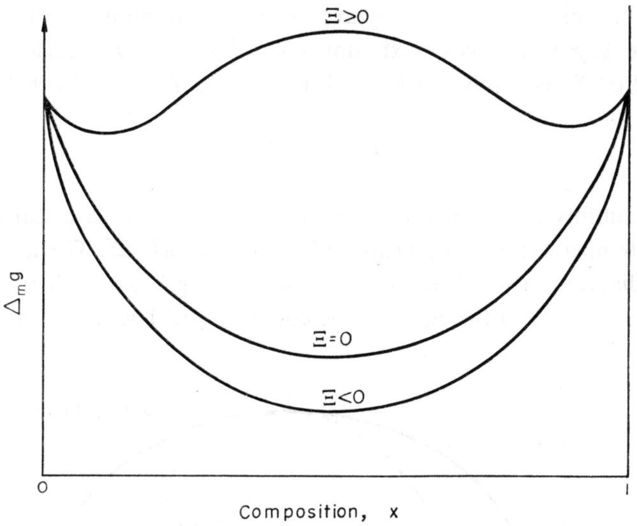

FIG. 6.7. Free energy of mixing vs. composition curves according to (23.5).

From eqns. (23.5) and (23.6),

$$\left(\frac{1}{2x-1}\right)\ln\left(\frac{x}{1-x}\right) = \frac{z\Xi}{kT}. \tag{23.7}$$

At temperatures where the curve of $\Delta_m g$ has the form of Fig. 6.1, there are three solutions to (23.7). The solution at $x = \frac{1}{2}$ gives a maximum value to $\Delta_m g$; the other two solutions, x_1 and x_2, give minima, and are symmetrical about $x = \frac{1}{2}$. For dilute solutions $(x \ll 1)$, we write as an approximation $(1-x) \approx (1-2x) \approx 1$, and the variation of the solubility limit x_1 is then

$$x_1 = \exp-(z\Xi/kT), \tag{23.8}$$

so that the solubility approximates to an exponential curve if the solubility is small, i.e. if $z\Xi/kT$ is large.

The similarity of eqn. (23.7) to Freedman and Nowick's expression eqn. (22.27) should be noted, but the underlying physical assumptions are different. Equation (22.27) applies to any solubility limit if the solubility is small, and includes the effects of thermal entropy. Equation (23.7) applies only to the artificial regular solution model, but within this limitation it is valid for all values of x. The limiting forms of the two equations in very dilute solutions (eqns. (22.23) and (23.8)) differ insofar as a thermal entropy contribution from each solute atom is present in a real solution, but not in the model.

Equation (23.7) is sometimes written in the form

$$(2/(2x-1))\tanh^{-1}(2x-1) = z\Xi/kT. \tag{23.9}$$

As the temperature is raised, the two roots x_1 and x_2 both approach the value $x = \frac{1}{2}$ until, finally, they coincide at this value, and the curve of $\Delta_m g$ assumes a simple U-form. At all temperatures above this critical temperature, there is a continuous solid solution from pure A to pure B. The $\Delta_m g$ curve has a maximum at $x = \frac{1}{2}$ for $T < T_c$, and a minimum at $x = \frac{1}{2}$ for $T > T_c$, so that T_c is given by the condition $(\partial^2 \Delta_m g / \partial x^2)_{x=1/2} = 0$. Using eqn. (23.5),

$$T_c = z\Xi/2\boldsymbol{k}. \tag{23.10}$$

The complete solubility curve for the hypothetical assembly we are considering is plotted in Fig. 6.8, the temperatures being measured in units of $\boldsymbol{k}T/z\Xi$. The assumptions lead to an equilibrium diagram in which there is a solubility gap below T_c, the boundaries of the gap being symmetrical about the equi-atomic composition. The solubilities are zero at 0 K,

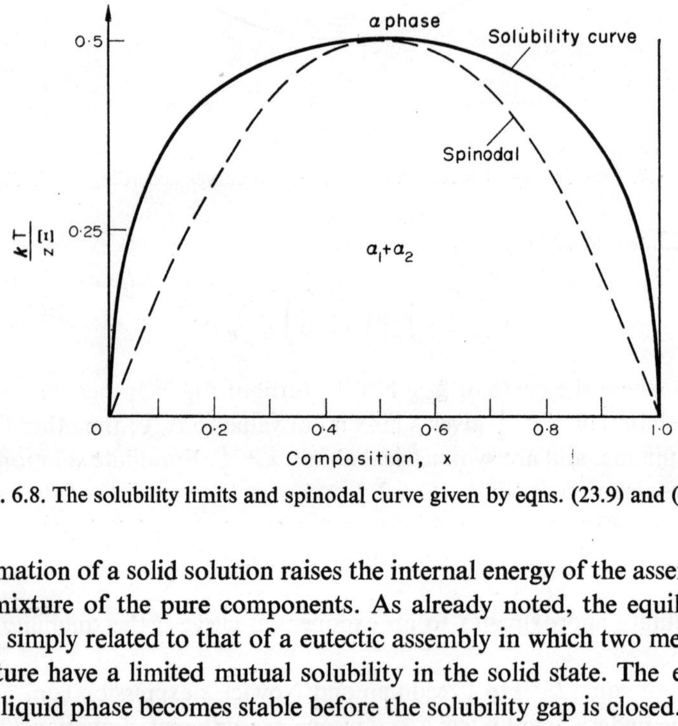

FIG. 6.8. The solubility limits and spinodal curve given by eqns. (23.9) and (23.11).

since the formation of a solid solution raises the internal energy of the assembly above that of a phase mixture of the pure components. As already noted, the equilibrium diagram of Fig. 6.8 is simply related to that of a eutectic assembly in which two metals of the same crystal structure have a limited mutual solubility in the solid state. The eutectic reaction occurs if the liquid phase becomes stable before the solubility gap is closed.

In addition to the boundaries of the two-phase region we are also interested in the chemical spinodal [eqn. (22.42)] which is given by

$$x(1-x) = \boldsymbol{k}T/2z\Xi \tag{23.11}$$

in a regular solution. At all temperatures below $T_c = z\Xi/2\boldsymbol{k}$, the roots of this equation lie inside the roots x_1 and x_2 of eqn. (23.9). The two curves, which touch at $T = 0$ K and $T = T_c$, are shown in Fig. 6.8.

The chemical potential of A in the regular solution is given by

$$g_A - g_A^0 = z\Xi x^2 + \boldsymbol{k}T \ln(1-x)$$

from eqn. (23.5). The activity coefficient is thus

$$\gamma_A = \exp(z\Xi x^2/kT), \qquad (23.12)$$

and we have, similarly,

$$\gamma_B = \exp\{z\Xi(1-x)^2/kT\}.$$

24. THE QUASI-CHEMICAL THEORY: OTHER STATISTICAL THEORIES

We now consider the possibility of removing some of the very restrictive assumptions used in the zeroth order approximation of the treatment of regular solutions. The simple theory above is incorrect because of the neglect of thermal entropy factors, and in the assumption of random atomic arrangement. Both these factors may be important, but it is difficult to devise a model which will include them together in a satisfactory manner. In the quasi-chemical theory, which we shall describe first, emphasis is placed on improving the calculation of the configurational entropy, and the properties of the solution are again expressed in terms of the single parameter Ξ. In the usual application, Ξ is temperature independent, as assumed above, and the only contribution to $\Delta_e s$ comes from configurational terms. Thermal entropy terms can be formally included if Ξ is allowed to vary with temperature.

In a real solution, it is physically obvious that the atom distribution will be random only at very high temperatures, and the configurational entropy of mixing must therefore be a function of temperature. Since a random distribution gives the maximum number of distinguishable arrangements which the assembly can possess, a more accurate calculation of the configurational entropy term must lead to a negative excess entropy of mixing if the thermal entropy of mixing is ignored. Consider the assembly in a particular macroscopic configuration which we may specify by the number of nearest neighbour $A-B$ pairs zN_{AB}. To find the equilibrium state, we have to calculate the number of microscopic states leading to this .configuration, and the energy of the configuration. We can then take the sum over all values of N_{AB}, and thus obtain the partition function.

The total energy of the assembly is $U = -z(N_A\Xi_{AA}+N_B\Xi_{BB}-N_{AB}\Xi)$ (using eqn. (23.2)), and the heat of mixing is thus

$$\Delta_m H = \Delta_m U = z\overline{N_{AB}}\Xi, \qquad (24.1)$$

where $\overline{N_{AB}}$ is the equilibrium value of N_{AB}. Denoting the total number of arrangements for given N_{AB} by $\Omega(N_A, N_B, N_{AB})$, we have the partition function

$$Q = \sum_{N_{AB}} \Omega(N_A, N_B, N_{AB}) \exp(-U/kT). \qquad (24.2)$$

This partition function has not been evaluated exactly for a three-dimensional lattice. An exact solution in two dimensions was obtained by Onsager (1944) using a complex mathematical method, and his results have been extended by other workers using simpler methods.

In three dimensions, the value of $\Omega(N_A, N_B, N_{AB})$ may be evaluated if we assume as an approximation that the various kinds of pair do not interfere with each other, and may be treated as separate entities. We note first of all that

$$\Sigma\Omega(N_A, N_B, N_{AB}) = N!/(N_A)!\,(N_B)! \tag{24.3}$$

since this is merely the expression for the total number of arrangements of the assembly. We are going to be concerned with the logarithm of the function Ω, and we may then replace the sum on the left of this expression by the maximum term; this procedure is very commonly used in statistical thermodynamics. Suppose the maximum value of $\Omega(N_A, N_B, N_{AB})$ is obtained for some particular value N^{θ}_{AB}. We then have

$$\Omega(N_A, N_B, N^{\theta}_{AB}) = N!/(N_A)!\,(N_B)!.$$

The assumption that the total number of arrangements for given N_{AB} is obtained by treating the various pairs as independent entities requires that $\Omega(N_A, N_B, N_{AB})$ is proportional to

$$\frac{(\frac{1}{2}zN)!}{\{\frac{1}{2}z(N_A-N_{AB})\}!\,\{\frac{1}{2}z(N_B-N_{AB})\}!\,(\frac{1}{2}zN_{AB})!\,(\frac{1}{2}zN_{BA})!}$$

in which we have allowed for the orientation of the sites, which effectively distinguishes an $A-B$ pair from a $B-A$ pair. The number of arrangements is only proportional to the above expression, not equal to it, since the total number of arrangements must satisfy (24.3). We can achieve this by introducing a normalization factor, so that from (24.3) we find

$$\Omega(N_A, N_B, N_{AB}) = \frac{N!}{N_A!\,N_B!}\,\frac{\{\frac{1}{2}z(N_A-N^{\theta}_{AB}\}!\,\{\frac{1}{2}z(N_B-N^{\theta}_{AB})\}!\,\{\frac{1}{2}zN^{\theta}_{AB}\}!\,\{\frac{1}{2}zN^{\theta}_{AB}\}!}{\{\frac{1}{2}z(N_A-N_{AB})\}!\,\{\frac{1}{2}z(N_B-N_{AB})\}!\,\{\frac{1}{2}zN_{AB}\}!\,\{\frac{1}{2}zN_{AB}\}!}. \tag{24.4}$$

Also, by differentiating expression (24.4) with respect to N_{AB}, we obtain the maximum value of N_{AB} which is given by

$$0 = \partial\Omega(N_A, N_B, N_{AB})/\partial N_{AB} = \partial \ln \Omega(N_A, N_B, N_{AB})/\partial N_{AB}.$$

On taking logarithms and using Stirling's theorem, this reduces to

$$N^{\theta}_{AB} = N_A N_B/N = Nx(1-x), \tag{24.5}$$

which is the value of N_{AB} for a completely random arrangement, as anticipated.

We can now rewrite eqn. (24.2) as

$$Q = \Sigma\Omega(N_A, N_B, N_{AB})\exp\{-z(N_A\Xi_{AA}+N_B\Xi_{BB}+N_{AB}\Xi)/kT\}.$$

and we now have an explicit value for Ω. Once again we replace the sum by its maximum term; this corresponds to N_{AB} having its equilibrium value $\overline{N_{AB}}$. Thus

$$Q = \Omega(N_A, N_B, \overline{N_{AB}})\exp\{-z(N_A\Xi_{AA}+N_B\Xi_{BB}+\overline{N_{AB}}\Xi)/kT\} \tag{24.6}$$

and $\partial \ln Q/\partial N_{AB} = 0$ for $N_{AB} = \overline{N_{AB}}$. We have, therefore, to substitute eqn. (24.4) into (24.6) and differentiate in order to find $\overline{N_{AB}}$. Proceeding as we did for N^{θ}_{AB}, we find that

$$(\overline{N_{AB}})^2 = (N_A-\overline{N_{AB}})\,(N_B-\overline{N_{AB}})\exp(-2\Xi/kT). \tag{24.7}$$

This is the basic formula of the present approximation, and was originally derived by Guggenheim by a different method. He treated the interchange of an A atom and a B atom as a chemical process in which $A-A$ bonds and $B-B$ bonds "react" to form $A-B$ bonds, and vice versa:

$$AA+BB = 2AB.$$

If the law of mass action is applied to this symbolic equation, we obtain the result (24.7). For this reason, the method is often called the quasi-chemical theory. Guggenheim has tlso shown it is equivalent to the method which Bethe developed for order–disorder reacaions (see p. 219).

Equation (24.7) is a rather unpleasant quadratic. The algebra is simplified by writing the solution in the form

$$\overline{N_{AB}} = 2N_{AB}^{\theta}/(\beta+1) = 2Nx(1-x)/(\beta+1). \tag{24.8}$$

Substituting this into (24.7) shows that β is the positive root of the equation

$$\beta^2-(1-2x) = 4x(1-x)\exp(2\Xi/kT). \tag{24.9}$$

Equation (24.8) shows that $\beta = 1$ in the zeroth approximation, and $\beta > 1$ in the present treatment. It follows from (24.1) and (24.8) that the heat of mixing per atom is

$$\Delta_m h = \Delta_m u = 2zx(1-x)\Xi/(\beta+1). \tag{24.10}$$

We can now deduce the other thermodynamic properties of the assembly, using the relation

$$G = F = -kT\ln Q = -kT\ln \Omega(N_A, N_B, \overline{N_{AB}})+U.$$

From eqn. (24.6) it follows that $\partial F/\partial \overline{N_{AB}} = 0$, and we already know that $\partial \ln \Omega/\partial N_{AB}^{\theta} = 0$. Using these relations we obtain the chemical potential

$$g_A-g_A^0 = kT\left[\ln x_A+\tfrac{1}{2}z\ln\{(N_A-\overline{N_{AB}})/(N_A-N_{AB}^{\theta})\}\right] \tag{24.11}$$

and hence the activity coefficient

$$\gamma_A = \{(N_A-\overline{N_{AB}})/(N_A-N_{AB}^{\theta})\}^{z/2} \tag{24.12}$$

with similar expressions for g_B and γ_B.

We may also write:

$$\Delta_m g = x_A(g_A-g_A^0)+x_B(g_B-g_B^0) = kT\big[(1-x)\ln(1-x)+x\ln x$$
$$+\tfrac{1}{2}z\big\{x_A\ln\big((N_A-\overline{N_{AB}})/(N_A-N_{AB}^{\theta})\big)+x_B\ln\big((N_B-\overline{N_{AB}})/(N_B-N_{AB}^{\theta})\big)\big\}\big]. \tag{24.13}$$

This equation is simple and symmetrical. The quantities $\overline{N_{AB}}$ and N_{AB}^{θ} are defined in eqn. (24.7) and (24.5); as already noted, they have simple physical interpretations. N_{AB}^{θ} is the number of $A-B$ bonds for a random mixture, and $\overline{N_{AB}}$ is the mean number of $A-B$ bonds for the actual solution. If we wish to eliminate the numbers N_A, N_B, $\overline{N_{AB}}$, we may write the equation in the form

$$\Delta_m g = kT\Big[(1-x)\ln(1-x)+x\ln x$$
$$+\frac{1}{2}z\Big\{(1-x)\ln\frac{\beta+1-2x}{(1-x)(\beta+1)}+x\ln\frac{\beta-1+2x}{x(\beta+1)}\Big\}\Big]. \qquad (24.14)$$

From eqns. (24.10) and (24.14) an expression for the excess entropy of mixing may be obtained.

To find the equilibrium state, we must equate $\partial\Delta_m g/\partial x = 0$. As in the zeroth approximation, we find $x = \frac{1}{2}$ is always a solution, but that this gives a maximum value to $\Delta_m g$ below some critical temperature, and a minimum value above this temperature. Below the critical temperature, there are two other solutions symmetrically disposed about $x = \frac{1}{2}$.

The differentiation of eqn. (24.14) involves some rather lengthy algebra because β is a function of x (eqn. (24.9). Equating the first differential coefficient to zero, we obtain

$$\ln\frac{x}{1-x}+\frac{z}{2}\ln\frac{(\beta-1+2x)(1-x)}{(\beta+1-2x)x} = 0$$

or

$$\left\{\frac{x}{1-x}\right\}^{(z-2)/z} = \frac{\beta-1+2x}{\beta+1-2x} = C_5. \qquad (24.15)$$

This is the equivalent of eqn. (23.7) of the zeroth approximation, and the roots give the equilibrium compositions of the coexisting phases at any temperature. The algebra involved in differentiating may be avoided if we note that for the equilibrium compositions $(1-x)\gamma_A = x\gamma_B$. The use of eqn. (24.12) then leads direct to (24.15). It is possible to eliminate β from (24.15) and thus obtain a more satisfactory expression for the solubility limits. By manipulation of (24.15)

$$\beta^2-(1-2x)^2 = 4C_5(1-2x)^2/(1-C_5)^2$$

and eliminating β using (24.9)

$$\exp(2\Xi/kT) = C_5(1-2x)^2/x(1-x)(1-C_5)^2$$

or

$$\exp(\Xi/kT) = \frac{1-x/(1-x)}{[x/(1-x)]^{1/z}-[x/1-x)]^{1-1/z}}. \qquad (24.16)$$

The equilibrium conditions are thus given in terms of the ratio $x/(1-x)$. The solutions are symmetrical; if one root is x_1, the other is $1-x_1$. This is a necessary consequence of the nearest neighbour model.

At the critical temperature, the two roots coincide at $x = \frac{1}{2}$. Substituting $x = \frac{1}{2}$ into eqn. (24.16) gives an indeterminate equation, so we write $x/(1-x) = 1+\delta$, and by expanding in terms of δ and letting $\delta \to 0$,

$$\exp(\Xi/kT_c) = z/(z-2). \qquad (24.17)$$

For a f.c.c. structure, $z = 12$ and

$$T_c = \Xi/\boldsymbol{k} \ln 1{\cdot}2 = z\Xi/2{\cdot}19\boldsymbol{k},$$

whilst for a b.c.c. structure

$$T_c = \Xi/\boldsymbol{k} \ln 1{\cdot}33 = z\Xi/2{\cdot}3\boldsymbol{k}.$$

These values may be compared with eqn. (23.10) for the zeroth approximation. If we let $z \to \infty$ in eqn. (24.17), we re-obtain the value $T_c = z\Xi/2\boldsymbol{k}$. Guggenheim has emphasized that this is a general result, and any formula of the quasi-chemical method in the first approximation can be converted into the corresponding zeroth order approximation by letting $z \to \infty$.

Equation (24.13) is a much more satisfactory expression for the free energy of a solid solution than is (23.5), especially as it includes a temperature-dependent energy term. In practical applications, however, eqns. (23.7) and (23.10) are usually used in preference to (24.16) and (24.17) because of their greater simplicity. In Fig. 6.9, the predicted solubility

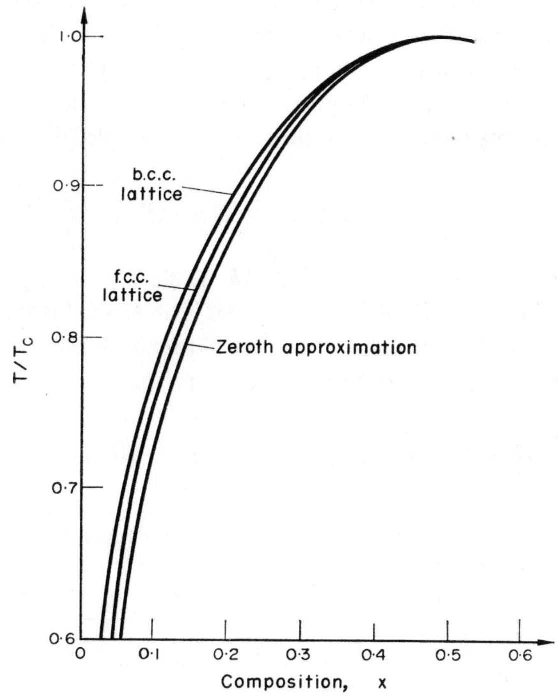

FIG. 6.9. Solubility limits according to the zeroth and first approximations of the quasi-chemical theory.

imits for the first and zeroth approximations are compared by plotting them as a function of T/T_c. The curve for the first approximation is drawn for both the f.c.c. and the b.c.c. structure. We have assumed identical values of T_c rather than of Ξ, since this most clearly reveals the shape of the curves. In practice, T_c is measured and Ξ is the disposable parameter.

The basic assumption of the method of calculation used in this section is the non-interference of atom pairs. This is a very artificial hypothesis since each atom belongs simultaneously to several pairs. Higher approximations may be obtained by considering the non-interference of interacting groups of larger numbers of sites. All such treatments may be classed as examples of the "cluster variation" method (Kurata, *et al.*, 1953), and the three-dimensional partition function of the Ising model is found approximately by a procedure similar to that outlined above for pairs of sites. A small group of lattice sites is chosen as the basic unit, the energy of this cluster is computed exactly for each way in which the sites may be occupied, and the number of microscopic configurations of the crystal is calculated approximately. The inconsistencies in the calculation of the number of configurations decrease as the size of the basic cluster increases, but the mathematics become correspondingly more complex. Most calculations have been made with triangular clusters of three sites, or with tetrahedral clusters of four sites; in some problems, involving superlattice formation, the choice of the basic cluster is governed by the crystallography in these higher approximations.

We shall not consider further the mathematical development of the Ising model; although the higher approximations lead to better solutions, they are mainly of value in order–disorder problems, where the use of pair interactions sometimes leads to difficulties (see p. 222). The results obtained do not generally differ greatly from those of the zeroth and first approximations; if systems of three sites are considered, for example, the critical temperature is given by

$$T_c = \Xi/\boldsymbol{k} \ln\{(z+1)/(z-3)\}^{1/2},$$

and for a f.c.c. structure, this gives $T_c = z\Xi/2{\cdot}21\boldsymbol{k}$. Table V gives some comparisons of the predictions of the various cluster approximations; it is abstracted from more complete data given by Guggenheim (1952). The value of z in the cluster approximations is taken to be 12. The values of the excess entropy are negative and very small.

TABLE V. SUMMARY OF PREDICTIONS OF QUASI-CHEMICAL THEORY FOR THERMODYNAMIC FUNCTIONS AT $T = T_c$

x	Function	Approximation			
		Zeroth	Pairs	Triplets	Quadruplets
0·1		0·15	0·13	0·13	0·13
0·3	$-\Delta_m g/\boldsymbol{k}T$	0·19	0·17	0·17	0·16
0·5		0·19	0·17	0·17	0·17
0·1		0·18	0·19	0·19	0·19
0·3	$\Delta_m h/\boldsymbol{k}T$	0·42	0·42	0·42	0·42
0·5		0·5000	0·50	0·50	0·50
0·1		0	0·014	0·014	0·015
0·3	$-\Delta_e s/\boldsymbol{k}$	0	0·018	0·021	0·023
0·5		0	0·025	0·028	0·033
—	$z\Xi/\boldsymbol{k}T_c$	2	2·19	2·21	2·23

We have already noted that exact solutions of the Ising model have been obtained in two dimensions. For a square lattice, the closed solution is

$$\sinh(z\Xi/4\boldsymbol{k}T_c) = 1 \tag{24.18}$$

which gives

$$T_c = z\Xi/3{\cdot}526\boldsymbol{k}.$$

This formula has been used in Chapter 5, Section 19. Exact solutions in three dimensions have not been found in closed form, but quite early in the development of the theory Kirkwood pointed out that an exact solution for the partition function may be written as a power series in $\Xi/\boldsymbol{k}T$. Unfortunately, this series converges very slowly, and since we are more interested in obtaining a reasonable physical model than in an expression for T_c, we shall not investigate this further.

In all the above derivations we have assumed that Ξ is a constant energy. The non-dependence of Ξ on the surrounding configuration of atoms is a fundamental postulate of the quasi-chemical theory, but Guggenheim (1948) has pointed out that there is no need to assume that Ξ is independent of temperature. An alternative approach is to regard Ξ as a quantity defined by eqn. (24.7). If we do this, 2Ξ is the free energy increase when an $A-A$ and a $B-B$ pair are destroyed to produce two $A-B$ pairs. This co-operative free energy may now include thermal entropy terms, insofar as these are affected by the interchange, and the restriction to solutions obeying the Kopp–Neumann rule has been removed. Since Ξ is now a function of temperature, we may write formally

$$\Xi' = \Xi - T(\mathrm{d}\Xi/\mathrm{d}T), \tag{24.19}$$

where Ξ' corresponds to our old definition of Ξ as an interchange energy. The equations for the thermodynamic quantities at any one temperature still contain only one parameter, which may be fixed empirically. On the other hand, the equations for the variation of solubility with temperature are no longer valid, and we cannot deduce Ξ or Ξ' from experimental data of this kind unless some assumption is made about $\mathrm{d}\Xi'/\mathrm{d}T$. Guggenheim suggested that the simplest assumption, Ξ' independent of temperature, may be adequate.

We now turn to a brief comparison of the predictions of the quasi-chemical theory with experiment. The comparison may be made in various ways; for example, values of Ξ can be determined from independently measured quantities and examined for consistency. If the theory is applicable, Ξ should also be independent of composition, and the composition dependence of experimentally measured activity coefficients and pair probability functions will show whether or not this is so. Detailed tests of this kind are possible in some assemblies, but are not needed in others, where the signs or the magnitudes of the excess thermodynamic quantities are sufficient to establish that the theory is unsatisfactory.

We note first of all that the formation of a metallic solid solution from its constituents produces fairly small changes in the thermodynamic functions, since the atomic interactions are not greatly changed. Thus, the experimentally measured excess functions are small; the heat of mixing per atom, $\Delta_m g$, usually lies within the limits $\pm\boldsymbol{k}T$, and the excess entropy of mixing per atom, $\Delta_e s$, usually lies within the limits $\pm\boldsymbol{k}$. Nevertheless, the magnitude of of $\Delta_e s$ provides one of the most convenient tests of the theory, and especially of the assumption that the thermal entropy may be neglected. According to the quasi-chemical theory, $\Delta_e s$ is always negative, and it is readily seen that $|\Delta_e s/\boldsymbol{k}|$ is unlikely to exceed 0·05 (see Table V). Data collected by Oriani (1959) for a large number of liquid and solid solutions (mainly at $x = \frac{1}{2}$) show that this is rarely true. The experimental values range from $\Delta_e s/\boldsymbol{k} \approx -1$ for liquid $Mg_{1/2}\ Bi_{1/2}$ to $\Delta_e s/\boldsymbol{k} \approx +0{\cdot}7$ for solid $Au_{1/2}\ Ni_{1/2}$.

A related test of the assumption about thermal entropy is to consider the equilibrium solubility curves for dilute solid solutions, instead of measured thermodynamic functions. This has been done by Freedman and Nowick (1958) using eqn. (22.27). The partial excess entropies obtained in this way were all positive (with one doubtful exception) and of appreciable magnitude; this is true for assemblies giving $\Delta_e s_B$ as well as for assemblies in which $\Delta_f s_B$ (related to $\Delta_e s_B$ by eqn. (22.26)) is obtained from the equilibrium diagram.

It follows from these results that the assumption of zero thermal entropy of mixing is a serious defect of the theory in most assemblies, and is more important than the error in the configurational term. This means that a theory with Ξ constant is not good enough, and the predictions of the model in respect of temperature variation are inadequate. However, if Ξ is allowed to vary with temperature, the quasi-chemical theory may still be a reasonable representation of the situation in some alloys at constant temperature, and it is, indeed, usually more successful in dealing with the heat of mixing than with the excess entropy.

We end this section by referring briefly to the attempts which have been made to develop better statistical models of solutions. The first of these is the theory of conformal solutions (Longuet-Higgins, 1951), which is a generalization of the pair interaction model, but is free from any structural assumptions. This makes it particularly suitable for dealing with liquid solutions.† A conformal solution is defined by certain assumptions about the form of the potential energy of the assembly, expressed as the sum of a number of bimolecular terms. These assumptions, which derive essentially from a law of corresponding states, are that the interaction energies between $A-A$, $B-B$, and $A-B$ pairs depend only on the nature and separation of the two atoms or molecules considered, and have the same functional dependence on the separation for all pairs. Thus the interaction energy of an $A-A$ pair can be expressed in the form $\Xi_{AA}(r) = f_{AA}\Xi_0(g_{AA}r)$, with equivalent expressions for $\Xi_{BB}(r)$ and $\Xi_{AB}(r)$. The factors f_{AA}, f_{AB}, g_{AA}, etc., are all constants, and $-\Xi_0(r)$ is a universal function giving the potential energy of two atoms or molecules in a reference state as a function of their separation r. If the reference state is taken to be pure A, for example, $f_{AA} = g_{AA} = 1$.

In developing the theory, it is necessary to assume that $1-f_{AA}$, $1-g_{AA}$ are both small, so that all the forces have approximately the same magnitude, and it is also assumed that $g_{AB} = \frac{1}{2}(g_{AA}+g_{BB})$. The advantage of this approach is that the pair interaction energies vary with composition, since the mean separation is a function of composition, and the difficulties inherent in the usual quasi-chemical assumption of a temperature-independent Ξ are also avoided. The results of the theory are obtained in a series of successive approximations. The averaging procedure in the first approximation leads to a random distribution

† The quasi-chemical method is applicable in principle to liquids, and is often used for liquid solutions, but it does carry the implication of a "lattice model" of the liquid.

of the components, and is equivalent to the zeroth approximation of the theory of regular solutions; the higher approximations lead to statistical terms which are very difficult to evaluate (Brown and Longuet-Higgins, 1951).

Prigogine *et al.* (1957) have developed a theory of solutions which we may regard as based on the theory of conformal solutions and on the cell model. They use average interactions between a molecule or atom and its neighbours, and assume composition-dependent potentials of the Lennard-Jones (6−12) type. The interaction potentials are thus written $-\Xi_{AA}(r) = -\Xi^*_{AA}\varphi(r/r^*_{AA})$, etc., where Ξ^*_{AA}, etc., are characteristic energies, and r^*_{AA}, r^*_{AB}, etc., are characteristic lengths. The function φ is a universal function specifying the type of interaction assumed; for Lennard-Jones forces, $\varphi(x) = x^{-12}-2x^{-6}$. The average potential model assumes random atomic distribution, and the expressions for the thermodynamic functions are obtained in terms of three parameters which are defined by the equations

$$\left.\begin{aligned} \varrho &= (r^*_{BB}/r^*_{AA})-1, \\ \delta &= (\Xi^*_{BB}/\Xi^*_{AA})-1, \\ \theta &= (\Xi^*_{AA}+\Xi^*_{BB}-2\Xi^*_{AB})/\Xi^*_{AA}. \end{aligned}\right\} \qquad (24.20)$$

These parameters are essentially "reduced" quantities, and the theory gives expressions for $\Delta_m h$ and $\Delta_e s$ which can be simplified considerably when it is justifiable to neglect all products of ϱ, δ, and θ and all powers of higher order than δ, θ, and ϱ^2 respectively. Under these conditions, which obviously correspond to those mentioned above in the theory of conformal solutions, we have

$$\left.\begin{aligned} \Delta_m h &\approx 1{\cdot}435(\theta+4{\cdot}5\varrho^2)z\Xi^*_{AA}x(1-x), \\ \Delta_e s &\approx 78{\cdot}2k\varrho^2 x(1-x), \end{aligned}\right\} \qquad (24.21)$$

where the factor 1·435 is a result of using Lennard-Jones forces for interactions between atoms which are not nearest neighbours; this factor is unity if only nearest neighbour interactions are considered. The results then reduce to the zeroth approximation of the quasi-chemical theory when $\varrho = 0$, and it will be seen that, according to this model, the positive excess entropy which is usually observed is attributable to the size disparity measured by ϱ.

Shimoji and Niwa (1957) have modified this theory by using generalized interatomic potentials of the Morse type in place of the 6–12 potential. They obtain an expression which is equivalent to Prigogine's equation for $\Delta_m h$ if the Morse potential is chosen to correspond to Lennard-Jones forces, but they suggest that for liquid metal solutions, a more appropriate choice of potential gives

$$\Delta_m h \approx (\theta+0{\cdot}5\varrho^2)\, z\Xi^*_{AA}x(1-x). \qquad (24.22)$$

The predictions of the average potential model have been compared with measured thermodynamic parameters by Oriani (1959). The procedure adopted was to assume $r^*_{AB} = \frac{1}{2}(r^*_{AA}+r^*_{BB})$ and to evaluate θ from the measured $\Delta_m g$. The values of $\Delta_m h$ and $\Delta_e s$ were then calculated. The agreement found is rather poor, especially as the theory is only being used to decide how the excess free energy of mixing is distributed between the heat and excess entropy of mixing. The degree of agreement between the experimental and theoretical $\Delta_m h$ is, moreover, little affected by whether or not the size disparity is taken into account.

The provisional conclusion is thus that the improved statistical theories are little better than the quasi-chemical theory in explaining the properties of solid metallic solutions. The main reason for this situation is that already emphasized on p. 109. Metals and metallic solutions do not obey a law of corresponding states, and the assumption of central force interactions of any type is a poor approximation.

25. MISFIT ENERGY IN SOLID SOLUTIONS: SUB-REGULAR SOLUTIONS

We have emphasized the restriction of the quasi-chemical theory to solutions in which the constituents have nearly equal atomic volumes. The failure of the theory in many cases in which large heats of mixing are associated with a considerable difference in the radii of the component atoms is frequently ascribed to "strain energy". This term is seldom precisely defined, but it is usually implied that the energy is a mechanical energy of long-range character (i.e. distributed throughout the crystal), and is somehow separable from any other (chemical) terms in the energy of mixing.

Theories of solid solution which utilize the notion of strain energy attempt to calculate the heat of mixing, or rather the partial excess free energy of mixing, from more fundamental parameters such as elastic constants. All treatments use the approximation of an elastic continuum to represent the solid, and then consider the mechanical strain energy when a solute atom in inserted into this continuum. The model used is one in which a sphere of specified radius is forced into a hole of smaller or larger radius, the surfaces welded together, and the body allowed to relax in its self-stressed state. Each solute atom is thus regarded as a centre of dilatation.

This model seems quite specific, but in fact the calculation has been applied in various ways. The most realistic physical assumption would seem to be to regard the atomic displacements as Hookean outside some limiting radius r_0 from the centre of dilatation, and thus to calculate the strain energy in the region of crystal for which $r > r_0$. The total (free) energy of the solute atom, apart from the configurational term, would then be obtained by adding a "core energy" to the elastic energy; this is the approach which is normally used in dislocation theory (Chapter 7). A serious difficulty in the misfitting sphere model is that the elastic energy varies inversely as the cube of r_0, and so is very sensitive to the choice of this core radius. Different authors have put r_0 equal to the radius of the solute atom itself, the radius of a cluster of twelve nearest neighbours, and the radius of the third nearest neighbour shell.

We should emphasize here that providing the processes envisaged above are isothermal, the calculation gives an estimate of the free energy† change associated with the introduction of the solute atom in a particular place, and not the internal energy or the enthalpy. Thus for an isolated solute atom, we are attempting to calculate $\Delta_e g_B$, and differentiation with

† Strictly, the elastic energy gives the change in the Helmholtz free energy, and the Gibbs free energy is obtained by adding the (negligible) work done against external forces in changing the volume of the assembly. The usual configurational term has, of course, to be added to the elastic energy to give the total free energy change associated with the solute atom.

respect to temperature will give $\Delta_e s_B$. This has been mentioned already in connection with Zener's conclusion about the sign of the excess (thermal) entropy of a point defect (p. 127). In some papers it is wrongly implied that the elastic energy is an internal energy.

The usual way of avoiding, or rather concealing, the difficulty of defining r_0 is to apply elastic theory both to the surrounding matrix and to the solute atom and its immediate neighbours. Two energy terms are calculated, one being the work necessary to expand (contract) a hole in the matrix from an initial radius to a new final radius, and the other the work needed to compress (expand) a sphere from its initial radius to the final radius of the hole. The sum of these two contributions is then minimized to give the equilibrium final radius. The initial radii are commonly taken as those of a solvent and solute atom respectively, although a larger cluster may be used with equal justification.

Oriani (1956) has pointed out that this calculation is very unsatisfactory. The minimization procedure of the mechanical model has no real significance in the problem of determining the equilibrium "radius" of the solute atom. It is easy to see that the primary role in determining the distances of the first few coordination shells must be played by electronic interactions (including effects such as charge transfer and screening) with the immediate neighbours of the atom concerned. In a liquid solution, these interactions must entirely determine the "size" of the atom, since there are no long-range elastic stresses. For solid solutions, the same interactions must be present, and will still be the main factor determining the local configuration, but there will be a perturbation resulting from the imposition of lattice periodicity.

In a careful discussion of the significance of misfit energy, Oriani concludes that the term has no operational significance in liquid solutions, and any theoretical concept which might be invoked relates only to hypothetical situations which cannot be obtained physically. (For example, a theoretical definition of misfit energy might be the difference in energy of a real liquid solution and a hypothetical solution in which the components had the same ionic radii; this clearly has no operational significance and the quantity could never be measured.) The idea of size effects and misfit energy should thus be used very cautiously, if at all, for liquid solutions. For solid solutions it is possible to give an operational definition of misfit energy based on the changes produced by the imposition of long-range lattice periodicity. Only that part of the misfit energy outside some (unknown) radius r_0 is susceptible to calculation by elastic theory, and might thus fairly be called strain energy.

These are severe criticisms of the elastic model, and we shall see later that they are to some extent supported by the experimental evidence. The model is capable of giving heats of mixing of the correct magnitude, and entropies of mixing of the correct sign and magnitude, but more detailed tests seem to show that $\Delta_e g$ cannot really be ascribed to the elastic strain energy in the matrix. Nevertheless, we shall now give the elastic calculation in some detail. The reasons for this are twofold. First, similar calculations are still frequently used in many applications to both solid solutions and lattice defects, and as Eshelby (1956) has remarked, the limitations of this model are perhaps more immediately obvious than the equally serious defects of other approximations used in solid state theory. Second, and more important, we shall require the identical calculation in a later chapter when we consider the energy of a precipitate growing in a matrix. The minimization of the mechanical strain energy then has much greater validity, and in fact the procedure was first used for this problem by Mott and Nabarro (1940).

In the present application, we consider a solid solution formed from two components of atomic volumes v_A^0 and v_B^0 in the respective pure states. Consider first the effect of introducing a single B atom into a lattice of pure A. According to the misfitting sphere model, we treat the B atom as an elastic sphere inserted into a hole of volume v_A^0 in an isotropic elastic continuum.

Let the atomic radii of the atoms in the pure components be r_A^0 and $r_B^0 = (1+\varepsilon)r_A^0$, where ε may be positive or negative. We suppose first that the B atom is introduced into an infinite A crystal, where its volume becomes v_B and its effective radius is $r_B = (1+C_6\varepsilon)r_A^0$. Use an origin at the centre of the B atom. From the symmetry of the problem it follows that the displacements $\mathbf{w}$ are all radial, and $|\mathbf{w}|$ is a function only of the radius vector r. In polar coordinates, the strain components are

$$e_{rr} = \partial w/\partial r, \quad e_{\theta\theta} = e_{\varphi\varphi} = w/r,$$

and hence

$$\Delta = (\partial w/\partial r) + 2w/r, \quad \boldsymbol{\omega} = \text{curl } \mathbf{w} = 0. \tag{25.1}$$

The equations of equilibrium in the form (11.30) thus reduce to

$$(\lambda+2\mu)\{(\partial^2 w/\partial r^2)+(2/r)(\partial w/\partial r)-2w/r^2\} = 0,$$

and the general solution is

$$w = A_1 r + A_2/r^2.$$

The displacements in any spherically symmetrical elastic problem must be of this type. In our infinite medium, we have to satisfy the boundary conditions

$$w(r_A^0) = C_6\varepsilon r_A^0, \quad w(\infty) = w(0) = 0, \tag{25.2}$$

if the initial state is taken to be the introduction of the compressed B atom into the hole of radius r_A^0. The displacements are thus

$$\left.\begin{aligned} w'_B &= C_6\varepsilon r & (r < r_A^0), \\ w_A &= C_6\varepsilon(r_A^0)^3/r^2 & (r > r_A^0). \end{aligned}\right\} \tag{25.3}$$

The vector $\mathbf{w}_A = C_6\varepsilon\{(r_A^0)^3/r^3\}\mathbf{r}$ gives the displacements in the A crystal relative to the unstressed state. The chosen initial state for the B atom, however, is when it has been compressed or extended from radius $(1+\varepsilon)r_A^0$ to r_A^0. The displacements w'_B refer to this state, and relative to the unstressed B sphere (before its introduction into the hole), we have the displacements

$$w_B = \varepsilon(C_6-1)r,$$

so that within this sphere $e_{rr} = e_{\theta\theta} = \Delta_B/3 = (C_6-1)\varepsilon$. The state of stress is a uniform hydrostatic pressure (or tension) given by

$$p = 3K_B(C_6-1)\varepsilon, \tag{25.4}$$

where K_B is the effective bulk modulus of the solute atom B. It is usual in this model to take K_B as the bulk modulus of the component B. This is admittedly a poor approximation because of the changes in the electronic state of a strongly deformed atom.

The introduction of the B atom leaves the whole assembly in a state of self-stress, and the stress components in the surrounding matrix are X_{rr} and $X_{\theta\theta} = X_{\varphi\varphi}$. From (25.1) and (25.3) we see that $\Delta_A = 0$, so that only shear strains are present, and the stresses may be written entirely in terms of the shear modulus μ_A of A. Using (11.26) the stress components are

$$\left.\begin{aligned} X_{rr} &= 2\mu_A(\partial w/\partial r) = -4\mu_A C_6 \varepsilon (r_A^0)^3/r^3, \\ X_{\theta\theta} &= 2\mu_A w/r = 2\mu_A C_6 \varepsilon (r_A^0)^3/r^3. \end{aligned}\right\} \tag{25.5}$$

Since the matrix and inserted sphere are in equilibrium, X_{rr} must equal p (eqn. (25.4)) when $r = r_B \approx r_A^0$. This gives an expression for C_6:

$$C_6 = 3K_B/(3K_B + 4\mu_A). \tag{25.6}$$

If the B atom is incompressible ($K_B = \infty$), $C_6 = 1$ and $v_B = v_B^0$.

Let S be any closed surface in the assembly which totally encloses the B atom. An element $\mathrm{d}O$ of this surface, having unit vector normal $\mathbf{n}$, moves when the B atom is introduced so as to sweep out a volume $\mathbf{w}_A \cdot \mathbf{n}\, \mathrm{d}O$. The volume within S thus increases by

$$\left.\begin{aligned} \Delta v^\infty &= C_6 \varepsilon (r_A^0)^3 \int_S (\mathbf{r} \cdot \mathbf{n}/r^3)\, \mathrm{d}O \\ &= 4\pi C_6 \varepsilon (r_A^0)^3 \end{aligned}\right\} \tag{25.7}$$

since the integral is just the total solid angle subtended by S at the origin. We write the change of volume Δv^∞ to emphasize that this equation is only valid in an infinite crystal. Note that Δv^∞ is equal to the increase in volume of the hole, and is independent of S, as is necessary since there is no dilatation outside the B atom. If we write the difference in atomic volumes of the two pure components as $\Delta v_{AB} = v_B^0 - v_A^0 = 4\pi\varepsilon (r_A^0)^3$, we also have

$$\Delta v^\infty = C_6 \Delta v_{AB}. \tag{25.8}$$

Now consider what happens in a finite bounded crystal of A. The boundary conditions (25.2) are no longer applicable, and must be modified so that the external surface is free of traction. This may be accomplished by superimposing on the displacements (25.3) a second set of "image" displacements, caused by surface tractions which will just annul those given by (25.5). The work of Eshelby (1954, 1956) has shown that these image effects are often of surprising importance in the continuum theory of lattice defects.

Suppose the external surface is S. Then the image displacements are due to tractions $-X_{ij}n_j$ distributed over S. If S is a sphere of radius R, the surface traction to be applied is a uniform hydrostatic tension (or pressure) of $4\mu_A C_6 \varepsilon (r_A^0)^3/R^3$. The corresponding image displacements represent a uniform dilatation of the large sphere provided we neglect the small perturbation caused by the different elastic constants of the enclosed B atom. This dilatation is

$$\Delta_A = 4\mu_A C_6 \varepsilon (r_A^0)^3/(K_A R^3), \tag{25.9}$$

and the increase in volume of the whole assembly is

$$\left.\begin{aligned} \Delta v^i &= 16\pi\mu_A C_6 \varepsilon (r_A^0)^3/3K_A \\ &= (4\mu_A/3K_A)\,\Delta v^\infty \end{aligned}\right\} \tag{25.10}$$

Since the dilatation varies as $1/R^3$, Δv^i is independent of R. Thus for any sphere, the total change in volume is

$$\left.\begin{aligned} \Delta v = \Delta v^\infty + \Delta v^i &= 4\pi C_6 \varepsilon (r_A^0)^3/C_6' \\ &= (C_6/C_6')\,\Delta v_{AB}, \end{aligned}\right\} \tag{25.11}$$

where $C_6' = 3K_A/(3K_A+4\mu_A)$.

If the bulk moduli of the matrix and the B atom are equal, $C_6 = C_6'$ and $\Delta v = \Delta v_{AB}$. This illustrates the importance of including the image effects, since this result cannot be otherwise obtained.

A solution for the image displacements is possible only when the surface S has certain simple forms. However, it is readily proved (Eshelby, 1956) that the result for the change in volume is valid for any external surface. In a sphere with a uniform distribution of solute atoms, c_B per unit volume, it follows from symmetry that the shape of the assembly is unaltered, apart from small ripples in the surface, of magnitude determined by the mean separation of B atoms. The fractional change in the volume of the assembly will be

$$\Delta V/V = c_B(C_6/C_6')\,\Delta v_{AB}. \tag{25.12}$$

Eshelby has shown that this result remains true for a body of arbitrary shape which contains a uniform distribution of defects.

Returning to the single B atom in otherwise pure A, we now calculate the strain energy. Within the compressed or extended B sphere, the strain energy density, $\frac{1}{2}p\Delta$, is constant, and the total strain energy is thus

$$W_B = 9v_B K_B (C_6 - 1)^2\,\varepsilon^2/2. \tag{25.13}$$

If the atom is in an infinite matrix, the strain energy density at any point is

$$\tfrac{1}{2}X_{rr}e_{rr} + X_{\theta\theta}e_{\theta\theta} = 6\mu_A C_6^2 \varepsilon^2 (r_A^0)^6/r^6,$$

and the total strain energy by integration is

$$W_A = 6\mu_A C_6^2 \varepsilon^2 v_B \tag{25.14}$$

to the first order in which all these equations are valid. The same result for W_A is obtained by considering the strain energy density in a finite medium and integrating from r_B to R.

The strain energy of the whole assembly is thus

$$W_s = W_A + W_B = 6\mu_A C_6 \varepsilon^2 v_B \tag{25.15}$$

using the result of (25.6). Since $\varepsilon = \Delta v_{AB}/3v_B$, we may also write the strain energy in the form

$$W_s = 2\mu_A C_6 (\Delta v_{AB})^2/3v_B. \tag{25.16}$$

Note that the volume v_B in eqns. (25.13), (25.14), and (25.16) might equally well be written v_A^0 or v_B^0 to the order of approximation of linear elasticity theory, and it would perhaps be better to write these equations with some mean volume v.

We can now consider the properties of a solid solution containing N_A A atoms and N_B B atoms. If N_B is small, the volume of the assembly will be $V+N_B\Delta v$, where Δv is given by eqn. (25.11). The mean volume per atom is thus

$$\overline{v_{AB}} = v_A^0 + x\Delta v,$$

where x is the atomic concentration of B atoms. Substituting $\Delta v = (C_6/C_6')(v_B^0 - v_A^0)$, we find

$$\overline{v_{AB}} = (1-x)v_A^0 + xv_B^0 + x(C_6/C_6' - 1)(v_B^0 - v_A^0). \tag{25.17}$$

When $K_A = K_B$, this leads to an additive law for atomic volumes, or on the linear approximation to an additive law for atomic radii

$$\overline{r_{AB}} = (1-x)r_A^0 + xr_B^0 = (1+\varepsilon x)r_A^0, \tag{25.18}$$

so that ε is the fractional rate of change of lattice constant $\partial r/\partial x$. Equation (25.18) is usually known as Vegard's law, and according to the elastic model the radii are larger or smaller than their Vegard's law values according to whether the last term in (25.17) is positive or negative. The sign of the deviation is thus dependent on the sign of

$$(K_B - K_A)(r_B^0 - r_A^0).$$

Friedel (1955) has shown that this rule accounts for observed qualitative deviations from the law (using X-ray measurements), although quantitative agreement, as expected, is not very good.

In deriving eqn. (25.17) we have assumed that Δv is a constant independent of x, and we have also neglected the variation of elastic constants with x. It might strictly be more logical to assume that the B atom is forced into a hole of size $v_A^0(1+\Delta_A)$, where Δ_A is the dilatation produced in the matrix by all the preceding B atoms. However, the misfitting sphere model is probably too uncertain in its details for this kind of modification. Eshelby has pointed out that there is no very convincing reason to take the volumes of the hole and misfitting sphere as v_A^0 and v_B^0 respectively; they might, for example, have been chosen so that their radii were equal to nearest neighbour distances in the pure components. Experimental results which show that Vegard's law is approximately valid in many alloys are the best justification for the assumption we have made.

We are now able to derive the strain energy of the solid solution. When the composition is x, the addition of a further B atom changes the energy first by the self-energy W_s of this atom, and, secondly, by the interaction energy of the new solute atom with all the preceding B atoms. This interaction energy is due to the presence of a hydrostatic tension or compression in the matrix (the image stress), and as this helps the required expansion or contraction of the hole, the total extra energy is less than W_s. Although the interaction energy of any two B atoms is negligible in comparison with W_s, the fact that each B atom interacts with all the other B atoms means that the total interaction energy is comparable with the total self-energy.

When the concentration is c_B, the dilatation is $c_B \Delta v^i = (x/v_A^0)\Delta v^i$. The interaction energy when another B atom is added is thus $K_A(x/v_A^0)\Delta v^i \Delta v = 2W_s(C_6/C_6')x \simeq 2W_s(K_B/K_A)x$. The net change in energy on adding the B atom is thus

$$\delta W = W_s\{1-2(K_B/K_A)x\},$$

and this corresponds to a change of composition $\delta x = 1/N$. The total strain energy is thus

$$W = NW_s x\{1-(K_B/K_A)x\} \qquad (x \ll 1). \tag{25.19}$$

If chemical interaction terms are small, this strain energy is the atomic part of the free energy of solution, as pointed out on p. 198, and the total free energy of solution is obtained by adding the configurational term for random mixing to give

$$\Delta_m g = W_s x\{1-(K_B/K_A)x\}+kT\{x \ln x+(1-x)\ln(1-x)\} \qquad (x \ll 1). \tag{25.20}$$

Except for the factor (K_B/K_A), this equation is identical in form with eqn. (23.5) for a regular solution. However, we have already emphasized that W is a free energy, so that the temperature variation of the elastic constants gives rise to an appreciable temperature variation of W, and hence to an entropy

$$\Delta_e s = -(\partial W/\partial T). \tag{25.21}$$

From eqn. (25.16) we see that the temperature variation of W is governed by the term $\partial\mu/\partial T$, and since this is always negative, the elastic model always gives a positive value to $\Delta_e s$ (Zener, 1951). Friedel (1955) has compared the values of $\Delta_e s$ and $\Delta_m h$ obtained from eqns. (25.21) and (25.19) with experimental data on gold–nickel alloys, and obtained good agreement. In a survey of limiting solubility data for a number of binary alloys, using the method of analysis described on p. 196, Freedman and Nowick (1958) found that $\Delta_e s$ is always positive, in accordance with the predictions of the model. The model actually permits calculation of $\Delta_e s$, but, as pointed out earlier, it is too sensitive to a choice of integration limits for this calculation to have much significance. The values obtained are certainly of the correct magnitude, but it is also reasonable to expect some proportionality between the observed $\Delta_e s$ and that calculated from the size factor and temperature variation of the shear modulus. Freedman and Nowick showed that such a proportionality does not exist, and concluded that a quantitative interpretation of $\Delta_e s$ cannot be given by the elastic model. Such an interpretation almost certainly requires detailed consideration of nearest neighbour interactions, as in Huntington's calculations for point defects (see p. 128). Oriani (1959) reaches the same conclusion, which he bases partly on the observed experimental correlation between heats of solution in the liquid and solid phases of alloys with a size disparity. The elastic model is not applicable to a liquid solution, so there should be no correlation if it is valid.

Within the limitations of the model, the equations derived above are strictly valid only for very dilute solutions. Some attempts have been made to treat more concentrated solutions by supposing that each atomic site is occupied by a positive or negative strain centre acting in a mean lattice. We shall not describe these theories because of their complexity; some attention has been paid to the X-ray diffraction effects expected from a model of this type. The simplest treatment of more concentrated solutions comes from noting that eqn. (25.20) will be nearly identical for both A-rich and B-rich solutions if the components have nearly identical elastic properties, and as a first approximation may be applied to intermediate compositions, using mean values for the parameters. Comparison with the previous quasi-chemical treatment then shows that the critical temperature below which phase segregation occurs is given by

$$T_c = 3\bar{C}_6\varepsilon^2\bar{v}\bar{\mu}/k,$$

and if this is to be below the melting point of the alloy,

$$\varepsilon^2 < (kT^{sl}/3\bar{C}_6\bar{v}\bar{\mu}).$$

Reasonable values of the constants give $\varepsilon \sim 0{\cdot}15$, in agreement with the empirical Hume-Rothery rule that wide solid solution is possible only when the size factor ε is less than about 15%. In view of the reservations above, it is doubtful if this is anything other than a coincidence.

An early attempt at a semi-empirical equation for the strain energy was made by Lawson (1947). He used the expression for the self-energy of a B atom in an A lattice, but replaced Δv_{AB} by $v_B^0 - v_{AB}$. This means that the hole into which the B atom is forced is equal in volume to the mean volume per atom in the assembly. Lawson neglected interaction energies, and his equation for the free energy may be expressed in the form

$$\Delta_m g = A_1 x(1-x)^2 + A_2 x^2(1-x) + kT\{x \ln x + (1-x)\ln(1-x)\}. \tag{25.22}$$

The assumptions made in deriving the equation are too sweeping for the expressions for A_1 and A_2 to have any real value, but the equation itself may be useful as a representation of the properties of some real solutions. Its real advantage as an empirical equation is that if effectively replaces the constant Ξ of the quasi-chemical theory by a composition dependent exchange energy, and the assumed independence of the quantities Ξ_{AA}, Ξ_{AB}, etc. of composition is a major weakness of the quasi-chemical theory.

The formal properties of eqn. (25.22) have been investigated by Hardy (1953), who used it to define a "sub-regular solution". For the present, we assume that A_1 and A_2 are independent of temperature and composition, and the chemical potentials are then

$$g_A - g_A^0 = x^2(2A_1 - A_2) + 2x^3(A_2 - A_1) + kT\ln(1-x). \tag{25.23}$$

The compositions of the solubility curve at any temperature are found by differentiating (25.22) and applying the condition of eqn. (22.5). This gives

$$kT\ln\left\{\frac{x_2(1-x_1)}{x_1(1-x_2)}\right\} + (x_2 - x_1)\{2(A_2 - 2A_1) + 3(A_1 - A_2)(x_1 + x_2)\} = 0, \tag{25.24}$$

where, as before, x_1, x_2 are the limits of solubility of B in A and A in B respectively. The equation of the spinodal is similarly found to be

$$2(A_2 - 2A_1) + 6x(A_1 - A_2) + kT/x(1-x) = 0, \tag{25.25}$$

and at the critical point

$$\partial^3\Delta_m g/\partial x^3 = 6(A_1 - A_2) + (2x-1)\,kT/x^2(1-x)^2 = 0. \qquad (25.26)$$

From eqns. (25.25) and (25.26) the critical temperature T_c is related to the critical composition at which the solubility gap closes, x_c, by the equations

$$\left.\begin{aligned} A_1 &= kT_c(-9x_c^2+8x_c-1)/6x_c^2(1-x_c)^2\,. \\ A_2 &= kT_c(-9x_c^2+10x_c-2)/6x_c^2(1-x_c)^2\,. \end{aligned}\right\} \qquad (25.27)$$

From eqns. (25.27) we see that if $A_1 = A_2$, $x_c = \frac{1}{2}$ and $T_c = A_1/2k$. The model is then formally equivalent to the quasi-chemical model. As $A_1 - A_2$ increases, the composition of the maximum in the solubility gap moves further away from the equi-atomic composition, and the asymmetry of the solubility gap increases.

Instead of using eqn. (22.5), we could have obtained (25.24) from the equivalent thermodynamic conditions

$$g_A(x_1) = g_A(x_2), \quad g_B(x_1) = g_B(x_2).$$

If we multiply the first of these equations by $2-(x_1+x_2)$ and the second by (x_1+x_2) and then add, we obtain

$$(2-x_1-x_2)\,kT\ln(\{1-x_1\}/(1-x_2)\}+(x_1+x_2)\,kT\ln(x_1/x_2) = -(A_1-A_2)\,(x_2-x_1)^3. \quad (25.28)$$

In a regular solution, both sides of this equation are zero. Hardy used (25.28) to test the sub-regular solution model for a number of alloy assemblies. Provided $(A_1 - A_2)$ is independent of temperature, a straight line may be obtained by plotting the left-hand side of the equation (obtained from the experimental equilibrium diagram) against $(x_2-x_1)^3$. In this way, the alloys Ag–Cu, Ag–Pt, Al–Zn, and Au–Pt were shown to behave approximately as sub-regular solutions. The simple model could not be fitted to other alloys, e.g. Au–Fe, Au–Co, and Au–Ni, where $A_1 - A_2$ is not constant and A_1 and A_2 are apparently functions of composition. In the alloys which may be represented as sub-regular solutions, A_1 and A_2 may increase with temperature, although the difference $A_1 - A_2$ is nearly independent of temperature.

The theory of sub-regular solutions does not rest on any firm basis, and has very limited utility. It supplies an empirical equation for the solubility curve in terms of a single parameter $(A_1 - A_2)$ (eqn. (25.28)), and is thus useful in representing the properties of alloys with equilibrium diagrams which are not symmetrical about the equi-atomic composition.

26. ORDERED STRUCTURES IN ALLOYS

In the previous sections we have concentrated mainly on deriving the equilibrium conditions for an assembly with a positive heat of mixing, i.e. an assembly in which Ξ is positive according to the nearest neighbour model. As already noted, a negative $\Delta_m h$ (negative Ξ) implies that atoms of opposite kinds attract each other. At low temperatures, each A atom will thus surround itself with as many B atoms as possible; such a structure is called a superlattice. A completely ordered structure of this kind resembles a chemical compound, and is possible only when the atomic fractions of the different components are small integral numbers.

The term superlattice occasionally causes some confusion. As described on p. 169, we do not need to distinguish between the different kinds of atom in a substantially disordered solid solution, since they occupy the available sites approximately at random. When the structure is ordered, the atoms segregate in such a way that the atoms of one kind occupy one or more sets of sites, and atoms of another kind occupy different sets of sites. A distinction must then be made between vectors specifying the positions of A atoms and those specifying the positions of B atoms, and the size of the primitive unit cell of the Bravais lattice has to be increased. This is the origin of the term superlattice. The conventional unit cell may remain the same size, as happens, for example, in the simplest type of superlattice, which is formed from a disordered b.c.c. structure. If there are equal numbers of A and B atoms, they may arrange themselves so that all the A atoms are at cube corners and all the B atoms at cube centres. The superlattice unit cell is the same as the cubic unit cell of the b.c.c. structure, but the structure is now simple cubic because of the non-equivalence of the corner and centre atoms. The cubic unit cell is the smallest possible for the superlattice structure, but the disordered structure has a smaller rhombohedral unit cell. In other superlattices, more complex changes take place, and the formation of the superlattice may be accompanied by a lowering of lattice symmetry.

It would be inappropriate in this book to discuss the structural features of all the known superlattices, but it is convenient to summarize the main types. In binary alloys, most superlattices have one of five structures, two being derived from each of the common cubic structures and one from the h.c.p. structure. In addition, there are a number of superlattice structures of which only one example is known, and some structures which have large unit cells, corresponding to a modification of one of the basic types.

The superlattices derived from the f.c.c. structure ($A1$) are known as $L1_0$ and $L1_2$ in the Strukturbericht notation. The unit cell of each contains the four atoms found in the cubic unit cell of the disordered $A1$ structure, but $L1_0$ has tetragonal symmetry. In this structure, which occurs at equi-atomic positions, the A atoms are at points [000] and $[\frac{1}{2}\frac{1}{2}0]$ of the unit cell, and the B atoms at points $[\frac{1}{2}0\frac{1}{2}]$ and $[0\frac{1}{2}\frac{1}{2}]$. The structure consists of alternate layers of A and B atoms parallel to the (001) planes. The attraction between A and B atoms results in slightly smaller interatomic distances between nearest neighbours in adjacent layers, so the structure is tetragonal with c/a slightly smaller than unity. Each atom has four nearest neighbours of its own type in the same (001) layer, and eight nearest neighbours of opposite type in the two adjacent (001) layers. This contrasts with the completely disordered structure, where each atom on the average has six like and six unlike nearest neighbours.

The $L1_2$ structure corresponds to the ideal composition A_3B. The B atoms are in the [000] positions, and the A atoms in the remaining positions of the conventional unit cell of the f.c.c. structure. In the superlattice, the B atoms each have twelve unlike nearest neighbours, compared with an average of three like and nine unlike nearest neighbours in a random f.c.c. solid solution. There is also a single known example of the $L1_1$ structure, in which alternate (111) planes are composed entirely of A and B atoms respectively.

The simplest superlattice ($B2$ or $L2_0$) derived from the b.c.c. ($A2$) structure is found in equiatomic alloys, and was described above. The symmetry is simple cubic, and the unit cell contains the two atoms of the conventional b.c.c. unit cell. The A atoms are at the corners [000] and the B atoms at the body centres $[\frac{1}{2}\frac{1}{2}\frac{1}{2}]$, or vice versa; this is sometimes called the caesium chloride structure. Each atom has eight unlike nearest neighbours in the disordered state compared with an average of four like and four unlike neighbours in the completely disordered state. Another superlattice derived from the b.c.c. structure is more complex; it is known as the DO_3 type. This structure has a cubic unit cell formed from eight conventional b.c.c. unit cells, and thus containing sixteen atoms; the ideal composition is A_3B. The corner positions of the small cubic cells are occupied by equal numbers of the atoms of each kind, each set being arranged on tetrahedral groups of sites. The body centred positions of the small unit cells are occupied entirely by A atoms. Each B atom has eight unlike nearest neighbours in the superlattice, compared with an average of two like and six unlike nearest neighbours in the substantially disordered solid solution.

From the h.c.p. structure ($A3$), a superlattice is made by stacking together four unit cells of two atoms to give a larger unit cell of sides $2a$, $2a$, c containing eight atoms. The composition is again A_3B, and each close packed layer contains three times as many A atoms as B atoms. The B atoms form a hexagonal network of side $2a$, and the A atoms occupy the remaining sites of the hexagonal network of side a. Each B atom is surrounded by twelve A atoms in the superlattice compared with an average of three like and nine unlike atoms in the random solution.

The relations we have just described are summarized in Table VI, which also lists some of the known binary assemblies having superlattices of these kinds. In some cases, the

TABLE VI. COMMON SUPERLATTICE STRUCTURES

Disordered structure	Superlattice type	Composition	Atom positions	Examples
f.c.c.	$L1_0$ (tetragonal)	AB	$2A$ in $(000; \frac{1}{2}\frac{1}{2}0)$ $2B$ in $(\frac{1}{2}0\frac{1}{2}; 0\frac{1}{2}\frac{1}{2})$	AuCu, CoPt, MgIn MnNi, NiPt, FePd, FePt
f.c.c.	$L1_2$ (cubic)	A_3B	$3A$ in $(0\frac{1}{2}\frac{1}{2}; \frac{1}{2}0\frac{1}{2}; \frac{1}{2}\frac{1}{2}0)$ $1B$ in (000)	Cu_3Au, Au_3Cu, Pt_3Co, Fe_3Pt, Pt_3Fe, Cu_3Pt, Ni_3Mn, etc.
b.c.c.	$B2$ ($L2_0$) (cubic)	AB	$1A$ in (000) $1B$ in $(\frac{1}{2}\frac{1}{2}\frac{1}{2})$	CuZn, CuPd, AgCd, AgZn, CoFe
b.c.c.	DO_3 (cubic, face-centred)	A_3B	$(000; 0\frac{1}{2}\frac{1}{2}; \frac{1}{2}0\frac{1}{2}; \frac{1}{2}\frac{1}{2}0)$ $+4B$ in (000) $4A$ in $(\frac{1}{2}\frac{1}{2}\frac{1}{2})$ $8A$ in $(\frac{1}{4}\frac{1}{4}\frac{1}{4}; \frac{3}{4}\frac{3}{4}\frac{3}{4})$	Fe_3Al, Fe_3Si, Mg_3Li, Cu_3Al
h.c.p.	DO_{19} (hexagonal)	A_3B	$6A$ in $(\frac{1}{2}00; 0\frac{1}{2}0; \frac{1}{2}\frac{1}{2}0;$ $\frac{1}{6}\frac{1}{3}\frac{1}{2}; \frac{1}{6}\frac{5}{6}\frac{1}{2}; \frac{2}{3}\frac{5}{6}\frac{1}{2})$ $2B$ in $(000, \frac{2}{3}\frac{1}{3}\frac{1}{2})$	Ag_3In, Mn_3Ge, Mg_3Cd, Cd_3Mg. Ni_3Sn

nearest neighbour model with negative Ξ predicts that for a given crystal structure and composition the lowest energy state will correspond to an observed superlattice type. However, this is not always valid, and the model may be ambiguous; for example, the energy of a b.c.c. solution is reduced by the formation of a phase mixture of the $L2_0$ superlattice and pure A, so that all B atoms have only A atoms as nearest neighbours. For an AB_3 alloy it follows that the energy of this phase mixture is identical with that of a DO_3 superlattice, so that the nearest neighbour model is not able to predict which is the equilibrium state. It then becomes necessary to consider higher neighbour interactions; this procedure is not so artificial as first appears since, as previously noted, the interchange energies can be given operational definitions (Clapp and Moss, 1968).

Several early calculations of the lowest energy states with second and sometimes third nearest neighbour interactions taken into consideration were restricted either to stoichiometric compositions or to one-phase states. A systematic investigation for cubic alloys with first and second neighbour interactions was first made by Richards and Cahn (1971) and was supplemented by Allen and Cahn (1972). The results are expressed in terms of the ratio $\xi = \Xi_2/\Xi_1$ (see eqn. (23.1a)) which is unrestricted in sign and magnitude; Richards and Cahn considered only $\Xi_1 < 0$, but Allen and Cahn also treated $\Xi_1 > 0$. Because of symmetry, it is only necessary to consider explicitly the composition range $0 \leqslant x \leqslant \frac{1}{2}$.

The internal energy of any configuration may now be expressed in an obvious extension of eqn. (24.1) as

$$\left.\begin{aligned} \Delta_m U &= z_1 N_{AB,\,1}\Xi_1 + z_2 N_{AB,\,2}\Xi_2 \\ &= \tfrac{1}{2}N\Xi_1[z_{AB,\,1} + z_{AB,\,2}\,\xi], \end{aligned}\right\} \tag{26.1}$$

where $z_{AB,i}$ now denotes the average number per atom of ith neighbours of opposite type. Thus for Ξ_1 negative the equilibrium state at 0 K (ground state) is found by maximizing the quantity in square brackets, whilst for Ξ_1 positive the energy will be minimized by minimizing this quantity. There are limits on $z_{AB,1}$ and $z_{AB,2}$, given for $0 \leqslant x \leqslant \frac{1}{2}$ by

$$0 \leqslant z_{AB,\,i} \leqslant 2z_i x. \tag{26.2}$$

The lower limit corresponds to all B atoms having only B atoms for ith neighbours; one way of attaining this is to postulate a two-phase mixture of the two components in which $z_{AB,i} = 0$ for all i. At the upper limit, each B atom has only A atoms for ith neighbours, assuming B to be the minority component.

Clearly, if $z_{AB,1}$ and $z_{AB,2}$ can have the upper or lower limiting values (depending on the signs of Ξ_1 and Ξ_2) which will minimize $\Delta_m U$, a ground state has been found. Thus if both Ξ_1 and Ξ_2 are positive, minimum energy is given by the phase mixture ($\Delta_m U = 0$) as already concluded. However, the crystal structure does not always allow the upper limits of (26.2) to be attained; for example $z_1 = 12$ for a f.c.c. structure, but the maximum value of $z_{AB,1}$ for an equi-atomic superlattice is 8. This arises because in this structure two nearest neighbours of a given atom may also be nearest neighbours of each other, and the upper limit of (26.2) is in fact only possible for $0 \leqslant x \leqslant 0{\cdot}25$. In addition to this difficulty, $z_{AB,1}$ and $z_{AB,2}$ may not be independent of each other, so that it may be impossible to find a configuration which produces the appropriate limits simultaneously, even though both may be individually obtained.

In cases where $B-B$ bonds cannot be avoided for negative Ξ_i, or $A-B$ bonds cannot be avoided for positive Ξ_i, the equilibrium structure(s) can only be found by calculating the energies of various possible configurations. Richards and Cahn used an empirical procedure of examining superlattices with translational vectors that coincide as much as possible with energetically favourable distances for like atoms. The results of this procedure seem intuitively to be correct, but no proof could be given that all possible low energy configurations had been considered. Allen and Cahn used a more rigorous technique, the cluster method, that includes consideration of all possible structures but also sometimes includes cluster combinations which represent impossible "structures".

In the cluster method, a motif of M adjacent lattice points (atoms) is defined by a specific three-dimensional circuit and the 2^M clusters which correspond to the various occupancies of these sites by A and B atoms are enumerated. An energy can now be assigned to each cluster and the total energy of any arrangement of A and B atoms on a lattice can then be expressed as a linear sum of each such energy multiplied by the number of clusters of that type. The problem of minimizing this energy function subject to the constraints imposed by the overall composition can be solved by a mathematical technique known as linear programming and yields as a result the fractional numbers of clusters of each type in the whole arrangement. This gives an absolute minimum to the energy since all possible structures (single and multiphase) are represented. However, whilst the fractions of each cluster in any arrangement can always be specified, it is not always possible to construct an arrangement corresponding to specified fractions. Thus the numbers given by the technique may represent an "imaginary" structure, and the method then gives only a lower limit to the energy unless a true minimum energy structure can be found by selection of a different motif.

In the case of the b.c.c. structure, the results obtained by the cluster method for negative Ξ_1 coincide with those found by Richards and Cahn, and thus constitute a proof that their ground state diagram is correct. The equilibrium phases at 0 K are shown schematically in Fig. 6. 10 in which the phase fields are plotted as functions of $\xi = \Xi_2/\Xi_1$ and of atomic per cent of B. The only structure not listed in Table VI is the $B32$ (NaTl) type. For ξ negative a two-phase state is stable over the whole composition range, but for ξ positive the situation is more complex. For $0 \leq \xi \leq \frac{2}{3}$ and 25–50 atomic % B, for example the same energy is obtained either by a phase mixture of the two stoichiometric phases, $D0_3$ and $B2$, or by a gradual change in the occupancy of some of the sites. In the latter case, there is a continuous single phase region which includes the stoichiometric compositions. This degeneracy in the ground state is presumably a property of the particular model, and would disappear in a better approximation. Richards and Cahn point out that the configurational entropies of the degenerate states are not the same, and this will then determine the equilibrium state at finite low temperatures. On this basis, the fields of type $D0_3 \rightarrow B2$ are perhaps better regarded as single phase.

It is particularly noteworthy that the ground state is multiply degenerate in the nearest neighbour model ($\xi = 0$), except for $x = \frac{1}{2}$. At compositions near $x = \frac{1}{4}$, for example, it is possible for $z_{AB,1}$ to have its maximum value of 8 (i.e. for there to be no $B-B$ nearest neighbour bonds) in a configuration consisting of (i) $A+B2$, (ii) DO_3, (iii) single phase non-stoichiometric $B2$, or (iv) an almost random structure. When second neighbour interactions are taken into account, this degeneracy disappears and either (i) or (ii) becomes stable even when $|\xi|$ is very small. This is thus a strong argument for the inclusion of second neighbour interactions in some special situations. There are also degeneracies for $\xi = \frac{2}{3}$ and $0{\cdot}25 \leqslant x \leqslant 0{\cdot}5$.

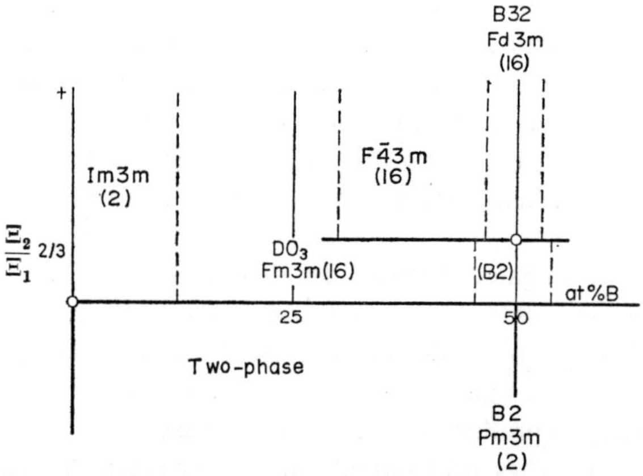

FIG. 6.10. Body-centred cubic ground state diagram showing the state of lowest energy as a function of composition and of the ratio of second to first neighbour interaction energies. Superlattices are identified by Struhturbericht and space-group symbols, with the number of atoms in the unit cell in brackets. Heavy lines show the limits of two-phase fields, and thin lines mark stoichiometric compositions which may be extended into single-phase fields as indicated by the broken lines (after Richards and Cahn, 1971).

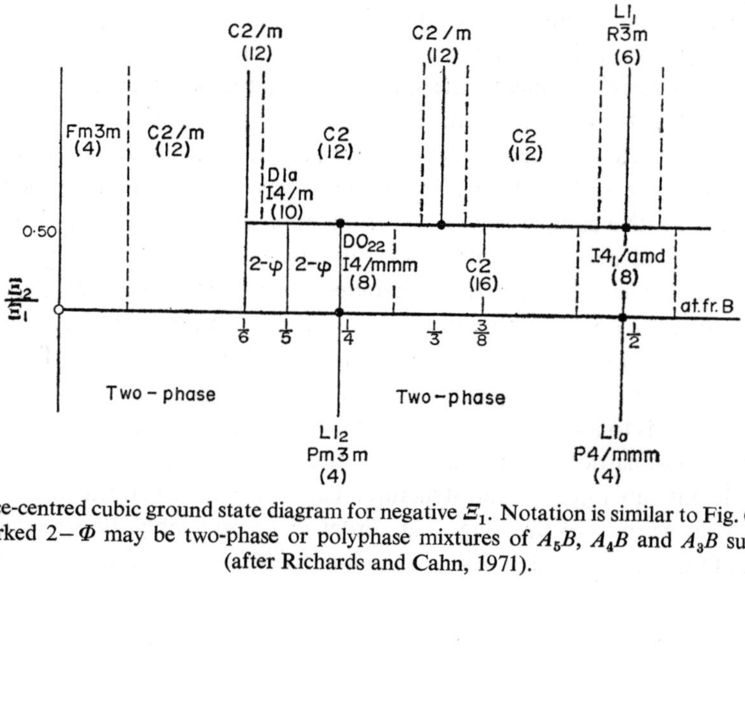

FIG. 6.11. Face-centred cubic ground state diagram for negative Ξ_1. Notation is similar to Fig. 6.10; the two regions marked $2-\Phi$ may be two-phase or polyphase mixtures of A_5B, A_4B and A_3B superlattices (after Richards and Cahn, 1971).

The ground-state structures for negative ξ in the more complex f.c.c. diagram (Fig. 6.11) are again two phase, and the Cu_3Au and CuAu superlattices are stable. Among the many superlattices predicted for positive ξ are the known types $D1a$ (Ni_4Mo), $D0_{22}$ (Al_3Ti) and $L1_1$ (CuPt), an A_5B structure with no known example, and at $x \sim 0{\cdot}5$ an eight-atom cell formed from two "antiphase" $L1_0$ cells which thus resembles the CuAu II structure (p. 225). The predictions were all confirmed by Allen and Cahn except for the region $0{\cdot}25 < x < 0{\cdot}5$ and $0 < \xi < 0{\cdot}5$, where the cluster method yielded impossible structures.

For large negative values of ξ ($<-\frac{4}{3}$) the energy of mixing of the disordered b.c.c. solution is positive even though the ground state is an ordered phase. This also applies to f.c.c. solutions and is contrary to the simple idea that low-temperature ordering is associated with negative $\Delta_m U$ and low-temperature phase separation with positive $\Delta_m U$. However, the second neighbour interchange energy is unlikely to be larger than the nearest neighbour interchange energy, so the usual association of ordering with the sign of $\Delta_m U$ is not unreasonable.

Allen and Cahn also investigated the ground state for positive Ξ_1, and in b.c.c. structures found it to be pure A + pure B for all $\xi > -\frac{2}{3}$ and a mixture of pure A and the $B32$ (NaTl) structure for $\xi < -\frac{2}{3}$. Similarly for f.c.c. structures, $A+B$ mixtures are stable for $\xi > -1$ and mixtures of A + the CuPt superlattice for $\xi < -1$. Very large second neighbour interactions are thus necessary to produce superlattices when Ξ_1 is positive.

Structures not included in Figs. 6.10 and 6.11 may also be stable because of higher neighbour interactions or non-central forces. Clapp and Moss (1968) assumed that the observed diffuse X-ray maxima due to short-range ordering (see below) indicate minima of the **k**-space potential (p. 111), and they deduced the stable structures for stoichiometric compositions with up to third neighbour real-space interactions. Khachaturyan (1962, 1973) uses a similar approach based on the symmetry rules of Landau and Lifshitz (1958) for second-order transitions, and especially on the Lifshitz criterion that minima of the harmonic part of the **k**-space potential occur at the special points of the Brillouin zone where symmetry elements intersect. He thus considers only structures generated by composition waves with wave vectors corresponding to the special points, and this excludes some f.c.c. superlattices in Fig. 6.11. On the other hand, Khachaturyan includes some b.c.c. structures which are not ground states, at least in a near-neighbour model, and which therefore do not appear in the Richards–Allen–Cahn scheme. An elegant discussion of symmetry rules for order–disorder reactions with a comparison of the various approaches is given by de Fontaine (1975).

The use of second neighbour interactions in calculations of partitition functions and entropies is very difficult and for the remainder of this chapter we shall use the nearest neighbour model. We now turn to a discussion of the transition from the disordered to the ordered structure, and we begin by distinguishing between the concepts of long-range and short-range order. In most binary superlattices, the tendency of the like atoms to separate and the unlike atoms to attract each other results in all the A atoms occupying sites comprising one or more sublattices of the whole structure, and the B atoms occupying sites which make up different sublattices of the structure. The extent to which this ideal arrangement is achieved is a measure of the long-range order of the assembly. Short-range order is a description of the atomic configuration in the immediate vicinity of an atom; if the average number of unlike neighbours of an A atom is higher than would be expected for random distribution, the alloy possesses short-range order. Both concepts may be defined in terms of the probability $P_{AB}(\mathbf{r})$ introduced in Section 23. The alloy possesses short-range order if $P_{AB}(\mathbf{r})$ is greater than the random value $2x(1-x)$ for nearest neighbour values of $\mathbf{r}$. (Negative short-range order, or clustering, corresponds to $P_{AB}(\mathbf{r}) < 2x(1-x)$; for convenience, any deviation from the random value is often referred to as short-range order, as discussed in the next section.) The alloy possesses long-range order if $P_{AB}(\mathbf{r})$ for large $\mathbf{r}$ approaches a limiting value which is not identical with the random value. It is perfectly possible and usual for there to be appreciable short-range order, but no long-range order.

The distinction between long-range and short-range order is sometimes obscured by the possible existence of antiphase domains. These are regions of perfect or nearly perfect order, at the boundaries of which the roles of two or more of the sets of sites are interchanged. Figure 6.12 shows a simple illustration of the effect in a two-dimensional AB super-

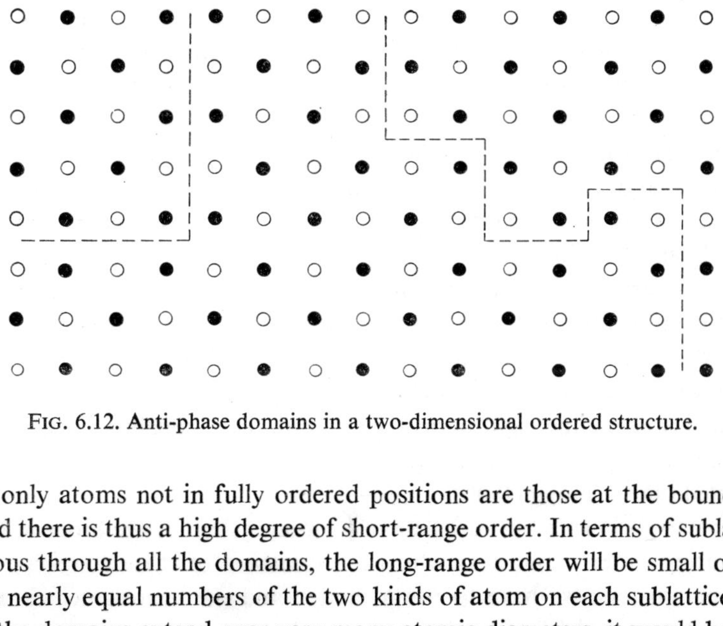

FIG. 6.12. Anti-phase domains in a two-dimensional ordered structure.

lattice. The only atoms not in fully ordered positions are those at the boundaries of the domains, and there is thus a high degree of short-range order. In terms of sublattices which are continuous through all the domains, the long-range order will be small or zero, since there will be nearly equal numbers of the two kinds of atom on each sublattice. Obviously, however, if the domains extend over very many atomic diameters, it would be undesirable to conclude that the structure has no long-range order, and the difficulty can be avoided by considering only a single domain.

Some review articles give a misleading impression that there is no real distinction between a structure of antiphase domains and one possessing short-range order. In fact the domain aggregate may be shown to be thermodynamically unstable, since it involves an increase in energy (at the boundaries) with very little extra entropy to compensate. A structure with genuine short-range order is often the stable configuration because of its high entropy. It is true, of course, that if the size of the antiphase domains continually decreases, the resultant configuration is eventually a state of short-range order only, but at this stage, the domains have lost their individual identity.

A structure of highly ordered antiphase domains is analogous in many ways to a polycrystalline aggregate of strain free grains. The domain boundaries will vanish when true equilibrium is attained, but it is possible in some circumstances to produce a metastable structure in which the domains persist for long periods. From any initial configuration of small domains, the approach to equilibrium occurs in two stages. The domain boundaries first contract, so as to reduce their area, and this is a relatively rapid process, during which the energy decreases continually. It results eventually in an array of planar or nearly planar boundaries, meeting along lines in groups of three at mutual angles of about 120°, and at corners in groups of four. Such an array of either domains or crystal grains is often compared to the structure of a soap froth, and referred to as a foam structure. When it is attained, small displacements of any of the boundaries may lead to an increase in energy.† At this stage, the structure changes more slowly, reductions in energy being obtained by movements of boundaries over distances of the order of the domain diameter. As a result of these movements, the smaller domains are progressively eliminated, and the larger domains grow.

There is an important distinction between an array of grains and an array of antiphase domains in a single grain. A grain boundary may separate two crystals of any orientations, but there is a very limited number of different antiphase domains possible with any superlattice structure. Thus in the $L2_0$ structure, there are only two domains, the corner atoms of the unit cell being occupied by either A or B atoms. In an A_3B superlattice with the $L1_0$ structure, the A atoms may occupy any of the four positions of the cubic unit cell, thus giving four different domains.

The foam structure requires the meeting of four different domains at a corner, and hence it is not possible for the $L2_0$ structure to form such a metastable array. This structure has only one type of antiphase domain boundary, and each such boundary must begin and end on the surface of the grain. The boundaries will straighten rapidly, giving a structure of a few large domains extending right across the crystal. This conclusion that superlattices which have only two types of antiphase domain cannot exist as aggregates of small domains was first pointed out by Bragg (1940).

We have now to formulate our ideas about long- and short-range order in more precise terms, introducing order parameters which are measures of these concepts. In order to evaluate the parameters, we have to use a model, and the nearest neighbour, or Ising, model is the only one which has been found tractable to mathematical analysis. It is useful to note the close analogy between the formation of a phase mixture and the formation of a superlattice, and between the presence of small local aggregates in a solid solution and the presence of short-range order. The analogy is particularly useful for an AB superlattice, and all results of order–disorder theory in successive approximations may be derived from the corresponding results of the regular solution theory (for a detailed discussion of this point, see Guggenheim, 1952, pp. 111–14). The simplest assumption in the theory of regular solutions does not consider clustering on an atomic scale. The corresponding approxi-

† In a fuller discussion, in Section 35, we show that this condition can be satisfied exactly in two dimensions but not in three dimensions. If the structure arises from randomly nucleated centres, the completely metastable array will not be formed, even in two dimensions.

mation in the theory of order–disorder transformations is similarly a theory of long-range order only. This treatment, which we describe first, is usually called the Bragg–Williams theory. These workers originally developed it in a rather different (but equivalent) manner from that which we shall use (Bragg and Williams, 1934, 1935.)

Consider a crystal of a binary alloy having N sites, of which Nx_A sites (the A sites) may be distinguished in some way from the remaining Nx_B sites (the B sites). When the alloy is fully ordered, these two sets of sites are completely occupied by A atoms and B atoms respectively. The completely disordered state is taken to be the random distribution, in which there will be Nx_A^2 A atoms and Nx_Ax_B B atoms on the A sites, and Nx_B^2 B atoms and Nx_Ax_B A atoms on the B sites. We define the partially ordered state by the number of "wrong" A atoms (i.e. A atoms on B sites), which we write Nx_w; this is also equal to the number of "wrong" B atoms. The probability that an A site is occupied by a "wrong" (i.e. B) atom is written $w_A = x_w/x_A$, and the probability of its being occupied by a "right" atom is $r_A = 1-w_A$. The corresponding probabilities for the B sites are r_B, w_B; note that $r_A \neq r_B$ except for an equi-atomic alloy. The definition of long-range order introduced by Bragg and Williams, and now generally used, is

$$L = \frac{r_A - x_A}{1-x_A} = \frac{r_B - x_B}{1-x_B} = 1 - \frac{x_w}{x_A x_B}. \tag{26.3}$$

This has a maximum value of $L = 1$ for $x_w = 0$, and $L = 0$ when x_w has the random value. For $x_w > x_Ax_B$, L becomes negative, but the long-range order must increase again, since this corresponds only to a different labelling of the sites. For example, in an $L2_0$ superlattice, the situation when all the A atoms are on the B sites ($L = -1$) is physically indistinguishable from the fully ordered state of $L = +1$. Hence only the magnitude of L is significant, and we need only consider $0 \leqslant x_w \leqslant x_Ax_B$.

Let us now consider the equi-atomic superlattice in more detail, confining our attention to the simplest condition (type $L2_0$) in which the atomic sites are situated on two equivalent interpenetrating lattices. Each site has z nearest neighbours, all of which are situated on the other lattice. As there are $\frac{1}{2}N$ sites of each type, $w = 2x_w$ and $L = 2r-1$, where the subscripts for w, r have been dropped, since $w_A = w_B$. In the simplest approximation, we assume that the $Nr/2$ A atoms and the $Nw/2$ B atoms are distributed completely randomly amongst the A positions, and similarly for the $Nw/2$ A atoms and the $Nr/2$ B atoms on the B positions. As on p. 186, this enables us to calculate the number, zN_{AB}, of nearest neighbour pairs of unlike atoms, and thus to obtain an expression for the free energy. A given atom of the first lattice will have on average zr B neighbours and zw A neighbours. The numbers of $A-A$, $B-B$, and $A-B$ pairs are thus $Nzrw/2$, $Nzrw/2$ and $Nz(r^2+w^2)/2$, and the internal energy may be written

$$\begin{aligned} U &= -Nzr(1-r)(\Xi_{AA}+\Xi_{BB}) - Nz\{r^2+(1-r)^2\}\Xi_{AB} \\ &= -Nz\Xi_{AB} - Nzr(1-r)\Xi. \end{aligned} \tag{26.4}$$

We are assuming random distributions of the atoms on each set of sites, and since any arrangement on the A sites is independent of the arrangement on the B sites, the number of distinguishable ways of arranging the atoms is

$$\frac{(N/2)!}{(Nr/2)!\,(Nw/2)!}\cdot$$

After using Stirling's theorem in the usual manner, we obtain for the configurational free energy

$$G \approx F = -Nz\{\Xi_{AB}+r(1-r)\Xi\}+NkT\{r \ln r+(1-r)\ln(1-r)\}. \qquad (26.5)$$

To find the equilibrium value of r, we equate $\partial G/\partial r = 0$, and obtain

$$(1/(2r-1))\ln\{r/(1-r)\} = z\Xi/\boldsymbol{k}T \qquad (26.6)$$

or

$$(2/L)\tanh^{-1} L = z\Xi/\boldsymbol{k}T. \qquad (26.7)$$

Equations (26.6) and (26.7) are identical in form with eqns. (23.7) and (23.9), and they have the same properties. For all temperatures, a possible solution is $r = \frac{1}{2}$ $(L = 0)$; this gives a maximum free energy below a critical temperature $T_\lambda = -z\Xi/2\boldsymbol{k}$, and a minimum free energy above this temperature. The degree of long-range order is thus zero above T_λ and increases with falling temperature below T_λ. In the latter range, there are two further roots of eqn. (26.6) symmetrically disposed about $r = \frac{1}{2}$. These roots minimize the free energy, and either of them gives the equilibrium value of L at a given temperature.

In Fig. 6.13 the equilibrium value of L is plotted as a function of T/T_λ. The curve is nearly

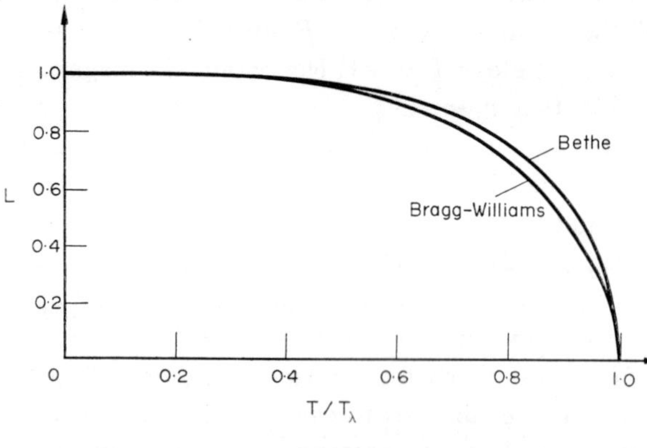

FIG. 6.13. Equilibrium long-range order vs. temperature curve for an *AB* superlattice according to the Bragg–Williams and Bethe approximations.

horizontal at first, and then falls more and more steeply as the degree of disorder increases. This behaviour is characteristic of co-operative phenomena, in which the resistance to further disordering (using the term order in a general sense) decreases with decreasing order. The best-known example of such a co-operative phenomenon is the alignment of elementary atomic spins, which leads to ferromagnetism; the Bragg–Williams theory of long-range order is formally analogous to the Weiss theory of ferromagnetism.

The temperature T_λ is often referred to as a λ point in order–disorder and similar theories, the name being taken from the shape of the specific heat anomaly in the neighbourhood of T_λ. From eqn. (26.4), the internal energy may be written

$$U(L) = -Nz\{\Xi_{AB}+(1-L^2)\Xi/4\} \tag{26.8}$$

and the total change in internal energy associated with the ordering process is thus $-Nz\Xi/4 = NkT_\lambda/2$. This value is in fair agreement with experiment. The curve of U as a function of temperature, obtained by combining (26.7) and (26.8), is shown in Fig. 6.14; there is a discontinuity at $T = T_\lambda$. The gradient of U gives the configurational contribution to the specific heat, i.e. the excess specific heat caused by the transformation. This is plotted in Fig. 6.15, together with the experimental curve for the ordering of β-brass, which is an

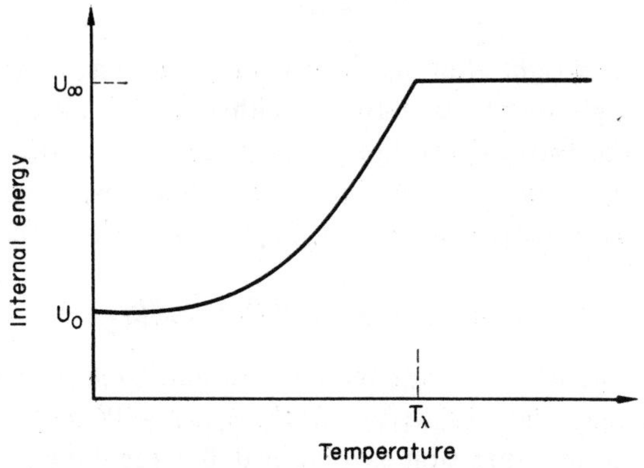

FIG. 6.14. Internal energy of an *AB* superlattice according to the Bragg–Williams theory.

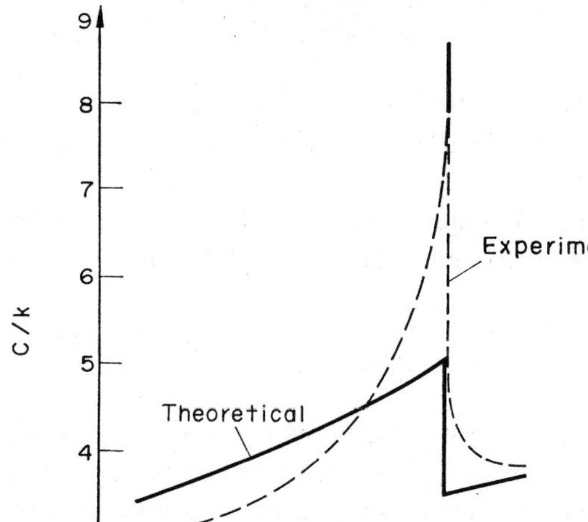

FIG. 6.15. Comparison of the Bragg–Williams prediction of the atomic specific heat with the experimental measurements of Sykes and Wilkinson (1937) on β-brass.

order–disorder transformation of the $L2_0-A2$ type. In this approximation, the specific heat remains finite at all temperatures, and the ordering reaction is not a first-order phase change.

The original treatment of this theory of long-range order did not refer specifically to the nearest neighbour model, and preceded Bethe's development of this model for alloys. Bragg and Williams considered the change in energy W associated with the interchange of an A and a B atom, both of which were originally in "right" positions, and, finally, in "wrong" positions. Use of Boltzmann's equation then gives an expression for r_A/w_A in terms of W; alternatively this expression may be derived by considering the kinetic balance of the rates at which right atoms interchange to give wrong atoms, and vice versa. These two processes may be written symbolically as an equation (analogous to a chemical equation)

$$A^r+B^r \rightleftharpoons A^w+B^w.$$

Note that interchanges of right A atoms with wrong B atoms, or vice versa, are without significance, since the right or wrong status of neither atom is affected by the interchange.

The rate at which the forward reaction takes place is proportional to the number of right A atoms, $N(x_A-x_w)$, and to the number of right B atoms, $N(x_B-x_w)$, whilst the rate of the back reaction is proportional to $(Nx_w)^2$. At equilibrium, these two rates may be equated, and

$$(x_A-x_w)(x_B-x_w)/x_w^2 = k_w/k_r, \tag{26.9}$$

where k_w, k_r are the rate constants for the forward and back reactions respectively. The equilibrium constant may then be written $k_w/k_r = \exp(-W/\boldsymbol{k}T)$.† On substituting from eqn. (26.3), the following expression is obtained for the long-range order parameter

$$L = 1-\frac{[4x_Ax_B\{\exp(W/\boldsymbol{k}T)-1\}+1]^{1/2}-1}{2x_Ax_B\{\exp(W/\boldsymbol{k}T)-1\}}. \tag{26.10}$$

The energy W introduced into this theory is not a constant, but depends upon L. Clearly W is zero in a state of complete disorder, since there is then no energy change on interchanging any A and B atoms,‡ and has a maximum value W_0 in the completely ordered state. Bragg and Williams made the simple assumption that there is a linear relation $W = W_0L$ For the equi-atomic superlattice, eqn. (26.10) simplifies to

$$L = 1-\frac{2\exp(W/2\boldsymbol{k}T)-1}{\exp(W/\boldsymbol{k}T)-1} = \tanh(W/4\boldsymbol{k}T). \tag{26.11}$$

Comparing this with eqn. (26.7), we see that the expressions are identical if $W = W_0L = -2z\Xi L$. It may readily be seen that this equivalence applies to the more general expression (26.10) and the corresponding equation derived by the zeroth approximation of the quasi-chemical method. The two theories are in fact identical if the substitution $W_0 = -2z\Xi$ is

† Note that this derivation shows the implicit assumption is being made that the entropy of activation is equal for the two processes. For a further discussion, see Part II, Chapter 18.

‡ More correctly, the energy change is equally likely to be positive or negative.

made, and the Bragg–Williams assumption that $W = W_0 L$ is equivalent to the quasi-chemical assumption of random mixing in the two sets of positions. This serves to emphasise that although the zeroth approximation is developed in terms of the nearest neighbour model, it makes no real use of the short-range character of the binding forces.

We next consider the nature of the changes introduced by removing the restrictive approximation that the atoms are arranged randomly on the two sets of sites. Clearly, it is possible to develop a treatment or order–disorder phenomena based on the assumption of independent nearest neighbour interaction, and this will be exactly analogous to the first approximation of the quasi-chemical method. Such a theory of order–disorder reactions was first given by Bethe (1935). Bethe's analysis was more complex than that given above, but the equivalence of the two methods was proved by Rushbrooke (1938) and Fowler and Guggenheim (1940). We shall not repeat the derivations, but simply assume that we can transcribe the appropriate formulae from Section 24. In particular, we find for the critical temperature of long-range order in an $L2_0$ superlattice

$$T_\lambda = -\Xi/\boldsymbol{k}\ \ln(z/(z-2)). \tag{26.12}$$

The equilibrium roots of r below $T = T_\lambda$ are given by

$$\frac{1-[r/(1-r)]}{[r/(1-r)]^{1/z}-[r/(1-r)]^{(z-1)/z}} = \exp(-\Xi/\boldsymbol{k}T), \tag{26.13}$$

which is the equivalent of (24.16). In Fig. 6.13 we compare the curve for r, and hence for $L = 2r-1$, given by eqn. (26.13) for an $L2_0$ superlattice with that given by eqn. (26.6). Both equations predict that the long-range order falls to zero at a critical temperature (λ point), but the Bethe theory shows that the long-range order decreases more slowly than is predicted by the Bragg–Williams approximation. It follows from the general correspondence of the methods, that the order–disorder curves of Fig. 6.13 are the same as the solubility limit curves of Fig. 6.9.

The configurational internal energy will be given by

$$U = -\tfrac{1}{2}Nz(\Xi_{AA}+\Xi_{BB}-2\overline{N_{AB}}\Xi/N) \tag{26.14}$$

(see p. 189), where $\overline{N_{AB}}$ is a function of temperature and is determined by eqn. (24.7) with $N_A = N_B = \tfrac{1}{2}N$. As $T \to 0$ K, the superlattice structure becomes perfect, and $\overline{N_{AB}} \to \tfrac{1}{2}N$. Thus at 0 K

$$U_0 = -\tfrac{1}{2}Nz(\Xi_{AA}+\Xi_{BB}-\Xi). \tag{26.15}$$

At very high temperatures, $\overline{N_{AB}}$ is equal to the value N^θ_{AB} corresponding to $r = \tfrac{1}{2}$ and random mixing on each of the sublattices, i.e. to $N/4$. We then have

$$U_\infty = -\tfrac{1}{2}Nz(\Xi_{AA}+\Xi_{BB}-\tfrac{1}{2}\Xi), \tag{26.16}$$

and the total energy of disordering is $-Nz\Xi/4$, as in the zeroth approximation. The present treatment shows, however, that only part of the configurational energy of ordering disappears below the λ point. From eqns. (26.14), (26.15), and (26.16)

$$(U_T-U_0)/(U_\infty-U_0) = 2(1-2\overline{N_{AB}}/N). \tag{26.17}$$

This equation gives implicitly the value of the configurational energy U_T as a function of T or of the order parameter L. At any temperature, $\overline{N_{AB}}$ may be found from the following equation, which is analogous to eqn. (24.7),

$$\xi^2 = (r-\xi)(1-r-\xi)\exp(2\Xi/kT),$$

where $2\xi = 1-2\overline{N_{AB}}/N$. The solution may be written in the form (24.8) as

$$\xi = 2\xi/(\beta+1) = 2r(1-r)/(\beta+1),$$

and β is given by eqn. (24.9) with Ξ replaced by $-\Xi$ and x replaced by r. The value of r in these equations is given by (26.6). At $T = T_\lambda$ and higher temperatures, $r = \frac{1}{2}$ and $\beta = \exp(-\Xi/kT)$. This gives

$$\overline{N_{AB}} = \frac{N\exp(-\Xi/kT)}{2[1+\exp(-\Xi/kT)]}. \qquad (26.18)$$

Above the critical point, the curve of U_T against T is represented by the equation

$$U_T - U_0 = \frac{-Nz\Xi}{2[1+\exp(-\Xi/kT)]}. \qquad (26.19)$$

Below the critical point, it is not possible to write a simple analytical expression for $U_T - U_0$. The curve of $U_T - U_0$ is plotted against T/T_λ in Fig. 6.16, together with the corres-

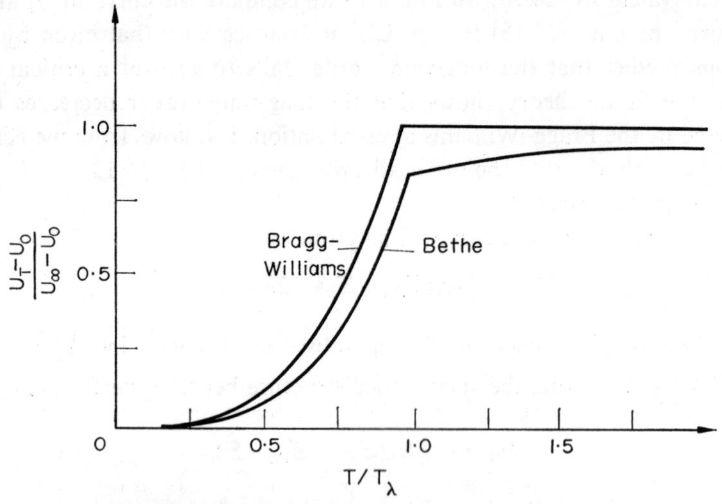

FIG. 6.16. Comparison of the internal energy vs. temperature curves given by the zeroth (Bragg–Williams) and first (Bethe) approximations for an *AB* superlattice.

ponding curve for the zeroth approximation. We may also calculate from (26.19) that

$$(U_\lambda - U_0)/(U_\infty - U_0) = (z-2)/(z-1).$$

For a b.c.c. structure, sixth-sevenths of the total ordering energy is destroyed below the critical temperature according to this approximation, and the remaining one-seventh disappears more gradually as the temperature is raised above T_λ. The gradient of U_T gives the configurational specific heat, i.e. the excess specific heat above the Debye curve. This is shown in Fig. 6.17.

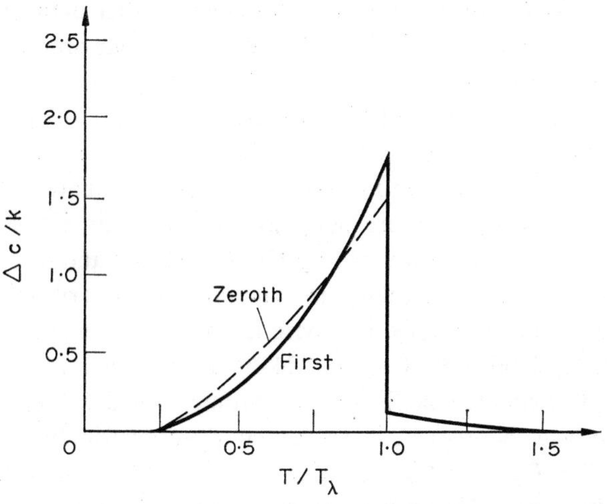

FIG. 6.17. Comparison of the excess atomic specific heat curves given by the zeroth and first approximations.

The principal advantage of the Bethe treatment over the Bragg–Williams approximation is in this prediction of an excess specific heat above the disordering temperature, since this is always observed experimentally. The physical interpretation of the effect is in terms of the concept of short-range order. Although the long-range order becomes zero at the critical temperature, the number of unlike nearest neighbour pairs remains larger than would be found in a random distribution, so that short-range order persists above the critical point. The destruction of long-range order has been frequently compared to the melting of a solid; local crystallinity persists in the liquid, and local order in the disordered solution.

Bethe introduced a short-range order parameter σ, which is defined so that $\sigma = 1$ for perfect long-range order, and $\sigma = 0$ for a completely random distribution. In these respects, σ is equivalent to L, but the definition is directly in terms of the nearest neighbours of an atom rather than of the segregation into different lattices. The general definition of the short-range order parameter is

$$\sigma = (\overline{N_{AB}} - N^{\theta}_{AB})/(N^{0}_{AB} - N^{\theta}_{AB}), \tag{26.20}$$

where $\overline{N_{AB}}$, N^{θ}_{AB} are the actual value of N_{AB} and the value for a purely random arrangement, as before, and N^{0}_{AB} is the value of N_{AB} for the fully ordered state. In this simple case we have been considering, $N^{0}_{AB} = N/2$, and $N^{\theta}_{AB} = N/4$, so that

$$\sigma = (4\overline{N_{AB}}/N) - 1. \tag{26.21}$$

Comparing this with eqn. (26.17), we see

$$(U_T - U_0)/(U_\infty - U_0) = 1 - \sigma. \tag{26.22}$$

With this definition of ordering, the internal energy of ordering is proportional to the degree of short-range order. At T_λ, for example, the short-range order parameter $\sigma = \frac{1}{7}$.

Although Bethe's theory is successful in explaining short-range order, detailed comparison of experimental and theoretical results shows rather poor agreement. Attempts at more exact theories usually take the form of higher cluster approximations or series expansions (see Takagi, 1941). We shall not consider these here, but it is interesting to refer briefly again to the exact solutions obtained in two dimensions. A square lattice forms a superlattice of the type described above, there being no *AA* or *BB* pairs in the fully ordered state. The rather surprising result, first obtained by Onsager (1944) and discussed more fully by Wannier (1945), is that the specific heat of the structure has a logarithmic infinity at $T = T_\lambda$. The heat content remains continuous, i.e. there is no latent heat, and the change is thermodynamically of the second order (see pp. 224–6). This result is at variance with the predictions of all the approximate calculations, according to which there is a finite discontinuity in the specific heat. Moreover, in the exact two-dimensional solution, the curve is symmetrical about $T = T_\lambda$, so that as much of the energy of disordering is required above T_λ as below it. Exact solutions in three dimensions are unlikely to be obtained; it is probable that such a solution would still contain a logarithmic infinity in the specific heat, but would not be symmetrical about T_λ.

The above discussion has been confined to the simplest equi-atomic superlattice. The next most important example is the structure of type $L1_2$, formed in A_3B alloys. We shall not develop the theory of this superlattice, but shall merely describe the results. We now have $w_B = 3w_A = 4x_w$, and the Bragg–Williams definition of long-range order becomes

$$L = (4r_A - 1)/3 = 1 - 16x_w/3. \tag{26.23}$$

Attempts to apply the first approximation, or Bethe theory, to the $L1_2$ superlattice result in a contradiction; the hypothesis of the non-interference of nearest neighbour pairs does not lead to equilibrium long-range order at all. The reason for this difficulty seems to be that at least four atom sites are needed to define a unit cell of the superlattice. It is thus necessary to use a higher approximation of the quasi-chemical theory (cluster variation method), embodying the hypothesis of the non-interference of tetrahedral groups of atoms. The stability of the $L1_2$ structure at the A_3B composition was first shown in this way by Yang (1945), and the treatment was considerably extended in later papers by Yang and Li (1947) and Li (1949). These papers also considered the more difficult problems associated with non-stoichiometric compositions. When x_B is a variable, the use of the quasi-chemical method ensures that all properties are symmetrical about $x_B = \frac{1}{2}$, so, in this treatment, superlattices based on both A_3B and AB_3 necessarily appear in the same binary alloys.

The results of the Bragg–Williams approximation and of the tetrahedral cluster method are shown in Fig. 6.18, which gives the equilibrium degree of long-range order as a function of temperature. In contrast to the $L2_0$ type of superlattice, L drops discontinuously to zero at the critical temperature. The transformation thus requires a latent heat, and is correctly

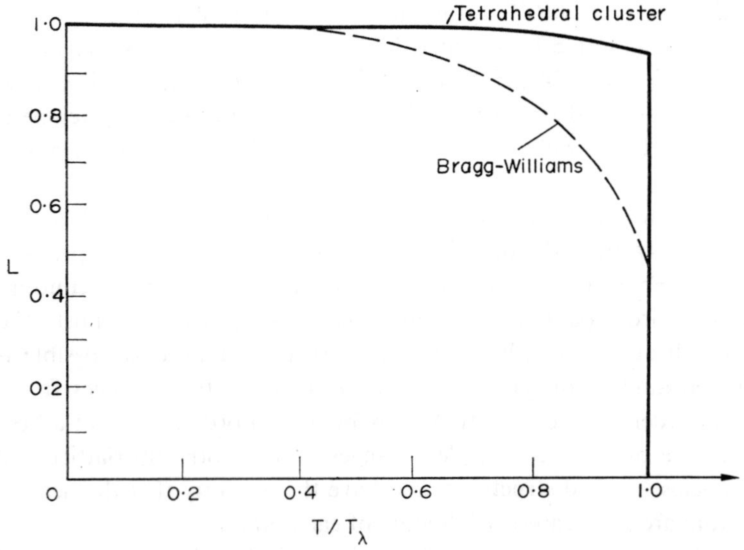

FIG. 6.18. Equilibrium long-range order vs. temperature curve for an A_3B superlattice of type $L1_2$ according to the Bragg–Williams and tetrahedral cluster approximations (after Guggenheim, 1952).

described as a thermodynamic phase change of the first order. The transformation temperature, at which the superlattice phase with a finite degree of long-range order is in equilibrium with the solution of zero long-range order, is given by

$$\left.\begin{aligned} T &= -z\Xi/7{\cdot}3k \qquad \text{(Bragg–Williams)},\\ T &= -z\Xi/14{\cdot}6k \qquad \text{(tetrahedral cluster)}. \end{aligned}\right\} \tag{26.24}$$

These values are obtained by numerical solution of the equations, no simple analytical expressions being obtained.

An approximate theory of order in alloys due to Cowley (1950 a, b) differs in some respects from the cluster variation methods. Cowley considers a set of short-range order coefficients which effectively define the probability $P_{AB}(\mathbf{r})$ for each of the different shells (successively larger values of r) around an atom. The internal energy is defined in terms of the interactions of pairs of atoms in different shells, so that the theory is not confined to nearest neighbour interactions. It is assumed as an approximation that the order coefficients are independent of each other, this corresponding approximately to the independence of pairs in the first approximation of the quasi-chemical theory.

Cowley's theory has the advantage of being related rather more closely to the experimental X-ray measurements of the $P_{AB}(\mathbf{r})$ quantities than are the above descriptions. The results of work on the diffuse X-ray scattering from solid solutions are frequently expressed in terms of the order coefficients used by Cowley (positive or negative for short-range order and clustering; see Section 27), together with size coefficients representing the displacements of the atoms from the ideal sites of a structure with the measured lattice parameter. This work will not be described here; reviews have been given by Averbach (1956) and

Sivertsen and Nicholson (1961). From the mean pair probabilities for Cu_3Au, Cowley deduced that the values of Ξ for the first three shells (sets of nearest neighbours) are given by $\Xi_1/\boldsymbol{k} = 358°$, $\Xi_2/\boldsymbol{k} = -34°$, $\Xi_3/\boldsymbol{k} = -19°$. For the $L2_0$ superlattice, Cowley's theory gives the same results as the Bragg–Williams theory; the disordering temperature for the other types of superlattice differs slightly from the predictions of the various theories mentioned above.

The question of whether superlattice changes are first- or second-order thermodynamic transformations has attracted much discussion. As we have seen, the approximate solutions of the nearest neighbour model suggest that some transitions involve a continuous decrease of the long-range order parameter to zero (second-order change), whilst others predict a discontinuous fall at $T = T_c$ (first-order change). Insofar as it is possible to generalize from the limited number of types of transition, it seems that changes from $L2_0$ or DO_3 structures to disordered b.c.c. structures may be second order, and that all others are first order. The feature of the second-order changes is that both superlattice and disordered phase are non-close-packed structures, and have the property that the nearest neighbours of any one atom are not nearest neighbours of each other.

The thermodynamic classification of transitions is based on the order of the lowest derivative of G which shows a discontinuity at the transition temperature. If the values of this function are G^α and G^β for the two forms concerned, the transition temperature is defined by the condition $G^\alpha = G^\beta$. In a normal first-order transition, there are discontinuities in the derivatives of G with respect to temperature (entropy) and with respect to pressure (volume). In a second-order transition, the entropy and volume are continuous, but their derivatives (e.g. the specific heat and the compressibility) are not. It follows that the enthalpy is also continuous, and the transition has an anomalous specific heat but no latent heat.

A well-known difficulty in the theory of even-order transitions arises if the difference in the free energies of the two phases $\Delta G^{\alpha\beta}$ is expanded as a Taylor series in terms of the difference in temperature $\Delta T = T - T^{\alpha\beta}$. This gives

$$\Delta G^{\alpha\beta} = -(\Delta S^{\alpha\beta})\,\Delta T + \tfrac{1}{2}(\partial^2 G/\partial T^2)\,\Delta T^2 + \tfrac{1}{6}(\partial^3 G/\partial T^3)\,\Delta T^3 + \ldots$$

The sign of $\Delta G^{\alpha\beta}$ is determined by the sign of the first non-vanishing term on the right, and this is the nth term for an nth-order transition. For a first- or third-order transition, $\Delta G^{\alpha\beta}$ changes sign with ΔT, but for an even-order transition, it has the same sign above and below $T^{\alpha\beta}$. Thus if attempts are made to draw G^α and G^β curves, they do not intersect at the transition temperature, but only touch, with one curve always below the other. This apparently results in either α or β always being stable.

This difficulty is thought to arise because of the attempt to extrapolate properties of the assembly into regions in which they do not exist. We cannot really draw separate G^α and G^β curves for a second-order superlattice transition, but only a single curve which represents an ordered structure below the transition temperature and a substantially disordered structure above it. The existence of the singularity in $\partial^2 G/\partial T^2$ renders the analytic expansion above invalid.

A second-order transition in a solid solution implies not only a vanishing of the latent heat of transition, but also a continuity of composition. The transition temperature is marked by a single line on the phase diagram, and the coexistence of ordered and disordered phases is not possible at equilibrium. This provides one of the most convenient experimental tests of the order of a transition, it being very difficult in practice to distinguish between latent heats and anomalous specific heats. In the early stages of order–disorder theory, there was a tendency to regard the disordering of almost all superlattices as second-order transitions, but it has become increasingly evident that many of these changes are first order. It seems obvious that the transition must be first order when there is a change in lattice symmetry, as in the $L1_0$ (tetragonal) superlattice formed from the f.c.c. structure, and the most convincing evidence of the existence of equilibrium two phase regions comes from alloys undergoing this structural change.†

Rhines and Newkirk (1953) have suggested that all superlattice transitions are first-order phase changes, and they presented some evidence, based on electrical resistivity measurements, to show that this is true even for the $L2_0-A2$ transition in copper–zinc alloys. In general, this suggestion has not been substantiated by later workers; for example, a two-phase region was sought but not found in a very careful investigation of copper–zinc by Beck and Smith (1952). Moreover, measurements of long-range order by X-ray methods tend to support the division of superlattice changes into first- and second-order types. For the $L2_0$ structures in copper–zinc and silver–zinc alloys, the long-range order decreases continually with temperature (Chipman and Warren, 1950; Muldawer, 1951), the decrease becoming more and more rapid as the critical temperature is approached. Whilst the experimental techniques do not allow the definite conclusion that there is no discontinuity in L, it is certain that the magnitude of any discontinuous drop is much smaller than that for the transitions from $L1_0$ or $L1_2$ superlattices in copper–gold and copper–platinum alloys. In these alloys, the amount of long-range order decreases only slightly as the temperature is increased within the superlattice region, and then drops discontinuously to zero (Keating and Warren, 1951; Roberts, 1954; Walker, 1952). Thus these measurements tend to confirm the qualitative predictions of the nearest neighbour model for all kinds of superlattice.

We have assumed throughout this section that the ordering tendency is provided by the chemical interactions between nearest neighbour atoms. Suggestions are often made that an atomic size disparity is an important factor in lowering the energy of the superlattice relative to that of a random arrangement, a view first proposed by Hume-Rothery and Powell (1935). It is not entirely clear whether this distinction is meaningful (see p. 199), but it is obvious that a strain energy calculation of the type used in Section 25 has no relevance to the ordering energy, which must arise from more localized interactions. At the same time, there are a number of superlattices known to have large unit cells, and the formation of these structures cannot be explained on the basis of purely nearest neighbour interactions.

The best-known example of a superlattice with a large unit cell is the structure CuAu II. At low temperatures, equi-atomic copper–gold alloys form a tetragonal superlattice of the $L1_0$ form (CuAu I) and, at high temperatures, the structure is disordered f.c.c. However,

† The Bragg–Williams theory predicts a second-order transition, but the first approximation of the quasi-chemical theory correctly predicts a first-order transition (Guggenheim, 1952).

it was found in 1936 that in an intermediate temperature range, now known to be ~ 380°C to ~ 405°C, the equilibrium structure has a more complex unit cell with dimensions $10a$, a, and c, where c and a are the tetragonal parameters (Johannson and Linde, 1936). The structure may be considered to be formed by introducing antiphase domain boundaries parallel to (100) planes at every five lattice planes of the tetragonal structure, so that the repeat unit has to contain two opposite "domains". The complete diffraction effects, which are rather complex, also indicate a small expansion across the antiphase domain boundaries. Direct evidence for the existence of these domains has been obtained by transmission electron-microscopy of thin films (Pashley and Presland 1958–9; Glossop and Pashley, 1959). The structure consists of regions within which the antiphase domain boundaries are parallel planes; two kinds of region are observed, corresponding to the two perpendicular a directions of the tetragonal cell.

Other examples of structures with large unit cells are now known, and are formed by a similar disturbance of the structure of a simple superlattice, but the spacing of the antiphase boundaries is not always constant, sometimes varying with temperature and composition (Schubert *et al.*, 1954, 1955). From this point of view, it seems attractive to regard the structures as superlattices containing some kind of fault. Nevertheless, and in spite of our description of the CuAu II structure as a tetragonal superlattice with regularly spaced antiphase domain boundaries, it is quite clear that, in these alloys at least, the large superlattice is the thermodynamically stable form. This cannot be explained by the nearest neighbour model because an A atom has the same number of nearest neighbour B atoms in both the CuAu I and the CuAu II structures if the slight deviations from cubic symmetry are ignored.

The thermodynamics of the transitions CuAu I → CuAu II → disordered f.c.c. have been investigated by Oriani and Murphy (1958). The latent heats for the two changes are 888 and 1590 J g atom^{-1} respectively; the former value seems remarkably large in view of the slight structural rearrangement involved. The results show other disagreements with the quasi-chemical theory. For example, the total disordering energy per g-atom for the transition CuAu I → disordered f.c.c. can be evaluated, and the temperature T_λ for this transition in the absence of the CuAu II structure can be estimated. The ratio of the disordering energy to $\boldsymbol{R}T_\lambda$ is found to be 0·73 compared with 0·5 of the zeroth approximation and 1·37 given by Li's treatment of the $L1_2-A1$ change using the tetrahedral cluster method. The heats of formation of both ordered and disordered phases were measured by Oriani and Murphy, and give independent estimates of Ξ. These values are 1570 and 1940 J g atom^{-1} respectively. It thus seems clear that the quasi-chemical treatment is extremely unsatisfactory for the CuAu superlattices.

27. FLUCTUATIONS IN SOLID SOLUTIONS: SHORT-RANGE ORDER AND CLUSTERING

We have emphasized that the pair probability functions will generally not have the values corresponding to a completely random arrangement, even in a substantially disordered solid solution. These functions, or related order parameters, may be measured by X-ray methods, and give quantitative information on the amount of short-range order or clustering. In general, short-range order, for which $P_{AB}(\mathbf{r}) > 2x(1-x)$ when $\mathbf{r}$ represents a nearest neighbour vector, is expected if the solid solution undergoes a superlattice transition at lower temperatures. Clustering (negative short-range order), for which $P_{AB}(\mathbf{r}) < 2x(1-x)$ for nearest neighbour vectors, is expected if there is a solubility gap at lower temperatures.†

The X-ray measurements of short-range order give the average number of neighbours of given kinds possessed by any atom. We now examine the nature of the qualification expressed by the word "average". An instantaneous picture of a solid solution would obviously show that many atoms had a surrounding configuration quite different from the average. This is because the average configuration is obtained by a dynamic averaging process over all the available configurations. In postulating this, we are assuming some mechanism for changing from one configuration to another; this is achieved in practice by atomic diffusion.

For kinetic applications, we require a quantitative estimate of how the arrangement of a small region of the assembly may momentarily differ from the average arrangement. This is provided by the theory of fluctuations, and we express the probable state of the assembly in terms of the root mean square deviation from the average arrangement. We begin by considering a solid solution in which the atomic arrangement is completely random. The statistical properties of such a solution are especially easy to derive, since all arrangements have equal probability.

We consider in all N atoms, each having z nearest neighbours. For each atom, we fix attention first on a particular neighbour. The probability that this neighbour is a B atom is x, and hence there will be Nx of the neighbours we are considering which are B atoms. Now consider another neighbour of each of our N atoms. Since the probabilities are independent, there will again be a chance x that each such neighbour is a B atom. The number of $B-B$ pairs among the N sets of two neighbours will thus be proportional to Nx^2, whereas the number of $A-B$ pairs will be $Nx(1-x)$. Similarly, the number of $A-A$ pairs and of $A-B$ pairs will be proportional to $N(1-x)^2$ and $Nx(1-x)$. We thus have that for $z = 2$, the numbers of atoms having two A atoms, an A atom and a B atom, and two B atoms as nearest neighbours are $N(1-x)^2$, $2Nx(1-x)$ and Nx^2 respectively. Exactly similar reasoning shows that for $z = 3$, the numbers of atoms with three A atoms, two A atoms and one B atom, one A atom and two B atoms, and three B atoms are proportional to $N(1-x)^3$, $3N(1-x)^2x$, $3N(1-x)x^2$ and Nx^3 respectively. The relative numbers of neighbours of the different types are given by the terms of the expansions $\{(1-x)+x\}^2$ and $\{(1-x)+x\}^3$. We conclude by extension of this reasoning that the relative numbers of atoms which have as nearest neighbours zA atoms, $(z-1)\,A$ atoms and one B atom, $(z-2)\,A$ atoms and two B atoms, ..., one A atom and $(z-1)\,B$ atoms, and $z\,B$ atoms are given by the terms of the binomial expansion

$$\{(1-x)+x\}^z.$$

† For many years, gold–nickel alloys were quoted as an extreme example of the failure of the quasi-chemical theory, since they have a solubility gap at low temperatures, and X-ray measurements (Flinn, *et al.*, 1953) indicated positive short-range order at high temperatures. Later work by Munster and Sagel (1959) has shown that the interpretation of the earlier X-ray results may have been incorrect, and there may actually be clustering above the solubility gap. It is possible that this is a necessary result, and is independent of the assumptions of any particular model.

This expression is not a continuous mathematical function, i.e. there is no chance of finding an atom with a fractional number of A atoms among its nearest neighbours. There is a distinction between the average number of neighbours of a given type and the most probable number. Thus if $x = 0{\cdot}7$ and $z = 4$, each atom has on the average 1·2 A atoms and 2·8 B atoms as nearest neighbours. The most probable neighbours for any atom are one A atom and three B atoms.

The above result is not confined to nearest neighbour probabilities, since we have simply calculated the probability of having a given number of B atoms amongst the z atoms which happen to be neighbours of any atom. We are often interested in rather larger clusters of (say) n atoms. The probabilities of our finding 0, 1, 2, 3, ..., $n-1$, n B atoms in such a cluster when the average composition is x are given by the terms of the expansion $\{(1-x)+x\}^n$, and the general term, i.e. the probability of the group containing (say) m B atoms is

$$\binom{n}{m}(1-x)^{n-m}x^m. \tag{27.1}$$

This formula may be used as a first approximation in the problem of nucleus formation in a solid solution.

If the solution is very dilute, the above law may be simplified. We let $x \to 0$, increasing n at the same time, so that the average number of solute atoms in the group, nx, remains constant. The general term of the binomial expansion can now be written as

$$\frac{n!}{m!(n-m)!}\left(\frac{nx}{n}\right)^m\left(1-\frac{nx}{n}\right)^{n-m} = \frac{(nx)^m}{m!}\left(1-\frac{nx}{n}\right)^n\left[\frac{n!}{(n-m)!\,n^m(1-nx/n)^m}\right].$$

By using Stirling's theorem, the term in square brackets reduces to

$$\left(1-\frac{nx}{n}\right)^{-m}\left(1-\frac{m}{n}\right)^{-n+m-\frac{1}{2}}\exp(-m)$$

and as $n \to \infty$ with nx constant, this may be replaced by unity. At the same time, the term $(1-nx/n)^n$ tends to $\exp(-nx)$. We thus have, finally, for the probability of finding mB atoms amongst n atoms:

$$\{(nx)^m/m!\}\exp(-nx). \tag{27.2}$$

This is Poisson's distribution. The expression has a maximum value when $m = nx$, i.e. when the number of solute atoms in the group equals the average number, and the concentration fluctuation is zero. The fluctuation is most conveniently expressed by the root mean square deviation from the average value; this is the conventional mean deviation of statistical theory.

Since $(m-nx)^2 = m^2-(nx)^2+2nx(nx-m)$, and the mean deviation of $(nx-m)$ is zero, we have

$$\overline{(m-nx)^2} = \sum_{m=0}^{n}[m^2-(nx)^2]\{(nx)^m/m!\}\exp(-nx).$$

The first part of this expression is

$$\sum_{m=0}^{n} (m^2/m!)\,(nx)^m \exp(-nx) = nx(1+nx),$$

so that $\overline{(m-nx)^2} = nx$, and the mean deviation is $(nx)^{1/2}$. Expressed as a composition fluctuation, we write $m = n(x+\Delta x)$, and the mean deviation is given by

$$\overline{(\Delta x)^2} = x/n. \tag{27.3}$$

The concentration fluctuations thus increase as the number of atoms in the group diminishes. Provided n is large enough (or, equivalently, x is small enough), to justify the use of Poisson's equation, the mean deviation is proportional to $n^{-1/2}$.

A similar method may be used to find the mean deviation of the more exact binomial expansion. We then obtain

$$\left.\begin{aligned} \overline{(m-nx)^2} &= nx(1-x) \\ \text{or} \qquad \overline{(\Delta x)^2} &= x(1-x)/n. \end{aligned}\right\} \tag{27.4}$$

Obviously, this reduces to (27.3) when x is small.

In the above discussion, we have not been concerned with the effects of temperature at all. This is because all arrangements of the ideal solution have the same energy. Temperature has no effect on the magnitude of the fluctuations, although it will determine the rate of fluctuation. In a real solution, we must consider the change in free energy which is associated with a fluctuation. The probability of the fluctuation is then related to this change in energy by the Boltzmann factor.

Consider a region of solid solution which has initially a concentration of B atoms equal to the average or equilibrium concentration, x. As a result of a fluctuation, one half of this region may be supposed to contain a fraction $x+\Delta x$ of B atoms, whilst the other half contains a fraction $x-\Delta x$ of B atoms. If the free energy per atom is $g(x)$, we have

$$\Delta g = \tfrac{1}{2}\{g(x+\Delta x)+g(x-\Delta x)\}-g(x).$$

And after expanding in a Taylor series, we obtain

$$\Delta g = \tfrac{1}{2}(\partial^2 g/\partial x^2)\,(\Delta x)^2$$

provided that Δx is small. If the original volume element contained n atoms, the total change in the free energy of the assembly is $n\,\Delta g$. Since the average value of $n\,\Delta g$ must be $\tfrac{1}{2}kT$, the mean square fluctuation is

$$\overline{(\Delta x)^2} = kT/n(\partial^2 g/\partial x^2) \tag{27.5}$$

and this equation replaces (27.4) when the energy of the assembly is a function of the arrangement. The value of $\partial^2 g/\partial x^2$ for the zeroth approximation of the quasi-chemical theory is derived from eqn. (23.5). Substituting this into (27.5),

$$\overline{(\Delta x)^2} = \left(\frac{n}{x(1-x)} - \frac{2z\Xi}{kT}\,n\right)^{-1}. \tag{27.6}$$

This equation reduces to (27.4) when $\Xi \to 0$, this being the condition for the solution to be ideal. A similar result is obtained when T becomes very large, so that, at high temperatures, the fluctuations approach those expected in a random solid solution, as is physically obvious. At lower temperatures, the fluctuations depend on the sign of Ξ. If Ξ is positive, there is a tendency to clustering, and the fluctuations are greater than those in an ideal solution. As the temperature is lowered, the fluctuations become larger, and according to (27.5) are infinite when $\partial^2 g/\partial x^2 = 0$. This equation is, of course, only valid for small x, but it is clear that large fluctuations may be expected under these conditions since there is no first-order dependence of the free energy on the atomic arrangement. Thus when the spinodal is reached, all concentrations have the same probability. When Ξ is negative (tendency to ordering) the fluctuations are always smaller than would be found in an ideal solution.

The above treatment, though giving a better approximation to the fluctuations in a solution than do the purely statistical expressions, is logically rather unsatisfactory. We have used the zeroth approximation for the free energy, and so have assumed that the equilibrium configuration is completely random, even though the energy depends on this configuration. In a more accurate treatment, $\partial^2 g/\partial x^2$ would be calculated from the first approximation, and used in eqn. (27.5). Unfortunately, this does not lead to any simple analytical expression. We should note, however, the difference between the value of $(\Delta x)^2$ given by eqn. (27.6) and the mean number of neighbours of a given kind possessed by any atom. Thus above the critical temperature, the equilibrium, or mean, composition in any region of n atoms

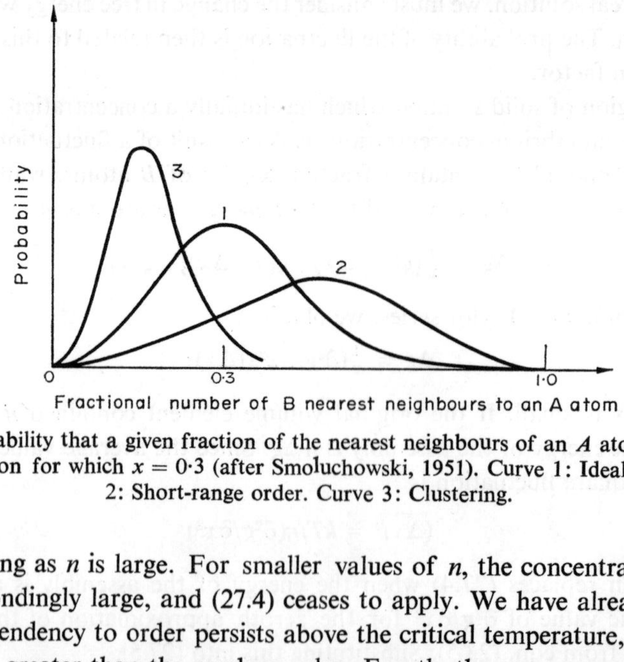

FIG. 6.19. Probability that a given fraction of the nearest neighbours of an A atom are B atoms in a solid solution for which $x = 0{\cdot}3$ (after Smoluchowski, 1951). Curve 1: Ideal solution, Curve 2: Short-range order. Curve 3: Clustering.

is always x so long as n is large. For smaller values of n, the concentration fluctuations become correspondingly large, and (27.4) ceases to apply. We have already seen that for Ξ negative, the tendency to order persists above the critical temperature, and the number of $A-B$ pairs is greater than the random value. Exactly the same reasoning holds for Ξ positive; we expect a finite configurational specific heat corresponding to the atomic clusters which gradually disappear with increasing temperature. This clustering or short-range ordering is quite distinct from the fluctuations described above, and expresses the fact that if we view on a sufficiently fine scale, the average distribution is no longer uniform. For Ξ positive we have a short-range phase mixture, just as we have short-range order for Ξ negative. Thermodynamically, the solution remains one phase, since thermodynamic concepts only have meaning for large numbers of atoms. Figure 6.19 shows qualitatively the number of nearest neighbours of type $A-B$ for $x = 0{\cdot}3$.

REFERENCES

ALLEN, S. M. and CAHN, J. W. (1972) *Acta Metall*, **20**, 423.
AVERBACH, B. L. (1956) *Theory of Alloy Phases*, p. 301, American Society for Metals, Cleveland, Ohio.
BECK, L. H. and SMITH, C. S. (1952) *Trans. Am. Inst. Min (Metall.) Engrs.* **194**, 1079.
BETHE, H. (1935) *Proc. R. Soc.* A, **150**, 552.
BRAGG, W. L. (1940) *Proc. Phys. Soc.* **52**, 105.
BRAGG, W. L. and WILLIAMS, E. J. (1934) *Proc. R. Soc.* A, **145**, 699; (1935) *Ibid.* **151**, 540.
BROWN, W. B. and LONGUET-HIGGINS, H. C. (1951) *Proc. R. Soc.* A, **209**, 416.
CAHN, J. W. (1961) *Acta Metall.* **9**, 795; (1969) *The Mechanism of Phase Transformations in Crystalline Solids*, p. 1, Institute of Metals, London.
CAHN, J. W. and HILLIARD, J. E. (1958) *J. Chem. Phys.* **28**, 258; (1971) *Acta Metall.* **19**, 151.
CHIPMAN, W. B. and WARREN, B. E. (1950) *J. Appl. Phys.* **21**, 696.
CLAPP, P. C. and MOSS, S. C. (1968) *Phys. Rev.* **171**, 754.
COTTRELL, A. H. (1955) *Theoretical Structural Metallurgy*, Arnold, London.
COWLEY, J. M. (1950a) *J. Appl. Phys.* **21**, 24; (1950b) *Phys. Rev.* **77**, 664.
DE FONTAINE, D. (1975) *Acta Metall.* (in press).
ESHELBY, J. D. (1954) *J. Appl. Phys.* **25**, 255; (1956) *Solid State Phys.* **3**, 79.
FLINN, P. A., AVERBACH, B. L., and COHEN, M. (1953) *Acta Metall.* **1**, 664.
FOWLER, R. H. and GUGGENHEIM, E. A. (1940) *Proc. R. Soc.* A, **174**, 189.
FREEDMAN, J. F. and NOWICK, A. S. (1958) *Acta Metall.* **6**, 176.
FREUNDLICH, H. (1926) *Colloid and Capillary Chemistry* (transl. of 3rd German edition).
FRIEDEL, J. (1955) *Phil. Mag.*, **46**, 514.
GIBBS, J. W. (1878) *Trans. Conn. Academy* **3**, 439; *Collected Works*, I, p. 219, Yale University Press, New Haven, Connecticut.
GLOSSOP, A. B. and PASHLEY, D. W. (1959) *Proc. Roy. Soc.* A, **250**, 132.
GUGGENHEIM, E. A. (1948) *Trans. Faraday Soc.* **44**, 1007; (1952) *Mixtures*, Clarendon Press, Oxford.
GUINIER, A. (1959) *Solid State Phys.* **9**, 293.
HARDY, A. H. (1953) *Acta Metall*, **1**, 202.
HILLERT, M. (1970) *Phase Transformations*, p. 181, American Society for Metals, Metals Park, Ohio.
HUME-ROTHERY, W. and POWELL, H. M. (1935) *Z. Kristallogr.* **91**, 23.
HUME-ROTHERY, W. and RAYNOR, G. V. (1954) *The Structure of Metals and Alloys*, Institute of Metals, London.
JOHANNSON, C. H. and LINDE, J. O. (1936) *Ann. Phys.* **25**, 1.
KAUFMAN, L. (1967) *Phase Stability of Metals and Alloys*, p. 125, McGraw-Hill, New York; (1969) *Prog. Mater. Sci.* **14**, 57.
KAUFMAN, L. and BERNSTEIN, H. (1970) *Computer Calculation of Phase Diagrams*, Academic Press, New York.
KEATING, D. T. and WARREN, B. E. (1951) *J. Appl. Phys.* **22**, 286.
KHACHATURYAN, A. G. (1962) *Phys. Met. and Metallg.* **13**, 493; (1973) *Phys. Stat. Sol.* **60**, 9.
KIRKWOOD, J. G. (1938) *J. Chem. Phys.* **6**, 70.
KURATA, M., KIKUCHI, R., and WATARI, T. (1953) *J. Chem. Phys.* **21**, 434.
LANDAU, L. D. and LIFSHITZ, E. M. (1958) *Statistical Physics*, Addison-Wesley, Reading, Mass.
LAWSON, A. W. (1947) *J. Chem. Phys.* **15**, 831.
LI, Y. Y. (1949) *Phys. Rev.* **76**, 972.
LONGUET-HIGGINS, H. C. (1951) *Proc. R. Soc.* A, **205**, 247.
MOTT, N. F. and NABARRO, F. R. N. (1940) *Proc. Phys. Soc.* **52**, 86.
MULDAWER, L. (1951) *J. Appl. Phys.* **22**, 663.

MUNSTER, A. and SAGEL, K. (1959) *The Physical Chemistry of Metallic Solutions and Intermetallic Compounds*, paper 2D, HMSO, London.
ONSAGER, L. (1944) *Phys. Rev.* **65,** 117.
ORIANI, R. A. (1956) *Acta Metall.* **4,** 15; (1959) *The Physical Chemistry of Metallic Solutions and Intermetallic Compounds*, paper 2A, HMSO, London.
ORIANI, R. A. and MURPHY, W. K. (1958) *J. Phys. Chem. Solids* **6,** 277.
PASHLEY, D. W. and PRESLAND, A. E. B. (1958–9) *J. Inst. Metals* **87,** 419.
PRIGOGINE, I., BELLEMANS, A., and MATHOT, V. (1957) *The Molecular Theory of Solutions*, North-Holland, Amsterdam.
RHINES, F. N. and NEWKIRK, J. B. (1953) *Trans. Am. Soc. Metals* **45,** 1029.
RICHARDS, M. J. and CAHN, J. W. (1971) *Acta Metall.* **19,** 1263.
ROBERTS, B. W. (1954) *Acta Metall.* **2,** 597.
RUSHBROOKE, G. S. (1938) *Proc. R. Soc.* A, **166,** 296.
SCHUBERT, K., KIEFER, B., and WILKENS, M. (1954) *Z. Naturforsch.* **9***a*, 987.
SCHUBERT, K., KIEFER, B., WILKENS, M., and HAUFLER, R. (1955) *Z. Metallk.* **46,** 692.
SHIMOJI, M. and NIWA, K. (1957) *Acta Metall.* **5,** 496.
SIVERTSEN, J. M. and NICHOLSON, M. E. (1961) *Prog. Mater. Sci.* **9,** 303.
SMOLUCHOWSKI, R. (1951) *Phase Transformation in Solids*, p. 149, Wiley, New York.
SWAN, E. and URQUHART, A. R. (1927) *J. Phys. Chem.* **21,** 251.
SYKES, C. and WILKINSON, H. (1937) *J. Inst. Metals* **61,** 223.
TAKAGI, Y. (1941) *Proc. Phys. Math. Soc. Japan* **23,** 44.
THOMSON, J. J. (1888) *Application of Dynamics to Physics and Chemistry*.
TILLER, W. A., POUND, G. M., and HIRTH, J. P. (1970) *Acta Metall.* **18,** 225.
TRIVEDI, R. (1970) *Metall. Trans.* **1,** 921.
WALKER, C. B. (1952) *J. Appl. Phys.* **23,** 118.
WANNIER, G. H. (1945) *Rev. Mod. Phys.* **17,** 50.
YANG, C. N. (1945) *J. Chem. Phys.* **13,** 66.
YANG, C. N., and LI, Y. Y. (1947) *Chin. J. Phys.* **7,** 59.
ZENER, C. (1951) *J. Appl. Phys.* **22,** 372.

CHAPTER 7

The Theory of Dislocations

28. INTRODUCTION: EDGE AND SCREW DISLOCATIONS

The line defect which we call a dislocation is of great importance in the description of almost all solid-state phenomena. For many years the existence of dislocations was inferred rather than observed, and the theory of dislocations was developed during this time into a reasonably well-ordered body of knowledge. More recently, experimental techniques have been devised by means of which individual dislocations may seen and their properties studied directly, and the most important advances are now being made in this way.

We cannot hope to cover the theory of dislocations adequately in a single chapter, but we shall summarize the main results. The postulate that real crystals contain dislocations was originally made in order to develop a satisfactory theory of plastic deformation, and the majority of the applications of the theory have been made in this field. We shall not discuss these applications in any detail, but we find it convenient to follow the historical order and introduce the dislocation first by considering the deformation process. This is not strictly necessary, since dislocations can be discussed *ab initio* as line defects possessing certain topological properties (see p. 242), but their glide motion is so important that the deformation approach seems most natural.

Experimental results show that deformation usually occurs by the slipping or gliding of close-packed atomic planes over one another. Before this gliding can begin, the component of the applied shear stress acting across the glide plane and resolved along a close-packed direction of this plane must exceed a certain value. This critical resolved shear stress is characteristic of the state of the material, and is influenced by its thermal and mechanical history. It follows that the atoms in the glide plane maintain a highly ordered (crystallographic) arrangement, and slip does not involve the formation of a locally melted layer, as once supposed. The coupling of the atoms is elastic, not rigid, and it is therefore inconceivable that all the atoms in the glide plane should move simultaneously over the plane beneath. We see then that at a given time, the portion of the crystal on one side of a glide plane may have slipped over the remainder by different amounts in different regions. The simplest definition of a dislocation is that it is a line discontinuity separating two such regions.† From the definition it follows at once that the dislocation must either begin and

† We give a formal and more complete definition on p. 242.

end on the surface of the crystal or must form a closed line or part of a network in the interior.

A dislocation is characterized principally by its Burgers vector, the scalar magnitude of which is also called the strength of the dislocation. This is the difference in slip, i.e. in relative atomic positions, produced by crossing the dislocation line from one region to another. The Burgers vector **b** is usually written in the form $[b_1 b_2 b_3]$, and for cubic structures this may be expressed as $c[u_1 u_2 u_3]$, where the u_i give the crystallographic direction of the displacement, and $cu_i = b_i$. Usually, we expect **b** to be equal to the interatomic vector in the glide plane, or at least to a small lattice vector. Dislocations of this kind are called perfect or lattice dislocations. In certain cases, it is also possible to have **b** equal to a fraction of a lattice repeat vector. The discontinuity is then an imperfect or partial dislocation, and the original lattice structure is not preserved when the dislocation line is crossed.

The Burgers vector is constant along any dislocation line, but this invariant characteristic is not sufficient to specify completely the properties of the discontinuity. The strain field, and hence the detailed atomic arrangement, depends on the relation between the Burgers vector and the direction of the line itself. This latter direction need not be fixed; that is, the line may be curved so that its local direction changes continuously in the slip plane. In the early development of the theory, it was assumed that a dislocation line is straight, and that there is a special relation between its own direction and that of its Burgers vector. The assumption of such a relation enables the structure of two fundamental types, edge and screw dislocations, to be discussed in detail. The procedure is useful, since any element of dislocation lying in an arbitrary direction may be resolved into edge and screw components.

In the edge dislocation, introduced by Taylor (1934), the Burgers vector is perpendicular to the line of the dislocation. Such a discontinuity is illustrated in Fig. 7.1. The upper part of the crystal block is gliding over the lower part along the plane *ABCD* in the direction Ox_1. The gliding motion has been completed over the region *ABPQ* of the glide plane, and this is separated from the unslipped portion by the dislocation line *PQ*. As a result of

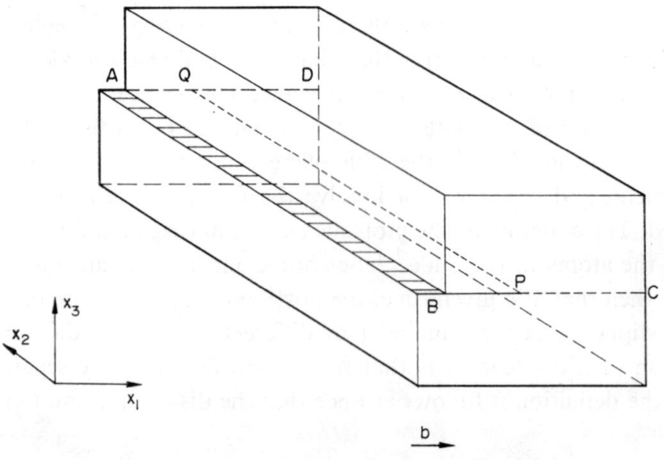

FIG. 7.1. Simple illustration of edge dislocation.

the slip, the atoms are displaced through a distance $|\mathbf{b}|$ in the direction Ox_1. It is obvious that in the region of the dislocation line, the atoms on one side of the glide plane are compressed, and those on the other side are extended. This state of strain is characteristic of an edge dislocation.

In the early work, the lattice was assumed to be simple cubic, since this enables the structure of the dislocation line to be readily visualized. For such a lattice, we take $\mathbf{b} = a[100]$, so that the block of material shown in Fig. 7.1 has edges parallel to those of the unit cube. If we consider the structure to be made from rows of atoms parallel to *PQ*, we see that $(n+1)$ such rows in the atom plane above the glide plane will be opposite only n such rows in the atom plane below the glide plane. There is thus an extra plane of atoms in the region of crystal above *PQ*. This is illustrated in Fig. 7.2 which shows the approximate atomic

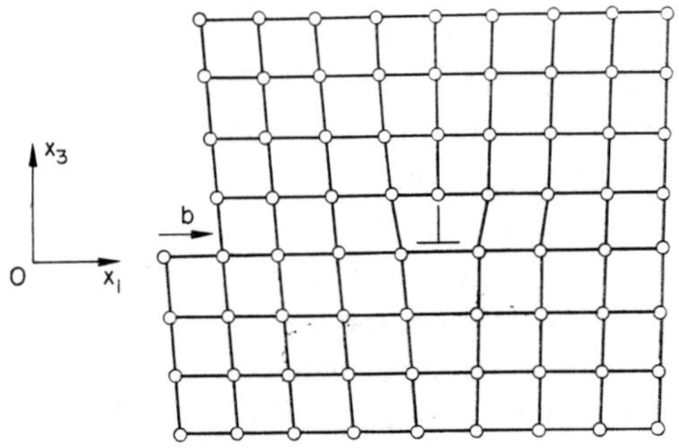

FIG. 7.2. Approximate atomic structure of an edge dislocation in a simple cubic crystal.

structure in a plane perpendicular to *PQ*, looking in the direction from *P* to *Q*. From this figure, we see that instead of the slipping motion discussed above, we could make an edge dislocation by the equivalent hypothetical process of cutting the crystal above *PQ* and inserting an extra half plane of atoms; the dislocation line lies at the edge of this extra half-plane. It is formally equally valid to regard the dislocation as situated at the edge of an extra half-plane of vacant lattice sites inserted below *PQ*, i.e. of an extra half-plane of atoms removed from below *PQ*. In diagrams, the edge dislocation is conventionally represented by the symbol ⊥ or ⊤ where the vertical line indicates the extra half-plane of atoms, and the horizontal line the glide plane.

When the crystal structure is not simple cubic, there is an extra half-plane of material of thickness $|\mathbf{b}|$ on one side of the glide plane, but this does not necessarily consist of a planar arrangement of atoms. Fig. 7.2 no longer shows the atomic arrangement, but it is illustrative of the strain pattern of the edge dislocation; it would be obtained, for example, by scribing a square reference net of lines on the surface prior to introducing the dislocation. It is also clear from this figure that although we consider dislocations to be line defects, each has a finite width and consists of a roughly cylindrical region of bad crystal. A convenient definition of the width of a dislocation is the distance in the slip plane over which the relative displacements of the atoms above and below this plane are more than half their maximum values. The width is determined by the nature of the atomic binding forces, and many properties of dislocations are sensitive to its exact value. We shall also find in Section 32 that in many close-packed structures each dislocation is dissociated into two parallel (imperfect) dislocation lines, separated by a region of stacking fault. When this happens, the whole dislocation is a planar, rather than a linear defect.

We see from Fig. 7.1 that slip is produced by the migration of the dislocation *PQ* from *BA* towards *CD*. The passage of *PQ* right through the crystal from one edge of the slip plane to the other would leave a perfect lattice slipped through a distance equal to the Burgers vector. The formation of a dislocation requires a high energy and will be discussed later; once formed, however, a dislocation in a close-packed structure can move under the action of a small shear stress. This is shown qualitatively by diagrams of the type of Fig. 7.2. If the configurations representing the dislocation in two adjacent atomic positions of the type shown in that figure are compared, it is seen that the movement of the dislocation through this distance has been achieved by displacements of the atoms in the neighbourhood of the dislocation through very small fractions of an interatomic distance. To a first approximation, the energy of the dislocation may be assumed to be constant as this movement is made, so that the dislocation will glide under a vanishingly small shear stress. More accurately we see that when the atomic configuration is symmetrical, as in Fig. 7.2, the dislocation will have a lower energy than it has in an arbitrary position. There is thus a constraining force tending to anchor the dislocation in the lattice; its magnitude depends on the crystal structure, but it is quite small in close-packed metallic structures. The configuration of Fig. 7.2 represents a stable position of the dislocation line; there is also a symmetrical stable configuration halfway between it and the next configuration of the same type. When moving from one stable configuration to another, the dislocation must increase its energy, either by thermal agitation, or because a finite shear stress is applied. The problem is considered further on p. 274.

An edge dislocation can glide only in its slip plane, defined by the dislocation itself and its Burgers vector. It is also geometrically possible for the dislocation to move out of this plane by "climbing" along the perpendicular direction Ox_3 (Fig. 7.1). If the dislocation moves in the $+x_3$ direction, the extra half-plane of atoms shrinks, or equivalently the extra half-plane of vacancies grows. Such a motion thus requires the removal of atoms from their normal sites into interstitial sites, and their subsequent diffusion away through the lattice, or else the absorption of vacant lattice sites, which diffuse to the dislocation from the surrounding lattice. The reverse is true for motion along $-x_3$. The motion of an edge dislocation in a direction out of the slip plane is thus a slow process, depending on the diffusion of vacancies or interstitial atoms.

The absorption or emission of individual point defects implies that the dislocation climbs one atom at a time. Thus the dislocation will not lie entirely in one slip plane but will contain steps where it moves from one such plane to an adjacent plane. These steps are called jogs (Fig. 7.3). We should expect point defects to be absorbed or emitted at the jogs in the dislocation line, the jog moving along the line as this happens. If the local point defect concentration is in excess of the equilibrium value, there will be a decrease in free energy when a defect disappears at the dislocation, this being the binding energy of the defect to the dislocation. Clearly, the absorption of a point defect in a straight region of dislocation

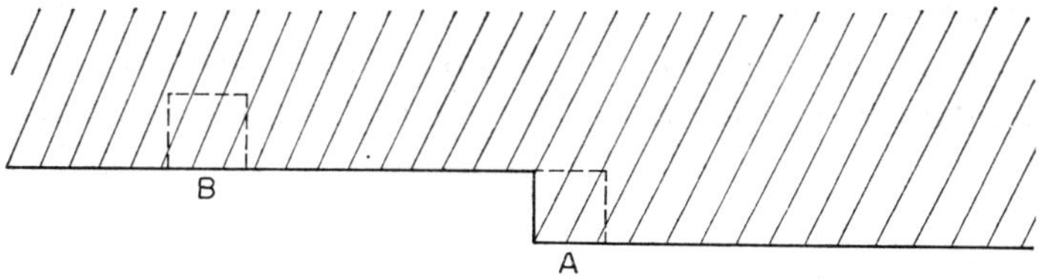

FIG. 7.3. A jog in an edge dislocation. Absorption of a vacancy at the jog *A* does not change the configuration. A higher energy state results from absorption of a vacancy at *B*.

is much less probable, since this would create a new double jog (Fig. 7.3). The energy of this double jog is probably greater than the binding energy, so that if such a configuration forms, it will tend to disappear again. The point defect must thus migrate along the dislocation line until it finds a jog at which it can be absorbed; such movement may be very rapid because of the distorted structure of the line.

The efficiency of a dislocation line as a source or sink for point defects thus depends, at least partially, on the number of jogs which it contains. There will always be a certain jog density maintained in thermal equilibrium, since a dislocation line with a few jogs has a higher entropy to compensate for its higher internal energy. In addition, non-equilibrium jog concentrations may be produced by motion of the dislocation line (see p. 250). We should like to emphasize that this type of dislocation climb is presented here solely as a geometrical possibility; we discuss later the conditions under which it may or may not take place.

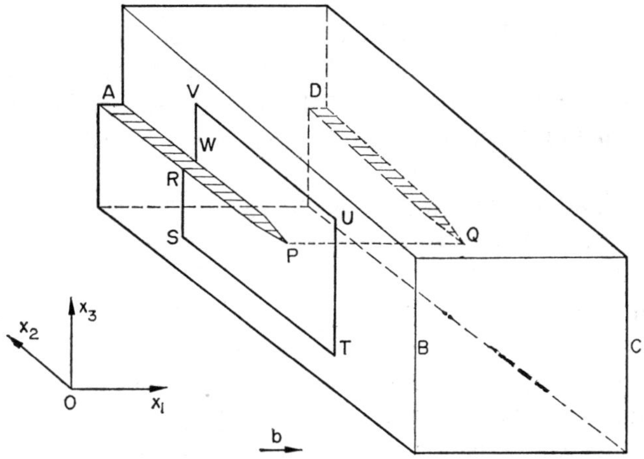

FIG. 7.4. Simple illustration of a screw dislocation.

We now consider the other fundamental type of dislocation, which was introduced by Burgers (1939), and is illustrated in Fig. 7.4. Slip is again occurring along the plane *ABCD* in the direction Ox_1, and *PQ* is the dislocation line. *PQ* is now parallel to Ox_1, and slip has been completed over the region *APQD*. Migration of *PQ* from *AD* to *BC* produces the same resultant slip as migration of the dislocation in Fig. 7.1.

The structure of the screw dislocation in the simple cubic structure, or the strain pattern in any structure, is shown by projecting the atomic positions on to the slip plane, as in Fig. 7.5. The filled circles represent atoms immediately below the glide plane, the unfilled

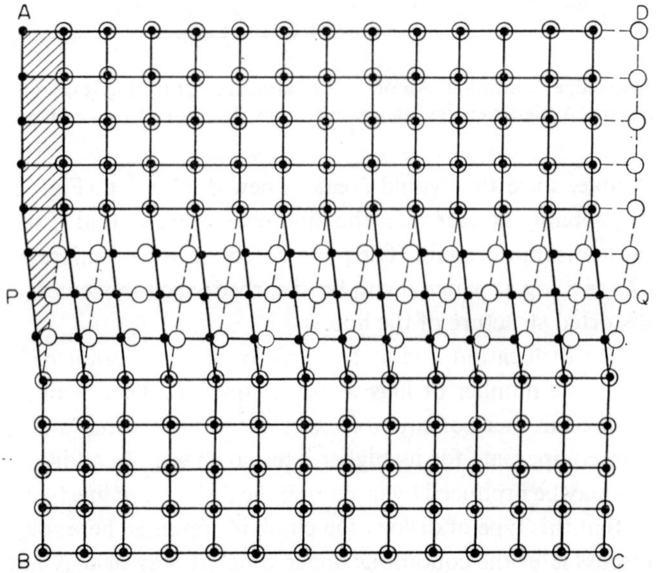

FIG. 7.5. Approximate atomic structure of a screw dislocation in a simple cubic crystal.

circles atoms above the glide plane. The crystal does not contain an extra half-plane of atoms; instead the whole crystal is one continuously connected atomic plane. Consider, for example, the circuit *RSTUVW* in Fig. 7.4, which is made entirely in the Ox_2x_3 plane of the crystal. Because of the dislocation, this circuit is not closed; its end-points are displaced by **b**.

A point which continually encircles the dislocation line, following the atomic bonds (i.e. making nearest neighbour jumps) describes a helical path, like the motion of a point on a screw thread.

The dislocation line *PQ* of Fig. 7.4 will migrate in the direction $-Ox_2$ under the action of a small shear stress applied in the direction Ox_1. For the motion of the screw dislocation, however, there is clearly no distinction between the directions Ox_2 and Ox_3, and the dislocation can also readily glide in the Ox_3 direction. This is a rapid motion and does not require thermal activation energy. The screw dislocation cannot move slowly by absorption or emission of vacant lattice sites. For the edge dislocation, we saw that the glide plane was the plane containing both the Burgers vector and the dislocation line. If the direction of the dislocation line is given by the unit vector **i**, the glide plane has unit normal $\mathbf{b}\wedge\mathbf{i}/|\mathbf{b}\wedge\mathbf{i}|$, and all atomic planes containing **b** are possible glide planes for the screw dislocation.

Figure 7.4 shows that where the screw dislocation emerges at a crystal face there is a step running from the dislocation end to the edge of the face. Movement of the dislocation results in a gradual increase or decrease in the length of this step. In contrast, a step extending from one edge of a crystal face to another suddenly appears or disappears when an edge dislocation emerges at that face.

The edge dislocation of Fig. 7.1 was obtained by inserting an extra plane of atoms above *PQ*. Equally we could have obtained an edge dislocation by inserting an extra plane of atoms below *PQ*, and the passage of such a dislocation from *CD* to *BA* would produce the same resultant slip as the passage of *PQ* from *BA* to *CD*. One description can in fact be converted into the other simply by inverting the crystal. The distinction is less trivial than this implies, since dislocations exert forces on one another. These forces are attractive for dislocations with opposing Burgers vectors and repulsive if the Burgers vectors are parallel. It is convenient to introduce the terms positive and negative dislocations, a positive edge dislocation conventionally having the extra half-plane of atoms in the upper half of the crystal.

Two opposite edge dislocations in the same slip plane are shown in Fig. 7.6. Separation of these lines to the edges of the crystal results in the upper half of the crystal slipping to the right over the lower half. This process is equivalent to moving a single dislocation

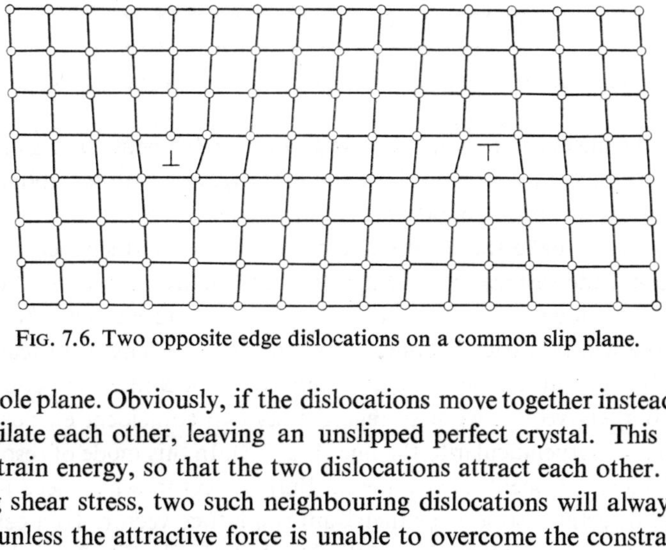

FIG. 7.6. Two opposite edge dislocations on a common slip plane.

through the whole plane. Obviously, if the dislocations move together instead of separating, they will annihilate each other, leaving an unslipped perfect crystal. This results in a lowering of the strain energy, so that the two dislocations attract each other. In the absence of an opposing shear stress, two such neighbouring dislocations will always run together and disappear unless the attractive force is unable to overcome the constraining forces of the lattice. Now consider two opposite dislocations on parallel, adjacent slip planes. There will again be an attractive force, and if the two dislocations run together, they will disappear with the formation of a line of point defects (vacancies or interstitials, depending on whether the extra half-planes do not meet or overlap).

In a similar way, we may form right- and left-hand screw dislocations which are the opposites of each other. The structure of a right-handed screw dislocation is the mirror image of that of a left-handed one. Two such dislocations on a common glide plane can run together, leaving a region of perfect crystal.

The screw and edge dislocations described above are straight lines. One way of forming a more general dislocation in the glide plane is to combine together edge and screw elements. As stated above, a dislocation cannot end within the crystal. The helical surface of a screw dislocation, for example, can only be terminated in the extra half-plane of an edge dislocation, and vice versa. Thus each time we end a dislocation within the crystal we introduce another dislocation perpendicular to the first but having the same Burgers vector. This is illustrated in Fig. 7.7, which shows a slip plane in which the slipped region

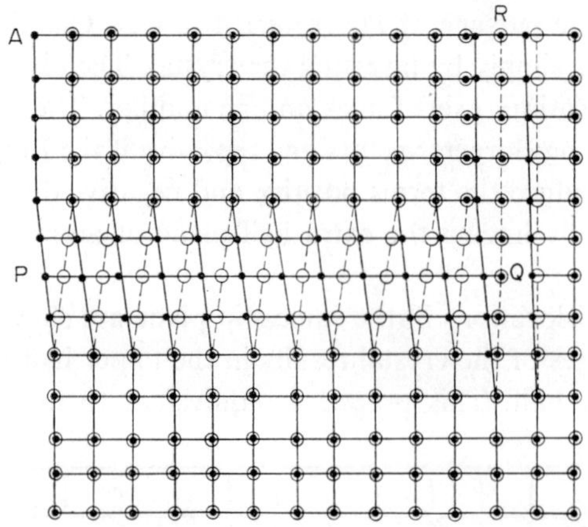

FIG. 7.7. The dislocation line *PQR* consists of an edge segment *QR* and a screw segment *PQ*.

is divided from the unslipped region by screw dislocation segments such as *PQ* and edge dislocation segments such as *QR*. Note how the extra plane of atoms above *QR* terminates in the helical surface wrapped round *PQ*.

In Fig. 7.7 we may make the separate elements smaller and smaller, and as the length of each is reduced, the dislocation line approximates more nearly to a curve. If we continue the reduction down to an atomic scale, the separate elements begin to lose their identity, since the width of each becomes comparable to its length. Instead of regarding a small length of curved dislocation line as composed of separate, non-coincident elements with identical Burgers vector, we may regard it as two coincident elements with Burgers vectors parallel and perpendicular to the line (Fig. 7.8). In this mode of resolution, the component edge and screw elements have different Burgers vectors which need not have rational crystallographic directions so long as their sum is a lattice vector. To a certain extent, the manner of resolution is a matter of convenience, but clearly Fig. 7.8 only gives a correct

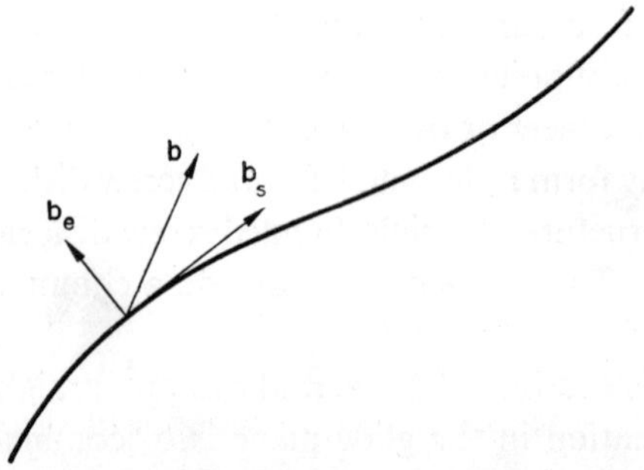

FIG. 7.8. Resolution of the Burgers vector into edge and screw components.

picture if the dislocation line can be represented as a curve, rather than as a zigzag, even on an atomic scale, and if this is possible, the type of resolution shown in Fig. 7.7 is misleading, since the elements are arbitrary. Calculations by Mott and Nabarro (1948) show that a dislocation prefers to be as straight as possible, and the energy rises if the zigzag configuration is adopted. By analogy with surface tension we introduce the idea of the line tension of a dislocation, which may thus be compared to a slightly stretched piece of elastic. The concept of curved, flexible dislocation lines is needed in many applications of the theory.

We see from the above discussion that the general dislocation line in the slip plane is a curve of arbitrary shape, and the edge and screw dislocations described above are best regarded as special orientations of the general line. Each small section of the line may be resolved into edge and screw components; when one of these components is zero, the line is locally a screw or edge dislocation. Of particular interest is the formation of a closed loop of dislocation, isolating a region of local slip. In the absence of a stress, the loop will tend to disappear and leave a perfect crystal. If an external shear stress is applied, a loop of sufficient size will expand, spreading slip over the plane in which it lies.

Although we have only discussed dislocations in a single slip-plane, we have already noted that a dislocation line can move from one slip-plane to another. The most general form of dislocation may have any orientation in the crystal, and Fig. 7.9 shows an elemen-

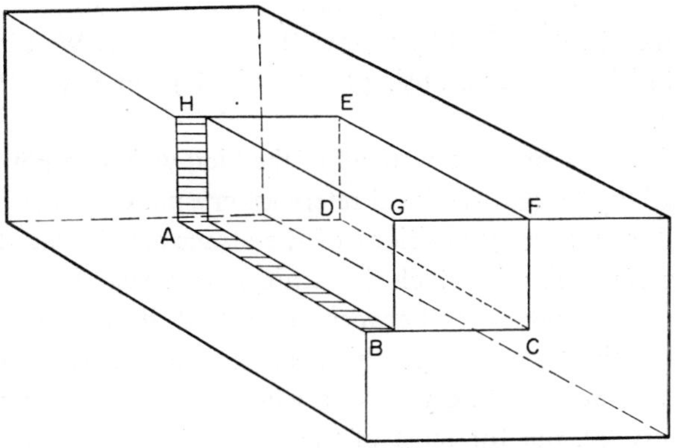

FIG. 7.9. The dislocation line *CDE* consists of two perpendicular edge segments.

tary example of a discontinuity in two slip-planes. The region of crystal *ABCDEFGH* has slipped through one interatomic distance with respect to the rest of the crystal. This slip has taken place on the planes *ABCD* and *ADEH*, and its limit is marked by the edge dislocation elements *CD* and *DE*, making up the complete dislocation line *CDE*. By combining perpendicular edge elements of this kind with screw elements in the two slip-planes, we may build up a dislocation line which is curved in three dimensions. More generally, a three-dimensional dislocation may lie in any number of slip planes, which are defined whenever a segment of the line has edge character; these slip planes have a common zone axis which is the direction of the Burgers vector.

The two edge elements of Fig. 7.9 may be formed by making an imaginary cut over the area *FCDE* and inserting an extra plane of atoms. Obviously a closed dislocation loop in the x_2x_3 plane can be formed by inserting an extra plane of atoms over any arbitrary area of this plane. The loop will have Burgers vector along Ox_1, i.e. it is composed entirely of edge elements, and forms the limit of a prismatic region of slipped crystal. The term prismatic dislocation is often used for any closed loop of dislocation line with a Burgers vector inclined to the plane of the loop. A prismatic dislocation of opposite sign could be formed by removing part of an atomic plane (inserting an area of vacancies). An excess concentration of vacancies in a region of perfect crystal could conceivably lower the free energy by condensing to form a spherical cavity or a flat disc of atomic thickness. In the latter case, the disc would collapse inwards to leave a prismatic dislocation loop, and this kind of mechanism was postulated by Seitz (1952) as one way in which dislocations might be introduced into crystals. There is now convincing electron-microscope evidence that dislocations are formed in this way in specimens quenched from high temperatures to produce supersaturations of vacancies.

29. GEOMETRICAL PROPERTIES OF DISLOCATIONS

On p. 239, we defined the sign of an edge dislocation. A closed dislocation loop contains edge and screw elements of opposite sign, but it is convenient to adopt a convention in which such a loop may be described by a single Burgers vector. We shall first give a more formal definition of dislocation lines, using the terminology of Frank (1951) already introduced in Chapter 5.

The defects described in Chapter 5 were all of the kind which do not disturb the correlation between the real crystal and the ideal reference crystal. Consider a crystal containing internal strains, so that the mean positions of the atoms are displaced from those of the reference crystal. In any region in which the strains are sufficiently small, there will again be an unambiguous correspondence between the actual atomic positions and the ideal positions, but if the strains become large, this unambiguous correspondence will disappear. When large strains are present, the crystal as a whole can no longer be compared directly with the reference crystal, but for most of the atoms local correspondence will remain. The exceptional regions of "bad crystal" are dislocations.

In a region of good crystal, any four atoms which are arranged in approximately tetrahedral fashion may be associated with four similarly arranged atoms of the reference crystal. A neighbouring atom of the real crystal can then be correspondingly associated with a reference atom, in the sense that its position relative to the other four atoms is almost identical with that of the reference atom with respect to the other four reference atoms. By repeating this process we may trace a path through the real crystal by successive atom jumps and associate with this path a corresponding reference path. The path in the real lattice must be made entirely through good crystal; when such a path closes on itself, it is called a Burgers circuit. If a Burgers circuit encloses regions of bad crystal, the reference path is not necessarily closed, and the displacement needed to close it must be a lattice vector of the reference crystal. This displacement is defined as the resultant Burgers vector of all the dislocation lines enclosed by the path in the real crystal. An example of a Burgers circuit was shown in Fig. 7.4; if the circuit *RSTUVWR* were plotted in the reference crystal, the point corresponding to *W* would coincide with the starting point corresponding to *R*, and the displacement corresponding to *RW* would thus give the Burgers vector. The vector is independent of the starting point of the circuit, and is the same for all circuits which are separated only by good crystal.

Notice that in this definition, the Burgers vector is strictly defined in an ideal reference lattice. Clearly it is equally valid to complete a circuit in the reference crystal, and to define the Burgers vector as the closure failure of the corresponding circuit in the real crystal. The Burgers vector will then connect two lattice points of the real crystal, and since both points are in regions of good crystal, it will differ only very slightly from a lattice vector of the ideal crystal. For most purposes, it is quite immaterial which circuit is regarded as closed, but in the formal theory of continuous distributions of dislocations (p. 314) and similar applications, this point can be important.

The dislocations defined by a closed circuit made entirely through good crystal must be "perfect" since the translations preserve the original structure. Perfect dislocations have Burgers vectors determined entirely by the lattice vectors **u** of eqn. (5.8); the vectors ξ are not directly related to dislocations, which are a property of the lattice rather than of the crystal structure. Imperfect dislocations are discussed in Section 32. A Burgers circuit which encloses more than one dislocation line may be subdivided by joining points on the circuit through regions of good crystal. Each circuit which can no longer be subdivided encloses a single dislocation.

We may now use this rather formal definition to formulate a sign convention. We choose the positive direction of the dislocation arbitrarily, and describe the Burgers circuit in a clockwise (right-handed) direction when looking along this line. The displacement required to close the reference path then gives both the magnitude and sign of the Burgers vector. Bilby (1951) has given an elementary discussion of this convention, which is sometimes described as FS/RH (Bilby *et al.*, 1955). This identifies the vector with the displacement from the finish F to the start S of the reference path when the right-handed circuit (S^1F^1) of the real lattice is closed. The sign of the Burgers vector is changed by taking the closure failure SF, by making a left-handed circuit, or by closing the reference circuit and measuring F^1S^1. Thus $FS/RH = SF/LH = F^1S^1/LH = S^1F^1/RH = -F^1S^1/RH$, etc. There is no general agreement on which convention should be used.

If we have a closed dislocation loop, we may define the positive direction of the line by specifying that the loop is to be traversed in a certain sense, and we then have the same Burgers vector at all points on the loop. Consider a loop lying in a single glide plane. At two points along the loop, the strain field will be of the type we previously associated with positive and negative edge dislocations, and at two other points the local arrangement will correspond to right-and left-handed screw dislocations. Whereas we should previously have stated that the two edge components have equal and opposite Burgers vectors, we should now state that they have the same Burgers vector but opposite directions. The two screw elements are still describable as right- and left-handed (simple rotation can never convert a right-handed thread into a left-handed thread) but are now regarded as having opposite directions, so that they possess the same Burgers vector. We thus see that there is always an ambiguity in the sign of the Burgers vector until the direction of the line itself is specified in some way. For three-dimensional dislocations, or closed twodimensional loops, it is often more convenient to regard elements of the line which are opposites in the sense that they can mutually annihilate as having opposite directions rather than opposite Burgers vectors.

Suppose the Burgers circuit in the real crystal encloses two separate regions of bad crystal. If we displace the circuit in the general direction of these dislocations (i.e. along the dislocations), it is geometrically possible for the two bad regions to coalesce into a single bad region. Two perfect dislocations may thus unite to form a single dislocation or, conversely, a dislocation may split into two dislocations in the interior of the crystal. Since the Burgers circuits enclosing the dislocation or its branched pair are entirely in good crystal, the resultant Burgers vector is unchanged and

$$\mathbf{b}_1+\mathbf{b}_2 = \mathbf{b}_3. \tag{29.1}$$

The meeting of three or more dislocation lines in the interior of a crystal is called a node. Clearly, in a node of three lines we are free to regard any two of them as resulting from the decomposition of the third. The relation between their Burgers vectors is then expressed more symmetrically if we define the positive direction of each line as the direction looking outward from the node. By describing right-handed Burgers circuits round each of the lines, we then find

$$\mathbf{b}_1+\mathbf{b}_2+\mathbf{b}_3 = 0,$$

or more generally if i dislocation lines meet at a node,

$$\sum_i \mathbf{b}_i = 0. \tag{29.2}$$

Equation (29.1) may also be considered as a dislocation reaction in which (by analogy with chemical reactions) the two dislocations $\mathbf{b}_1$ and $\mathbf{b}_2$ combine to form a new dislocation $\mathbf{b}_3$. In theory, any number of dislocations may contribute to a reaction. Non-parallel dislocations in the same glide plane may glide together, uniting over part of their lengths, at the ends of which nodes are formed. The positions of the nodes then change continuously as the reaction proceeds.

Consider any element of dislocation line, lying along a direction specified by the unit vector $\mathbf{i}$ and having Burgers vector $\mathbf{b}$. If we resolve into edge and screw components, as described on p. 240, the edge component of the Burgers vector lies in the slip plane specified by $\mathbf{b}$ and $\mathbf{i}$, and is perpendicular to $\mathbf{i}$, and the screw component is parallel to $\mathbf{i}$. The screw component is thus given by

$$\mathbf{b}_s = (\mathbf{b}\cdot\mathbf{i})\mathbf{i} \tag{29.3}$$

and the edge component by

$$\mathbf{b}_e = \mathbf{b}-\mathbf{b}_s = (\mathbf{b}\wedge\mathbf{i})\wedge\mathbf{i}. \tag{29.4}$$

We introduced a dislocation as a line discontinuity in the glide plane, but we have already seen that for a general dislocation loop there is no unique glide plane. However, if we draw generators from the dislocation parallel to the Burgers vector, we obtain a prismatic surface which may be called the virtual glide surface for the dislocation (Read and Shockley, 1952). From the previous consideration of slipping movements, it follows that any movements of parts of the dislocation loop along this surface are slipping motions; such motions have the property that the projected area of the loop on any plane normal to **b** remains unchanged. Conversely, any motion of the loop which changes this projected area requires the movement of edge elements out of their slip planes, and is a diffusive or climbing motion. It is also obvious that an element of loop which has pure screw character may move off the virtual glide surface without diffusion, so long as this does not change the projected area.

We should also note which kinds of motion are likely under the action of applied stress. In Fig. 7.10 the virtual glide surface has been drawn as a cylinder for simplicity. The ini-

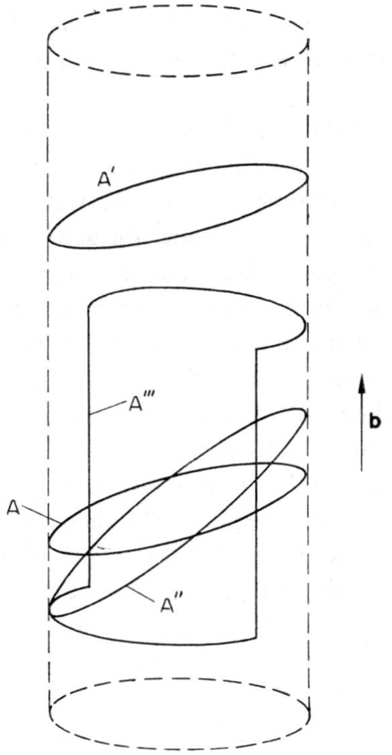

FIG. 7.10. Possible movements of a dislocation loop on its virtual glide surface. An external shear stress may displace the loop A to positions such as A'' and A'''.

tial dislocation loop A lying on this surface may be imagined to move in two ways: bodily along the surface without change in length, or by gliding to a new position A'' in which its length is greatly increased or decreased. Slipping motions of the second kind are found when the loop is subjected to a shear stress, and are the three-dimensional analogues of the simple expansion of a glide plane loop described on p. 241. In slipping motions of the first kind there is a displacement of the material inside the virtual glide surface relative to that outside the surface. Such a motion cannot relieve any externally applied shear stress, but it may be responsible for the phenomenon observed in prismatic punching experiments. The possible dislocation motions just discussed refer only to perfect dislocations; partial dislocations are subjected to additional constraints and have more restricted possibilities of motion.

Early observations of the surface of deformed materials by electron-microscopy indicated that in some circumstances large amounts of slip may occur on a single glide-plane. These slip steps represent dislocations which have escaped from the material, but it is also well established that the density of dislocation lines inside the crystal (see p. 311) increases by several powers of ten during work hardening. We thus require a mechanism for the production of large numbers of perfect dislocations during plastic deformation, and we shall prove later that such dislocations cannot be nucleated spontaneously by the combined effects of the external stress and the thermal energy of the lattice. An important development in the theory of dislocations was the recognition by Frank and Read (1950) that a process which is in principle purely geometrical can be used to produce new dislocations from old dislocations.

Consider a crystal which contains a dislocation line *CDE* composed of two edge elements, exactly as in Fig. 7.9. Suppose that when an external shear stress is applied to the crystal, the part *DE* is prevented from gliding in some way, so that the horizontal planes of the figure are the only active glide planes. Under the action of a suitable stress, the portion of the line *CD* will move in its glide plane, and as it does so the crystal above this plane glides over the crystal below it. The point *D* must remain fixed, so the line *CD* rotates about *D*. The crystal is slipped through one Burgers vector for each complete rotation. Actually, *CD* will not remain radial, since the parts of the line *CD* will glide initially with constant linear velocity (except very close to *D*). The angular velocity of the interior parts of the line will thus be greater than that of the outer parts, and the line will wind up into a spiral shape. Once an equilibrium spiral has been formed, it will continue to rotate with constant angular velocity, increasing the slip in each region of the glide plane by **b** for each complete revolution. We shall not calculate the shape of this equilibrium spiral, but an analogous problem in the theory of crystal growth will be treated in Part II, Chapter 13.

In the example just discussed, the dislocation line *CDE* only leaves the active glide plane at one point *D*, and *CD* extends to the surface of the crystal. This is improbable, even for the low dislocation densities of a well-annealed crystal, unless the whole configuration is close to the surface. As we shall describe later, the dislocations in an annealed crystal are largely concentrated into two-dimensional networks (sub-boundaries), with a few dislocations forming a three-dimensional network inside the subgrains. In the interior of a crystal we thus expect a limited length of any one dislocation line to lie in a given glide plane, the line leaving the glide plane at two points. This gives the geometrical configuration necessary for the operation of a double-ended Frank–Read source as opposed to the single-ended source described above.

We suppose that the dislocation line is anchored at the points of emergence from the slip plane, but is free to glide in this plane. Under the action of an applied stress, it expands in the glide plane, forming a loop which eventually winds round and joins together again. When this happens, a closed loop of dislocation is formed, and this spreads out over the glide-plane, whilst the original length of line repeats the process. This action is illustrated at various stages of the expansion in Fig. 7.11. The extra energy of the dislocation line which has been created has to come from the work done by the external stress. The operation of the source is opposed by the line tension of the dislocation, which tends to straighten it, and the critical stage is that shown at (c) in Fig. 7.11 where the dislocation line reaches its minimum radius of curvature. The greater the distance between the fixed points A and B, the lower the stress needed to activate the source. A single-ended source should become active at about half the stress needed for a double-ended source of the same length, since only one end of the gliding dislocation segment is anchored.

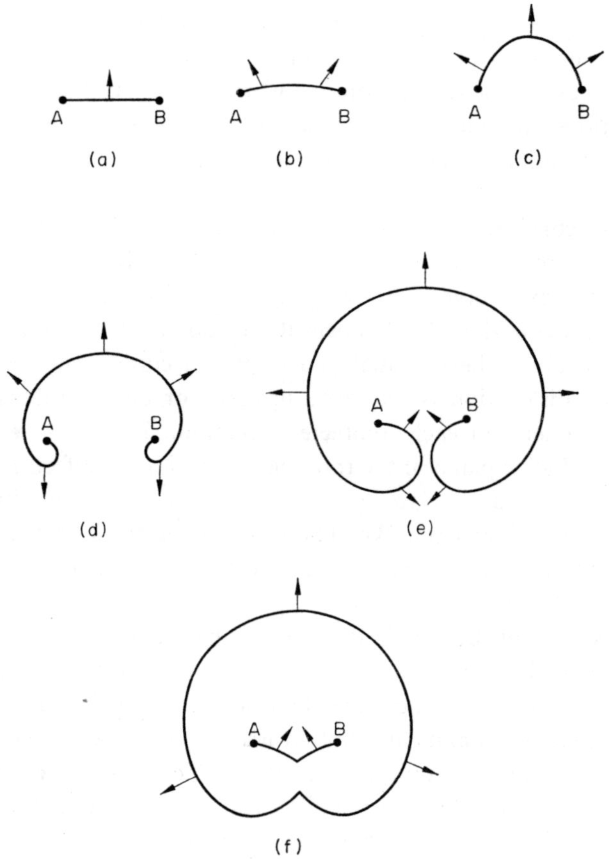

FIG. 7.11. Successive stages in the operation of a double-ended Frank–Read source.

The points A and B of Fig. 7.11 could represent places where the dislocation line merely bends out of the glide plane, as in Fig. 7.9, but it is perhaps more likely that nodes are formed at these places. The geometrical action of the source is unchanged provided the new dislocations formed at the nodes do not lie in the glide plane but have Burgers vectors which do. Nodes may be a method of ensuring firm anchoring of the ends of the source, since the dislocations produced by the dissociation of the glide dislocations may be types which cannot glide readily.

The theory of Frank–Read sources received early experimental support from the work of Dash (1956), who revealed such a configuration in a silicon crystal by precipitating copper on the dislocation lines which were then visible when the crystal was examined in infrared light. Observations of source operation are rather rare, however, and are mainly confined to very lightly deformed crystals (e.g. Miltat and Bowen, 1970). Whilst Frank–Read sources are undoubtedly operative in certain circumstances, there is also abundant evidence that in some materials glide dislocations originate from inclusions, from boundaries, or from sources near a free surface. Dislocation multiplication also often occurs by spreading of slip from one plane to a small group of neighbouring planes. This can happen by cross-slipping of screw segments or by interaction with point defects to form "super-jogs"; in either case, the part of the dislocation line which is displaced acts as a source, sending out a few dislocations on a parallel plane, before the cross-slip or glide again takes place. This dynamic modification of the Frank–Read mechanism is often described as a double cross-slip source.

We now turn to consider an interesting topological property of a double-ended source in which nodes are formed at the points of emergence. There is no necessity for all the dislocations at a node to have Burgers vectors lying in the glide plane so long as eqn. (29.2) is satisfied, and we now consider what happens if they do not. If the Burgers vectors of (say) two emerging dislocations have (equal and opposite) components normal to the glide plane, the rotating dislocation is displaced upwards or downwards through a distance equal to this component for each complete revolution. This could lead to uniform slip on a series of parallel glide planes if the rotating dislocation is perfect, but for our purpose, the more interesting application is to mechanical twinning and martensitic transformations, when this dislocation is imperfect. We shall give a detailed description of this modified form of Frank–Read source, originally suggested by Cottrell and Bilby (1951) in Section 32 and Part II, Chapter 20.

Another modification of the Frank–Read source mechanism is its application to dislocation climb. In Fig. 7.11 we now assume that the dislocation *AB* is a pure edge dislocation with its Burgers vector normal to the plane of the paper. If there is a large excess of vacancies in the region of the dislocation, it can climb by addition of these vacancies, and in so doing it increases its length and takes up a curved shape. The successive curves of Fig. 7.11 are now successive stages in the climb of the dislocation line, and eventually a closed loop of prismatic dislocation is formed, together with an edge element which can repeat the process. The closed loop at the edge of a disc of vacancies can grow outwards, removing an atom plane from the crystal as it does so. The configuration can thus act as a continuous sink for vacancies (source of interstitials), or by climbing in the opposite direction as a source of vacancies (sink for interstitials). This is usually described as a Bardeen–Herring source (Bardeen and Herring, 1952), and was first applied to the Kirkendall effect in diffusion (see p. 400).

Figure 7.12 illustrates the climb of a dislocation line fixed at *A* and *B* which is of the mixed type. The Burgers vector is parallel to *A′A*, so that the glide plane is *ABA′* and the configuration could act as an ordinary Frank–Read source in this plane. Suppose that the dislocation climbs. The line *AB* then becomes curved and lies along the surface of some cylinder with generators parallel to **b**; this surface is now the new virtual glide surface for the line *AB*. The area *F* between *A'B* and the new projection of the dislocation line on the plane

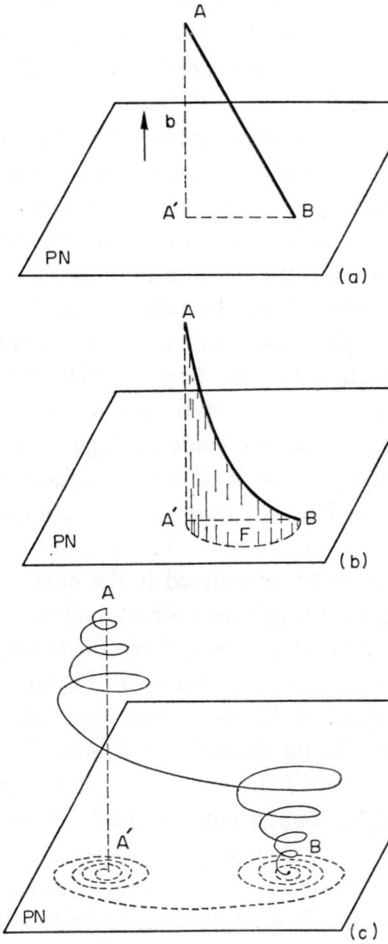

FIG. 7.12. To illustrate the climb of a general dislocation line *AB* into a spiral configuration (after Amelinckx *et al.*, 1957).

normal to **b** specifies the amount of material added or removed (i.e. the number of vacancies emitted or absorbed). Further climb displaces the dislocation line in a direction normal to the virtual glide surface at each point, and the whole line may change into a single spiral, or into a double spiral as shown in Fig. 7.12(c). The successive turns of the spiral (which has a radius of curvature decreasing towards the fixed points) will tend to repel each other by glide along the virtual glide surface. If *A'A* is large in comparison with *A'B* and with the spiral radius, the outer turns of the spiral will glide towards the centre, and a simple helix will be formed. Two such helices meeting in opposite directions produce a series of closed loops of dislocation line.

In the limit, helical dislocations can be produced by the interaction of point defects and pure screw dislocations. The point defects collapse in discs to give pure edge loops normal to the screw dislocation line, and the interaction of the screw with these prismatic loops then gives a helix. The earlier conclusion that pure screw dislocations cannot climb thus needs modification. Dislocation lines of the general form just discussed have been called spiral prismatic dislocations (Seitz, 1952).

It is necessary to make the same kind of reservation about Bardeen–Herring sources as we made about Frank–Read sources. Experimental evidence on dislocation climb appears rather confusing, insofar as there are well established cases in which dislocations do move by absorption or emission of point defects, and other equally well authenticated circumstances in which dislocations are ineffective as sources or sinks. Kuhlmann-Wilsdorf *et al.* (1962) suggest that all these experiments can be rationalized by making a distinction between old dislocations and new dislocations. New dislocations have been formed or have moved at fairly low temperatures, and have not been subjected to subsequent ageing treatment. Old dislocations, in contrast, have been stationary in the structure whilst it has been heated to moderately high temperature. The evidence suggests that new dislocations can climb, but old dislocations cannot do so, at least in f.c.c. structures. A possible reason for this is that the jogs are poisoned by impurity atoms after heating to temperatures at which such atoms can move.

The final geometrical property to be considered is the effect of two dislocation lines, lying in different planes, which glide through each other. Suppose we have two dislocations, *A* and *B*, in intersecting planes, and that *A* glides and so cuts through *B*. The motion moves the part of the crystal on one side of the glide plane of *A* by the vector $\mathbf{b}_A$ relative to the part on the other side. The two parts of the dislocation line *B* separated by the glide plane of *A* must therefore suffer this relative displacement, and since dislocations cannot end within the crystal, the two original parts of *B* are joined by a short length of line (i.e. by a jog), equal in length and magnitude to $\mathbf{b}_A$. Equivalent arguments apply to the line *A* which acquires a jog equal to $\mathbf{b}_B$, since we could equally well suppose that *B* moved through *A*. Thus whenever two dislocations move through each other, they both acquire a jog.

Some simple geometrical configurations which arise when two lines are of simple types and are perpendicular to each other are of interest. A gliding dislocation which acquires a jog can always continue to move easily if its slip plane contains the line of the jog, i.e. if it contains the Burgers vector of any dislocation which it has intersected. If this is not true, only pure edge dislocations can continue to move conservatively together with a fixed jog produced by intersection. When both dislocations are of edge type, the two jogs produced may both be pure screws, and can be eliminated completely by the continuing glide motion, or else one dislocation may acquire an edge jog which continues to glide readily, whilst the other acquires a jog parallel to itself (i.e. increases its length slightly). If an edge dislocation glides through a screw, it acquires an edge jog which can continue to glide, and the screw also acquires an edge jog. Two screw dislocations gliding through each other both acquire edge jogs.

An edge jog in a pure screw dislocation is able to glide only along the dislocation. Thus if the screw dislocation itself is moved in any direction (other than that parallel to the jog), and the jog is carried along with it, a line of vacant lattice sites or interstitials must be left behind by the climb of the edge jog. Seitz (1952) first suggested that during plastic deformation there is a large increase in the concentration of point defects because of this and similar processes. The generation of such point defects would greatly hinder the movement of a jogged screw dislocation.

A jog which has edge character in a general dislocation line is also unable to move conservatively if it is fixed in position along the dislocation line. Thus motion of the line will also lead to the generation of point defects, though in lower density if the jog is not pure edge. However, Seeger (1955a) pointed out that the jog can move conservatively in a slip plane defined by its own direction and that of the Burgers vector, that is, in a direction inclined to the normal to the dislocation line. This should be possible if the jog moves along the line as the line itself moves forward. Although occasional point defects might still be produced, for example, if the jog is temporarily halted for some reason, this type of conservative motion should be much more probable than the non-conservative motion.

The possibility of conservative jog motion may reduce the importance of moving jogs as sources of point defects, but there are numerous other ways in which defects may be created by moving dislocations. Figure 7.13 illustrates the production of an edge dislocation dipole

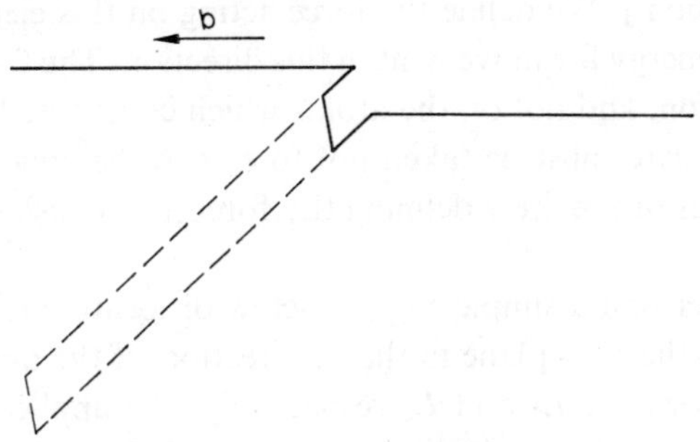

FIG. 7.13. Production of point defects by "pinching off" of edge dislocation dipole.

when a jog in a pure screw is held back by the effect just discussed, the rest of the dislocation continuing to glide. If the jog is only one or two Burgers vectors in length, the pair of edge dislocations are equivalent to a line of point defects, which may be pinched off by the linking together again of the two screw parts of the dislocation line. If the jog is rather longer than this, eventual pinching off will leave a loop of prismatic dislocation. Electron-microscope observations have shown that at sufficiently high temperatures this will break up into a number of approximately circular dislocation loops, and these will eventually disappear by climb. Finally, if the jog is a long "super-jog", produced by cross-slip of part of the original screw, the parts of the dislocation line may behave independently, and may act as single-ended dislocation sources by spiralling around the super-jog. These effects have all been observed in the electron-microscope (e.g. Low and Turkalo, 1962), and there is increasing evidence that the production of "debris" in the form of dislocation dipoles is important in the deformation of many materials.

In crystals where the dislocations are extended (see p. 285), the equilibrium structure of a jog is very difficult to calculate. Detailed consideration of the possible configurations leads to the conclusion (Hirsch, 1962) that vacancy-producing jogs may not be able to move sideways along the dislocation line, and so cannot glide conservatively with a dislocation. Conservative motion of interstitial producing jogs is shown to be much easier, although this motion is thermally activated and may require an increasing stress at very low temperatures. Whatever the mechanism, there is good evidence for some metals that vacancies but not interstitials are produced in considerable numbers during plastic deformation. This is important in some low-temperature transformation processes.

30. DISLOCATIONS IN AN ELASTIC MEDIUM

No real metals have simple cubic structures, and some extra factors have to be considered when applying the results of the last section. Before doing this, we shall briefly review some of the more mathematical aspects of the theory. The results of this section are independent of the actual crystal structure, and are, indeed, obtained by ignoring the crystalline nature of the material.

When a dislocation is present in an external stress field, the energy of the whole assembly may be reduced by movement of the dislocation line. Suppose a small element of the line is displaced in the direction **j**. We define the force acting on this element in the **j** direction as the rate of change of energy for movement in this direction. The force acts on the dislocation line as a configuration, and not on the atoms which constitute the core of the dislocation at any one instant; care must be taken not to extend the concept beyond the formal definition. The difficulties of precisely defining the force on a dislocation have been discussed by Peach (1951).

For simplicity, consider first a simple edge or screw dislocation (Fig. 7.1 or 7.4) with an applied stress X_{31} across the x_1x_2 plane in the x_1 direction. If the dimensions of the crystal in the x_1 and x_2 directions are L_1 and L_2 respectively, the applied force is $X_{31}L_1L_2$. By moving the dislocation across the whole slip plane, the two halves of the crystal are displaced a relative distance b in the x_1 direction, so that the work done is $bX_{31}L_1L_2$. If the force on the dislocation line has magnitude f per unit length, the total force on the edge dislocation is fL_2, and in moving it across the crystal the work done is fL_2L_1. Similarly, the screw dislocation has a force fL_1 acting on it, and the work done is again fL_1L_2. Equating the expressions for work done, we find for both edge and screw dislocations

$$f = bX_{31}. \tag{30.1}$$

The force is a vector perpendicular to the length of the line, i.e. in the Ox_1, Ox_2 directions for edges and screws respectively.

The above derivation is obviously approximate; the general expression was first given by Peach and Koehler (1950). Consider any element of dislocation line $\mathbf{dl}$ present in an external stress field. The field is completely represented by the symmetrical stress tensor $\mathbf{X}$, the components of which are functions of position. The forces acting across an arbitrary plane at any point in the stress field are $\mathbf{X}\,\mathrm{d}\mathbf{O}$, where $\mathrm{d}\mathbf{O}$ is the vector area of the plane.

Now let the stress field move the dislocation line a distance **dx**, where the direction of **dx** specifies the direction of the resultant force on d**d**. The area swept out by d**d** during the movement is d**x**∧d**d**, and hence the force acting across this area is **X**(d**x**∧d**d**). As a result of the motion, the two parts of the crystal on either side of the swept area are displaced relative to each other by the Burgers vector **b**. Hence the work done during the displacement is **b**•**X**(d**x**∧d**d**), and since **X** is a symmetric tensor, this may also be written **Xb**•(d**x**∧d**d**). If **f** is the force per unit length acting on the element of dislocation line during the displacement, it does work |d**d**| **f**•d**x**. Equating the two expressions, we find

$$\mathbf{f} = \mathbf{Xb}\wedge\mathbf{i}, \tag{30.2}$$

where **i** is the unit vector in the direction of the dislocation element (d**d**) at the point considered.

Peach and Koehler emphasised that this result is perfectly general, and especially that the stress field **X** may arise in any way from externally applied forces or from interactions with dislocations and other irregularities in isotropic or anisotropic cyrstals. However, Weertman (1965) pointed out that eqn. (30.2) appears to give wrong results in certain cases, and he suggested that **X** should be replaced by the "deviator" stress **X**′, where

$$X'_{ij} = X_{ij} - (\tfrac{1}{3})\delta_{ij}X_{kk}.$$

The difficulties arise because the division of forces into mechanical, chemical, and other terms is not necessarily unique when climb processes are involved. According to Lewthwaite (1966) when a dislocation loop is formed in the presence of an applied stress **X** in a body in which material is conserved, the change in energy is given by

$$W_d = -\int_O \mathbf{Xb}\cdot d\mathbf{S} - \int_A \mathbf{Xw}\cdot d\mathbf{A}, \tag{30.3}$$

where O is the cut surface used to introduce the dislocation, A is the external surface, and **w** is a displacement of that surface caused by rearrangement of material after the dislocation is created. A virtual displacement of the dislocation element d**d** in any direction d**x** now gives

$$\delta W_d = -\mathbf{Xb}(d\mathbf{x}\wedge d\mathbf{d}) - \int_A \mathbf{X}\,\delta\mathbf{w}\cdot d\mathbf{A}. \tag{30.4}$$

If this equation is used to define the force per unit length **f** on the element d**d** through the identity

$$\delta W_d = |d\mathbf{d}|\,\mathbf{f}\cdot d\mathbf{x},$$

it follows on equating the two expressions that the first term in (30.4) gives the Peach–Koehler formula (30.2), whilst the second term represents modifications caused by material transfer, and can only be given explicitly when the boundary conditions for this transfer have been specified. Thus the climb of an edge dislocation in a plane normal to the end faces of a cylindrical crystal subjected to a uniform tensile or compressive stress X_{33}, involves the emission or absorption of vacancies at the dislocation line, and the effective force depends on the sink or source of vacancies. For vacancies absorbed or emitted uniformly at the external surface of the crystal, the force has magnitude $f = 2bX_{33}/3 = bX'_{33}$, whilst if the vacancies all appear or disappear at the cylindrical surfaces, $f = bX_{33}$. If the vacancies are retained within the lattice, the force is $bX_{33}(1-(v-v_{\square})/3v)$ where $v_{\square}$ is the volume of a vacancy. Weertman wrote this last expression as bX'_{33} by incorporating the term in $v_{\square}$ in the chemical potential of the lattice.

The above equations were derived on the assumption that the dislocation moves in the direction of the resultant force. Nabarro has pointed out that we are usually interested only in the effective force causing conservative motion in the glide plane, and has shown how this leads to the law of constant resolved shear stress mentioned on p. 233. The normal to the glide plane is the unit vector **n** defined by

$$\mathbf{n} = (\mathbf{b}\wedge\mathbf{i})/|\mathbf{b}\wedge\mathbf{i}|$$

The force **f** may be resolved into a component $\mathbf{f}_g$ in the glide plane and a component $(\mathbf{f}\cdot\mathbf{n})\mathbf{n}$ normal to the glide plane, where

$$\mathbf{f}_g = \mathbf{f}-(\mathbf{f}\cdot\mathbf{n})\mathbf{n} = -\mathbf{n}\wedge(\mathbf{n}\wedge\mathbf{f})$$

using a standard equation for a vector triple product. Substituting for **f** from eqn. (30.2),

$$\mathbf{n}\wedge\mathbf{f} = \mathbf{n}\wedge(\mathbf{Xb}\wedge\mathbf{i}) = (\mathbf{n}\cdot\mathbf{i})\mathbf{Xb}-(\mathbf{Xb}\cdot\mathbf{n})\mathbf{i}$$
$$= -(\mathbf{Xb}\cdot\mathbf{n})\mathbf{i}$$

and, finally,

$$\mathbf{f}_g = (\mathbf{Xb}\cdot\mathbf{n})\mathbf{n}\wedge\mathbf{i} = (\mathbf{Xb}\cdot\mathbf{n})\mathbf{j}, \tag{30.5}$$

where **j** is the unit vector in the glide plane perpendicular to **i**. The force across unit area of the glide plane is **Xn**, and hence the force per unit length of dislocation tending to produce glide, $\mathbf{Xb}\cdot\mathbf{n} = \mathbf{Xn}\cdot\mathbf{b}$, is equal to $|\mathbf{b}|$ times the component of the applied stress across the glide plane resolved in the direction of **b**. If dislocations begin to move at a critical value of $\mathbf{f}_g$, or if their mean velocity is determined by $\mathbf{f}_g$ alone, eqn. (30.5) provides a theoretical justification for the experimental Schmid law that yielding begins at a critical resolved shear stress.

In some crystalline materials, e.g. b.c.c. metals at low temperatures, the Schmid law is not observed, and it must be concluded that the force in the glide plane is not the only factor controlling motion of the dislocation line. Moreover, there is an asymmetry in the observed behaviour such that reversal of all the components of **X** does not necessarily lead to a reversal of the motion. Not only is the critical stress for slip different in tension and compression, but the operative glide-plane of the crystal may change when the stress is reversed. Such effects, although best documented in b.c.c. metals (Christian, 1970), may well exist in other materials in which the core structure of the dislocation is an important factor in determining its mobility (see Section 31).

A single dislocation is highly mobile, and in the presence of a stress field will tend to move in the direction of $\mathbf{f}_g$. In problems of interest, the stress field includes contributions from other dislocations in the structure, each of which is the centre of an internal stress field. In order to calculate the stress distribution, we may disregard the atomic structure of the material, and treat it as an elastic continuum; this procedure is justified so long as we are interested in stresses which vary only slowly over distances of atomic dimensions.

We may form a dislocation in a continuum in the following way. A slit is cut over part of a plane, and the two surfaces of the cut are moved over each other a distance **b** and then rejoined; in order to do this, it may be necessary to add or remove thin layers of material over the cut surfaces. If **b** is perpendicular to the edge of the slit, this edge forms an edge dislocation; if **b** is parallel to the edge of the slit, it forms a screw dislocation. In a true continuum, the stresses would be infinite at the dislocation line, and we have therefore to imagine a narrow cylinder of material a few atomic spacings in diameter cut out along the edge. Physically, the significance of this is that the condition of slowly varying stresses is no longer satisfied near the dislocation line, and the continuum approximation is invalid there. The actual displacements are finite everywhere, but in the region of bad crystal, the stresses vary too rapidly with distance to be treated as continuous.

A general dislocation loop may be formed in a continuum in the same manner. We cut the material over any surface which has as its limit the dislocation line. The two cut surfaces are then displaced by the vector **b** and rejoined. As before, it is necessary to remove a thin cylinder of material along the dislocation core, and also to remove or add thin layers of material wherever the displacement is not parallel to the surface. The dislocation is determined entirely by the limiting line of the cut surface and the vector displacement; the shape of the cut surface is without significance.

The removal of the thin cylinder along the dislocation core signifies mathematically that we can only introduce dislocations into a multiply connected body. The theory of dislocations in an elastic body was formulated long before they were introduced into the physical theory of crystals, and dislocations of more general types were considered. Consider a doubly connected body obtained by boring a cylindrical hole through a solid body. Now cut the body over any surface which terminates on the hole, as in the above procedure, and give the two cut surfaces any elastic deformation and displacements, adding or removing material as necessary, before rejoining them. We have now produced the most general type of dislocation, known as a Somigliana dislocation. A more restricted class of elastic dislocations is obtained if we only permit rigid displacements of the cut surfaces, and these are known as Volterra dislocations. There are six fundamental types of Volterra dislocation, two being (equivalent) edge dislocations and one a screw dislocation obtained by vector translations of the cut surfaces normal or parallel to the edge of the cut. The other three types of Volterra dislocation correspond to rotations of the two cut surfaces with respect to each other, and do not represent single dislocations in crystals.

Following a suggestion by Frank, the rotational Volterra dislocations are now termed "disclinations" (Nabarro, 1967). A dislocation line is characterized by the net displacement of a point when carried around the line, and a disclination line is similarly characterized by the net rotation of a vector carried around a circuit enclosing the line. Clearly, if the lattice structure is to be preserved except in the immediate vicinity of the line, the rotation angle must be $2\pi/n$ when the axis of rotation is an n-fold symmetry axis. Lattice disclinations must thus have large strengths, and hence do not occur in three-dimensional crystals since the elastic strains are large at large distances from the singularity. However, disclinations occur in liquid crystals (Kléman and Friedel, 1969), and it is also possible to envisage disclinations of small strength in internally stressed crystals of finite size. In the latter case, the disclination must form the edge of a planar surface across which there is a rotational discontinuity which may now be made indefinitely small; by analogy with the corresponding (translational) dislocation (see p. 284), this configuration may be called a partial disclination. An example is a low-angle grain boundary, which terminates inside the crystal, so that its edge is topologically a disclination. Such a boundary may be regarded as an array of line dislocations (see Chapter 8), and a disclination is thus a discontinuity in the regular dislocation array.

The three translational Volterra dislocations are characterized by constant Burgers vectors, and hence correspond to real dislocations in crystals. The more general dislocations have varying Burgers vectors, and hence can be used to represent any arbitrary collection of real crystal dislocations. Any network of dislocation lines can then be regarded as a suitable Somigliana dislocation, and this has been useful in some problems. Centres of compression or dilatation (point defects) may also be regarded as Somigliana dislocations. We shall generally use the word dislocation to mean only the translational Volterra dislocations of constant Burgers vector, although we shall also introduce the idea of a "surface dislocation" in Section 34.

To calculate the stress field of an edge dislocation, we suppose that it lies along the axis of a long cylinder of infinite radius. We take the **b** direction as Ox_1 and the axis of the cylinder as Ox_3. Over most of the cylinder, the displacements in the Ox_3 direction are zero, and the problem is one of plane strain (see p. 58). The stresses may be found by means of a suitable stress function, or alternatively we may first solve the elastic equations to find the strains, and then deduce the stresses. We assume that the external surface of the cylinder (at infinity) is free from stress. Solutions have been given by many authors, especially Burgers (1939) and Koehler (1941). The three components of the displacements may be shown to be

$$\left.\begin{aligned} w_1 &= \{b/4\pi(1-\nu)\}\{2(1-\nu)\tan^{-1}(x_2/x_1)+x_1x_2/(x_1^2+x_2^2)\}, \\ w_2 &= -\{b/4\pi(1-\nu)\}\{(1-2\nu)\ln(x_1^2+x_2^2)-x_1^2/(x_1^2+x_2^2)\}, \\ w_3 &= 0, \end{aligned}\right\} \tag{30.6}$$

where ν is Poisson's ratio. The displacement w_1 is a continuous but not a single valued function; every time we describe a circuit round the dislocation line, w_1 increases by b. This characteristic of all dislocation lines is a consequence of the body not being singly connected, and it was utilized in the formal definition on p. 242–3.

Differentiation of eqns. (30.6) gives us the components of the strain tensor, and the stresses are then obtained by Hooke's law. Since we are assuming the material to be elastically isotropic, there are only two independent constants in the stress tensor. From eqn. (11.26)

$$\left.\begin{aligned} X_{11} &= -B_e x_2(3x_1^2+x_2^2)/(x_1^2+x_2^2)^2, \\ X_{22} &= B_e x_2(x_1^2-x_2^2)/(x_1^2+x_2^2)^2, \\ X_{33} &= -2B_e\nu x_2/(x_1^2+x_2^2), \\ X_{12} &= X_{21} = B_e x_1(x_1^2-x_2^2)/(x_1^2+x_2^2)^2, \end{aligned}\right\} \tag{30.7}$$

where we have written B_e in place of $\mu b/2\pi(1-\nu)$. Alternatively, if we use cylindrical coordinates, r, θ, x_3, the stress components are

$$\left.\begin{aligned} X_{rr} &= X_{\theta\theta} = -(B_e \sin\theta)/r, \\ X_{r\theta} &= (B_e \cos\theta)/r, \quad \text{and} \quad X_{33} \text{ as above.} \end{aligned}\right\} \tag{30.8}$$

The stresses thus decrease with distance from the dislocation line, becoming zero at infinity in accordance with the assumed boundary conditions. For a dislocation in a real crystal, however, a more realistic boundary condition is that the external surface at some finite value of r shall be free from traction. This is achieved if we superimpose a second set of displacements which are single-valued and continuous, and which just cancel the stresses X_{rr} and $X_{r\theta}$ at $r = r_e$, the external radius. Since these displacements are single-valued, they do not affect the dislocation. The necessary conditions are obtained if we superimpose the stresses

$$\left.\begin{aligned} X_{rr} &= (B_e/r_e^2)r\sin\theta, \\ X_{\theta\theta} &= (3B_e/r_e^2)r\sin\theta, \\ X_{r\theta} &= -(B_e/r_e^2)r\cos\theta, \end{aligned}\right\} \tag{30.9}$$

on to those of eqns. (30.8). The resultant stress $X_{\theta\theta}$ given by these two sets of equations does not become zero at $r = r_e$, so that there is a tangential stress in the external surface caused by the edge dislocation in the material. Provided r_e is large, the stresses given by (30.9) are small in the region of the dislocation line, and may thus be neglected in comparison with those of (30.8).

The stresses on the internal surface of radius r_i (the radius of the central excluded region in the continuum approach) do not disappear when the stress fields of (30.8) and (30.9) are taken together. This corresponds to the true situation in a crystal, since this is not a real surface in the crystal, and the radius r_i is arbitrarily chosen. If r_i is very small, the displacements are large, and the stresses given by the continuum theory are much larger than the real stresses in a crystal; this is certainly true if we put $r_i = b$, so in general the excluded region must be a few interatomic distances wide. If we wish, we can free the internal surface from tractions by combining (30.8) with a stress function

$$\left.\begin{aligned} X_{rr} &= -X_{\theta\theta} = (B_e r_i^2 \sin\theta)/r^3, \\ X_{r\theta} &= -(B_e r_i^2 \cos\theta)/r^3. \end{aligned}\right\} \tag{30.10}$$

These stresses fall off rapidly with r and are only appreciable in the vicinity of $r = r_i$. Physically, we expect that the stresses at $r = r_i$ will be intermediate between those given by (30.8) alone, and those zero values of X_{rr} and $X_{r\theta}$ obtained by combining (30.8), and (30.10), but they should certainly be closer to the former assumption.

If we add together the three stress functions specified by eqns. (30.8), (30.9), and (30.10), we obtain a solution in which both internal and external surfaces are free from traction. Strictly, the constants in (30.9) and (30.10) have to be adjusted to give this result, but since the effect of (30.9) is negligible when $r = r_i$, and of (30.10) is negligible when $r = r_e$, the changes are only by factors of order $(1-(r_i^2/r_e^2))$, which can be ignored.

Similarly, we may imagine a screw dislocation lying along the x_3 axis of the cylinder, so that the Burgers vector is now parallel to x_3. The strain field is not now a plane deformation, but in fact the problem is simpler, since the displacement w_3 is constant along the dislocation line, i.e. is independent of x_3, and this is the only component of the displacement, which is specified by

$$w_1 = w_2 = 0, \quad w_3 = (b/2\pi)\tan^{-1}(x_2/x_1), \tag{30.11}$$

and the corresponding stress components are

$$\left.\begin{aligned} X_{13} &= X_{31} = -B_s x_2/(x_1^2+x_2^2), \\ X_{23} &= X_{32} = B_s x_1/(x_1^2+x_2^2), \end{aligned}\right\} \tag{30.12}$$

where $B_s = \mu b/2\pi$, and all other components of **X** are zero.

The mathematical theory of screw dislocations is simpler than that of edge dislocations, because in an elastically isotropic medium only shear stresses and strains are involved. In cylindrical coordinates we have

$$w_3 = b\theta/2\pi, \quad X_{\theta 3} = \mu b/2\pi r, \tag{30.13}$$

and all other components of stress are zero. The stress field thus has radial symmetry, and depends only on r. This is related to the absence of a unique slip plane for the screw dislocation, so that the axes x_1 and x_2 are completely arbitrary. The multi-valued component of displacement w_3 is independent of r.

We note that for a screw dislocation there is no traction across the outside surface of the cylinder of finite radius r_e. However, there is a couple on the end faces of a finite cylinder, so that to obtain correct boundary conditions in a finite specimen, we must superimpose stresses which cancel this couple and give single-valued displacements. The necessary stress field is

$$X_{\theta 3} = -\mu b r/\pi(r_e^2+r_i^2), \tag{30.14}$$

where r_e is the external radius, as before, and r_i is the internal radius of the toroidal cylinder, and can be neglected in comparison with r_e. As in the case of edge dislocations, the additional stress field of (30.14) can be neglected reasonably close to the dislocation line, provided r_e is large.

The stress field of an arbitrary dislocation line will be more complex than that of the two simple types of Volterra dislocation which we have considered. More general formulae have been given by Peach and Koehler (1950) and other workers. The stresses all tend to zero as x_1 or x_2 tends to infinity, but the stress field extends throughout the material unless it is limited by other dislocations (see below).

Since we are interested in dislocations in real crystals, which are elastically anisotropic, we should really consider Volterra dislocations introduced into a continuum with anisotropic elastic properties. This theory is much more complex; it no longer follows that all the displacements produced by a screw dislocation must be parallel to the dislocation line, or that those produced by an edge dislocation must be normal to it. When all three components of displacement are non-zero, there is little advantage in treating edges and screws separately, since the displacements due to an arbitrary element of line are not then separable in this way. However, the components of stress in the x_1x_2 plane can be expressed in terms of the derivatives of w_1 and w_2 only, and the component of stress in the x_3 direction in terms of derivatives of w_3 only, provided the x_1x_2 plane is a symmetry plane of the lattice. Several particular examples of dislocations in cubic structures have been solved in this way; the general theory and some important particular solutions were given by Eshelby *et al.* (1953).

The elastic theory of dislocations in an anisotropic medium is algebraically complex and cannot be described here in detail. Many numerical results are changed from the corresponding isotropic values by only $\sim$20–50%, which may not be significant in view of the approximations. However, anisotropic theory also leads to some results which are not predicted at all if isotropic theory is used; for example, in some metals straight dislocations in certain orientations are found to be unstable. For cubic structures the anisotropy constant (Table IV, p. 74) is usually a good measure of the deviations expected, and for highly anisotropic metals such as the alkali metals or β-brass these may be large. Although there are five independent elastic constants in a hexagonal material, there is, nevertheless, a large class of problems for which exact solutions may be obtained in hexagonal symmetry but not in cubic symmetry (Kröner, 1953; Eshelby, 1956a).

Eshelby *et al.* (1953) used a complex variable method to find the stress and strain fields of an infinite straight dislocation, and a slightly different method was developed by Stroh (1958). Except for certain symmetrical orientations (as noted above), the isotropic conditions $w_1 = w_2 = 0$ for a pure screw dislocation and $w_3 = 0$ for a pure edge dislocation do not apply, and a set of linear equations together with a sixth-order algebraic equation have to be solved. In general, analytical solutions may be obtained for dislocations parallel to symmetry axes and in planes normal to even-fold symmetry axes, and a number of solutions have been published. Anisotropic solutions for the elastic problem of a general dislocation configuration have been obtained more recently.

We may now calculate the energy of the dislocation in an otherwise perfect, elastically isotropic lattice. The energy is the sum of the elastic strain energy in the material, and the energy of the core of the dislocation, which cannot be treated by elastic theory. From eqn. (11.16), we see that the strain energy per unit volume at the point $x_1x_2x_3$ is given by

$$\mathrm{d}W_s/\mathrm{d}v = B_e b/4\pi(x_1^2+x_2^2) = B_e b/4\pi r^2,$$

for an edge dislocation, using the displacements and stresses of eqns. (30.6) and (30.7). Taking an annular volume element, and integrating from $r = r_i$ to $r = r_e$, the internal and external radii of the cylinder,

$$^{L}W_s = (B_e b/2)\ln(r_e/r_i), \tag{30.15}$$

as the strain energy per unit length of dislocation line.

More logically, we should use not the stresses of eqn. (30.8) alone, but those from (30.8), (30.9), and (30.10) together. This gives the slightly modified expression

$$^{L}W_s = (B_e b/2)\{\ln(r_e/r_i)-1\}. \tag{30.16}$$

The total energy of the dislocation line is obtained by adding the energy of the core of the dislocation to the elastic energy of (30.15) or (30.16). We thus have

$$ {}^{L}W_d = {}^{L}W_s + {}^{L}W_c \tag{30.17} $$

as the self-energy of the dislocation. ${}^{L}W_c$, the core energy, has been estimated in various ways, and shown to be between 0·5 and 2 eV per atom plane of the dislocation.

The same calculation applied to a screw dislocation gives for the strain energy per unit length of dislocation line

$$ {}^{L}W_s = (B_s b/2)\{\ln(r_e/r_i) - 1\}, \tag{30.18} $$

which differs from (30.16) only by a factor $(1-\nu)$, which is between two-thirds and three-quarters for most metals.

Instead of integrating the elastic energy density, the energy of a dislocation line may alternatively be obtained directly as the work done on the surfaces of the cut (p. 255) whilst the displacement is increased from 0 to **b**. Bullough and Foreman (1964) pointed out that this method requires inclusion of the work done by the tractions on the core surface (e.g. eqn. (30.10)), since otherwise inconsistent results are obtained from different orientations of the cut. However, the constant terms in eqns. (30.16) and (30.18) are normally unimportant, since the core radius is not defined in any case.

It follows from (30.16) and (30.18) that the strain energy of any straight dislocation line may be written

$$ {}^{L}W_s = \tfrac{1}{2}Bb\ln(r_e/r_i), \tag{30.19} $$

where B varies between limits B_e and B_s. If the Burgers vector makes an angle φ with the line, it may be regarded as coincident screw and edge dislocations of vectors $b\cos\varphi$ and $b\sin\varphi$ respectively, so that

$$ B = B_e(1-\nu\cos^2\varphi). \tag{30.20} $$

Also since the shear stress on the slip plane is B_e/r and B_s/r for edges and screw respectively (eqns. (30.8) and (30.13)) it also follows that

$$ \tau_s = B/r \tag{30.21} $$

is the effective shear stress on the slip plane produced by any straight dislocation.

In an anisotropic medium eqn. (30.19) is modified (Foreman, 1955; Stroh, 1958), and becomes

$$ {}^{L}W_s = (Kb^2/4\pi)\ln(r_e/r_i), \tag{30.22} $$

where K is a function of the elastic constants and of the orientations of both the dislocation line and its Burgers vector. For symmetrical orientations, K may be expressed in analytical form; and in the general case it may be written as a Fourier series

$$ K = \sum_{n=0}^{\infty}(\alpha_n\cos n\varphi + \beta_n\sin n\varphi). \tag{30.23} $$

Equations (30.19) and (30.22) show that in an infinite crystal the strain energy of a single dislocation diverges logarithmically to infinity. In a real crystal there is a finite strain energy which is about 10 eV per atom plane if r_e is taken as 10^{-2} m and we assume typical values for elastic constants and lattice parameters. Actually the effective radius of the elastic field of each dislocation in a real crystal is limited by the presence of other dislocations, and the integration should probably be cut off at a separation $r_e = r_d$, where r_d is the mean separation of dislocations. A typical value of r_d is 1 μm, and the strain energy of the dislocation is then rather less than half the value quoted above, but is still several electron volts per atom plane.

It was formerly assumed that a random distribution of dislocations would correspond to a cut-off radius r_d, but Wilkens (1967, 1968) showed that the mean interaction energy of any two dislocations vanishes in a random arrangement, so that the energy of each dislocation diverges logarithmically with a radius of the order of the specimen size. This implies that a random arrangement is prevented by the dislocation interactions and the real arrangement will probably be such that the effective r_e is not appreciably larger than r_d.

Equation (30.17) represents the internal or self-energy of the dislocation. In order to calculate the free energy, it is necessary to include the effects of entropy terms, and as with the defects previously considered, these may be separated into the effects of the dislocation on the vibrational spectrum of the crystal (thermal entropy), and the configurational entropy of the dislocation itself. Cottrell (1953) has shown that both these effects are negligible in comparison with strain energy, even if the dislocation is assumed to be completely flexible on an atomic scale. We thus conclude that the expression for the self-energy of a dislocation also gives its free energy to a close approximation. Since this energy is positive, and of magnitude very much larger than kT, the dislocation cannot exist as a thermodynamically stable defect. This is in contrast to the point defects considered in Chapter 5. It follows from the form of eqns. (30.16) and (30.18) that to a good approximation the free energy of any dislocation may be assumed to be proportional to b^2. There is a small dependence on the orientation of the line, which may be neglected since the factor B in the energy equation in nearly isotropic crystals only varies between about $\frac{2}{3}$ and 1. The dependence on crystal anisotropy, is important, however, in some special cases. The dependence of the dislocation energy on b^2 is important in the theory of dislocation reactions.

When two dislocations are present in a crystal, the energy of the stress field can be formally represented as the sum of the self-energies of the two dislocations plus a term representing the energy of interaction. If the two dislocation have equal and opposite Burgers vectors, the strains produced will cancel at distances large compared with their separation,† and the energy density thus tends to zero except in the region around the dislocations. Two equal and opposite dislocations have finite energy even in an infinite crystal, and the same conclusion applies to a closed dislocation loop.

The problem of two dislocations can be treated by the principle of superposition (p. 62). The integration of the part of the strain energy which represents the interaction of the dislocations is, however, difficult, and may be avoided by means of a method due to Cottrell

† More correctly, the displacements will fall off as r^{-2}, and the strain energy integral is thus convergent.

(1948, 1953). We imagine one dislocation to be already present in the crystal, and consider the work done when a second dislocation is introduced. This work may be divided into work done against the elastic resistance of the material, which represents the self-energy of the second dislocation, and the work done on or by the stress field of the first dislocation, which represents the interaction energy.

Consider first a positive edge dislocation lying along the x_3 axis with Burgers vector along Ox_1 and the extra plane of atoms Ox_2x_3. Let a second, parallel dislocation be brought into the crystal so that it lies along a line distant r from the x_3 axis and having equation $x_1 = r\cos\theta$, $x_2 = r\sin\theta$. This dislocation may be formed by making a cut parallel to the Ox_2x_3 plane and inserting an extra plane of atoms or vacancies, the boundaries of the cut being displaced by b in the x_1 direction. Taking a unit length of this dislocation parallel to Ox_3, the forces due to the first dislocation acting across an element of area dx_2 of the Ox_2x_3 plane are $X_{11}\,dx_2$, and when a positive dislocation is formed, the work done by the forces acting on this area is $X_1 b\,dx_2$. The total gain in energy is thus obtained by integrating the negative of this expression between limits $x_2 = r\sin\theta$ and $x_2 = r_e\sin\theta$. After substituting for X_{11} from eqn. (30.7), the interaction energy is thus obtained as

$$^{L}W_i = B_e b[\ln r_e - \ln r + \cos^2\theta], \tag{30.24}$$

where $(x_1/r_e\sin\theta)^2$ has been neglected since $r_e \gg x_1$.

The force between the dislocations may now be obtained by differentiating this expression. We find

$$\left.\begin{aligned} f_r &= \partial\, ^{L}W_i/\partial r = B_e b/r, \\ f_\theta &= -(1/r)\,\partial\, ^{L}W_i/\partial\theta = (B_e b/r)\sin 2\theta, \end{aligned}\right\} \tag{30.25}$$

as the radial and tangential forces per unit length of either dislocation. The radial force is positive (repulsive) for dislocations of the same sign, but the existence of the f_θ component shows that it is not central. For opposite dislocations, the radial force is attractive.

Since the dislocations are edge types, they can glide only in a particular plane. The component of force providing glide motion is

$$\left.\begin{aligned} f_1 &= f_r\cos\theta - f_\theta\sin\theta \\ &= (B_e b/r)\cos\theta\cos 2\theta \\ &= (B_e b/4y)\sin 4\theta, \end{aligned}\right\} \tag{30.26}$$

where y is the normal separation between the glide planes. For dislocations of like sign, the force f_1 as a function of $\cot\theta$ is shown in Fig. 7.14. We see that the dislocations attract each other (i.e. the force acts so as to increase θ) for $\theta > \pi/4$ and repel one another for $\theta < \pi/4$. The force is zero at $\theta = \pi/2$ and $\theta = \pi/4$, but whereas the former equilibrium is stable, the latter is unstable, a small displacement to higher or lower angles leading to repulsive or attractive forces respectively. A set of edge dislocations of like sign on parallel slip planes are thus in stable equilibrium when they line up in the direction perpendicular to the slip planes; this is further discussed in Chapter 8. It also follows that two dislocations of unlike sign on different parallel slip planes are in stable equilibrium when $\theta = \pi/4$.

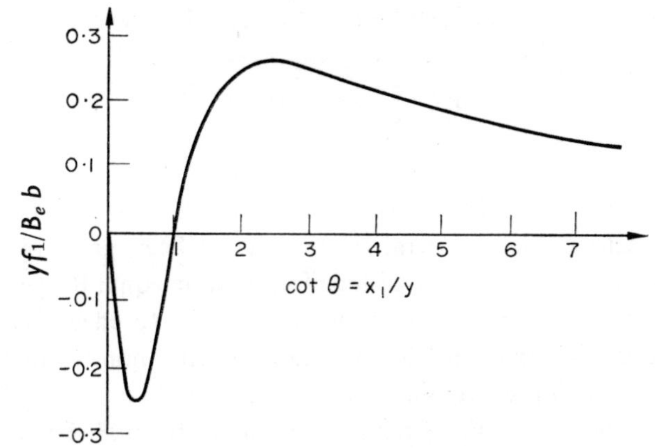

FIG. 7.14. Component of force acting in the glide plane between two like edge dislocations moving in glide planes with a separation of y in the x_2 direction.

We can perform an identical calculation for two parallel screw dislocations, to obtain

$$^{L}W_i = B_s b[\ln r_e - \ln r], \tag{30.27}$$

and the corresponding forces are

$$f_r = B_s b/r, \quad f_\theta = 0. \tag{30.28}$$

The radial forces between two parallel dislocations with parallel Burgers vectors thus vary only between the limits represented by the factor B_e or B_s. For the edge dislocations, the tangential force f_θ disappears when the dislocations have a common slip plane. This condition is automatically satisfied for two parallel screw dislocations, and there is thus no tangential component of force.

The total strain energy for two parallel dislocations of opposite sign is obtained by adding the expression for the self-energies (eqns. (30.15)–(30.18)) to the interaction energy (eqns. (30.24) and (30.27)). Thus for unit length of an edge dislocation pair, or dipole,

$$^{L}W = B_e b[\ln (r/r_i) - \cos^2 \theta] + 2\,^{L}W_c, \tag{30.29}$$

and there is a similar expression for a screw dislocation dipole. Since r_e does not appear in eqn. (30.29), the energy per unit length of such a dipole is finite even in an infinite crystal, as anticipated above. In this approximation, the binding energy of the edge dipole (with respect to dissociation by glide) is obtained by putting $\theta = \pi/4$ in eqn. (30.24), and is thus approximately $B_e b \ln(r_e/y)$, where y is the separation of the glide planes. This is a large energy, comparable with the self-energy when $y \leqslant 100b$.

Edge dipoles and multipoles (see p. 313) are often observed as a principal feature of the dislocation structure in the early stages of deformation of single crystals. Screw dipoles are observed much less frequently, presumably because the component dislocations can often annihilate each other by cross-slip. A closed loop may be regarded as composed of infinitesimal dipole elements, and so its energy will also be finite. In first approximation, the energy is $(Bb/2)\ln(r/r_i)$ per unit length of the loop, so that the energy of a loop of radius r is

$$W_d = \tfrac{1}{2}\mu b^2 r \ln(r/r_i). \tag{30.30}$$

A more accurate expression is given below.

The above expressions were derived for parallel dislocations with equal, or equal and opposite, Burgers vectors. More general relations have been given by Bilby (1950) and Nabarro (1952). For two parallel screw dislocations with unequal Burgers vectors we have obviously merely to replace b^2 by b_1b_2, and the same applies to edge dislocations with collinear Burgers vectors. Two parallel edge dislocations with non-parallel Burgers vectors $\mathbf{b}_1$ and $\mathbf{b}_2$ give an interaction energy in which the term $b^2 \ln r$ in eqn. (30.24) is replaced by $\mathbf{b}_1\cdot\mathbf{b}_2 \ln r$, and the term $b^2\cos^2\theta$ is replaced by $(\mathbf{b}_1\cdot\mathbf{r})(\mathbf{b}_2\cdot\mathbf{r})/r^2$. The latter term equals $b_1b_2\cos^2\theta$ when the two Burgers vectors are parallel, and $\frac{1}{2}b_1b_2\sin 2\theta$ when they are perpendicular to each other.

More generally, two parallel dislocation lines lying along a direction $\mathbf{i}$ and having edge and screw elements $\mathbf{b}_{e1}\mathbf{b}_{e2}\mathbf{b}_{s1}\mathbf{b}_{s2}$ (see p. 240) have an energy of interaction given by

$${}^{L}W_i = \frac{\mu}{2\pi}\left[b_{s1}b_{s2}\ln\frac{r_e}{r} + \frac{1}{1-\nu}\left\{\mathbf{b}_{e1}\cdot\mathbf{b}_{e2}\ln\frac{r_e}{r} + \frac{(\mathbf{b}_{e1}\cdot\mathbf{r})(\mathbf{b}_{e2}\cdot\mathbf{r})}{r^2}\right\}\right].$$

The force between the dislocations is obtained by differentiating this expression.

Parallel dislocations exert a force on each other which has a constant magnitude per unit length. Perpendicular dislocation lines, in contrast, only interact strongly in the region where they approach closest to each other. The relevant formulae have been summarized by Nabarro (1952). The is no interaction between perpendicular dislocations unless the Burgers vector of each is parallel to the line of the other; when this condition is satisfied there is a finite interaction which does not lead to glide. The more general interaction between two nonparallel dislocations on intersecting glide planes is very complex; it is important in some theories of work hardening (Nabarro *et al.*, 1964). Dislocations which pass through each other produce jogs, as described on p. 250.

The displacement field of a general dislocation loop in isotropic approximation was first derived by Burgers (1939) as an integral over the cut surface of the loop; it may alternatively be found by a Green's function method (Seeger, 1955a, b; de Wit, 1960). In either case, the algebra is unwieldly, although numerous particular solutions have been found. For a circular loop in the glide plane, Kröner (1958) found the stress field in the form of complete elliptical integrals, and the corresponding strain energy is

$$W_d = \tfrac{1}{2}\pi(2-\nu)B_e br\{\ln(4r/r_i)-1\}. \tag{30.31}$$

This is insignificantly different from the approximate expression (30.30). The stress field of a circular pure prismatic loop is obtained as definite integrals of Bessel functions (Kroupa, 1960; Bullough and Newman, 1960), and the energy is

$$W_d = \pi B_e br\{\ln(8r/r_0)-1\}. \tag{30.32}$$

A general method for the calculation of self-energies and interaction energies in which any dislocation line is constructed from angular dislocations was developed by Yoffe (1960), and a similar approach in which the dislocation is treated as piecewise straight is due to Jøssang, *et al.* (1965), de Wit (1967), and others. The energies of some complex loop shapes have been determined analytically by these methods; these calculations were reviewed by Kroupa (1966).

An anisotropic calculation of the elastic field and associated energy of a general dislocation configuration has only recently become possible. Indenbom and Orlov (1967) solved the elastic problems of a finite dislocation segment by generalizing approaches due to Lothe (1967) and Brown (1967). Their solution utilizes the derivatives with respect to orientation of the distortions produced by an infinite straight dislocation. Analytical solutions which do not require solution of a set of linear or algebraic equations have been derived by Willis (1970), also by means of a Green's function method. Willis gives for the infinite straight dislocation explicit (though complex) expressions which are easier to apply than the formulations of Eshelby *et al.* (1953) and Stroh (1958), and he also derives equations for the distortions produced by a dislocation segment and by any planar dislocation loop. These expressions do not require the analytical solutions for infinite, straight dislocations, and can be used directly for the calculation of the stress and strain fields or energies of any dislocation configuration. A further simplification of the computing procedure has been suggested by Barnett and Swanger (1971), Barnett *et al.* (1972), and Barnett (1972), who extended Brown's method and gave formulae for the line energy ${}^{L}W_s$ and its first and second derivatives with respect to orientation. Their procedure avoids the necessity for solving the sextic algebraic equation by using a numerical integration method based on a Fourier transform. Alternative numerical procedures are given by Malen (1970) and Malen and Lothe (1970). These important techniques have already been applied to the elliptical loop (Willis, 1970) and the rhombus-shaped loop (Bacon *et al.*, 1970) and are being further developed, but details cannot be given here because of the complexity of the expressions.

We next consider briefly the concept of the line tension of a dislocation which was introduced by Mott and Nabarro (1948) who used the taut-string analogy mentioned above and defined the line tension as the increase in energy for a unit increase in line length. The line tension T thus has the dimensions of force and for a straight dislocation in isotropic approximation is given from eqn. (30.19) as

$$T = {}^{L}W_s = \tfrac{1}{2}Bb \ln(r_e/r_i). \tag{30.33}$$

However, the Mott and Nabarro definition gives a line tension which depends on the orientation and shape of the dislocation; obviously a slightly different expression is obtained by differentiating (30.31) to obtain the line tension in a circular loop. De Witt and Koehler (1959) first suggested that any change in the orientation of the dislocation line should be included in the line tension, which would thus be defined as

$$T = {}^{L}W_s + (\partial^2\, {}^{L}W_s/\partial\varphi^2), \tag{30.34}$$

where φ is the angle between **b** and the direction of the line element. A more detailed study of this problem was made by Brown (1964), who introduced the concept of self-stress, i.e.

the stress on an element of dislocation due to the dislocation itself. This stress varies along the dislocation with the local orientation, and also depends on the shape of the whole line; it is expressed as a rather complex line integral. If the curvature is everywhere small, the self-stress is well-approximated by a simple tension, as assumed above, but if this condition is not satisfied it is necessary to integrate numerically.

Further calculations of the self-stress have been made by Jøssang *et al.* (1965), Jossang (1968), and other workers who treated a bowed dislocation as composed of piecewise straight segments, and by Brailsford (1965) who approximated a curved dislocation by a sequence of infinitesimal kinks. These results are important in certain problems in dislocation theory, e.g. the use of measurements on extended nodes to estimate stacking fault energies, but for many purposes it is adequate to use the rough approximation $T = \mu b^2$ which follows from (30.33) if $r_e \sim 10^4 b$ and $r_i \sim b$ (Friedel, 1964).

Line tension may be used to estimate the equilibrium shape of a dislocation with fixed ends in an external or internal stress field. An element $\mathbf{dd}$ which subtends an angle $d\psi$ at its centre of curvature is in equilibrium under the force per unit length f from the stress field and its own line tension T if

$$T d\psi = f dd, \tag{30.35}$$

and hence the curvature is

$$\frac{1}{r} = d\psi/dd = f/T. \tag{30.36}$$

For a Frank–Read source of length d, the critical condition is when the semicircular con figuration is reached, so that $r = d/2$. If the resolved shear stress in the direction of b is τ the source will operate if

$$r \geqslant 2T/db. \tag{30.37}$$

More accurately, the dislocation will not adopt a semicircular configuration, even in isotropic approximation, and the equilibrium shape must be determined by the condition that the total force (from imposed stresses, internal stresses, and self-stress) on each element of the line must be zero (Brown, 1964; Bacon, 1967; Foreman, 1967). Moreover, when anisotropic theory is used a new result appears, namely that it is possible for the line tension of a straight dislocation line to be negative (de Wit and Koehler, 1959). This means that the straight line is unstable, and if its mean direction is fixed it will adopt some kind of zigzag configuration. The conditions leading to negative line tension in several metals have been calculated by Head (1967) and Barnett *et al.* (1972) and are futher discussed in Section 33.

In the above discussion we have ignored any internal or external surfaces, except insofar as the latter determine r_e. However, the stress field of a dislocation near an interface or a free surface is modified by the requirement that the shear and normal components must be continuous across the boundary so that there is an interaction of the dislocation with the interface. Solutions to this elastic problem can often be obtained by the method of images which is exactly analogous to that used in electrostatics. Imaginary sources of stress, e.g. other dislocations, are introduced so that the superimposed stress fields can be made to satisfy the boundary conditions. The stress field of an image source appears as an external stress field to the real dislocation, and the resultant force on the dislocation is called an image force.

For a screw dislocation in a medium of shear modulus μ_1 near a planar boundary with a medium of shear modulus μ_2, the image dislocation has a Burgers vector

$$\mathbf{b}_I = \{(\mu_2 - \mu_1)/(\mu_2 + \mu_1)\}\mathbf{b} \tag{30.38}$$

(Head, 1953), and the dislocation experiences a force normal to the boundary of magnitude

$$f = B_s b_I / 2r, \tag{30.39}$$

where r is the distance of the dislocation from the boundary. The dislocation is attracted to the boundary if $\mu_2 < \mu_1$ and is repelled if $\mu_2 > \mu_1$; in particular it is attracted to a free surface ($\mu_2 = 0$). Similar calculations have been made for an edge dislocation (Head, 1953), a dislocation inclined to the surface (Lothe, 1967), and a general dislocation loop (Groves and Bacon, 1970). Image forces are obviously of great importance in considering dislocation configurations observed by transmission electron-microscopy of thin foils.

In addition to the forces which they exert on each other, dislocations will also interact with the stress fields of other types of lattice defect. Of particular interest is the interaction between a dislocation and a solute atom, which may be dissolved substitutionally or interstitially in the lattice, or between a dislocation and a point defect. We have already seen that an edge dislocation produces an expanded region on one side of its slip plane, and a compressed region on the other. If the solute atom is larger than the solvent atom, or if it is dissolved interstitially, we expect it to be repelled from the compressed side and attracted into the expanded side. Conversely, a small solute atom or a lattice vacancy will be attracted to the compressed region; the vacancy may disappear completely at a jog, as discussed earlier, or may be merely bound to the dislocation line in the same way as a solute atom.

A simple expression for the energy of interaction of a dislocation line and a solute atom was first given by Cottrell (1948). He used the misfitting sphere model of the solute atom which we have already discussed in Section 25. The self-energy of the solute atom in this model is equal to the work done against the elastic resistance of the crystal in forcing a spherical atom into a spherical hole of different size. In calculating this energy previously, we made the implicit assumption that the crystal is free from internal stress fields. If an internal stress field is present, however, additional work will be done on or by the forces of this field, and this work may be termed the energy of interaction. The method is clearly identical with that used above for finding the energy of interaction of two dislocation lines.

Suppose the change in volume is Δv when a single solute atom is introduced. Then if the stress field is assumed uniform over the region occupied by the atom, and not to change during the expansion, the work done against the stress field is $p\,\Delta v$, where p is the hydrostatic pressure in the region of the atom. We thus have

$$W_i = p\,\Delta v = K\Delta\,\Delta v, \tag{30.40}$$

where K is the bulk modulus of elasticity, and Δ is the cubic dilatation near the atom due to the stress field of the dislocation. Eshelby (1954, 1956a) has shown that in this expression Δv is the change in volume of the whole crystal (including "image" effects) when the point defect is introduced. Using the result that the hydrostatic pressure in a stress field $\mathbf{X}$ is $p = -X_{ii}/3$ we have from eqn. (30.8)

$$W_i = \frac{\mu b}{3\pi} \frac{1+\nu}{1-\nu} \frac{\sin\theta}{r} \Delta v. \tag{30.41}$$

For a solute atom which expands the lattice, this is positive when $0 > \theta > \pi$, i.e. when the atom is above the slip plane, and negative when the atom is below the slip plane. The atom is thus expelled from the compressed region and attracted into the expanded region.

If the model of a solute atom as a compressible sphere in a misfitting hole is used, the interaction energy may also be written (Bilby, 1950) as

$$W_i = \frac{\mu b}{\pi} \frac{\sin\theta}{r} \Delta v^{\infty}. \tag{30.42}$$

where Δv^{∞} (see p. 201) is the change in volume of a region enclosed by a surface immediately surrounding the solute atom. Equation (25.8) gives the relation between Δv^{∞} and the difference Δv_{AB} in atomic volumes of the two components. In principle, W_i may be obtained from an estimate of Δv_{AB}, but in practice it is preferable to use measurements of the change of lattice parameter with composition to obtain the mean strain $\varepsilon = \mathrm{d} \ln a/\mathrm{d}x$ (see p. 183) and then to substitute $\Delta v = 3v\varepsilon$ directly into (30.25).

The existence of an interaction energy implies that there is an effective force between an edge dislocation and a solute atom (or a vacancy) at any point. The force components $f_r = -\partial W_i/\partial r$ and $f_\theta = -(1/r)(\partial W_i/\partial\theta)$ may be evaluated from (30.41) or (30.42).

Clearly this type of calculation suffers from all the defects mentioned in Section 25, but it is likely to be qualitatively correct. Even if the elastic model for the solute atom is accepted, however, the calculation is only applicable in the region where the internal stress field of the dislocation is given by linear elastic theory. The expression should thus not be applied directly to solute atoms which are very close to the dislocation axis. When Δv is positive, it is clear that the strongest binding (lowest value of W_i) is obtained at $\theta = 3\pi/2$ and at some finite distance r which will be about one interatomic distance. The change in the stress field of the dislocation due to the presence of solute atoms near the dislocation is also not considered in this simple form of calculation.

Cottrell supposed the strain field of the solute atom to be a pure dilatation, so that it interacts only with hydrostatic stress. When elastic anisotropy is taken into account, this conclusion is no longer valid, but for symmetrical solute atoms (or point defects) the main energy will still arise from the dilatational component of the dislocation field. An interesting result which follows from the different geometries is that a screw dislocation has a dilatational field in b.c.c. structures, but not in f.c.c. or h.c.p. structures (Chang, 1962; Chou, 1965). Thus even in the linear elastic approximation, screw dislocations interact with centres of dilatation in b.c.c. crystals. However, a more important effect may be the presence of solutes in sites of lower than cubic symmetry.

Point defects which do not possess the spherical (or cubical) symmetry required for Cottrell's theory have a long-range elastic interaction with all dislocations. Tetragonal defects are of especial interest since they represent interstitial solutes in b.c.c. crystals and various other defects in f.c.c., ionic or diamond cubic structures. In many b.c.c. metals, interstitial solutes such as carbon, nitrogen, or oxygen are believed to occupy octahedral sites, that is equivalent sites at the centres of {100} faces and ⟨100⟩ edges of the unit cube. The resultant shear strains can interact with the shear stress field of a screw dislocation to give an appreciable energy (Crussard, 1950; Cochardt *et al.*, 1955). If the defect is specified by principal strains $e_{11} = e_1$ and $e_{22} = e_{33} = e_2$, the interaction energy depends on the shear strain $(e_1 - e_2)$ and according to Cochardt *et al.* it is given by

$$W_i = 8^{\frac{1}{2}} B_s(e_1 - e_2)v \sin \theta / r, \tag{30.43}$$

where θ is now the smallest angle between the radius vector from the dislocation to the solute atom and a ⟨110⟩ direction normal to the dislocation line. The volume v with which the strains e_1 and e_2 are associated is effectively an ellipsoid of revolution, but is of ill-defined size. However, if the values of e_1 and e_2 are derived from experimental measurements of the lattice parameters of aligned defects in dilute solution (as is possible for carbon or nitrogen martensites), the choice of v is immaterial. The lattice parameter measurements are extrapolated to one defect per volume v and the product $(e_1 - e_2)v$ is then independent of v. Some confusion has arisen because Cochardt *et al.* chose to regard the volume v as the cubic cell volume a^3. This implies rather large values for the principal strains, and Hirth and Cohen (1969a, b) argued that a more reasonable choice of v is $\frac{1}{2}a^3$, which effectively doubles these large strains. These authors therefore argued that the lattice parameter extrapolation procedure is invalid and that relative displacements of nearest neighbours obtained by an atomistic calculations should be used in (30.43). Schoeck (1969) and Bacon (1969) have shown that this is incorrect and that (30.43) is exact within the limitations of continuum theory.

The interaction of a tetragonal defect with an edge dislocation is given by

$$W_i = \{\tfrac{1}{3}(e_1 + 2e_2)(1 + \nu + 2\cos^2\theta) - e_2 \cos 2\theta + e_1 \nu\}(B_e v / r) \sin \theta. \tag{30.44}$$

The interactions with edge and screw dislocations are of comparable magnitude.

The above interactions all involve energies falling off as $1/r$ and may be described as long-range. Shorter-range interactions, decreasing as $1/r^2$, arise from a model in which the solute atom is treated as an elastic inhomogenity, i.e. as a finite volume of different elastic properties. For example, an atom of slightly different shear modulus has an interaction energy with a screw dislocation (Saxl, 1964),

$$W_i = e_\mu v\, {}^vW = (\mu / 8\pi^2)\, e_\mu (b/r)^2. \tag{30.45}$$

where ${}^vW$ is the strain energy density of the dislocation and e_μ is the fractional rate of change of μ with atomic fraction of solute. A similar equation for an edge dislocation involves the rate of change of both bulk and shear moduli.

Finally, there is a second-order volume expansion around a screw dislocation of magnitude $Cb^2/4\pi^2r^2$, where $0{\cdot}3 \leqslant C \leqslant 1$ (Stehle and Seeger, 1956). This gives a short-range interaction

$$W_i = \left\{\frac{C(1+\nu)}{2\pi^2(1-2\nu)}\right\} \mu v e\left(\frac{b}{r}\right)^2. \tag{30.46}$$

The change of volume arises from anharmonic vibrations, and Lomer (1957) pointed out that C may be derived from the Grüneisen relation between thermal expansion and specific heat.

A full discussion of the interaction energy between a line dislocation and a point defect should include electrical and chemical interactions as well as elastic interactions. Generally, the electrical interaction is small for metals in comparison with the elastic interaction, though this is not necessarily true for polar crystals. The chemical interaction arises mainly when the dislocation is extended into a ribbon of stacking fault, since a lowering of energy can result from a change of composition in the stacking fault region. This effect was first discussed by Suzuki (1952).

The maximum value of the interaction energy (the binding energy) of a solute atom and a dislocation may be estimated theoretically or from experimental measurements. The calculations of Cochardt *et al.* suggest a binding energy of about 0·75 eV for a carbon atom in iron for both screw and edge dislocations. This is almost certainly too high, and various other estimates fall in the region 0·3–0·5 eV, but the results are significant in showing that for b.c.c. metals with interstitial solutes, the binding energies are comparable for all types of dislocations. For f.c.c. metals, where solutions are usually substitutional, this is not true, and edge dislocations are bound more strongly than screw dislocations, at least if chemical binding is ignored for the present. Typical binding energies for substitutional solutes are probably about 0·02 eV, and this should also be about the order of magnitude of the interaction between a dislocation and a vacancy.

Provided the necessary diffusion of solute atoms is possible, we may expect the interactions to lead to a segregation of solute atoms to the vicinity of a dislocation. The equilibrium state of a crystal containing both dislocations and impurities will thus be one in which each dislocation is surrounded by a cloud of impurity atoms; this has been called by Cottrell (1948) a dislocation "atmosphere". Under suitable conditions, the equilibrium solute distribution around a dislocation may be given by a Maxwellian formula of the type

$$c = c_0 \exp(W_i/kT), \tag{30.47}$$

where c_0 is the mean solute concentration. This equation will be valid so long as the change in concentration which it predicts at any point is small, i.e. the atmosphere is "dilute". The necessary conditions for this are that c_0 is small and $W_i \ll kT$. If these conditions are not satisfied, the atmosphere will condense, probably into a row of solute atoms parallel to the dislocation and situated at the points of maximum binding. A rough condition for the condensation of the atmosphere is that the dilatations due to the solute atoms just cancel those due to the dislocation, so that the hydrostatic stress is completely relaxed. This happens approximately when $(c-c_0)\,\Delta v$ is equal to the maximum value of the dilatation near the dislocation centre. In the case of interstitial solutes in b.c.c. metals, the interaction is so strong that $W_i \gg kT$ at room temperature, thus leading to condensed or saturated atmospheres. As we have already seen, this strong interaction does not arise primarily from purely hydrostatic stresses.

Even for a saturated atmosphere, there should be only one interstitial atom per atom plane of the dislocation within a radial distance of a few Ångstrom from the dislocation core. Clearly, this distribution is not described by eqn. (30.47); Louat (1956) has suggested instead the use of a distribution law of the Fermi type. We should also note here that there is much evidence that at low-temperatures dislocation lines in metals like iron acquire a very large number of interstitial atoms per atom plane of dislocation. This is, in fact, a precipitation phenomenon, the dislocation line acting as a centre on which the new phase precipitates. The theory of the segregation of solute atoms to dislocations is obviously closely related to that of precipitation proper in certain circumstances, and we shall discuss this in more detail in Part II, Chapter 16.

31. FORMATION AND PROPERTIES OF CRYSTAL DISLOCATIONS

In the last section we obtained explicit expressions for the stress field of a dislocation by treating the solid as an elastic continuum. This procedure is only valid if the region around the dislocation line is excluded from consideration. The stress field at the centre of a dislocation is infinite in an elastic continuum, but in a crystalline material, the shear stress between the planes which glide over each other must be a periodic function of **b**. In the model of a dislocation used by Peierls (1940) and extended by Nabarro (1947), the half crystals on

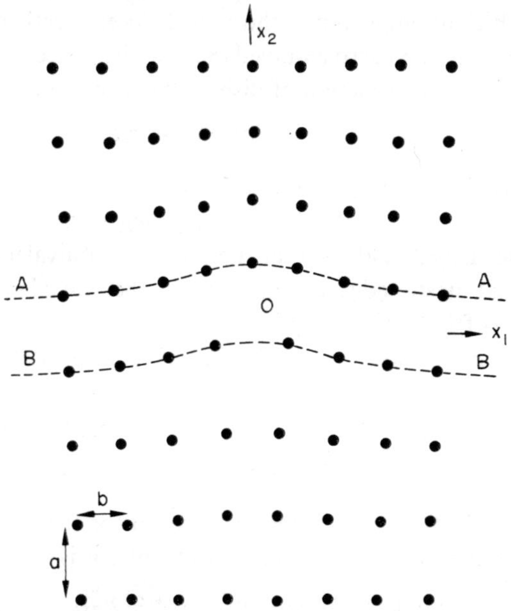

FIG. 7.15. The Peierls–Nabarro model of a dislocation (after Nabarro 1952).

either side of the glide plane are treated as elastic continua, but the shear stress between two atomic planes is assumed to be a sinusoidal function of the relative displacement of the atoms.

Figure 7.15 shows a section through a straight-edge dislocation lying along the x_3 axis. The two planes A and B are curved by the dislocation, but in the model used there is no normal stress X_{22}. If the interatomic distances in the x_1 and x_2 directions are b and a respectively, the Burgers vector has magnitude b; for a simple cubic crystal as used in most of the calculations, $a = b$. Let the distance along the x_1 direction between neighbouring atoms in the planes A and B be written

$$\Phi_1(x_1) = \begin{cases} 2w_1(x_1)+\frac{1}{2}b & (x_1 > 0), \\ 2w_1(x_1)-\frac{1}{2}b & (x_1 < 0), \end{cases} \tag{31.1}$$

where $w_1, -w_1$ are the components in the x_1 direction of the displacements of the atoms in A and B respectively, measured from an initial configuration in which the atoms in the two planes are all out of registry by $\frac{1}{2}b$ but the planes themselves are undistorted. Any components of displacement w_2 are assumed to be equal in A and B. The choice of origin for $\mathbf{w}$ means that the displacements tend to zero near the centre of the dislocation line if the core has the symmetrical configuration shown in Fig. 7.15. There is also a boundary condition $w_1(\infty) = -w_1(-\infty) = -b/4$ which ensures that the atoms are in registry ($\Phi_1 = 0$) far from the centre.

The width of the dislocation is the region over which the disregistry is appreciable, and is determined by the balance between two opposing forces. The atoms in A are compressed in the x_1 direction, and try to reduce their energy by spreading this compression uniformly along the plane; this force is calculated by elastic theory. The opposing force is due to the extended atoms in B which attempt to pull those in A closer together, and thus reduce the dislocation width. Peierls and Nabarro assumed that the shear stress X_{12} between the planes can be expressed as a sinusoidal function of displacement of the form

$$X_{12} = (-\mu b/2\pi a)\sin(4\pi w/b), \tag{31.2}$$

where the constant is fixed by requiring Hooke's law to be valid in the limit of small shear.

The elastic shear stress in A due to the displacements $w_1(x_1)$ may be calculated directly or, alternatively, by noting (Eshelby, 1949) that the model is equivalent to a continuous distribution of infinitesimal elastic edge dislocations on the slip plane with Burgers vector between x_1 and x_1+dx_1 of $-2\,dw_1(x_1)$. By considering the shear stress at x_1 due to the infinitesimal dislocations lying between $x_1 = x'$, $x'+dx'$ (eqn. 30.7) and integrating,

$$X_{12}(x_1, 0) = \frac{\mu}{\pi(1-\nu)} \int_{-\infty}^{\infty} \left\{ \frac{[dw_1(x_1)/dx_1]_{x_1=x'}}{x_1-x'} \right\} dx'. \tag{31.3}$$

Equations (31.2) and (31.3) give the integral equation for the displacements w_1 and the solution which satisfies the boundary conditions at infinity is

$$w_1 = -(b/2\pi)\tan^{-1}\{x_1+\alpha b)/\zeta\}, \tag{31.4}$$

where

$$\zeta = a/2(1-\nu) \tag{31.5}$$

and $0 \leqslant \alpha \leqslant 1$. The disregistry Φ_1 falls from its maximum value to one-half of this value in a distance ζ, so that 2ζ is a measure of the width of the core. The parameter α specifies the symmetry of the core, and is such that αb is the distance of the centre of the dislocation from the symmetrical configuration shown in Fig. 7.15. In this configuration ($\alpha = 0$), the additional half-plane of atoms above A is midway between two half-planes below B; in the simple cubic lattice there is a second symmetrical configuration ($\alpha = \frac{1}{2}$) in which one half-plane on the expanded side, below B, is midway between two half-planes on the compressed side, above A.

Equations (31.1) and (31.4) enable the shear stress in the slip plane to be written as

$$X_{12}(x_1, 0) = B_e z_1/(z_1^2+\zeta^2), \tag{31.6}$$

where $z_1 = x_1+\alpha b$. It also follows that the density of Burgers vector along the x_1 axis in the continuous distribution model is $b\zeta/\pi(z_1^2+\zeta^2)$, and this provides one method of finding the stress field X_{ij} throughout the crystal by integration of the field of an individual elastic dislocation over this distribution. This gives for the shear stress component, for example,

$$X_{12}(x_1, x_2) = \frac{B_e z_1(z_1^2+z_2^2-2x_2z_2)}{(z_1^2+z_2^2)^2}, \tag{31.7}$$

where $z_2 = x_2+\zeta$ for $x_2 > 0$ and $z_2-\zeta$ for $x_2 < 0$. Comparing with eqn. (30.7) we see that this is almost identical with the field of a single elastic dislocation with its centre at $x_1 = -\alpha b$, $x_2 = \mp\zeta$ for the upper and lower half-crystals respectively.

The energy of the dislocation in this model consists of the elastic energy stored in the two parts of the crystal plus the misfit energy across the plane $x_2 = 0$, which is the effective core energy. The elastic energy per unit length is given by the integral $\int X_{12}(x_1, 0)\, u_1(x_1)\, dx_1$ and within a cylinder of large radius r_e, this becomes

$$^{L}W_s = \tfrac{1}{2}B_e b \ln(r_e/2\zeta). \tag{31.8}$$

For a small ribbon of unit length and width dx_1, the misfit energy is $\left[\int_0^{\Phi_1} -X_{12}(x_1,0)\, d\Phi_1\right] dx_1 = (\mu b^2/4\pi a)\{1+\cos(4\pi w_1/b)\}\, dx_1$, and on integrating this between limits $x_1 = \pm\infty$, the total misfit energy is found to be

$$^{L}W_c = \tfrac{1}{2}B_e b. \tag{31.9}$$

Nabarro's expression for the self-energy of an edge dislocation is thus

$$^{L}W_d = \tfrac{1}{2}B_e b\{\ln(r_e/2\zeta)+1\}. \tag{31.10}$$

Comparing this with eqn. (30.15) we see that the energy is identical with that of an elastic dislocation in an isotropic continuum if the integration is cut off at a distance $r_i = 2\zeta/\exp(1)$ from the centre of the dislocation. Alternatively, we may regard eqn. (31.8) as giving the elastic energy with the integration cut off at a core radius $r_i = 2\zeta$, and to $^{L}W_s$ is added the core energy of eqn. (31.9). It is perhaps more logical to make the external surface of the cylinder of radius r_e stress free, as before, and when the work done by the additional external forces on the cylinder is included, an energy $B_e b/4$ has to be subtracted from the right-hand sides of eqns. (31.8) and (31.10). Comparing now with eqn. (30.16), we find the equivalent cut-off radius is $r_i = 2\zeta/[\exp(1)]^{3/2}$.

The same procedure may be used for a screw dislocation, although it does not then seem immediately obvious that it is correct to choose a particular plane on which to calculate the misfit energy. Eshelby (1949) showed, however, that if this assumption is made, identical results are obtained except that $\zeta = \frac{1}{2}a$ and B_e is replaced by B_s. The results are independent of the choice of misfit plane.

The width of the dislocation given by eqn. (31.5) is very small and this implies a contradiction; the narrow core is a consequence of the sinusoidal approximation, but the approximation is valid only if the misfit is spread over many atoms. A more realistic law of force would correspond to the same initial slope but a smaller maximum force, and Foreman *et al.* (1951) developed a parametric solution for a family of force laws of more general type than (31.2). If the maximum force is reduced, so also is the energy of the regions of severe misfit in comparison with other regions, and the core therefore becomes wider. The form of the solution is similar to eqn. (31.4) but allows the dislocation width to be arbitrary; the product of the width and the maximum force is almost constant.

On p. 236 we referred to the constraining force tending to keep a dislocation in an equilibrium position in the lattice. The barrier to dislocation motion arises because of the periodic variation in the core energy of the dislocation as it is displaced, and the stress needed to move the dislocation over this barrier is generally called the Peierls stress. In the model we have just described, the effects of the discrete lattice are smoothed by the integration leading to eqn. (31.9), and in order to find the energy as a function of displacement (i.e. of the parameter α) it is necessary to sum the misfit energy over the atom rows in the cut. In the method used by Nabarro, this sum is evaluated separately over the top and bottom rows, and gives for the core energy

$$^{L}W_c(\alpha) = \tfrac{1}{2}B_e b\{1+2\cos(4\pi\alpha)\exp(-4\pi\zeta/b)\}. \tag{31.11}$$

The difference between the maximum and minimum values of this expression is the Peierls energy

$$^{L}W_P = 2B_e b\exp(-4\pi\zeta/b) \tag{31.12}$$

and from the maximum slope of this expression, the corresponding Peierls stress is

$$X_P = (2\pi/b^2)^{L}W_P. \tag{31.13}$$

Using typical values for the elastic constants, eqn. (31.13) gives a stress of $\sim 2-4\times10^{-4}\,\mu$, i.e. $\sim 10^7$ N m^{-2} or 1 kg mm^{-2}. This is about 10^3 times smaller than the theoretical stress to shear a perfect lattice, but is much larger than the observed low-temperature yield stresses of well-annealed single crystals of soft metals. However, it follows from (31.12) that the Peierls energy and stress are both very sensitive to dislocation width and the above values will be reduced when a is greater than b (as will usually be the case for slip on a close-packed plane) or when the force law is such as to give a greater width than that predicted by eqn.

(31.5). Foreman *et al.* confirmed that the exponential dependences of ${}^{L}W_P$ and X_P on ζ remain in their more general treatment.

Criticisms of the above calculation, and attempts to improve it, have been made by several authors. Huntington (1955) noted that the two symmetrical configurations of the dislocations have equal energies according to the above calculation, so that the wavelength of the Peierls force is $\frac{1}{2}b$ instead of b. He pointed out that when the atoms are almost in registry in A and B, the forces on the atoms are nevertheless defined in terms of the large displacements (approaching $\pm b/4$) from the reference positions, and he developed an alternative treatment in which the forces depend on the actual positions of the atoms. This gives a similar exponential dependence on the width of the dislocation, but now has the correct periodicity b; the predicted Peierls stress has higher numerical values of $\sim 10^{-2}$–$10^{-3}\,\mu$. Kuhlmann–Wilsdorf (1960) pointed out that the Peierls energy given by eqn. (31.12) is only a very small percentage of the core energy, and she suggested that the first-order effect has been lost in the Peierls–Nabarro calculation because of the equal energy of the two symmetrical configurations. She attempted to estimate the Peierls stress directly by summing the tangential forces on the atoms without calculating the energy, and obtained a result in terms of the core radius r_i

$$X_P = (b/8r_i)X_{\text{crit}}, \tag{31.14}$$

where X_{crit} is the critical stress to cause slip in a perfect crystal. The exponential dependence on core width does not appear in this equation, but the numerical values of the Peierls stress are comparable with those obtained by Huntington.

Equation (31.12) is unsatisfactory not only in giving equal energies to the $\alpha = 0$ and $\alpha = \frac{1}{2}$ configurations, but also because these are the *maximum* energy configurations, whilst the asymmetric configuration ($\alpha = \frac{1}{4}$) gives the minimum energy. This occurs because of the independent summations over the two planes A and B, and it seems physically more realistic to sum the interaction energy between pairs of atom rows in the two surfaces. This has been done numerically by Vítek (see Christian and Vítek, 1970) and produces the result that the minimum and maximum energy configurations now occur at $\alpha = 0$ and $\alpha = \frac{1}{2}$ respectively. The stress again decreases rapidly with the width of the core.

In recent years there have been many attempts to overcome the limitations of the Peierls–Nabarro model by the use of more realistic interatomic forces and by considering the atomic structure of a block of material and not just of the slip plane. In early work a discrete elastic approach (linear force laws between the atoms) was often used, and solutions for edge dislocations in a simple cubic lattice (Babuska *et al.*, 1960) and for screw dislocations in simple cubic (Maradudin, 1958) and diamond cubic (Celli, 1961) structures were obtained. However, discrete elastic models can be expected to be a good approximation only for relatively small deviations of the atomic structure from the defect-free equilibrium configuration, and thus do not work well for dislocation cores. Non-linear force laws, of the types discussed in Chapter 5, are thus required, and many computer calculations of the core structure of particular dislocations have now been published. One of the earliest of these calculations was made for edge and screw dislocations in sodium chloride (Huntington *et al.*, 1955, 1959) using electrostatic attractive and Born–Mayer repulsive forces; in this work, rows of ions were treated as solid rods, but were not restricted in motion parallel to their length. The width of the edge dislocation was found to be similar to that given by the Peierls–Nabarro model.

Morse potentials were used by Cotterill and Doyama (1966) and Doyama and Cotterill (1966) to compute the structure of edge and screw dislocations in f.c.c. metals. The potentials were fitted to the measured elastic constants of copper and truncated to 179 neighbours. The dislocation cores were found to be undissociated when relaxation was allowed only parallel to b, but split into partials on $\{111\}$ planes (see Sections 32 and 33) when full relaxation was allowed. Chang and Graham (1966) used a Johnson potential (eqn. 16.12) truncated to second neighbour interactions to calculate the core structure of an edge dislocation in a b.c.c. structure, and Yamaguchi and Vítek (1973) have calculated the structures of several non-screw dislocations in this structure, using general potentials of the Johnson type. These results all show the cores to be rather narrow and planar, as assumed in the Peierls–Nabarro model. However, most attention has been paid recently to core calculations for screw dislocations in b.c.c. structures, which are of especial interest since for reasons connected with their symmetry they have a non-planar structure (see pp. 309). Various calculations for the screw dislocation (Suzuki, 1968; Chang, 1967; Bullough and Perrin, 1968; Gehlen, 1970; Vítek *et al.*, 1970; Basinski *et al.*, 1970) are not in complete agreement, but some of the discrepancies arise not so much from the results themselves as from the difficulties of interpretation. The calculations have been extensively reviewed by Vítek (1974), but since they require a discussion of the geometry of the lattice, we defer further description to Section 33.

In an atomistic calculation it is not possible to follow the Peierls–Nabarro procedure of displacing the centre of the dislocation from its equilibrium configuration and calculating the corresponding energy. This is because each configuration is obtained by allowing the atomic positions to relax, so that only the energies corresponding to either stable or metastable configurations may be determined. The Peierls energy can thus not be calculated, since the configuration at which the energy is a (saddle-point) maximum is necessarily unstable. However, a correct procedure for the calculation of the Peierls stress is to apply external forces to the crystal block in small increments and to calculate the equilibrium configuration at each stage until a stress is reached at which the dislocation runs away. Such calculations were first made successfully for sodium under the action of shear stresses (Basinski *et al.*, 1971) and uniaxial stresses (Basinski *et al.*, 1972), using the potential described previously (eqn. 16.10). Under a pure shear stress the Peierls stress for a screw dislocation was found to depend on the orientation of the stress axis but to have a high minimum value of $\sim 0.01\mu$, whereas the corresponding stress in the h.c.p. lattice with the same interatomic potential was at least 25 times smaller. When uniaxial stresses were applied, the minimum resolved shear stress to move the b.c.c. screw dislocation was reduced, but only to $0.007\,\mu$. The results provide direct evidence that the crystal structure itself may be an important factor in fixing the magnitude of the Peierls stress. More recently, the distortion of the core and the motion of the screw dislocation in a b.c.c. structure under an applied shear stress has been calculated for a wide variety of effective interionic potentials of the type described by eqn. (16.12) (Duesbery *et al.*, 1973). The behaviour is very complex and is believed to account for some of the anomalous results obtained from studies of the deformation of b.c.c. metals at low temperatures (Christian, 1970).

In b.c.c. metals and in many semi-conductors and non-metals, there is good experimental evidence for a high Peierls force, at least at low temperatures, but in close-packed structures the stress to move dislocations may be as low as 10^{-5}–10^{-6} μ. If the theoretical calculations are accepted, it is thus necessary to consider what additional effects may allow deformation to occur at such low stresses. One possibility considered by Dietze (1952) is that the Peierls stress is rather strongly temperature-dependent because of the effects of thermal vibrations. Unfortunately there is no evidence of a steep rise in stress at very low temperatures for the soft materials, and in the materials which have a strong temperature dependence of stress there is good evidence that this arises from thermally activated processes, i.e. from fluctuations rather than from an intrinsic temperature dependence of the barrier.

The effect considered by Dietze takes account of the fact that the atoms are not fixed points, but the dislocation line itself is essentially considered to be fixed in position. Kuhlmann-Wilsdorf (1960) pointed out that because of the atomic oscillations the exact position of the dislocation axis cannot be defined since certain modes of oscillation correspond to a displacement of the axis, and these modes can scarcely be correlated over long distances along the dislocation line. Her calculation suggested that this uncertainty in the position of the dislocation is much more important than the diffuseness in the positions of individual atoms, and that it may reduce the Peierls force to a very low value in f.c.c. metals, even at low temperatures. In materials such as diamond, this effect will not be operative until high temperatures are reached, whilst b.c.c. structures represent an intermediate case. In principle, direct temperature effects of this type (as distinct from problems involving thermal activation) may be calculated with atomistic models by adding kinetic energy to the atoms (see Gehlen, 1972), but the difficulties are considerable. A criticism of the Kuhlmann-Wilsdorf calculation is that over short lengths of dislocation line it requires coordinated motions of the atoms in opposite senses above and below the slip plane, and the amplitude ascribed to this mode is probably much too large.

Calculations of the Peierls force for straight edge and screw dislocations may be rather misleading, since the restraining force is much smaller when the dislocation does not lie along a close-packed direction of the lattice, and is zero for a completely arbitrary (irrational) direction. Shockley (1947) pointed out that we should thus expect restraining forces from the lattice only for straight dislocations of simple type, and he suggested that the force is probably zero even for these dislocations since the atomic configuration along a macroscopically straight line need not be constant. At sufficiently high temperatures, such a dislocation line will contain kinks at which it moves from one close-packed row in the slip plane to a neighbouring row. These kinks are analogous to the kinks in the steps on a crystal surface described in Section 18, and the dislocation line can move by sideways glide of the kinks along the line at a very much lower stress than is required to move the line forward as a whole. For a line with a macroscopic direction deviating slightly from a close-packed direction, there will be a certain concentration of geometrically necessary kinks; under the action of an applied shear stress, these will glide to the end of the line, which will thus be turned into the close-packed direction. The continued motion of the line at finite temperatures then depends on whether thermal kinks form sufficiently rapidly, and the situation is again analogous to the growth of a crystal surface by addition of atoms at a step. Shockley suggested that thermal activation will provide kinks even at quite low temperatures, so that the resistance of the lattice to dislocation motion will be negligible except near 0 K. However, detailed consideration shows that the energy to produce kinks can be quite large when the Peierls barrier is large, and whilst Shockley's suggestion is probably true for close-packed metals at most temperatures, present evidence is that there is appreciable lattice resistance at quite high temperatures for b.c.c. metals and materials such as silicon and germanium.

Thermal energy may assist a dislocation line which lies in an equilibrium position to move forward to the next equilibrium position by providing an energy fluctuation which enables the applied stress to push a small portion of the line over the barrier. The critical condition thus involves the production of a double kink in an initially straight line, and processes of this kind were first discussed by Seeger *et al.* (1957) in connection with a low-temperature internal friction peak (Bordoni peak) in close-packed structures. For b.c.c. metals, measurements of the temperature and strain rate sensitivity of the flow stress, first made by Basinski and Christian (1960) and Conrad and Schoeck (1960), showed that the gliding dislocations are overcoming some obstacle to motion by means of thermal activation, and in these and many subsequent papers it has been suggested that the barrier is the lattice itself. The application of the rate theory of Chapter 3 to dislocation processes is very complex since stress is an independent variable, and a detailed description cannot be given here. The theory has been developed by many authors, and is reviewed, for example, by Hirth and Nix (1969), Evans and Rawlings (1969), and Christian and Vítek (1970). It is now generally accepted that dislocation motion in b.c.c. metals is thermally activated, but some controversy has existed for many years about whether the resistance is due to the lattice periodicity or to dispersed interstitial impurities which interact strongly with dislocations (see Section 30). Stein and Low (1966) and others have provided experimental evidence in support of the impurity theory, but in a critical survey Christian (1968) concluded that it is very difficult to explain the low-temperature strength in this way. Whilst it is now clear that there are important impurity effects, present evidence indicates that the sessile equilibrium configuration of the screw dislocation (which is a form of Peierls–Nabarro force) is responsible for many of the low-temperature properties of b.c.c. metals. This conclusion is consistent with many electron-microscope investigations which show the low-temperature dislocation structure to consist of long, apparently immobile, screw dislocations.

In both the elastic solution and the Peierls–Nabarro approximation, the energy of an isolated dislocation line diverges, but the energy of a closed loop is finite. This suggests the possibility that the energy of formation of a loop within a perfect lattice might be supplied by thermal fluctuations assisted by an external stress. In view of our previous conclusion that the free energy of a dislocation is too high for it to be in thermal equilibrium with the lattice, however, it seems unlikely that this process is feasible. A detailed examination of the possibility of nucleating an opposite dislocation pair was made by Nabarro

(1947), using the Peierls–Nabarro expression for the self-energy of an edge dislocation pair, which differs from the elastic eqn. (30.29) only in the assumption that the normal stress X_{22} is zero. It seems preferable to treat this problem directly in terms of the nucleation of a closed loop of dislocation, using the approximate expression (30.30) (Cottrell, 1953). The increase in energy when a small loop is formed in the interior of a perfect crystal will be given by the dislocation energy (30.30) less the work done by the external shear stress X in forming the loop, i.e.

$$W_d = \tfrac{1}{2}\mu b^2 r \ln(r/r_i) - \pi X b r^2. \tag{31.15}$$

As the radius of the loop increases, the energy first increases and then decreases again. There is thus a critical size of loop given by

$$r_c = (\mu b/4\pi X)\{\ln(r_c/r_i)+1\} \tag{31.16}$$

at which the energy is a maximum

$$W_c = (\mu b^2 r_c/4)\{\ln(r_c/r_i)-1\}. \tag{31.17}$$

In terms of the theory of Chapter 3, W_c is now the activation energy for the formation of the loop, and the chance of a favourable thermal fluctuation is proportional to $\exp(-W_c/\boldsymbol{k}T)$. Using typical figures, we find that with a critical radius of about 40 atom diameters, the applied stress is $10^{-2}\mu$ and the activation energy about 20 eV. Thus even with this very high stress, the probability of spontaneously nucleating a dislocation loop is virtually zero. For still smaller loops, the activation energy decreases but the required external stress increases, and the general conclusion is that spontaneous formation of a slip dislocation in a perfect lattice (without stress raisers) will not take place unless the applied stress approaches the theoretical shear strength of the lattice. This conclusion does not necessarily apply to the formation of small loops of imperfect dislocation, e.g. at a twin interface, since the dependence of the activation energy on b^2 makes thermal activation much more effective when b is small.

We now turn to consider the motion of a dislocation line in a crystal. In many circumstances the average velocity of a dislocation is limited by obstacles to its motion such as intersecting ("forest") dislocations, impurity atoms, point defects, or the frictional drag of the lattice itself. When these obstacles are overcome with the aid of thermal fluctuations, the effective velocity is determined by the mean time of stay at each obstacle and not by the speed of the dislocation moving through the perfect crystal between obstacles. In a region of perfect crystal, however, we may expect the velocity of a dislocation to rise rapidly once it begins to move under an external stress, provided the lattice resistance to motion is small.

In the absence of energy dissipation, the work done by the stress field in moving the dislocation is equal to its gain in kinetic energy (we can use the concept of kinetic energy of the dislocation, which is a configuration of atoms, in much the same way as we use the concept of the force on a dislocation). It is obvious physically that the speed of sound sets an upper limit to the velocity of the dislocation, since motion of the latter corresponds to the propagation of an elastic discontinuity. We may thus anticipate that the energy of a moving dislocation will tend to infinity as its velocity approaches that of the speed of sound. Actually, moving dislocations lose energy to the surrounding lattice, and this dissipation effectively produces a viscous drag, so that a limiting velocity is obtained under given stress conditions.

We begin by neglecting energy dissipation and consider a simple screw dislocation lying along the x_3 axis, so that the strain field is given by eqn. (30.11). We wish to determine the energy of the dislocation when it moves with velocity u in the x_1 direction. The simplest assumption is that the stress field of the dislocation is unaltered, so that if we use a set of coordinate axes which move with the dislocation line, the displacements are again given by eqn. (30.11). Referred to the usual stationary axes, we have

$$w_3 = (b/2\pi)\tan^{-1}\{x_2/(x_1-ut)\}. \tag{31.18}$$

If we now calculate the energy of the moving dislocation, we find that it increases continually with u and is equal to twice its rest energy when $u = \boldsymbol{c}$, the velocity of sound in the crystal. The solution is thus physically inadmissible since it permits dislocations to move with velocities greater than $\boldsymbol{c}$. The error lies in the assumption that the strain field of a moving dislocation is identical with that of a stationary dislocation. As the motion proceeds, the material particles of the crystal are accelerated and then decelerated, and the inertial forces thereby introduced have to be included in the elastic equations of equilibrium. The situation is very closely analogous to the problem of a moving particle in the special theory of relativity; in deriving eqn. (31.18), we have applied a classical (Galilean) transformation instead of a Lorentz transformation.

Consider a small unit volume of material in the field of the moving screw dislocation. If the density is ϱ, the inertial forces are equal to $\varrho(\partial^2 w_3/\partial t^2)$, and the equations of elastic motion (11.7) become

$$\frac{\partial^2 w_3}{\partial x_1^2}+\frac{\partial^2 w_3}{\partial x_2^2} - \frac{1}{\boldsymbol{c}^2}\frac{\partial^2 w_3}{\partial t^2},$$

where $\boldsymbol{c} = (\mu/\varrho)^{1/2}$ is the velocity of transverse sound waves. We now introduce the new variables y_i where

$$y_1 = (x_1-ut)/[1-(u^2/\boldsymbol{c}^2)]^{1/2}, \quad y_2 = x_2, \quad y_3 = x_3. \tag{31.19}$$

The equations of equilibrium then have solution

$$w_3 = (b/2\pi)\tan^{-1}(y_2/y_1). \tag{31.20}$$

Comparing with eqn. (31.18), we see that the only differences lies in the term $[1-(u^2/\boldsymbol{c}^2)]^{1/2}$. This corresponds to a reduction in the width of the strain field in the x_1 direction, and may be compared to the Lorentz contraction of a moving measuring rod.

We have anticipated that the energy of the dislocation will tend to infinity as $u \to \boldsymbol{c}$. Consider first the elastic energy. For a stationary dislocation, the elastic energy density in a small volume $dx_1\,dx_2\,dx_3$ is given by

$$dW_s = \tfrac{1}{2}\mu\{(\partial w_3/\partial x_1)^2+(\partial w_3/\partial x_2)^2\} \tag{31.21}$$

and the total elastic energy is

$$^{L}W_s = (\mu b^2/4\pi)\ln(r_e/r_i) \tag{31.22}$$

per unit length as in eqn. (30.18). For the moving dislocation, (31.21) still gives the elastic energy density, and in terms of the new variables y_i we have

$$^{L}W_s' = \iiint \tfrac{1}{2}\mu\{(\partial w_3/\partial y_1)^2[1-(u^2/c^2)]^{-1/2}+(\partial w_3/\partial y_2)^2[1-(u^2/c^2)]^{1/2}\}\cdot dy_1\,dy_2\,dy_3.$$

This integral may be evaluated since the displacements are now the same function of y_i as they were of x_i for a stationary dislocation. By symmetry, each term in (31.21) contributes equally to the integral in (31.22); it follows that we may write

$$^{L}W_s = \tfrac{1}{2}\,{}^{L}W_s[1-(u^2/c^2)]^{-1/2}+\tfrac{1}{2}\,{}^{L}W_s[1-(u^2/c^2)]^{1/2} = {}^{L}W_s[1-(\tfrac{1}{2}u^2/c^2)][1-(u^2/c^2)]^{-1/2}. \tag{31.23}$$

In addition to the elastic energy, the moving dislocation now possesses kinetic energy. The energy of a small volume dv is $\frac{1}{2}\varrho(\partial w_3/\partial t)^2\,dv$, so that we have

$$^{L}W_k' = \iiint \tfrac{1}{2}\varrho(\partial w_3/\partial y_1)^2 u^2[1-(u^2/c^2)]^{-1/2}\cdot dy_1\,dy_2\,dy_3 = {}^{L}W_s(u^2/2c^2)[1-(u^2/c^2)]^{-1/2} \tag{31.24}$$

substituting $\varrho/\mu = 1/c^2$. Adding together eqns. (31.23) and (31.24), the total energy of the dislocation per unit length is found to be

$$^{L}W_d' = {}^{L}W_s[1-(u^2/c^2)]^{-1/2}, \tag{31.25}$$

where $^{L}W_s$ is the elastic energy of the stationary dislocation. This result is exactly parallel to the formula for the energy of a moving particle in the theory of relativity. As $u \to c$, the energy approaches infinity.

The simple analogy between the formulae of the special theory of relativity and those applicable to a moving screw dislocation is a little misleading. For edge dislocations, the situation is more complicated since the displacements are not all perpendicular to the direction of motion, and the velocities of transverse and longitudinal sound waves have both to be considered. Although the equations are more complex, however, the general results remain valid. Eshelby (1949) showed that c is also the limiting velocity for a uniformly moving edge dislocation, but the energy diverges as $\{1-(u^2/c^2)\}^{-3/2}$. An interesting result is that the shear stress in the glide plane changes sign for $c_R < u < c$, where $c_R \simeq 0{\cdot}9c$ is the velocity of Rayleigh waves. Thus a very fast edge dislocation attracts other edge dislocations of like sign and repels dislocations of opposite sign.

For velocities $u \ll c$, an expansion of eqn. (31.25) enables the energy to be written

$$^{L}W_d' = {}^{L}W_s+\tfrac{1}{2}m_{0s}u^2, \tag{31.26}$$

where

$$m_{0s} = {}^{L}W_s/c^2 = (\varrho b^2/4\pi)\ln(r_e/r_i) \tag{31.27}$$

may be regarded as the "rest mass" of unit length of a screw dislocation. The corresponding expression for the rest mass of an edge dislocation includes a factor $\{1+(\boldsymbol{c}/\boldsymbol{c}_1)^4\}$, where $\boldsymbol{c}_1$ is the longitudinal sound velocity.

From eqn. (31.25) we see that dislocations moving with velocities which are an appreciable fraction of the velocity of sound possess energies which are much larger than the self-energy of a stationary dislocation. When $u \approx 0{\cdot}8\boldsymbol{c}$, for example, the rest energy is doubled, but for $u \approx 0{\cdot}1\boldsymbol{c}$, the difference is very small. A distinction has sometimes been made between dislocations which possess additional energy of motion comparable to their rest energies (fast dislocations), and those which do not (slow dislocations). The division is fairly sharp because of the form of equation (31.25). In principle, fast dislocations should be able to use their extra energy to overcome obstacles, and possibly to create new dislocations, and this was the basis of an early theory of dynamic multiplication of dislocations during mechanical deformation (Frank, 1951). The available experimental evidence at present is that dislocations do not attain velocities of the order of half the velocity of sound, either because of the presence of obstacles of various kinds, or because of the energy dissipation from a moving dislocation. However, the effective stress on an interface dislocation (see pp. 289, 361) during a phase transformation can be very high, and it is not yet certain that fast dislocations do not play a role in transformation phenomena.

Although the speed of sound appears as a limiting velocity in the above equations, Eshelby (1956b) has investigated the formal possibility of supersonic dislocations. Instead of (31.19), the transformation

$$y_1 = (x_1 - ut)/[(u^2/\boldsymbol{c}^2)-1]^{1/2} \tag{31.28}$$

is used and leads to the hyperbolic form

$$\frac{\partial^2 w_3}{\partial y_1^2} - \frac{\partial^2 w_3}{\partial y_2^2} = 0 \tag{31.29}$$

with the general solution

$$w_3 = \Omega_a(y_1+y_2)+\Omega_b(y_1-y_2). \tag{31.30}$$

Here Ω_a, Ω_b are plane waves of velocity $\boldsymbol{c}$ inclined to the x_1 axis at angles of $\tan^{-1}\pm[(u^2/\boldsymbol{c}^2)-1]^{1/2}$. They represent outward radiation of energy from the moving disturbance and are similar to supersonic shock waves.

Eshelby showed that steady supersonic motion is only possible if energy is continuously supplied by atomic readjustment in the glide plane. In the Peierls–Nabarro model, for example, the solution implies that $\alpha = 0$ before the disturbance has passed and $\alpha = \frac{1}{4}$ after it has passed; i.e. the misfit energy across the glide plane is a maximum before passage of the supersonic dislocation and a minimum after it. In detailed studies, Weertman (1967) has confirmed that a dislocation moving on a glide plane which cannot contribute energy by atomic readjustments is unable to move supersonically. Clearly supersonic motion can not arise in ordinary glide, but it may occur for the special kind of dislocations which are essentially steps in twin boundary or martensitic interfaces (see Section 32). It may be significant that audible clicks are often produced by the formation of twins or martensite plates.

We now briefly consider some of the ways in which energy may be lost from a moving dislocation. The least important, at least until quite high velocities are attained, is thermoelastic damping. As a dislocation moves, the hydrostatic stresses at any point in the material are altered, and these changes produce corresponding changes in the local temperature. Heat flow takes place as a result of the small temperature gradients thus created, and this is one source of energy loss. Estimates show that the effect is small at most velocities. A second source of loss is the interaction of a dislocation with the natural vibrations of the lattice. In effect, the dislocation scatters sound waves which impinge upon it, so that these waves offer a resistance to the motion of the dislocation. Leibfried (1950) considered this problem and concluded that the limiting velocity of a dislocation would be only 0·07c for an applied stress of $5\times10^6\,\mathrm{Nm}^{-2}$. Later work by Nabarro (1951) showed that Leibfried had incorrectly calculated the scattering cross-section, the contribution of long elastic waves to the resistance being overestimated. A good estimate of this effect is very difficult to obtain, but Nabarro's tentative conclusion was that fast dislocations might be produced at very high stresses ($\sim 10^{-3}\,\mu$), but it is improbable that there will be any fast dislocations at the much lower stresses required to deform annealed single crystals.

A third source of energy dissipation arises from the periodic nature of the strain field of the dislocation. The variation in the form of the dislocation as it moves from one equilibrium position in the lattice to the next has already been considered; this variation leads to the radiation of elastic waves out from the dislocation. Nabarro also estimated the magnitude of this radiation damping; it is smaller than the scattering of sound waves, but not negligible. His calculations suggest that an applied stress ten times greater than the yield stress of soft single crystals is probably required to overcome this resistance sufficiently to produce fast dislocations.

The dynamic properties of dislocations form one of the most difficult parts of the formal theory, and many other mechanisms of energy dissipation have also been suggested (see Nabarro, 1967, and Hirth and Lothe, 1968, for further description). The magnitudes of these effects are very difficult to estimate, and as Eshelby has remarked, it is not always clear that they are separate from one another. Some experimental observations which were formerly thought to require fast dislocations have been explained in other ways, and it remains uncertain whether or not fast dislocations are encountered in normal deformation conditions.

We conclude this section by briefly discussing experimental measurements of individual dislocation velocities which were first made by Johnston and Gilman (1959) and Stein and Low (1960) who used an etch-pit technique to reveal the positions of a dislocation before and after a stress pulse of known duration. The results are usually fitted to empirical equations of form

$$u = u_0(\tau/\tau_0)^m \qquad (31.31)$$

or

$$u = u_0 \exp(-\tau_0/\tau), \qquad (31.32)$$

where τ is shear stress; the parameters in these equations are probably functions of temperature and possibly also of stress, i.e. they are not representative of the correct functional relations. The first equation is usually applied only over the range of velocities which correspond to obstacle-limited motion, and the true functional relation between velocity and stress is then probably better represented by the rate equation

$$u = \nu d \exp(-\Delta_a G/kT), \tag{31.33}$$

where ν is the attempt frequency of the dislocation held up at an obstacle, d is the distance moved found after a successful activation and $\Delta_a G$ is the free energy of activation (cf. Chapter 3). These velocity measurements are usually consistent with those deduced from macroscopic measurements of the stress and temperature sensitivity of the strain rate, which is given by

$$\dot{\varepsilon} = \varrho b u, \tag{31.34}$$

where ϱ is the total length of mobile dislocation in unit volume (see p. 311). We have already mentioned the rate theory analysis of flow stress data (p. 278), but we should emphasize that the applicability of rate theory to either individual velocities or to macroscopic strain rates is not universally accepted.

Equation (31.32) was first used by Gilman (1960) and was developed on the basis of a rather specific and oversimplified rate theory model. As an empirical equation, however, it has been claimed to represent the relation between velocity and stress over most of the range which can be measured (in the case of LiF from 10^{-8} to $10^{+2.5}$ m s^{-1}). At the highest velocities, which correspond to one-tenth of the velocity of sound, the motion is probably limited by energy dissipation rather than by the time of stay at obstacles, and such velocities correspond to very high strain rates in macroscopic experiments.

32. PARTIAL DISLOCATIONS, TWINNING DISLOCATIONS, AND TRANSFORMATION DISLOCATIONS

Dislocations which are "perfect" are surrounded entirely by good crystal. We now consider whether any physical significance can be assigned to dislocations with Burgers vectors not equal to repeat vectors of the lattice. In principle, such dislocations can be made by the virtual process described on p. 255; we cut a perfect crystal along some surface having the dislocation line as its limit, move the two cut faces by a non-lattice vector (adding or removing thin layers of material as necessary), and then rejoin. An important difference is at once apparent. After displacement, the lattices of the two cut portions will no longer match, and when rejoined, the cut surface will generally be a region of high energy. An imperfect, or partial, dislocation is thus associated with a surface of misfit, and the possible virtual processes leading to the same imperfect dislocation are not equivalent, since the energy of the misfit surface will depend on its orientation relative to the lattice. In most, and possibly all, orientations, this energy will be very large, and the existence of such irregularities within a crystal is most improbable.

Imperfect dislocations are thus to be expected only when misfit surfaces of low energy can exist. Such surfaces are provided by the stacking faults or translational twin interfaces described on pp. 120–4. Unlike ordinary twin boundaries, a translational twin interface may end within the crystal, and the edge of the fault is then an imperfect dislocation. If there is no marked discontinuity in the elastic properties at the stacking fault, there will be no strong asymmetry in the stress field of the imperfect dislocation, and the previous elastic theory may be used. Since the stacking fault is planar, the imperfect dislocation is properly regarded as a two-dimensional fault.

The Burgers vector of an imperfect dislocation may be defined formally in exactly the same way as for a perfect dislocation (p. 242) with one or two additional conditions (Frank, 1951). The stacking fault is strictly a region of bad crystal, and to avoid ambiguity the Burgers circuit must begin and end on the fault and not at an arbitrary point. This gives an unambiguous path in the reference lattice for intrinsic faults (p. 120), since the path in the real crystal is entirely in good crystal. For extrinsic faults, we must also specify that the interpolated layer is crossed normally, and in the reference lattice we take an equal distance normal to the corresponding plane.

Suppose we have a strip or a loop of stacking fault, bounded by imperfect dislocations on either side. The sum of the Burgers vectors of opposite bounding dislocations must clearly be a lattice vector in order to leave the structure undisturbed outside the fault. If this sum be zero, the partial dislocations are opposites and may annihilate each other by moving together. This is analogous to the formation of a loop of perfect dislocation by local slip, and we may assign a single Burgers vector to the whole dislocation by specifying that the loop is to be traversed in a certain sense (see p. 243). When the two edges of a stacking fault are not opposites, their Burgers vectors add to give a non-zero lattice vector. If the two vectors have a component in the same direction, the lines will repel each other and the edges of the stacking fault move apart. As they do so, the elastic energy of the partial dislocation decreases, but the area of the fault increases, so that there is an equilibrium separation at which the total energy (free energy) is minimized. This separation may be estimated by balancing the elastic repulsive force between the partials and the attractive force which results from the surface tension of the stacking fault. In metals and alloys of low stacking fault energy, ordinary lattice dislocations are dissociated into ribbons of stacking fault in this way whenever they lie in low fault energy planes. The combination of two partial dislocations and a connecting ribbon of stacking fault is described as an extended dislocation.

The two-dimensional nature of the fault attached to an imperfect dislocation imposes additional restrictions on the motion of the dislocation. Suppose the Burgers vector lies in the composition plane of the stacking fault. Gliding motions of the dislocation in this plane are then possible; as they take place, the area of stacking fault increases or decreases, and the corresponding change in free energy imposes an additional force on the dislocation. Gliding motions on any other plane are not possible, even if the dislocation assumes screw orientation, since the surface of misfit would then be extended out of the composition plane and so would have high energy. An exception to this rule occurs if the Burgers vector of an imperfect dislocation is common to two fault-planes. A dislocation lying in one plane can then assume screw character over part of its length, and thus extend into the second plane. The necessary condition is obtained in the b.c.c. structure, and may be important in connection with the formation of mechanical twins. In the close-packed lattices, however, the possible imperfect dislocations bounding stacking faults each lie in only one of the composition planes, and their gliding motions are thus confined to these planes.† Climbing motions of the dislocation line outside its composition plane would also lead to the production of a high energy interface, and so are also impossible.

Now consider the dislocation line to have a Burgers vector which does not lie in the composition plane of the low energy stacking fault which it produces. Gliding motions are then not possible in the composition plane, since the glide plane must contain the Burgers vector, and gliding motions out of the composition plane are forbidden as described above. Such imperfect dislocations are thus unable to glide at all; they are called "sessile" dislocations, in contrast to normal dislocations which are "glissile". Sessile dislocations may climb by atomic diffusion, provided they remain in the same composition plane. All imperfect dislocations are thus constrained to move only in the composition plane of the low energy fault; the motion requires atomic diffusion if the Burgers vector does not lie in this composition plane, and occurs by glide if it is in this plane.

Imperfect dislocations exist only at the edges of low energy stacking faults. It is possible to give a description of any internal misfit surface in dislocation terms, though in some cases this has formal rather than physical significance. We shall consider internal boundaries in the next chapter, but it is convenient here to describe the properties of a defect closely related to imperfect dislocations. We have seen in Chapter 2 that mechanical twinning may be described as a simple shear of the lattice along a composition plane K_1. Figure 7.16 shows diagramatically a twin of type I in which the simple shear is produced by the lattice planes each slipping a certain distance over the plane beneath. On the plane *AB*, this slipping motion has extended over part of the plane only, so that the actual boundary between the twin and the parent crystal extends from *A* to *P* and then steps on to the next lattice plane.

It is clear that around the step at *P*, which runs perpendicular to the figure, there is a region of high strain energy. This step is the limit of a region of slip, and hence has all the properties normally associated with a dislocation. It is called a twinning dislocation, since it exists only in a twin boundary and not in the interior of a crystal. The Burgers vector of a twinning dislocation is not a lattice vector of the parent or twinned lattice, but is equal to the distance moved by each lattice plane in the twinning shear. Thus for a shear **s**, the Burgers vector of the twinning dislocation is

$$\mathbf{b} = d\mathbf{s}, \tag{32.1}$$

where d is the step height. The minimum Burgers vector is fixed by the smallest repeat distance between lattice planes, and for a given K_1 plane we expect all the twinning dislocations to have minimum Burgers vectors, since multiple steps will tend to dissociate into single steps. However, this conclusion is only valid if the atomic configuration at the interface is identical in the two adjacent lattice K_1 planes connected by a step of minimum height. As mentioned on p. 53, the crystallography of twinning is sometimes such that the lattice deformation **S** does not relate all of the lattice vectors of the two structures, and this implies

† This does not mean that two stacking faults on different planes of a f.c.c. structure, for example, may not meet along the line of intersection of the two planes. The difference is that the line of intersection would itself be an imperfect dislocation line, whereas the line of intersection is not a dislocation if the imperfect dislocation can glide in both planes.

that the interface structure is repeated only at distances corresponding to an integral multiple of the spacing of K_1 planes. A detailed discussion of this complication is given in Part II, Chapter 20; if the differing interface configurations have very different energies, the actual step heights may correspond to the distance at which the lowest energy configuration repeats. Such a step is called a zonal twinning dislocation (Kronberg, 1959, 1961; Westlake, 1961).

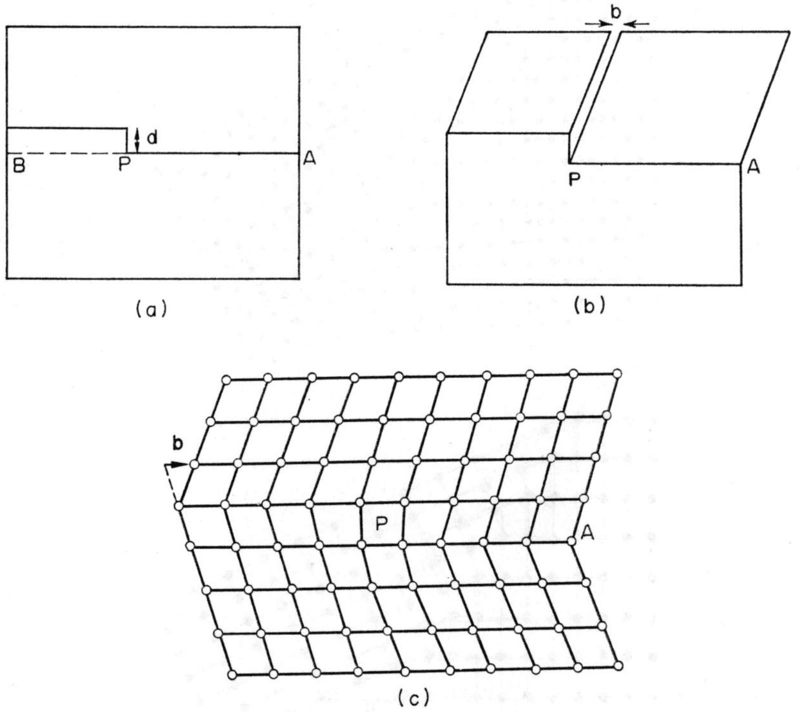

FIG. 7.16. To illustrate the formation of a twinning dislocation (after Bilby and Christian, 1956). When the upper half of the crystal shown in (a) is sheared into twin orientation, a fissure appears because of the step at *P*. When this is rewelded to produce a twin boundary with a step, the discontinuity at *P* has the characteristics of an edge dislocation. The approximate atomic configuration in a simple structure is shown in (c).

The Burgers vector of the twinning dislocation is always smaller than a lattice vector. The dislocation can glide in its composition plane (the K_1 plane), and as it does so, the amount of one orientation grows at the expense of the other. Macroscopic growth of a twin could obviously result from a mechanism which allows a twinning dislocation to move through a whole series of parallel lattice planes. Gliding motions of the twinning dislocation outside the K_1 plane, or climbing motions, produce a high energy interface (an incoherent twin boundary), and are thus impossible under normal stresses.

The step in the twin boundary of Fig. 7.16 is perpendicular to the η_1 direction, and is thus an edge-type twinning dislocation. Obviously dislocations of screw and mixed types are also possible, and closed dislocation loops may be formed. Figure 7.17 illustrates both edge- and screw-type twinning dislocations in a face-centred tetragonal structure with a {101} twinning plane. Mechanical twins often have a lenticular shape, and the progressive twinning may be described formally in terms of closed loops of twinning dislocation with decreasing radius as we move outwards from the centre plane of the lens.

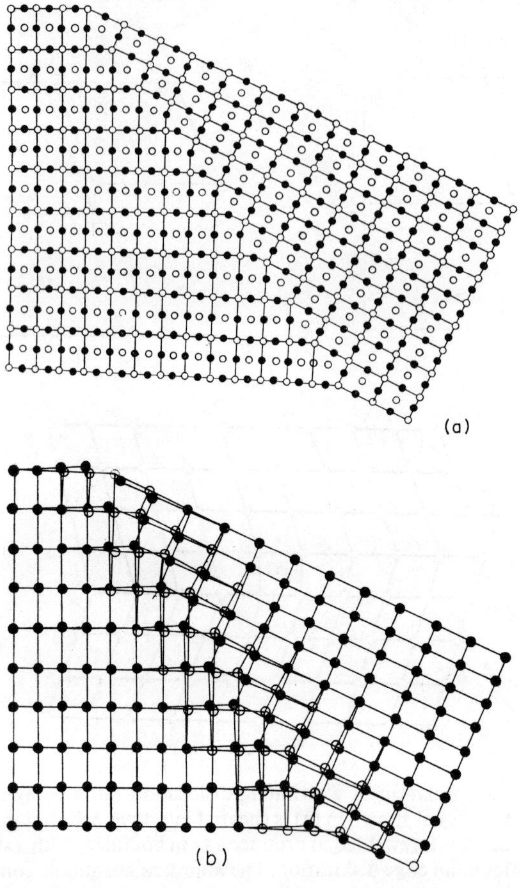

FIG. 7.17. Two edge-type twinning dislocations (a) and one screw-type twinning dislocation (b) in the ⟨101⟩ twin boundary of a tetragonal structure (after Basinski and Christian, 1954).

The plane of the figures is {010}. Open and filled circles represent two successive atom layers projected on this plane. To avoid confusion, (b) is drawn for a simple tetragonal structure, but may readily be related to the face-centred tetragonal structure of (a).

The Burgers vector of a twinning dislocation may be defined formally by using a reference lattice which represents not a single crystal but a crystal and its coherent twin meeting on a K_1 plane. The reference lattice is then multi-valued; it consists of two interpenetrating lattices since the composition plane can be any one of the infinite series of lattice planes parallel to K_1. If the Burgers circuit is made round a step in the real boundary, the corresponding circuit in the reference lattice will have a closure failure which will be a vector parallel to **s** and connecting a point of one of the interpenetrating lattices to a neighbouring point of the other. This is the formal version of the definition of a twinning dislocation used above, but we note especially that it is valid whether or not the twin has formed by a physical deformation. A step in a twin boundary has the characteristics of a dislocation, even if the twin itself were formed by individual thermally activated atomic movements, and not by shear.

In Section 9 we generalized the concept of simple shear to that of invariant plane strain, which produces a new lattice. We may now repeat the above arguments by imagining two lattices which are related by an invariant plane strain, and which meet along some composition plane which we shall assume, for the present, to be rational. If the composition plane contains a step where it leaves one lattice plane and moves to a neighbouring plane, this discontinuity will have properties very similar to those of the twinning dislocation described above; it is described as a transformation dislocation (Bilby, 1953). As a transformation dislocation glides in its composition plane, atoms are transferred from one phase to another. We saw on p. 58 that the displacements in an invariant plane strain are equivalent to the combination of a simple shear on the composition plane and a uniaxial expansion or contraction normal to this plane. The Burgers vector of the transformation dislocation, defined formally by using a multivalued reference lattice comprising the interpenetrating lattices of the structures, will obviously be of this form, and will be equal to the vector $\mathbf{e}$ of eqn. (9.10) multiplied by the height of the step. We may also envisage the situation when the invariant plane strain $\mathbf{S}$ relates only some fractions of the lattice points in the two structures; this leads to the concept of multiple step heights or "zonal transformation dislocations".

According to the above descriptions, the transformation dislocation moves in a fixed composition plane which need not contain its Burgers vector. The usual restriction that glide motion of a dislocation is only possible in a plane containing its Burgers vector is relaxed because the dislocation lies in the boundary between two lattices which possess identical but differently spaced composition planes.

We next consider what happens when a dislocation lying in the parent lattice meets a coherent twin boundary. If this boundary were the free surface of the crystal, the dislocation would produce a step of height equal to the component of the Burgers vector normal to the boundary, and running from one end of the dislocation to the edge of the surface, or to the end of another dislocation. We may treat the twin by making an imaginary cut along the coherent composition plane and introducing the dislocation into the parent crystal, thus producing the step. Clearly the parent and twin can only be rejoined if there is also an incomplete step on the twin surface of the composition plane, and there must therefore be a dislocation line in the twin crystal. There is thus a node in the interface formed from (at least) one dislocation line in each crystal and a step, or twinning dislocation, lying in the interface. If the node contains three lines, as above, conservation of the Burgers vectors requires that the dislocations in the two lattices have equal components normal to the twinning plane (equal and opposite if the convention of eqn. (29.2) is used), and this component gives the height of the step. It is, in fact, geometrically obvious that the Burgers vector of the dislocation in the twin is produced from that of the dislocation in the parent by operation of the twinning shear; if we start with a single crystal containing a dislocation and convert part of it into twin orientation by means of a simple shear, a node of the above type is formed naturally at the point where the dislocation crosses from the sheared to the unsheared portion. The only exception is for a dislocation with its Burgers vector parallel to the composition plane. The vector is then undistorted by the twinning law, there is no step in the composition plane, and the dislocation simply bends in passing from one lattice to the other.

Provided the twinning law is such that a simple shear **S** can be found to relate all the lattice sites of the two crystals, it follows that if the dislocation in one lattice has a Burgers vector which is a repeat vector of that lattice, so also does the dislocation of the other lattice. However, if **S** converts only some fraction of the lattice sites of one crystal to those of its twin, this condition may or may not be fulfilled. In the general case, both dislocations will have lattice Burgers vectors if the component of Burgers vector normal to the interface corresponds to the step height of a zonal twinning dislocation, rather than to that of an elementary twinning dislocation (Saxl, 1968); this is discussed further in Part II, Chapter 20. This result has an immediate application to the problem of glide of a dislocation across a coherent twin boundary. Unless the dislocation crosses in pure screw orientation, it must leave a step of height equal to the normal component of its Burgers vector along the line on which it crosses, with an interface node at each end of this line. If the step height is a repeat distance of the interface periodicity, the dislocation can continue to glide in a slip plane defined by its new Burgers vector and by the line of intersection, or step.

In other cases, the dislocation cannot glide into the new lattice without leaving a fault, but a group of dislocations in one lattice may form a single dislocation which glides into the other. It should be noted that anomalous slip planes and directions may be produced by this mechanism, as has been confirmed experimentally for slip across twin boundaries in zinc (Tomsett and Bevis, 1969). The new Burgers vector is related to the net Burgers vector of the group of dislocations by **S**, and the two slip-planes meet edge to edge along the line of the zonal twinning dislocation.

The above results may again be generalized to two lattices related by an invariant plane strain. In this case, the dislocations in the lattices will produce steps of unequal height on the composition plane, the difference being the expansion or contraction normal to the composition plane which is produced by the normal component of the Burgers vector of the transformation dislocation. A formal proof of the possibility of constructing this kind of node is as follows. If the relation between the lattices is represented by $\mathbf{v} = \mathbf{S}\mathbf{u}$, it was shown on p. 58 that for an invariant plane strain, the tensor **S** must have matrix representation $\mathsf{I} + \mathsf{e}\nu'$ where ν is the normal to the invariant plane, and **e** is a constant vector which gives the direction of movement of the lattice points. It follows that we can construct a node of dislocation lines with Burgers vectors having representations u, $\mathsf{e}\,\nu'\,\mathsf{u}$, and $-\mathsf{v}$, since their vector sum is necessarily zero. Now if **S** relates all the lattices points and u represents the Burgers vector of a perfect dislocation in the first crystal, $-\mathsf{v}$ is a lattice vector of the second crystal and so represents the Burgers vector of a perfect dislocation in that crystal, whilst the vector $\mathsf{e}\,\nu'\,\mathsf{u}$ represents a transformation dislocation. In the more general case, both dislocations in the lattices will have lattice Burgers vectors if the step represents a zonal transformation dislocation, and it is easily seen that all nodes between lattices are of the above form. If two lattices are not related by an invariant plane strain, they do not possess a low-energy composition plane, and the concept of transformation dislocation is then not so useful. In some circumstances it may be possible to define a step in a semi-coherent interface as a transformation dislocation, but it is probably not possible to define unambiguously what happens to a dislocation line when it meets an incoherent boundary between two structures. The formation of a generating node, as discussed above, is illustrated in Fig. 7.18.

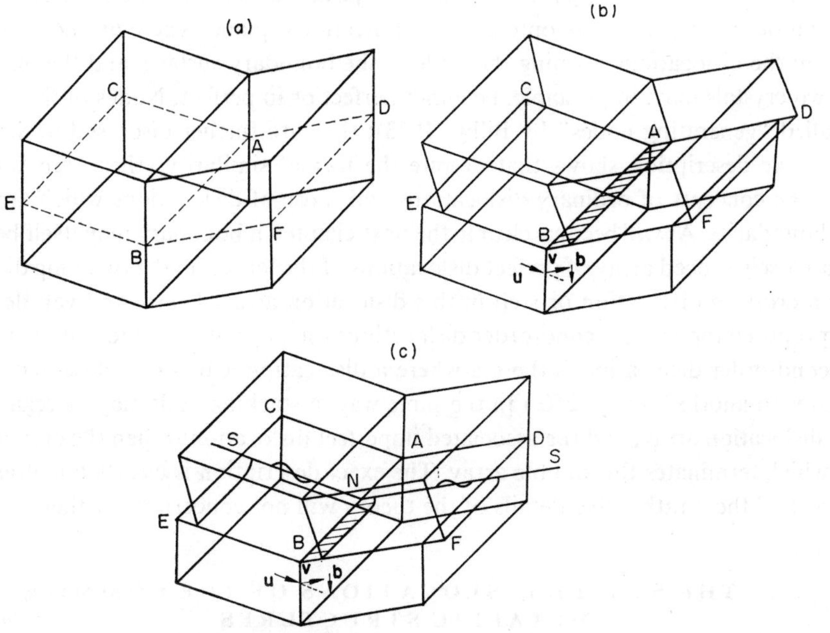

FIG. 7.18. To illustrate the formation of a generating node between two lattices (after Bilby and Christian, 1956). (a) Two coherent lattices (related by an invariant plane strain) with corresponding slip planes outlined. (b) Effect of slipping upper half of both crystals over lower half. The slip vector **v** of one crystal is produced from the corresponding vector **u** of the other crystal by the invariant plane strain. A transformation dislocation of Burgers vector $\mathbf{b} = \mathbf{v} - \mathbf{u}$ is formed along *AB*. (c) The slip now extends only from *EBF* to *SNS*, which is a screw dislocation in both crystals. The transformation dislocation extends from *B* to *N* where *SS* intersects the boundary. The boundary surface is now helicoidal, and rotation of *NB* about *S N S* generates one lattice from the other.

The concepts of twinning dislocation and transformation dislocation were introduced above only for rational composition planes, where the atomic arrangement in the neighbourhood of the composition plane is unambiguous and a clear meaning can be given to a step connecting two lattice planes. We should note, however, that a step produced by any dislocation, even the shortest perfect dislocation, crosses many lattice planes of type $(h_1 h_2 h_3)$ if the indices h_i are notsm all. As the indices increase, the number of lattice planes crossed by the step increases and becomes infinity for the limiting case of irrational planes. The position of the interface is then no longer clearly defined, but the above arguments show that the concepts of twinning and transformation dislocations may still be useful.

If a dislocation in one lattice meets such an interface, e.g. in twinning of the second kind, it will produce a step equal to the component of the Burgers vector normal to the irrational plane, and this step is still a dislocation in the sense used above. The definition is such that if a given lattice point is in or near the composition plane, a lattice point related to it by the Burgers vector of the dislocation producing the step will lie in or near the new position of the composition plane (extended if necessary). A rather similar discussion has been given by Cahn (1960) of the existence of steps on the (probably diffuse) liquid–solid interface (see p. 164). In this further extension of the concept of a dislocation, we have now included discontinuities which can glide only in a given irrational plane. There are no further conditions on the dislocations forming the node in the boundary surface, and the dislocations in the two crystals may, in principle, be either perfect or imperfect. Nodes of this type have been called "generating nodes" by Bilby (1953); they are further discussed in Section 87.

The above description shows that despite the formal similarity, there are differences between the concepts of ordinary dislocations and those of dislocations which are present only in boundaries. As will become clear in the next chapter, a boundary may itself be regarded as a closely spaced array of perfect dislocations of the lattice, and a twinning dislocation is then a crossing dislocation or step in this dislocation array. Frank and van der Merwe (1949) proposed the name second-order dislocations for steps of this kind; another example of a second-order dislocation is the jog where a dislocation leaves one close-packed plane and moves to another (see p. 250.) In the same way, a stacking fault may be regarded as a double dislocation array, and the associated imperfect dislocation is then the crossing dislocation which terminates this double array. The exact description is clearly a matter of convenience, and these rather fine details of the theory will not concern us further.

33. THE STABLE DISLOCATIONS OF THE COMMON METALLIC STRUCTURES

The Burgers vector of a perfect dislocation may be any of the vectors **u** of eqn. (5.1), so that in principle there is an infinity of possible dislocations for any crystal structure. However, the strain energy increases rapidly with the length of the Burgers vector, and large dislocations will thus tend to dissociate spontaneously into lines with smaller Burgers vectors. By examining the change in energy resulting from a given dissociation, we can determine whether or not it will occur spontaneously, and in this way we arrive at a finite number of dislocations which are stable in each lattice.

According to the formulae of Sections 30 and 31, the elastic energy per unit length of dislocation is proportional to b^2, and to a small factor between 1 and $(1-\nu)^{-1}$ which depends on the orientation of the line itself. In real crystals, the anisotropy of elastic properties and the variation of core structure may increase the orientation dependence, and in extreme cases this leads to the result that some orientations of a dislocation line cannot exist. We consider only effects due to elastic anisotropy, since the variation of core energy with orientation is generally unknown. The concept of line tension shows that in the absence of a stress field, a dislocation fixed at two points of an elastically isotropic crystal will take the form of a straight line joining these points. However, if the orientation dependence of the line energy ${}^{L}W_s$ is sufficiently large, the line tension as defined by eqn. (30.34) may become negative, and the straight dislocation line may then lower its energy by adopting some kind of zigzag configuration. This effect was first predicted for lithium by de Wit and Koehler (1959).

The equilibrium shape of a closed dislocation loop is given by a two-dimensional Wulff-plot type of construction (see p. 153), as first pointed out by Mullins (see Friedel, 1964). Head (1967) has used Frank's inverse Wulff plot (i.e. in the dislocation case a plot of $\{{}^{L}W_s(\varphi)\}^{-1}$ vs. φ) to examine the problem of dislocation line stability in detail. If there is any region in which part of this plot lies inside a common tangent to the plot, the corresponding orientations of the dislocation line are unstable and may change spontaneously to a V-shaped, Z-shaped, or more complex zigzag in which the component directions of the line correspond to the points of contact of the common tangent. This is a more general condition than that of negative line tension, and predicts wider ranges of instability.

Except for certain simple orientations, the energy ${}^{L}W_s$ must be calculated numerically. Head has shown that in cubic crystals the instability ranges depend on the usual elastic anisotropy factor (Table IV, p. 74) and also on a second factor $(c_{11}+2c_{12})/c_{44}$. The experimental values are such that ordinary lattice dislocations, but not necessarily partial dislocations, are stable in all orientations of f.c.c. metals, but there are forbidden dislocation orientations for b.c.c. metals. However, we must first determine what are the stable Burgers vectors in the common metallic structures, and for this purpose we assume the validity of the simple criterion that the energy varies as b^2.

In a dislocation reaction

$$\mathbf{b}_3 = \mathbf{b}_1 + \mathbf{b}_2$$

the dislocation with Burgers vector $\mathbf{b}_3$ will be unstable if $b_3^2 > b_1^2 + b_2^2$. Since $b_3^2 = b_1^2 + b_2^2 + \mathbf{b}_1 \cdot \mathbf{b}_2$, the dislocation will dissociate spontaneously if $\mathbf{b}_1 \cdot \mathbf{b}_2$ is positive (angle between $\mathbf{b}_1$ and $\mathbf{b}_2$ less than 90°) and will be stable if $\mathbf{b}_1 \cdot \mathbf{b}_2$ is negative (angle greater than 90°). There is no first-order change in the energy if $\mathbf{b}_1$ and $\mathbf{b}_2$ are perpendicular, and Frank and Nicholas (1953) refer to such dislocations ($\mathbf{b}_3$) as doubtfully stable.

In addition to the perfect dislocations, we must consider the possibility of dissociation into allowable imperfect dislocations. The Burgers vectors of the possible imperfect dislocations are obtained by adding any lattice vector $\mathbf{u}$ (including, of course, $\mathbf{u} = 0$) to a non-lattice vector which will generate a low energy stacking fault. The energy of the imperfect dislocation is again supposed to be characterized by b^2, but in addition we must include the energy of the stacking fault which is associated with the imperfect dislocation.

We now consider the various common lattices and describe their stable dislocations and the important dislocation reactions. In the simple cubic lattice, the shortest lattice vectors joining nearest neighbours are of the form $a\langle 100\rangle$, and dislocations with Burgers vectors equal to these are always stable. Vectors joining second and third nearest neighbours are of types $a\langle 110\rangle$ and $a\langle 111\rangle$ respectively, and the corresponding dislocations are of doubtful stability with respect to dislocations $a\langle 100\rangle$ All other perfect dislocations are unstable, and there are no imperfect dislocations.

The shortest lattice vectors of the f.c.c. structure again join nearest atomic neighbours, and are of the form $\frac{1}{2}a\langle 110\rangle$ when referred to the conventional unit cell; three non-coplanar lattice vectors of this type define the rhombohedral unit cell of the primitive lattice. The twelve dislocations of this type are stable with respect to other perfect dislocations. There are six second nearest neighbour vectors of type $a\langle 100\rangle$ and the corresponding dislocations are doubtfully stable; all other perfect dislocations are unstable.

The possible stacking faults in the f.c.c. lattice were described on p. 121. The operations involved in making intrinsic (1△) or extrinsic (2△) faults by inserting or removing close-packed planes correspond to displacements of $(a/3)\langle 111\rangle$ normal to these planes. Faults of both types may thus be bounded by imperfect dislocations having Burgers vectors equal to these displacements; as discussed previously, the sign of the Burgers vector is ambiguous until some convention is formulated. All possible imperfect dislocations of the f.c.c. structure are thus obtained by adding $(a/3)\langle 111\rangle$ to any lattice vector. Two faults of the same type may be bounded by different imperfect dislocations, just as different perfect dislocations may exist in the same lattice.

Frank and Nicholas (1953) have used a notation which gives a description of both dislocation and associated fault. They suppose the [111] direction to be positive upwards, and regard the left- and right-hand edges of the stacking fault as having the same direction, so that for a fault produced by adding or removing part of a plane these edges have opposite Burgers vectors. For (1△) faults, the two edges are called L and R, for (2△) faults λ and ϱ, and the four imperfect dislocations are then

$$-\frac{a}{3}[111]L, \quad \frac{a}{3}[111]R, \quad \frac{a}{3}[111]\lambda, \quad -\frac{a}{3}[111]\varrho.$$

The possible imperfect dislocations are now tested for stability against dissociation, as above. A dislocation, say of type L, may dissociate into a perfect dislocation and another L dislocation, or into two λ dislocations, and there are corresponding rules for the other types. This leads to the following L dislocations bounding (1△) stacking faults on a (111) plane.

$$\left.\begin{array}{ll}(1) & -\frac{a}{3}[111]L,\\ (2) & \frac{a}{6}[11\bar{2}]L, \quad \frac{a}{6}[1\bar{2}1]L, \quad \frac{a}{6}[\bar{2}11]L,\\ (3) & \frac{a}{6}[411]L, \quad \frac{a}{6}[141]L, \quad \frac{a}{6}[114]L.\end{array}\right\} \qquad (33.1)$$

Dislocations of types (1) and (2) are stable. Type (1) is the original sessile dislocation, first introduced by Frank (1949) and sometimes called a Frank partial. Type (2) has its Burgers vector in the plane of the stacking fault, which may be regarded as produced by slip on the (111) plane in a $\langle 11\bar{2}\rangle$ direction (see deformation stacking faults, p. 122. These dislocations were introduced by Heidenreich and Shockley (1948) and are called half-dislocations or Shockley partials. Fig. 7.19 shows Frank and Shockley partials bounding

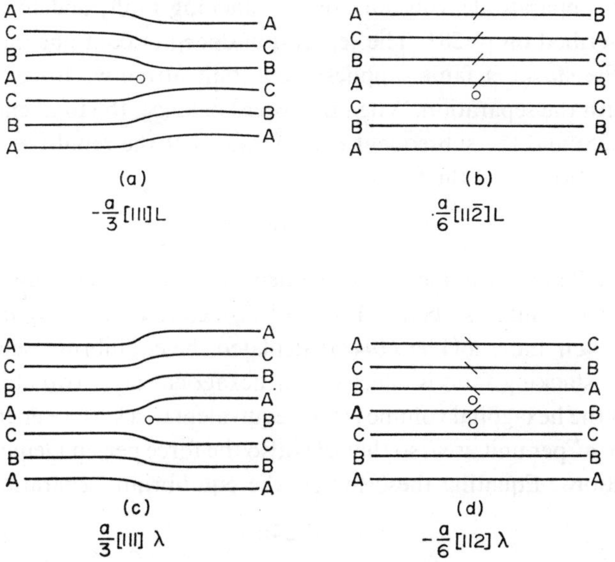

FIG. 7.19. Partial dislocations and stacking faults on a {111} plane of a f.c.c. structure (after Frank and Nicholas, 1953). o: Position of dislocation. Sloping lines indicate where changes must be made in the nomenclature of the planes.

the left-hand side of (1△) and (2△) faults respectively. Note that the (1△) stacking fault is a one-layer twin on the tensile side of the sessile (edge) dislocation, whereas the (2△) fault is a two-layer twin on the compressive side. Dislocations of type (3) are sessile; they are doubtfully stable against dissociations of type

$$\frac{a}{6}[411]L = \frac{a}{6}[2\bar{1}\bar{1}]\lambda + \frac{a}{3}[111]\lambda. \qquad (33.2)$$

Since there are four sets of {111} planes, each of which may contain L, R, λ, and ϱ dislocations of the above types, there are in all 64 stable dislocations of types (1) and (2), and 48 doubtfully stable dislocations of type (3). Frank and Nicholas use the notation *a*, *b*, *c*, *d* to denote the planes $(1\bar{1}\bar{1})$, $(\bar{1}1\bar{1})$, $(\bar{1}\bar{1}1)$, and (111) respectively. Then any of the above imperfect dislocations is represented by the Burgers vector combined with a symbol such as D_i, where D stands for one of the quantities L, R, λ, ϱ, and i for one of *a*, *b*, *c*, *d*. The rules for obtaining all the dislocations from those of the L_d set described above are obvious.

The existence of imperfect dislocations may allow perfect dislocations to lower their energies by dissociation. Consider a dislocation $\frac{1}{2}a[1\bar{1}0]$. When it lies in a (111) plane, the dissociation

$$\frac{a}{2}[1\bar{1}0] = \frac{a}{6}[1\bar{2}1]L_d + \frac{a}{6}[2\bar{1}\bar{1}]R_d \qquad (33.3)$$

is possible and leads to a decrease in energy. The two partial dislocations have parallel components in the $[1\bar{1}0]$ direction, and thus repel each other. Since they are both glissile in (111), they will separate, leaving an area of stacking fault, and forming an extended dislocation as described on p. 285. The separation does not continue indefinitely since the finite energy of the stacking fault supplies a constant attractive force, and the repulsive force decreases with the separation. When the forces balance, the total energy is minimized.

From eqn. (30.25) and the subsequent text, we see that the repulsive force between two parallel half-dislocations is of the form

$$f_r = \mu \mathbf{b}_1 \cdot \mathbf{b}_2 / 2\pi r \tag{33.4}$$

leaving out the small numerical factors which depend on the orientation of the dislocation line in relation to its Burgers vector. For half-dislocations, $\mathbf{b}_1 \cdot \mathbf{b}_2 = a^2/12$, so that $f_r = \mu a^2/24\pi r$. Heidenreich and Shockley (1948) calculated the equilibrium separation on the assumption that the stacking fault is effectively a hexagonal layer two atoms thick. The excess free energy of the hexagonal volume is then equivalent to a surface energy of the stacking fault of magnitude σ^f per unit area, so that σ^f is also the force per unit length of the bounding imperfect dislocations. Equating these forces, the equilibrium separation is

$$r = \mu a^2/24\pi\sigma^f. \tag{33.5}$$

In the approximation used by Heidenreich and Shockley, a free energy difference between the phases of 420 J g atom^{-1} corresponds to a surface energy of the fault of ~0·02 J m^{-2} (20 erg cm^{-2}) and r is of the order of 20a. The separation of the half-dislocations is thus small but much greater than the effective width of each. It is important to note that in this approximation the fault energy, and hence r, can be functions of temperature. In a metal which undergoes a transition from a f.c.c. to a h.c.p. phase, for example, the fault energy will become very small, and the partials will therefore tend to separate spontaneously as the transition temperature is approached. Actually, of course, the justification for using the macroscopic free energy difference between bulk phases in order to estimate the stacking fault energy is very slender, and it is rather surprising that in fact there seems to be a good empirical correlation between the occurrence of low fault energies and phase transitions between f.c.c. and h.c.p. phases. Another early method of estimating the fault energy is to regard it as a monolayer twin, so that the energy of the fault is equal to twice the experimental surface energy of a coherent twin interface. According to this approximation, intrinsic and extrinsic faults have the same energy. Estimates of this kind have been largely superseded by experimental methods which have been developed for measuring fault energies.

Seeger and Schoeck (1953) considered the formation of extended dislocations in anisotropic f.c.c. materials, making use of a method due to Leibfried and Dietze. Their results differ in important respects from those of the isotropic theory. In particular, when the extended dislocation has an edge orientation (line parallel to ⟨112⟩) the separation is much greater than when it has a screw orientation (line parallel to ⟨110⟩). For surface energies of 0·2, 0·04, and 0·02 J m^{-2} (then thought to be the appropriate energies for aluminium, copper, and cobalt respectively), the equilibrium separations in interatomic distances were estimated as 1·6, 12, and 50 for edge dislocations, and 1, 5, and 7 for screw dislocations. The effect of the high stacking fault energy is that the dislocations in aluminium are completely unextended, the calculated widths being smaller than the core radii.

The development of high resolution techniques such as weak-beam dark-field electron-microscopy (Cockayne *et al.*, 1969; Cockayne, 1972) has enabled the separation of the partial dislocations to be observed and measured with relatively high accuracy, even for metals such as silver and copper where the separations are small. Stobbs and Sworn (1971) and Cockayne *et al.* (1971) have measured the variation of the separation with orientation for copper, and the latter authors have obtained similar results for silver. In the case of copper, the observed spacing is 4 nm (40 Å) for edge orientations and 2 nm (20 Å) for screw orientations, and the edge spacing corresponds to a stacking fault energy of 0·041 J m^{-2} (41 erg cm^{-2}). There is very good agreement between the two sets of measurements, but it appears that the separations cannot be calculated solely from anisotropic elastic theory and an assumed constant value of σ^f, at least at the smaller separations. The ratio of the edge width to the screw width should be ~3·2 for copper, but is observed experimentally to be ~2·1. Thus the effects of the dislocation cores and/or non-linear elastic interactions between the partials has to be included to explain the observed orientation dependence. This is illustrated also by calculations made by Cockayne *et al.* by the method of Seeger and Schoeck which, as mentioned above, is based on the Peierls model. This gives a stacking fault energy of 0·032 J m^{-2}, which is more than 20% smaller than that obtained from anisotropic elasticity alone. The corresponding values obtained for silver are 0·016 and 0·014 J m^{-2} on the basis of anisotropic theory and of the Seeger–Schoeck model respectively. The "best" values for the stacking fault energies of these metals are thus still a little uncertain, but the results now obtained are in reasonably good agreement with those obtained from the only other direct technique (measurement of the curvature of an extended dislocation line at a threefold node, see p. 302 and the spread compares very favourably with previous estimates by indirect techniques which, for copper, ranged up to 0·163 J m^{-2} (Seeger *et al.*, 1959).

Instead of the dissociation represented by eqn. (33.3) it is also possible for the following reaction to occur:

$$\frac{a}{2}[1\bar{1}0] = \frac{a}{6}[1\bar{2}1]\varrho_d + \frac{a}{6}[2\bar{1}\bar{1}]\lambda_d. \qquad (33.6)$$

The energy of this pair of half-dislocations is expected to be slightly greater than that of the pair bounding the intrinsic fault, but it is possible that both types of dissociation occur. Loop annealing methods for comparing fault energies (see p. 303) suggest that for some metals at least, the energies of (1△) and (2△) faults are comparable, so that the equilibrium separations corresponding to equations (33.3) and (33.6) may be similar.

Each perfect dislocation of type $\frac{1}{2}a\langle 110\rangle$ may thus reduce its energy by extending in either of the two {111} planes which contain its Burgers vector. In a well-annealed f.c.c. crystal there may be a three-dimensional network of extended dislocations of this type, forming ribbons of stacking fault on all the close-packed planes, although, as we shall see, it is more common for most of the dislocations to be concentrated into planar nets. Whenever an extended dislocation turns out of a particular {111} plane, a node may be formed and the partial dislocations come together, but this need not happen (see below).

When a $\frac{1}{2}a\langle 110\rangle$ dislocation lies in a {111} plane which does not contain its Burgers vector, it cannot glide. Nevertheless a dissociation may occur; for example in the plane $(1\bar{1}\bar{1})$, a $\frac{1}{2}a[1\bar{1}0]$ dislocation may split as follows

$$\frac{a}{2}[1\bar{1}0] = \frac{a}{6}[1\bar{1}2]L_a + \frac{a}{3}[1\bar{1}\bar{1}]R_a. \tag{33.7}$$

This reaction, first considered by Cottrell and Bilby (1951), produces no first-order change in elastic energy, and thus occurs only under the influence of stress, since the energy of the stacking fault must be supplied externally. The R_a dislocation is sessile and remains in the site of the original dislocation, but the L_a dislocation may glide away, leaving a region of (1△) stacking fault. Once again, the analogous dissociation into λ_a and ϱ_a dislocations is also possible. We thus see that each $\frac{1}{2}a\langle 110\rangle$ dislocation may dissociate in eight ways, four of them in the two $\{111\}$ planes in which it can glide, and four in the other two $\{111\}$ planes.

Kuhlmann-Wilsdorf (1958) has pointed out that in appropriate circumstances, we may imagine the reaction of eqn. (33.7) to proceed from right to left. Consider the formation of a dislocation by the collapse of a disc of vacancies aggregated on a $\{111\}$ plane in the manner described on p. 242. As already discussed, this will produce the R_a dislocation of eqn. (33.7) as a circular loop enclosing a disc of stacking fault. However, if the stacking fault energy is sufficiently high, the lattice might not tolerate this situation, and a loop of Shockley partial may be spontaneously nucleated to combine with the Frank partial and give the resultant perfect dislocation loop. The vacancies have then condensed to give a prismatic loop of lattice dislocation, which has a higher strain energy than the Frank partial it has replaced, but a lower total energy because of the elimination of the stacking fault.

Calculations show that it is energetically favourable to form a faulted loop rather than an unfaulted loop on initial collapse of a small vacancy disc, even for quite high values of σ^f. A faulted loop may continue to grow by climb if there is a vacancy supersaturation, and it will eventually become metastable relative to an unfaulted loop. However, the Shockley partial required to remove the fault must now be nucleated on the periphery of the fault, and the activation energy for this event is very large ($\sim$10 eV), so that once a faulted loop has formed it is difficult for it to be converted (Saada, 1962; Saada and Washburn, 1963). In metals quenched to produce high vacancy supersaturations, both faulted and unfaulted loops have been observed under different circumstances; faulted loops are observed in aluminium, for example, only with very pure specimens quenched from relatively low temperatures. High temperatures or impurities were formerly believed to assist the thermal nucleation of Shockley partials, but it appears that the transition depends mainly on stress-assisted nucleation (Edington and Smallman, 1965).

The imperfect dislocations we have considered have all separated stacking faults from good crystal. It is also possible to have two different stacking faults meeting on the same plane, and the line of separation is then another type of imperfect dislocation. To define its Burgers vector we make two Burgers circuits, one beginning on the first fault and one on the second fault, and take the difference of the two closure failures. These dislocations are thus combinations of the simple imperfect dislocations already described; from the nature of the definition, they result from the combination of the right-hand side of one fault with the left-hand side of the other, and are thus of the form $L_i\varrho_i$ or $R_i\lambda_i$. A typical dislocation of type $L_d\varrho_d$ is $(a/3)$ [111]. and all the Burgers vectors of dislocations of this group are the same as those of the R_d group. Equally all those of the $R_d\lambda_d$ group are equivalent to L_d vectors. Corresponding to the existence of these imperfect dislocations, we may have more complex extended dislocations in which translation twin surfaces of different types are successively joined together. The total Burgers vectors of these extended dislocations are rather large.

Two non-parallel {111} planes meet in a $\langle 110 \rangle$ direction. If each of the planes contains a stacking fault, the line of intersection of the two stacking faults will be an imperfect dislocation. This imperfect dislocation may be obtained by combining two previous imperfect dislocations lying on separate planes; there is a threefold node of imperfect dislocation lines at each end of the line of intersection of the stacking faults. Thompson (1953), following a suggestion of Nabarro, has called these stacking fault intersections "stair-rod dislocations". Frank and Nicholas show that there are in fact 96 different sets of the form $D_iD'_j$, where i and j are different and D and D' may represent the same or different symbols. However, all of these dislocations may be derived from the simple forms L_aL_b $(= -R_aR_b)$ in which the included angle between the two stacking faults is obtuse and L_aR_b $(= -R_aL_b)$ in which the included angle is acute. The vector $(a/3)$ $[00\bar{1}]$ L_aL_b gives the only stable type of dislocation of the first group. In the second group, there are stable Burgers vectors of type $(a/6)$ $[\bar{1}10]L_aR_b$ and $(a/3)$ $[1\bar{1}0]L_aR_b$; all other possible Burgers vectors are unstable. Frank and Nicholas give a table for deriving the correct representation of any dislocation from the type vectors of the two basic groups.

When two extended dislocations on intersecting planes meet in a stair-rod dislocation, a reduction in energy is possible if the partials bounding the stacking faults curve inwards towards the line of intersection. The increase in energy due to the increased interactions of the partial dislocations is then compensated by the reduction in length of the stair-rod dislocation. In the limit, there is zero length of stair-rod dislocation, each extended dislocation being fully constricted to give a fourfold node or partial dislocations. This situation was once assumed to exist in all circumstances, but is really a limiting case and need not give the lowest energy. A related problem is the structure of a jog in an extended dislocation, which may or may not be fully constricted. According to Hirsch (1962) the jog configurations may be analysed into rather complex arrangements of partial dislocations.

Stair-rod dislocations are necessarily edge and sessile. Since they exist only at the intersection of two stacking faults, they are in fact "supersessile", and can neither glide nor climb without first dissociating into component partials. However, if both stacking faults arise from extended dislocations which move in their respective glide planes, it is possible that the stair-rod dislocation is moved along parallel to its length. What really happens in such motion is that the leading two partials (one on each plane) create additional length of stair-rod as they glide, and the trailing partials remove some of the existing stair-rod.

Most attention has been paid to the stair-rod dislocation of type $(a/6)$ $[\bar{1}10]L_aR_b$. This was first described by Cottrell (1952) following a dislocation reaction suggested by Lomer (1951). Lomer supposed two perfect dislocations, $\frac{1}{2}a[10\bar{1}]$ and $\frac{1}{2}a[011]$, the first lying in the

(111) or d plane and the second in the $(\bar{1}\bar{1}1)$ or c plane. Then the following reaction leads to a reduction in energy

$$\frac{a}{2}[10\bar{1}]+\frac{a}{2}[011] = \frac{a}{2}[110]. \tag{33.8}$$

The $\frac{1}{2}a[110]$ perfect dislocation is an ordinary stable dislocation of the structure, but it must lie along a $[1\bar{1}0]$ line which is the intersection of the two glide planes. The dislocation is thus an edge, and can glide only in the plane (001) which is not a usual glide plane in the f.c.c. structure. Cottrell pointed out that the energy can be further reduced by the dissociation

$$\frac{a}{2}[110] = \frac{a}{6}[11\bar{2}]+\frac{a}{6}[112]+\frac{a}{6}[110]. \tag{33.9}$$

The $(a/6)\langle 112\rangle$ dislocations move away from the $[1\bar{1}0]$ line in the two glide planes, leaving stacking faults in these planes, and the stair-rod type dislocation $(a/6)[110]$ is left at the line of intersection. Clearly, the same result is obtained by supposing the original dislocations to be dissociated into Shockley partials, so that in the extended notation

$$\frac{a}{2}[10\bar{1}] = \frac{a}{6}[11\bar{2}]L_d+\frac{a}{6}[2\bar{1}\bar{1}]R_d,$$

$$\frac{a}{2}[011] = \frac{a}{6}[112]R_c+\frac{a}{6}[\bar{1}21]L_c,$$

and the reaction is then

$$\frac{a}{6}[2\bar{1}\bar{1}]R_d+\frac{a}{6}[\bar{1}21]L_c = \frac{a}{6}[110]L_cR_d. \tag{33.10}$$

The group of three partial dislocations is very important in some theories of work hardening, which assume that Lomer–Cottrell barriers are stable obstacles to glide dislocations. However, calculations by Stroh (1956) indicate that even an infinite barrier of this type is much weaker than once believed, and will support only a small number of dislocations at the stress levels attained in plastic deformation. A real barrier is expected to be finite in extent, ending at nodes, and this should give way by "unzipping" from the ends. Hence many of the more recent work-hardening theories do not attribute any very special significance to Lomer–Cottrell barriers. Experimental observations in the electron-microscope have revealed the early formation of dislocation networks by interactions of the kind discussed above, but although Lomer–Cottrell barriers are observed (e.g. Basinski, 1964), similar interactions between glide and forest dislocations which do not lead to sessile configurations seem to be equally important.

The above description does not quite exhaust the imperfect dislocations of the f.c.c. structure. It is possible to have three stacking faults meeting in a line, two (of intrinsic and extrinsic types respectively) lying in one plane and the third in an intersecting plane. Similarly, if two intersecting planes each have two types of fault, all four may meet in the line of intersection. In the first case, we get dislocation combinations of types $L_a\varrho_aL_b$ and $L_a\varrho_aR_b$, and these have vectors identical with those of R_aL_b and R_aR_b respectively. Moreover, the stable dislocations have Burgers vectors identical with those of the stable stair-rod dislocations. The four dislocation combinations, e.g. $L_a\varrho_a L_b\varrho_b$ or $L_a\varrho_a R_b\lambda_b$, have Burgers vectors corresponding to those of the stable R_aR_b and R_aL_b sets.

The notation used in the above description of the f.c.c. structure is rather clumsy because of the necessity of specifying the type of stacking fault as well as the Burgers vector of the partial. If the stacking fault specification is omitted, or if these are assumed to be always of the (1△) type, a considerable simplification is possible, especially if a geometrical representation due to Thompson (1953) is used. The four sets of {111} planes correspond to the faces of a regular tetrahedron, shown opened out in Fig. 7.20. The vertices opposite

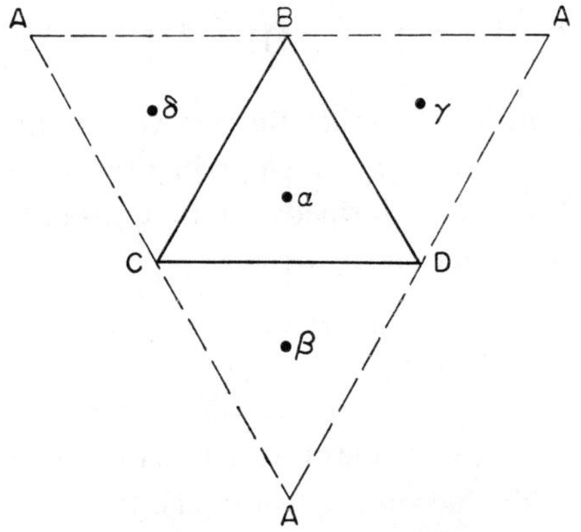

FIG. 7.20. The Thompson tetrahedron.

to the faces *a*, *b*, *c*, *d* are marked *A*, *B*, *C*, *D*, and the mid-points of these faces by α, β, γ, δ. The stable dislocations of the f.c.c. structure then have Burgers vectors with the following representations:

(1) Perfect dislocations of type $\frac{1}{2}a\langle 110\rangle$ are given by the edges *AB*, etc., the letters and their order covering the twelve cases.

(2) Perfect dislocations (doubtfully stable) of type $a\langle 100\rangle$ are represented by symbols *AB*/*CD* and permutations in which no attention is paid to the order of the grouped letters (i.e. $AB/CD = BA/CD = -CD/AB$). The symbol means a vector equal to twice the join of the mid-points of *AB* and *CD*, and this is equal to the vector sum of *AC* and *BD*.

(3) Imperfect dislocations of type $(a/3)\langle 111\rangle$ are represented by lines αA, etc. No distinction is possible between L and λ dislocations. Dislocations of type $R(\varrho)$ have the same symbols as the $L(\lambda)$ dislocations with the order of the letters reversed.

(4) Imperfect dislocations of type $(a/6)\langle 211\rangle$ are represented by symbols αB, αC, etc.

(5) Doubtfully stable imperfect dislocations of type $(a/6)\langle 411\rangle$ are represented by DS/AB and permutations.

(6) Stair-rod dislocations of type L_aL_b are given by symbols $\alpha\beta/CD$ etc.

(7) Stair-rod dislocations of type $(a/6)\langle 110\rangle$ L_aR_b are given by $\alpha\beta$, etc., and those of types $(a/3)\langle 110\rangle$ L_aR_b by $A\alpha/B\beta$, etc.

Thompson's notation has considerable attractions, since the geometrical relations between the Burgers vectors and the close-packed planes are readily visualized, and it has been used in the literature much more than that of Frank and Nicholas. Only when the possibility of extrinsic stacking faults is being specifically considered will it be necessary to use the more elaborate method. As an example of the simplicity of the tetrahedron representation, consider eqn. (33.7). This becomes

$$BA = B\alpha + \alpha A \tag{33.11}$$

Another example arises from the dislocation reaction

$$\tfrac{1}{2}a[1\bar{1}0] + \tfrac{1}{2}a[01\bar{1}] = \tfrac{1}{2}a[10\bar{1}] \tag{33.12}$$

in which two dislocations with non-parallel Burgers vectors in the same (111) slip plane combine to form a new dislocation with corresponding reduction in energy. If the reaction occurs over a limited length, a node is formed and, in terms of the Thompson tetrahedron, (33.12) may be written more symmetrically as

$$BA + AC + CB = 0. \tag{33.13}$$

Nodes of this kind occur in networks, and are very stable when the dislocation lines are mutually at 120°. By means of the Thompson tetrahedron it may now be shown that if the dislocations are all extended according to (33.11), the nodes in a hexagonal network are alternately extended and constricted provided only that all faults are of the same type. This is, indeed, observed in many metals, and following a suggestion by Whelan (1959) measurements of the radius of curvature at an extended node have been used by several authors to estimate the stacking fault energy. The method depends on the equilibrium between the mutual and self-stresses of the dislocations (see p. 265) and the tension due to the fault, and is now capable of giving reasonable accurate results for materials with moderately low fault energies (see Gallagher, 1968, and Gallagher and Liu, 1969, for experimental values).

If the extrinsic fault energy is sufficiently low, all threefold nodes are extended, but enclose stacking faults which are alternately of 1△ and 2△ types with cross-over points (twofold constrictions) between the nodes. In principle it is now possible to determine the ratio of the energies of the two types of fault although there are difficulties caused by the energy of the twofold constrictions. Gallagher (1966) found the ratio of the extrinsic to the intrinsic fault energies in two silver–indium alloys to be $\sim 1{\cdot}09$ and $\sim 1{\cdot}03$.

Yet another example of the utility of the Thompson tetrahedron is provided by an analysis of the possible structures of long jogs in extended dislocations and of related defects such as extended dipoles. If a jog lowers its energy by dissociating in a plane inclined to the slip planes of the extended dislocations, the resulting configurations (Thompson, 1955; Hirsch, 1962) can be quite complex. The motion of dislocations containing dissociated long jogs can lead to the trailing of faulted dislocation dipoles (i.e. long strips of stacking fault on an inclined {111} plane, bounded by Frank partials), and in Hirsch's theory these dipole loops are always of vacancy kind since jogs which would trail interstitial-type dipoles either move conservatively or leave unfaulted dipoles. This asymmetry disappears if the dissociation allows 2△ faults to form as readily as 1△ faults. Faulted dipoles can also occur by the dissociation of unfaulted dipoles, obtained, for example, from interactions of dislocations from different sources. Seeger (1964) pointed out that further dissociation of a faulted dipole in a {111} plane could occur by dissociation of the opposite Frank partials into stair-rods and Shockley partials, the latter dislocations moving off in parallel slip planes to give a final configuration which is Z-shaped in section if all three faults are of 1△ type, and S-shaped if the middle fault is 2△. Faulted dipoles appear in electron-micrographs as thin straight lines, and careful contrast experiments are needed to distinguish them from similarly oriented Lomer–Cottrell dislocations. Seeger proposed that observations on these defects would provide a method for measurement of stacking fault energies and following developments by several authors, Steeds (1967) formulated an anisotropic elastic theory of the interactions, and a detailed theory of the fault contrast. He used his experimental measurements on copper, silver, and gold to estimate values of the fault energies of these materials, but the values obtained are rather high in comparison with those given by other direct methods.

Elementary steps between stacking faults on adjacent close-packed planes will also have complex configurations when the dislocations are extended, and are sometimes known as jog-lines (Thompson, 1955). Jog-lines are dipole-like defects corresponding to the overlapping of the cores of two partial dislocations, and there are four basic kinds which are equivalent geometrically to lines of one-third or two-thirds vacancies or one-third or two-thirds interstitials. The lowest energy configuration is the one-third vacancy type, which corresponds formally to a $\gamma\delta+\delta\gamma$ dipole. Hirsch applied arguments similar to those applied to long jogs to discuss the motion of dislocations containing elementary jogs, and used this as the basis of a theory of work-hardening.

We have already referred to planar Frank loops formed by vacancy condensation, but more complex loops which involve stepped stacking faults have also been detected (Clarebrough *et al.*, 1966; Morton and Clarebrough, 1969). Although some of these could arise from two independently nucleated planar loops, some observations require the concept of the climb of the fault, as first considered by Escaig (1963) and Schapink and de Jong (1964). These authors showed that a row of four vacancies absorbed on a fault may be changed by a simple atomic shift into a triangular array of one-third vacancy jog-lines. Subsequent vacancies can then be absorbed one by one, each vacancy leading to an increase in the length of the triangular fault by one atom distance. This process progressively transfers the stacking fault on to the adjacent plane, and is more probable in materials of very low fault energy where addition of partials at the edge of a growing fault (i.e. climb of the Frank partial rather than of the fault) may be inhibited because of the dissociation (33.14) which is considered below.

A different situation is the formation of multiple loops by condensation of vacancies on successive {111} planes, first observed by Westmacott *et al.* (1961). In later work, two-, three- and four-layer defects have been observed (see Edington and West, 1966, 1967) in quenched aluminium and aluminium alloys. The observations allow the deduction that the extrinsic fault energy is less than twice the intrinsic energy; the observation of four-layer defects, however, suggests that heterogeneous nucleation on some impurity has occurred.

In many quenched f.c.c. metals of low fault energy, a characteristic defect observed in the electron-microscope is a more or less regular tetrahedron of stacking fault, the edges of the tetrahedron being stair-rod dislocations. The Thompson model provides a particularly simple way of discussing this fault.

Suppose a flat disc of vacancies collapses to give a Frank sessile dislocation which has a triangular shape, the edges being parallel to the $\langle 110 \rangle$ directions in the $\{111\}$ plane of the disc. If the stacking fault energy is sufficiently high, the Kuhlmann-Wilsdorf reaction (the reverse of eqn. (33.7)) will take place to give a prismatic loop of perfect dislocation, eliminating the fault at the expense of some elastic energy. If the fault energy is sufficiently low, however, there will be a reverse tendency to lower the elastic energy at the expense of some increase in the area of fault. The Frank partial lying along a close-packed direction may dissociate into a low-energy stair-rod dislocation and a Shockley partial dislocation. The Shockley partial cannot glide in the plane of the original vacancy disc, but it can glide in the plane which intersects this in the direction of the Frank partial considered. A typical such reaction in the d plane, for example, is

$$\frac{a}{3}[111] = \frac{a}{6}[101] + \frac{a}{6}[121], \tag{33.14}$$

or, in terms of the Thompson model,

$$\delta D = \delta\beta + \beta D. \tag{33.15}$$

The $\beta D = (a/6)[121]$ dislocation can glide in the b plane, leaving the $\delta\beta$ stair-rod dislocation at the intersection of the two stacking faults which meet at an acute angle. This reaction is favourable so far as the elastic energy of the partial dislocations is concerned.

Now consider three such dissociations to take place, corresponding to the three orientations of the triangular Frank partial. Using a partial in the a plane, these would be

$$\left.\begin{aligned} \alpha A &= \alpha\beta + \beta A, \\ \alpha A &= \alpha\gamma + \gamma A, \\ \alpha A &= \alpha\delta + \delta A. \end{aligned}\right\} \tag{33.16}$$

The Shockley partials will begin to bow out on the three $\{111\}$ planes which intersect the a plane along the sides of the original Frank dislocation. If we use the sign convention suggested on p. 243, looking outwards from each vertex of the original triangle, the Shockley partials will have opposite signs at their two ends. These partials attract one another in pairs, according to equations of type

$$\left.\begin{aligned} \beta A + A\gamma &= \beta\gamma, \\ \gamma A + A\delta &= \gamma\delta, \\ \delta A + A\beta &= \delta\beta, \end{aligned}\right\} \tag{33.17}$$

so that the four lines of dislocation radiating from each corner of the triangle become three lines, and the nodes where the pairs of Shockley partials meet move away from the original *a* plane until they all join together in a single nodal point. The reactions (33.17) are all of the type (33.10), and hence also lead to a reduction of energy. The final result is the defect we described above, a tetrahedron of stacking fault with stair-rod dislocations as its edges. The tetrahedron may be identified with the Thompson tetrahedron, and the edges *AB*, *BC*, *CD*, *DA* have Burgers vectors $\gamma\delta$, $\delta\alpha$, $\alpha\beta$, $\beta\gamma$ which form a tetrahedron inverse to that of the dislocation lines themselves. The above theory is due to Silcox and Hirsch (1959) who also made the first observations of these defects in gold. Clearly the final defect is highly symmetrical, and an alternative to the two-stage nucleation process just described is the direct collapse of small three-dimensional vacancy clusters into tetrahedra which subsequently grow by further vacancy addition (de Jong and Koehler, 1963). Both these processes probably occur in quenched metals but, in addition, tetrahedra appear to form by a dislocation reaction during deformation (Loretto *et al.*, 1965).

The final defect consists of six partial dislocation lines, the elastic energy of each of which is approximately proportional to $(a^2/18)$ per unit length, whereas the equivalent loop of Frank partial consists of three dislocation lines of the same length, and elastic energy of $(a^2/3)$ per unit length. The elastic energy of the tetrahedral defect is thus only about one-third of that of the corresponding loop of Frank partial formed by an equivalent number of vacancies condensing. On the other hand, the finite stacking fault energy sets an upper limit to the size of the tetrahedral regions which may be formed; an approximate calculation suggests that with a fault energy of 0.033 J m^{-2}, the edge of the tetrahedron will be about 43 nm.

Under conditions of vacancy supersaturation, tetrahedra increase in size by further vacancy condensation. Each face could climb by the nucleation of a high-energy jog-line along the edge of the tetrahedron or by the formation of a low-energy (one-third vacancy) jog-line at the apex; the latter process is much more probable (Kuhlmann-Wilsdorf, 1965). Shrinkage of a tetrahedron by vacancy emission is more difficult to initiate, since the step nucleus is either a short length of high-energy jog-line at the apex, or a long low-energy jog-line. In practice, tetrahedra are found to be very stable against dissolution, and persist in gold up to $\sim$800°C.

Since triangular Frank faults and tetrahedra are sometimes observed in the same specimens, considerable attention has been paid to the conditions governing the transition. Simple theory, as outlined above, shows that the tetrahedron has approximately four times the fault energy and one-third the elastic self-energy of the equivalent Frank loop, so that as its size increases the tetrahedron eventually becomes less stable than the loop. However, a full calculation must include the elastic interaction energies of all the partials and is difficult even in isotropic approximation. The transition from tetrahedron to planar loop essentially involves the inverse of the Silcox–Hirsch mechanism, and at an intermediate stage the defect has the form of a truncated tetrahedron. After various calculations of the energy of this defect, Humble and Forewood (1968a, b) concluded that there is essentially no energy barrier to the transition and that experimental measurements of the largest tetrahedron and smallest Frank loop in a plastically deformed material give a reliable indication of the stacking fault energy.

The kinetics of growth or shrinkage of prismatic loops, faulted loops, or tetrahedra may be followed by *in situ* experiments in the electron-microscope. Most of these observations have been made on aluminium, and it appears that the rate-limiting process is usually point defect diffusion to or from the foil surface (Seidman and Balluffi, 1966; Dobson *et al.*, 1967). Faulted loops anneal at a different rate from unfaulted loops because of the dominating effect of the fault energy on the effective climb force. Estimates of fault energy may be made from the climb rates but are subject to some uncertainty.

The aggregation of point defects in quenched or irradiated f.c.c. metals is intrinsically important, but also has great relevance, especially in aluminium alloys, to precipitation and age-hardening processes. Although the existence of vacancy-solute interactions introduces some important modifications into the processes described above, we defer further consideration to Part II, Chapter 16. It remains finally to consider whether any dislocations in the f.c.c. structure are unstable in particular orientations.

Since the Peierls force is believed to be small for f.c.c. metals any appreciable variation in ${}^LW_a$ with φ must come from ${}^LW_s$. The calculations made by Head show that elastic instability for lattice dislocations arises only when both the elastic factors mentioned on p. 393 are rather large, and this excludes most common metals, but regions of instability are predicted for elastically anisotropic alloys such as indium–thallium. The individual Shockley partials of an extended dislocation are unstable under less rigid conditions; for example, over a range $\varphi \simeq 79$–$101°$ for copper. Experimental observations on copper–aluminium alloys show that one of the partials of an extended dislocation adopts a zigzag configuration as predicted by theory (Clarebrough and Head, 1969).

We now turn to consider the h.c.p. structure, which is the only common metallic double lattice structure. This means that not all nearest neighbour atomic translations are possible Burgers vectors of perfect dislocations. There are, in fact, six stable perfect dislocations of type $(a/3)\langle 11\bar{2}0\rangle$ which represent nearest neighbour translations within the close-packed planes. Perfect dislocations of type $c\langle 0001\rangle$ are also stable, this being the next smallest lattice vector. Finally, there are possible Burgers vectors of type $\left\langle \frac{a}{3}\,\frac{a}{3}\,\frac{2a}{3}\,c \right\rangle$, but these dislocations are doubtfully stable against dissociation into one each of the above two types.

Vectors in the h.c.p. structure have been written above (see p. 37) in terms of the conventional four-axis reference system, which is adopted to ensure that equivalent planes and directions have similar indices. This advantage, however, is much reduced by the difficulty of handling four indices, and it is often more convenient to refer vectors to the three axes of the conventional h.c.p. cell, or of the orthohexagonal cell. With the conventional three-axis system, the above dislocations have Burgers vectors $a\langle 100\rangle$ and $a\langle 110\rangle$ for the first type, $c\langle 001\rangle$ for the second, and $\langle a0c\rangle$ and $\langle aac\rangle$ for the third type.

The possible low-energy stacking faults of the h.c.p. structure were discussed in Section 17, and it remains to enumerate the Burgers vectors of the dislocations which bound them. A typical displacement producing a $(1\triangle)$ fault is $\left[\frac{2a}{3}\,\frac{a}{3}\,\frac{c}{2}\right]$ (expressed in the three-axis system), i.e. a nearest neighbour displacement between atoms not in the same basal plane. With a suitable choice of axes, the Burgers vector of either the left-hand side of a $(1\triangle)$

fault or the right-hand side of a (1▽) fault is given by this displacement, and there are five other equivalent vectors. The $L(1\triangle)$ or $R(1\triangledown)$ set may thus be

$$\left[\frac{2a}{3}\;\frac{a}{3}\pm\frac{c}{2}\right],\quad\left[-\frac{a}{3}-\frac{2a}{3}\pm\frac{c}{2}\right],\quad\left[-\frac{a}{3}\;\frac{a}{3}\pm\frac{c}{2}\right],\tag{33.18}$$

and the negative of these give the $L(1\triangledown)$ or $R(1\triangle)$ sets. As mentioned previously, the distinction between △ and ▽ depends only on the choice of axes in the basal plane, and the two sets are interchanged by a rotation of 60°.

The deformation fault (2△) corresponds to slip on the basal planes, and the dislocations which bound it thus correspond to the half-dislocations of the f.c.c. structure. With the above axes, the $L(2\triangle)$ and the $R(2\triangledown)$ dislocations have Burgers vectors

$$\left[\frac{a}{3}-\frac{a}{3}\,0\right],\quad\left[\frac{a}{3}\;\frac{2a}{3}\,0\right],\quad\left[-\frac{2a}{3}-\frac{a}{3}\,0\right],\tag{33.19}$$

and the $L(2\triangledown)$ and the $R(2\triangle)$ dislocations have opposite Burgers vectors.

The extrinsic fault (3△) may be produced by a displacement of $c/2$ perpendicular to the close-packed planes, and the corresponding dislocations are thus rather similar to sessile dislocations of the f.c.c. structure. The two possible Burgers vectors $[0\;0\;\mp\frac{1}{2}c]$ may be associated with either edge of (3△) or (3▽) faults.

Application of the stability rule shows that the dislocations bounding (2△) and (3△) faults are stable, but that those bounding (1△) faults are only doubtfully stable against dissociations of the type

$$\left[\frac{2a}{3}\;\frac{a}{3}\;\frac{c}{2}\right]L(1\triangle)=\left[\frac{2a}{3}\;\frac{a}{3}\,0\right]L(2\triangle)+\left[0\,0\,\frac{c}{2}\right]L(3\triangle).\tag{33.20}$$

The dislocations bounding (1△) faults and (3△) faults are sessile, but those bounding (2△) faults are glissile, so that the above reaction could occur if energetically favoured. In addition to the vectors listed above, there are three sets of imperfect dislocations which may bound (2△) faults, but all of these are doubtfully stable against dissociation into a perfect dislocation and a (2△) dislocation of the above type.

The existence of low energy-faults allows stable perfect dislocations which lie in the close-packed planes to lower their energies by dissociation into imperfect dislocations. The reaction analogous to eqn. (33.3) is

$$a[100]=\frac{a}{3}[1\bar{1}0]\,L(2\triangle)+\frac{a}{3}[210]\,R(2\triangle).\tag{33.21}$$

Since the (2△) dislocations are glissile, the reaction will occur spontaneously in the basal planes, and the equilibrium separation will depend on the stacking fault energy, as in the f.c.c. case. Calculations by Seeger (1955b) suggest that for zinc, cadmium, and magnesium, the stacking fault energy is high and the width of the extended dislocations less than 0·7 nm (7 Å). On the other hand, h.c.p. cobalt has a very low stacking fault energy.

The other types of perfect dislocation may also reduce their energies by dissociation in the basal planes, e.g.

$$\left.\begin{aligned} c[001] &= \left[\frac{2a}{3}\ \frac{a}{3}\ \frac{c}{2}\right] L(1\triangle)+\left[-\frac{2a}{3}-\frac{a}{3}\ \frac{c}{2}\right] R(1\triangle), \\ [a0c] &= \left[\frac{2a}{3}\ \frac{a}{3}\ \frac{c}{2}\right] L(1\triangle)+\left[\frac{a}{3}-\frac{a}{3}\ \frac{c}{2}\right] R(1\triangle). \end{aligned}\right\} \quad (33.22)$$

The $(1\triangle)$ dislocations are sessile, and these separations may thus be achieved only by diffusive processes (dislocation climb), and are much less probable. The extended dislocation would, of course, be sessile.

As with f.c.c. structures, there is no limit to the number of geometrically possible extended dislocations, formed by joining together stacking faults of different types to give rather large resultant Burgers vectors. However, if only basal plane faults are possible, all faults are parallel in this structure, so there is no equivalent to the f.c.c. stair-rod dislocations. In some h.c.p. metals in which the axial ratio is close to the ideal value for close-packing, or is less than this, the most prominent slip planes may be prismatic $\{10\bar{1}0\}$ or pyramidal $\{10\bar{1}1\}$ planes. Atomistic calculations with Morse potentials (Schwartzkopff, 1969) show that stacking faults on these planes may be metastable. However, the computed energies are very high, and there is no experimental evidence to support the hypothesis.

The collapse of a monolayer disc of vacancies on a basal plane may give either a $3\triangle$ fault with a low-energy partial dislocation, or a $1\triangle$ fault with a higher-energy partial dislocation. This is analogous to the production of a faulted loop with a Frank partial or an unfaulted loop with a prismatic dislocation in a f.c.c. material, and provided the $3\triangle$ fault does indeed have the higher energy, the $1\triangle$ loop should be stable above some critical radius. Conversion of the $3\triangle$ to a $1\triangle$ fault by the inverse of the dissociation (33.20) requires the nucleation of a "Shockley partial" (as in the analogous f.c.c. reaction), but it is probable that in many metals vacancies condense directly to $1\triangle$ loops (e.g. Lally and Partridge, 1966).

Precipitation of vacancies on two successive (001) planes will give unfaulted prismatic loops of types $[00c]$ or $[a0c]$. In principle loops of the latter type may lower their energies by dissociation into $[00c]$ loops + $[a00]$ loops which then collapse. Also both loops may dissociate by climb according to (33.22); at an intermediate stage the 1Δ fault is contained between two concentric loops, the inner of which is shrinking and the outer growing by diffusion across the annulus until a single faulted loop enclosing twice the area of the original loop is obtained. In conditions of high vacancy supersaturation the reverse process has been observed, so that a second layer of vacancies nucleates on a faulted loop, and both of the concentric dislocation loops then grow outwards. Measurements of the kinetics of loop climb in foils have been used to estimate fault energies in some hexagonal metals by methods similar to those used for f.c.c. metals (Harris and Masters, 1966a, b; Dobson and Smallmann, 1966; Hales *et al.*, 1968).

Finally we consider the b.c.c. structure. Using the axes of the conventional unit cell as the reference system, we find there are two distinct types of stable perfect dislocation. There are eight vectors of type $\frac{1}{2}a\langle 111\rangle$ which define the displacements from a lattice point to its nearest neighbour lattice points, and six vectors of type $a\langle 100\rangle$ which represent the next nearest neighbour displacements. All the other possible perfect dislocations are unstable. Perfect dislocations of type $\frac{1}{2}a\langle 111\rangle$ gliding on intersecting $\{110\}$ slip planes will attract each other and form an edge type $a\langle 100\rangle$ dislocation with reduction in elastic energy. This is one suggested mechanism for crack nucleation in b.c.c. structures, which are usually brittle at low temperatures, since the width of the $\langle 100\rangle$ dislocation may be so narrow that it can be regarded as an incipient crack (Cottrell, 1958). There is some evidence for crack nucleation from intersecting slip bands in non-metallic materials such as magnesium oxide, but many b.c.c. brittle fractures begin from twin intersections (Hull, 1960).

As explained on p. 124, there is no firm evidence for the existence of monolayer stacking faults and partial dislocations in b.c.c. metals except under anomalous conditions, and theoretical calculations lead to the conclusion that single-layer faults are not mechanically stable on either $\{112\}$ or $\{110\}$ planes. Prior to this work, there was much speculation about possible dissociations of lattice dislocations with the production of faults on these planes; for example, a monolayer twin fault on (112) could have a left-hand edge bounded by a partial dislocation with Burgers vector either $[-a/6, -a/6, a/6]L$ or $[a/3, a/3, -a/3]L$ and the right-hand edge bounded by a partial dislocation with the negatives of these faults. Thus there would be a dissociation

$$\frac{a}{2}[11\bar{1}] = \frac{a}{3}[11\bar{1}]L + \frac{a}{6}[11\bar{1}]R, \tag{33.23}$$

which leads to a reduction of elastic energy. A greater reduction of energy is represented by the dissociation of a screw dislocation

$$\frac{a}{2}[11\bar{1}] = \frac{a}{6}[11\bar{1}] + \frac{a}{6}[11\bar{1}] + \frac{a}{6}[11\bar{1}] \tag{33.24}$$

in which there are either three stacking faults meeting along a common line of no fault (Hirsch, 1960) or two stacking faults meeting along the central partial dislocation (Sleeswyk, 1963). These and other models formerly proposed to account for the immobility of the screw dislocation are reviewed, for example, by Hirsch (1968) and Christian and Vítek (1970), but they have been largely superseded by atomistic calculations of the core structure.

The results of various calculations all show that the screw dislocation core in a b.c.c. metal has a three-dimensional structure with threefold symmetry. The centre of the core must lie along one of the two threefold screw axes of symmetry (Suzuki, 1968), and this has the effect of either removing or reversing the spiral stacking sequence of neighbouring $\langle 111\rangle$ atom rows, depending on the relation between the site and the sign of the Burgers vector. The lowest-energy configurations of the stress-free dislocation are in those sites where this spiral order is reversed, and there are then two possible configurations for each dislocation which are related by a $\langle 110\rangle$ diad axis. The configurations have been displayed in various ways of which the most useful seems to be a differential displacement map (Vítek *et al.*, 1970) and a stress field representation (Basinski *et al.*, 1971). The differential displacement map (Fig. 7.21) shows that the largest displacements are along the three $\{110\}$ planes intersecting the centre of the dislocation, but in each case the displacements are

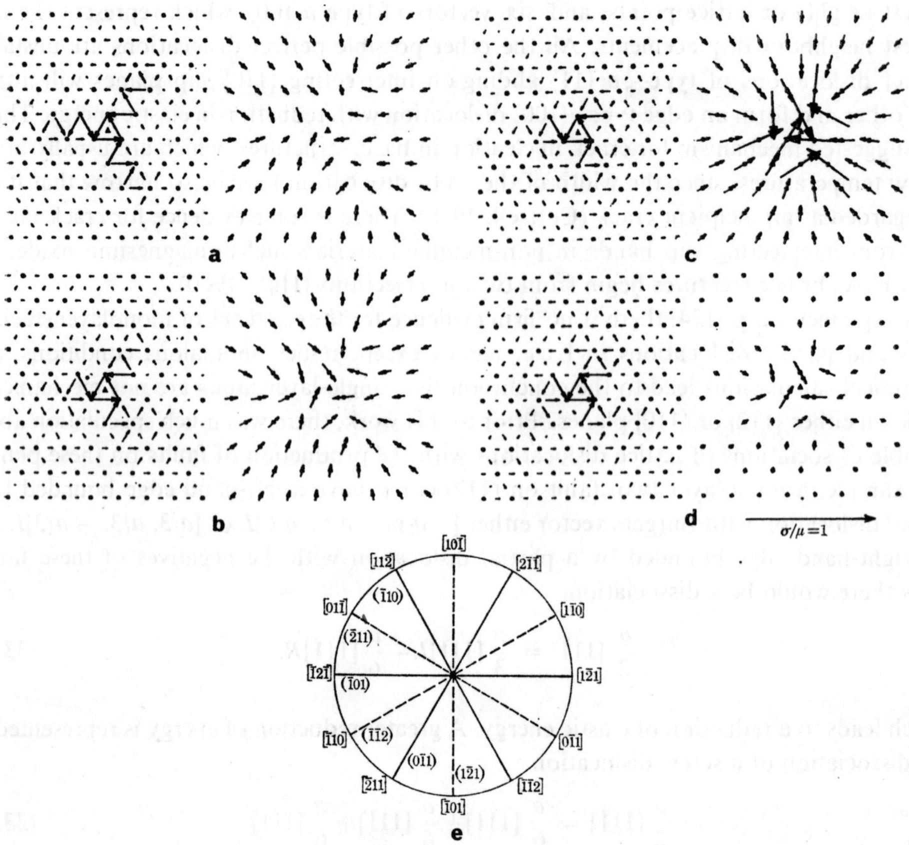

FIG. 7.21. Asymmetric core configurations computed for the unstressed a/2 [111] screw dislocation in the b.c.c. structure.

Displacement maps (on the left) and stress fields are showns in (a) – (c) for three different Johnson-type potentials and in (d) for the sodium potential of Fig. 5.5. The atomic structure in three successive (111) planes is projected along [111] (see (e)) and the arrows in the displacement maps are centred about the mid-points of atom pairs on successive planes and indicate the change in the [111] separations of these pairs caused by the introduction of the dislocation. The arrows are scaled so that the largest differential displacements of $(a/6)$ [111] are represented by arrows which join two projected atom positions. The stress field map shows in the same projection the magnitude and the plane of the maximum shear stress σ in the [111] direction acting at each atom; the scale of this representation is given by the length of the $\sigma/\mu = 1$ arrow.

large on only one side of the dislocation. The centre of the dislocation can be loosely regarded as the intersection of three stacking faults of generalized type (non-constant fault vector). This is also shown by the stress field map, which represents the plane and magnitude of the maximum shear stress at each atom. The stress field (Fig. 7.21) would be radial in the elastic solution, but it can be seen that with the exception of the three nearest neighbour atom rows, the arrows appear to radiate from three symmetrically placed sites away from the centre of the dislocation.

These observations on the core structure are closely connected with the asymmetry of the flow stress, which on {112} planes is smaller in the sense of shear which corresponds to a twinning shear than in the opposite sense, and with other anomalous slip observations in b.c.c. metals at low temperatures. The behaviour of the dislocation under stress is complex; see Basinski *et al.* (1971, 1972) and Duesbery *et al.* (1973).

Finally, we must recognize that many b.c.c. metals are elastically quite anisotropic with values of the Zener constant (Table IV. p. 74) which are especially large for the alkali metals and for β-brass. Head (1967) calculated, for example, that in β-brass dislocations with [111] Burgers vectors in a $(1\bar{1}0)$ plane should have two instability ranges of φ, namely -26 to $+33°$ and 88–116° from pure screw orientation, whilst there are three instability ranges for [111] Burgers vectors in $(11\bar{2})$ and for [001] Burgers vectors in (010). Electron-microscopy shows that there are many V- or Z-shaped dislocations in this ordered b.c.c. structure. For less anisotropic metals like iron, the ordinary slip dislocations are stable in all orientations on $(1\bar{1}0)$ and $(11\bar{2})$, but [001] dislocations have instability ranges on $(1\bar{1}0)$ and (010). Identification of Burgers vectors by electron-miscroscopy in b.c.c. metals is often difficult because of complex diffraction conditions (France and Lorretto, 1968).

34. DISTRIBUTIONS OF DISLOCATIONS

In discussing phenomena involving point defects, we need to specify the density or atomic concentration of each defect, which may vary from point to point within the crystal. Clearly, we need a similar measure of the total amount of dislocation line present in each small volume of a crystal as a first stage in providing a full description of the dislocated state. Since a dislocation is a line defect, a specification of a scalar density alone will generally be insufficient to determine the properties of the crystal, which will depend on the types of dislocation and the ways in which they are arranged. A more complete description thus requires the specification of a tensor dislocation density at each point.

The scalar dislocation density is usually defined as the total length of all dislocation lines contained in unit volume of the crystal, and has units of m^{-2}. If the dislocations are all parallel straight lines, this density is simply the number of dislocations cutting unit area normal to the lines, and this was the definition used in the early days of dislocation theory. For dislocations which are arranged randomly, the dislocation density as defined above is about three times the number of dislocations per unit area threading through any randomly orientated planar surface in the material. When dislocations are concentrated into "walls", or nearly planar regions, it is sometimes convenient to define a line density as the number of dislocations crossing unit length of a line in the plane of the dislocations. We discuss this further in Chapter 8 which deals with crystal boundaries.

Many experimental methods are available for estimating dislocation densities, ranging from direct counts of dislocations seen in the transmission electron-microscopy of thin films, or of dislocation nucleated etch pits on the surface, to deductions from X-ray misorientations. All of these methods give reasonably self-consistent values, the densities in well-annealed pure metals being usually in the range 10^4–10^6 lines mm^{-2}. In exceptional circumstances much lower densities may be obtained, especially in semi-conductors such as silicon and germanium grown under carefully controlled conditions. Large crystals of germanium have been obtained virtually free of dislocations, except possibly for very small dislocation loops. Metals in "whisker" form can also be obtained substantially free from dislocations, or with a single axial screw dislocation, and macroscopic crystals of metals have been grown with densities as low as $1-10$ dislocations mm^{-2} (Young, 1962; Wittels *et al.*, 1962).

A dislocation density of 10^6 lines mm^{-2} means that on the average about one dislocation threads through an area of about one square micron. The average dislocation separation, or the mesh size of a dislocation network, is thus of the order of 1 μm in a well-annealed metal. It follows that dislocation effects are generally likely to be unimportant in phenomena on a scale much finer than this, since the chances are that there will be no dislocations in the region considered. This means, for example, that very fine precipitate particles, in the 10^{-2}–10^{-1} μm (100–1000 Å) size range will not generally contain dislocations, and will have the properties of ideal materials.

As we have already noted, quenched metals may contain much higher defect concentrations than annealed and slowly cooled metals. Consider, for example, the tetrahedral defects formed in quenched f.c.c. metals of low stacking fault energy, which were described on p. 304. The experimental results for gold showed a typical volume density of tetrahedra of about 5×10^{11} mm^{-3}, each tetrahedron corresponding to the condensation of about 7400 vacancies. This gives a total vacancy concentration at the quenching temperature of about 6×10^{-5}, in agreement with the estimates in Section 17. The equivalent dislocation density of the stair-rod dislocations in the fault tetrahedra is about 10^8 mm^{-2}, compared with a dislocation density of 5×10^7 mm^{-2} which would have resulted if the vacancies had condensed as Frank sessile loops. In the same specimen, the density of dislocations other than those formed by vacancy condensation was estimated at 5×10^6 mm^{-2}.

Direct counts have been attempted in a few instances for the much larger dislocation densities found in heavily deformed materials, but the work is tedious and the results rather uncertain. Dislocation densities in such specimens are usually obtained from less reliable indirect methods, based on the measurement of such physical properties as the stored energy of cold work, the extra thermal or electrical resistivity produced by deformation, or the breadths of X-ray diffraction lines. The limiting densities in severely worked metals are generally found to be $\sim10^9$–10^{10} lines mm^{-2}, the upper limit corresponding to about one atom in a thousand being at the centre of the core of a dislocation line.

If a well-annealed single crystal attained thermodynamic equilibrium, it would contain virtually no dislocation lines. Once these lines have been introduced as inevitable accidents of growth, a mechanism for their complete removal does not exist, so a true equilibrium state is not attained, even at temperatures near to the melting point. However, if the dislocations are able to move, we expect them to adjust their positions so as to reduce their energy as much as possible. The dislocations will be in metastable equilibrium with the structure if they have adopted a configuration which gives a relative minimum to the free energy of the crystal, so that any further small displacements are resisted. This may lead to the formation of a three-dimensional dislocation network with stable nodes, as described on p. 246, or it may lead to the assembly of dislocations into stable two-dimensional networks, which then form the boundaries between slightly misaligned sub-grains. There is experimental evidence for both these situations in different materials or different circumstances, but the commonest configuration is found to be that in which most dislocations are in two-dimensional arrays, with a smaller number of dislocations forming a three-dimensional (Frank) network distributed across the subgrains. This result was deduced first from X-ray observations, but has been amply confirmed by electron-microscopy.

When a metal is deformed, the distribution of dislocations depends upon the type of deformation, the metal concerned, and the temperature, and very many variations are possible. Experimental observations lead to the conclusion that the distributions are qualitatively similar in the majority of close-packed metals with the exception of those having very low stacking fault energy. A deformed metal does not usually contain a uniform distribution of dislocations, and those dislocations which are observed, e.g. by thin foil electron-microscopy, are often not well concentrated on to particular slip planes. In the easy glide region of single crystal deformation, the predominant structures observed are dipoles and multipole clusters of primary dislocations in mainly edge orientations. Later these clusters build up into tangles which include dislocations of several secondary systems, and eventually carpets and walls of dislocations link together in a pronounced substructure in which cells of relatively low dislocation density are separated from each other by regions of very high dislocation density. In metals deformed at sufficiently low temperatures, these dense tangles are thick dislocation walls rather than genuine sub-boundaries, but the cells on either side of a particular tangle are misorientated with respect to each other. As the temperature of deformation is raised, or if a deformed metal is heated to a higher temperature, two effects can be noticed. The first is a general cleaning of the dislocation structure in the regions of low density, and the second is a sharpening of the tangled regions until ultimately they correspond to effectively two-dimensional distributions representing sharp sub-boundaries. We shall discuss these changes further in Part II, Chapter 19; for the present, we note that the sharpness of the substructure seems to be correlated with the ease of cross-slip in the material. Sharp sub-structures are found in aluminium, which has a high stacking fault energy, allowing dislocations to cross-slip readily, even after deformation at quite low temperatures. The sub-structures in the noble metals with lower fault energies are not so sharp unless the material is annealed at a high temperature, and the scale of the sub-structure is usually too small to be detected by X-ray diffraction methods.

Alloys of very low stacking fault energy show extended dislocations lined up on slip planes in the early stages of deformation, and this is the only case in which the observations made in the ordinary way approximate to the classical "pile-ups" assumed in many theories. However, it has often been suggested that pile-ups disappear through relaxation processes on removal of stress or thinning of the foil, and in order to overcome this effect Mughrabi (1968) deformed copper crystals at 77 K and then irradiated them at 20 K without removal of stress in order to pin the dislocations in position. Subsequent electron-microscopy showed piled-up groups of dislocations containing 10–20 dislocations of the same sign.

Body-centred cubic metals develop dislocation substructures similar to those of f.c.c. metals after deformation at moderately high temperatures (e.g. room temperature for pure crystals of most of the high melting point transition metals). At low temperatures, however, very different structures are developed. The predominant features are now long screw dislocations which are fairly uniformly distributed and tend to alternate in type, so that the long-range stress field is minimized. These screw dislocations often contain long jogs, and are accompanied by small dipoles or loop debris. This deformation substructure is believed to be associated with a much lower mobility of screws in comparison with non-screw dislocations at low temperatures.

The theory of equilibrium networks of dislocations has been developed by many workers (e.g. Frank, 1955; Ball and Hirsch, 1955; Amelinckx, 1957). We shall discuss the theory of dislocation boundaries in the next chapter, but we shall not analyse the numerous possibilities for particular types of boundary built up from square or hexagonal arrays of dislocations. Many beautiful examples of such networks in transparent crystals have been studied experimentally by precipitation techniques (Hedges and Mitchell, 1953; Amelinckx, 1957, 1958; etc.) and in foils by thin-film electron-microscopy (e.g. Carrington *et al.* 1960), and these have been correlated with theoretical predictions.

We now turn to a more mathematical description of the dislocation condition of a crystal, and we shall give a very sketchy outline of the theory of continuous distributions of dislocations, which includes the definition of a tensor dislocation density. When discussing the stress field of an individual dislocation, we made the approximate assumption that the integration can be cut off at some effective radius of the order of a mean dislocation spacing. In general, however, an arbitrary distribution of dislocations will give stresses which are not zero when averaged over distances much larger than the mean spacing, and the attainment of a configuration which reduces the free energy will correspond to a reduction in the far-reaching stresses. This leads us to consider the dislocation arrangements which are compatible with given imposed restraints, e.g. in the way in which a crystal is deformed. The first analysis of this kind was due to Nye (1953) who assumed that there is no accumulating long-range stress, i.e. that the above average is zero. Nye's work has been greatly extended by Bilby and co-workers, by Kondo, and by Kröner, who have independently and on slightly different lines developed formal theories of the compatibility relations between the overall (shape) deformation, the lattice deformation and the deformation caused by movement of dislocation lines.

We begin with a discussion of the tensor dislocation density, first introduced by Nye. Consider the dislocation lines threading through a unit area perpendicular to the x_j axis, and let the sum of the components of the Burgers vectors of these dislocation lines in the x_i direction be written A_{ij}. The quantity A_{ij} is a dislocation tensor, and it follows that the resultant Burgers vector of the dislocation lines threading a planar area O with a unit normal $\mathbf{n}$ is given by

$$b_i = A_{ij}O_j, \tag{34.1}$$

where the vector area $\mathbf{O}$ has components $O_i = On_i$. Now let C be a closed curve forming the limit of the area O, and let S be any cap ending on C. All the dislocations threading through O must also thread through S (they may have combined or dissociated, but this is immaterial), so that we have the more general expression for the resultant Burgers vector

$$b_i = \iint_S A_{ij}\, dS_j \tag{34.2}$$

and b_i is independent of the choice of S.

Clearly in a real crystal if the curve C is large, we obtain the average value of the dislocation density over a region of material, whereas if it is small, A_{ij} must show large fluctuations since the dislocations are discrete. It is convenient mathemati callyto treat the dislocation distribution as continuous by considering an element of the crystal containing several dislocation lines, and then allowing the number of dislocation lines of each type to tend to infinity whilst the Burgers vector of each tends to zero in such a way that the product remains finite. We thus arrive at the concept of the continuously dislocated state.

When we have a continuous distribution of dislocations, we expect to be able to define a tensor dislocation density at each point, although there are no individual dislocation lines. There is now a difficulty in finding the resultant Burgers vector by using the closure failure of a circuit in the reference crystal (p. 242), since there is no good crystal in which to make the closed circuit of the real crystal, and in the strict sense there is no lattice. Bilby *et al.* (1955) have shown how this difficulty may be overcome by making use of a local correspondence between the real crystal and the reference crystal.

At any point of the real crystal, choose three independent basic vectors, $\mathbf{e}_i$, which correspond to a set of basic vectors, $\mathbf{a}_i$, of the reference crystal in the sense that an identification of each $\mathbf{e}_i$ with a corresponding $\mathbf{a}_i$ may be made. This choice must be made continuously at each point of the real crystal, so that $\mathbf{e}_i$ are always the same crystallographic vectors, and define the local crystal lattice. To an observer moving in the real crystal, the local $\mathbf{e}_i$ vectors are everywhere "parallel", and any two "parallel" vectors (defined by reference to the local lattice) have the same $\mathbf{e}_i$ components. Bilby and Smith (1956) illustrate this point graphically by the example of a number of aircraft each flying due north along a great circle of longitude. The pilots of these aircraft, using their local systems of reference, will consider that they are flying parallel to each other, but this will not be the view of an outside observer, using a Euclidean orthonormal reference system.

The local vectors $\mathbf{e}_i$ may be regarded as generated from the reference vectors $\mathbf{a}_i$ at each point of the crystal by a deformation

$$\mathbf{e}_i = D_{ij}\mathbf{a}_j, \tag{34.3}$$

where the components D_{ij} of the matrix D vary from point to point in the crystal. Let us now introduce a fixed orthonormal Cartesian coordinate system x_i, which is defined by a set of orthonormal vectors $\mathbf{i}_i$ and which we shall use to discuss displacements in the real crystal.† The set of base vectors $\mathbf{i}_i$ may be directly related to the lattice vectors $\mathbf{a}_i$ of the reference crystal; in the case of simple cubic systems, the two bases will coincide. A displacement in the real crystal from x_i to $x_i + dx_i$ can thus be written as a vector $dx_i\mathbf{i}_i$. If C is now a small closed circuit in the real crystal, we have

$$\int_C \mathbf{i}_i\, dx_i = 0. \tag{34.4}$$

† The following development could be carried out in a coordinate system defined by the lattice base, $\mathbf{a}_i$, or in generalised curvilinear coordinates, as shown by Bilby *et al.*, but the use of an orthonormal basis is simpler for our purpose.

Let the vectors $\mathbf{j}_i$ be the local equivalent of the orthonormal set $\mathbf{i}_i$, that is let them be obtained from $\mathbf{i}_i$ by the local deformation $\mathbf{D}$ of eqn. (34.3). Then eqn. (34.4) becomes

$$\int_C D_{ij}^{-1}\mathbf{j}_j \, dx_i = 0. \tag{34.5}$$

We now require the closure failure of the corresponding circuit in the reference lattice. Each vector $\mathbf{j}_k$ of the real lattice is replaced by its corresponding vector $\mathbf{i}_k$ of the reference lattice, so that the closure failure giving the net Burgers vector of the distribution encircled by C is the negative of the vector sum of the reference lattice displacements, i.e.

$$\mathbf{b} = -\int_C D_{ij}^{-1}\mathbf{i}_j \, dx_i . \tag{34.6}$$

The sign of this equation has been written so that the Burgers vector is the vector needed to complete the circuit when this is traversed in a right-handed sense; this is obviously just a matter of defining a convention (see p. 243). In order to compare directly with eqn. (34.2), we now use Stoke's theorem to transform the line integral over C into a surface integral over any cap S having C as its limit. This gives

$$\mathbf{b} = -\iint_S \varepsilon_{ikl} \frac{\partial D_{ij}^{-1}}{\partial x_l} \mathbf{i}_j \, dS_k , \tag{34.7}$$

where ε_{ikl} is $+1$ or -1 according to whether i, k, l is an even or odd permutation of 1, 2, 3, and zero otherwise. Comparing with (34.2), we see that

$$A_{ij} = -\varepsilon_{jkl}(\partial D_{li}^{-1}/\partial x_k). \tag{34.8}$$

We may think of the matrix D in two equivalent ways. It establishes a one-to-one correspondence betweeen vectors of the reference lattice and the local lattice vectors drawn from a point, so that one local region may be mapped in another. Alternatively, we may regard the lattice in an infinitesimal region of real crystal around a point at which D_{ij} is given as formed by a deformation of a corresponding region of perfect lattice. The quantities D_{ij} have thus been called both lattice correspondence functions and generating deformations. They are not, of course, correspondence matrices of the type discussed on p. 56; since the base vectors are related by the deformation D, the correspondence matrix is I. This only means that corresponding vectors in the real and reference crystals have the same components in coordinate systems based on $\mathbf{e}_i$ and $\mathbf{a}_i$ (or $\mathbf{j}_i$ and $\mathbf{i}_i$) respectively.

Small lattice vectors about a point in the real crystal may be written as $dy_k\mathbf{j}_k$, where dy_i is a system of local coordinates based on the vectors $\mathbf{j}_i$. It follows from (34.3) that the relation between the local and the reference coordinates may be written

$$dy_i = D_{ji}^{-1} dx_j , \quad dx_i = D_{ji} \, dy_j \tag{34.9}$$

(cf. eqns. (6.1) and (6.4).

The Burgers vector of the circuit C could equally well have been written $b_i = -\int_C (\mathrm{d}y_i - \mathrm{d}x_i)$, so that

$$b_j = -\int_C (D_{ij}^{-1} - \delta_{ji})\,\mathrm{d}x_i$$
$$= -\int_C (\delta_{ji} - D_{ij})\,\mathrm{d}y_i.$$

On using Stokes' theorem, the first expression on the right gives eqn. (34.7) again, whilst the second gives

$$b_j = -\iint_S \varepsilon_{ikl} \frac{\partial D_{ij}}{\partial y_l}\,\mathrm{d}S_k, \qquad (34.10)$$

which corresponds to

$$A_{ij} = -\varepsilon_{jkl}(\partial D_{li}/\partial y_k). \qquad (34.11)$$

Equation (34.8) was derived first in the form (34.11) by Bilby (1955).

The relations (34.9) (known as Pfaffian forms) imply that the values of y_i at any point Q may be found from their values at any other point P by integrating along some path from P to Q. In general, however, the integral depends upon the path, so that y_i are not functions of x_i. This is an expression of the fact that we can only define a local correspondence in a dislocated crystal (see p. 242); we cannot generate the whole of the dislocated lattice from the reference lattice by applying a continuous deformation for which a displacement function exists. Imagine the reference crystal cut into small volumes, each of which is then given a local deformation specified by D_{ij}. If the separate elements can then be fitted together again so as to form a continuous lattice in ordinary space, the deformations are said to be compatible, and a continuous deformation exists. The condition for this is simply that

$$(\partial D_{li}^{-1}/\partial x_k) = (\partial D_{ki}^{-1}/\partial x_l), \qquad (34.12)$$

and we see from (34.7) or (34.10) that this corresponds to zero dislocation density.

The condition for compatibility, or for a continuous mapping of the reference lattice in the dislocated crystal, is thus the absence of dislocations. Conversely, when the local deformations of the separate elements imagined above are not compatible, dislocations will be needed to fit these elements together in real space.

Equation (34.7) gives the Burgers vector associated with the circuit element terminating the surface S, and should be distinguished from the *local Burgers vector*, which is the vector of the real lattice corresponding to the vector $\mathbf{b}$ of the reference crystal. The local Burgers vector could be obtained, for example, by completing a circuit in the reference crystal and taking the negative of the closure failure of the corresponding circuit in the real crystal (see p. 243). By definition, the local Burgers vector will have the same components in the $\mathbf{j}_k$ system as the Burgers vector has in the $\mathbf{i}_k$ system, so that it may be written

$$\mathbf{l} = \iint_S A_{ij} D_{ik} \mathbf{i}_k\,\mathrm{d}S_j$$
$$= -\iint_S \varepsilon_{jkl} D_{im} \frac{\partial D_{li}^{-1}}{\partial x_k} \mathbf{i}_m\,\mathrm{d}S_j. \qquad (34.13)$$

The geometrical significance of the local Burgers vector is that if a dislocation line moves in the real crystal by glide or climb so as to intersect the closed circuit C, the parts of the circuit on each side of the point of intersection suffer a relative displacement of $\mathbf{dl}$, where $\mathbf{dl}$ is the change in the local Burgers vector at the point of intersection. Clearly the distinction between $\mathbf{b}$ and $\mathbf{l}$ is unimportant when $\mathbf{D}$ represents a small deformation; it becomes important when there are large discontinuities in the tensor dislocation density, as in the theory of surface dislocations (p. 363).

Although we have written the area element as a vector in the above equations, it is strictly an axial vector, that is an asymmetric second order tensor. The tensor element of area is related to the (pseudo) vector element of area by the relation

$$\left.\begin{aligned} dS_k &= \tfrac{1}{2}\varepsilon_{kmn}\, dS_{mn}, \\ dS_{mn} &= \varepsilon_{kmn}\, dS_k. \end{aligned}\right\} \tag{34.14}$$

On substituting into (34.13), we find that the local Burgers vector can be written in the form

$$l_p = T_{mnp}\, dS_{mn}, \tag{34.15}$$

where

$$T_{mnp} = \frac{1}{2} D_{jp}\left\{\frac{\partial D^{-1}_{mj}}{\partial x_n} - \frac{\partial D^{-1}_{nj}}{\partial x_m}\right\}. \tag{34.16}$$

Alternatively, we may define a local dislocation tensor density ${}^{l}A_{ij}$ where

$$l_j = {}^{l}A_{ij}\, dS_j \tag{34.17}$$

and

$${}^{l}A_{ij} = \varepsilon_{jkl} T_{ikl}. \tag{34.18}$$

The quantity T_{ijk} is called the torsion tensor.

In the situation considered by Nye, in which there are no far reaching stresses, the relation between $\mathbf{i}_i$ and $\mathbf{j}_i$ is everywhere a pure rotation, so that the real lattice although continuously curved is unstrained. The general theory has been applied to this case, and to many other specific examples (for review see Bilby, 1960), but we shall not consider them in detail. The further development of the theory makes extensive use of the geometry of generalized spaces in which the lattice generating deformations are compatible. We recall that the individual elements of a reference crystal, given arbitrary deformations, will not fit together without the introduction of further deformations or the insertion of dislocations between the elements. However, the elements may be "fitted together" in a general space with appropriate geometry, so that the continuously dislocated crystal may be associated with such a space. The advantage of this formalism is that the previously developed methods of differential geometry for such a space may be used to solve dislocation problems.

The analogy with the theory of generalized spaces may be introduced by considering equivalent lattice vectors at two neighbouring points P and Q. Clearly, two vectors are crystallographically equivalent if they have the same components relative to the local basis ($\mathbf{e}_i$ or $\mathbf{j}_i$). Let two such vectors at P and Q be written $p_i\mathbf{i}_i = p_k D^{-1}_{ki}(P)\mathbf{j}_i$ and $(p_i + dp_i)\mathbf{i}_i =$

$(p_k + dp_k)\, D_{ki}^{-1}(Q)\mathbf{j}_i$, where we have found it necessary to distinguish the values of D_{ki}^{-1} at P and Q. Since the components are identical in the $\mathbf{j}_i$ system, we have

$$p_k D_{ki}^{-1}(P) = (p_k + dp_k)\, D_{ki}^{-1}(Q)$$

or, since P and Q are neighbouring points,

$$dp_k D_{ki}^{-1} = -p_k \frac{\partial D_{ki}^{-1}}{\partial x_l}\, dx_l .$$

This may be written

$$dp_m = -D_{im} \frac{\partial D_{ki}^{-1}}{\partial x_l} p_k\, dx_l = -L_{klm} p_k\, dx_l . \tag{34.19}$$

In differential geometry, a relation like (34.19) is described as a linear connection; it prescribes which vectors of the generalized space are equivalent or "parallel". The quantities L_{klm} are called the coefficients of connection, and we see that they may be written

$$L_{klm} = D_{im} \frac{\partial D_{ki}^{-1}}{\partial x_l} . \tag{34.20}$$

The important geometrical property of this connection is that L_{klm} is antisymmetric in k and l, i.e. $L_{klm} \neq L_{lkm}$. For a Euclidean geometry, $L_{klm} = 0$, and in a more general (Riemannian) space, L_{klm} is still symmetric. A space for which L_{klm} is not symmetric is said to possess torsion, and its geometry is described as non-Riemannian. The quantities L_{klm} may be written as the sum of a symmetric and an antisymmetric part; by comparing with eqn. (34.16), we see that the antisymmetric part gives the torsion tensor associated with the local Burgers vector since

$$T_{klm} = \tfrac{1}{2}(L_{klm} - L_{lkm}). \tag{34.21}$$

The unusual geometrical properties of a space with torsion are just those properties which we associate with a crystal containing dislocations if we imagine ourselves inside the crystal with only the lattice lines to guide us. For example, an infinitesimal closed parallelogram does not exist in a space with torsion. If we draw two non-parallel small vectors PQ and PR from P, and then draw QS equal to and parallel with PR and RS' equal to and parallel with PQ, in general S does not coincide with S'. This is just the situation in a dislocated crystal, the vector SS' being, of course, the local Burgers vector encircled by the circuit.

We conclude this section by referring again to the problem of the generation of a dislocated crystal from a reference crystal. We have emphasized that the local lattice generating deformations are not compatible in the sense that they will not fit together in ordinary space without the introduction of discontinuities (dislocations). We find it convenient to distinguish between these deformations and the *shape deformation*, which could be measured in principle by the distortion of a network of lines scribed on the crystal before the introduction of the dislocations. The shape deformation describes the behaviour of large vectors, and is the deformation usually considered in the macroscopic theory of plasticity. Since the final crystal may be regarded as obtained by the deformation of a reference crystal in ordinary space, the shape deformation is necessarily compatible. This means that the deformation is continuous and the displacements are the derivatives of a displacement function.

In most of this section we have been concerned with the geometry of an existing dislocated crystal, and for this reason we have not considered the shape deformation. In such a crystal the lattice strain is clearly related to the dislocation density, as we have seen. However, when we consider the production of the dislocated crystal from the reference crystal, it is obvious that there may be changes of shape which do not affect the lattice at all. Such a change, for example, is produced by the generation of a dislocation line which passes right through the crystal and disappears again at the surface. In general, all dislocation motion by glide or climb will produce changes in the external shape of the body.

We may thus regard the shape deformation as made up of two parts, one due to changes in the lattice (the *lattice deformation*), and one due to the introduction and movement of dislocations without affecting the lattice. This second part has been called the *lattice invariant deformation* (Bilby and Christian, 1956) or the *dislocation deformation* (Bilby and Smith, 1956). In terms of our previous discussion, we apply separate lattice deformations to individual elements of the reference crystal, and we now make these elements fit together in real space. We can do this by causing slip to occur in the elements and by adding or removing lattice planes to the elements. Since these operations will differ from element to element, lattice dislocations will appear when the elements are fitted together. The operations required to make the elements fit together in ordinary space constitute the lattice invariant deformation. We pass from the discrete (physically real) dislocation model to the continuous (mathematically convenient) model by making the elements infinitesimal. The torsion calculated from D_{ij} then gives just the dislocation distribution required to make the resultant shape changes of the elements compatible with each other.

We have regarded the shape change as continuous in the sense that two points which are initially very close together remain very close together. However, some of the more important applications of the above theory correspond to surfaces of discontinuity in the dislocation field, where the shape and lattice deformations suddenly change. Such a change represents the boundary between two crystals of different structure and orientation, and when a lattice correspondence exists, the dislocation distribution at the interface ensures compatibility of the two shape deformations on each side of the interface. The dislocation distribution† thus ensures that macroscopic regions of the two crystals will fit smoothly together, even although the lattices will not. We discuss the application of the theory of continuous distributions of dislocations to two-dimensional sheets ("surface dislocations") in Section 38. A fuller description of the relation between lattice and shape deformations in the theory of martensite is given in Part II, Chapters 21 and 22.

REFERENCES

AMELINCKX, S. (1957) *Dislocations and Mechanical Properties of Crystals*, p. 2, Wiley New York; (1958) *Acta metall.* **6,** 34.

AMELINCKX, S., BONTINCK, W., DEKEYSER, W., and SEITZ, F. (1957) *Phil. Mag.* **2,** 355.

BABUSKA, I., VITÁSEK, E., and KROUPA, F. (1960) *Czech. J. Phys. B*, **10,** 419, 488.

† Or the equivalent effect produced by alternating fine twins: see Part II, Chapter 22.

BACON, D. J. (1967) *Phys. Stat. Sol.* **23,** 527; (1969) *Scripta metall.* **3,** 735.
BACON, D. J., BULLOUGH, R., and WILLIS, J. R. (1970) *Phil. Mag.* **22,** 31.
BALL, C. J. and HIRSCH, P. B. (1955) *Phil. Mag.* **46,** 1343.
BARDEEN, J. and HERRING, C. (1952) *Imperfections in nearly Perfect Crystals*, p. 261, Wiley, New York.
BARNETT, D. M. (1972) *Phys. stat. sol.* (b), **49,** 741.
BARNETT, D. M., and SWANGER, L. A. (1971) *Phys. stat. sol.* **48,** 419.
BARNETT, D. M., ASARO, R. J., BACON, D. J., GAVAZZA, S. D., and SCATTERGOOD, R. O. (1972) *J. Phys.* (F) **2,** 854.
BASINSKI, Z. S. (1959) *Phil. Mag.* **4,** 393; (1964) *Disc. Foraday Soc.* **38,** 93.
BASINSKI, Z. S. and CHRISTIAN, J. W. (1954) *Acta metall.* **2,** 101; (1960) *Aust. J. Phys.* **13,** 299.
BASINSKI, Z. S., DUESBERY, M. S., and TAYLOR, R. (1970) *Phil. Mag.* **21,** 1201; (1971) *Can. J. Phys.* **49,** 2160; (1972) *Interatomic Potentials and Simulation of Lattice Defects*, p. 537, Plenum Press, New York.
BILBY, B. A. (1950) *Proc. Phys. Soc.* A, **63,** 191; (1951) *Research* **4,** 389; (1953) *Phil. Mag.* **44,** 782; (1955) *Report of the Conference on Defects in Crystallic Solids*, p. 123, Physical Society, London; (1960) *Prog. Solid Mechan.* **1,** 329.
BILBY, B. A. and CHRISTIAN, J. W. (1956) *The Mechanisms of Phase Transformations in Metals*, p. 121, Institute of Metals, London.
BILBY, B. A. and SMITH, E. (1956) *Proc. R. Soc.* A, **236,** 481.
BILBY, B. A., BULLOUGH, R., and SMITH, E. (1955) *Proc. R. Soc.* A, **231,** 263.
BRAILSFORD, A. D. (1965) *Phys. Rev.* **139,** A1813.
BROWN, L. M. (1964) *Phil. Mag.* **10,** 441; (1967) *Ibid.* **15,** 363.
BULLOUGH, R. and FOREMAN, A. J. E. (1964) *Phil. Mag.* **9,** 315.
BULLOUGH, R. and NEWMAN, R. C. (1960) *Phil. Mag.* **5,** 921.
BULLOUGH, R. and PERRIN, R. C. (1968) *Proc. R. Soc.* A, **305,** 541.
BURGERS, J. M. (1939) *Proc. K. ned. Acad. Wet.* **42,** 293, 378.
CAHN, J. W. (1960) *Acta metall.* **8,** 554.
CARRINGTON, W., HALE, K. F., and MACLEAN, D. (1960) *Proc. R. Soc.* A, **259,** 203.
CELLI, V. (1961) *J. Phys. Chem. Solids* **19,** 100.
CHANG, R. (1962) *Acta metall.* **10,** 951; (1967) *Phil. Mag.* **16,** 1021.
CHANG, R., and GRAHAM, L. J., (1966) *Phys. stat. sol.*, **18,** 99.
CHOU, Y. T. (1965) *Acta metall.* **13,** 251, 779.
CHRISTIAN, J. W. (1968) *The Interaction between Dislocations and Point Defects*, p. 604, HMSO, London.; (1970) *Proceedings of the 2nd International Conference on Strength of Metals and Alloys*, p. 31, American Society for Metals, Ohio; (1973) *Int. Met. Rev.* **18,** 24.
CHRISTIAN, J. W. and SWANN, P. R. (1965) *Alloying Behaviour and Effects of Concentrated Solid Solutions*, Gordon and Breach, New York. p. 105.
CHRISTIAN, J. W. and VÍTEK, V. (1970) *Rep. Progr. Phys.* **33,** 307.
CLAREBROUGH, L. M. and HEAD, A. K. (1969) *Phys. stat. sol.* **33,** 431.
CLAREBROUGH, L. M., SEGALL, R. L., and LORETTO, M. H. (1966) *Phil. Mag.* **13,** 1285.
COCHARDT, A. W., SCHOECK, G., and WIEDERSICH, H. (1955) *Acta metall.* **3,** 533.
COCKAYNE, D. J. H. (1972) *Z. Naturforsch.* **27***a*, 452.
COCKAYNE, D. J. H., JENKINS, M. L., and RAY, I. L. F. (1971) *Phil. Mag.* **24,** 1383.
COCKAYNE, D. J. H., RAY, I. L. F., and WHELAN, M. J. (1969) *Phil. Mag.* **20**, 1265.
CONRAD, H. (1963) *The Relation between the Structure and Mechanical Properties of Metals*, p. 474, HMSO, London.
CONRAD, H. and SCHOECK, G. (1960) *Acta metall.* **8,** 791.
COTTERILL, R. M. J. and DOYAMA, M. (1966) *Phys. Rev.* **145,** 465.
COTTRELL, A. H. (1948) *Conference on Strength of Solids*, p. 30, The Physical Society, London; (1952) *Phil. Mag.* **42,** 1327; (1953) *Dislocations and Plastic Flow in Crystals*, Clarendon Press, Oxford; (1958) *Trans. Metall. Soc. Am. Inst. Min. (Metall.) Engrs.* **212,** 192.
COTTRELL, A. H. and BILBY, B. A. (1951) *Phil. Mag.* **42,** 573.
CRUSSARD, C. (1950) *Métaux Corros.* **25,** 203.
DASH, W. C. (1956) *Dislocations and Mechanical Properties of Crystals* p. 57, Wiley, New York.
DE JONG, M. and KOEHLER, J. S. (1963) *Phys. Rev.* **129,** 49.
DE WIT, R (1960) *Solid State Physics.* **10**, 249; (1967) *Phys. stat. sol.* **20**, 575.
DÉ WIT, R. and KOEHLER, J. S. (1959) *Phys. Rev.*, **116**, 1113.
DIETZE, H. D. (1952) *Z. Physik* **132,** 107.

DOBSON, P. S. and SMALLMAN, R. E. (1966) *Proc. R. Soc.* A, **293,** 423.
DOBSON, P. S., GOODREW, P. J., and SMALLMAN, R. E. (1967) *Phil. Mag.* **16,** 9.
DOYAMA, M. S. and COTTERILL, R. M. J. (1966) *Phys. Rev.* **150,** 448.
DUESBERY, M. S., VÍTEK, V., and BOWEN, D. K. (1973) *Proc. R. Soc.* A, **332,** 85.
EDINGTON, J. W. and SMALLMAN, R. E. (1965) *Phil. Mag.* **11,** 1109.
EDINGTON, J. W. and WEST, D. R. (1966) *Phil. Mag.* **14,** 603; (1967) *Ibid.* **15,** 229.
ESCAIG, B. (1963) *Acta metall.* **11,** 595.
ESHELBY, J. D. (1949) *Proc. Phys. Soc.* A, **62,** 307; (1954) *J. Appl. Phys.* **25,** 255; (1956a) *Solid State Phys.* **3,** 79; (1956b) *Proc. Phys. Soc.* B, **69,** 1013.
ESHELBY, J. D., READ, W. T., and SHOCKLEY, W. (1953) *Acta metall.* **1,** 251.
EVANS, A. G. and RAWLINGS, R. D. (1969) *Phys. stat. sol.* **34,** 9.
FOREMAN, A. J. E. (1955) *Acta metall.* **3,** 322; (1967) *Phil. Mag.* **15,** 1011.
FOREMAN, A. J. E., JASWON, M. A., and WOOD, J. K. (1951) *Proc. Phys. Soc.* A, **64,** 156.
FRANCE, L. K. and LORETTO, M. H. (1968) *Proc. R. Soc.* A, **307,** 83.
FRANK, F. C. (1949) *Proc. Phys. Soc.* A, **62,** 202; (1951) *Phil. Mag.* **42,** 809; (1955) *Conference on Defects in Crystalline Solids*, p. 159, The Physical Society, London.
FRANK, F. C. and NICHOLAS, J. F. (1953) *Phil. Mag.* **44,** 1213.
FRANK, F. C. and READ, W. T. (1950) *Phys. Rev.* **79,** 722.
FRANK, F. C. and VAN DER MERWE, J. H. (1949) *Proc. R. Soc.* A, **198,** 205.
FRIEDEL, J. (1964) *Dislocations*, Pergamon Press, Oxford.
GALLAGHER P. C. J. (1966) *Phys. stat. sol.* **16,** 695; (1968) *J. Appl. Phys.* **39,** 160.
GALLAGHER, P. C. J. and LIU, Y. C. (1969) *Acta metall.* **17,** 127.
GEHLEN, P. C. (1970) *J. Appl. Phys.* **41,** 5165; (1972) *Interatomic Potentials and Simulation of Lattice Defects*, p. 475, Plenum Press, New York.
GILMAN, J. J. (1960) *Aust. J. Phys.* **13,** 327.
GROVES, P. P. and BACON, D. J. (1970) *Phil. Mag.* **22,** 83.
HAASEN, P. (1957) *Acta metall.* **5,** 598.
HALES, R., SMALLMAN, R. E., and DOBSON, P. S. (1968) *Proc. R. Soc.* A, **307,** 71.
HARRIS, J. E. and MASTERS, B. C. (1966a) *Phil. Mag.* **13,** 963; (1966b) *Proc. R. Soc.* A, **292,** 240.
HEAD, A. K. (1953) *Phil. Mag.* **44,** 92; (1967) *Phys. stat. sol.* **19,** 185.
HEDGES, J. M. and MITCHELL, J. W. (1953) *Phil. Mag.* **44,** 223, 357.
HEIDENREICH, R. D. and SHOCKLEY, W. (1948) *Report of the Conference on Strength of Solids*, p. 57, The Physical Society, London.
HIRSCH, P. B. (1959a) *Conference on Internal Stresses and Fatigue*, p. 139, Elsevier, Amsterdam; (1959b) *Metall. Rev.* **4,** 113; (1960) *Fifth International Conference in Crystallography, Cambridge* (oral discussion); (1962) *Phil. Mag.* **7,** 67; (1968) *Trans. Jap. Inst. Metals*, Suppl. **9,** XXX.
HIRSCH, P. B. and WARRINGTON, D. H. (1961) *Phil. Mag.* **6,** 735.
HIRTH, J. P. and COHEN, M. (1969a) *Scripta metall.* **3,** 107; (1969b) *Ibid.* **3,** 311.
HIRTH, J. P. and LOTHE, J. (1968) *Theory of Dislocations*, McGraw-Hill, New York.
HIRTH, J. P. and NIX, W. D. (1969) *Phys. stat. sol.* **35,** 177.
HULL, D. (1958) *Phil. Mag.* **3,** 1468; (1960) *Acta metall.* **8,** 11.
HUMBLE, P. and FORWOOD, C. T. (1968a) *Phys. stat. sol.* **29,** 99; (1968b) *Aust. J. Phys.* **21,** 941.
HUNTINGTON, H. B. (1955) *Proc. Phys. Soc.* B, **68,** 1043.
HUNTINGTON, H. B., DICKEY, J. E., and THOMSON, R. (1955) *Phys. Rev.* **100,** 1117; (1959) *Ibid.* **113,** 1696.
INDENBOM, V. L. and ORLOV, S. S. (1967) *Kristallografiya* **12,** 971 (*Soviet Phys. Crystallogr.* **12,** 849).
JOHNSTON, W. G. and GILMAN, J. J. (1959) *J. Appl. Phys.* **30,** 129.
JØSSANG, T. (1968) *Phys. stat. sol.* **27,** 579.
JØSSANG, T., LOTHE, J., and SKYLSTAD, K. (1965) *Acta metall.* **13,** 271.
JØSSANG, T., STOWELL, M. J., HIRTH, J. P., and LOTHE, J. (1965) *Acta metall.* **13,** 279.
KLÉMAN, M. and FRIEDEL, J. (1969) *J. Phys. Paris* **30,** C4–43.
KOEHLER, J. S. (1941) *Phys. Rev.* **60,** 397.
KRONBERG, M. L. (1959) *J. Nucl. Mater.* **1,** 85; (1961) *Acta metall.* **9,** 970.
KRÖNER, E. (1953) *Z. Phys.* **136,** 402; (1958) Kontinuums theorie der Versetzungen und Eigenspannungen, *Ergebn. Angew. Math.* **5,** 1.
KROUPA, F. (1960) *Czech. J. Phys.* B, **10,** 284; (1966) *Theory of Crystal Defects*, p. 275, Academia, Prague.
KUHLMANN-WILSDORF, D. (1958) *Phil. Mag.* **3,** 125; (1960) *Phys. Rev.* **120,** 773; (1962) *Strengthening Mecha-*

nisms in Solids, p. 137, American Society for Metals, Cleveland, Ohio; (1962) *Trans. metall. Soc. AIME* **224,** 1047; (1965) *Acta metall.* **13,** 257.
LALLY, J. S. and PARTRIDGE, P. G. (1966) *Phil. Mag.* **13,** 9.
LEIBFRIED, G. (1950) *Z. Phys.* **127,** 344.
LEWTHWAITE, G. W. (1966) *Phil. Mag.* **13,** 437.
LOMER, W. M. (1951) *Phil. Mag.* **42,** 1327; (1957) *Ibid.* **2,** 1053.
LORETTO, M. H., CLAREBROUGH, L. M., and SEGALL, R. L. (1965) *Phil. Mag.* **11,** 459.
LOTHE, J. (1967) *Phil. Mag.* **15,** 353.
LOUAT, N., (1956) *Proc. Phys. Soc.* B, **69,** 459.
LOW, J. R. and TURKALO, A. M. (1962) *Acta metall.* **3,** 215.
MALEN, K. (1970) *Phys. stat. sol.* **38,** 259.
MALEN, K. and LOTHE, J. (1970) *Phys. stat. sol.* **39,** 287.
MARADUDIN, A. A. (1958) *J. Phys. Chem. Solids* **9,** 1.
MASTERS, B. C. and CHRISTIAN, J. W. (1964) *Proc. R. Soc.* A, **281,** 223.
MILTAT, J. E. A. and BOWEN, D. K. (1970) *Phys. stat. sol.* (a) **3,** 431.
MORTON, A. J. and CLAREBROUGH, L. M. (1969) *Aust. J. Phys.* **22,** 393.
MOTT, N. F. and NABARRO, F. R. N. (1948) *Report of the Conference on Strength of Solids*, p. 38, The Physical Society, London.
MUGHRABI, H. (1968) *Phil. Mag.* **18,** 1211.
NABARRO, F.R.N. (1947) *Proc. Phys. Soc.* **59,** 256; (1951) *Proc. R. Soc.* A, **209,** 278; (1952) *Adv. Phys.* **1,** 319; (1967) *Theory of Crystal Dislocations*, Clarendon Press, Oxford.
NABARRO, F. R. N., BASINSKI, Z. S. and HOLT, D. B. (1964) *Adv. in Phys* **13,** 193.
NYE, J. F. (1953) *Acta metall.* **1,** 153.
PEACH, M. O. (1951) *J. Appl. Phys.* **22,** 1359.
PEACH, M. O. and KOEHLER, J. S. (1950) *Phys. Rev.* **80,** 436.
PEIERLS, P., (1940) *Proc. Phys. Soc.* **52,** 34.
READ, W. T. and SHOCKLEY, W. (1952) *Imperfections in Nearly Perfect Crystals*, p. 77, Wiley, New York.
SAADA, G. (1962) *Acta metall.* **10,** 551.
SAADA, G. and WASHBURN, J. (1963) *J. Phys. Soc. Japan* **18,** Suppl. **3,** 43.
SAXL, I. (1964) *Czech. J. Phys.* B, **14,** 381; (1968) *Czech. J. Phys.* B, **18,** 39.
SCHAPINK, F. W. and DE JONG, M. (1964) *Acta metall.* **12,** 756.
SCHOECK, G. (1969) *Scripta metall.* **3,** 239.
SCHWARTZKOPFF, K. (1969) *Acta metall.* **17,** 345.
SEEGER, A. (1955a) *Phil. Mag.* **46,** 1194; (1955b) *Conference on Defects in Crystalline Solids*, p. 328, The Physical Society, London; (1964) *Discuss. Faraday Soc.* **38,** 82.
SEEGER, A., BERNER, R., and WOLF, H. (1959) *Z. Phys.* **155,** 247.
SEEGER, A., DONTH, H., and PFAFF, F. (1957) *Discuss. Faraday Soc.* **23,** 19.
SEEGER, A. and SCHOECK, G. (1953) *Acta metall.* **1,** 519.
SEIDMAN, D. N. and BALLUFFI, R. W. (1966) *Phys. stat. sol.* **17,** 531.
SEITZ, F. (1952) *Adv. Phys.* **1,** 43.
SHOCKLEY, W. (1947) Quoted by Mott and Nabarro (1948).
SILCOX, J. and HIRSCH, P. B. (1959) *Phil. Mag.* **4,** 72.
SLEESWYK, A. W. (1963) *Phil. Mag.* **8,** 1467.
STEEDS, J. W. (1967) *Phil. Mag.* **16,** 771, 785.
STEHLE, H. and SEEGER, A. (1956) *Z. Phys.* **146,** 217.
STEIN, D. F. and LOW, J. R. (1960) *J. Appl. Phys.* **31,** 362; (1966) *Acta metall.* **14,** 1183.
STOBBS, W. M. and SWORN, C. H. (1971) *Phil. Mag.* **24,** 1365.
STROH, A. N. (1956) *Phil. Mag.* **1,** 489; (1958) *Ibid.* **3,** 625.
SUZUKI, H. (1952) *Sci. Rep. Res. Inst. Tohôku Univ.* **A4,** 455; (1968) *Dislocation Dynamics*, p. 679, McGraw-Hill, New York.
TAYLOR, G. I. (1934) *Proc. R. Soc.* A, **145,** 362.
THOMPSON, N. (1953) *Proc. Phys. Soc.* B, **66,** 481; (1955) *Report of the Conference on Defects in Crystalline Solids*, p. 153, The Physical Society, London.
TOMSETT, D. I. and BEVIS, M. (1969) *Phil. Mag.* **19,** 129.
VÍTEK, V. (1974) *Crystal Lattice Defects* **5,** 1.
VÍTEK, V., PERRIN, R. C., and BOWEN, D. K. (1970) *Phil. Mag.* **21,** 1049.

WEERTMAN, J. (1965) *Phil. Mag.* **11,** 1217; (1967) *J. Appl. Phys.* **38,** 5293.
WESTLAKE, D. G. (1961) *Acta metall.* **9,** 327.
WESTMACOTT, K. H., BARNES, R. S., HULL, D., and SMALLMAN, R. E. (1961) *Phil. Mag.* **6,** 929.
WHELAN, M. J. (1959) *Proc. R. Soc.* A, **249,** 114.
WILKENS, M. (1967) *Acta metall.* **15,** 1417; (1968) *Scripta metall.* **2,** 299.
WILLIS, J. R. (1970) *Phil. Mag.* **21,** 931.
WITTELS, M. C., SHERRILL, F. A., and YOUNG, F. W., (1962) *Appl. Phys. Letters* **1**, 22.
YAMAGUCHI, M. and VÍTEK, V. (1973) *F. Phys. (F)* **3**, 523. 537.
YOUNG, F. W. (1962) *Bull. Am. Phys. Soc.* **7**, 215.
YOFFE, E. H. (1960) *Phil. Mag.* **5,** 161.

CHAPTER 8

Polycrystalline Aggregates

35. MACROSCOPIC THEORY

The description of the solid state in Chapters 5 and 6 refers only to single crystals of a metal or one-phase alloy. Very large single crystals can often be grown by a suitable technique, but macroscopic specimens usually consist of a compact polycrystalline mass. The crystals are allotriomorphic. i.e. their limiting surfaces are not regular and do not display the symmetry of the internal structure. In a single-phase assembly, neighbouring crystals differ only in the orientations of their respective lattices: it is then usual to refer to the individual crystals as the grains of the structure and the regions over which the lattice orientation changes are the grain boundaries. We shall restrict the use of the term grain boundary to this sense, and regions of discontinuity separating crystals having different structures or chemical compositions will be referred to as crystal boundaries or interphase boundaries.

A one-phase assembly is in true thermodynamical equilibrium only when it forms a single crystal, the exterior surface of which has the shape giving the minimum energy. In practice the energies of the free surface and the grain boundaries of a polycrystalline specimen are very small, and there is a negligible rate of approach to this equilibrium except at temperatures near the melting point. Even at these temperatures, the structure will usually remain polycrystalline indefinitely, unless there has been prior mechanical deformation. When calculating the equilibrium state of a solid metal or alloy, we have, therefore, to accept the various crystals as frozen into the structure.† In exactly the same way, we have to accept the presence of isolated dislocations in the structure, even though they can never be in thermal equilibrium with the lattice.

In a one-phase assembly, the orientations of the grains may be completely random, or there may be a preferred orientation induced by mechanical deformation or thermal treatment. In a drawn wire, for example, there is a tendency for some particular crystallographic axis to lie along the wire axis, the various grains having random rotations about this direction. This is described as a fibre structure. The grains in a cold-rolled sheet have a preferred crystal direction in the direction of rolling, and a crystal plane in the plane of rolling. The

† Speculations have sometimes been made, e.g. Born (1946), that there may be a natural or equilibrium limit to the size of a region of perfect lattice. It seems most improbable that this can be true.

preferred orientations are never perfect, and are represented statistically by "pole figures" plotted in stereographic projection.

The thickness of the transition region between two lattices is always small, and grain boundaries may be treated macroscopically as surfaces of discontinuity. A general grain boundary has five degrees of freedom, of which three degrees are associated with the relative orientations of the two lattices on either side of the boundary, and two are required to specify the local inclination of the boundary surface relative to one of these lattices. The energy of the boundary will be a function of all these variables, but the relative orientation of the two lattices proves to be much more important than that of the boundary itself, except in certain special cases. The atoms in a grain boundary region are acted upon by forces tending to move them into positions corresponding to the two competing orientations. In general, this implies a considerable distortion of the periodic structure, and part or all of the grain boundary must be bad crystal. As we have seen, the atoms in a free surface are distorted little from equilibrium positions; nevertheless, the energy of a grain boundary is less than that of a free surface, since there are fewer unsatisfied atomic bonds in the former.

It is sometimes suggested, e.g. Chalmers and Gleiter (1971), that in addition to the above five parameters, three further quantities are required to specify the components of any relative translation **t** of the atoms at large distances from the boundary in the two structures (cf. eqn. (9.1)), thus making eight parameters, and that a ninth may be necessary to specify the position of the boundary itself. However, this appears to confuse internal parameters which may be utilized in an economic description of the boundary structure with macroscopic degrees of freedom: only five parameters may be varied independently of the others. As discussed later in this chapter, the translation **t** must be a vector from a lattice point of one structure to an internal point of a unit cell which has that lattice point at a corner and, together with any localized atomic adjustments or relaxations, the eight parameters fix the positions of all the atoms on both sides of the interface. There is thus no necessity separately to specify the position of the interface. Once the five degrees of freedom are fixed, there are in principle $3N$ additional but dependent parameters to specify the positions of the N atoms; these may be reduced approximately to three parameters in many models, but they are still not independent.

The free energy of an internal or external surface is a useful macroscopic concept, which we have hitherto used without detailed justification. Such a free energy may be defined quite generally as the change in the free energy of the whole assembly per unit area of boundary formed when a geometrical boundary is introduced by some virtual process. It is important to realize that this energy is a property of the whole solid, and is not necessarily localized in the immediate vicinity of the boundary region. When the boundary is plane, the free energy is independent of the exact location of the supposed geometrical boundary within the transition region but, in common with other macroscopic surface parameters, the free energy per unit area of a curved boundary depends slightly on the way in which the boundary is defined. The important result, however, is that the thermodynamic relations between these parameters are independent of the position of the geometrical surface; a detailed discussion has been given by Herring (1953). For specimens having dimensions large compared with the effective thickness of the transition region, the variation of specific surface energy with curvature may be neglected, and the extra energy associated with the presence of edges or corners may similarly be ignored.

Another useful macroscopic concept is that of surface stress. This may be defined by considering the difference in the forces acting across a small area normal to the boundary before and after introduction of the boundary. The surface stress in a liquid is a uniform tensile force acting normally to any line in the liquid surface, and numerically equal to the surface free energy of the liquid. This surface tension is just another measure of the tendency of the liquid to attain a minimum area surface, and may be introduced by considering the work done in increasing the area by a small amount. In general, however, the surface stress has three independent components, since solids can withstand elastic shearing forces, and the numerical equivalence of the surface free energy and the surface stress disappears. Indeed, the surface free energy is always positive, but the non-shear component of the surface stress may have either sign. This complication has led to some confusion in the terminology, and particularly in the use of "surface tension". Shuttleworth (1950) defined the surface tension as one half of the sum of the principal surface stresses, by analogy with the three-dimensional definition of hydrostatic pressure. Herring and other workers have continued to use the term surface tension to mean the surface free energy defined for a particular choice of the geometrical interface which gives exactly the classical relation between excess pressure and curvature. To avoid confusion, we shall not refer to surface tension, except for liquids.

The distinction between the surface free energy and the surface stress is that the former measures the work required to create new boundary, while the latter is a measure of the work required to increase the boundary area by deformation, without changing the number of atoms in the boundary. Brooks (1952) has shown that for internal boundaries, there are two surface stress tensors, one for each side of the boundary. Even with an external surface, there are complications arising from situations intermediate between those just envisaged. Thus, in general, the application of a stress will result in changes both in the surface area per atom and in the number of surface atoms, and the work done may be used to define an effective surface stress tensor which is intermediate between the purely elastic and the completely liquid-like cases. Couchman *et al.* (1972) have discussed this complication in some detail. They find it possible to give a generalized discussion of the surface stress in terms of the linear components of the area per surface atom, and thus to avoid non-linear elastic terms which appear when the stress is defined in terms of the components of an area element of surface; this is analogous to the use of Eulerian rather than Lagrangian components in finite elasticity. However, this surface stress is not simply related to any readily defined surface strain, and the condition of zero surface strain can not be unambiguously defined.

Although complete equilibrium can seldom be attained, a polycrystalline mass at high temperatures will be able to reduce its grain boundary energy, and hence its total free energy, by atomic movements over relatively short distances. The grains will usually have originated from random centres, and there will thus be a statistical distribution of different polyhedral crystals. In order that these polyhedra shall completely fill space, certain purely mathematical conditions must be satisfied; when these are combined with the conditions for local surface energy equilibrium, deductions may be made about the (metastable) equilibrium state, towards which the assembly will tend to move. This topological theory has been extensively investigated by C. S. Smith (1952).

It is instructive to consider first a two-dimensional analogy in which we have an assembly of polygons meeting along lines and at corners. We suppose that each separating edge possesses a certain free energy per unit length, and for simplicity we take this to be a constant, independent of the two polygons which it separates. The problem of minimizing the total free energy, subject to the restrictive condition that there are a given number of polygons present, reduces to that of finding the arrangement which gives minimum line (edge) length. Consider first corners which are the junctions of three edges. Then by considering small displacements of the edges, it is at once evident that the energy is minimized if the three edges make equal angles of 120° with each other. Similarly, any four-junction corner will be unstable, since a small re-adjustment will allow it to split into two three-junction corners of 120° as shown in Fig. 8.1. It follows that the arrangement of lowest energy is

FIG. 8.1. Unstable configuration (a) changes rapidly to (b) during two-dimensional growth.

one in which all corners are equi-angled three-edge junctions, and this is possible with straight edges if the whole two-dimensional structure consists of an array of regular hexagons. Such an array would be almost indefinitely metastable, since any displacement of any of the boundaries would result in an increase of energy.

We now suppose that our two-dimensional array has arisen by growth from random centres. This means that inevitably there will be some polygons with more than six edges and some with less. The approach to equilibrium will now take place in stages. Corners formed from four or more edges will still be very unstable and will dissociate into three-edge corners by small boundary displacements. Similarly, all three-edge corners will adjust their angles to the equilibrium 120° array. Since the polygons do not all have six edges, these adjustments mean that many of the edges must become curved. Now while a straight edge is indefinitely metastable in two dimensions, a curved edge is not. Provided material transfer across the boundary is possible, the edge will migrate slowly towards its centre of curvature, decreasing its area, and thus the total energy, as it does so. In a constrained two-dimensional assembly, where the edges meet at corners at 120°, not all of them will be able to migrate towards centres of curvature, but on the average this will be the dominating movement, and will result in the growth of some polygons at the expense of others. In this way, successive polygons may be eliminated entirely from the structure. The important conclusion is that this process will continue indefinitely until the whole assembly consists of polygons separated by straight boundaries passsing from one edge of the assembly to the other, unless, by chance, the stable hexagonal array is achieved. Any actual random arrangement of polygons thus tends to change, not into the metastable hexagonal array, but into a few very large polygons.

The two stages of the above process correspond to the attainment of local equilibrium over small regions of the boundaries, and of general equilibrium over the whole assembly. The whole process is a two-dimensional analogue of the grain growth of a metal, described in Part II, Chapter 19, and corresponds rather closely to actual observations of microstructural changes. However, the polygons in a microsection are sections of three-dimensional polyhedral grains, so we must further consider the equilibrium of a three-dimensional stack of such polyhedra.

We first consider the attainment of local surface energy equilibrium, which may be achieved at sufficiently high temperatures by the migration of atoms over short distances. Suppose the grain boundary surfaces of three grains of fixed orientations meet in a common line. Figure 8.2 shows a section through this line at a point O, where the direction of the line is

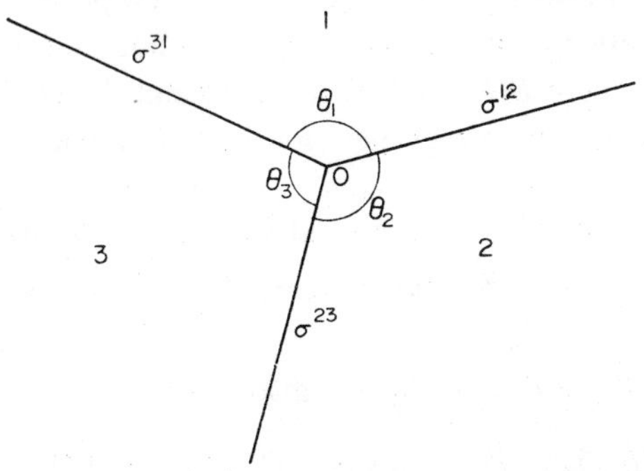

FIG. 8.2. Section through a three-grain junction.

normal to the plane of the paper. We assume that the grain boundary free energies, σ^{12}, σ^{23}, σ^{31} are constants determined by the orientations of the lattices concerned. The configuration attains local equilibrium when a virtual displacement of the boundary gives no first-order change in the free energy, and the condition for this is readily seen to be

$$\frac{\sigma^{23}}{\sin\theta_1} = \frac{\sigma^{31}}{\sin\theta_2} = \frac{\sigma^{12}}{\sin\theta_3}, \qquad (35.1)$$

where θ_i are the dihedral angles. In the special case where the surface stresses contain no shearing components, the surface free energies are numerically equal to tension forces in the boundaries, and the above equation gives the familiar triangle of forces condition. If the surface free energy is independent even of the lattice orientations, the free energies are all equal, and the dihedral angles are all 120°. As we shall see below, this is approximately true for large orientation differences.

The more general problem of the equilibrium of three such grains when the energies σ vary with the position of the boundary has been treated by Herring (1951). Figure 8.3 shows the same three grains as Fig. 8.2, and we consider a change resulting from the

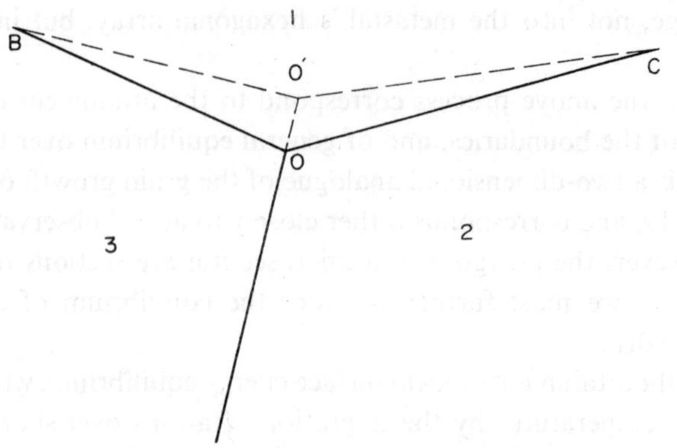

FIG. 8.3. To illustrate Herring's conditions for surface free energy equilibrium.

displacement of the line of intersection through O to a new position O' in the plane of the boundary between orientations 2 and 3. The other two boundaries are supposed to acquire slight kinks at B and C, where BO', CO' are much greater than OO', but still infinitesimal. The first-order change of surface energy per unit length of line normal to the plane of the figure is then

$$\delta G = [\sigma^{23} - \sigma^{31}\cos(\pi - \theta_3) - \sigma^{12}\cos(\pi - \theta_2)]OO' \\ - O'B(\partial\sigma^{31}/\partial\theta_3)\,\delta\theta_3 - O'C(\partial\sigma^{12}/\partial\theta_2)\,\delta\theta_2 .$$

Since $\delta\theta_3 = \widehat{BO'O} - \theta_3$, $\delta\theta_2 = \widehat{CO'O} - \theta_2$, we also have

$$O'B\,\delta\theta_3 = O'O\sin\theta_3, \quad O'C\,\delta\theta_2 = O'O\sin\theta_2$$

and equating $\delta G = 0$,

$$\sigma^{23} + \sigma^{31}\cos\theta_3 + \sigma^{12}\cos\theta_2 - (\sin\theta_3)(\partial\sigma^{31}/\partial\theta_3) - (\sin\theta_2)(\partial\sigma^{12}/\partial\theta_2) = 0. \qquad (35.2)$$

In this equation, the derivatives $\partial\sigma^{31}/\partial\theta_3$, $\partial\sigma^{12}/\partial\theta_2$ are measured in opposite directions of rotation. If derivatives with respect to θ are evaluated in the same sense as the labelling of the phases (in this case in the clockwise direction of rotation), we may write alternatively

$$\sigma^{23} + \sigma^{31}\cos\theta_3 + \sigma^{12}\cos\theta_2 - (\sin\theta_3)(\partial\sigma^{31}/\partial\theta) + (\sin\theta_2)(\partial\sigma^{12}/\partial\theta) = 0. \qquad (35.3)$$

Equations (35.2) or (35.3), together with two similar relations which may be written down by inspection, are Herring's conditions for equilibrium when three grains meet along a line. The set of equations, like the simpler eqns. (35.1), contains only two independent relations. If the boundary free energies are constants, there are only three unknowns, and experimental determinations of the dihedral angles enable the ratios of the three energies to be determined. In the general case, there are five unknown ratios of the energies and their derivatives, so that dihedral angle measurements are unable to establish the grain boundary parameters.

The arguments used for the two-dimensional array may now be extended. Again, a random array will never achieve a metastable equilibrium, since the requirements of local equilibrium will inevitably result in grains with curved faces. In principle, such grains will migrate under surface energy forces until only a few large grains are left, although the variation of boundary energy with orientation may introduce some modifications in the simple picture. Thus a few boundaries (coherent twin boundaries) may have energies which vary so rapidly with orientation of the boundary surface that virtually no movement of the surface is possible, and these will be frozen into the structure indefinitely. Most boundaries, however, may be supposed to become mobile at sufficiently high temperatures, so that grain growth should always occur.

In two dimensions, there is a theoretical possibility of achieving a stable array by regular packing. This is no longer true in three dimensions. In the simplest example (constant grain boundary energy), the requirements of local surface energy equilibrium are that all surfaces should meet each other in groups of three at angles of 120° along lines which themselves meet in groups of four at angles of $\cos^{-1}(-\frac{1}{3}) = 109°28'$. As in two dimensions, junctions at which four or more grains meet along a line will rapidly separate into two junctions of three grains. However, there is no regular polyhedron with plane faces having edges meeting at the correct angle, and although a pentagonal dodecahedron is a close approximation, this figure lacks the symmetry required to form the basis of a repeating pattern filling all space. The problem was considered long ago by Lord Kelvin, who showed that if a b.c.c. stack of regular tetrakaidecahedra (truncated octahedra), shown in Fig. 8.4, is modified

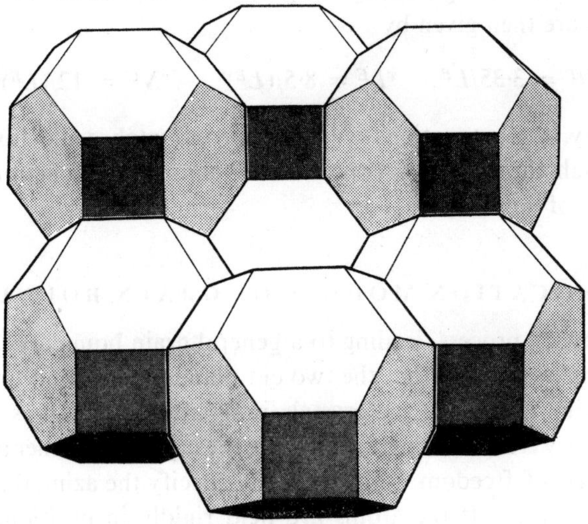

FIG. 8.4. A stack of regular tetrakaidecahedra (after Smith, 1952).

by introducing double curvatures into the hexagonal faces, the local surface tension requirements can all be satisfied. The curvatures, as we have pointed out, prevent the array from achieving a stable configuration.

Attempts to find an equilibrium stack require a combination of surface energy conditions with the purely mathematical conditions which ensure that the stack shall fill all space. In two dimensions, the maximum number of corners per polygon cannot exceed two. An array of uniform hexagons, which possesses this maximum sharing of corners, also gives the shortest edge length for a given number of cells. Attempts were made by Smith (1952) to invoke similar topological principles for three-dimensional arrays. He stated that the maximum number of corners per polyhedron in a space filling array is six, and that minimum interfacial area occurs when there is a maximum sharing of faces and corners. These statements were retracted in a later paper (1953), where processes which lead to an apparently limitless increase in the number of corners, edges and faces per polyhedron were described. This later paper also withdraws a former "proof" that for minimum total interfacial area, the average number of corners on each face is $5\frac{1}{7}$, although reasons for believing that this statement is correct are given. In a later note, Meijering (1953) showed that the number of corners per polyhedron can exceed six, even when all faces are plane. Purely topological arguments thus do not seem to lead to the same sort of useful result in three dimensions as they do in two dimensions, although many interesting relations are pointed out in Smith's papers.

In discussing problems which depend upon grain size, it is often useful to have expressions for the mean boundary area, mean edge length, and mean number of corners per unit volume in terms of the mean grain diameter. Probably the most realistic simple assumption is to consider all grains to be equal tetrakaidecahedra, as above. If the separation of square faces is L^B, the edge length is $L^B/2\sqrt{2}$ and the number of grains per unit volume is $2/(L^B)^3$. The grain boundary area, ${}^vO^B$, grain edge length, ${}^vL^E$, and number of grain corners, ${}^vN^C$, all per unit volume, are then given by

$$ {}^vO^B = 3{\cdot}35/L^B, \quad {}^vL^E = 8{\cdot}5/(L^B)^2, \quad {}^vN^C = 12/(L^B)^3. \tag{35.4} $$

In a real assembly, these equations will be approximately valid for some mean grain diameter L^B. Although the numerical constants will change, the functional dependence will be correct for grains of any shape.

36. DISLOCATION MODELS OF GRAIN BOUNDARIES

Consider the following process, leading to a general grain boundary. Two pieces of crystal are each cut along a plane, so that the two cut planes correspond to the orientations of the required boundary relative to the crystals it separates. This uses up four degrees of freedom. The boundary is now made by cementing the crystals together along the cut planes, the remaining degree of freedom being used to specify the azimuthal orientation about the normal to these planes. If the atoms are held rigidly in position, it is obvious that the atomic bonding across the boundary will be quite different from that in the interior of the crystals. The changes in configuration and in interatomic distances will result in the boundary energy per atom being much larger than that in the interior of the crystals, and the whole of this extra misfit energy may be considered to be localized in the immediate vicinity of the interface. In many cases, however, the misfit may be reduced considerably by readjustment in the positions of the atoms near the boundary. These readjustments, in turn, disturb the ideal arrangement in regions a little further from the boundary, and thus cause further small atomic displacements. The displacements evidently fall off rapidly with distance from the boundary, and at sufficient distances are equivalent to elastic strains of the two lattices. The boundary energy now contains two terms, a misfit energy localized at the boundary, and an elastic energy which in principle extends throughout the crystal but which is largely concentrated in a small volume near the boundary. This illustrates the importance of the general definition of boundary energy given in the last section.

The actual atomic configuration and the relative importance of the two terms will be such as to minimize the total boundary free energy. In general, the greater the amount of misfit when rigid structures are joined, the less effective will be readjustments of atomic position in lowering the energy. Thus grain boundaries separating lattices with large relative orientation have high energies localized at the interface, whereas boundaries separating lattices of nearly the same orientation have lower energies which are mainly spread through a larger volume. We have the apparently anomalous result that the smaller the misorientation, the greater is the distance over which atomic adjustments are made, and hence the "wider" is the boundary. For a small misorientation, the boundary region does not correspond to uniform disregistry of the atoms on the two sides, but is rather to be pictured as a surface over which most atoms are in almost perfect registry, and a few atoms are not in registry. This is a special example of a type of boundary in the solid state which we shall describe as semi-coherent. The centres of strain are dislocations, and a surface array of dislocation lines forms a suitable model for such a boundary.

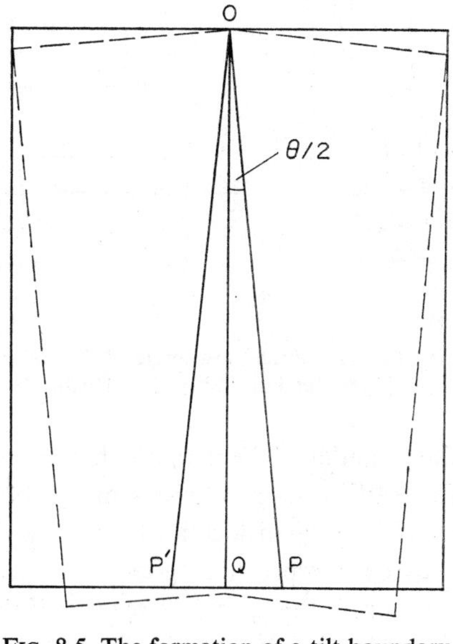

FIG. 8.5. The formation of a tilt boundary.

In the simplest type of interface, the two lattices are connected by a rotation about an axis lying in the boundary. This may be constructed macroscopically by removing a wedge of material *POP′* (Fig. 8.5) from a perfect crystal and rejoining along *OP*, *OP′*. When $\widehat{OPP'} = \widehat{OP'P}$, the joining surface is equally inclined to the two lattices, and we have a symmetrical tilt boundary. A dislocation model for this configuration was first suggested by Burgers (1939) and Bragg (1940). On an atomic scale, the surfaces *OP*, *OP'* are not plane but are stepped, as shown in Fig. 8.6(a). Each step corresponds to the ending of a plane of atoms, or more generally to a lattice plane, now running towards the boundary from left or right, but originally parallel to the x_2 axis. By elastic distortion of the crystals, the two surfaces may be fitted together as shown in Fig. 8.6(b). Most atoms are then in registry except near the steps. As the steps are the edges of incomplete atomic planes, they may be considered

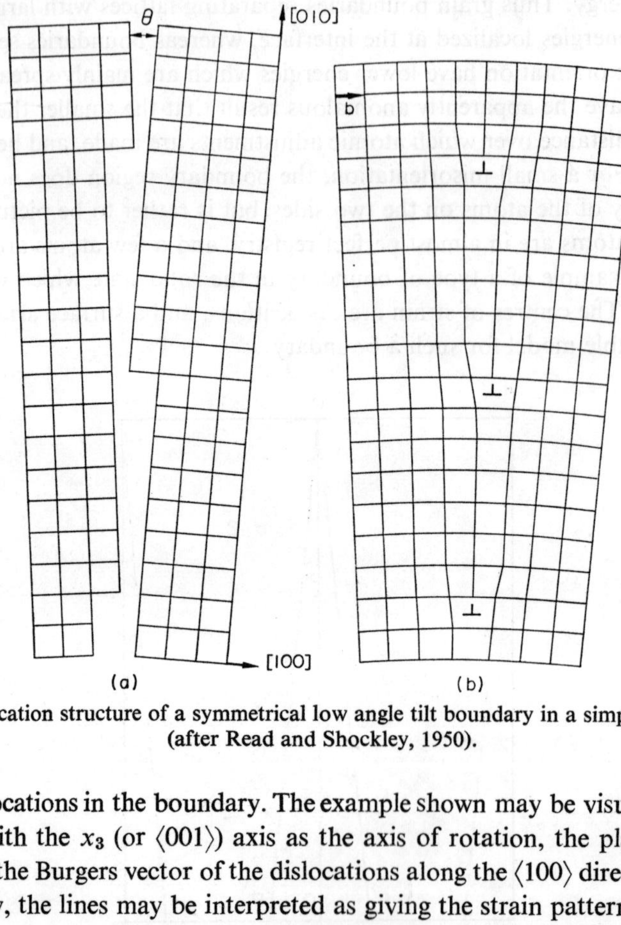

FIG. 8.6. Dislocation structure of a symmetrical low angle tilt boundary in a simple cubic structure (after Read and Shockley, 1950).

to be edge dislocations in the boundary. The example shown may be visualized as a simple cubic lattice, with the x_3 (or $\langle 001 \rangle$) axis as the axis of rotation, the plane $x_1 = 0$ as the boundary, and the Burgers vector of the dislocations along the $\langle 100 \rangle$ direction of one of the lattices. Equally, the lines may be interpreted as giving the strain pattern of a symmetrical tilt boundary in any structure. From the figure, we see that if the spacing of the dislocation lines (measured along the boundary) is Y, and the Burgers vector of each is **b**, the total rotation between the lattices is

$$\theta = 2 \sin^{-1}(b/2Y) \approx b/Y, \tag{36.1}$$

where the approximation is valid only for small tilts.

We notice that although the steps were assumed to belong to one or other half-crystal, the edge dislocations are present in the boundary itself rather than in one or other lattice. The quantity b in eqn. (36.1) is the magnitude of the Burgers vector defined in a reference crystal in the usual manner, and the equation is independent of the orientation of this reference crystal. Dislocations in boundaries clearly differ in some important respects from dislocations in the interior of crystals, one obvious difference being that the Burgers vector of a boundary dislocation is not fixed in space. In discussing the definition of the Burgers vector, we have always tacitly assumed that the reference crystal has the same mean orientation as the real crystal. For a dislocation in a boundary, we may choose the reference crystal to correspond in orientation to either of the real crystals, or to some intermediate situation. In principle, the boundary dislocation may move into either lattice, the Burgers vector appropriate to this movement being defined by using a reference crystal in the orientation of the lattice into which it moves. The local Burgers vector suffers a sharp discontinuity as we pass through the boundary. These geometrical relations are obvious, but are emphasized here because care is sometimes needed before dislocations in boundaries can be assigned the properties normally associated with lattice dislocations.

Equation (36.1) may also be derived from Fig. 8.5 since PP' is the total Burgers vector of the dislocations in a length OP of boundary. The equation is sometimes presented in the alternative form

$$\theta = 2\tan^{-1}(b/2Y'), \tag{36.1a}$$

where Y' is the dislocation spacing measured normal to the slip planes in one of the crystal lattices, i.e. parallel to OQ in Fig. 8.5 rather than to the interface OP. Although the dislocations are equally spaced in Fig. 8.6, this cannot be arranged for arbitrary θ since the actual separation of dislocations parallel to OQ must be an integral multiple of the interplanar spacing d. There is thus generally some variability in the individual separations, and eqns. (36.1) and (36.1a) refer to mean values Y, Y' of these separations. Uniform spacing is possible only at discrete values of θ, where $(2\tan\frac{1}{2}\theta)^{-1}$ is an integral multiple of d/b, or for small misorientations when $\theta = b/nd$, where n is an integer. According to the strict definition of a twin, all uniform spacings correspond to coherent twin boundaries inasmuch as the boundary plane must be rational and the lattices on each side are mirror images (Brooks, 1952). However, the low angle boundaries of this type have very high indices, and would not be classed as twins in normal circumstances. The more important twin boundaries, with low indices, correspond to small integral values of Y'/d; they are often represented as coincidence site boundaries (see below).

A tilt boundary may be formed in an originally single but imperfect crystal if a number of edge dislocations of the same sign move by glide and/or climb processes until they are aligned parallel to each other and normal to the slip planes. As each dislocation moves, the lattice and shape deformations (p. 320) change, and when the dislocations are aligned, the lattice dislocations have added together so as to produce a small rotation. We recall that an isolated edge dislocation in a finite crystal produces a rotation of the lattice planes (p. 272), this being the tilt boundary of minimum angle. The shape deformation produced by the lining up of the dislocations is discussed below; the array of edge dislocations provides one of the simplest examples of the compatibility conditions discussed on p. 319.

Once formed, the tilt boundary is stable against dissociation by loss of individual dislocations into the crystals on each side. This is because the force between parallel dislocations on parallel, non-identical, glide planes pulls them into alignment (eqn. 30.26). At distances from the boundary greater than the mean dislocation separation, the effects of the dislocations will be nearly identical with those produced by removing a uniform wedge. The stress field of the dislocation array thus extends only for an effective distance of the order of Y into the crystals, since the displacements cancel at larger distances. The expression for the energy of each dislocation in a small angle boundary will thus contain a term in ln Y to replace the ln r_e of eqn. (30.15). Note that the smaller the misorientation the greater the extent of the elastic field, as we remarked above. However, when the dislocation spacing is non-uniform, there are additional, longe-range terms in the expression for the elastic energy.

Equation (36.1) in its exact form is valid in principle even for large misorientations. As θ increases, the dislocations become closer together, until eventually Y is of the order of the core radius, r_i. It is clear that the boundary must be considered as a whole for separations much smaller than this, and its analysis in terms of individual dislocations has only formal significance. The dislocation interpretation is nevertheless still extremely useful in considering compatibility relations, since continuous distributions may be assumed; the physical structure of such boundaries is considered later in this section. Lomer and Nye (1952) used the bubble model to study the change in boundary structure as θ increases; this is of value since the model gives a reasonable approximation to the interatomic forces in a metal like copper. They found that when the spacing of the dislocations becomes of the order of their original widths (see p. 272), this width begins to decrease. Eventually, the dislocations coalesce, so that the defects resemble closely spaced vacancies, and the whole boundary is incoherent.

The symmetrical tilt boundary has only one degree of freedom. An extra degree of freedom is obtained if we allow the boundary to rotate about the direction common to the two lattices, i.e. if in Fig. 8.5, $\widehat{OPP'} \neq \widehat{OP'P}$. The crystal spacings along the x_2 direction are no longer equal, and some of the {100} planes must also end in the boundary. The asymmetrical tilt boundary thus contains dislocations of two types, as shown in Fig. 8.7. The angles which the boundary plane makes with the [010] directions are now $\frac{1}{2}\theta+\varphi$ and $\frac{1}{2}\theta-\varphi$, where $\varphi = 0$ for the symmetrical boundary. By considering the number of {010} planes which end on unit length of the boundary from the two sides, it is readily seen that for small misorientations the spacing of [100] dislocations is given by

$$Y_{100} = b/\theta \cos \varphi. \tag{36.2}$$

Similarly, the spacing of the [010] dislocations is obtained by finding the number of {100} planes which approach the boundary from the two sides, and is

$$Y_{010} = b/\theta \sin \varphi. \tag{36.3}$$

When $\varphi = 0$, only the [100] set are present; when $\varphi = 90°$, only the [010] set. The latter case represents a symmetrical tilt boundary with axis of rotation again [001], but the boundary surface perpendicular to [010].

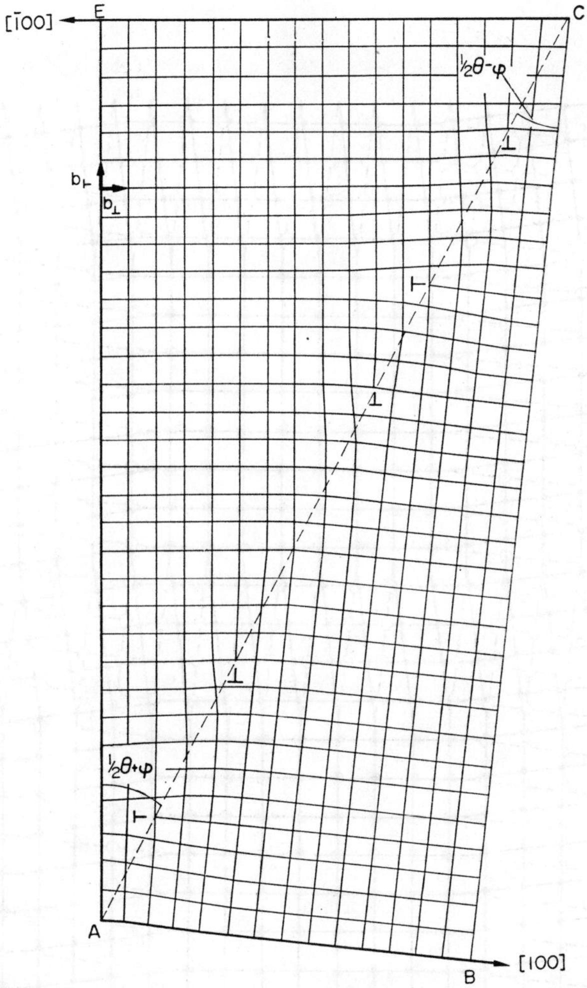

FIG. 8.7. Structure of an asymmetrical tilt boundary (after Read, 1953).

In the examples we have been considering, the tilt boundaries have been constructed from edge dislocations alone. It will only be possible to do this with allowable lattice dislocations, however, when the axis of rotation is perpendicular to a possible Burgers vector, or to two such vectors for an asymmetric boundary. In other cases, the boundary could be constructed of straight dislocation lines of mixed edge and screw type parallel to the axis of rotation. The Burgers vectors of these dislocations would have to be such that the edge components are equal and give the required tilt, while the screw components alternate in sign, and do not contribute to the misorientation. The existence of the screw component does, however, increase the boundary free energy.

We next consider how further degrees of freedom in the relative orientation of the two lattices may be obtained. In Fig. 8.6 we may give the two crystals an independent rotation about the x_2 axis by inserting a second set of dislocations, perpendicular to the first, but

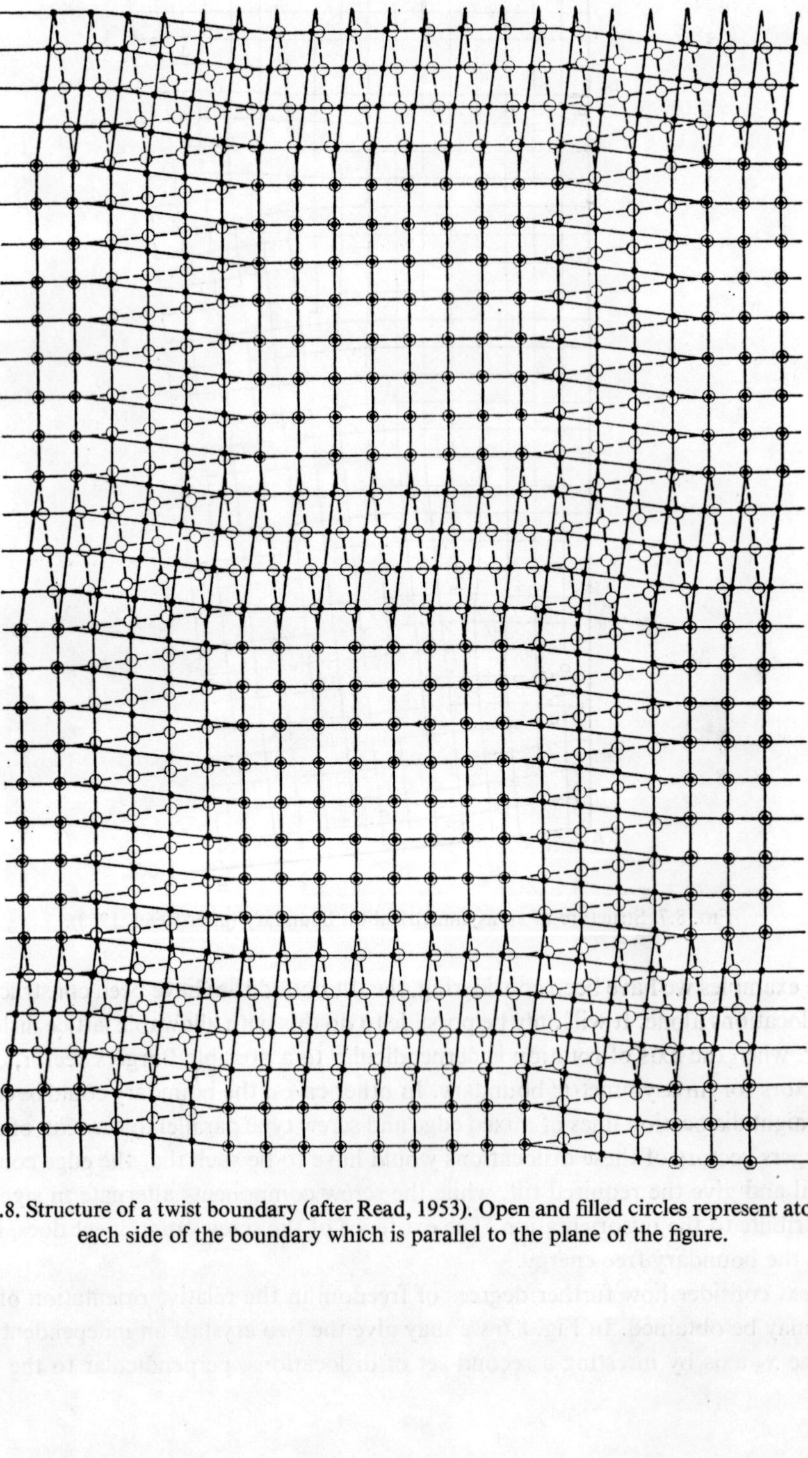

FIG. 8.8. Structure of a twist boundary (after Read, 1953). Open and filled circles represent atoms on each side of the boundary which is parallel to the plane of the figure.

having the same Burgers vector. For the most general relation, we also require a rotation about x_1, i.e. about an axis perpendicular to the boundary, and we shall now show that this may be achieved by arrays of screw dislocations. When the only misorientation is a relative rotation about this normal, we have a simple twist boundary.

In the simple tilt boundary, atomic planes from the two crystals meet the boundary along parallel lines but with different densities. In a twist boundary, the densities of atomic planes are equal on both sides, but corresponding planes do not meet in parallel directions. The function of the screw dislocations is to provide this change of direction whilst allowing most atoms at the interface to remain in registry. In Fig. 8.8 we show a {100} twist boundary in a simple cubic structure. The boundary is in the plane of the figure, open circles representing the atoms on one side of the boundary, and filled circles the atoms on the other side. The [010] and [001] rows of atoms have different mean directions on the two sides of the boundary, but the introduction of the screw dislocations enables these rows at the interface to remain parallel over most of their lengths. The relative rotation of the [010] rows, for example, is corrected by a set of parallel screw dislocations of spacing $Y = \frac{1}{2}b \operatorname{cosec}(\theta/2)$, where θ is the angle of twist between the two lattices. This set of screw dislocations alone would produce a long-range shear in the lattices, and the [001] rows on the two sides of the boundary would still not be parallel. The insertion of a second set of screw dislocations, perpendicular to the first and having the same spacing, brings the [001] rows into coincidence over most of the boundary, and removes the long-range shear. The stress field now extends only for a distance of order Y from the boundary, and the [100] rows of atoms are continuous across the interface (except near the dislocations). A crossed grid of screw dislocations thus gives a stable twist boundary, the misorientation again being given by eqn. (36.1).

Twist boundaries can only be made from pure screw dislocations if the boundary surface contains two perpendicular directions corresponding to allowable Burgers vectors, as does, for example, a f.c.c. {100} boundary. In general, dislocations of mixed type must be used, the edge components being arranged to cancel one another. In the f.c.c. structure, a simple hexagonal net of slip dislocations will give a twist boundary on a {111} plane (Frank, 1955). There are thus many possibilities for low angle dislocation boundaries in real structures, but most of them are based on simple rectangular or hexagonal nets. Slight deviations in the orientation of the boundary plane, or of the orientation relations from the simplest types, are then achieved by having local "faults" in the mesh.

The simple boundaries are specified by only one or two parameters, and it is usually easy to find the correct dislocation description. Models of more general boundaries will clearly involve rather complex arrays of dislocations of several Burgers vectors. Even for small misorientations, the representation of such a boundary by a particular array is not unique, and it is not immediately obvious which of the possible descriptions best corresponds to the actual configuration. Experimental studies of the dislocation structure of many low angle boundaries have been made, using precipitation techniques in transparent crystals, or thin film electron microscopy in metals. The boundaries usually approximate to the simple symmetrical types (for details see, e.g., Frank, 1955; Amelinckx, 1957, *et al.* 1958; Carrington, 1960), but we shall not discuss then here.

For a general boundary we may specify a resultant or net dislocation content, as first shown by Frank (1950). The following method for finding this net content is equivalent to that originally given by Frank, except that we use rotation matrices instead of vector notation. As we shall see in the next section, Frank's formula is a particular case of a more general expression for the dislocation content of a boundary between two crystals of different structure and orientation.

Consider a reference lattice with origin O. Two crystals of different orientations may be constructed from this reference lattice by giving it arbitrary rotations represented by $\mathbf{R}_+$ and $\mathbf{R}_-$ respectively. We consider these two crystals to meet along a boundary surface with normal $\boldsymbol{\nu}$, and we now show how this configuration can be achieved by introducing dislocations into the surface $\boldsymbol{\nu}$.

Let $OP = \mathbf{p}$ be a large vector in the boundary, and consider a right-handed Burgers circuit PA_+OA_-P (Fig. 8.9(a)) where A_+ is any point of the +lattice and A_- is any point

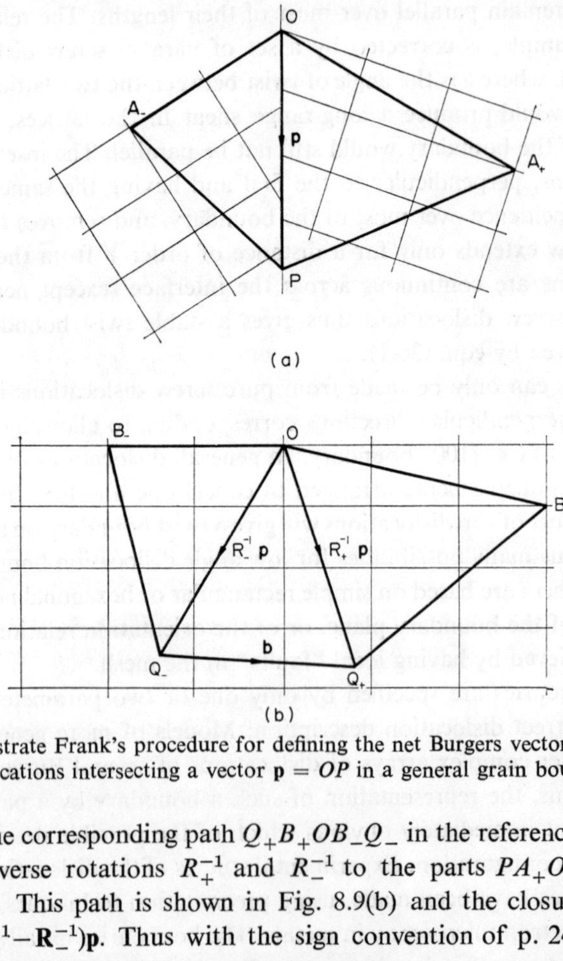

FIG. 8.9. To illustrate Frank's procedure for defining the net Burgers vector of the boundary dislocations intersecting a vector $\mathbf{p} = OP$ in a general grain boundary.

of the $-$ lattice. The corresponding path $Q_+B_+OB_-Q_-$ in the reference lattice is obtained by applying the inverse rotations R_+^{-1} and R_-^{-1} to the parts PA_+O and OA_-P of the circuit respectively. This path is shown in Fig. 8.9(b) and the closure failure $Q_-Q_+ = OQ_+ - OQ_- = (\mathbf{R}_+^{-1} - \mathbf{R}_-^{-1})\mathbf{p}$. Thus with the sign convention of p. 243, the net Burgers vector is given by

$$\mathbf{b} = (\mathbf{R}_+^{-1} - \mathbf{R}_-^{-1})\mathbf{p}. \tag{36.4}$$

Equation (36.4) gives the total Burgers vector of the dislocations in the surface crossing any vector **p** in the surface, referred to the reference lattice specified above. Frank gave the equivalent vector expression, taking $\mathbf{R}_+$ to represent a rotation of θ_+ about a unit vector $\mathbf{l}_+$, and $\mathbf{R}_-$ to represent a rotation of θ_- about a unit vector $\mathbf{l}_-$. This is

$$\mathbf{b} = (\cos\theta_- - \cos\theta_+)\mathbf{p} + \mathbf{p}\wedge(\mathbf{l}_+ \sin\theta_+ - \mathbf{l}_- \sin\theta_-)$$
$$+(\mathbf{l}_+\cdot\mathbf{p})(1-\cos\theta_+)\,\mathbf{l}_+ - (\mathbf{l}_-\cdot\mathbf{p})(1-\cos\theta_-)\mathbf{l}_-. \qquad (36.5)$$

It will be noted that the procedure we have used to deduce eqn. (36.4) is essentially identical with that used on p. 316 for finding the dislocation tensor associated with a continuous distribution of dislocations. We may obtain eqn. (36.4) in fact by assuming that we have a continuous distribution of dislocations confined to a thin sheet (Bilby, 1955), as discussed in Section 38.

The net dislocation content is a geometrical property of the boundary considered, but it is not a unique description since the relations between the two lattices can be expressed in an infinity of different ways, each of which leads to a different **b**. This problem is considered below; for the present we note that for low angle boundaries a particular description has physical significance and **b** in eqn. (36.4) may then be considered to be the sum of the contributions from the individual dislocations forming the boundary. We have seen that in three dimensions the appropriate dislocation density is given by the number of dislocations of any type crossing a unit area of any orientation. Similarly, in a two-dimensional distribution an appropriate density is the number of dislocations crossing unit length of line of any orientation in the boundary. Suppose now that **p** is a unit vector. Then if n_i dislocations, each having Burgers vector $\mathbf{b}_i$, cross **p**, the net Burgers vector **b** is given by the sum

$$\mathbf{b} = n_i\mathbf{b}_i. \qquad (36.6)$$

The number of possible terms on the right-hand side of eqn. (36.6) is restricted to the number of stable dislocations of the lattice, at least in cases where a sub-boundary is formed physically by amalgamation of lattice dislocations. If there are only three non-coplanar Burgers vectors to be considered, the resolution of **b** into its components is unique. When, as is more usual, there are more than three Burgers vectors of lowest energy, **b** may be compounded into component dislocations in several ways. Since **b** is directly proportional to **p**, and $\mathbf{b}_i$ is fixed, n_i must be proportional to $|\mathbf{p}|$. The dislocations of type i are thus (statistically) uniformly spaced along any direction **p** in the boundary, and must all be straight lines. Determination of n_i for two different boundary directions uniquely specifies the direction and spacing of the dislocation lines with Burgers vector $\mathbf{b}_i$. Thus any grain boundary may be defined in a reference lattice by arrays of dislocation lines, having as a minimum three different sets of non-coplanar vectors. When several different models are possible, they represent different ways of making the atomic transition from one grain to the other compatible, and they are no longer distinguishable when the dislocation distribution is regarded as continuous. Nevertheless, the lowest energy will generally correspond to the model in which the (discrete) dislocation density is lowest. The above remarks apply in principle to all boundaries, but as already remarked, the dislocation model begins to lose its physical significance when the relative orientations of the grains becomes large.

We have not said anything about the choice of reference lattice orientation in which the Burgers vectors are defined. The most convenient reference lattice is usually in an orientation equivalent to that of one of the crystals, or else is the median lattice from which the two real lattices are obtained by equal and opposite rotations. The first choice is particularly useful when we consider the motion of an interface by glide motions of the dislocations, as in the theory of martensite or the stress induced migration of a symmetrical low angle tilt boundary. For example, if we use the (−) lattice as the reference crystal, we have, from (36.4),

$$\mathbf{b} = (\mathbf{R}^{-1} - \mathbf{I})\mathbf{p}, \tag{36.7}$$

where **R** is the rotation giving the relative orientations of the two crystals.

When the median lattice is used as the reference lattice, the dislocation density takes a particularly simple form. The two real lattices are now generated by rotations $\theta_+ = -\theta_- = \frac{1}{2}\theta$ about a common direction **l**, and eqn. (36.5) becomes

$$\mathbf{b} = 2 \sin \tfrac{1}{2}\theta \, \mathbf{p} \wedge \mathbf{l}. \tag{36.8}$$

The more general expression (36.4) or (36.5) need be retained only when it is desired to have a common reference lattice for a number of different boundaries. This occasionally has some advantages; for example, the total Burgers vectors are conserved at a "surface node" where three or more boundaries meet.

We now consider the atomic structures of high-angle grain boundaries where the above dislocation analysis has only formal significance and the misfit is concentrated mainly in the immediate vicinity of the boundary, which has a width of only a few atoms. One possible assumption is that there is practically no correlation of atomic positions across the boundary, and that the positions of the atoms in the boundary region may only be described statistically as in a two-dimensional disordered or liquid-like array. However, there is now strong evidence, both from structural studies and from measurements of physical properties such as grain boundary diffusion, that high-angle boundaries have some form of periodic structure. An early suggestion of this type (Mott, 1948) was that the boundary structure consists of "islands" of good atomic matching, separated by regions where the matching is poor. The islands are small groups of atoms, and the disordered regions have a structure rather like that of a liquid. This is similar to a dislocation model except that the bad matching regions cannot be pictured as isolated dislocations. As the misorientation of the grains for a boundary of given type is increased, the structure of the boundary might be expected to change from the low-angle dislocation model to Mott's model, and eventually to a completely disordered boundary. Such a proposal was actually made by Smoluchowski (1952) for tilt boundaries in connection with some early results on grain boundary diffusion. The misfit regions were assumed to be formed originally by the coalescence of groups of individual regions when the tilt angle exceeds $\sim 15°$.

Smoluchowski's proposal is similar to some recent descriptions of grain boundaries in terms of disclinations rather than dislocations (e.g. Li, 1972). As pointed out in Section 30, a disclination of arbitrary strength is topologically equivalent to the internal termination of a grain boundary, and in particular a wedge disclination represents the termination of a tilt boundary. It follows that a wedge disclination dipole is formally equivalent to an edge dislocation when the separation of the opposite disclinations is of the order of the atomic distance, and to a finite wall of edge dislocations when the separation is larger. Now consider a symmetrical tilt boundary formed by edge dislocations and suppose that as the tilt angle and hence the density of dislocations increase the dislocations begin to cluster into closely spaced walls separated by larger distances. When the walls become very dense, the individual dislocations lose their identities and the walls may be considered to have condensed into disclination dipoles (Mott's bad regions) separated by regions of good fit. The relation between the usual dislocation model and this disclination model is shown in Fig. 8.10. In the figure, each disclination dipole replaces a single dislocation, so that the

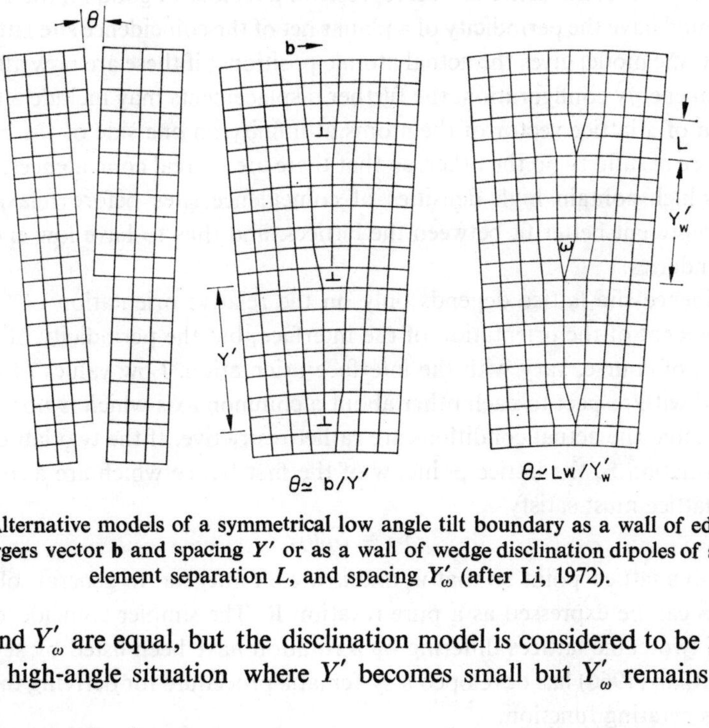

FIG. 8.10. Alternative models of a symmetrical low angle tilt boundary as a wall of edge dislocations of Burgers vector b and spacing Y' or as a wall of wedge disclination dipoles of strength ω, element separation L, and spacing Y'_ω (after Li, 1972).

spacings Y' and Y'_ω are equal, but the disclination model is considered to be more appropriate to the high-angle situation where Y' becomes small but Y'_ω remains appreciably larger than b.

Other recent models of grain boundary structure also embody Mott's concept, but are differently formulated. These theories arose from the discovery that certain high-angle boundaries have special properties which suggest that their structures are simpler and their energies lower than random boundaries. The most obvious example of a special boundary is a coherent twin interface, but there are many other boundaries of unusually good atomic fit. These boundaries appear to have higher mobilities than random boundaries, at least when impurities are present in solution, as will be fully discussed in Part II, Chapter 19. In discussing results on secondary recrystallization, Kronberg and Wilson (1949) plotted the positions of the atoms in the two grains that meet at a boundary of high mobility. They found that if the lattices are assumed to interpenetrate, an appreciable fraction of the lattice sites are common to both structures, and thus form a "coincidence site lattice". The concept of the coincidence lattice has had an important influence in the development of theories of high-angle boundaries.

Let two lattices interpenetrate to fill all space, and assume that there is a common lattice point. In general, there may then be other lattice points which coincide, and the set of such points forms the coincidence site lattice. In some special situations, the reciprocal density Σ of coincidence sites relative to ordinary lattice sites may be small, but for an arbitrary orientation Σ may tend to infinity and there is then effectively no coincidence site lattice. A model of a grain boundary is introduced by fixing the interface somewhere in the space and placing atoms on the appropriate sites of one lattice on one side of the interface and of the other lattice on the opposite side of the interface. If this boundary contains a high density of sites of the coincidence lattice, representing regions of good fit, the structure of the boundary should have the periodicity of a planar net of the coincidence site lattice. This does not mean that the model gives the actual atomic positions; if these are now allowed to relax to a minimum energy configuration, the further displacements may include a translation by some fraction of a lattice vector of the atoms at infinity on one side of the boundary with respect to those at infinity on the other, so that there are no real coincidences. Nevertheless, boundaries which contain high densities of coincidence sites before relaxation may be expected to represent better fit between the lattices, and thus to have low energies relative to other boundaries.

The coincidence site lattice depends only on the relative orientation of the two grains and is independent of the orientation of the interface, but the periodicity of the boundary structure will, of course, vary with the interface orientation. Low values of Σ result from grains rotated with respect to each other about a common axis which is normal to a close-packed plane, and the actual conditions are rather restrictive. If the two lattices are related by a transformation **S**, the lattice points **w** of the first lattice which are also points of the coincidence lattice must satisfy

$$w_i = S_{ij}u_j, \tag{36.9}$$

where **u** is also a lattice point so that w_i and u_i are all integers. In general solutions do not exist unless **S** can be expressed as a pure rotation **R**. The simpler coincidence site lattices formed from two cubic lattices differing by a rotation have been listed by several authors, and Ranganathan (1966) has developed a systematic procedure for deriving these lattices by means of a generating function.

Consider a planar rectangular net of axial ratio γ which is rotated about the normal to the plane which passes through a lattice point of the net. If the net is used to define an orthogonal coordinate system, each rotation of a point with coordinates $x, -y$ into a new position x, y will give rise to a coincidence lattice formed from the original and rotated nets. The angle of this rotation is

$$\theta = 2\tan^{-1}(\gamma y/x) \tag{36.10}$$

and the area ratio of the two-dimensional cell of the coincidence and original lattices is

$$\Sigma' = x^2+\gamma^2 y^2. \tag{36.11}$$

The lattice sites in any plane of a cubic crystal may be represented as a rectangular net of axial ratio $\gamma = (h_i h_i)^{1/2}$, or as a series of interpenetrating nets of this type, and it follows that the above function (36.11) can be used to generate the coincidence site lattices for rotations about the normal to the plane **h**. However, it may be necessary to verify by inspection that all lattice sites are brought into coincidence by this operation, since otherwise eqn. (36.11) gives a sub-multiple of the correct reciprocal density of coincidence sites, and also that there are no additional coincidences, in which case it gives a multiple. Additional coincidences always occur if Σ' is even, and the true density of coincidence sites is then twice, or more generally 2^n times, that given by eqn. (36.11). Otherwise, Ranganathan states that the factors leading to the above possibilities tend to cancel, so that the equation usually gives the true multiplicity. Thus all that is necessary in most cases is to assign relatively prime, integral values of x and y in the generating function $x^2 + h_i h_i y^2$ and to divide even values of Σ' repeatedly by two until an odd value corresponding to Σ is obtained.

Obviously when some coincidence lattice relations are known, others may be generated by combining these together. Thus with two rotations θ_1 and θ_2 about the same axis corresponding to relatively prime reciprocal densities Σ_1 and Σ_2, a new lattice of reciprocal density $\Sigma_1\Sigma_2$ will represent rotations of $\theta_1 \pm \theta_2$ about this axis. It should also be noted that any orientation represented by an axis-angle pair **l**, θ can be represented in a number of different ways by making use of the symmetry properties of the lattice. Thus for any rotation matrix R there are in the cubic system up to 23 non-equivalent alternative matrices

$$\mathsf{R} = \mathsf{U}_i \mathsf{R} \mathsf{U}_j, \qquad (36.12)$$

where U_i, U_j are orthogonal matrices corresponding to any of the 24 possible symmetry operations of the point group 432. Suppose R is known for one representation of a coincidence lattice with reciprocal density Σ. Then it follows that R may be written in the form

$$\mathsf{R} = \frac{1}{\Sigma}\begin{pmatrix} u_1 & v_1 & w_1 \\ u_2 & v_2 & w_2 \\ u_3 & v_3 & w_3 \end{pmatrix}, \qquad (36.13)$$

where the matrix elements u_i, v_i, w_i are all integers and have no common factor with Σ. The vectors **u**, **v**, **w** give the directions into which the base vectors of the first lattice are rotated by R. It may now be seen that R represents a rotation about a unit vector **l** through an angle θ where

$$l_1 : l_2 : l_3 : = w_2 - v_3 : u_3 - w_1 : v_1 - u_2 \qquad (36.14)$$

and

$$\cos\theta = (u_1 + v_2 + w_3 - \Sigma)/2\Sigma. \qquad (36.15)$$

Warrington and Bufalini (1971) have used eqns. (36.12)–(36.15) to give a computer printout of the main coincidence lattices, their possible **l**, θ representations, and the corresponding matrices R.

Table VII lists some of the orientation relations which give the highest density of coincidence sites in cubic crystals; $\Sigma = 1$ implies $\theta = 0°$, so that the lowest value of Σ is 3 and this corresponds to the simplest type of twin relation. In general an arbitrary boundary

TABLE VII. SOME COINCIDENCE SITE RELATIONS IN CUBIC CRYSTALS

Reciprocal density of coincidence sites. Σ	Minimum angle and axis of rotation. $\theta°$	Minimum angle and axis of rotation. **l**	Twin axes	Most densely packed planes of coincidence site lattice b.c.c.	Most densely packed planes of coincidence site lattice f.c.c.	Area per lattice point in coincidence in units of b^2 b.c.c.	Area per lattice point in coincidence in units of b^2 f.c.c.
3	60·0	⟨111⟩	⟨111⟩ ⟨112⟩	{112}	{111}	1·6	0·87
5	36·9	⟨100⟩	⟨012⟩ ⟨013⟩	{013}	{012}	2·1	2·25
7	38·2	⟨111⟩	⟨123⟩	{123}	{135}	2·5	2·95
9	38·9	⟨110⟩	⟨122⟩ ⟨114⟩	{114}	{115}	2·9	2·6
					{111} (3)†		2·6
11	50·5	⟨110⟩	⟨113⟩ ⟨233⟩	{233}	{113}	3·1	1·65
13a	22·6	⟨100⟩	⟨023⟩ ⟨015⟩	{015}	{023}	3·4	3·6
13b	27·8	⟨110⟩	⟨134⟩	{134}	{139}	3·4	4·8
15	48·2	⟨210⟩	⟨125⟩	{125}	{157}	3·55	4·3
					{111} (5)†		4·3
17a	28·1	⟨100⟩	⟨014⟩ ⟨035⟩	{035}	{014}	3·9	4·1
17b	61·9	⟨221⟩	⟨223⟩ ⟨334⟩	{334}	{155}	3·9	3·6
19a	26·5	⟨110⟩	⟨133⟩ ⟨116⟩	{116}	{133}	4·1	2·2
19b	46·8	⟨111⟩	⟨235⟩	{235}	{235}	4·1	6·15
21a	21·8	⟨111⟩	⟨124⟩	{124}	{124}	6·1	4·6
21b	44·4	⟨211⟩	⟨145⟩	{145}	{111} (7)†	4·3	6·05

† These figures give the reciprocal density of lattice sites which are in coincidence in the close-packed planes. In all other cases, all sites in the close-packed planes are in coincidence.

will have the same ratio of normal sites to coincidence sites as the ratio Σ for the whole lattice, but there are also some special orientations of the boundary, parallel to densely packed planes of the coincidence lattice, which have a higher proportion of coincidence sites. Thus it is possible for all the lattice sites of a given plane to be coincidences, and there are then $\Sigma - 1$ parallel planes which contain no coincidences. Symmetrical tilt boundaries with rational indices are always of this type, and as noted on p. 335, they may be considered to be fully coherent twin boundaries. The corresponding orientation relation may alternatively be regarded as a special case of eqns. (36.10) and (36.11) for which $x = 0$, $y = 1$, $\theta = 180°$, and the rotation axis is normal to the plane of exact coincidence. When the indices of this plane satisfy the condition

$$h_i h_i = \Sigma' \quad (\Sigma' \text{ odd}) \tag{36.16}$$

the number of lattice planes traversed by a primitive lattice vector normal to the planes is either Σ' or $2\Sigma'$ (see Tables II and III, pp. 36–7) but in both cases there are $\Sigma' - 1$ planes of no coincidences in every group of Σ' planes so that $\Sigma = \Sigma'$. Alternatively, for a net with normal k rotated by 180°, where

$$k_i k_i = \Sigma' \quad (\Sigma' \text{ even}), \tag{36.17}$$

the number of lattice planes traversed by the primitive lattice vector parallel to **k** is always Σ', and the value of Σ is then $\Sigma = \frac{1}{2}\Sigma'$, where Σ is necessarily odd as k_i are co-prime.

Equations (36.16) and (36.17) show that in general a coincidence site lattice with fixed Σ may be represented by rotations of π about two different axes but for some values of Σ (7, 15, 23, ...) satisfying $\Sigma = 4^p(8n+7)$, where p and n are integers, there is no solution to (36.16) and there is then only one type of twin axis. Another factor is that either (36.16) or (36.17) may have more than one solution, and these solutions give rise to independent coincidence lattices which are denoted by adding *a*, *b*, ..., to the value of Σ. If one of the equations has two solutions and the other has only one solution, one of the two orientation relations will have two representations as a 180° rotation (e.g. 13*a*) and the other only one (e.g. 13*b*).

Although a twin orientation always implies that there is a coincidence site lattice, the reverse statement is not valid. Fortes (1972) showed that a coincidence site lattice in a cubic system generated by Ranganathan's procedure will be equivalent to a 180° rotation about some other axis if either (a) the generating axis is parallel to a plane of type {100} or {110} or (b) the ratio x/y in eqn. (36.11) is equal to any of the relatively prime indices of the rotation axis or to the sum of all three, allowing independent choices of sign for each index. The lowest index direction which does not satisfy (a) is ⟨123⟩, and for rotations about this axis conditions (b) show that the smallest value of Σ which does not represent a twin relation is $\Sigma = 39$.

Planes in which every site may be a coincidence site may be expected to represent boundaries of low energy, and the most densely packed planes of this type are usually, but not invariably, one of the twin planes for the given orientation relation. However, the closest packed planes of the coincidence lattice are often relatively high index planes of the crystal lattice, so that the absolute density of lattice points is small and atoms on several of the closely spaced neighbouring parallel planes are also in the interface region. When comparing the energies of high-angle boundaries with different crystal orientations, the most important criterion is likely to be a high absolute density of coincidence sites, or equivalently a small size for the minimum period at which the boundary structure is repeated. We have therefore also included in the table the size of the minimum repeat unit for the f.c.c. and b.c.c. lattices in terms of the magnitude of the shortest lattice vector **b**; this corresponds, of course, to the nearest neighbour atom separation r_1 in the case of the f.c.c. and b.c.c. structures.

The multiple descriptions of the orientation relation given by eqn. (36.12) lead to a corresponding multiplicity of values for the net Burgers vector in the interface given by eqns. (36.7) or (36.8). For any description, the magnitude of the Burgers vector $|\mathbf{b}|$ has maximum and minimum values given by

$$|\mathbf{b}^{\max}| = 2|\mathbf{p}| \sin \tfrac{1}{2}\theta \tag{36.18}$$

and

$$|\mathbf{b}^{\min}| = 2|\mathbf{p}| \cos \varphi \sin \tfrac{1}{2}\theta, \tag{36.19}$$

where φ is the angle between the interface normal and the rotation axis **l**. If the assumption is made that the description which is physically most significant is that for which $|\mathbf{b}^{\max}|$ is smallest, the choice of the appropriate rotation matrix is straightforward. The angle θ_U corresponding to any representation R_U of eqn. (36.12) is readily obtained from the trace of the matrix which is

$$(\mathsf{R}_U)_{ii} = 1+2\cos\theta_U$$

and the minimum value of θ_U substituted into eqns. (36.18) and (36.19) then gives the dislocation density of the interface.

The criterion of minimum $|\mathbf{b}^{max}|$ is, not unexpectedly, equivalent to choosing the axis-angle representation of minimum rotation angle. However, the lattice relations need not be expressed as a simple rotation and if shears are also permitted the matrices U of eqn. (36.12) are any unimodular matrices with integral elements. As will be shown in Section 38, this permits descriptions in which $|\mathbf{b}|$ is further reduced, and in particular any coherent twin boundary in which the lattices are related by a simple shear may be assigned zero dislocation density. This is not a physically very useful representation when the value of Σ is large, so that the lattice planes parallel to the interface are sparsely populated.

Figure 8.11, which is due originally to Frank, shows the atom positions in the common $\{111\}$ planes of two f.c.c. crystals which are rotated by 38° about the $\langle 111 \rangle$ direction to produce a $\Sigma = 7$ coincidence site lattice. The atoms are assumed to remain in coincidence site relation, and the figure illustrates the atomic configuration for two tilt boundaries of

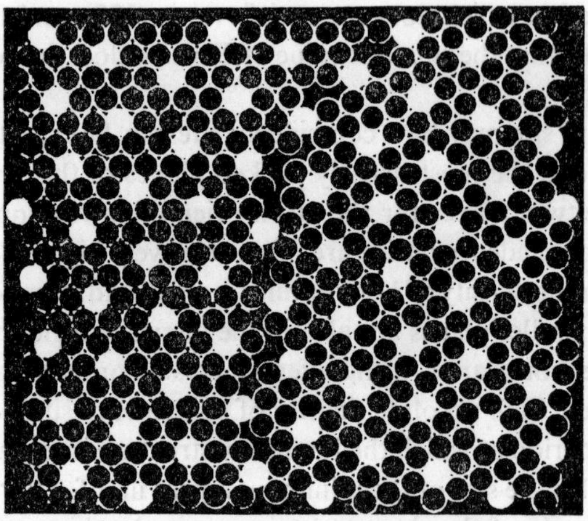

FIG. 8.11. Frank's model of a grain boundary between two f.c.c. crystals rotated 38° about a common $\langle 111 \rangle$ axis (after Aust and Rutter, 1959). (Blocked-in circles represent "coincidence sites".)

different orientations. In the hard sphere model shown, it is impossible for two atoms to approach closer than the normal nearest neighbour separation, and as a result there is a considerable amount of empty volume in the vicinity of the interface. In some other models, atoms are inserted into holes of radii greater than $0{\cdot}9r_1$, or equivalently an atom is removed only when two atoms "overlap" by more than a critical amount. However, it is clear that a still lower energy may be achieved by translation away from the coincidence site position, and the structure will then depend on the details of the interatomic forces. Atomistic calculations of this type have been made for a few boundaries, and are expected to become more important. Before discussing them, however, we take up the problem of describing the structure of a boundary which deviates by a relatively small amount from an ideal coincidence site relationship.

We find it convenient to discuss separately two kinds of deviation. The first occurs when the relations between the two lattices are such that a good fit, low-energy boundary might exist but the actual boundary is rotated slightly from this ideal orientation. The real interface can then often be considered to consist of stepped planar sections of ideal low-energy interface. The steps are discontinuities in this interface, and for symmetrical tilt boundaries they may be described as twinning dislocations. This is a simple extension of the model described in Section 32 to the generalized coherent twin boundaries introduced above. It is clear that the equivalent Burgers vectors will be smaller than a lattice vector.

The other type of deviation occurs when the orientation relation between two lattices is close to but different from a coincidence lattice relationship. The interface dislocations in the ordinary model of a low-angle boundary may be regarded as discontinuities which compensate for the discrepancy between the ideal orientation relation ($\theta = 0°$, $\Sigma = 1$) and the actual orientation relation; we call these "primary" interface dislocations and note that their Burgers vectors must be repeat vectors of the reference lattice. In an analogous fashion, a high-angle grain boundary of good fit may be treated as an entity, and the boundary structure may be modelled by introducing "secondary" interface dislocations which accommodate the deviation from the ideal orientation relations. This concept was first formulated by Read and Shockley (1950) who used a description in terms of two arrays of dislocations. The dense array gives the mathematical description of a coincidence site, double tilt boundary in terms of a uniformly spaced set of primary interface dislocations, whilst the deviations are represented by a low-density dislocation array.

The secondary dislocation description was developed in terms of the coincidence site model by Brandon *et al.* (1964) and Brandon (1966) who linked it to field-ion microscope observations of grain boundary structures, and by Bishop and Chalmers (1968, 1971), who emphasized that boundary coincidences can be maintained even when three-dimensional coincidences vanish. The most general mathematical theory has been developed by Bollmann (1967, 1970), who has introduced a concept known as the *O*-lattice. Bollmann proposed that regions of good fit in a boundary should not be identified solely by coincidences of points of the two lattices, but also by coincidences of interior cell points which are defined so that if for any cell of one lattice the interior coordinates of a point (expressed as fractions of the cell edges ξ_i as in eqn. (5.8)) are identical with the interior coordinates of that same point measured relative to a cell of the other lattice, the point is considered to be a coincidence. The *O*-lattice is the totality of such coincidences, and, unlike the coincidence site lattice, it changes continuously as the orientation is varied away from a coincidence lattice relation.

The *O*-lattice is used to discuss the primary and secondary dislocation content of any boundary, and is especially powerful in the case of interphase as distinct from grain boundaries. We thus find it convenient to defer a fuller discussion to Section 38. However, we may usefully summarize the type of dislocation model required for various grain boundary situations of interest. The simplest case to consider is when the axis of misorientation corresponds exactly to that required for a coincidence site relation, but the angular misorientation differs slightly from an exact coincidence value. The structure can then be described as that of a coincidence boundary with a superimposed sub-boundary network of dislocations corresponding to a low angle rotation about this axis. Thus for a tilt boundary of this type, the secondary dislocations are pure edges parallel to the tilt axis, but as shown below, their Burgers vectors do not correspond to lattice vectors of either real lattice.

Deviations in the axis of misorientation **r**, rather than in θ, may be accommodated by a superimposed array in which the sub-boundary misorientation axis is perpendicular to the coincidence site misorientation axis. This means, for example, that a small change from a coincidence site tilt boundary is now accommodated by a secondary dislocation twist boundary. In general, deviations may involve changes in both **r** and θ, so that the axis of misorientation of the secondary array will lie at an arbitrary angle to the axis of the reference coincidence site relation.

Even under the most favourable conditions, these dislocation models of grain boundaries can only be regarded as approximations, and there is a major difficulty in the choice of the reference situation for secondary dislocations which will be further discussed in Section 38. Some attempts have been made to calculate the actual atomic positions in the boundary region using the interatomic force laws discussed in Section 16. There are, however, some special difficulties in making such calculations for grain boundaries.

Consider a planar grain boundary of unit normal $\boldsymbol{\nu}$ so that the positions of the atoms in in the two crystals in the regions remote from the interface are known, apart from a possible small rigid translation **t** (see eqn. (9.1)). The interface structure may be calculated in principle by placing the atoms in a zone around the interface in arbitrary positions which are then relaxed by an iterative displacement of each atom in turn until an energy minimum is attained. The size of the crystal block used for the computer calculation must be such that periodic boundary conditions can be applied parallel to the interface, i.e. the structure of the interface must be doubly periodic in two independent directions, and the area used for the calculation must include at least one complete unit of pattern. This restricts calculations to those interfaces in which there is a reasonably sized two-dimensional net which is a section of a coincidence site lattice (ignoring any translation **t** which may move the sites out of coincidence but does not change the periodicity). Even then, it is not obvious that a single pattern repeat unit is adequate; the results will appear more convincing if a larger area is used, and the shorter periodicity should then be revealed directly in the computed atomic positions.

The fixed positions of the atoms at large distances from the interface may be used as a boundary condition for the relaxation procedure, but themselves have to be determined because of the translation **t**. Thus if one lattice is regarded as fixed, the atoms in the other lattice remote from the interface may be in positions corresponding exactly to a coincidence site relation or they may be translated away from these positions by any vector which joins the corner of a unit cell to an interior point of the cell. Weins (1972) has described a procedure in which the energy of the crystal is first minimized with respect to this translation, keeping all the atoms on normal sites, and then the iterative relaxation of the atoms in the boundary zone is used to find the lower energy configuration. Presumably in a complete calculation it would be necessary to iterate the displacement and atomic relaxation stages more than once, but in Weins's results the relaxations from normal lattice sites were found to be very small.

The necessity for the two-stage calculation arises from an assumed boundary condition in which the atoms remote from the interface are regarded as fixed. A possible alternative procedure is to fix the atoms at infinity on one side of the interface only and to carry out the atomic relaxation plane by plane. Any components of displacement which are common to all the atoms in a plane are immediately applied to all the succeeding planes, so that the translation at infinity is computed as the sum of the individual translations near the interface. Vítek (1970) used this method to calculate the structure of a coherent twin boundary, but its application to a general boundary is obviously more difficult.

Unlike most defect problems, it is not known at the beginning of the calculation how many atoms should be placed in the region around the interface to be relaxed. Dahl *et al.* (1972) used rigid boundary conditions and rejected overlapping atoms, i.e. those atoms with geometrical centres closer than the nearest neighbour distance, whilst Weins used a similar but opposite procedure in which the initial structure contains voids, and atoms are inserted into any holes with radii greater than 90% of the nearest neighbour distance. Hasson *et al.* (1972) tackle this problem systematically by starting with a number of atoms which is certainly too high, relaxing completely to minimum energy, and then removing the atom which has the highest energy. The structure is then relaxed again to produce a still lower energy, and the process is repeated until removal of an atom causes the energy to rise. Once again, in a full calculation it would seem necessary to iterate this process with any separate calculation of the translation vector.

There are obviously many variables in these calculations, and the best procedures have probably still to be determined; in particular it cannot be certain that the calculations do not give structures which are metastable rather than stable. However, some of the results appear to confirm the dislocation models of low-angle boundaries and the significance of periodic structures in some high-angle boundaries. They also show that the actual coincidence of boundary sites is not the important criterion in periodic boundaries since in general the lowest energy configuration corresponds to a relative translation away from the coincidence site position.

The simplest type of boundary is the fully coherent twin boundary, and one calculation for a {112} boundary in a b.c.c. structure has been made by Vítek (1970) using empirical, Johnson-type potentials. The work arose from the discovery (see Chapter 5) that wide monolayer twins, i.e. intrinsic stacking faults, on {112} planes of b.c.c. structures are mechanically unstable. Vítek therefore investigated multilayer faults, and found that a three-layer fault is metastable, but the successive displacements of the layers are not all $(a/6)\langle\bar{1}\bar{1}1\rangle$, as expected, but are approximately $(a/12)\langle\bar{1}\bar{1}1\rangle$, $(a/6)\langle\bar{1}\bar{1}1\rangle$ and $(a/12)\langle\bar{1}\bar{1}1\rangle$. The same situation was found to characterize the boundary of the infinite twin where the first atomic layer may be regarded as shifted by $(a/12)\langle\bar{1}\bar{1}1\rangle$ with respect to the matrix and the remaining layers are each shifted by $(a/6)\langle\bar{1}\bar{1}1\rangle$ with respect to each other. The calculated structure, Fig. 8.12, is dislocation free as expected, but there is no plane of atoms in coincidence site positions common to the two lattices. The computed structure may be derived from the coincidence site structure by a relative translation of the lattices through $(a/12)\langle 11\bar{1}\rangle$; it is also rather syimmetrical and the atomic structure appears identical when viewed from the two sides of the interface. The calculations also show that the reason for the stability

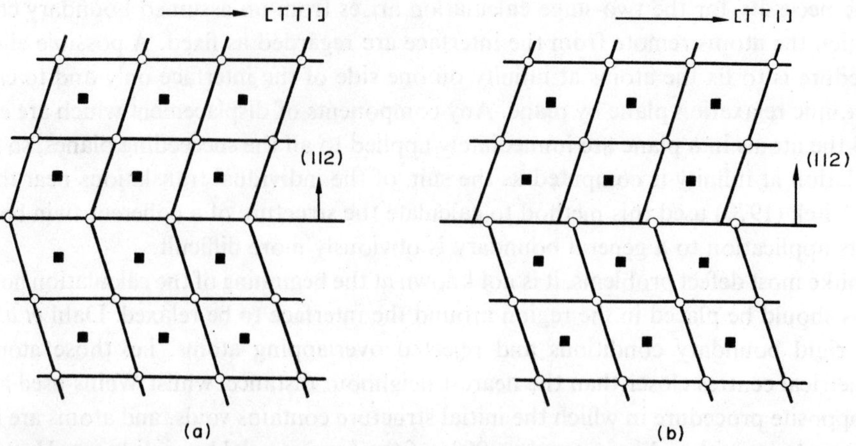

FIG. 8.12. Possible atomic structures of a {112} twin boundary in a b.c.c. material. (a) Coincidence site boundary. (b) Boundary calculated by Vítek (1970). Atoms represented by circles and squares lie in successive {1̄10} planes parallel to the plane of projection (after Christian and Crocker, 1974).

of the structure of Fig. 8.12(b) lies in the strong repulsive interaction, and Fig. 8.12(a) could represent the stable structure only for very "soft" atoms.

Calculations of the structures and energies of various tilt boundaries have been summarized by Hasson *et al.* (1972), Weins (1972), and Dahl *et al.* (1972). The first two groups used Morse potentials adjusted to fit certain physical parameters of aluminium or copper, and Weins also used Lennard-Jones potentials of 6–12 and 4–7 types; Dahl *et al.* worked with a Johnson potential applicable to iron. Weins and Hasson *et al.* both calculated the structures of symmetrical coincidence site ⟨100⟩ tilt boundaries, and the latter workers also studied assymetrical tilt boundaries, low-angle boundaries, and ⟨011⟩ tilt boundaries.

Figure 8.13(a) shows the computed structure for a low angle (10°27′) symmetrical tilt about [100], corresponding to a boundary plane (0, 1, 11) and a Σ value of 133. The relation to the dislocation model is immediately obvious, and the regions of severe misfit extend only for one or two atomic distances around the dislocation cores. A higher angle (36°52′) tilt boundary, for which the boundary plane is (013) and $\Sigma = 5$, is shown in Fig. 8.13(b), and here the boundary misfit is almost continuous although the configurations corresponding to dislocations can still be identified. Rotation of the boundary of Fig. 8.13(b) to an asymmetrical position gives the configuration shown in Fig. 8.13(c) in which the structure is much less regular, although dislocation-like groupings of atoms similar to those in the symmetrical boundaries can still be recognized.

Goux has stated that the lattices in Fig. 8.13 are not in coincidence site relations, but it is clear that any relative translation must be fairly small. Rather different results were obtained by Weins (1972) who found quite large relative translations of the two lattices, the details of which depend on the assumed potentials. In some [100] tilt boundaries, for example, the (100) planes of the two crystals do not remain in registry across the interface. Subsequent atomistic relaxation results in very small displacements of the atoms from

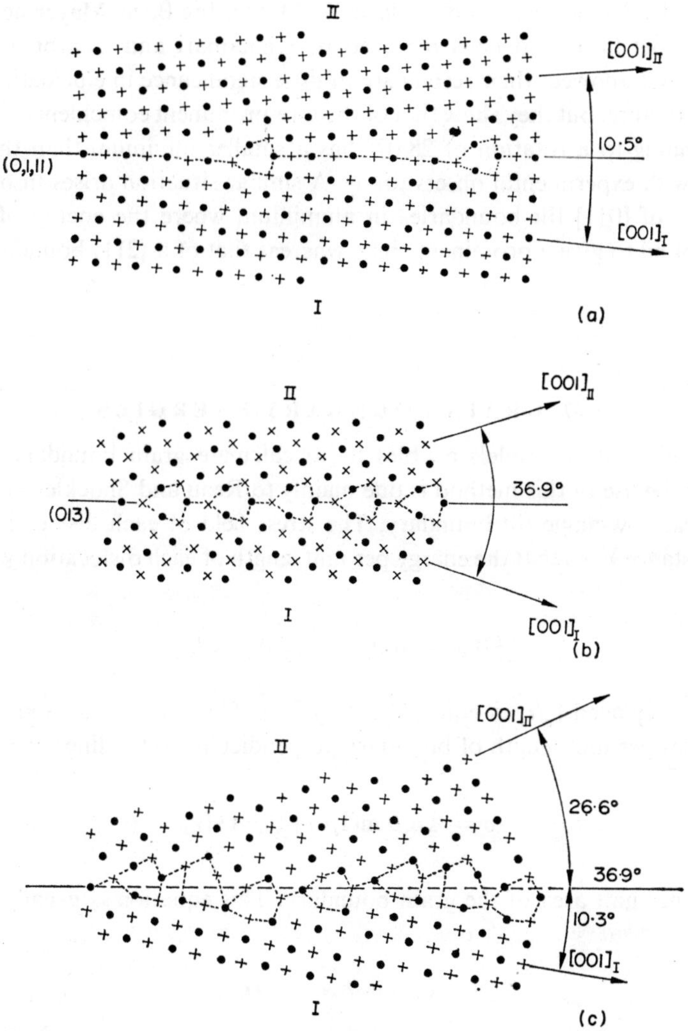

FIG. 8.13. Simulated atomic structures of some [100] tilt boundaries in aluminium. (a) Symmetric (0, 1, 11) interface, $\Sigma = 133$, $\theta = 10{\cdot}5°$ (b) Symmetric (013) interface, $\Sigma = 5$, $\theta = 36{\cdot}9°$ (c) Asymmetric $\Sigma = 5$ interface rotated 8·1° from the symmetric configuration.

Atoms represented by circles and crosses lie in successive (100) planes parallel to the plane of projection (after Hasson *et al.*, 1972).

their lattice sites, so that the relative translation from the coincidence site orientation is the major factor determining the boundary structure according to these calculations.

Weins *et al.* (1972) report one calculation for a [100] twist boundary of 36°52′; the only non-zero component of translation was parallel to the twist axis. Some calculations of the energies of [100] coincidence twist boundaries in MgO using Born–Mayer potentials, and a rigid lattice model have also been reported by Chaudhari and Charbnau (1972). Since relaxation was not allowed, their results are of little significance in elucidating the problem of boundary structure, but they show that of the four prominent coincidence site boundaries, that corresponding to a rotation of 28·07° has a smaller minimum than the other three, in agreement with experimental observations. A similar situation arises in one calculation of the energies of [011] tilt boundaries in aluminium where the energy of a {311} twin boundary is found to give a prominent cusp, whereas that of a {211} boundary does not.

37. GRAIN BOUNDARY ENERGIES

The use of dislocation models enables us to calculate grain boundary energies from elastic theory; the use of this method is due mainly to Read and Shockley (1950). Consider the symmetrical, low-angle tilt boundary. The stress field of each dislocation extends for an effective distance Y, so that the energy per unit length of each dislocation will be approximately

$$^{L}W_d = (B_e b/2)\ln(Y/b) + {^{L}W_c}, \tag{37.1}$$

where we have replaced r_e/r_i in eqn. (30.13) by Y/b. Since $Y = b/\theta$, and there are $1/Y$ dislocation lines per unit length of boundary perpendicular to the lines, we have

$$\sigma = \tfrac{1}{2}B_e\theta\ln(1/\theta) + (\theta/b)\,{^{L}W_c} \tag{37.2}$$

as the energy per unit area of the grain boundary. This equation is usually written in the orm

$$\sigma = \sigma_0\theta(A - \ln\theta), \tag{37.3}$$

where σ_0 and A do not depend on θ. The parameter σ_0 may be calculated from elastic theory, but A contains the unknown core energy of the dislocations.

This derivation is intuitive, and we have given no formal justification for the assumption that the extent of the stress field is about equal to Y. In their original paper, Read and Shockley summed the stress field of all the dislocations to find the shear stress acting on the slip plane of a particular dislocation in the array. They thus found the energy of this dislocation from the work required to create it by a virtual process. We shall follow a later, simpler method, which may be illustrated by reference to Fig. 8.14.

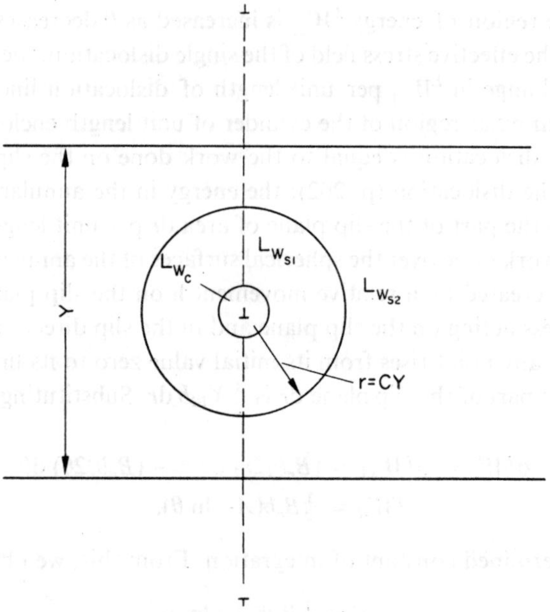

FIG. 8.14. To illustrate the calculation of grain boundary energy (after Read and Shockley, 1950).

We divide the crystal into uniform parallel strips, each containing one dislocation and having energy ^LW_d. In calculating ^LW_d, we first write, as before,

$$^LW_d = {}^LW_s + {}^LW_c,$$

where ^LW_c is the core energy of the bad crystal at the centre of the dislocation line. We then subdivide ^LW_s into the elastic energy in a cylindrical region around the dislocation line ($^LW_{s1}$) and the energy in the volume of the strip outside this region ($^LW_{s2}$). The radius of this cylindrical region is $r = CY$, where the constant $C < 1$ is not precisely defined, but must be large enough to ensure $r \gg b$ and small enough for the stress field inside the cylinder to be approximately equal to that of the enclosed dislocation alone.

Read and Shockley now consider a small change in which the boundary angle decreases by $d\theta$, leading to increases in Y and r given by

$$-d\theta/\theta = dY/Y = dr/r.$$

The energy ^LW_c is obviously unaffected by the change in spacing, so the change in energy is given by the sum of d^LW_{s1} and d^LW_{s2}. Consider first d^LW_{s2}. The volume of this region increases, but the energy in a small unit volume decreases because the dislocations move further apart. The volumes of corresponding small cylindrical elements with axes parallel to the dislocation line vary as Y^2 as θ changes. A dimensional argument shows that the energy density in each such volume varies as b^2/Y^2, so that the strain energies of corresponding cylindrical elements are unchanged. The two effects thus exactly balance and $d^LW_{s2} = 0$.

The volume of the region of energy ${}^LW_{s1}$ is increased as θ decreases. Since this region is chosen to lie within the effective stress field of the single dislocation, the strain energy density is unchanged. The change in ${}^LW_{s1}$ per unit length of dislocation line is thus equal to the strain energy in the annular region of the cylinder of unit length enclosed by radii r, $r+dr$. The self-energy of a dislocation is equal to the work done on the slip plane in the virtual process of creating the dislocation (p. 262); the energy in the annular region is thus equal to the work done on the part of the slip plane of area dr per unit length of line. This result follows because the work done over the spherical surfaces of the annular region vanishes.

The dislocation is created by a relative movement **b** on the slip plane, work being done against the shear stress acting on the slip plane and in the slip direction. During the virtual process, the stress at any point rises from its initial value zero to its final value X_{12}, so that the work done on the part of the slip plane dr is $\frac{1}{2}X_{12}b\,dr$. Substituting from eqn. (30.7) for X_{12} we have†

$$d^LW_d = d^LW_{s1} = (B_eb/2r)\,dr = -(B_eb/2\theta)\,d\theta$$

or

$${}^LW_d = \tfrac{1}{2}B_eb(A-\ln\theta), \tag{37.4}$$

where A is an undetermined constant of integration. From this, we obtain as before

$$\sigma = \tfrac{1}{2}B_e\theta(A-\ln\theta), \tag{37.5}$$

which is identical with (37.3).

As previously mentioned, the constant term includes the unknown core energies of the dislocations, but the form of the above equation has been deduced from elastic theory alone. The derivation shows that A is determined not only by the core energy W_c, but also by the constant parts of the elastic energy, such as that contained in W_{s2}. In eqn. (37.3), we have used the shear stress calculated from isotropic elasticity, but Read and Shockley have shown that the shear stress on the slip plane at a distance r from a straight dislocation line can always be written

$$X_{12} = \tau_0\frac{b}{r}+\tau_1\frac{b^2}{r^2}+\ \ldots,$$

where τ_0 is uniquely determined by the anisotropic elastic constants. The coefficients $\tau_1, \ldots,$ are determined by the interaction of good material near the core with the bad material of the core itself, and may thus be regarded as corrections when linear elasticity theory is no longer applicable. These terms contribute negligibly to the shear stress at large r, and the extra energy which they represent may thus be included in ${}^LW_c$ since it also is localized near the dislocation. The general expression for the energy of a small angle tilt boundary is then

$$\sigma = \frac{b\tau_0}{2}\,\theta(A-\ln\theta). \tag{37.6}$$

An estimate of the value of A was made by van der Merwe (1950) who developed a more rigorous theory of the grain boundary energy in which he used the Peierls–Nabarro approxi-

† Note that r, θ as used here have different meanings from the cylindrical coordinates r, θ of (30.8).

mation (p. 271). With typical values of the elastic constants, he obtained $A \simeq 0{\cdot}5$. Experimental results seem to fit the equation quite well for $A \simeq 0{\cdot}2$, in very good agreement with this.

Equation (37.3) shows that σ approaches zero as $\theta \to 0$, as is physically obvious. The energy per dislocation, $b\sigma/\theta$, however, becomes infinite, in agreement with the previous conclusion that the energy of a single dislocation in an infinitely large crystal is infinite. The derivation assumes that the spacing of dislocations is uniform, but as discussed above this is only possible at a number of discrete angles. For intermediate angles, the irregularities in the array cause extra energy terms. For a small deviation $\delta\theta$ from the nearest rational angle $\theta = 1/m$, the extra energy is of order

$$-\frac{\sigma_0\,\delta\theta}{m}\ln\delta\theta.$$

As $\delta\theta \to 0$, the slope of the σ–θ curve thus becomes infinite, and the rational angles $\theta = 1/m$ represent cusps in the true curve of σ–θ. Equation (37.3) gives the locus of these cusps (Fig. 8.15). The point $\theta = 0$ is one such cusp, and the slope of the energy curve is

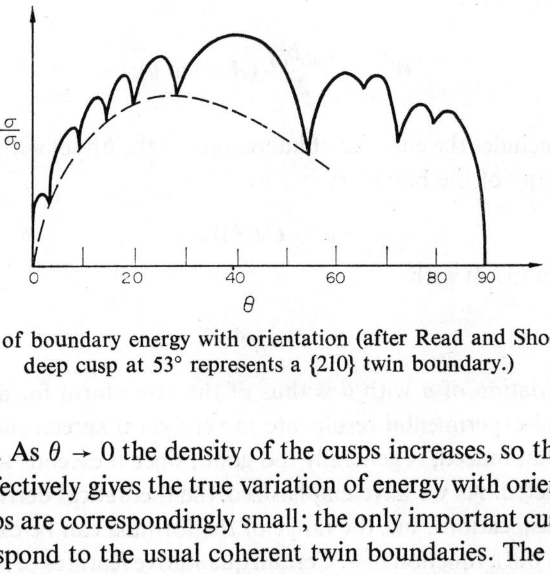

FIG. 8.15. Variation of boundary energy with orientation (after Read and Shockley, 1950). (The deep cusp at 53° represents a {210} twin boundary.)

infinite at the origin. As $\theta \to 0$ the density of the cusps increases, so that in the low-angle region eqn. (37.3) effectively gives the true variation of energy with orientation. The depths of the low-angle cusps are correspondingly small; the only important cusps are those of low indices, which correspond to the usual coherent twin boundaries. The relative rotation of the lattice is then too large for the energy formula to apply in any case.

In the above we have discussed the energy of a tilt boundary with only one degree of freedom. If we next consider the asymmetrical tilt boundary of Fig. 8.7, we have to sum the energies of two sets of dislocations. This energy may be divided into three parts, namely an energy per dislocation ${}^{L}W_{100}$ which the first array would possess in the absence of the second array, a similar energy ${}^{L}W_{010}$ for the second array, and an energy per dislocation which results from the interactions of the stress field of one array with the strain field of the other. The previous method of calculation, however, shows that the interaction energy can only contribute to the terms ${}^{L}W_{s2}$. The argument above may thus be repeated separately for each set of dislocations, and the interaction energy absorbed in the constants of integration. This gives

$$\sigma = ({}^{L}W_{100}/Y_{100}) + ({}^{L}W_{010}/Y_{010}) = \sigma_0\theta(A - \ln\theta) \tag{37.7}$$

as before. The term σ_0 now depends explicitly on φ; using eqns. (36.2) and (36.3)

$$\sigma_0 = \frac{b\tau_0}{2}(\cos\varphi + \sin\varphi), \tag{37.8}$$

and σ_0 is proportional to the total density of dislocations. This is true only if the two sets have equal strengths b and equal values of τ_0. The constant of integration A is now also a function of φ, and formulae have been given for its variation (Read and Shockley, 1950).

The argument may be further extended to the general small-angle grain boundary, in which the two lattices are related by a rotation θ about an arbitrary axis **l**. In the last section we described how this boundary could be described by three or more sets of dislocation lines. The spacing of each set of lines Y_i was shown to be inversely proportional to θ; we write $Y_i = 1/C_i\theta$ so that $C_i\theta$ gives the density of lines of Burgers vector $\mathbf{b}_i$. The energy per dislocation of this set is then

$${}^{L}W_i = \frac{\tau_{0(i)}b_{(i)}^2}{2}(A_{(i)} - \ln\theta), \tag{37.9}$$

where the term A_i includes the energies of interaction of the ith set with all the other dislocations. The total energy of the boundary is now

$$\sigma = C_i\theta\, {}^{L}W_i \tag{37.10}$$

which is of the form (37.3) with

$$\sigma_0 = \tfrac{1}{2}\tau_{0i}C_i b_i^2. \tag{37.11}$$

The predicted variation of σ with θ is thus of the same form for all types of low-angle grain boundary, and experimental results are in very good agreement with the theoretical curve. The agreement, indeed, is generally too good, since it extends to much higher angles than would be expected. As we have emphasized, the theoretical derivations are only valid for widely spaced dislocations, but the range of the formula can be extended by supposing A to be not constant but a function of θ. Certain qualitative features of high-angle boundaries have already been mentioned, namely the existence of deep energy cusps, corresponding to coherent twin boundaries. It should also be mentioned that different choices of the reference lattice, which lead to different dislocation descriptions of a boundary, will give different apparent energies if eqn. (36.7) is used. Thus two grain boundaries related by a relative rotation matrix R may also be described as related by R_1RR_2, where R_1, R_2 are rotations which leave the lattice in an identical orientation. Since these descriptions represent the same physical situation, the energies must actually be equal. The anomaly arises because at least one of the descriptions must be in terms of a large rotation and the values of A must thus be such as to give the same energy.

Although dislocation calculations of grain boundary energies are valid only for low-angle boundaries, they may also be applied in principle to secondary dislocation distributions, in which case they give the excess energy of the boundary over that of a reference (coincidence site) boundary. However, the difficulties in assigning specific Burgers vectors to these dislocations make this calculation very uncertain, and Li (1972) has proposed a slightly different approach based on the disclination model of Fig. 8.10. From that figure, we see that

$$\tan \tfrac{1}{2}\theta = b/2Y' = L\omega/Y'_{\omega}, \tag{37.12}$$

where b, Y' as before are the Burgers vector and projected spacing of the dislocations, and ω, L, Y'_{ω} are respectively the disclination strength, the separation of the dipole elements, and the separation of the disclination dipoles. The strain energy of an isolated disclination dipole is the same as that of the equivalent dislocation wall, and may be expressed (Li, 1960) as

$$^{L}W_{\omega} = \{\mu\omega^2 L^2/4\pi(1-\nu)\} \ln(r_e/L). \tag{37.13}$$

Using a model in which infinite dislocation walls of identical spacing are added together to form a high-angle boundary in such a way that the dislocations cluster into disclinations, Li obtains an expression for the variation of energy with misorientation for situations intermediate between two low-energy configurations. The energy appears as a sum and the disclinations are approximated as walls containing M dislocations with separations of N Burgers vectors. The angle of the boundary may be varied by varying L or Y'_{ω} or the two simultaneously, eqn. (37.12), and the excess energy is then zero for $M = 1$ and $M/N = 1$, and reaches a maximum for $M/N \sim \frac{1}{2}$. Further work is needed to test the validity of this approach.

38. INTERPHASE BOUNDARIES: SURFACE DISLOCATIONS

The boundary between two crystals of different structure also has five degrees of freedom in the general case, and its structure and behaviour is similar in many ways to that of a grain boundary. Three degrees of freedom are needed to specify the relative orientations of the two lattices, but of course it is no longer possible to carry one lattice into the other by means of a pure rotation. As discussed in Section 9, the lattices are now related by the general linear transformation **RP**, and may also be described by any other linear transformation $\mathbf{U}_i\mathbf{RPU}_j$, where $\mathbf{U}_i$ and $\mathbf{U}_j$ as in eqn. (36.12) represent either symmetry operations of the lattice or other lattice-preserving deformations.

In discussing the physical structure of a boundary, we shall find it convenient to distinguish three types of interface which we describe as incoherent, semi-coherent, and coherent respectively. The incoherent phase interface is the analogue of the high-angle grain boundary. There are no continuity conditions for lattice vectors or planes to be satisfied across the interface, the structure of which is relatively disordered. Although a formal description of this boundary may be given in terms of dislocation theory, this has no physical significance, and there is no lattice correspondence when the boundary moves.

The fully coherent interface is to be compared with the meeting of two twins along their composition (K_1) plane. The lattices match exactly at the interface, and "corresponding" lattice planes and directions are continuous across the interface, although they change direction as they pass from one structure to another. We emphasize here that a coherent interface in the sense in which we use the term does not necessarily imply a rational interface; in mechanical twinning, for example, the coherent interface may be rational or irrational, depending on the type of twinning. As we have seen in Section 9, it is not generally possible to find a coherent interface between two arbitrary structures. A necessary condition for a plane of exact matching to exist is that the deformation relating selected unit cells of the two structures should have a zero value of one principal strain, and since this implies a relation between the cell parameters it can only be satisfied coincidentally. The best-known examples of nearly exact matching are those involving f.c.c. and h.c.p. structures with virtually identical interatomic distances in the octahedral and basal planes respectively. If the matching condition is nearly satisfied, two phases may be forced elastically into coherence across a planar boundary. This is usually possible only when one (included) crystal is very small, since the stress at any point increases with the size of the crystal. Forced elastic coherence of this kind may exist at the nucleation or early growth stage of a transformation, the elastic strains near the interface being possibly much larger than the normal elastic limit.

We now turn to the semi-coherent interface, the prototype of which is the low-angle grain boundary. Such an interface consists of regions in which the two structures may be regarded as being in forced elastic coherence, separated by regions of misfit. A fully coherent interface is possible only when the deformations generating the two lattices from some reference lattice are compatible in ordinary space, in the sense discussed in Section 34. When these deformations are not compatible, we must specify an additional (lattice invariant) deformation to enable us to fit the regions together in ordinary space. If we use the theory of the continuous distribution of dislocations, we can do this for any two lattices, but in many cases the resultant surface density of dislocations will be extremely high. In cases where the density is low enough for individual dislocations in the interface to accomplish the lattice matching, the description acquires a physical significance, and we may have a semi-coherent boundary. When such a boundary exists, we cannot specify an overall correspondence (lattice generating deformation) relating one lattice to the other, but a local correspondence of this kind exists. This is exactly the analogy of the relation between the reference lattice and the dislocated lattice discussed on p. 315.

Note that in the sense used here, a semi-coherent boundary is a particular type of plane interface. A curved boundary, as in lenticular plates, necessarily contains additional misfit regions. If the curvature is small, these may be considered as steps of atomic height, and correspond to the twinning dislocations and transformation dislocations already discussed in Section 32. These discontinuities, which can glide in the composition or habit plane, are quite distinct from the discontinuities needed to correct the mismatch of the two lattices.

During a phase transition, one crystal grows at the expense of the other by the migration of the boundary interface. It is clear that if the structures are fully coherent a lattice correspondence is implicit in this process, so that labelled rows or planes of lattice sites in one crystal become rows or planes in the other. We write lattice sites rather than atoms since only lattices and not structures are related by affine transformations; in the simplest cases, there will be a correspondence of atomic sites. The motion of a coherent boundary thus produces a macroscopic change of shape which is specified by the relation between the lattices.

If the boundary is incoherent, there is no correspondence and no shape change when it moves. Thus the growth of a mechanical twin changes the shape of the specimen, but the motion of a grain boundary does not. A semi-coherent boundary requires more detailed discussion. Since the lattice deformations are not compatible, the motion of the boundary cannot produce a change of shape given by the local lattice correspondence. In fact, the shape change produced is obtained by combining the lattice change with the change due to the migration of the discontinuities in the boundary (lattice invariant deformation) exactly as on p. 320. It follows that for both fully coherent and semi-coherent boundaries the shape change produced by motion of the boundary is an invariant plane strain.

We shall now discuss a simple model of a semi-coherent boundary, before giving a more general treatment of the structure of such a boundary. Suppose first that the boundary contains a single set of parallel dislocations with a common Burgers vector. If this vector does not lie in the interface, or if the dislocations are pure screws, we may suppose that the corresponding glide planes for the dislocations in the two structures meet edge to edge in the interface. The line of intersection, which is parallel to the dislocations, is an invariant line of the lattice deformation (see p. 61). With a single set of parallel dislocations, the lattice invariant deformation produced by motion of the dislocation array is a simple shear, and the boundary we have just described is the usual dislocation model of a martensitic type interface. If the dislocations are pure edges, the lattice deformation reduces to a pure rotation, and the interface is a symmetrical low-angle tilt boundary, as already discussed.

If the direction of the Burgers vector is in the interface plane and is not parallel to the dislocation lines, we have a different type of boundary. The habit plane is now the glide plane, so that as the boundary is displaced the dislocations in it must climb. The motion is thus non-conservative in the sense that if the boundary sweeps through a region of one crystal, the number of atoms incorporated into the other crystal will differ from the number originally present in the swept region. The lattices again match in the direction of the dislocation lines, but the density of equivalent lattice points† in planes parallel to the habit plane is different in the two structures. The energy of a boundary of this type was calculated by van der Merwe (1950) and Brooks (1952); it may be called "epitaxial" since the structure envisaged is similar to that postulated for epitaxial layers deposited on crystals.

Figure 8.16 shows a simple epitaxial semi-coherent boundary in which the dislocations are of edge type. As the boundary moves towards the top of the figure, the region *ABCD* of the upper crystal is transformed into a region of lower crystal with shape *D'C'BA*. This macroscopic shape change is an invariant plane strain, and results from a combination of the shape change of a unit cell with the uniaxial contraction produced by motion of the dislocation array. However, we may note an important difference between the motion of this type of semi-coherent boundary and that of boundaries of the martensitic type. The

† We define equivalent lattice points as lattice points outlining unit cells in the two structures containing the same number of atoms.

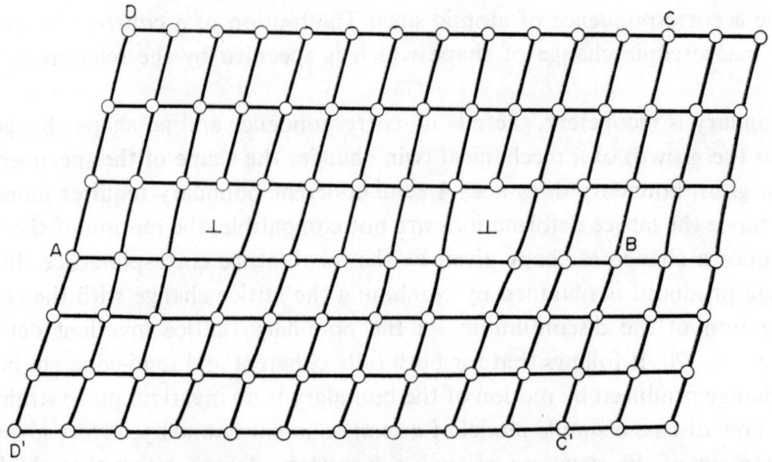

FIG. 8.16. To illustrate a semi-coherent interphase boundary containing discontinuities of edge-dislocation type.

dislocation climb associated with the motion of the boundary in Fig. 8.16 requires atomic displacements through distances of the order of the distance moved by the boundary. The motion will thus be relatively slow, and may be accompanied by a transfer of atoms from one side of the growing crystal to the other, thus effectively destroying the correspondence and its associated shape change. In the particular case of Fig. 8.16 the shape change could be minimized by a net flow of atoms from right to left as the boundary moves upwards, sites being destroyed on the right of the growing crystal and created on the left. The extra strain energy associated with the shape change within a constraining matrix should provide a driving force for this motion (Christian, 1962). Thus if epitaxial semi-coherent boundaries exist, it is rather improbable that their motion will produce observable shape changes in the same way as the motion of martensitic semi-coherent boundaries.

We have considered only a single set of parallel dislocations, but certain other boundaries may be envisaged in which the dislocation array is equivalent to a single set of dislocations. These either contain parallel sets of dislocations with differing Burgers vectors, or sets of dislocations in different directions but with a common Burgers vector. If the individual sets of dislocations are all able to glide in planes of the two lattices which intersect the habit plane, the interface will be of the martensitic type. The lattice invariant deformation produced by the glide motion of the dislocations is then again a simple shear, although the shear direction (in the first case) or the invariant plane of the shear (in the second case) will generally be irrational in both lattices.

Dislocation arrays which reduce to a single array with a Burgers vector in the interface again represent an epitaxial boundary with a single direction of lattice fit. A more general boundary of this type is obtained by combining two or more sets of non-parallel interface dislocations, each set having a Burgers vector in the interface. The whole dislocation content of the boundary may then be replaced by a crossed grid of edge dislocations, compensating for the different densities of the equivalent lattice points in two mutually perpendicular directions. This is a simple extension of the boundary shown in Fig. 8.16, and van der Merwe showed that its energy is similar to that of a low-angle grain boundary. The shape deformation is again an invariant plane strain, but the habit plane no longer contains an invariant line of the lattice deformation.

In the above discussion, we have emphasized the distinction between a boundary in which the dislocations can glide and one in which they must climb as the boundary moves. More generally, we may distinguish between boundaries which are dependent on thermally activated processes for their motion, and boundaries which are mobile under suitable driving forces even at very low temperatures. The latter type of boundary may be called "glissile". A necessary condition for a boundary to be glissile is that its motion be conservative.

We now turn to the general dislocation description of the boundary between any two lattices in given orientation. As on p. 340, we may specify a net dislocation content, although its resolution into individual dislocation lines is not necessarily unique. The dislocation content may be found from first principles, or by utilizing the theory of continuous distributions of dislocations.

Consider first the argument we used on p. 340. We may repeat this exactly for the case in which the two real crystals have different structures if we replace $\mathbf{R}_+$ by $\mathbf{S}_+$, the deformation carrying the reference lattice into the lattice of one crystal in its final orientation, and $\mathbf{R}_-$ by $\mathbf{S}_-$, with a similar interpretation. We thus have for the total Burgers vector of the dislocations crossing any unit vector $\mathbf{p}$ in the interface

$$\mathbf{b} = (\mathbf{S}_+^{-1} - \mathbf{S}_-^{-1})\mathbf{p}. \tag{38.1}$$

In particular, if we take the reference crystal to have the structure and orientation of the (−) crystal, we have

$$\mathbf{b} = (\mathbf{S}^{-1} - \mathbf{I})\mathbf{p} = (\mathbf{P}^{-1}\mathbf{R}^{-1} - \mathbf{I})\mathbf{p}, \tag{38.2}$$

where $\mathbf{S} = \mathbf{RP}$ is the lattice deformation relating the two crystals, and $\mathbf{P}$ and $\mathbf{R}$ are a pure deformation and a pure rotation respectively.

Alternatively, suppose we have a continuous distribution of dislocations specified by the tensor A_{ij} (p. 314), and concentrate all the dislocations into a shell of thickness t so that A_{ij} vanishes outside this shell. Now let $t \to 0$ and $A_{ij} \to \infty$ in such a way that the product tA_{ij} remains finite and tends to B_{ij}. We describe B_{ij} as the surface dislocation tensor. By applying Stokes's theorem to a small circuit intersected by the dislocation shell, we find (Bilby, 1955)

$$B_{ij} = -\varepsilon_{jkl}\nu_k(S_{+il}^{-1} - S_{-il}^{-1}), \tag{38.3}$$

where $\boldsymbol{\nu}$ is the positive normal to the interface plane. It should be noted that in accordance with our usual practice in Chapter 2 and elsewhere, $\mathbf{S}_+$ and $\mathbf{S}_-$ in eqns. (38.1) and (38.3) (when written as matrices) transform the components of a reference lattice vector into those of a vector in the real lattices. They are thus not directly identifiable with $\mathbf{D}$ of eqn. (34.3), which gave the relation between the lattice bases, and in terms of the notation used in Section 34 we should have $\mathbf{S}^{-1} = \mathbf{D}'$, i.e. $S_{+il}^{-1} = D_{+li}$ where $\mathbf{D}_+$ and $\mathbf{D}_-$ are the values of $\mathbf{D}$ in the two crystals.

Now consider the dislocations which cross a small area of the shell defined by the vector **p** and the vector thickness $t\boldsymbol{\nu}$. The normal to this area has components $dS_j = t\varepsilon_{jmn}p_m\nu_n$, and using eqn. (34.2) and putting $tA_{ij} = B_{ij}$ in the limit,

$$b_i = B_{ij}\varepsilon_{jmn}p_m\nu_n \tag{38.4}$$

for the resultant Burgers vector. Since we have now allowed t to tend to zero, this is the net Burgers vector of the dislocation lines which cross **p**. Substituting from (38.3), and using the relation

$$\varepsilon_{jkl}\varepsilon_{jmn} = \delta_{km}\delta_{ln} - \delta_{kn}\delta_{lm},$$

we obtain finally

$$b_i = (S^{-1}_{+il} - S^{-1}_{-il})p_l, \tag{38.5}$$

which is identical with (38.1).

The surface dislocation tensor B_{ij} gives the ith component of the resultant Burgers vector of the dislocation lines cutting unit length of line in the surface perpendicular to the jth direction. When the dislocation content of the surface is smoothed out (continuous), we may think of the whole array as a single defect, called by Bilby a "surface dislocation". The tensor B_{ij} is the analogue for a surface dislocation of the Burgers vector of a line dislocation.

We have emphasized that for a fully coherent boundary, the lattice deformation must be an invariant plane strain. Two lattices are related by such a strain if $\mathbf{S}_+$ has the form of eqn. (9.10) and $\mathbf{S}_- = \mathbf{I}$. Substituting into (38.3) we find that B_{ij} is identically zero, proving again that no dislocations are required in the interface when the lattice deformations are compatible (Basinski and Christian, 1956; Bilby and Smith, 1958). This applies, of course, to any twin boundary, but difficulties arise in the case of symmetrical low-angle tilt boundaries which may be regarded as twin boundaries for coincidence site relations of high Σ (Section 36). Although the lattice points may be formally related by a simple shear, the dislocation model obtained by treating **S** as a rotation is physically more significant than a zero-dislocation model (see p. 348). In fact it is very difficult to formulate rules for the specification of the dislocation content of any boundary unless restrictions are placed on **S**. Figure 8.17, for example, shows the same grain boundary as Fig. 8.9, but the deformation **S** is no longer taken to be a pure rotation. The net Burgers vector defined by eqn. (38.1) in the same reference lattice as that shown in Fig. 8.9 has been changed, although the positions of all lattice points are identical; in the notation used by Hirth and Balluffi (1973) and some others, the boundary is composed of intrinsic grain boundary dislocations (IGBDs) and its Burgers vector content cannot be made specific.

It might seem obvious that we should find the minimum value of $|\mathbf{b}|$ for all possible descriptions of the lattice relation, just as we found it for all descriptions which are rotations in Section 36. This is closely related to a procedure suggested by Bollmann for choosing the "correct" representation **S**, and although we shall find that this is not valid without further restrictions, it is convenient to develop further the elegant formalism of his *O*-lattice theory, which has already been mentioned on p. 349.

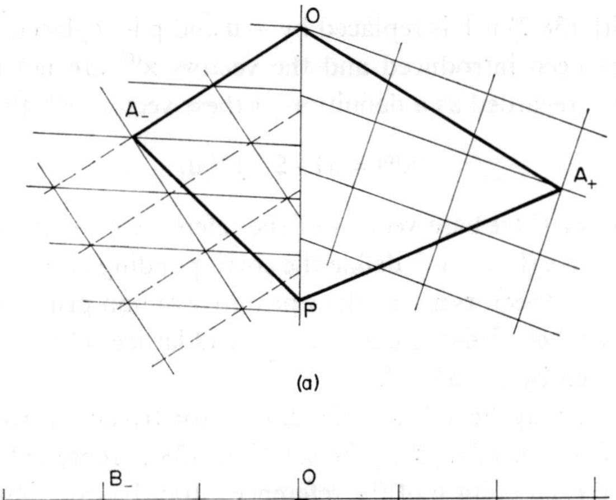

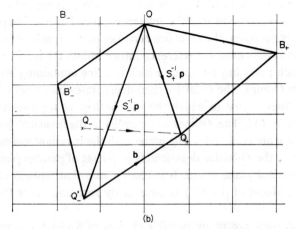

FIG. 8.17. To illustrate how a redefinition of the lattice relations in the grain boundary of Fig. 8.9 gives a different result for the net Burgers vector of the dislocations crossing OP. The + part of the circuit shown in (b) is identical with that in Fig. 8.9 but the reference unit cell is now related to a different cell of the − lattice (shown by the unbroken lines in (a)), so that $\mathbf{S}_-$ in (38.1) is no longer a pure rotation. The closure failure in (b) is now Q'_-Q_+; the previous closure failure Q_-Q_+ is shown by a broken line.

For two lattices related by a general linear transformation $\mathsf{y} = \mathsf{S}\mathsf{x}$, a point $\mathsf{x}^{(0)}$ of the O-lattice is defined by the simultaneous conditions

$$\mathsf{S}\mathsf{x} = \mathsf{x}+\mathsf{u} = \mathsf{y} = \mathsf{x}^{(0)}, \tag{38.6}$$

where $\mathbf{x} = \mathbf{w}+\boldsymbol{\xi}$ is any vector and $\mathbf{w}$ and $\mathbf{u}$ are lattice vectors of the first (or −) lattice which is used as the reference lattice. The conditions ensure that the end point of the vector $\mathbf{x}^{(0)}$ has the same internal coordinates ξ_i in the original basis and in the "natural" basis of the second lattice which is related though $\mathbf{S}$ to the original basis.

Eliminating $\mathbf{x}$ from eqn. (38.6) gives

$$\mathsf{u} = \mathsf{y}-\mathsf{x} = (\mathsf{I}-\mathsf{S}^{-1})\mathsf{x}^{(0)}, \tag{38.7a}$$

which is identical with (38.2) if b is replaced by $-\mathbf{u}$ and p is replaced by $\mathbf{x}^{(0)}$. However, the interface has not yet been introduced and the vectors $\mathbf{x}^{(0)}$ are not restricted to a plane; in fact, the equation is regarded as a definition of these vectors $\mathbf{x}^{(0)}$ through

$$\mathbf{x}^{(0)} = (\mathsf{I}-\mathsf{S}^{-1})^{-1}\mathbf{u}. \qquad (38.7\text{b})$$

By substituting for u the three base vectors of the reference (−) lattice in turn, we see that the columns of the matrix $(\mathsf{I}-\mathsf{S}^{-1})^{-1}$ define the corresponding base vectors of the *O*-lattice. Any non-coplanar triad of vectors **u** specifies a primitive or non-primitive cell of the reference lattice and eqn. (38.7b) then gives the corresponding *O*-lattice cell; the ratios of the volumes of the two cells is given by $(\mathsf{I}-\mathsf{S}^{-1})^{-1}$.

The matrix $(\mathsf{I}-\mathsf{S}^{-1})$ may be of rank 3, 2, or 1 (or trivially zero when the + and − lattices coincide). When it is of rank 3, the solutions (38.7) represent a lattice of points in three dimensions, and each point u of the reference lattice has an "image" which is a point of the *O*-lattice. In other cases, the *O*-lattice becomes partly continuous. Thus if the matrix is of rank 2, solutions only exist for reference lattice points u lying in a plane through the origin, and each such point has an image which is a line containing an infinite number of *O*-lattice points (an *O*-line). The *O*-lattice then degenerates into a set of parallel lines. The direction of these lines is one in which the two lattices fit together exactly; i.e. S is an invariant line strain (p. 61). The *O*-lines are parallel to the invariant line and the independent components of u are confined to the plane with the invariant normal. Similarly, if the rank of $(\mathsf{I}-\mathsf{S}^{-1})$ is 1, the *O*-lattice degenerates into a set of parallel planes and the relation S represents an invariant plane strain. It is readily seen that when *S* has the form (9.10), each 2×2 sub-determinant of $(\mathsf{I}-\mathsf{S}^{-1})$ is necessarily zero and hence the rank of this matrix is 1.

As already noted, there are many possible choices of S which will relate two lattices in fixed relative orientations. Bollmann postulated that the correct choice is that which maximizes the three-, two- or one-dimensional unit cell of the *O*-lattice; this means that the distance between corresponding points y and x of eqn. (38.6) is minimized, and that S relates nearest neighbours in the region round the origin. (Note that whereas a coincidence site lattice for fixed real lattices is independent of the choice of S, the *O*-lattice varies with S.) Since the highest non-zero determinant contained in $(\mathsf{I}-\mathsf{S}^{-1})$ gives the ratio of the volume, area, etc., of the unit cell of the reference lattice to that of the *O*-lattice, the condition is equivalent to finding the smallest absolute value of this determinant.

All rotations are invariant line strains and hence give *O*-line lattices. In a plane normal to the rotation axis, the rotation has a simple two-dimensional representation A, and in an orthonormal coordinate system

$$(\mathsf{I}-\mathsf{A}^{-1})^{-1} = \begin{pmatrix} \frac{1}{2} & \frac{1}{2}\cot\frac{1}{2}\theta \\ -\frac{1}{2}\cot\frac{1}{2}\theta & \frac{1}{2} \end{pmatrix}. \qquad (38.8)$$

Figure 8.18 shows the orientations of two cubic lattices rotated about a cube edge through $\theta = 2\arctan\left(\frac{3}{20}\right)$ and the corresponding planar section of the *O*-line lattice. The figure illustrates an important geometric property of the *O*-lattice, namely that each *O*-point may act as an origin for the relation $\mathbf{y} = \mathsf{S}\mathbf{x}$ which then relates nearest neighbour points

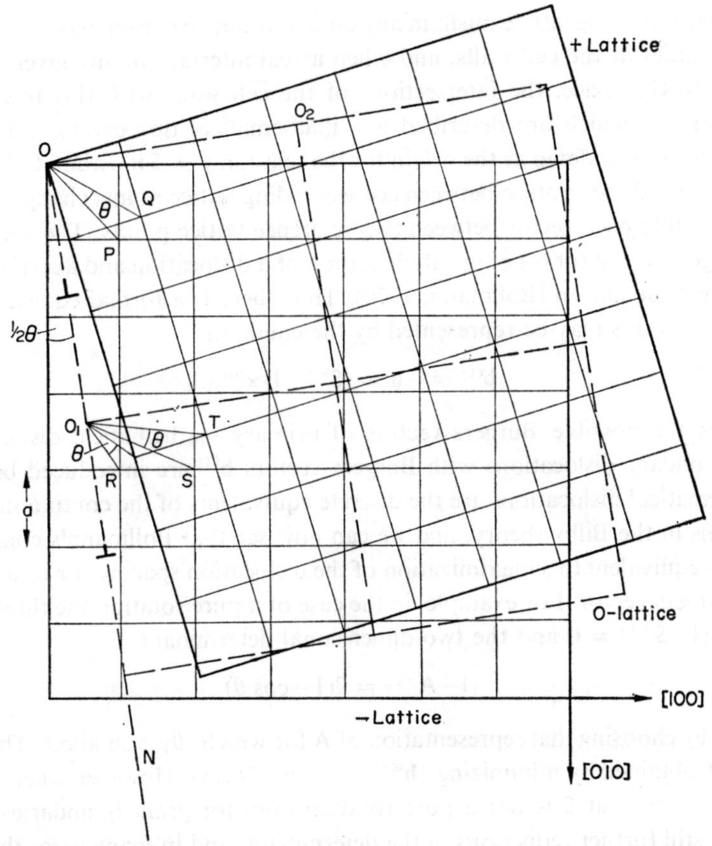

FIG. 8.18. (001) section through the O-line lattice for two cubic crystals rotated about [001] through $\theta = 2 \arctan(\frac{3}{20})$ All points of the + lattice may be obtained from those of the − lattice by a rotation of θ about the origin O; e.g. this carries OP into OQ and OR into OT. Nearest neighbour points of the two lattices are related by a rotation of θ about the nearest point of the O-lattice; e.g. OP into OQ and O_1S into O_1T.

The positions of the mathematical dislocations in a symmetric tilt boundary OO_1N are also shown.

of the two lattices in the vicinity of the O-point. This may be shown analytically by considering the relations between the two sets of lattice points when the origin is shifted to the O-lattice point $\mathbf{x}^{(0)}$. Using eqn. (38.7), the relation

$$\mathbf{y} - \mathbf{x}^{(0)} = \mathbf{S}\mathbf{x} - \mathbf{x}^{(0)}$$

may be put in the form

$$\mathbf{y} - \mathbf{x}^{(0)} = \mathbf{S}(\mathbf{x} + \mathbf{u} - \mathbf{x}^{(0)}) \tag{38.9}$$

which is of the original form $\mathbf{y} = \mathbf{S}\mathbf{x}$ but corresponding points of the two crystals are changed. Thus point $\mathbf{y}$ of the + crystal was originally derived from point $\mathbf{x}$ of the − crystal, but is now derived from the point $\mathbf{x}+\mathbf{u}$ (measured from the first origin).

We can now make the transition from the theory of the O-lattice to a dislocation model of a boundary by separating the different O-elements (points, lines, or planes) by "cell walls". The accumulating lattice misfit in any direction can be considered to be concentrated into discontinuities at the cell walls, and when a real interface in any given orientation is introduced into the space, the intersections of the cell walls with this interface become line discontinuities which are described as "mathematical dislocations". Each *O*-lattice point may now be considered as the origin for the relation $\mathbf{y} = \mathbf{S}\mathbf{x}$ within its own cell; when a cell wall is crossed the relation between corresponding lattice points changes by the vector $\mathbf{u}$ which is the difference vector between two reference lattice points. The vector $\mathbf{u}$ satisfies both the Burgers circuit (or topological) definition of a dislocation and also the kinematical, or displacement, definition (Bollmann, 1970). Thus there is a formal equivalence with the Bilby theory, and this may be represented by the equation

$$\mathbf{b}^{(1)} = -\mathbf{u} = (\mathbf{S}^{-1}-\mathbf{I})\,\mathbf{x}^{(0)}, \tag{38.10}$$

where $\mathbf{b}^{(1)}$ are the possible Burgers vectors of primary mathematical dislocations in the interface. Secondary dislocations with Burgers vectors $\mathbf{b}^{(2)}$ are introduced below.

The mathematical dislocations are the discrete equivalents of the continuous distribution of dislocations in the Bilby theory, and we can now see that Bollmann's condition for the choice of $\mathbf{S}$ is equivalent to a maximization of the dislocation spacings or to a minimization of the net Burgers vector. For example, in the case of a pure rotation the three dimensional determinant $|\mathbf{I}-\mathbf{S}^{-1}| = 0$ and the two-dimensional determinant

$$|\mathbf{I}-\mathbf{A}^{-1}| = 2(1-\cos\theta) \tag{38.11}$$

is minimized by choosing that representation of $\mathbf{A}$ for which $|\theta|$ is smallest. This is the same result as that obtained by minimizing $|\mathbf{b}^{max}|$ in eqn. (36.18). However when more general forms are used, so that $\mathbf{S}$ is not a pure rotation even for grain boundaries, there is the possibility of still further reductions in the determinant, and in many cases the two-dimensional determinant may be made zero. The Bollmann condition is then inapplicable without additional arbitrary assumptions (Christian and Crocker 1974).

We have noted above that if $\mathbf{S}$ represents an invariant plane strain, the *O*-lattice degenerates into an *O*-plane lattice. The cell walls of this lattice are parallel planes midway between the *O*-elements and an interface parallel to the *O*-planes thus intersects no cell walls and hence contains no mathematical dislocations. This description is self-consistent since the interface is fully coherent, i.e. it is the invariant plane of the deformation $\mathbf{S}$.

The difficulty which now arises is that unless some restrictions are placed on $\mathbf{S}$ there are many zero dislocation descriptions. As previously noted, for example, all rational symmetrical tilt boundaries in cubic crystals may be represented as coherent twin boundaries and hence, by appropriate choice of $\mathbf{S}$, as dislocation-free boundaries. If the Miller indices of the boundary are allowed to increase without limit, the region near $\theta = 0°$ becomes more and more populated with "coherent" twin relations. However, for low-angle boundaries with low densities of coincidence sites and very small spacing of lattice planes parallel to the boundary, the usual description (Section 36) of the boundary as an array of edge dislocations, which is obtained by considering $\mathbf{S}$ as a rotation, is physically more realistic. Thus the Bollmann criterion does not work in this case: the difficulty is exactly that mentioned on p. 348.

For high-angle grain boundaries, more general expressions for S are often appropriate, as is clearly shown by the existence of deformation twins where the simple shear S describes a physical correspondence, rather than just one of many possible mathematical representations. Even here however the Bollmann condition can only be used if combined with an assumption about which correspondences (or lattice reproducing shears U, eqn. (36.12)) are permissible: the decision is equivalent to a choice of which coincidence site relations have special properties. Thus Bollmann considered $[1\bar{1}0]$ tilt boundaries in b.c.c. lattices and used two forms for A, one representing a rotation of θ and the other relating corresponding unit cells as in the (112) twin. The second form gives

$$|\mathsf{I}-\mathsf{A}^{-1}| = 2-\cos\theta-(5\sqrt{2}/4)\sin\theta, \qquad (38.12)$$

and has a smaller absolute value than (38.11) for $\theta \geqslant 29{\cdot}5°$. Moreover, the determinant in (38.12) has zero values at $\sim 50{\cdot}5°$ and $70{\cdot}5°$, corresponding to (332) ($\Sigma = 11$) and (112) ($\Sigma = 3$) twin boundaries. However, there is also a (114) ($\Sigma = 9$) twin boundary at $\sim 39°$, and there seems no reason why this also should not be regarded as a full coherent, zero dislocation boundary, although Bollmann chose not to do so.

For interphase boundaries, the possibilities of zero dislocation descriptions are remote because of the restrictive conditions needed to enable S to be described as an invariant plane strain (Section 9). However, there is an equivalent difficulty, since by choosing correspondences involving large shears and/or shuffles, the value of the Bollmann determinant may be reduced.

The element of arbitrariness in the primary dislocation description is also relevant to the secondary dislocation models of boundaries which deviate from ideal conditions. However, we first have to consider some other aspects of the primary model of discrete mathematical dislocations. The theory completely specifies their spacing in any given direction, but there is some ambiguity about the definitions of their positions. Bollmann assumes that the cell wall between, for example, the origin and the *O*-lattice point $\mathsf{x}^{(0)}$ is defined by the condition that the magnitude of the relative displacement of corresponding points y and x is smaller within the cell around the origin than the relative displacement of new corresponding points y and x+u produced by a change of origin to $\mathsf{x}^{(0)}$, but is larger outside this cell. This means that points on the cell wall may be defined by the condition

$$|\mathsf{y}-\mathsf{x}|^2-|\mathsf{y}-\mathsf{x}-\mathsf{u}|^2 = 0 \qquad (38.13)$$

or after some algebraic manipulation

$$\mathsf{u}'\,\mathsf{G}\,(\mathsf{y}-\mathsf{x})-\tfrac{1}{2}\mathsf{u}'\,\mathsf{G}\,\mathsf{u} = 0, \qquad (38.14)$$

where G is the metric tensor. Finally, since $\mathsf{y}-\mathsf{x} = (\mathsf{I}-\mathsf{S}^{-1})\mathsf{y}$, the equation giving any point y on the cell wall may be expressed as

$$\mathsf{u}'\,\mathsf{G}\,(\mathsf{I}-\mathsf{S}^{-1})\,\mathsf{y}-\tfrac{1}{2}\mathsf{u}'\,\mathsf{G}\,\mathsf{u} = 0. \qquad (38.15)$$

It follows from eqn. (37.7a) that eqn. (38.15) is satisfied by the point $\mathsf{y} = \tfrac{1}{2}\mathsf{x}^{(0)}$; in general, the cell wall bisects, but is not necessarily perpendicular to, the line joining the two *O*-elements which it separates. The whole procedure is equivalent to the mapping of the Wigner–Seitz–Slater type cells of the reference lattice into the real lattice.

When the O-lattice cell is large compared with those of the real lattices, the cell walls are spaced at many atom distances, and the structure of the interface, obtained by relaxation from the initial state we have just described, may correspond to arrays of physical dislocations. Unlike the mathematical dislocations, the positions of which are defined in the O-lattice, the physical dislocations cannot in general be uniformly spaced (cf p. 335). Thus in Fig. 8.18, mathematical edge dislocations corresponding to a symmetrical tilt boundary along $OO_1N = (20, \bar{3}, 0)$ are at positions $[\pm(n+\frac{1}{2})x_1^{(0)}, 0]$, where n is any integer, and their mutual separation is $\frac{1}{2}a$ cosec $\frac{1}{2}\theta$ along the interface or $\frac{1}{2}a \cot \frac{1}{2}\theta = 10a/3$ projected along the [010] direction of the reference lattice. The physical dislocations may be placed on the (010) planes of the reference (−) lattice if they are spaced at repeating intervals of $3a$, $3a$, and $4a$. The centre of the physical dislocation should lie on the lattice plane nearest to a cell wall. For small O-lattice cells it may no longer be possible to describe the structure by physical dislocations. However, for both large and small O-lattice cells, any boundaries which contain a relatively high density of O-lattice points are likely to have lower energies than other boundaries. When the O-lattice becomes a lattice of O-lines, the cell walls consist of planes parallel to the invariant line, and any boundary which contains the invariant line will intersect only ony type of cell wall. Thus the structure of an interface of this type may consist of a single set of parallel mathematical dislocations. This applies to all tilt boundaries and also to all martensitic interfaces. The formal theory of martensite crystallography includes a necessary condition that $\mathbf{S}$ represents an invariant line strain and that this invariant line lies in the interface.

Some complications arise when the plane with the invariant normal is not rational and thus contains few or no points of the reference lattice. This plane may be considered as a sub-space of the reference lattice, within which each lattice point gives rise to an image as an O-line. As the density of lattice points in this sub-space decreases, the distances between the O-lines increases and the Burgers vectors of the mathematical dislocations becomes very large. This scarcely seems a plausible physical model, and to overcome its limitations the effect of a relative translation of the two lattices may be considered. Bollmann shows that if the − lattice is displaced relative to the + lattice, the reference and O-lattice are also displaced. Let the displacement of the reference lattice be $\mathbf{t}$; then the displacement of the − lattice is $\mathbf{S}\mathbf{t}$ and of the O-lattice is

$$\mathbf{t}^{(0)} = (\mathbf{I}-\mathbf{S}^{-1})^{-1}\,\mathbf{t}. \tag{38.16}$$

The geometrical structure of a three-dimensional O-lattice is preserved by this translation, but when the rank of $(\mathbf{I}-\mathbf{S}^{-1})$ is only 2, the lattice of O-lines may disappear completely if $\mathbf{t}$ has a component normal to the plane of the reference lattice which has an invariant normal. Now consider again the situation when the plane with the invariant normal contains few lattice points. By allowing relative translations of the + and − lattices, these points of the reference lattice which are close to this plane may be projected on to it and a relatively dense lattice of O-lines may thus result. This means that the irrational plane of the reference lattice which generates the lattice of O-lines is replaced by a stepped series of rational planes, which together generate the *O*-lattice. Physically, the transformation around each *O*-line is still described by S but in addition the lattices are allowed a small relative translation which may not be the same around different *O*-lines. This inhomogeneous transformation thus permits the periodicity of the interface structure to become much smaller, and the discontinuities less drastic, and thus it may be assumed that it is likely to give a lower energy.

In a similar way, for an *O*-plane lattice it is necessary to realize that the *O*-planes are the images of reference lattice points which lie along a line. If this line is not a rational lattice direction of the reference lattice adjacent lattice points may be projected on to it in order to give a reasonably dense lattice of *O*-planes.

The *O*-lattice theory may be regarded as a quantized version of the formal Bilby theory since it adds to this theory the requirement that the net Burgers vector **b** of eqn. (38.2) is obtained by summation of individual mathematical or physical dislocation lines with lattice Burgers vectors. For the special case of grain boundaries we have also introduced the concept of secondary dislocations which represent the deviation from an ideal or coincidence site situation. In the same way an ideal or low energy interface between two lattices (if such exists) may in principle be used to give a secondary dislocation model of an interface which deviates from this ideal condition.

The theory of secondary dislocation arrays has been expressed in general form by Bollman, who introduced the concept of the "*O*-2" lattice, which is an *O*-lattice formed between two *O*-lattices rather than between two real lattices. The two *O*-lattices correspond to the actual transformation S and the ideal or reference relation S_i, and are both defined in the usual way by eqn. (38.7). The relation between the two *O*-lattices is then

$$B = S^{-1} S_i, \tag{38.17}$$

where B specifies the deviation to be described by the secondary dislocation array. The *O*-2 lattice is now defined by an equation analogous to (38.7), namely

$$x^{(02)} = (I - B^{-1})^{-1} u^{(0)}, \tag{38.18}$$

where $u^{(0)}$ represents possible translations (Burgers vectors) in the ideal or reference *O*-lattice. Clearly, the translation repeat vectors of the *O*-lattice are possible vectors $u^{(0)}$, just as crystal lattice repeat vectors are possible vectors **u** in eqn. (38.7). However, the repeat vectors of the *O*-lattice only form a subset of the possible vectors $u^{(0)}$, as will now be shown.

We pointed out in Section 36 that when a coincidence lattice exists there are a finite number of different internal coordinates ξ_i at the *O*-lattice points. The arrangement of crystal lattice points within one *O*-cell is referred to as a pattern element, and the combination of different pattern elements gives the complete pattern of lattice points. If the complete pattern is periodic, it consists of a finite number of pattern elements, this number being the same as the number of different values of ξ. Bollmann introduced the idea of the reduced *O*-lattice, which is defined as a single cell of lattice *A* within which the ξ_i values of the *O*-lattice points are plotted. Now consider a shift of the origin of the *O*-lattice to one of these internal points, and correspondingly a relative shift of the two crystal lattices. Such a displacement will obviously lead to conservation of the pattern as a whole since the displacement gives another pattern element. The lattice of all such displacements is called the "complete pattern shift lattice" or "DSC lattice"; it is formed by an infinite repetition of the reduced *O*-lattice. This lattice contains as a superlattice the *O*-lattice itself and also the lattice of all displacements of the primary origin, here supposed to be a coincidence site, to any other lattice site of that lattice.

We have assumed that the existence of a periodic pattern implies a coincidence site lattice, but this follows only if the primary origin for the relation between the two crystal lattices has been taken to be at a lattice site. However, the pattern will remain periodic if this origin is not at a crystal lattice site, and there is then no coincidence site lattice. When considering best fit between lattices at an interface, it is the existence of a periodic pattern which is important, and the actual positions of the two sets of atoms may be translated away from a coincidence lattice situation. Any such translation may be represented as a displacement of the *O*-lattice by a vector which is not a repeat vector of the DSC lattice; this displacement modifies the pattern in such a way that the reduced *O*-lattice is translated within a unit cell of lattice *A*, and there is no longer a reduced *O*-lattice point at the origin. The number of points in the reduced *O*-lattice, however, remains constant, so that the number of pattern elements and the periodicity of the pattern are conserved. Clearly, the complete pattern displacement lattice is also unchanged, and any translation through a vector of this lattice will reproduce the new pattern.

It is now clear that the vectors $\mathbf{u}^{(0)}$ of eqn. (38.18) may be any vectors of the DSC lattice since linear discontinuities with this displacement (Burgers vector) in an interface will reproduce the interface structure. However, the vectors $\mathbf{u}^{(0)}$ are referred to the *O*-lattice of the ideal interface, and it is necessary to transfer them to the reference (−) lattice. This may be done through eqn. (38.16) which relates displacements in the reference and *O*-lattices. Thus the possible Burgers vectors $\mathbf{b}^{(2)}$, referred to the − lattice, are

$$\mathbf{b}^{(2)} = (\mathbf{S}^{-1}-\mathbf{I})\,\mathbf{u}^{(0)} \tag{38.19}$$

which may be compared with (38.10). Combining (38.15) with (38.18) gives a new basic equation for the *O*-2 lattice which replaces eqn. (38.7) for the *O*-lattice as

$$(\mathbf{S}^{-1}-\mathbf{I})(\mathbf{I}-\mathbf{B}^{-1})\,\mathbf{x}^{(02)} = \mathbf{b}^{(2)}. \tag{38.20}$$

The *O*-2 lattice may be subdivided by cell walls in an exactly analogous manner to the *O*-lattice, and the intersection of each such wall with the interface then represents a mathematical "secondary" dislocation with a Burgers vector $\mathbf{b}^{(2)}$. The positions of the cell walls are defined by an equation equivalent to (38.15), namely

$$\mathbf{b}^{(2)\prime}\,\mathbf{G}\,(\mathbf{I}-\mathbf{S}^{-1})(\mathbf{I}-\mathbf{B}^{-1})\,\mathbf{y}-\tfrac{1}{2}\,\mathbf{b}^{(2)\prime}\,\mathbf{G}\,\mathbf{b}^{(2)} = 0. \tag{38.21}$$

As already noted, the Burgers vectors $\mathbf{b}^{(2)}$ of the secondary dislocations can be markedly smaller than those of the primary dislocations.

Experimental observations of secondary dislocations have been made by several authors, and especially by Schober and Balluffi (1971) and Balluffi *et al.* (1972) using thin film bicrystals of gold made by welding single crystal films together. The grain boundary dislocation

(GBD) structures were observed by transmission electron microscopy. The authors make a distinction between the geometrically necessary dislocations which in accordance with the above theory are part of the equilibrium structure of a boundary and which they term IGBDs and additional or "extrinsic" GBDs, later called by Hirth and Balluffi (1973) just GBDs. The non-intrinsic dislocations are present because of accidents of growth, plastic deformation, etc., but the distinction cannot always be maintained without ambiguity. For several types of boundary close to coincidence site conditions, arrays of IGBDs as predicted by theory have been found, but in all cases these were also some extrinsic GBDs which sometimes had complex interactions with the IGBDs. Extrinsic GBDs also occasionally arise from Bardeen–Herring type defect sources and sinks in the boundaries.

The most general type of semi-coherent boundary might contain sets of non-parallel dislocations with non-parallel Burgers vectors. Motion of such a boundary could involve the climb of some dislocations and the glide of others; it would then be a combination of the simple martensitic and epitaxial types discussed above. The concept of surface dislocations provides a convenient way of discussing the average structure of such a boundary. Clearly a surface dislocation, like a line dislocation, cannot end within a crystal. Three or more surface dislocations may, however, meet in a common line, which Bilby (1955) has called a surface dislocation node. There is then a compatibility condition on the individual surface dislocation tensors which may be compared with the equation for the conservation of Burgers vectors. Examples of surface dislocation nodes occur in the deformation patterns produced by indenting zinc single crystals (Li *et al.*, 1953), and in the structures produced by martensitic transformation in indium–thallium alloys (Basinski and Christian, 1956).

We have discussed only boundaries between two single crystal regions. As we shall discuss in detail in Part II, Chapter 22, martensite plates often contain fine parallel twins which accommodate the misfit of their individual lattices with the parent lattice in a manner similar to that which could be achieved by having glide dislocations with a Burgers vector in the twinning direction. The formal theory of this situation shows that it is equivalent to a dislocation boundary, but of course the energy will be different in the two cases. Twinning cannot change the density of equivalent lattice points, so that it need not be considered in relation to epitaxial semi-coherence. There are other situations in which fine lamellae of two different structures have a common boundary with a third structure, for example in eutectoidal reactions or discontinuous precipitation, but the boundary is then incoherent.

REFERENCES

AMELINCKX, S. (1957) *Dislocations and Mechanical Properties of Crystals*, p. 3, Wiley, New York; (1958) *Phil. Mag.* **6**, 34.

AUST, K. T. and RUTTER, J. W. (1959) *Trans. Am. Min. (Metall.) Engrs.* **215**, 820.

BALLUFFI, R. W., KOMEN, Y., and SCHOBER, T. (1972) *Surf. Sci.* **31**, 68.

BASINSKI, Z. S. and CHRISTIAN, J. W. (1956) *Acta metall.* **4**, 371.

BILBY, B. A. (1955) *Report of the Conference on Defects in Crystalline Solids*, p. 123, The Physical Society, London.

BILBY, B. A. and SMITH, E. (1958) *Proc. Roy. Soc.* A, **236**, 481.

BISHOP, G. H. and CHALMERS, B. (1968) *Scripta metall.* **2**, 133; (1971) *Phil. Mag.* **24**, 515.

BOLLMANN, W. (1967) *Phil. Mag.* **16**, 363, 383; (1970) *Crystal Defects and Crystalline Interfaces*, Springer-Verlag, Berlin.

BORN, M. (1946) *Proc. Math. Phys. Soc. Egypt* **3,** 35.
BRAGG, W. L. (1940) *Proc. Phys. Soc.* **52,** 54.
BRANDON, D. G. (1966) *Acta metall.* **14,** 479.
BRANDON, D. G., RALPH, B., RANGANATHAN, S., and WALD, M. S. (1964) *Acta metall.* **12,** 813.
BROOKS, H. (1952) *Metal Interfaces,* p. 20, American Society for Metals, Cleveland, Ohio.
BURGERS, J. M. (1939) *Proc. Akad. Sci. Amst.* **42,** 293.
CARRINGTON, W., HALE, K. F. and MCLEAN, D. (1960) *Proc. Roy. Soc.* A, **259,** 203.
CHALMERS, B. and GLEITER, H. (1971) *Phil. Mag.* **23,** 1541.
CHAUDHARI, P. and CHARBNAU, H. (1972) *Surf. Sci.* **31,** 104.
CHRISTIAN, J. W. (1962) *The Decomposition of Austenite by Diffusional Processes,* p. 371, Wiley, New York.
CHRISTIAN, J. W. and CROCKER, A. G. (1974) In *Dislocation Theory: A Collective Treatise,* (ed. F. R. N. Nabarro), Dekker, New York (in press).
COUCHMAN, P. R., JESSER, W. A., and KUHLMANN-WILSDORF, D. L. (1972) *Surf. Sci.* **33,** 429.
DAHL, R. E., Jr., BEELER, J. R., and BOURQUIN, R. D. (1972) *Interatomic Potentials and Simulation of Lattice Defects,* p. 673, Plenum Press, New York and London.
FORTES, M. A. (1972) *Revista de fisica Quimica e Engenharia* **4A,** 7.
FRANK, F. C. (1950) *Symposium on the Plastic Deformation of Crystalline Solids,* p. 150, Office of Naval Research, Pittsburgh Pennsylvania; (1955) *Report of the Conference on Defects in Crystalline Solids,* p. 159, The Physical Society, London.
HASSON, G., BOOS, J. Y., HERBEUVAL, I., BISCONDI, M., and GOUX, C. (1972) *Surf. Sci.* **31,** 115.
HERRING, C. (1951) *The Physics of Powder Metallurgy,* p. 143, McGraw-Hill, New York; (1953) *Structure and Properties of Solid Surfaces,* p. 5, University of Chicago, Chicago, Illinois.
HIRTH, J. P. and BALLUFFI, R. W. (1973) *Acta metall.* **7,** 929.
KRONBERG, M. L. and WILSON, F. H. (1949) *Trans. Am. Inst. Min. (Metall.) Engrs.* **185,** 50.
LI, J. C. M. (1972) *Surf. Sci.* **31,** 12.
LI, C. H., EDWARDS, E. H., WASHBURN, J., and PARKER, E. R. (1953) *Acta metall.* **1,** 223.
LOMER, W. M. and NYE, J. F. (1952) *Proc. R. Soc.* A, **212,** 516.
MEIJERING, J. L. (1953) *Acta metall.* **1,** 607.
MOTT, N. F. (1948) *Proc. Phys. Soc.* **60,** 391.
RANGANATHAN, S. (1966) *Acta Crystallogr.* **21,** 197.
READ, W. T. (1953) *Dislocations in Crystals,* McGraw-Hill, New York.
READ, W. T. and SHOCKLEY, W. (1950) *Phys. Rev.* **78,** 275; (1952) *Imperfections in Nearly Perfect Crystals,* p. 77, Wiley, New York.
SCHOBER, T. and BALLUFFI, R. W. (1971) *Phil. Mag.* **24,** 165, 469.
SHOCKLEY, W. and READ, W. T. (1949) *Phys. Rev.* **75,** 692.
SHUTTLEWORTH, R. (1950) *Proc. Phys. Soc.* A, **63,** 444.
SMITH, C. S. (1952) *Metal Interfaces,* p. 65, American Society for Metals, Cleveland, Ohio; (1953) *Acta metall.* **1,** 244.
SMOLUCHOWSKI, R. (1952) *Phys. Rev.* **87,** 482.
VAN DER MERWE, J. H. (1950) *Proc. Phys. Soc.* A, **63,** 613.
VÍTEK, V. (1970) *Scripta metall,* **4,** 725.
WARRINGTON, D. H. and BUFALINI, P. (1971) *Scripta metall.* **5,** 771.
WEINS, M. J. (1972) *Surf. Sci.* **31,** 138.

CHAPTER 9

Diffusion in the Solid State

39. MECHANISM OF ATOMIC MIGRATION

It is usual to begin a discussion of diffusion with a statement of Fick's law, which is the analogue for material flow of Fourier's law for heat flow by conduction. Although it is not strictly valid, and has had to be modified in the formal theory, Fick's equation retains its significance since experimental results are almost always presented in the form of diffusion coefficients defined by reference to it. We shall find it simpler, however, to approach the subject in a more fundamental manner by considering first the ways in which atoms may move through the crystal. A general account of diffusion can be developed without reference to the actual mechanism of atomic migration, but the more detailed theories depend on this mechanism, and may readily be linked to our previous discussion of point defects.

Consider an assembly containing two or more kinds of different atoms. "Chemical" diffusion takes place when this assembly is not in equilibrium with respect to distribution of the different atomic species, so that there is, for example, a displacement of B atoms relative to A atoms. "Tracer" diffusion presupposes a situation in which the assembly is in equilibrium, except for the distribution of two components which differ from each other in an insignificant manner (so far as the diffusion problem is concerned). These two components must thus be virtually identical in chemical properties, but must be "labelled" in some way, e.g. by their nuclear properties. Tracer diffusion is, in fact, generally investigated by setting up a concentration gradient of a radioactive isotope of one of the components of an otherwise homogeneous assembly. Self-diffusion usually means tracer diffusion of an isotope of a pure element, or of the solvent atoms of a dilute solution.

Consider the chemical diffusion of an atomic species, B. The B atoms may share a set of sites with the other components, forming a substitutional solid solution, or they may partially occupy a separate set of sites, so that B is an interstitial solute. In the latter case, the mechanism by which B atoms diffuse is fairly certain, and needs little comment. In all interstitial solutions, the fraction of possible solute sites which are actually occupied by atoms is quite small, and most of these sites are vacant. An interstitial atom can then migrate by jumping from one such site to a neighbouring vacant site. This process is usually treated by the chemical rate theory of Chapter 3, there being a saddle point of free energy at some intermediate configuration which is critical for the attempt at a jump to succeed. Interstitial diffusion is shown diagrammatically in Fig. 9.1(a). We note that the A atoms form

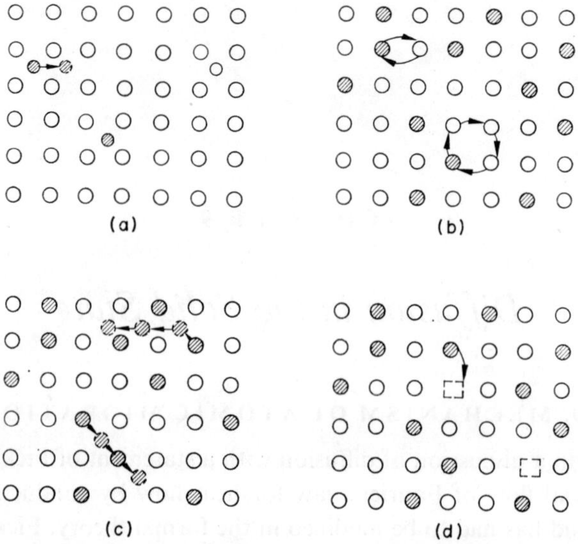

FIG. 9.1. Possible atomic mechanisms for diffusion. *A* atoms; open circles. *B* atoms; filled circles. (a) Solute diffusion in interstitial solid solution. (b) Place exchange and ring diffusion. (c) Interstitial and interstitialcy diffusion. (d) Vacancy diffusion.

a rigid frame of reference, relative to which the migration of the *B* atoms may be studied.

For substitutional solid solutions, the atomic mechanism is less certain. There are three basic possibilities for the relative displacement of *A* and *B* atoms occupying a common set of sites. The first of these is shown in Fig. 9.1(b), and involves the direct interchange of two atoms on neighbouring sites. If this interchange is imagined to occur by the rotation of the two atoms about their common centre of separation, it is obvious that considerable distortion of the surrounding structure will be required. For this reason, the probability of such a direct interchange is believed to be very low in metallic crystals. As a result of the process, both *A* and *B* atoms are displaced (in opposite directions) relative to the remaining atoms.

The second basic mechanism is the movement of atoms through the interstices of the lattice, as for interstitial solid solutions. An individual atom migrates by leaving a normal site and successively occupying a series of interstitial sites, until it meets a vacancy and returns to the normal sites. Such migration will be energetically unfavourable unless the structure is very "open". The interstitial mechanism is illustrated in Fig. 9.1(c) which also shows a variation suggested by Seitz (1950a, b). In Seitz's "interstitialcy" mechanism, the interstitial atom does not "squeeze past" the atoms on normal sites, but occupies the next normal site, "pushing" the atom on this site into an interstitial position, and so on. We have already discussed this mechanism in Section 17 for the case of self-interstitials.

Figure 9.1(d) shows the third mechanism for atomic migration, which depends on the presence of vacancies in the structure. An atom next to a vacant site may jump into that site, thus in effect changing places with the vacancy. The relative motion of *A* and *B* atoms results from the migration of vacancies through the lattice, and diffusion by vacancies is in some of its effects intermediate between the interstitial and place exchange mechanisms. If only B atoms could change places with a vacancy, the motion of B would take place with reference to an unchanging lattice frame of A atoms. On the other hand, if the probabilities of A and B atoms jumping into a neighbouring vacant site were equal, the average displacement of the two kinds of atom would be equal and opposite, as in place exchange.

The above suggestions were once thought to cover all possibilities for the atomic mechanism of diffusion in the solid state. This is true so long as processes involving one or two atoms are considered, but it is now recognized that the co-operative motion of a larger number of atoms is also possible. The first type of co-operative process, suggested by Zener (1950) is an extension of the place exchange mechanism, and is called ring diffusion. A small number of atoms n move together in a ring, thus enabling the distortion to be spread over a larger volume, and reducing the energy. As the number of atoms in the ring increases, the activation energy first falls steeply, and then rises slowly; the minimum for a b.c.c. structure probably corresponds to $n = 4$. Ring diffusion is illustrated in Fig. 9.1(b).

On p. 125, we considered the possibility that an interstitial defect may spread its distortion along a line to form a crowdion. This suggestion was first made by Paneth (1950) as a possible diffusion mechanism in the alkali metals, and the motion of a crowdion may be regarded as a co-operative version of the interstitial mechanism. In the b.c.c. structure, crowdions will form (if at all) along $\langle 111 \rangle$ directions. In general, crowdions should be able to migrate readily along nearest neighbour axes in the crystal, but cannot easily turn corners.

An additional mechanism for diffusion depends on the migration of vacancy pairs rather than of single vacancies. As discussed in Section 17, the divacancies migrate more readily than single vacancies, but under conditions of thermal equilibrium the concentration of divacancies is much smaller than that of single vacancies. This implies that divacancies contribute relatively little to the equilibrium diffusion rate of most metals, although this conclusion depends sensitively on the value of the divacancy binding energy $2\Delta h_{\square} - \Delta h_{\square\square}$. The divacancy contribution increases with increasing temperature, and with a high binding energy it may be $\sim 10\%$ of the total diffusion at temperatures near the melting point. Moreover, pair or higher cluster migration may become important when there are large numbers of excess vacancies present, e.g. in a quenched specimen. The divacancy binding energy is again important, since it determines the average number of jumps made by a divacancy before it dissociates again into single vacancies. It is perhaps worth noting here that non-equilibrium concentrations of point defects may be very effective in producing enhanced atomic migration over short distances, but are unlikely to contribute much to long-range diffusion effects, since the defects themselves anneal out so rapidly. Enhanced diffusion will be obtained, however, if a high defect concentration can be maintained throughout the diffusion anneal by external means.

A co-operative diffusion mechanism suggested by Nachtrieb and Handler (1954) is based on the idea (p. 125) that a "relaxation" or local melting of the structure occurs in the immediate vicinity of a vacancy. The disordering of a small number of atoms, amongst which the volume of the vacancy is distributed, leads to rapid rearrangements by processes equivalent to liquid diffusion. The disordered volume itself is supposed to move through the structure by local "melting" and "freezing" of atoms at its boundaries, one or two at a time. If its configuration remained constant during this migration, the whole process would be equivalent to vacancy diffusion, but it is supposed that rearrangements within the disordered volume occur at about the same rate as the movement of the volume itself. According to this theory, the various atomic mechanisms listed above may thus all be involved to some extent in the actual diffusion process.

In the above discussion, we have confined ourselves mainly to geometrical possibilities, and mentioned energies only occasionally. For any particular metal, we expect one of these possibilities to be favoured at the expense of the others, and thus to be the dominant mechanism for diffusion. We briefly repeat the reasons for this. The atoms, though vibrating about their equilibrium positions, spend most of their time in the vicinity of an atomic site, or minimum in the energy field. In order to move away from a given position, an atom must temporarily acquire an energy much greater than the mean thermal energy of vibration. The activation energy depends on the process concerned, and since the rate of the process is inversely proportional to the exponential of the corresponding free energy of activation, the mechanism with the lowest activation energy will probably be much more important than any other mechanism. Nevertheless, the accuracy of diffusion measurements is now often sufficient to permit contributions from the second most important process to be recognized and analysed.

The 'dominant' mechanism need not be that of lowest activation energy if different geometries exist to compensate for the different probabilites of individual atom movements. As we shall discuss later, for example, the activation energy for grain boundary migration is normally much lower than that for migration through the lattice. Nevertheless, the lattice makes the greater contribution to the overall diffusion rate at high temperatures because so few of the atoms are in grain boundary regions. As the temperature is lowered, the effect of the lower activation energy for boundary motion becomes progressively more important, and at sufficiently low temperatures virtually the whole of the diffusion may be along the boundaries.

Various methods have been used to deduce the principal mechanism of diffusion in particular metals or alloys. As discussed in Chapter 5, theoretical estimates of the activation energy are subject to severe limitations, but they may nevertheless sometimes enable meaningful distinctions to be made. Such estimates may also be compared with experimental activation energies, obtained from the measured variations of diffusion rates with temperature, and with the energies of formation and motion of the various atomic defects discussed in Section 17. A related method is to measure the pressure dependence of the diffusion coefficient, which defines an activation volume, and to compare this with predictions for the various models. For self-diffusion in pure metals, it has been suggested that the magnitude of the measured entropy of activation also provides a useful indication of mechanism.

Another group of methods depends on differences which arise from the geometry of the defects. For example, h.c.p. metals are anisotropic in diffusion properties and the ratio of the diffusion coefficients parallel and perpendicular to the basal plane is partly determined by entirely geometrical factors which differ with the mechanism. This has been used by Shirn *et al.* (1953) to deduce that vacancy diffusion is probably dominant in zinc, cadmium, and thallium. For cubic metals, we shall define later a quantity known as the correlation factor which has purely geometrical values in the case of tracer self-diffusion. This factor may be derived in various ways, e.g. from measurements of the self-diffusion coefficients of different isotopes of the same element, and the results used to differentiate between alternative possible mechanisms. An older method depends on the distinction made above between the interchange and interstitial mechanisms. If the lattice structure is regarded as a rigid framework, diffusion by an interchange method will not result in the displacement of the atoms relative to this framework. Diffusion by the other two methods can, however, lead to a mass flow of atoms, since there is no need for the motion of the *B* atoms to be balanced by an equal and opposite motion of *A* atoms. The experimental observation of this phenomenon is now known as the Kirkendall effect, and first became widely known after the work of Smigelskas and Kirkendall (1947) on diffusion in copper–zinc alloys. The mass motion was detected by inserting inert markers (molybdenum wires) at the original interfaces of a copper–brass–copper diffusion couple. At the end of the experiment, the separation of the wires was found to have decreased, and the magnitude of the change was too great for it to be attributed simply to the different atomic volumes of the diffusing species. If the wires, which take no part in the diffusion process, are assumed to be fixed to the lattice, the effect can only be explained by assuming that the number of zinc atoms which diffuse out of the brass is greater than the number of copper atoms which diffuse into it.

In a large number of subsequent investigations, it has been shown that the Kirkendall effect must usually be ascribed to unequal rates of diffusion of the two species through the lattice and not to secondary causes, such as transfer in the vapour phase. Thus, in the above experiment, the zinc atoms could diffuse interstitially from the brass, leaving vacancies, and take up normal sites in the copper. Alternatively, the zinc atoms may diffuse via vacancies, the net current of atoms being balanced by an opposite current of vacancies into the brass. The vacancies could coagulate together, forming macroscopic holes, or could diffuse to dislocations, with a resultant shrinking of the structure. The extra atoms in the copper would similarly cause an expansion of this region.

The results of both experimental and theoretical methods of determining the diffusion mechanism have been reviewed by many authors, e.g. Seitz (1950a, b), Huntington (1951), Lazarus (1960), Shewmon (1963), Howard and Lidiard (1964), Peterson (1968), and Seeger (1972). Seeger concludes that in all metals investigated in detail, the dominant mechanism of self-diffusion is the migration of monovacancies, and the migration of divacancies is the second most important process. The metals considered by Seeger were all f.c.c. and the mechanism in b.c.c. metals has always been more uncertain, although the existence of Kirkendall effects in many cases gives a strong indication that vacancy diffusion is dominant here also. However, there are certain anomalies in the diffusion behaviour of many b.c.c. metals which are not yet fully explained, and these will be described later. There are also some special situations; for example, solute diffusion of the noble metals in germanium, silicon, lead, tin, indium, and thallium appears to involve a combination of vacancy and interstitial mechanisms with the interstitial process dominant (Frank and Turnbull, 1956; Dyson *et al.*, 1966, 1967, 1968).

40. STATISTICAL BASIS OF DIFFUSION: FICK'S LAW

A typical diffusion experiment consists in bringing two homogeneous specimens into contact at an interface. The specimens may be different pure components, or may contain the same components in different concentrations. After a period of time, it is found that the sharp change of composition has been replaced by a more gradual change, extending for some distance on either side of the original interface. A diffusion process of this type, which tends to produce a homogeneous assembly, is easily visualized. If the probability of an atomic migration were independent of direction, i.e. of the neighbours of the atom which moves, then diffusion would be merely the statistical result of a large number of chance migrations. This is often referred to as a "random walk" problem The net statistical flow of any component would obviously be from regions of high concentration of that component to regions of low concentration.

In practice, the probability of atomic migration usually depends both on the type of atom and the direction of motion, and simple statistical flow does not occur. However, it is often convenient to divide the net atomic movements into a random flow and a superimposed drift velocity. The driving force for the random flow is the change in the configurational entropy of the assembly, whilst the drift velocity may be attributed to internal or external stresses, thermal gradients, or electrical fields, etc. A drift velocity in non-ideal solutions results from gradients of chemical potential.

The simple kinetic theory which we shall develop first in this section shows that random flow leads to a linear relation between flux $\mathbf{I}$ and concentration gradient ∇c, and this is used to define a diffusion coefficient. The drift velocity is treated separately unless it is also proportional to ∇c, in which case it may be combined with the statistical flow to define a different diffusion coefficient. An example is the important case of chemical diffusion in binary alloys. Since the terminology used in different books is not always identical, we find it convenient here to define the various diffusion coefficients by reference to experimental measurements. The three main types are tracer diffusion coefficients, intrinsic diffusion coefficients, and chemical or interdiffusion coefficients (Peterson, 1968).

Tracer-diffusion coefficients are measured from the atomic flux of an isotope which is present in low concentration in an otherwise homogeneous matrix. If the tracer is an isotope A_+ of the only or major component A, the coefficient D_{A+} is the tracer self-diffusion coefficient, whereas if the isotope B^+ is a minor component, D_{B+} is a tracer impurity or solute diffusion coefficient; in concentrated solutions these terms are not appropriate. Tracer self-diffusion by means of a vacancy mechanism is closely related to the motion of the vacancies, which may be used to define a vacancy diffusion coefficient. The vacancy motion in a pure element may be treated to a sufficient approximation as a random walk, but we shall show that the tracer jumps are not strictly random. The tracer self-diffusion coefficient is often called simply the self-diffusion coefficient, but experiments on the mass flow in a pure element subjected to a temperature gradient or an electrical field, etc., lead to the definition of another coefficient, the macroscopic or transport self-diffusion coefficient (Seeger, 1972). The atomic displacements in macroscopic self-diffusion are equivalent to a true random walk and the ratio of the tracer to the macroscopic self-diffusion coefficient is called the correlation factor (see Section 41).

The intrinsic diffusion coefficient D_A is used to describe the flux of A atoms in an alloy containing concentration gradients of A and hence of another component B. This is the type of diffusion which is much more difficult to discuss theoretically than is tracer diffusion. In a binary alloy, intrinsic diffusion coefficients D_A and D_B may be defined, but diffusion in a ternary alloy cannot be treated in this way unless the third component C is uniform in concentration. The chemical or interdiffusion coefficient D_{chem} also describes the diffusion process in a binary alloy with a concentration gradient, and it is used to describe the rate of mixing (or unmixing!) of the two species. It follows that D_{chem} may be expressed in terms of D_A and D_B.

The usual diffusion experiments involve net flow down the concentration gradient, even though this flux is not the result of purely random migrations. These investigations, in which both parts of the specimen are initially at equilibrium, do not correspond to the conditions which are often important in transformations, such as diffusion in a supersaturated solid solution. In certain circumstances, diffusion may take place against the direction of purely statistical flow, thus producing local regions which differ in concentration. This is sometimes called "uphill diffusion", and will occur when the probability of migration against the concentration gradient is large enough to affect the flow more than the greater number of atoms which are available for motion down this gradient.

The above description emphasizes the distinction between the individual migrations of the atoms and the net resultant flow. If the assembly as a whole is not in thermodynamic equilibrium, diffusion will redistribute the atoms in the direction of this equilibrium. A given atom moves from its position as a result of a fluctuation in the local thermal energy. There is no restriction on the direction in which it may move, and all that we can do is to calculate the relative probabilities of its moving in different directions. If these probabilities are not equal, the assembly is not in an equilibrium state, and the net result of a large number of atoms undergoing a large number of displacements will be to provide a net flow in some direction. As already stated, it is this net flow which we describe as diffusion, and it is misleading, and in general incorrect, to use the term for the motion of atoms over distances of the order of interatomic distances. Diffusion rates are often interpreted in terms of the velocities of individual atoms, expressed as the product of an average diffusion "force" and a "mobility" (compare the theory of Chapter 4). This formal approach can be very useful, but the true statistical nature of the atom movements must always be kept in mind.

Let us consider a simplified situation in which there is net flow in only one direction of a crystal of a binary alloy. We use the labels 1 and 2 for two neighbouring planes, distance d apart, perpendicular to the direction of flow, and we assume that the elementary diffusion process consists of atomic jumps between positions on planes of this type. The numbers of A atoms on unit areas of the two planes may be written $c_{A1}d$, $c_{A2}d$, where c_{A1}, c_{A2} are the concentrations of A atoms per unit volume at the two planes. If the probability per unit time that an A atom on plane 1 will jump to plane 2 is written $\pi_{A,12}$, and that of the reverse jump is $\pi_{A,21}$, the net flow of A atoms between the two planes is

$$I_A = c_{A1}d\pi_{A,12} - c_{A2}d\pi_{A,21}. \tag{40.1}$$

The frequency of jumping, π, depends on the chemical environment of the atom concerned. If all atoms are chemically identical, the frequency of jumping for a vacancy or interstitial mechanism may be assumed to be constant, and we write

$$\pi_{A,\,12} = \pi_{A,\,21} = \pi_A\,,$$

$$I_A = \pi_A d(c_{A1} - c_{A2}) = -\pi_A d^2 \frac{\partial c_A}{\partial x} = -D_A \frac{\partial c_A}{\partial x}\,, \tag{40.2}$$

where the x axis has been taken perpendicular to the planes 1 and 2. This is the simplest form of Fick's law, with a steady flow of atoms proportional to the gradient of concentration, expressed in atoms per unit volume. The diffusion coefficient, $D_A = \pi_A d^2$, is constant, and has units $m^2\ s^{-1}$ We shall also have

$$I_B = -\pi_B d^2 \frac{\partial c_B}{\partial x} = -D_B \frac{\partial c_B}{\partial x}\,. \tag{40.3}$$

Since $\pi_A = \pi_B$ if all atoms are equivalent, and $\partial c_A/\partial x = -\partial c_B/\partial x$, $I_A = -I_B$, and the diffusion coefficients are equal. This result applies, at least approximately, to chemical diffusion in an ideal solution, or to the diffusion of a radioactive isotope.

The frequency π_A is related to the frequency k_A with which an A atom in any site moves to a neighbouring vacant site. If there are z nearest sites to which an atom on plane 1 may jump, and a fraction $1/j$ of these lie on the plane 2, then

$$\pi_A = Pzk_A/j, \tag{40.4}$$

where P expresses the probability that a given site is vacant, and is thus effectively unity for interstitial diffusion, and equal to $x_\square$ for vacancy diffusion. The rate constant k_A is expressed in terms of an atomic frequency factor and a free energy of activation by means of the theory of Chapter 3.

The simple assumption that the atomic jumps take place between two neighbouring planes may readily be generalized to the case where the composition gradient has arbitrary direction and atomic jumps of different effective length along this direction are possible. Let $\Gamma_A = Pzk_A$ be the total number of jumps made by an A atom in unit time (see p. 135 for the analogous problem of vacancy jumps). If successive jumps represent vector displacements $\mathbf{s}_1$, $\mathbf{s}_2$, $\mathbf{s}_3$, etc., the total displacement after a large number of jumps $n = \Gamma_A t$ will be a vector $\mathbf{S} = \sum_{i=1}^{n} \mathbf{s}_i$. We cannot predict $\mathbf{S}$ for any given atom, but we can find a mean value, $\overline{S^2}$, of the square of the distance travelled by a large number of identical atoms. A simple kinetic argument, essentially equivalent to that above, then shows that the diffusion coefficient for a cubic material is given by

$$D_A = \overline{S^2}/6t. \tag{40.5}$$

This is sometimes called the Einstein–Smoluchowski relation and was derived by Einstein (1905).

Expanding $\mathbf{S} = \Sigma \mathbf{s}_i$, we find that the mean value of S^2 will be given by

$$\overline{S^2} = \sum_{i=1}^{n} \overline{s_i^2} + 2 \sum_{i=1}^{n-j} \sum_{j=1}^{n-1} \overline{\mathbf{s}_i \cdot \mathbf{s}_{i+}} . \tag{40.6}$$

We consider only the case when all elementary jumps are of the same length; this is believed to cover diffusion in both f.c.c. and b.c.c. structures. We let $|\mathbf{s}_i| = r_1$, the nearest neighbour distance for the sites concerned in the diffusion process. Then (40.6) becomes

$$\overline{S^2} = nr_1^2 + 2r_1^2 \sum_{j=1}^{n-1} (n-j) \overline{\cos \theta_j}, \tag{40.7}$$

where $\overline{\cos \theta_j}$ is the average value of the cosine of the angle between the directions of the ith and jth jumps. If the motion of the atoms considered is a true random walk, in which there is no correlation between the directions of the individual jumps made by an atom, $\overline{\cos \theta_j}$ must be zero for all values of j. We thus have the expression

$$D_A = \Gamma_A r_1^2/6, \tag{40.8}$$

which may be compared with that given by (40.2) and (40.4)

$$D_A = \Gamma_A d^2/f. \tag{40.9}$$

The two equations are equivalent if the factor $(1/f)$ is interpreted as the fractional number of jumps which are effectively along the composition gradient, and d is the average distance covered by each such jump. If the sites for the diffusing atoms have cubic symmetry, the flow of atoms depends only on the concentration gradient, and the diffusion coefficient is a constant. In crystals of lower symmetry, there will be non-equivalent sites to which an atom may move at different rates. The diffusion coefficient is then a second-order tensor, relating the two vector quantities $\mathbf{I}_A$ and grad c. Instead of eqn. (40.5) we have three equations of form

$$D_i = \overline{S_i^2}/2t$$

relating the principal values of the tensor $\mathbf{D}$ to the mean square displacements along the principal axes.

As we shall see in a later section (p. 393), the assumption that $\overline{\cos \theta_j} = 0$ is not valid, even for tracer self-diffusion by a vacancy mechanism. In practice, the most important example in which diffusion may be treated as a random walk problem is probably the diffusion of interstitial solute atoms. In the f.c.c. structure, the interstitial positions are at the body centre and the centres of the edges of the cubic unit cell, each site having twelve nearest neighbour sites. From eqns. (40.8) or (40.9) we find

$$D = (\tfrac{1}{12})\Gamma_A a^2 = k_A a^2, \tag{40.10}$$

where a is the cube edge. Similarly, the interstitial sites in the b.c.c. structure are at the centres of the faces and edges of the unit cube, each site having four nearest neighbours. In this case

$$D = (\tfrac{1}{24})\Gamma_A a^2 = (\tfrac{1}{6})k_A a^2 . \tag{40.11}$$

In deriving (40.10) and (40.11), we have assumed that all the interstitial sites next to an interstitial atom are vacant, so that P of eqn. (40.4) is unity. This will be true in dilute solution, and it is also the reason why we are able to treat interstitial diffusion as a random process, since the jump frequency will be almost constant throughout the specimen.

Returning to eqn. (40.1) we can assume that for a place exchange mechanism, the probability of an A atom jumping to plane 2 from plane 1 will be proportional to the number of B atoms on this plane, i.e. to $c_{B2} = c - c_{A2}$. We can thus write

$$\pi_{A,12} = \pi'_A(c - c_{A2})/c, \quad \pi_{A,21} = \pi'_A(c - c_{A1})/c,$$

and

$$I_A = \pi'_A d(c_{A1} - c_{A2}), \tag{40.12}$$

which is of the same form as (40.2) with the diffusion coefficient $D_A = \pi'_A d^2$. Once again, $I_A = -I_B$, and D is independent of composition.

In deriving these results we have considered only atoms on the two neighbouring planes. We therefore obtain the same equations if we replace the assumption that all atoms are chemically equivalent by the slightly less restrictive condition that the jump frequencies π depend only on the average concentration near the jumping atom. If we write $c_A = \frac{1}{2}(c_{A1} + c_{A2})$, the constants π_A, π'_A now become parameters $\pi_A(c_A), \pi'_A(c_A)$ which vary with composition. Fick's law now has to be written

$$I_A = -\partial(Dc_A)/\partial x. \tag{40.13}$$

In practice, D is nearly always a function of composition, and it is more realistic to express the difference between $\pi_{A,12}$ and $\pi_{A,21}$ in terms of the variation of π_A with c_{A1} and c_{A2}. We discuss this more satisfactory kinetic theory in Section 43.

The form of Fick's law given in eqn. (40.13) is inconvenient, since stationary flow can seldom be obtained in practice. The differential form of the law is obtained by considering the rate of accumulation of atoms in a small volume. For one-dimensional diffusion

$$\frac{\partial c_A}{\partial t} = \frac{\partial}{\partial x}\left\{D\frac{\partial c_A}{\partial x}\right\}$$

and the general equation for three-dimensional diffusion is

$$\frac{\partial c_A}{\partial t} = \frac{\partial}{\partial x_1}\left\{D_1\frac{\partial c_A}{\partial x_1}\right\} + \frac{\partial}{\partial x_2}\left\{D_2\frac{\partial c_A}{\partial x_2}\right\} + \frac{\partial}{\partial x_3}\left\{D_3\frac{\partial c_A}{\partial x_3}\right\}, \tag{40.14}$$

where the three diffusion coefficients allow for anisotropy of diffusion in non-cubic metals. The vector form of this equation is

$$\partial c/\partial t = \text{div}\,(\mathbf{D}\,\text{grad}\,c) = \nabla\cdot(\mathbf{D}\,\nabla c), \tag{40.15}$$

where $\mathbf{D}$ has a constant value for cubic crystals, and is a tensor with principal values D_1, D_2, D_3 for anisotropic crystals.

In particular diffusion problems, eqn. (40.14) has to be solved for given initial and boundary conditions. The equation cannot be integrated directly unless the diffusion coefficient is independent of position (i.e. of composition), but many solutions have been obtained subject to this restriction. Although some of these solutions are required in connection with the theory of diffusional growth described in Sections 54 and 55, we shall give here only an outline treatment of the one-dimensional problems which correspond to common experimental arrangements; for more detailed discussion, reference should be made to specialist texts such as Carslaw and Jaeger (1947) and Crank (1956).

In one dimension, the partial differential equation

$$(\partial c/\partial t) = D(\partial^2 c/\partial x^2) \tag{40.16}$$

has to be solved with given initial and boundary conditions, and the most appropriate form of solution is very dependent on these conditions. In a few experiments, such as the diffusion of a gas through a solid membrane or a thin-walled tube, the flux is constant so that the left-hand side of eqn. (40.16) is zero and the appropriate solution may be obtained directly by integration of (40.13). More general methods of solution include the use of Laplace transforms, the Boltzmann substitution $\lambda = x/t^{1/2}$, and the method of separation of variables, but we examine first the use of source functions. If a given component is initially present only as a very thin layer of mass (or number of atoms) M per unit area at one end of a long specimen, the concentration over a plane at a distance x from this end after time t is given by

$$c(x, t) = \{M/(\pi Dt)^{1/2}\} \exp(-x^2/4Dt), \tag{40.17}$$

whereas a thin sandwich of solute in the centre of a long specimen ($x = 0$) will give a distribution of half that of (40.16) in both positive and negative x directions. This result, which may be verified by inspection, can be used to solve problems in which the initial condition is a uniform or non-uniform extended distribution of solute. Thus a common experimental condition is to have two pure components meeting at an interface, so that the initial conditions are $c = c_0\,(x < 0)$ and $c = 0\,(x > 0)$. If the length of the bar is much greater than the mean diffusion distance from the original interface, the boundary conditions are $c(-\infty, t) = c_0$, $c(+\infty, t) = 0$. By regarding the initial distribution as a series of infinitesimal planar sources of strengths $M = c_0\,dx'$ situated at distances $\xi = (x-x')$ from the plane x where $0 \geqslant x' \geqslant -\infty$, it follows from (40.17) that

$$\begin{aligned} c(x, t) &= (c_0^2/4\pi Dt)^{1/2} \int_x^\infty \exp(-\xi^2/4Dt)\,d\xi \\ &= \tfrac{1}{2}c_0[1-\text{erf}\{x/2(Dt)^{1/2}\}], \end{aligned} \tag{40.18}$$

where the error function is defined by

$$\text{erf}\, z = (2/\pi)^{1/2} \int_0^z \exp(-\eta^2)\,d\eta. \tag{40.19}$$

The distribution may be written more compactly as

$$c(x, t) = \tfrac{1}{2}c_0\, \text{erfc}\{x/2(Dt)^{1/2}\}, \tag{40.20}$$

26

where the complementary error function is

$$\operatorname{erfc} z = 1-\operatorname{erf} z = (2/\pi)^{1/2}\int_z^\infty \exp(-\eta^2)\,d\eta. \tag{40.21}$$

Similarly, if we have an initial distribution of a component given by $c(-x, 0) = c_0$, $c(+x, 0) = c_1$, the final distribution becomes

$$\begin{aligned} c(x, t) &= \tfrac{1}{2}(c_0+c_1)+\tfrac{1}{2}(c_1-c_0)\operatorname{erf}\{x/2(Dt)^{1/2}\} \\ &= c_1+\tfrac{1}{2}(c_0-c_1)\operatorname{erfc}\{x/2(Dt)^{1/2}\}. \end{aligned} \tag{40.22}$$

A generalization of this solution is relevant to a growth problem to be discussed in Section 54. We note that in all cases, the quantity $(Dt)^{1/2}$, which has the dimensions of a length, is a measure of the distance travelled by an average diffusing atom, that is of the penetration depth. This is true for all geometrical situations, including initial point and line distributions. The penetration depth, defined as the distance at which the concentration of the diffusing component has decreased to a value of $1/e$ of its maximum value, is given by $2(Dt)^{1/2}$. It is frequently useful to remember this in order of magnitude calculations of diffusion effects.

The substitution $\lambda = x/t^{1/2}$ first used by Boltzmann leads directly to eqn. (40.18). Thus if $c(x, t) = f(\lambda)$, the diffusion equation (40.16) reduces to an ordinary differential equation with solution

$$f(\lambda) = C\int \exp(-\lambda^2/4D)\,d\lambda. \tag{40.23}$$

The solution is valid provided the boundary conditions can also be expressed in terms of λ alone; if one condition is given for $\lambda = 0$

$$f(\lambda)-f(0) = 2CD^{1/2}\int_{\eta=0}^{x/2(Dt)^{1/2}} \exp(-\eta^2)\,d\eta, \tag{40.24}$$

where $\eta = \lambda/2D^{1/2}$ is a dimensionless integration variable. In order to fix C, another condition is needed, for example for $\lambda = \pm\infty$. Thus the boundary conditions $f(0) = \frac{1}{2}(c_0+c_1)$, $f(-\infty) = c_0$, $f(\infty) = c_1$ gives (40.22) from (40.24). The Boltzmann substitution does not apply to the plane source problem (note $c(x, t)$ in eqn. (40.18) $\neq f(\lambda)$), but it is in general applicable for diffusion problems in infinite or semi-infinite media with uniform or zero initial concentrations. The solutions are usually in the form of error functions or related functions.

When the extent of x is comparable with the diffusion distance $(Dt)^{1/2}$, the above methods of solution become difficult to apply. Whilst it is possible to adapt infinite media solutions by considering repeated reflections at physical boundaries, it is generally preferable to try the method of separation of variables. Thus, if $c(x, t)$ is written as the product of two functions $F(x)$ and $G(t)$, eqn. (40.16) has a particular solution

$$c(x, t) = (A\sin kx+B\cos Kx)\exp(-k^2Dt)$$

and a general solution

$$c(x, t) = \sum_{n=1}^{\infty} (A_n\sin k_nx+B_n\cos k_nx)\exp(-k_n^2Dt). \tag{40.25}$$

This solution is most useful when only the first few terms, preferably only the first, need be retained, and this condition is usually satisfied for times $t \gtrsim x_t^2/D$, where x_t is the total range of x.

Since erf $x = -\text{erf}(-x)$, the concentration contours given by eqn. (40.18) should be symmetrical about the origin $x = 0$. Although this is sometimes approximately true, there is usually marked asymmetry. This may be handled in the formal theory by defining a concentration dependent diffusion coefficient, and using the Boltzmann substitution. The procedure has been especially developed by Matano (1933), who showed how to evaluate a concentration dependent diffusion coefficient from the equation

$$D(c_i) = \frac{1}{2t}\left(\frac{dx}{dc}\right)_{c_i} \int_{c_i}^{c_0} x \, dc. \tag{40.26}$$

All A atoms which have left the region $x < 0$ must have crossed the $x = 0$ interface and be contained in the region $x > 0$. Similarly, all B atoms which have left the original region $x > 0$ must be contained in the region $x < 0$. Thus we have the condition

$$\int_{c_0}^{0} x \, dc = 0, \tag{40.27}$$

and this may be used to define the position of the $x = 0$ interface. The meaning of the equation is illustrated in Fig. 9.2, which shows the original and final concentration contours

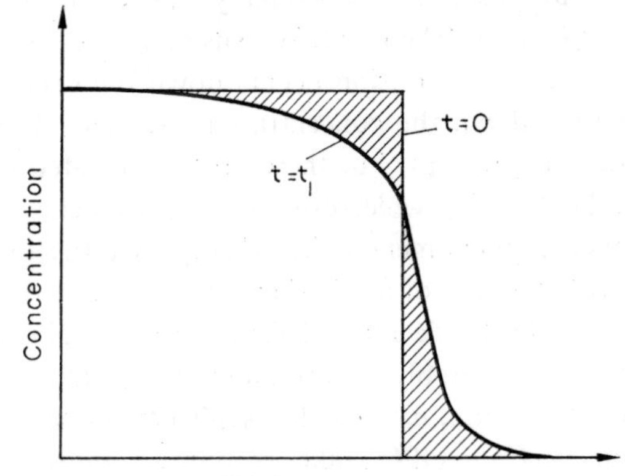

FIG. 9.2. Concentration contours in a typical diffusion experiment. The Matano interface is defined by the condition that the two shaded areas be equal.

of the A component. The $x = 0$ interface is defined by the condition that the two shaded areas be equal; it is usually called the Matano interface.

The existence of a Kirkendall effect means that if we plot the concentration contours of B instead of A, we shall again find that the shaded areas representing the changes in concentration are equal on each side of the Matano interface, but these areas are not the same as those of the A plot. Thus if inert markers are inserted at the original interface, these will move during the experiment relative to the Matano interface, or to the ends of the specimen. As we shall see in the next section, this situation may be described by assigning separate diffusion coefficients to the two components, the movement of the marker interface relative to the Matano interface then being proportional to the difference in these diffusion coefficients. In diffusion experiments of this kind, care should be used about the meaning given by different authors to the term "original" interface. Both the marker interface and the Matano interface may be regarded (in different senses) as the original interface. We should also note that if the diffusion couple consists of two alloys, the Matano interface is still defined by an equation of type (40.27), the limits of integration being from the initial concentration of a component on one side of the interface to its initial concentration on the other.

41. PHENOMENOLOGICAL THEORY OF DIFFUSION

In the original formulation of Fick's law it is assumed that diffusion takes place down the concentration gradient. This is certainly true if the equilibrium state of the assembly is single phase, since at equilibrium corresponding small volumes will contain equal numbers of atoms of a given kind. In any assembly which is not thermodynamically stable, however, the net flow of atoms will result in a lowering of the free energy, and it therefore seems more logical to write the diffusion equations in terms of free energies rather than compositions.

An assembly is in equilibrium with respect to a component i if the chemical potential g_i is the same for all parts of all phases of the assembly, and when this condition is attained there will be no net flow of i atoms. The natural extension of the statistical argument of the last section is thus to assume that diffusion occurs down the chemical potential (or free energy) gradient, rather than down the concentration gradient. This net flow is still the average effect of individual migrations, but these are so weighted that there is a greater probability of motion in the direction which evens out the gradient of free energy.

Darken (1948) developed a phenomenological theory of diffusion based on the above ideas and which included a treatment of the Kirkendall effect. A more general phenomenological theory has been given by Bardeen and Herring (1951, 1952), and has the advantage of being readily applied to an assembly of any number of components. Their treatment, which we shall follow in this section, includes Darken's equations as a special case. A still more general treatment due to Cahn (1961) takes account of additional energy terms due to inhomogeneities in composition, and is dissussed in Section 42.

We consider an alloy containing any number of component atoms, $A, B, C, \ldots$, with N equivalent sites, of which N_i are occupied by atoms of the ith component, and $N_\square$ are vacant. The phenomenological equations are obtained in effect by treating the vacancies as one of the components of the alloy. To minimize the possibility of confusion between our notation x_A, x_B, x, etc., for atomic fractions and the Cartesian coordinate x, we shall write most of our equations in this and the following section in three-dimensional vector form, so that x does not appear explicitly; moreover, we shall use concentrations c_A, c_B, etc., instead of atomic fractions wherever possible, and we shall not use the previous abbreviation $x = x_B$. Care is necessary if the vector equations are transposed into Cartesian form, but this difficulty is not encountered in other chapters.

Our assumption that the diffusion rate is proportional to the gradient of the free energy might lead us to write as the general expression for the diffusion current of the *i*th component in place of eqn. (40.2)

$$\mathbf{I}_i = -c_i M_i' \nabla g_i \tag{41.1}$$

where g_i is the chemical potential per atom and c_i is the number of *i* atoms per unit volume. M_i' is termed the mobility of *i*. The equation thus gives the diffusion current in terms of the average velocity of each atom, expressed as the product of a mobility and an average diffusion "force" on the atom, grad g_i. However, this is not the most general expression for the diffusion current. The flux of *i* atoms, although mainly dependent on the gradient of the chemical potential g_i, may also be influenced by the gradients of the chemical potentials of the other components, and of the vacancies if these are present. This concept is in accordance with the general theory of Chapter 4, and we must therefore write the general equation in the form

$$\mathbf{I}_i = -c_j M_{ij}' \nabla g_j - c_\square M_{i\square}' / \nabla g_\square , \tag{41.2}$$

the first term being summed over all values of *j*. In this equation, the most important term is M_{ii}', corresponding to M_i' of eqn. (41.1). The current $\mathbf{I}_i$ has units $m^{-2}\, s^{-1}$, and it is therefore convenient to introduce the mobilities per unit volume, M_{ij}, which are related to the mobilities per atom M_{ij}' by the equations

$$M_{ij} = c_{(j)} M_{i(j)}' , \tag{41.3}$$

where the brackets on the subscripts mean, as usual, the suspension of the summation convention. Equation (41.2) can thus be written

$$\mathbf{I}_i = -M_{ij} \nabla g_j - M_{i\square} \nabla g_\square , \tag{41.4}$$

In addition to the *i* equations giving the net flow of the various atoms in the *x* direction, there is obviously an identical equation giving the net vacancy flow $\mathbf{I}_\square$. The total number of sites will be assumed to remain constant, leading to a condition

$$\mathbf{I}_\square = -\sum_i \mathbf{I}_i$$

or

$$M_{\square j} = -\sum_i M_{ij} . \tag{41.5}$$

Equations (41.4) are now of the same form as the general equations (14.3), and it is usually assumed without further justification that the Onsager relations are valid, giving $M_{ij} = M_{ji}$ and $M_{i\square} = M_{\square i}$. In view of the discussion on p. 100, it will be realized that this assumption is logically unsatisfactory, and Coleman and Truesdell (1960) have empasized that it has never been shown that the above choice of conjugate forces and fluxes for diffusion satisfy eqns. (14.8) and (14.9). Nevertheless, we shall develop the theory of the rest of this section on the assumption that the Onsager relations are valid; some attempts have been made in nonmetallic systems to test this experimentally. It is interesting to note that the reciprocal relations were first applied to the corresponding phenomenological equations for diffusion in electrolytes (Onsager and Fuoss, 1932).

The theory presented above does not include the possibility of a marked association between one kind of atom and a vacancy. Such an association is sometimes called a "Johnson molecule" (see p. 401) and may be very important for diffusion in dilute solutions. The statistics of this kind of diffusion are very complex, and it is not at all obvious how the Onsager relations might be applied.

From eqn. (41.5) we obtain $M_{i\square} = -\sum_i M_{ij}$, and eqn. (41.4) can be written in the form

$$\mathbf{I}_i = -M_{ij} \nabla(g_j - g_\square). \tag{41.6}$$

This is the most convenient form for the diffusion equations in a multi-component assembly when vacancies are also present. By making certain assumptions, many more familiar but less general equations may be derived from (41.6). If the vacancies are everywhere maintained in local thermal equilibrium with the structure $g_\square = 0$ and

$$\mathbf{I}_i = -M_{ij} \nabla g_j. \tag{41.7}$$

Let us consider diffusion in a binary alloy containing A and B atoms. The atomic fluxes are given by

$$\left.\begin{aligned} \mathbf{I}_A &= -M_{AA} \nabla g_A - M_{AB} \nabla g_B, \\ \mathbf{I}_B &= -M_{AB} \nabla g_A - M_{BB} \nabla g_B. \end{aligned}\right\} \tag{41.8}$$

Now from the Gibbs–Duhem equation,

$$(\partial g_A / \partial \ln c_A) = (\partial g_B / \partial \ln c_B). \tag{41.9}$$

This result depends on the assumption that $g_\square = 0$, but will be approximately valid even if this is not true, since the concentration of vacancies is so small. Writing

$$\nabla g_A = (\partial g_A / \partial \ln c_A)(1/c_A) \nabla c_A$$

we have, from (41.8) and (41.9),

$$\begin{aligned} \mathbf{I}_A &= -\frac{\partial g_A}{\partial \ln c_A} \left\{ \frac{M_{AA}}{c_A} - \frac{M_{AB}}{c_B} \right\} \nabla c_A \\ &= -\frac{\partial g_A}{\partial \ln c_A} (M'_{AA} - M'_{AB}) \nabla c_A \\ &= -D_A \nabla c_A, \end{aligned} \tag{41.10}$$

where the intrinsic diffusion coefficient is given by

$$D_A = (\partial g_A / \partial \ln c_A)(M'_{AA} - M'_{AB}) \tag{41.11}$$

and, similarly,

$$D_B = (\partial g_B / \partial \ln c_B)(M'_{BB} - M'_{BA}).$$

Diffusion in a binary assembly may thus be described by two intrinsic diffusion coefficients D_A and D_B which are in general unequal, so that the components diffuse at different rates.

The difference between the components arises solely from the mobility terms, and the diffusion force on each atom is equal and opposite. For ternary and higher component assemblies, the Gibbs–Duhem relation does not simplify the equations in this way, so that the average diffusion force varies as well as the mobility. It is not then possible to specify the whole diffusion by means of a single diffusion coefficient for each component.

There will be a mass flow of atoms relative to the lattice, given by the negative of the vacancy current. This flow may be written

$$\mathbf{I}_A+\mathbf{I}_B = -(D_A-D_B)\,\nabla c_A \tag{41.12}$$

for a binary assembly. Instead of taking the lattice as our reference frame we may use a frame relative to which there is no mass flow. The flux of A atoms is then

$$\begin{aligned}\mathbf{I}'_A &= \mathbf{I}_A-x_A(\mathbf{I}_A+\mathbf{I}_B) = x_B\mathbf{I}_A-x_A\mathbf{I}_B\\ &= -(x_BD_A+x_AD_B)\,\nabla c_A\,.\end{aligned}$$

The quantity

$$D_{\text{chem}} = x_BD_A+x_AD_B \tag{41.13}$$

is the chemical diffusion coefficient as determined in ordinary experiments (e.g. eqn. (40.26)). Instead of specifying the flow by two diffusion coefficients (eqn. (41.11), we can therefore use a single coefficient and a velocity of mass flow eqns.((41.12) and (41.13)). The two methods are equivalent, although the first seems generally preferable. Equations (41.11)–(41.13) form Darken's phenomenological equations for binary diffusion when there is a Kirkendall effect; they depend only on the assumption that the vacancies are maintained in local thermal equilibrium. The validity of the assumption is discussed in Section 43. Seitz has shown that the equations actually depend on the vacancy concentration being a function only of the composition.

As stated on p. 379, the Kirkendall effect provides an experimental test to distinguish between place exchange mechanisms and vacancy or interstitial mechanisms. If diffusion is by some form of place exchange, then in place of eqn. (41.5) we have the condition $\sum_i \mathbf{I}_i = 0$. There is then no net flow of matter relative to the lattice, and

$$D_A = D_B = D_{\text{chem}}\,.$$

Darken also derived relations between the intrinsic diffusion coefficients and the tracer diffusion coefficients. Suppose we have an alloy in which the concentration of A and B atoms is uniform but in which the A atoms consists of two isotopes A and A_+ such that there is a concentration gradient. The currents of A and A_+ atoms will be given by (41.10) as

$$\left.\begin{aligned}\mathbf{I}_A &= -(\partial g_A/\partial \ln c_A)\,(M'_{AA}-M'_{AA_+})\,\nabla c_A\\ \mathbf{I}_{A_+} &= -(\partial g_{A_+}/\partial \ln c_{A_+})\,(M'_{A_+A_+}-M'_{A_+A})\,\nabla c_{A_+}\end{aligned}\right\} \tag{41.14}$$

Since the isotopes are chemically equivalent, the chemical potentials depend only on the entropy of mixing, and

$$\begin{gathered} g_A = kT \ln x_A = kT \ln(c_A/c), \quad g_{A_+} = kT \ln(c_{A_+}/c),\\ (\partial g_A/\partial \ln c_A) = (\partial g_{A_+}/\partial \ln c_{A_+}) = kT.\end{gathered}$$

The diffusion currents $\mathbf{I}_A$, $\mathbf{I}_{A+}$ must be equal and opposite. If we neglect the cross-terms, we thus have $M'_{AA} = M'_{A+A+}$, and the diffusion coefficient may be written

$$D_{A+} = M'_{AA}kT. \tag{41.15}$$

This relation between mobility and diffusion coefficient is known as Einstein's equation, and was first derived in 1905. It is valid for any component of an ideal solution provided the cross-terms can be neglected.

We also have from eqns. (41.11) and (41.15)

$$\frac{D_A}{D_{A+}} = \frac{\partial(g_A/kT)}{\partial \ln c_A} = \frac{D_B}{D_{B+}}. \tag{41.16}$$

This is Darken's thermodynamic relation, and also depends on the approximation of neglecting M_{AA+}, M_{AB}, etc. The quantities D_A, D_{A+} are respectively the intrinsic coefficient and the tracer diffusion coefficient of A in an alloy of A and B.

The extent of the approximation involved in neglecting the cross-terms was examined by Bardeen and Herring in the following way. Consider the simple case of radioactive diffusion in pure A. Equations (41.14) are valid, and we may write the diffusion coefficients

$$D_A = kT(M'_{AA} - M'_{AA+}) = D_{AA} - D_{AA+},$$
$$D_{A+} = kT(M'_{A+A+} - M'_{A+A}) = D_{A+A+} - D_{A+A}.$$

We also know that

$$M'_{AA+} = M_{AA+}/c_{A+}, \quad M'_{A+A} = M_{AA+}/c_A,$$

and if the tracer A_+ is present in very small concentration we may thus neglect M'_{A+A} in comparison with M'_{AA+}. Since $D_A = D_{A+}$, we therefore have

$$D_{A+} = D_{A+A+} = D_{AA} - D_{AA+}. \tag{41.17}$$

Now suppose that we have pure A with a concentration gradient of vacancies, so that $g_\square = kT \ln (c_\square/\bar{c}_\square)$, where $\bar{c}_\square$ is the equilibrium vacancy concentration. The vacancy current given by eqn. (41.4) is thus

$$\mathbf{I}_\square = -\mathbf{I}_A = M_{AA} \nabla g_\square = (kTM_{AA}/c_\square) \nabla c_\square.$$

The diffusion coefficient for vacancies is thus

$$D_\square = kTM_{AA}/c_\square = (c_A/c_\square) kTM_{AA} = D_{AA}/x_\square,$$

since $c_\square$ is small. Comparing this equation with (41.17),

$$D_{A+} = x_\square D_\square - D_{AA+}. \tag{41.18}$$

Now a simple kinetic argument has already been given on p. 382 to deduce $D_{A+} = x_\square D_\square$. If the motion of a tracer atom can be treated as a random walk, the probability per unit time that it will move to a neighbouring site is equal to the frequency with which a vacancy jumps, multiplied by the probability $x_\square$ that the site is vacant. The diffusion coefficients should thus be in this ratio. The defect in the reasoning lies in the assumption which we explicitly made on p. 383 that the successive jumps are uncorrelated. If an atom has just left a given site, there will be some time during which the probability of a reverse jump is greater than that of a further jump to another site. This is because the probability of the original site remaining vacant is greater than that of one of the remaining sites becoming vacant. This correlation results in D_{A+} being less than $x_{\square}D_{\square}$, and the difference is just the term D_{AA+}. For vacancy self-diffusion, we thus see that the correct physical interpretation of the vacancy cross-terms is the effect of one atomic movement on the probabilities of further movements.

Le Claire and Lidiard (1956) considered the correlation effects for various kinds of diffusion mechanism, using eqn. (40.7). For a vacancy mechanism, there is *direct* correlation only between successive jumps, i.e. the probability of a given jump being in any particular direction depends only on the direction of the jump immediately preceding the one considered. This means that

$$\overline{\cos \theta_j} = \overline{\cos \theta_{j-1}}\,\overline{\cos \theta_1} = (\overline{\cos \theta_1})^j. \tag{41.19}$$

Substituting into eqn. (40.7) and allowing n and t to become large then gives

$$\overline{S^2} = nr_1^2 f, \qquad \text{or} \qquad D = \tfrac{1}{6}\Gamma r_1^2 f \tag{41.20}$$

where the "correlation factor"

$$f = (1+\overline{\cos \theta_1})/(1-\overline{\cos \theta_1}). \tag{41.21}$$

Note that the magnitude of the cross-terms is given by $M'_{AA+} = 1-f$.

A rough estimate of $\overline{\cos \theta_1}$ is obtained by considering only the first jump of a vacancy with which a moving atom has just changed places. For a random walk of the vacancy, the probability is $1/z$ that it will re-exchange places with the same atom, and since $\theta_1 = -\pi$ for this jump $\overline{\cos \theta_1} = -1/z$ and

$$f \simeq 1-2/z. \tag{41.22}$$

For a more accurate computation, we let P_k be the probability that the next jump of the tracer atom will be along the direction $\theta_{(1)k}$ to the kth site around the atom. We must consider all possible paths which transfer the vacancy from site 1 to site k, and, in general,

$$P_k = \sum_{i=1}^{\infty} n_{ik} z^{-i}, \tag{41.23}$$

where n_{ik} is the number of paths along which the $(i-1)$th jump of the vacancy takes it to site k and z^{-i} is the cumulative probability of the vacancy moving along each such path and eventually changing places with the tracer atom. Then

$$\overline{\cos \theta_1} = \sum_{k=1}^{z} P_k \cos \theta_{(i)k}. \tag{41.24}$$

Equation (41.23) is based on the explicit assumption that the vacancy executes a random walk. This is not strictly valid even for tracer self-diffusion if the jump frequency of the isotope used as a tracer is not identical with that of the abundant isotope. Since the chemical properties are identical, changes in the jump frequency arise only because different isotopes have different masses. The correction to f from this *isotope effect* is very small, but may nevertheless be used to determine f experimentally (see below).

The difficulty in evaluating $\overline{\cos\theta_1}$ from eqn. (41.24) is that the series for P_k converges rather slowly. Bardeen and Herring used a numerical technique and Le Claire and Lidiard (1956) developed a matrix method which has also been discussed by Mullen (1961) and Manning (1968). An alternative method, based on an electrical analogue, was used by Compaan and Haven (1958) who give values of 0·5, 0·653, 0·727, and 0·781 for the correlation factor f in diamond cubic, simple cubic, b.c.c., and f.c.c. crystals respectively. Equation (41.22) is thus exact for the diamond structure but gives values which are slightly too high for the other structures.

Correlation effects do not occur for diffusion by the ring mechanism and simple interstitial mechanisms, but they are expected for interstitialcy diffusion. For self-diffusion in cubic crystals f is a numerical factor which, apart from the small isotope effect, depends only on the diffusion mechanism and crystal structure; its importance lies not so much in the magnitude of the correction but in the realization that eqn. (41.1) is not strictly valid even for one of the simplest diffusion problems. Moreover, measurements of the correlation factor may enable the diffusion mechanism to be deduced.

Equation (41.19) cannot be used for self-diffusion in non-cubic structures and for divacancy diffusion in any structure. The evaluation of the correlation factor is then much more difficult, but has been accomplished in several cases of interest by Mullen (1961), Howard (1966), Bakker (1971), and Mehrer (1972). Mehrer obtains $f = 0{\cdot}468$ for divacancy self-diffusion in f.c.c. metals. It is important to realize that f is not a purely geometrical factor in non-cubic crystals, since transitions with different jump frequencies are then possible. Expressions for the correlation factors for diffusion in different directions depend on ratios of jump frequencies, and may be temperature-dependent.

Correlation effects in chemical diffusion are often much more significant than in self-diffusion since the approximation of regarding the motion of a vacancy as a random walk is then invalid. We consider this further on p. 401.

42. UPHILL DIFFUSION

We shall now examine a little more fully the form of eqn. (41.11), giving the diffusion coefficient in a binary alloy. Since M'_{AA} is necessarily greater than M'_{AB}, the latter being neglected in first approximation, the term in brackets is necessarily positive. The condition for "uphill" diffusion, i.e. for D_A to be negative, is therefore

$$(\partial g_A/\partial c_A) < 0, \quad \text{i.e.} \quad (\partial g_A/\partial x_A) < 0. \qquad (42.1)$$

We have imposed no restrictions on the independent variables determining g_A, which may be a function not only of temperature and concentration but also of other external constraints and internal stress fields. One possible source of internal stress is the variation of lattice parameter with composition, and if this is included in the expression for the chemical potential is the inequality (42.1) gives a very general condition for uphill diffusion, which is valid for any imposed conditions of the assembly.

The inequality (42.1) governs uphill diffusion of one component, but it is generally more useful to consider uphill chemical diffusion, i.e. spontaneous segregation of two components, the condition for which is a negative value of D_{chem}. Applying eqn. (41.12) to the flux of B atoms, and neglecting the off diagonal terms in eqn. (41.8),

$$\begin{aligned}\mathbf{I}'_B &= \mathbf{I}_B - x_B(\mathbf{I}_A + \mathbf{I}_B) \\ &= -c_B x_A(M'_{BB} \nabla g_B - M'_{AA} \nabla g_A)\end{aligned} \tag{42.2}$$

since $c_B x_A = c_A x_B = x_A x_B/v$, where v is the atomic volume. This equation may also be written in the form

$$\mathbf{I}'_B = -M_{\text{chem}} \nabla(g_B - g_A) = -M_{\text{chem}} \nabla \varphi_D, \tag{42.3}$$

where φ_D may be regarded as the "driving force" for diffusion and

$$\begin{aligned}M_{\text{chem}} &= c_B x_A\{x_A M'_{BB} + x_B M'_{AA}\} \\ &= x_A^2 M_{BB} + x_B^2 M_{AA}\end{aligned} \tag{42.4}$$

is a mobility per unit volume; the remaining part of the right-hand side of (42.2) vanishes because of the Gibbs–Duhem equation. Now since $g_B - g_A = \partial g/\partial x_B$ (eqn. (22.7)), it follows that

$$D_{\text{chem}} = M_{\text{chem}} v(\partial^2 g/\partial x_B^2), \tag{42.5}$$

and for uphill diffusion,

$$\partial^2 g/\partial x_B^2 = v^{-2}(\partial^2 g/\partial c^2) < 0, \tag{42.6}$$

which means that the temperature and composition must lie inside the chemical spinodal (eqn. (22.42)). For condensed phases, (42.6) may be replaced by

$$\partial^2 f/\partial c^2 < 0, \tag{42.7}$$

where f is the Helmholtz free energy per atom. In the following development we use c rather than x_B as the composition variable (see p. 388), even though this introduces more terms involving the atomic volume than would otherwise be necessary.

Uphill diffusion inside the spinodal was discussed by Becker (1937) and later workers, but the inadequacy of the diffusion equation (42.3) was not realized until the work of Cahn (1961, 1962). Difficulties arise because composition gradients are established over fairly short distances during spinodal decomposition and the gradient and coherency energy terms discussed in Section 22 can then not be neglected. In the isotropic approximation, the free energy per atom of an inhomogeneous solid solution is given by

$$g_{ih} = \int [g(c) + v^3\{\varepsilon^2 Y'(c - c_0)^2 + K(\nabla c)^2\}] \,\mathrm{d}V, \tag{42.8}$$

where the integral is taken over the volume of the assembly and ε, Y', and K have the meanings defined on p. 183. For equilibrium of the non-homogeneous assembly this integral must be minimized subject to the requirement

$$\int (c-c_0)\,dV = 0, \tag{42.9}$$

which ensures that the average concentration c_0 is constant. The functional minimization by Euler's equation leads to the definition of a potential

$$\varphi_D' = v^{-1}(\partial g/\partial c)+v^2\{2\varepsilon^2 Y'(c-c_0)-2K\,\nabla^2 c-(\partial K/\partial c)(\nabla c)^2\}, \tag{42.10}$$

which has a constant value at equilibrium. The first term in φ_D' is φ_D of eqn. (42.3), and we may now generalize that equation by assuming the diffusion flux to be proportional to the gradient of φ_D' instead of φ_D. This gives

$$\mathbf{I}_B' = -M_{\text{chem}}\{v^{-1}(\partial^2 g/\partial c^2)+2\varepsilon^2 Y' v^2\}\,\nabla c+2M_{\text{chem}}Kv^2\,\nabla^3 c+M_{\text{chem}}v^2\,\nabla\{(\partial K/\partial c)(\nabla c)^2\} \tag{42.11}$$

and

$$(\partial c/\partial t) = M_{\text{chem}}\{v^{-1}(\partial^2 g/\partial c^2)+2\varepsilon^2 Y' 2v^2\}\,\nabla^2 c-2M_{\text{chem}}Kv^2\,\nabla^4 c$$

$$+\left[M_{\text{chem}}\frac{\partial}{\partial c}\{v^{-1}(\partial^2 g/\partial c^2)+2\varepsilon Y' v^2\}(\nabla c)^2+\ldots\right]. \tag{42.12}$$

The last term of eqn. (42.11) and the non-linear terms in square brackets of eqn. (42.12) may be neglected if $c-c_0$ is small. The two equations are then linear but differ from the classical forms of Fick's first and second laws because of the terms in $\nabla^3 c$ and $\nabla^4 c$ respectively.

A particular solution of the linear form of (42.12) may be written as a sine wave, and the general solution is obtained by superposition of all such solutions. This gives

$$c(\mathbf{r}, t)-c_0 = \iiint A(\boldsymbol{\beta}, t)\exp(-i\boldsymbol{\beta}\cdot\mathbf{r})\,d\boldsymbol{\beta}, \tag{42.13}$$

where $\mathbf{r}$ is the position vector, $\boldsymbol{\beta}$ is a wave vector representing a Fourier component of the composition variation with wavelength $\lambda = 2\pi/\beta$, and $A(\boldsymbol{\beta}, t)$ is the amplitude of this component at time t. By substituting back into the linearized diffusion equation, we find

$$A(\boldsymbol{\beta}, t) = A(\boldsymbol{\beta}, 0)\exp\{R(\boldsymbol{\beta})t\}, \tag{42.14}$$

where the initial fluctuation is specified by the amplitude $A(\boldsymbol{\beta}, 0)$ of the various components and $R(\boldsymbol{\beta})$ is called the amplification factor and is given by

$$R(\boldsymbol{\beta}) = -M_{\text{chem}}\beta^2\{v^{-1}(\partial^2 g/\partial c^2)+2\varepsilon^2 Y' v^2+2Kv^2\beta^2\}. \tag{42.15}$$

In this equation M is inherently positive and K is expected to be positive for a system with a spinodal; negative K occurs for ordering systems but cannot be handled by means of a continuum theory. Thus the condition for a particular Fourier component to increase in amplitude with time is that $R(\beta)$ is positive. This is only possible if the mean composition lines inside the curve

$$(\partial^2 g/\partial c^2) = -2\varepsilon^2 Y' v^3 \tag{42.16}$$

and if the wave vector has amplitude

$$\beta < \beta_c = -\{v^{-3}(\partial^2 g/\partial c^2) + 2\varepsilon^2 Y'\}/2K. \tag{42.17}$$

A positive amplification factor implies uphill diffusion, and we see that a necessary condition is that the composition and temperature lie inside the coherent spinodal since eqn. (42.16) is identical with (22.45). The coherency energy in eqn. (42.8) has two effects on diffusion processes; it restricts the region of negative diffusion to the area inside the coherent spinodal instead of that inside the chemical spinodal, and when an isotropic calculation is made it introduces directionality into diffusion, even in cubic crystals. The gradient energy does not restrict the temperature or composition range of negative diffusion, but ensures that it can only occur over distances greater than $\lambda_c = 2\pi/\beta_c$. A typical value for λ_c is 10 nm.

Cahn's generalized diffusion equation may be regarded as a correction for the assumption in Fick's law that the flux is proportional to the concentration gradient. However, the modifications he introduced have negligible effects in most situations where the diffusion distance is of the order of microns, and its application in practice tends to be restricted mainly to discussion of spinodal decomposition. We defer further development of this theory to Part II, Chapter 18, although we note that an atomistic (regular solution) model of spinodal decomposition was provided by Hillert (1961) prior to the work of Cahn. Hillert's development includes the gradient energy but not the coherency energy.

Despite the limitation mentioned above, it has proved possible to test the new diffusion equation in non-spinodal situations by preparing multi-layered sandwiches of two components by vapour deposition. Individual layers of 1–3 nm may be deposited successively, and the homogenization process on annealing can then be followed by suitable X-ray techniques. In work on gold–silver alloys (Cook and Hilliard, 1969) and copper–palladium (Philofsky and Hilliard, 1969) confirmation of both gradient energy and coherency energy effects was obtained and values of the gradient energy coefficient K were measured.

43. KINETIC THEORY OF VACANCY DIFFUSION

The elementary kinetic theory of Section 40 led to Fick's law; we shall now replace it by a more satisfactory theory which leads to the phenomenological equations of Section 41. This account is due initially to Seitz (1948, 1950a), and is important in showing the physical significance of the terms in the formal theory. We restrict the discussion to vacancy diffusion, although the method can readily be applied to the other possible mechanisms, and we again consider one-dimensional flow in a binary alloy.

The numbers of A atoms, B atoms, and vacancies on two neighbouring planes, distance d apart, may be written $c_{A1}d$, $c_{B1}d$, $c_{\square 1}d$, $c_{A2}d$, $c_{B2}d$, and $c_{\square 2}d$ respectively. Equation (40.1) gives the net current of A atoms, and the vacancy current is similarly

$$\mathbf{I}_{\square} = d(c_{\square 1}\pi_{\square 12} - c_{\square 2}\pi_{\square, 21}). \tag{43.1}$$

Since $c_\square$ is small, even though not necessarily constant, c_B is effectively determined by c_A. The probability per unit time that a vacancy will jump from a plane where the concentration of A atoms is ζ to a neighbouring plane where it is η is thus entirely determined by ζ and η, and may be written $\pi_\square(\zeta\eta)$. In particular, $\pi_{\square,12} \equiv \pi_\square(c_{A1}, c_{A2})$.

When the vacancy jumps from the region ζ to η, it may change places with either an A atom or a B atom. We write the frequency with which it changes places with an A atom as $\pi_{\square A}(\zeta\eta)$.† Clearly

$$\pi_\square(\zeta\eta) = \pi_{\square A}(\zeta\eta) + \pi_{\square B}(\zeta\eta). \tag{43.2}$$

In order to avoid confusion, we note that the terms in the bracket give the direction in which the vacancies move, and are opposite to the direction in which the atoms move.

Since c_{A1}, c_{A2} will be only slightly different on neighbouring planes, the two frequencies in eqn. (43.1) will both be nearly equal to $\pi_\square(c_A c_A)$, which is the frequency of vacancy jumps between two neighbouring planes both containing $c_A d = \frac{1}{2}(c_{A1}+c_{A2})d$ atoms of type A per unit area. We can thus express the difference between $\pi_\square(c_{A1}c_{A2})$ and $\pi_\square(c_{A2}c_{A1})$ by regarding $\pi_\square(\zeta\eta)$ as a continuous function of ζ and η, and expanding in a Taylor series about $\zeta = \eta = c_A$. This gives

$$\left.\begin{aligned} \pi_\square(c_{A1}c_{A2}) &= \pi_\square(c_A c_A) - \frac{d}{2}\frac{\partial \pi_\square}{\partial \zeta}\frac{\partial c_A}{\partial x} + \frac{d}{2}\frac{\partial \pi_\square}{\partial \eta}\frac{\partial c_A}{\partial x}, \\ \pi_\square(c_{A2}c_{A1}) &= \pi_\square(c_A c_A) + \frac{d}{2}\frac{\partial \pi_\square}{\partial \zeta}\frac{\partial c_A}{\partial x} - \frac{d}{2}\frac{\partial \pi_\square}{\partial \eta}\frac{\partial c_A}{\partial x}. \end{aligned}\right\} \tag{43.3}$$

The differentials $\partial\pi_\square/\partial\zeta$, $\partial\pi_\square/\partial\eta$ have, of course, to be evaluated at $\zeta = \eta = c_A$. Substituting into (43.1), we obtain the vacancy current

$$\mathbf{I}_\square = d\left[\pi_\square(c_A, c_A)\{c_{\square 1} - c_{\square 2}\} + \frac{d}{2}\{c_{\square 1} + c_{\square 2}\}\left\{-\frac{\partial \pi_\square}{\partial \zeta} + \frac{\partial \pi_\square}{\partial \eta}\right\}\frac{\partial c_A}{\partial x}\right]. \tag{43.4}$$

Purely for convenience, we now introduce the notation

$$\frac{\partial \pi_\square}{\partial c_A} = -\frac{\partial \pi_\square}{\partial \zeta} + \frac{\partial \pi_\square}{\partial \eta}. \tag{43.5}$$

Writing also $c_\square = \frac{1}{2}(c_{\square 1}+c_{\square 2})$,

$$I_\square = -d^2\pi_\square(c_A, c_A)\frac{\partial c_\square}{\partial x} + d^2 c_\square \frac{\partial \pi_\square}{\partial c_A}\frac{\partial c_A}{\partial x}. \tag{43.6}$$

There may thus be a net vacancy current, even when there is no concentration gradient of vacancies.

It should be noted that the term $\partial\pi_\square/\partial c_A$ is *defined* by (43.5); its use does not imply that $\pi_\square$ is a function of a single variable c_A. In some papers on diffusion theory, this latter

† It should be noted that $\pi_{\square A}(\zeta\eta)$ is not equal to $\pi_{A,21}$, used in Section 40; the two frequencies are in the ratio $c_{A2} : c_{\square 1}$.

assumption has in fact been made, and an equation of the same form as (43.6) is obtained. Seitz's theory is more general, insofar as it is recognized that $\partial\pi/\partial\zeta$ is not necessarily equal to $-\partial\pi/\partial\eta$.

We can now write the current of A atoms as

$$I_A = d[c_{\square 2}\pi_{\square A}(c_{A2}, c_{A1}) - c_{\square 1}\pi_{\square A}(c_{A1}, c_{A2})].$$

Expanding the $\pi_{\square A}$ in exactly the same way as we expanded the $\pi_\square$ terms, we obtain

$$I_A = d^2\pi_{\square A}(c_A, c_A)\frac{\partial c_\square}{\partial x} - d^2 c_\square \frac{\partial \pi_{\square A}}{\partial c_A}\frac{\partial c_A}{\partial x}, \tag{43.7}$$

where the term $\partial\pi_{\square A}/\partial c_A$ is defined by an equation of the type of (43.5). Similarly,

$$I_B = d^2\pi_{\square B}(c_A, c_A)\frac{\partial c_\square}{\partial x} - d^2 c_\square \frac{\partial \pi_{\square B}}{\partial c_A}\frac{\partial c_A}{\partial x}, \tag{43.8}$$

and in view of (43.2)

$$I_A = I_B + I_\square = 0, \tag{43.9}$$

which ensures the conservation of lattice sites.

Equations (43.6)–(43.8) correspond to the general phenomenological eqn. (41.6), in which no assumption is made about $g_\square$.

If we now assume that the concentration of vacancies is determined by the concentration of A atoms, we may write

$$\frac{\partial c_\square}{\partial x} = \frac{\partial c_\square}{\partial c_A}\frac{\partial c_A}{\partial x}. \tag{43.10}$$

On substituting into (43.7) and (43.8), we obtain the diffusion coefficients

$$\left.\begin{aligned} D_A &= -d^2\left[\pi_{\square A}(c_A, c_A)\frac{\partial c_\square}{\partial c_A} + c_\square\frac{\partial \pi_{\square A}}{\partial c_A}\right], \\ D_B &= d^2\left[\pi_{\square B}(c_A, c_A)\frac{\partial c_\square}{\partial c_A} + c_\square\frac{\partial \pi_{\square B}}{\partial c_A}\right], \end{aligned}\right\} \tag{43.11}$$

and from (43.6)

$$I_\square = -(D_A - D_B)\frac{\partial c_A}{\partial x}. \tag{43.12}$$

We thus obtain Darken's phenomenological equations if (43.10) is correct. This condition is slightly more general than the assumption of thermal equilibrium for the vacancies, used in deriving the same equations in Section 41.

Darken's thermodynamic relations (eqn. (41.16) were obtained by Bardeen on the assumption that the probability of a vacancy changing places with an A atom is proportional to the number of A atoms on the plane to which it jumps. This implies that each plane normal to the diffusion flow is in equilibrium, so that the probability of a given site being occupied by a given kind of atom is equal to the concentration of such atoms in the plane considered. This equilibrium must be established immediately after each atom has moved. In fact, as we have seen, correlation effects give a significantly higher probability to an immediate return jump, and in self-diffusion, the error in neglecting these terms is of the order $1/z$. Correlation effects in chemical diffusion are discussed below.

Plausible arguments were given by Bardeen and Herring (1952) to show that it is probable that the concentration of vacancies can be maintained close to its equilibrium value by the action of edge dislocations. If this is true, Darken's equations are valid, and lattice sites are created on one side of a diffusion couple and destroyed on the other, thus giving a net movement of the marker interface relative to the Matano interface. The mechanism envisaged for the maintenance of the equilibrium concentration of vacancies was the operation of dislocation sources in climb, as described on p. 248. However, there is now a large body of evidence to show that porosity is developed on one side of the original interface in nearly all Kirkendall effect experiments, and this must mean that local equilibrium of vacancy concentrations cannot be fully preserved. This has been demonstrated directly by Heumann and Walther (1957) who analysed the movements of inert markers distributed throughout the diffusion zone.

At first, it seems rather strange that vacancies in diffusion experiments should condense to form macroscopic spherical holes rather than flat discs which collapse into prismatic dislocations as in quenched specimens. The porosity is not a negligible effect; in typical experiments (e.g. Barnes, 1952) about one-half of the net vacancy current has been used in forming voids. Using arguments essentially equivalent to those of the classical theory of nucleation described in the next chapter, Seitz (1953) concluded that homogenous nucleation of pores by aggregation of single vacancies is most improbable, since this requires a relative excess vacancy concentration, defined as $(x_{\square}/\bar{x}_{\square})-1$, to be ~ 100. An analysis of the then available experimental data led to the conclusion that relative excess vacancy concentrations of ~ 1 are possible, and give a lifetime of $\sim 10^{11}$ vacancy jumps. However, if condensation on relatively large defects occurs, it is possible that the maximum excess concentration is much less than this. Resnick and Seigle (1957) have shown that oxide inclusions might act as nuclei for pore formation, and Barnes and Mazey (1958) demonstrated that an external pressure is able to prevent pore formation, whilst a larger pressure will collapse the voids produced by the diffusion process. These workers and others (e.g. Balluffi, 1954) also concluded that pore nucleation is heterogeneous, and that the vacancy supersaturation is of the order of 1%.

The full details of the processes leading to the Kirkendall effect are not yet certain; the whole phenomenon is rather complex, and various secondary reactions such as polygonization in the diffusion zone have been detected. The observed porosity accounts for only a fraction of the excess vacancies, generally about half, so that the previous conclusion that dislocations act as sources and sinks for vacancies in such experiments is still valid, even if the vacancy concentration does depart appreciably from its equilibrium value. There is also some experimental evidence that the climbing of dislocations to the free surface produces observable steps (Barnes, 1955).

In real crystal structures, the Bardeen–Herring climb mechanism needs modification since edge dislocations are not usually the boundaries of single atomic planes. The usual picture of a f.c.c. $\frac{1}{2}a\langle 110\rangle$ edge dislocation, for example, requires the insertion of two extra atomic $\{110\}$ planes above the $\{111\}$ glide plane. If the dislocation is undissociated in the glide plane, climb could take place by condensation of vacancies in a double layer, but if the dislocation dissociates into two Shockley partials, the extra atomic planes separate to opposite ends of the stacking fault, and the climb process is more complex. Analysis of results of thin film observations suggests that glide dislocations climb easily only when they are parallel to $\langle 110\rangle$, the vacancies being added as a single layer on the intersecting $\{111\}$ plane, and also when they are almost parallel to an intersecting $\{110\}$ plane, the vacancies being added as a double layer. This is contrary to the original concept that an edge or mixed dislocation just climbs normal to its slip plane.

The kinetic theory described in this section expresses the diffusion rate in terms of the vacancy concentration and of the different probabilities that an A atom and a B atom will change places with a neighbouring vacancy in unit time, but it does not include correlation effects. Let us now consider a dilute solution of B in A. It is found experimentally that tracers of the impurity (B) atoms may diffuse at rates which are considerably different from the self-diffusion rate of the solvent atoms, and which may in general be either greater or smaller. Part of the explanation for this effect will arise from the change in the energy of formation of a vacancy in the neighbourhood of an impurity atom, approximate treatments being possible in terms of electrostatic interactions between the impurity and the lattice or in terms of the misfitting sphere model. Another factor of importance is the correlation between the motion of the solute atom and that of the solvent atoms, the rate of jumping of the solute atom being limited either by the activation energy for its own motion, or that for the motion of the solvent atom. As first suggested by Johnson (1941), there may be a strong association between vacancies and impurity atoms. Instead of the random migration of vacancies, assumed in the treatment above, the solute-atom–vacancy pair ("Johnson molecule") will then move together through the lattice, thus increasing the diffusion coefficient. Correlation effects of this kind are of major importance in contrast to the minor corrections to the random walk theory required by correlation effects in the self-diffusion of a pure component.

For self-diffusion in a cubic pure metal, eqn. (40.8) reduces to $D = a^2Pk_A$. Le Claire and Lidiard (1956) treated the problem of solute diffusion by the Johnson mechanism in a f.c.c. structure, using the approach outlined on p. 393. We let P now be the probability of finding a vacancy as the nearest neighbour of a solute B atom. The rates of interchange of this vacancy with the B atom, with the four A atoms which are nearest neighbours to both vacancy and B atom ($A1$ sites) and with the seven A atoms which are neighbours of the vacancy, but not of the B atom ($A2$ sites) are assumed to be all different, and are denoted by k_B, k_{A1}, and k_{A2} respectively. The mean value of $\cos\theta_1$ in eqn. (41.20) is found by a method similar to that discussed previously and, the diffusion coefficient is expressed as

$$D_B = a^2Pk_B(k_{A1}+\tfrac{7}{2}k_{A2})/(k_B+k_{A1}+\tfrac{7}{2}k_{A2}). \tag{43.13}$$

When the rate of jump of the B atom is much greater than that of the solvent atoms, this expression takes the limiting form

$$D_B = a^2P(k_{A1}+\tfrac{7}{2}k_{A2}) \tag{43.14}$$

and is independent of k_B. The diffusion rate of the solute atom is then limited only by the jump frequency of the solvent atoms, but the solute will diffuse faster than the solvent. The other limiting form is when the B atoms take much longer to jump into neighbouring vacancies than do the A atoms. Equation (43.13) then has the limiting form

$$D_B = a^2 P k_B \tag{43.15}$$

and the diffusion rate of the solute is generally less than that of the solvent. An interesting reduction to the case of self-diffusion in a pure metal is obtained by writing $k_B = k_{A1} = k_{A2}$ in eqn. (43.13). This gives

$$D_A = \tfrac{9}{11} a^2 P k_A, \tag{43.16}$$

the factor (9/11) being the value of the correlation factor eqn. (41.21) previously derived by Bardeen and Herring.

The Le Claire–Lidiard calculation is based on the assumption that a vacancy which jumps to an $A2$ site is completely dissociated from the solute atom. In a more accurate treatment, Manning (1962, 1964) introduced another local frequency k_{A3} which describes an association jump (the reverse of the dissociation jump) and he further assumed that all jumps of the vacancy other than those described above take place with the frequency k_A which applies in pure solvent. He obtained for the correlation factor

$$f_B = (k_{A1} + \tfrac{7}{2} F k_{A2})/(k_B + k_{A1} + \tfrac{7}{2} F k_{A2}), \tag{43.17}$$

where F is a rather complex function of k_{A3}/k_A which reduces to $F = 1$ for $k_{A2} \ll k_A$, $F \simeq 0{\cdot}74$ for $k_{A2} = k_A$, and $F = \frac{2}{7}$ for $k_{A2} \gg k_A$.

Unless the atomic forces are very short-ranged, the assumption that there is a single frequency for vacancy jumps to each of the seven $A2$ positions is doubtful since these sites are variously second, third, or fourth neighbours of the solute atom. The error caused by this assumption is likely to be much more serious in b.c.c. structures where there are no $A1$ type sites but where the difference between first and second neighbour distances is not great (see p. 118). If the vacancy jump to a site which is the second neighbour of the solute atom has the same frequency as the jump to any other $A2$-type site, the correlation factor for solute diffusion in b.c.c. structures is

$$f_B = 7Fk_{A2}/(2k_B + 7Fk_{A2}), \tag{43.18}$$

where F is again a function of k_{A3}/k_A. This expression becomes more complicated if the $A2$ sites are divided into second neighbour and other sites. Note that for both (43.17) and (43.18), $0 \leqslant f \leqslant 1$, and f will in general be temperature dependent.

In the early days of diffusion theory a considerable difficulty presented by the vacancy mechanism was that the self-diffusion rate of the solvent in dilute solution is often found to be appreciably greater than that of the pure solvent; in some cases the tracer self-diffusion coefficient may be doubled by the addition of only 1% solute. Since dilute solutions behave ideally so far as the solvent is concerned, we should expect the diffusion coefficient to be given by $x_\square$ multiplied by the diffusion coefficient of vacancies, apart from small correlation effects, and since $c_{\square}$ should be nearly the same in dilute solutions as in the pure component, the self diffusion coefficients should be identical. The difficulty is removed by Johnson's proposal, since it is now possible for a faster moving impurity atom to enhance the rate of diffusion of the those solvent atoms which are close to it. This can happen either because P is greater near to the impurity than it is further away from it (association between the impurity and the vacancy), or because the rate of solvent jumping, k_{A1} and k_{A2}, is greater close to the impurity atom than it is in regions remote from it. Analyses of this rather difficult problem have been given by Lidiard (1960) and Howard and Manning (1967).

We finally consider correlation effects in concentrated solutions which have been treated approximately by Manning (1967) with the simplifying assumptions that the rates k_A, k_B at which tracer A and B atoms interchange with a vacancy are independent of all the surrounding atoms, and that non-tracer atoms have an average jump frequency of $x_A k_A + x_B k_B$. Provided the jump direction is at least an axis of twofold symmetry, he obtains

$$f_B = k_e/(2k_B + k_e), \tag{43.19}$$

where k_e is the effective frequency for escape of the vacancy from its position adjacent to the tracer B atom to a random position. Equation (43.19) may be compared with (43.13), (43.17), and (43.18) which contain explicit expressions for k_e for dilute solutions. In Manning's model, the vacancies are not bound to any particular atoms and

$$k_e = (x_A k_A + x_B k_B) f_{\square}\{2f_0(1-f_0)^{-1}\}, \tag{43.20}$$

where $f_{\square}$ is the correlation factor for vacancies and f_0 is the correlation factor for self-diffusion in a pure element. Thus the last factor in the equation has values (see p. 394) of 2, 3·77, 5·33, and 7·15 in diamond cubic, simple cubic, b.c.c., and f.c.c. crystals respectively. It remains to give an explicit expression for $f_{\square}$, which represents the deviation of the *vacancy* motion from a random walk. Manning gives

$$f_{\square} = \{x_A k_A (f_A/f_0) + x_B k_B (f_B/f_0)\}/(x_A k_A + x_B k_B) \tag{43.21}$$

so that

$$k_e = 2(x_A k_A f_A + x_B k_B f_B)(1-f_0)^{-1} \tag{43.22}$$

and

$$f_B = \frac{x_A k_A f_A + x_B k_B f_B}{k_B(1-f_0) + x_A k_A f_A + x_B k_B f_B}. \tag{43.23}$$

Note that the vacancies execute a random walk if $k_A = k_B$, in which case $f_A = f_B = f_0$; $f_{\square} = 1$ and (because of the assumptions of the model) $k_e = 2k_A f_0(1-f_0)^{-1}$.

Manning also calculated the effects of the correlations on the intrinsic diffusion coefficients D_A and D_B. The non-random vacancy motion gives rise to an additional vacancy flow term, but Darken's relation (41.12) remains valid. However, eqn. (41.16) must be replaced by the following relations between intrinsic and tracer diffusion coefficients:

$$D_A/D_{A+} = (1 + x_A B), \quad D_B/D_{B+} = 1 - x_B B, \tag{43.24}$$

where

$$B = \frac{(1-f_0)(D_{A+}-D_{B+})}{f_0(x_A D_{A+}+x_B D_{B+})}. \tag{43.25}$$

The vacancy flux may be expressed as

$$-(\mathbf{I}_A+\mathbf{I}_B) = (D_A^*-D_B^*)(\partial g_A/\partial \ln c_A) f_0^{-1} \nabla c_A. \tag{43.26}$$

Since $f_0^{-1} > 1$, the Kirkendall shift and the interdiffusion coefficient are larger than predicted by Darken's theory.

As discussed previously, the correlation effects may be represented formally by the off-diagonal elements of the mobility matrix (e.g. eqn. (41.4)). It is therefore interesting to note that with the assumptions made in Manning's theory, these off-diagonal terms obey the Onsager reciprocal relations (41.5).

44. THE THEORY OF DIFFUSION COEFFICIENTS

In this section we consider how the diffusion coefficient is related to the lattice energies. Consider the migration of an A atom from one position in the structure to a neighbouring vacant position. The initial and final conditions will both be solid solutions containing the same components, and will have the same standard free energies. The free energy will, however, rise to a maximum for some configuration where the atom is not on a regular site, and we may imagine an activated complex to be formed at this stage. The pseudo-thermodynamic theory of Chapter 3 may then be applied to relate the frequency of jumping to the difference in standard free energies of the complex and the equilibrium state. Most theories of diffusion coefficients use this approach, although attempts have been made to develop alternative dynamical theories.

If the solution behaves in a ideal manner, we can write the specific rate constant, k_A of eqn. (40.4), in the form

$$k_A = (\boldsymbol{k}T/\boldsymbol{h}) \exp(-\Delta g^+/\boldsymbol{k}T), \tag{44.1}$$

where the free energy of activation per atom Δg^+ is defined as in eqn. (13.9). From eqn. (41.20) the corresponding tracer diffusion coefficient is

$$D = (Pzr_1^2/6) f(\boldsymbol{k}T/\boldsymbol{h}) \exp(-\Delta g^+/\boldsymbol{k}T). \tag{44.2}$$

Many attempts have been made to apply the reaction rate theory to diffusion in non-ideal solutions. The less satisfactory use eqn. (13.9) and write expressions for $k_{A,12}$ and $k_{A,21}$ which include the activity coefficient, γ_A^+, of the activated state. By expanding about the values c_A, γ_A halfway between the compositions of the positions 1 and 2, essentially as in Section 43, an expression for the diffusion coefficient is obtained as

$$D/D_i = (1/\gamma_A^+)\{\gamma_A + c_A(\partial\gamma_A/\partial c_A)\}, \tag{44.3}$$

where D_i is the value of D given by eqn. (44.2), and we have made the implicit assumption that y_A^+ is constant. As already emphasized in Chapter 3, it is difficult to give a physical interpretation to quantities like $\gamma_A^{\ddagger}$ which appear in equations of this type. Le Claire (1949) has also emphasized that this treatment implies that the mobility of A depends only on the activity coefficient γ_A. If this is so, the experimental diffusion coefficient defined by Fick's law will vary with composition, but a diffusion coefficient defined in terms of activity gradients will be constant. In a few cases, the available experimental evidence suggests that this condition is nearly satisfied.

As already discussed, these difficulties may be avoided by means of the treatments due to Wert and Zener (1949), and to Vineyard (1957) who introduced only quantities with well-defined physical meanings. For interstitial and vacancy diffusion, the partition function when the diffusing atom is in an equilibrium position is factorized into a function Q_v representing vibration along the path taken in a unit jump process, and a partition function for the remaining degrees of freedom. The tracer diffusion coefficient is then given by

$$D = (Pzr_1^2/6)\, f(\boldsymbol{k}T/\boldsymbol{h})\,(1/Q_v)\, \exp(-\Delta_a g/\boldsymbol{k}T), \tag{44.4}$$

where $\Delta_a g$ has a clear meaning. It is the isothermal work required to transfer the diffusing atom slowly from its equilibrium position to its saddle-point position, constraining it always to remain in a plane normal to the diffusing path. At the temperatures used in most diffusion experiments, Q_v may be given its classical value $\boldsymbol{k}T/\boldsymbol{h}\nu$, where ν is some appropriate vibration frequency of the diffusing atom (see p. 88). We thus obtain

$$D = \nu(Pzr_1^2/6) f \exp(-\Delta_a g/\boldsymbol{k}T) = \nu(Pzr_1^2/6) f \exp(\Delta_a s/\boldsymbol{k}) \exp(-\Delta_a h/\boldsymbol{k}T) \tag{44.5}$$

The quantities ν and f which appear in eqn. (44.5) have hitherto been treated as constants, but it is now necessary to examine this assumption. If two different isotopes $A+$ and $\alpha+$ are used to measure separate tracer diffusion coefficients,

$$\Delta D/D_{\alpha+} = (\Delta k/k_{\alpha+})\{1+(k_{A+}/f_{\alpha+})(\Delta f/\Delta k)\} \simeq (\Delta k/k_{\alpha+})\{1+\partial \ln f/\partial \ln k\}, \tag{44.6}$$

where $\Delta D = D_{A+} - D_{\alpha+}$, etc. If we use the general form (43.19) for f and suppose that the non-tracer jumps of frequency k_e are independent of the isotope, the second bracket in (44.6) becomes simply f_{A+} and

$$(D_{A+}/D_{\alpha+}) - 1 = f_{A+}\{(k_{A+}/k_{\alpha+}) - 1\}. \tag{44.7}$$

If the activation energy is the same for both isotopes, the ratio of jump probabilities depends only on the attempt frequencies. Then provided the diffusion mechanism involves only one atom which is effectively decoupled from the surrounding atoms, we have (see eqn. (13.23))

$$(k_{A+}/k_{\alpha+}) - 1 = (\nu_{A+}/\nu_{\alpha+}) - 1 = (m_{\alpha+}/m_{A+})^{1/2} - 1. \tag{44.8}$$

This equation is clearly inapplicable to mechanisms such as ring diffusion or interstitialcy diffusion where $(n-1)$ atoms of the host matrix A move to new positions at the same time as the jump of the tracer atom. In this case, Vineyard (1957) proposed that eqn. (44.8) should be replaced by

$$(k_{A+}/k_{\alpha+}) - 1 = (\bar{m}_{\alpha+}/\bar{m}_{A+})^{1/2} - 1, \tag{44.9}$$

where

$$n\bar{m}_{A+} = (n-1)m_A + m_{A+}. \tag{44.10}$$

The assumption that the migrating atoms are decoupled from the remainder of the lattice is clearly not valid, and a correction must now be made for the participation of other atoms in the mode which leads to decomposition of the saddle-point configuration. This is expressed (eqn. (13.27)) as

$$(k_{A+}/k_{\alpha+}) - 1 = \Delta K\{(\bar{m}_{\alpha+}/\bar{m}_{A+})^{1/2} - 1\}, \tag{44.11}$$

where ΔK is the fraction of the translational kinetic energy in the unstable mode (Mullen, 1961; Le Claire, 1966). Equation (44.7) thus becomes in the general case

$$f_{A+}\Delta K = \{(D_{A+}/D_{\alpha+}) - 1\}\{(\bar{m}_{\alpha+}/\bar{m}_{A+})^{1/2} - 1\}^{-1}. \tag{44.12}$$

The experimental observation that $D_{A+} \neq D_{\alpha+}$ is known as the isotope effect and its measurement has assumed some importance because of the information it gives on diffusion mechanisms. The ratio of the diffusion coefficients is normally very small but may be measured rather accurately by diffusing both isotopes in the same specimen. Corrections to eqn. (44.12) arise mainly from two of the assumptions made above, namely that there is no mass effect on k_e and that the activation energies for migration of the two isotopes are equal; the latter is not valid when quantum effects are considered. These corrections have been considered by Le Claire (1966) and Ebisuzaki *et al.* (1967); using the treatment outlined in Section 13, they reached the conclusion that the equation is valid to within a few per cent. In order to extract the maximum information, experimental measurements have thus to be designed to reach this accuracy on the right-hand side.

Although it is not possible in general to separate the product $f_{A+}\Delta K$ into its separate factors, this may be done in appropriate cases. Thus for tracer self-diffusion in cubic structures, the values of f are known for the various possible diffusion mechanisms, and the knowledge that $\Delta K \leqslant 1$ is then sufficient in some cases to fix the mechanism from the measured values of $(D_{A+}/D_{\alpha+})$. For all close-packed metals investigated to date, it is possible in this way to eliminate all but vacancy mechanisms, and the values of ΔK range from ~0·75 to 1. On the other hand, the results for b.c.c. metals are consistent with all likely mechanisms except four-ring diffusion, and for sodium $\Delta K = 0{\cdot}5$ if the monovacancy mechanism is assumed to be operative.

Experimental results for diffusion coefficients are usually written in the form

$$D = D_0 \exp(-\varepsilon/kT), \tag{44.13}$$

and we now see (cf. p. 88) that if P is constant the experimental activation energy per atom may be identified with $\Delta_a h$. For interstitial diffusion, $P = f = 1$ and the other quantities in eqn. (44.5) have been evaluated (eqns. (40.10) and (40.11)). We thus have

$$\left.\begin{aligned} D_0 &= (a^2/6)\nu \exp(\Delta_a s/k) \quad \text{(f.c.c. structures),} \\ \text{and} \quad D_0 &= a^2\nu \exp(\Delta_a s/k) \quad \text{(b.c.c. structures).} \end{aligned}\right\} \tag{44.14}$$

There is a small correction for correlation effects ($f < 1$) if the mechanism is interstitialcy rather than interstitial.

We have already mentioned the proposal by Zener (1951) that $\Delta_a s$ is necessarily positive. His argument uses the thermodynamic relation

$$\Delta_a s = -(\partial \Delta_a g / \partial T)_p \tag{44.15}$$

to find a correlation between $\Delta_a s$ and $\Delta_a h$. If the free energy of activation is assumed to be due entirely to work done in straining the lattice at the activated complex, one may write $\Delta_a g / \Delta_a h = \mu / \mu_0$ where μ is some appropriate elastic modulus, and μ_0 is the value of this modulus at absolute zero. Hence

$$\Delta_a s = -\Delta_a h (1/\mu_0)(\partial \mu / \partial T) \tag{44.16}$$

and since $(\partial \mu / \partial T)$ is always negative, $\Delta_a s$ should always be positive. Note that if this model is correct, a non-linear dependence of μ on T implies that $\Delta_a s$, and hence $\Delta_a h$, are temperature dependent quantities.

The possible deficiencies of the elastic model used for calculations of this kind were stated on p. 127. Nevertheless, there is good experimental evidence of a correlation of the type indicated, experimental and theoretical values of $\Delta_a s$ agreeing well in most cases of interstitial diffusion (Wert, 1950). Zener suggested that where experimental results apparently indicated a negative entropy of activation, the results may be inaccurate because of the unsuspected presence of diffusion short circuits such as grain boundaries, sub-boundaries, or dislocation lines.

Turning now to substitutional diffusion by a monovacancy mechanism, we recall that the quantity P is equal to the probability $x_\square$ of a given site being vacant. Using eqn. (17.3), we thus find that the experimental activation energy for vacancy self diffusion should be given by

$$\varepsilon = \Delta h_\square + \Delta_a h_\square = \Delta h_1, \tag{44.17}$$

where $\Delta_a h$ in eqn. (44.5) has now been equated to the energy of motion of a vacancy $\Delta_a h_\square$. This relation is at least approximately valid for many f.c.c. metals, but the difficulties in making accurate measurements of $\Delta h_\square$ and $\Delta_a h_\square$ (see Section 17) limit its use as an experimental test for monovacancy diffusion more than was once realized.

Modern techniques have both improved the accuracy and extended the temperature range of diffusion measurements, and have thus permitted more detailed interpretation of $\ln D$ vs. $1/T$ plots. Although such plots are frequently almost linear for f.c.c. metals, some curvature can nevertheless be detected and there are strong indications that this is due, at least partially, to a divacancy contribution to self-diffusion. Thus it may be misleading to use a least squares linear fit to the experimental data in conjunction with eqn. (44.17) and we now consider how the measured values of D may be divided into monovacancy and divacancy contributions.

If diffusion involves monovacancies and divacancies, the self-diffusion coefficient may be written

$$D = D_{01} \exp(-\Delta h_1 / kT) + D_{02} \exp(-\Delta h_2 / kT), \tag{44.18}$$

where Δh_2 for divacancies is given by an expression similar to (44.17) and the pre-exponential factors (see eqn. (44.5)) are for f.c.c. structures.

$$D_{01} = a^2\nu_1 f_1 \exp\{(\Delta s_\square + \Delta_a s_\square)/k\} = a^2 \nu_1 f_1 \exp(\Delta s_1/k) \tag{44.19}$$

and

$$D_{02} = 4a^2\nu_2 f_2 \exp(\Delta s_2/k). \tag{44.20}$$

The numerical factors in this equation arise as follows. The Einstein–Smoluchowski factor eqn. (40.7) is equal to a^2 for monovacancy diffusion in f.c.c. (and b.c.c.) structures and is equal to $4a^2/6$ for divacancy diffusion in f.c.c. on the assumption that the moving atom is always a nearest neighbour of both vacancies; any other jump will decompose the divacancy. The probability P for divacancies contains a factor $z/2 = 6$ representing the number of orientations of the divacancy in the crystal (see p. 128). Note that the Einstein–Smoluchowski factor for diffusion of the divacancies themselves is $\frac{1}{6}$.

Eqn. (44.18) contains four disposable parameters and may thus be fitted to curved $\ln D$ vs. $1/T$ plots as was done for nickel by Seeger *et al.* (1965) and for gold by Wang *et al.* (1968). However, alternative interpretations may be based on temperature dependencies of the activation enthalpies and entropies, and it seems necessary to allow for this possibility. Seeger and Schumacher (1967) considered that it is sufficient to assume that the activation energy varies linearly with temperature so that

$$\Delta h_1 = \Delta h_{10} + C_7(T - T_0), \tag{44.21}$$

where Δh_{10} is the value at a reference temperature T_0 and C_7 is a constant which combines the separate temperature dependences of $\Delta h_\square$ and $\Delta_a h_\square$. Provided the temperature T_0 is above the Debye temperature θ_D, the second term in (44.14) is small in comparison with the first. The corresponding temperature dependence of the entropy is now

$$\Delta s_1 = \Delta s_{10} + C_7 \ln(T/T_0) \tag{44.22}$$

and eqn. (44.18) may be written in the form

$$\ln D = \ln D_{01} - (\Delta h_{10}/kT) + \ln\{1 + D_{02}/D_{01}) \exp(\Delta h_{10} - \Delta h_{20})\}/kT$$
$$+ C_7 \ln(T/T_0) - C_7(T - T_0)/kT, \tag{44.23}$$

where it has been assumed that the temperature dependences of Δh_2, Δs_2 are equivalent to those of Δh_1, Δs_1. The temperature variation of Δh_2 is a second-order correction and this assumption is unimportant.

Equation (44.23) contains five independent parameters, and since the observed deviation from linearity is small, an analysis of experimental results in terms of this equation alone cannot readily be made. However, additional information may also be employed; for example, experimental determinations of the correlation factor from the isotope effect or from a comparison of tracer diffusion with macroscopic thermal or electron diffusion will enable D_{02}/D_{01} to be deduced. This procedure is based on the knowledge that the correlation factor for single vacancies is $\sim 1{\cdot}67$ times larger than that for divacancies. The measured correlation factor is an effective factor which varies with the relative contributions of the two mechanisms, and hence with temperature. Mehrer and Seeger (1969) used the results of Rothman and Peterson (1969) on the isotope effect in copper to analyse the self-diffusion data which were originally represented by

$$D = 7{\cdot}8\times10^{-5}\exp(-2{\cdot}19\text{ eV}/kT)\text{ m}^2\text{ s}^{-1} \tag{44.24}$$

into a divacancy contribution and a single-vacancy contribution which is

$$D_1 = 1{\cdot}9\times10^{-5}\exp(-2{\cdot}09\text{ eV}/kT)\text{ m}^2\text{ s}^{-1} \tag{44.25}$$

at room temperature. When combined with results for $\Delta h_{\Box}$ obtained by Simmons and Balluffi (1963) (see Section 17), this gives

$$\Delta h_{\Box} = 1{\cdot}04\text{ eV},\quad \Delta_a h_{\Box} = 1{\cdot}05\text{ eV},$$

whilst eqn. (44.17) is not valid if ε is taken from (44.24).

A related procedure for analysing eqn. (44.23) is to make use of data on the pressure dependence of the diffusion rate which determines an effective activation volume at each temperature. Since the activation volumes for vacancies and divacancies are different, this again enables their relative contributions to D to be assessed (e.g. Mehrer and Seeger, 1972). The expressions for the effective activation volume are rather complex, but this method is possibly more accurate than the isotope effect procedure since the activation volume of a divacancy is approximately twice that of a monovacancy. Finally, nuclear magnetic resonance techniques enable diffusion coefficients to be measured at temperatures so low that the divacancy contribution is negligible; combination of these results with high temperature measurements then gives both the vacancy and divacancy parts of D.

We now turn to consider some of the special difficulties which are encountered in explaining diffusion measurements on b.c.c. metals. For most metals, self-diffusion appears to have the following characteristics (Le Claire, 1965); (1) the Arrhenius law (eqn. (44.13)) is approximately valid, (2) the activation energy ε is given within $\sim$20% by the empirical correlations

$$\varepsilon \simeq 34T^{sl} \simeq 16{\cdot}5\,\Delta h^{sl} \tag{44.26}$$

where T^{sl} and Δh^{sl} are respectively the melting temperature and the latent heat of melting' and (3) the pre-exponential factor D_0 is between $5{\cdot}10^{-6}$ and $5{\cdot}10^{-4}$ m^2 s^{-1}. Some b.c.c. metals conform to these general rules, albeit with rather larger curvatures in their Arrhenius plots than are usually found for f.c.c. metals, but other metals have markedly non-linear Arrhenius plots.

Analyses of the divacancy contribution to diffusion in b.c.c. metals are more complex than in f.c.c. metals because motion of a nearest neighbour divacancy by means of nearest neighbour atom jumps necessarily implies the temporary dissociation of the divacancy. There are, of course, other possibilities; for example, the stable divacancy configuration may be two next nearest neighbour vacant sites (p. 136), or atomic jumps may take place between next nearest neighbour sites. The possibility of associative and dissociative jumps means that the correlation factor f for divacancies is temperature-dependent in the b.c.c. structure, and Seeger (1972) has suggested that the most stable configurations and corresponding jump mechanisms may change from metal to metal. This could explain the appreciably greater variability in the experimental diffusion results for "normal" b.c.c. metals when compared with those for f.c.c. metals.

The "anomalous" b.c.c. metals include titanium, zirconium, hafnium and probably praesodymium, uranium, cerium, and plutonium, all of which have phase transitions on cooling from a b.c.c. structure to some other structure. When measurements are available over sufficiently large ranges of temperature, the resultant Arrhenius plots are very curved, and if these are approximated as two straight lines the lower temperature part of the curve indicates values of ε and D_0 much lower than are indicated by the above rules for normal metals. Although various explanations have been profferred in terms of diffusion short circuits or strong impurity–vacancy binding, the parameters required to fit the observations are implausible, and there also seems no reason why particular metals should be different in these respects from the other b.c.c. metals. An alternative suggestion (Seeger, 1972) is that the anomalous behaviour is linked to the phase transition, near to which there is an abnormally large temperature variation of the monovacancy parameters Δh_1 and Δs_1. The curvature of the Arrhenius plot is thus attributed mainly to this temperature dependence, although at high temperatures there will also be a divacancy contribution.

Vanadium and chromium were originally included in the anomalous metals, but although they exhibit rather marked curvatures in their $\ln D$ vs. $1/T$ plots, the deviation from normal behaviour is now at high temperatures, where D_0 and ε are apparently higher than usual. Seeger suggests that these metals differ only quantitatively from the normal b.c.c. metals in that they have a rather larger divacancy contribution which is characterized by an unusually large entropy factor.

The discussion in this section is mainly concerned with self-diffusion coefficients. Interstitial impurities usually diffuse at a much faster rate than the solvent atoms, but the impurity diffusion coefficients for substitutional impurities are often of the same magnitude as those of the host atoms. However, some substitutional impurities also diffuse at a very fast rate; these include noble metal solutes in silicon, germanium, lead, tin, thallium, etc. (Anthony, 1970). The explanation of this effect was first given by Frank and Turnbull (1956); the solute atoms dissolve both substitutionally and interstitially, and although the interstitial solubility may only be of the order of one-hundredth of the substitutional solubility, the interstitial atoms nevertheless make the biggest contribution to the diffusion rate. In effect, the diffusion mechanism involves the dissociation of a substitutional solute atom into an interstitial atom plus a vacancy.

45. STRUCTURE SENSITIVE DIFFUSION PROCESSES

In Sections 39–44 we have discussed diffusion through regions of crystal which are reasonably perfect except for small concentrations of defects. We expect that the activation energy for diffusion will be much lower in regions of bad crystal, so that diffusion should be a "structure sensitive" phenomenon. Under appropriate circumstances, we may define diffusion coefficients for the transfer of atoms along free surfaces, grain boundaries, or individual dislocation lines, and we expect that all of these will be larger than the lattice diffusion coefficient. Lattice diffusion is important only because the majority of the atoms in a crystal do not lie near regions of bad crystal, and so are unable to utilize the more rapid paths provided by such regions.

In principle, the measured diffusion coefficient should be a function of grain size since the amount of grain boundary per unit volume increases as the grain size decreases. This kind of effect is very difficult to detect, since the contribution of the grain boundaries to the overall diffusion rate is very small under the conditions of most diffusion anneals. Only at very low temperatures, where lattice diffusion virtually ceases, can grain boundary effects be detected by overall measurements of diffusion coefficients; this is in accordance with the general concept that processes of low activation energy become progressively more important as the temperature is lowered. Analyses of grain boundary diffusion thus have to be made by studying experimental concentration contours, which show greatest penetration along the boundaries.

A difference in diffusion behaviour is to be expected for low-angle and high-angle grain boundaries. In the disordered structure of a high-angle boundary, the activation energy for diffusion should be much lower than that for lattice diffusion, and the diffusion coefficient should be isotropic. For simple dislocation arrays, which form low-angle grain boundaries, the geometry of the dislocations will be expected to control boundary diffusion effects. In a symmetrical low-angle tilt boundary, for example, diffusion along the edge dislocations will be rapid, and a diffusion coefficient for each individual dislocation pipeline may be envisaged. For diffusion normal to the dislocation lines, the atoms must cross coherent regions of boundary, and the activation energy will not be very different from that for lattice diffusion. The boundary diffusion coefficient will thus be anisotropic in such a boundary.

Experimental measurements of grain boundary diffusion coefficients from observed concentration plots depend upon an analysis first given by Fisher (1951), who used the model shown in Fig. 9.3. The grain boundary is assumed to be a uniform slab of thickness δ^B having a diffusion coefficient D^B which is much greater than the lattice diffusion coefficient,

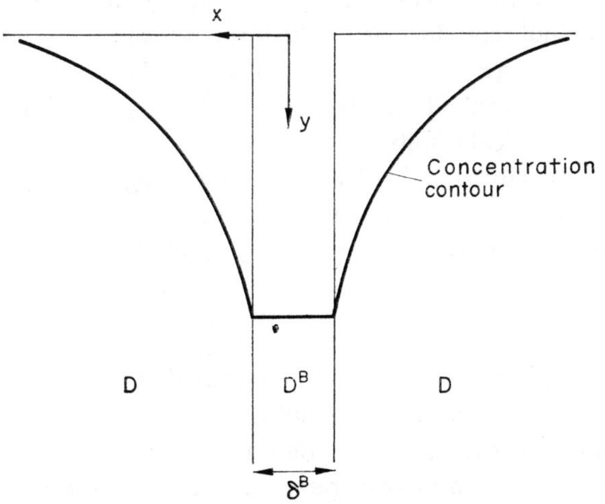

FIG. 9.3. To illustrate Fisher's model for grain boundary diffusion.

D. The concentration of diffusing substance at the original interface c_0 is assumed to remain constant with time, and the equilibrium concentration in the boundary is assumed to be the same as that in the lattice. The concentration in the boundary is also supposed to be uniform across its width, and both diffusion coefficients are regarded as independent of concentration. Finally, Fisher made the simplifying assumption that direct diffusion from the original interface into the lattice can be disregarded in comparison with diffusion along the boundary and then out sideways into the lattice. With these rather restrictive conditions, an approximate solution for the concentration at any point x, y (Fig. 9. 3) at time t can be written in the form

$$c = c_0 \exp\{-(2D/D^B\delta^B)^{1/2}/(\pi Dt)^{1/4}\}y \operatorname{erf}\{x/2(Dt)^{1/2}\}. \tag{45.1}$$

By integrating this equation, the following relation is obtained between the mean concentration of diffusing material in a thin slab at a distance y from the original interface and the distance y

$$\bar{c} \propto t^{1/2} \exp\{-2(D/D^B\delta^B)^{1/2}/(\pi Dt)^{1/4}\}y + \text{const}, \tag{45.2}$$

so that a plot of $\ln \bar{c}$ against y should give a straight line, from the slope of which $\delta^B D^B/D$ may be determined. Note that instead of the Gaussian penetration profile characteristic of volume diffusion (p. 385), the concentration falls off exponentially as the first power of the penetration distance y.

Fisher's assumption that the concentration change in the crystal due to direct diffusion through the lattice from the original interface may be neglected simplifies the problem, since it implies that lattice diffusion is everywhere normal to the grain boundary. It is clear, however, that this assumption is not really justified, and a more complete treatment of the Fisher model by Whipple (1954) manages to avoid it. The initial and boundary conditions of the model are $c(x, y, 0) = 0$ and $c(0, 0, t) = c_0$, and the diffusion equation has to be solved inside and outside the grain boundary slab in such a way that the concentration c and the diffusion flux $D(\partial c/\partial x)$ are both continuous at the surface of the slab. Whipple obtains a result for the concentration at any point $c(x, y)$ as a rather complex integral function which we shall not reproduce here. The form of the result is different from Fisher's inasmuch as there is no longer a linear relation between the mean concentration and the penetration distance, but Turnbull and Hoffman (1954) show that in a typical case two treatments give nearly identical results.

Le Claire (1951) suggested that a simpler way of obtaining the grain boundary diffusion coefficient is to measure the angle ψ at which a curve of constant composition meets the grain boundary. This angle is equal to $(dx/dy)_{x=0}$, and a simple differentiation of eqn. (45.1) gives

$$\delta^B D^B/D = 2(\pi Dt)^{1/2} \cot^2 \psi. \tag{45.3}$$

Although Whipple's solution does not give a simple analytical expression like this, he has computed an analogous relation. An important result emphasized by Whipple is that the depth of penetration of a particular concentration contour along a grain boundary is not markedly greater than the depth of penetration into the grain, even with large ratios of D^B/D. This is because most of the material which originally diffuses down the boundary is lost sideways into the lattice. Typical figures are that a ratio of D^B/D of about 10^5 is needed to produce a penetration depth along the boundary twice as large as that in the crystal in about 100 h. It will be realized from the above discussion that all grain boundary diffusion experiments depend upon $D^B\delta^B$; when it is desired to write values for D^B it is usual to assume a thickness of two or three atom diameters for the boundary region.

This analysis of grain boundary diffusion would be expected to apply most successfully to high angle boundaries, where the structure is very disordered, but Turnbull and Hoffman (1954) showed that it is also valid for the self-diffusion of silver along low-angle (dislocation) boundaries. Using symmetrical tilt boundaries, these workers found that the overall diffusivity varied with the misorientation θ as $\sin \frac{1}{2}\theta$. Since the density of dislocations in such a boundary is also proportional to $\sin \frac{1}{2}\theta$ (p. 334), the results imply that it is possible to define an intrinsic diffusion coefficient D^P along each dislocation pipe, the diffusion down one dislocation being unaffected by the other dislocations. The relation between the boundary diffusion coefficient and the dislocation diffusion coefficient is then

$$D^B = 2D^P a \sin \tfrac{1}{2}\theta \qquad (45.4)$$

for the simple boundary considered.

The activation energy for self-diffusion along dislocation pipes obtained in these and similar experiments, and the activation energy for diffusion in high-angle grain boundaries, are both found to be of the order of half the corresponding activation energy for lattice diffusion. This is not unexpected if the core of a dislocation and a high-angle boundary have similar structures, and several other experiments indicate that the diffusivity of a dislocation and a general grain boundary are similar. Results on chemical diffusion along grain boundaries are scarcer, but the activation energy is again probably of the order of half that for lattice diffusion. Chemical diffusion will be complicated, amongst other effects, by a tendency for the solute atoms to segregate preferentially either in or away from the boundaries.

In macroscopic diffusion experiments, the grain boundaries will make increasing contributions at low temperatures. The boundary mechanism will be dominant when the ratio of the diffusion coefficients exceeds the ratio of the cross-sectional areas of the respective paths, which is about 10^5 for a grain size of 0·1 mm. This gives a temperature of about one-half to three-quarters of the melting point temperature for the metals for which data are available. The transition will be gradual, and may account for some apparently incorrect early results on the temperature variation of diffusion coefficients. For greater accuracy in the determination of experimental activation enthalpies, it is normally recommended that the temperature range covered should be as large as possible. This is only true, of course, if the mechanism remains unchanged; unfortunately the scatter of diffusion data is such that apparently good Arrhenius type plots may be made even when the diffusion process changes with falling temperature from mainly lattice to mainly grain boundary. Many authors have emphasized the danger that incorrect activation energies and enthalpies may result from this kind of effect in polycrystalline specimens (see p. 407).

The enhanced diffusion rate along individual dislocation lines suggests the possibility that a grown-in three-dimensional dislocation network may make an appreciable contribution to the diffusion rate in a single crystal. This situation has been considered by Hart

(1957), using a random walk type of model for the diffusion process. The mean square displacement of an atom in a given time may be regarded as partly due to migration through the lattice and partly to much more rapid migration along dislocation lines. Provided the dislocation network has connecting lines of length short compared to the diffusion penetration distance, the penetration profile will be Gaussian as for true random volume diffusion, but the apparent diffusion coefficient is larger than the true volume diffusion coefficient.

Suppose an atom moves randomly through the crystal until it encounters a dislocation, and that it then remains for a time τ^D at the dislocation core, during which time it diffuses a mean square distance $\overline{P^2}$. If this procedure is repeated several times during a diffusion period t, the total mean square migration distance (averaged for many atoms) may be taken as the sum of the mean square migration distance ($\overline{Q^2}$) through the lattice in the time t, and the mean square migration distances along dislocation lines, giving

$$\overline{S^2} = m\overline{P^2} + \overline{Q^2}, \tag{45.5}$$

where m is the number of dislocations encountered during the time t. Correlation effects, which are ignored, should be small.

If the temperature and the time t are such that the mean migration time of an atom between two dislocations is much less than t, it is justifiable to assume that a diffusing atom spends a fraction x^D of the time diffusing along dislocations and a fraction $(1-x^D)$ of the time in the lattice, where x^D is the fractional concentration of atoms in dislocation core regions. This gives

$$m\tau^D = x^D t, \tag{45.6}$$

and the lattice diffusion coefficient D is given by

$$\overline{Q^2} = 6D(1-x^D)t \tag{45.7}$$

(see eqn. (40.5)).

From eqns. (45.5), (45.6), and (45.7) we now have the effective diffusion coefficient D' defined by†

$$\overline{S^2} = 6D't,$$

where

$$D' = (\overline{P^2}/6\tau^D)x^D + D(1-x^D). \tag{45.8}$$

The term $\left(\overline{P^2}/6\tau^D\right)$ is probably not much different from the diffusion coefficient for dislocation pipes in low angle boundaries. We have seen on p. 312 that a reasonable estimate of x^D is probably about 10^{-7} for well-annealed material. Thus if the grain boundary diffusion coefficient is 10^6 times larger than the bulk lattice diffusion coefficient, the dislocation network could make a 10% contribution to the overall diffusion flux. A ratio of this magnitude is quite possible in low-temperature diffusion experiments, and may explain deviations from the Arrhenius plot (i.e. discrepancies between high and low-temperature diffusion

† In Hart's paper, the factor 6 in eqns. (45.7) and (45.8) is replaced by a factor 2; this is because he considers the mean square displacements projected along a particular direction. This gives $\overline{S_i^2} = 2Dt$ for the diffusion coefficient in that direction, but reduces to (40.5) if the diffusion coefficient is isotropic (see p. 383).

results) even in single crystals. At sufficiently low temperatures, the dislocations may make much more significant contributions to diffusion that the lattice, but the overall diffusion rate is then so small that measurements are extremely difficult.

We now turn to consider the effects of non-equilibrium states on the rate of diffusion. We might expect enhanced diffusion in a severely deformed metal because of an increase in the number of dislocations or of vacancies, or in quenched or irradiated materials, because of increases in the vacancy concentration over the thermal equilibrium value. In normal diffusion anneals at reasonably high temperatures, the excess defect concentrations will be removed during the anneal, so that the amount of extra diffusion obtained will depend on the lifetime of the defects considered. More effective increases in diffusion rates should be obtained if defects are continuously created whilst the diffusion is taking place, e.g. by irradiating the specimen, or maintaining a stress on it. We now consider these effects more quantitatively.

For vacancy diffusion the diffusion coefficient is directly proportional to the atomic concentration $x_{\square}$ of vacancies. As we have seen in Section 17, this can be as high as 10^{-3} to 10^{-4} in severely quenched or damaged metals, so that at low temperatures the increase in diffusion coefficient can be a very large factor. The discussion on vacancy annealing in Section 17, however, shows that this is not an easy effect to detect in normal experiments. Thus Lomer (1958) points out that if a vacancy makes n jumps on the average before disappearing, the average number of jumps made by an atom is $x_{\square}n$ and this is equal to $6Dt/r_1^2$ (eqns. (40.5) and (40.7)). The distance $2(Dt)^{1/2}$ which defines the mean diffusion length is thus given by

$$2(Dt)^{1/2} = r_1(2nx_{\square}/3)^{1/2}, \tag{45.9}$$

and with typical values this is about one micron if n is taken as 10^{10} (see p. 137). Thus although the excess vacancies would be extremely effective in promoting diffusion over short distances, they anneal out too quickly to be important in macroscopic diffusion measurements. Effects of this kind, however, are probably extremely important in such transformation phenomena as the ageing of supersaturated solid solutions in which initial segregation over very short distances almost certainly utilizes the excess vacancies retained in the quenched specimen.

Similar arguments have been given by Lomer to show that the effect of extra vacancies produced by continuous irradiation is also rather small, expecially since the effective number of jumps made by a vacancy before disappearing is usually much less in irradiated specimens than in quenched specimens (see Section 17). The general conclusion is again that only at low temperatures do the contributions from the excess vacancies begin to outweigh those from the thermal vacancies which are always present, and that the excess vacancies are only important for fine scale phenomena.

Finally, we discuss the effect of plastic deformation on diffusion rate. A number of workers have reported very considerably enhanced diffusion rates in metals subjected to torsional or compressive strains during the diffusion process, the increase being of the order of a hundred or a thousand fold in some circumstances. Other workers have reported no change in diffusion rates, even for large total strains and high strain rates. This subject is thus rather confused and controversial and, in particular, there is disagreement about the magnitude of the effect in silver at about 800°C. It seems possible that genuine large increases are obtained in some circumstances, especially at lower temperatures, but the mechanism remains uncertain. Some workers have attributed the effect to point defects generated by moving dislocations, but the arguments above seem to suggest that very long vacancy lifetimes would be required for this model, and there is no available evidence to support such long lifetimes at high temperatures. However, the suggestion that "old" dislocations cannot absorb vacancies readily may be relevant here, since it suggests that sinks for vacancies will be less effective at diffusion temperatures. Annihilation of vacancies by interstitials will not be so important in deformation experiments as in irradiation, since present evidence is that moving dislocations generate vacancies but not interstitials. The alternative possibility is that enhanced diffusion is due to greatly increased sub-boundary diffusion, especially if the sub-boundaries are themselves migrating during the experiment.

REFERENCES

ANTHONY, T. R. (1970) *Vacancies and Interstitials in Metals*, p. 935, North-Holland, Amsterdam.
BAKKER, H. (1971) *Phys. stat. sol.* **44**, 369.
BALLUFFI, R. W. (1954) *Acta metall.* **2**, 194.
BARDEEN, J. and HERRING, C. (1951) *Atom Movements*, p. 87., American Society for Metals, Cleveland, Ohio; (1952) *Imperfections in Nearly Perfect Crystals*, p. 261, Wiley, New York.
BARNES, R. S. (1952) *Proc. Phys. Soc.* B, **65**, 512; (1955) *Defects in Crystalline Solids*, p. 359, The Physical Society, London.
BARNES, R. S. and MAZEY, D. J. (1958) *Acta metall.* **6**, 1.
BECKER, R. (1937) *Z. Metallk.* **29**, 245.
CAHN, J. W. (1961) *Acta metall.* **9**, 795; (1962) *Ibid.* **10**, 179.
CARSLAW, H. S. and JAEGER, J. C. (1947) *Conduction of Heat in Solids*, Clarendon Press, Oxford.
COLEMAN, B. D. and TRUESDELL, C. (1960) *J. Chem. Phys.* **33**, 28.
COMPAAN, K. and HAVEN, Y. (1958) *Trans. Faraday Soc.* **54**, 1498.
COOK, H. E. and HILLIARD, J. E. (1969) *J. Appl. Phys.* **40**, 2191.
CRANK, J. (1956) *The Mathematics of Diffusion*, Clarendon Press, Oxford.
DARKEN, L. S. (1948) *Trans. Am. Inst. Min. (Metall.) Engrs.* **175**, 184.
DYSON, B. F., ANTHONY, J., and TURNBULL, D. (1966) *J. Appl. Phys.* **37**, 2370; (1967) *Ibid.* **38**, 3408; (1968) *Ibid.* **39**, 1391.
EBISUZAKI, Y., KASS, W. J., and O'KEEFFE, M. (1967) *J. Chem. Phys.* **46**, 1373; *Phil. Mag.* **15**, 1071.
EINSTEIN, A. (1905) *Ann. Physik.* **14**, 549.
FISHER, J. C. (1951) *J. Appl. Phys.* **22**, 74.
FRANK, F. C. and TURNBULL, D. (1956) *Phys. Rev.* **104**, 617.
HART, E. W. (1957) *Acta metall.* **5**, 597.
HEUMANN, T. and WALTHER, G. (1957) *Z. Metallk.* **48**, 151.
HILLERT, M. (1961) *Acta metall.* **9**, 525.
HOWARD, R. E. (1966) *Phys. Rev.* **144**, 650.
HOWARD, R. E. and LIDIARD, A. B. (1964) *Reports Prog. Phys.* **27**, 161.
HOWARD, R. E. and MANNING, J. R. (1967) *Phys. Rev.* **154**, 561.
HUNTINGTON, H. B. (1951) *Atom Movements*, p. 69, American Society for Metals, Cleveland, Ohio.
JOHNSON, W. A. (1941) *Trans. Am. Inst. Min. (Metall.) Engrs.* **143**, 107.
KONOBEEVSKY, S. T. (1932) *J. Exp. Theor. Phys. USSR* **13**, 185; (1943) *Ibid.* **13**, 418.
LAZARUS, D. (1960) *Solid State Phys.* **10**, 71.
LE CLAIRE, A. D. (1949) *Prog. Metal Phys.* **1**, 306; (1951) *Phil. Mag.* **42**, 468; (1953a) *Prog. Metal Phys.* **4**, 265; (1953b) *Acta metall.* **1**, 438; (1965) *Diffusion in Body Centred Cubic Metals*, p. 3, American Society for Metals, Metals Park, Ohio; (1966) *Phil. Mag.* **14**, 1271.
LE CLAIRE, A. D. and LIDIARD, A. B. (1956) *Phil. Mag* **1**, 518.

LIDIARD, A. B. (1960) *Phil. Mag.* **5**, 1171.
LOMER, W. M. (1958) *Vacancies and Other Point Defects in Metals and Alloys*, p. 79, The Institute of Metals, London; (1959) *Prog. Metal Phys.* **8**, 255.
MANNING, J. R. (1962) *Phys. Rev.* **128**, 2169; (1964) *Ibid.* 136, A1758; (1967) *Acta metall.* **15**, 817; (1968) *Diffusion Kinetics for Atoms in Crystals*, Van Nostrand, Princeton, New Jersey.
MATANO, C. (1933) *Japan. J. Phys.* **8**, 109.
MEHRER, H. (1972) *J. Phys. (F)* **2**, L11.
MEHRER, H. and SEEGER, A. (1969) *Phys. stat. sol.* **35**, 313; (1972) *Crystal Lattice Defects* **3**, 1.
MULLEN, J. G. (1961) *Phys. Rev.* **121**, 1649.
NACHTRIEB, N. H. and HANDLER, G. S. (1954) *Acta Metall.* **2**, 797.
ONSAGER, L. and FUOSS, R. M. (1932) *J. Phys. Chem.* **36**, 2689.
PANETH, H. (1950) *Phys. Rev.* **80**, 708.
PETERSON, N. L. (1968) *Solid State Phys.* **22**, 409.
PHILOFSKY, E. M. and HILLIARD, J. E. (1969) *J. Appl. Phys.* **40**, 2198.
RESNICK, R. and SEIGLE, L. L. (1957) *Trans. Am. Inst. Min. (Metall.) Engr.* **209**, 87.
RICE, S. A. (1958) *Phys. Rev*, **112**, 804.
ROTHMAN, S. J. and PETERSON, N. L. (1969) *Phys. stat. sol.* **35**, 305.
SEEGER, A. (1972) *J. Less-common Metals* **28**, 387.
SEEGER, A. and SCHUMACHER, D. (1967) *Mater. Sci. Engr.* **2**, 31.
SEEGER, A., SCHOTTKY, G., and SCHUMACHER, D. (1965) *Phys. stat. sol.* **11**, 363.
SEITZ, F. (1948) *Phys. Rev.* **74**, 1513; (1950a) *Acta crystallogr.* **3**, 355; (1950b) *Phase Transformations in Solids*, p. 77, Wiley, New York; (1953) *Acta metall.* **1**, 355.
SHEWMON, P. G. (1963) *Diffusion in Solids*, McGraw-Hill, New York.
SHIRN, G. A., WAJDA, E. S., and HUNTINGTON, H. B. (1953) *Acta metall*, **1**, 513.
SIMMONS, R. O. and BALLUFFI, R. W. (1963) *Phys. Rev.* **129**, 1533.
SMIGELSKAS, A. D. and KIRKENDALL, E. O. (1947) *Trans. Am. Inst. Min. (Metall.) Engrs.* **171**, 130.
TURNBULL, D. and HOFFMAN, R. E. (1954) *Acta metall.* **2**, 419.
VINEYARD, G. H. (1957) *J. Phys. Chem. Solids* **3**, 121.
WANG, C. G., SEIDMAN, D. N., and SIEGEL, R. W. (1968) *Phys. Rev.* **169**, 553.
WERT, C. (1950) *Phys. Rev.* **79**, 601.
WERT, C. and ZENER, C. (1949) *Phys. Rev.* **76**, 1169.
WHIPPLE, R. T. P. (1954) *Phil. Mag.* **45**, 1225.
ZENER, C. (1950) *Acta crystallogr.* **3**, 346; (1951) *J. Appl. Phys.* **22**, 372; (1952) *Imperfections in Nearly Perfect Crystals*, p. 305, Wiley, New York.

CHAPTER 10

The Classical Theory of Nucleation

46. THE FORMATION OF NUCLEI OF A NEW PHASE

The kinetics of a heterogeneous reaction, as discussed in Section 4, can usually be described in terms of the separate nucleation and growth of the transformed regions. The classical theory of nucleation by random fluctuations in a metastable assembly is due mainly to Volmer and to Becker and Döring, but many other workers have made significant contributions. This theory was formulated first for the simplest nucleation process, the condensation of a pure vapour to form a liquid, and we shall find it convenient to study this change in some detail as a prelude to the more complex problems of nucleation in liquid and solid phases. Throughout this chapter we deal only with transformations which do not involve changes of composition; the complications which these introduce are considered in later chapters.

The essential driving force for a phase transformation is the difference in the free energies of the initial and final configurations of the assembly, but when small particles of the new phase are formed, the free energy rises at first. The increase is due to the considerable proportion of atoms in these particles which are situated in transition regions between the phases, where they do not have the environment characteristic of the new phase in bulk. The situation is conventionally described by assigning volume free energies to the bulk phases, and a surface free energy to the interface region; for sufficiently small particles of the new phase, the surface term is dominant. Additional factors have also to be considered in transformations in the solid state, where the volume change associated with the transformation, and the possible tendency for the two phases to remain coherent across the interphase boundary, may both produce considerable perturbations of the atomic arrangement. Each nucleus may then be regarded as a source of internal stress, and the resultant elastic strain energy has to be included in the overall free energy change of the assembly.

We begin with the vapour–liquid transformation where there are no strain energies to be considered. The usual theory is developed in terms of a model of a very small liquid droplet, which is certainly invalid; the justification for its use is partly mathematical, since we shall find that the assumptions are least important where they are most outrageous and partly empirical, since reasonable agreement with experiment is obtained. The essential assumption is that a small droplet may be treated in the same way as a large mass of liquid, and in particular may have its properties described by macroscopic thermodynamic parameters.

Thus we treat the transition region between vapour and liquid as a geometrical surface with a specific free energy σ per unit area. As discussed in Section 35, the value of σ for a curved boundary will vary with the position chosen for the geometrical surface, but no serious ambiguity arises unless the radius of curvature is of the order of the width of the transition region. Unfortunately, this is just the situation encountered in nucleation theory, where the rate of nucleation is commonly determined by the properties of droplets about twenty atomic diameters or less across. The concept of surface free energy is thus scarcely appropriate to the problem, and the value of σ for any particular choice of reference interface will not necessarily bear any simple relation to the usual macroscopic free energy σ_{∞} defined for a plane interface. Similar difficulties arise in connection with the other parameters used to define the droplet. In the transition zone, estimated at two to seven atomic or molecular† diameters, the density is intermediate between that of bulk liquid and vapour. The number of atoms within the droplet, and the mean free energy of the droplet per atom, are thus only approximate concepts, which cannot be rigidly defined for any choice of reference interface.

The classical theory of nucleation ignores these difficulties. A liquid droplet is regarded as a sphere having the density of bulk liquid right up to some limiting geometrical surface, and the interfacial energy σ is defined with reference to this surface. Tolman (1949), Kirkwood and Buff (1949) and others have shown that σ expressed in this way will not be constant but will decrease with decreasing drop size. The magnitude of the effect depends on the nature of the interatomic forces, and it is very small when only nearest neighbour interactions are considered. Buff (1951) has attempted to modify eqn. (46.2) for the free energy change given below to allow for the variation of σ; his theory suggests that values of σ calculated from experimental results using the classical nucleation theory will be substantially smaller than σ_{∞}. Assuming Lennard-Jones atomic forces, however, Benson and Shuttleworth (1951) have shown that for the extreme case of a "droplet" consisting of a close-packed cluster of thirteen atoms, σ is only 15% less than σ_{∞}.

Attempts at exact treatments which avoid the above inconsistencies have been made (see, for example, Reiss, 1952), but the statistical summations which replace the macroscopic parameters cannot usually be evaluated. Similar difficulties occur in some of the statistical theories of the liquid state (see Section 21). We shall accept the assumptions of the classical theory and shall not attempt to describe the more exact formulations of the problem. The variation of σ with droplet size, and the lack of precision in defining the size, are less serious than first appears, since the rate of nucleation is found to be determined principally by the properties of droplets in a restricted critical size range. Provided no formal attempt is made to identify σ with σ_{∞} (they are sometimes found *empirically* to be almost equal), the difficulties which we have described in detail do not seem to invalidate the results. However, in recent years it has become apparent that there are other difficulties in the quantitative formulation of the theory of nucleation which cannot be entirely avoided by

† Strictly, we should use the more general word "molecule", but it seems rather artificial to write of metallic molecules since almost all metallic vapours are monatomic. For the remainder of the chapter we write only of atoms, but the discussion in this and other chapters often applies to non-metals if "molecule" is substituted for "atom".

the macroscopic model and which may be attributed, at least partially, to our lack of knowledge of the detailed atomic configurations in the liquid state, and especially of the atomic motions which contribute to its thermal entropy. These difficulties have resulted in some confusion about the correct expression for the equilibrium distribution of very small liquid droplets in a vapour, and are discussed below.

Suppose then that the number of atoms in a liquid droplet and in the vapour phase are n and N^v respectively, and that the chemical potentials, or free energies per atom, in the bulk phases are g^l and g^v. The n atoms which have condensed will adopt a spherical configuration so as to minimize the excess surface energy, and the free energy of the assembly will thus change by an amount ΔG when the droplet is formed, where according to the classical theory and

$$\Delta G = \Delta G' \quad (46.1)$$

$$\Delta G' = (4\pi r^3/3v^l)(g^l - g^v) + 4\pi r^2 \sigma. \quad (46.2)$$

In this equation, r is the radius of the droplet, and v^l the volume per atom in the liquid state. It is often more convenient to focus attention on the number of atoms in the droplet, rather than on its radius, so the equation may be written in the alternative forms

$$\Delta G' = n(g^l - g^v) + O_n \sigma = n(g^l - g^v) + \eta n^{2/3} \sigma, \quad (46.3)$$

where O_n is the surface area of the droplet containing n atoms, and η is a shape factor. In the vapour–liquid transformation, $\eta = O_n/n^{2/3} = (36\pi)^{1/3}(v^l)^{2/3}$, but the same form of equation applies to other nucleation problems in which the nucleus is not spherical in shape. In solid transformations, the surface free energy is not isotropic, and a polyhedral nucleus may be formed. The change in free energy should then include a term in $n^{1/3}$ representing the edge free energy; if this is neglected, eqn. (46.3) can still be used, giving appropriate mean values to η and σ. The use of this equation implies, of course, that only surface energies need be considered.

We have now to consider what are the permitted values of n. The association of n vapour atoms to form a cluster can clearly be regarded as a liquid droplet with less and less plausibility as n decreases, and it makes the theory a little less artificial if we impose a restriction that a liquid droplet must contain a certain minimum number p of atoms. The value of p is, in fact, immaterial to the final result.

In the classical theory of nucleation, eqns. (46.2) and (46.3) are regarded as giving the free energy change of the whole assembly when a liquid droplet is formed from the vapour, and this is equivalent to the assumption that if a bulk liquid is separated into a vapour of droplets at the same temperature and pressure, the free energy change per droplet is given by the surface energy terms in these equations. Frenkel (1939, 1946) pointed out, however, that a liquid droplet in a vapour should be regarded as a kind of macromolecule, and a number of authors (Rodebush, 1952; Kuhrt, 1952; Lothe and Pound, 1962) independently recognized that this might appreciably affect the magnitude of ΔG. Any difference between ΔG and $\Delta G'$ essentially arises because a liquid droplet in a vapour differs from an equivalent region of bulk liquid not only in having a surface, but also in possessing much more freedom to translate and rotate independently of the other droplets. Thus, according to Lothe and Pound, a more correct expression for the free energy change would be

$$\Delta G = \Delta G' + \Delta G_{tr} + \Delta G_{rot} + \Delta G_{rep}, \tag{46.4}$$

where ΔG_{tr} and ΔG_{rot} are contributions to the energy which may be considered to arise from the "heating" of a stationary droplet until it acquires the kinetic energy appropriate to the state defined by the temperature T and pressure p of the vapour. The term ΔG_{rep} is the so-called replacement energy and arises because when the droplet acquires the gas-like translational and rotational motion it must lose six internal or liquid-like degrees of freedom. It is often stated that $\Delta G'$ represents the free energy of formation of a stationary droplet, but this is misleading; for example, its centre of mass is not stationary. A more correct statement might be that $\Delta G'$ represents the energy of a droplet confined as in a liquid rather than as in a vapour.

The terms ΔG_{tr} and ΔG_{rot} are both negative and may be calculated from the quantum-statistical contributions to the additional entropy of a droplet in the vapour, whilst ΔG_{rep} is correspondingly positive since it represents a loss of entropy. A considerable controversy exists over the magnitude of ΔG_{rep}, which at least partially reflects our ignorance of atomic motions in the liquid state. It is clear, for example, that if groups of atoms execute co-operative rotations in the liquid which are only slightly hindered, then one part of ΔG_{rep} must almost cancel ΔG_{rot}. Intuitively, it seems unlikely that the rotational component is cancelled in this way, but it is difficult to prove that it is not.

Since all the terms in ΔG except the first terms in (46.2) or (46.3) represent the difference in free energy between a "gas" of liquid droplets and the same liquid in bulk form, it would be possible in principle to define σ in (46.2) or (46.3) so as to include the additional terms of (46.4). This would have the advantage of retaining the classical formulation of the equations of nucleation theory, but would involve the unsatisfactory feature of a surface free energy σ dependent on the size of the assembly. An alternative is to write explicit expressions for the remaining terms of (46.4) and hope that σ then corresponds approximately to the macroscopic free energy, despite the difficulties, which have already been stressed, of using macroscopic concepts for very small regions. This latter course has generally been followed in recent work. Note that when the size of the droplet becomes very large, the difference between (46.1) and (46.4) tends to zero, since both terms in (46.3) increase indefinitely with the size of the embryo n, whereas the additional terms in (46.4) are nearly independent of n.

The translational partition function for the three-dimensional motion of an embryo of size n in the standard state "gas" of such embryos (cf. eqn. (13.2)) is

$$Q_{tr} = (2\pi mkT/h^2)^{3/2}\, n^{3/2} v^v, \tag{46.5}$$

where m is the mass of an individual atom and $v^v = kT/p$ is the molecular volume of the embryo. Clearly $v^v = V/N^v$ is also the volume per atom in the vapour phase.

The rotational partition function is similarly

$$Q_{rot} = \pi^{1/2}(8\pi^2 kTI)^{3/2}/h^3, \tag{46.6}$$

where I is the moment of inertia of the spherical droplet and no atomic symmetry (i.e. symmetry number = 1) is assumed. Certain difficulties clearly arise in the concept of rotation applied to a liquid droplet, but a detailed analysis by Nishioka *et al.* (1971) seems to show that the above formula is correct. For a special nucleus of radius r, $I = (\frac{2}{5})mnr^2$, and eqn. (46.6) can be put in the form

$$Q_{\text{rot}} = C_8(2\pi mkT/h^2)^{3/2}\, v^l n^{5/2}, \tag{46.7}$$

where $C_8 = (\frac{3}{4})(\frac{8}{5})^{3/2}\pi \simeq 4{\cdot}8$, and v^l is the volume per atom in the liquid phase. Hence

$$Q_{\text{tr}}Q_{\text{rot}} = C_8(2\pi mkT/h^2)^3\, v^v v^l n^4, \tag{46.8}$$

and if we write

$$\Delta G_{\text{tr}} + \Delta G_{\text{rot}} + \Delta G_{\text{rep}} = kT\ln(Q_{\text{tr}}Q_{\text{rot}}/Q_{\text{rep}}) \tag{46.9}$$

it remains only to evaluate Q_{rep}.

Frenkel assumed that Q_{rep} is small and may be neglected, but subsequent estimates have ranged from small values ($\sim 10^3$) to large values almost sufficient to cancel $Q_{\text{tr}}Q_{\text{rot}}$. Lothe and Pound originally considered the replacement partition function to be the entropy portion of the free energy of an atom in a liquid, and an argument for this in terms of a series of virtual processes was given by Feder *et al.* (1966). On the other hand, Kuhrt (1952), Dunning (1965, 1969), and others included the binding energy of the liquid atom in their estimate of Q_{rep}, and so obtained a much higher value. The Lothe–Pound expression is

$$Q^l_{\text{rep}} \simeq (2\pi n)^{1/2}\exp(s^l/k), \tag{46.10}$$

where s^l is the entropy per atom in the liquid. In later work, Lothe and Pound (1966) argued that Q_{rep} must correspond to six degrees of freedom for which the relative positions of the atoms in the region of bulk phase considered do not change. In the case of a solid, this corresponds to translational and torsional vibrations with

$$Q^s_{\text{rep}} = (kT/h\nu_D)^6, \tag{46.11}$$

where ν_D is the Debye frequency. The authors suggest, moreover, that $Q^l_{\text{rep}} \simeq Q^s_{\text{rep}}$. More recently, Nishioka *et al.* (1971) have calculated Q^s_{rep} for a small solid crystallite by a numerical normal mode analysis and obtained a result appreciably larger than that given by (46.11). The typical numerical value given by (46.10) or (46.11) is $\sim 10^2$–10^4, whereas the normal mode analysis gives $Q^s_{\text{rep}} \sim 10^8$. However, this discrepancy is stated to be largely due to the difference between the vibrational surface free energy of the bulk plane surface and that of the crystallite, which is included in eqn. (46.11) but excluded from the normal mode estimate.

If the vapour is supersaturated, the first of the two terms in (46.2) is negative, and the second is positive. Since these terms are proportional to r^3 and r^2 respectively, it follows that the influence of the second term will become less as r increases. The curve of ΔG against r (or n) will thus increase to a maximum and then decrease again. The position of the maximum is given by $\partial\Delta G'/\partial r = 0$, which leads to the critical radius of the droplet r_c as

$$r_c = 2\sigma v^l/(g^v - g^l). \tag{46.12}$$

The number of atoms in a drop of critical size is similarly

$$n_c = \left\{\frac{2\eta\sigma}{3(g^v - g^l)}\right\}^3 = \frac{32\pi\sigma^3(v^l)^2}{3(g^v - g^l)^3}. \tag{46.13}$$

A nucleus of radius r_c is in unstable equilibrium† with the vapour. Droplets of radius $r < r_c$ will tend to evaporate, since an increase in size leads to an increase in ΔG, whilst droplets of radius $r > r_c$ will tend to grow, since an increase in radius then decreases ΔG. Droplets with $r < r_c$ are often referred to as embryos, and those with $r > r_c$ as nuclei. The vapour is unsaturated for embryos, and supersaturated for nuclei.

By treating the vapour as an ideal gas, we can find a relation between p_r, the vapour pressure with which a drop of radius r is in equilibrium, and p_∞, the vapour pressure in equilibrium with a flat liquid surface. When the vapour pressure is p_∞, $g^v = g^l$, and from eqn. (46.12) when the vapour pressure is p_r, $g^v - g^l = 2\sigma v^l/r$. Differentiating this expression

$$(v^v - v^l)\,\mathrm{d}p = -(2\sigma v^l/r^2)\,\mathrm{d}r$$

since $\mathrm{d}g^v = v^v\,\mathrm{d}p$ and $\mathrm{d}g^l = v^l\,\mathrm{d}p$ at constant temperature, v^v and v^l being the volumes per atom in the vapour and liquid states. We neglect v^l in comparison with v^v, and write $v^v = kT/p$. Integrating from p_∞ to p_r while r varies from ∞ to r, we find

$$\left.\begin{aligned} kT\ln(p_r/p_\infty) &= 2\sigma v^l/r \\ \text{or} \qquad \ln i = \ln(p_r/p_\infty) &= 2\sigma v^l/rkT. \end{aligned}\right\} \qquad (46.14)$$

In this equation, the ratio $i = p_r/p_\infty$ is used to measure the supersaturation. A quantity more often used is the degree of supersaturation

$$j = (p_r - p_\infty)/p_\infty = i - 1.$$

The degree of supersaturation is often expressed as a percentage $100j$, instead of a fraction. For small supersaturations, $j = i - 1 \simeq \ln i$.

When the modified eqn. (46.4) is used for the free energy change, eqns. (46.13) and (46.14) become

$$n_c = [(\tfrac{2}{3}\eta\sigma\{g^v - g^l - (4kT/n_c)\}]^3, \qquad (46.15)$$

$$\ln(p_r/p_\infty) = (2\sigma v^2/rkT) - 4/n, \qquad (46.16)$$

respectively, where the $4/n$ factor comes from (46.8). With a typical value of $n_c \sim 100$, the corrected n_c is about 10% smaller than the "classical" value.

Equation (46.14) is another form of the Gibbs–Thomson or Thomson–Freundlich equation (see p. 182), and eqn. (46.16) shows that it remains valid to a very good approximation in the modified theory. It follows that the maximum value of the free energy increase ΔG_c is given sufficiently accurately by

$$\begin{aligned} \Delta G_c &= \Delta G'_c + \Delta G_{tr} + \Delta G_{rot} + \Delta G_{rep} \\ &= \Delta G'_c + kT\ln(Q_{tr}Q_{rot}/Q_{rep}), \end{aligned} \qquad (46.17)$$

where $\Delta G'_c$ is obtained by substituting (46.12) into (46.2) to give

$$\Delta G'_c = 4\pi\sigma r_c^2/3 \qquad (46.18)$$

† The fact that the equilibrium is unstable, which is physically obvious, is expressed mathematically by the condition that $\partial\Delta G/\partial r = 0$ gives a maximum ($\partial^2\Delta G/\partial r^2 < 0$), whereas for stable or metastable equilibrium $\partial\Delta G/\partial r = 0$ gives a minimum of ΔG.

so that the excess free energy is equal to one-third of the surface free energy of the critically sized nucleus. We may also write this expression in the equivalent forms

$$\Delta G_c' = \frac{16\pi(v^l)^2\sigma^3}{3(g^v - g^l)^2} = \frac{16\pi(v^l)^2\sigma^3}{3k^2T^2(\ln i)^2}, \tag{46.19}$$

$$\Delta G_c' = \sigma\eta n_c^{2/3}/3 = 4\sigma^3\eta^3/27(g^v - g^l)^2. \tag{46.20}$$

If $g^l > g^v$, ΔG does not have a maximum value, but increases rapidly with r. Under these conditions, the vapour is stable, and any liquid embryo which forms will quickly evaporate again. The relation between ΔG and r for an unsaturated and a supersaturated vapour is shown in Fig. 10.1.

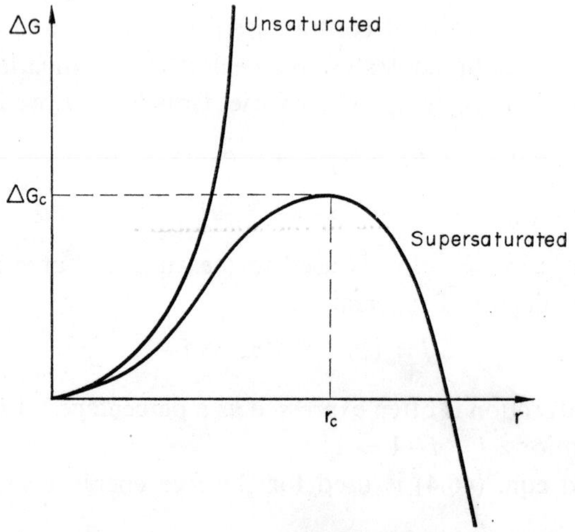

FIG. 10.1. Free energy of formation of a spherical liquid nucleus in a supersaturated and unsaturated vapour.

The above discussion assumes that liquid droplets form spontaneously in the interior of the vapour. In practice, it is well known that condensation usually occurs first on particles of dust or other impurities which may be present, or on the walls of the containing vessel. The impurities act by reducing the free energy barrier ΔG_c, and nucleation under these conditions is said to be heterogeneous in contrast to the homogeneous nucleation discussed above. Homogeneous nucleation can only be obtained if special precautions are taken, but we shall first develop the theory for this process and consider later the modifications necessary when nucleation is heterogeneous.

47. HETEROPHASE FLUCTUATIONS: VOLMER'S THEORY OF NUCLEATION

We see from the preceding section that the condensation of a supersaturated vapour requires the formation of nuclei of size $r > r_c$. The first satisfactory theory of nucleation was given by Volmer and Weber (1926) who assumed that there exists, effectively, a stationary distribution of embryos of size $r < r_c$. Frenkel (1939) generalized this theory to unsaturated vapour, and to other phase transformations, and we shall follow his treatment.

We have already seen that in a homogeneous phase there are local fluctuations of density, and of concentration if two or more components are present. These small fluctuations generally occur within the original phase, and their existence emphasizes the danger of applying the term homogeneous to very fine scale phenomena in macroscopic assemblies. In addition, we must consider the possibility of fluctuations which lead to local transitory phase transformations, so that small volumes can no longer be considered part of the original phase but have the atomic arrangement associated with a new phase. Fluctuations of this kind are called heterophase, and in a sense can be regarded as arising from a large homophase fluctuation. For example, a large density fluctuation in a small volume of vapour might lead to formation of a liquid droplet, or a large concentration fluctuation in a solid solution might lead to a rearrangement of atomic positions. Both of these are examples of heterophase fluctuations.

Heterophase fluctuations are, of course, responsible for the nucleation of phase transformations. Frenkel emphasized, however, that we must also regard them as existing in stable phases, the only difference being in the statistical distribution of the embryos. In a stable phase, the energy of an embryo of another phase (i.e. of a heterophase fluctuation) increases rapidly with its size, and the number of embryos present in equilibrium thus decreases extremely rapidly with size. For a metastable phase, the energy only increases initially, and then decreases again. If we attempt to find an equilibrium distribution of embryos in a metastable phase, we find that all the atoms are taking part in very large heterophase fluctuations; this is merely another way of expressing the condition that another phase has lower free energy. In nucleation problems of interest, we require the distribution of embryos in the initial stages of transformation, i.e. subject to the restrictive condition that almost all the atoms are present in the metastable state. The simplest assumption is that the distribution of embryos of size $n < n_c$ is the same as the equilibrium distribution if the phase were in fact stable. This is the basis of the Volmer theory of nucleation.

Fluctuations of density or concentration produce changes in the free energy. If a fluctuation of any kind produces a rise in the free energy ΔG, the probability of its occurrence in an equilibrium state of the assembly is proportional to $\exp(-\Delta G/kT)$. The number of embryos of size n is thus given by

$$N_n = N^v \exp(-\Delta G_n/kT), \tag{47.1}$$

where N_n is the statistical distribution function for embryos containing n atoms, N^v is the number of atoms in the vapour phase, and ΔG_n is the standard free energy change resulting from the conversion of vapour into embryos. We shall suppose that the assembly may be characterized by the distribution function N_n for the most probable type of embryo, so that ΔG_n is a function only of n. Strictly, this is a simplification of the problem, since we should also specify the shape of the embryos, but there will be a negligible error in limiting consideration to the shape corresponding to minimum surface energy for a given size. The procedure is essentially equivalent to replacing a summation by its largest term, which often has to be used in statistical mechanics. It then follows that ΔG_n may be identified with the free energy change of eqn. (46.4), and that the total number of atoms in the assembly is given by

$$N = N^v + \Sigma N_n n. \tag{47.2}$$

In the classical theory of nucleation, as developed in most books, the equilibrium distribution is represented by an equation similar to (47.1) but with ΔG_n replaced by $\Delta G_n'$ from eqn. (46.2). Feder *et al.* (1966) point out that if this were valid, the concentration N_n/V of embryos of size n in a fixed volume V would vary as p^{n+1}, which is contrary to the law of mass action. (One power of p comes from the pre-exponential factor (N^v/V) and the others from the exponentional term, since $n(g^v - g^l) = nkT \ln i$.) It follows from this, as well as from the preceeding discussion, that eqn. (47.1) cannot be correct unless the additional terms of (46.4) are included in ΔG_n. However, since these additional terms are almost independent of n, it is convenient to bring them into the pre-exponential and thus to write

$$N_n = N^v(Q_{tr}Q_{rot}/Q_{rep}) \exp(-\Delta G_n'/kT), \tag{47.3}$$

where the quantity in the first bracket represents the Lothe–Pound correction factor. In the form (47.3), the concentration (N_n/V) varies correctly as p^n (since one power of p has been cancelled by the factor $v^v = V/N^v$ in the translational partition function (eqn. (46.5)).

Frenkel gave a statistical-mechanical derivation of eqn. (47.1) for the case of homophase fluctuations, which was reproduced in the first edition of this book, but is here omitted for brevity. However, it is apparent that we have arrived at the equilibrium distribution by a rather tortuous path involving both thermodynamics and statistical mechanics, and it is possible to use the more direct procedure of minimizing the free energy of a vapour containing embryos with respect to the concentration of embryos of given class. This gives directly

$$N_n = Q_n \exp(ng_v/kT), \tag{47.4}$$

where Q_n is the partition function for an embryo of size n in the vapour phase. Reiss and Katz (1967) and Reiss *et al.* (1968) have attempted to evaluate Q_n using a method known as the standard phase integral, and they obtain a result

$$N_n = \frac{Q_{tr}^c}{Q_{tr}^d} \exp(-\Delta G'/kT)$$

$$= (V/v^d) \exp(-\Delta G'/kT), \tag{47.5}$$

where Q_{tr}^c is the translational partition function for a liquid cluster in the vapour phase and Q_{tr}^d is the corresponding partition function for a droplet in the bulk liquid. For Q_{tr}^c, the volume v^v of eqn. (46.5) is replaced by the volume of the whole assembly V, whilst for Q_{tr}^d it is replaced by the volume v^d in the bulk liquid over which the centre of mass of a droplet with fixed boundaries fluctuates. (In their papers a distinction is made between a cluster, which is a small liquid region in the vapour, and a droplet, which is an equivalent region in the bulk liquid.) Essentially the same result is obtained by Lin (1968) using the method of the grand canonical ensemble. Thus the correction to the classical distribution is given by Lothe and Pound as

$$\Gamma_{LP} = Q_{tr}Q_{rot}/Q_{rep} \tag{47.6}$$

and by Reiss *et al.* as

$$\Gamma_R = V/N^v v_d = v^v/v^d. \tag{47.7}$$

The rotational contribution does not appear in the Reiss equation, and according to Nishioka *et al.* (1971) this is because the authors suppose, incorrectly, that the effect of surroundings on a drop in bulk liquid may be completely simulated by enclosing the atoms in a fixed volume. Apart from the rotational contribution, however, the two equations should be equivalent, and the difficulty of calculating Q_{rep} has been replaced by that of calculating v^d. Reiss *et al.* believe that the correction (47.6) as used by Lothe and Pound is partially redundant because some states are being counted twice; Nishioka *et al.*, on the other hand, believe that v^d is much smaller than the volume used by Reiss *et al.*

We have emphasized these difficulties because the numbers involved are very large. For a critical nucleus size $n_c \sim 100$, applicable to water vapour for example, Lothe and Pound's original estimate for the factor (47.6) was 10^{17}, whilst Reiss *et al.* estimate from (47.7) a factor of 10^3–10^6. A rough estimate of v^d due to Nishioka *et al.* is

$$v^d = (8v^l/3)\,(3n+1)^{-2},$$

and this gives typically $\sim 5\times 10^8$ for the factor in (47.7). Thus the differences amount to a factor of 10^2–10^5 in the estimate of the translational contribution, and a disagreement about whether or not there is a rotational contribution.

Volmer assumed that an embryo is formed as a result of a large number of small scale (bimolecular) fluctuations, rather than by a sudden large fluctuation. If we use the symbol E_n to denote an embryo containing n atoms, and E_1 to denote a single atom, the process of formation may be written

$$\left.\begin{aligned} pE_1 &\rightleftharpoons E_p, \\ E_p \quad +E_1 &\rightleftharpoons E_{p+1}, \\ E_{p+1}+E_1 &\rightleftharpoons E_{p+2}, \\ \cdots\cdots&\cdots\cdots \\ E_{n-1}+E_1 &\rightleftharpoons E_n. \end{aligned}\right\} \tag{47.8}$$

The symbol E_p is used for the droplet containing p atoms, which we decide to regard as the smallest recognizable heterophase fluctuation (see p. 420). An embryo is thus assumed to grow or shrink by the addition or removal of individual atoms. When the primary phase is a vapour, the forward reactions of the above set will result from collisions between embryos and vapour atoms; the probability of embryo–embryo collisions is clearly negligible.

Consider now an unsaturated vapour ($i < 1$), in which the free energy of an embryo increases continually and rapidly with n. A state of dynamic equilibrium will be attained, in which the number of embryos of given size will remain effectively constant, although individual embryos are constantly growing or evaporating. Any non-stationary distribution will rapidly change towards the equilibrium distribution. Suppose, for example, that embryos E_n are being formed more rapidly than they disappear into either E_{n-1} or E_{n+1}. The concentration of E_n embryos will then increase with time, and this will increase the number decomposing until equilibrium is attained. The relaxation time for the attainment of the stationary distribution depends on kinetic features of the growth of embryos; it is possible that this has some importance in solid-state reactions (see pp. 442–8), but we shall neglect it in our discussion of the vapour–liquid change, since it will certainly be extremely small.

The fundamental condition for the equilibrium state can be expressed mathematically as follows. There will be a certain probability per unit time that an atom will condense on the surface of a E_n embryo, converting it into a E_{n+1} embryo. If each vapour atom which collides with the embryo condenses, the probability is given approximately by the product of its area and the kinetic theory collision factor $p/(2\pi mkT)^{1/2}$, where p is the pressure and m the atomic mass. More generally, however, we write this probability as q_0 per unit area per unit time, in order to preserve the same symbol for condensed state transformations. For vapour–liquid transformations, q_0 is given by the product of the collision factor and a condensation or accommodation coefficient $\alpha_c \ll 1$, and it is usual to assume that q_0 is independent of n. Actually, there may be a rather rapid variation of α_c, and hence of q_0, with n for very small embryos, but, as we shall show, this probably does not affect the final result. There will also be a probability q_n per unit area per unit time that an atom will evaporate from the surface of an embryo of n atoms converting it into a E_{n-1} embryo. In this case, q_n will not be independent of n, since the probability of evaporation is a function of the free energy of the embryo. Now consider the net transfer of atoms between embryos of size, n, $n+1$, as represented by the equations

$$\left.\begin{aligned} E_n \;\; + E_1 &= E_{n+1}, \\ E_{n+1} - E_1 &= E_n. \end{aligned}\right\} \tag{47.9}$$

The rate of the first process is $N_n O_n q_0$ and of the reverse process is $N_{n+1} O_{n+1} q_{n+1}$. In unit time, therefore, the net transfer is given by

$$I = N_n O_n q_0 - N_{n+1} O_{n+1} q_{n+1} = 0. \tag{47.10}$$

The expression is equated to zero, since the numbers, N_n, N_{n+1} are assumed to be the numbers of embryos in the equilibrium distribution. It then follows from the principle of detailed balancing (p. 81) that the net transfer of atoms in any separate process of the kind considered must be zero. Equation (47.10) is the fundamental expression of the equilibrium state; the corresponding equation for the metastable state (when $I \neq 0$) is the basis of the Becker–Döring theory of nucleation.†

† It should be remarked that in developing the Becker–Döring theory in his book *Kinetik der Phasenbildung*, Volmer gives this equation in different form. He assumes that a vapour atom effectively collides with an embryo when its centre passes through the surface of a sphere having radius greater by an atomic radius than the radius of the embryo itself. He writes the area of this effective collision sphere as O'_n, and he further assumes that an atom escapes from the $n+1$ embryo when its centre passes through the same limiting surface. Thus the quantities O_n, O_{n+1} in eqn. (47.10) are both replaced by O'_n. In view of the general inapplicability of macroscopic concepts to the embryos, it seems very doubtful whether refinements of this kind are worth while.

Now consider a supersaturated vapour ($i > 1$). Equation (47.1) then gives the result, as anticipated on p. 425, that N_n tends to infinity as n tends to infinity; all the atoms are thus present as large embryos, and there is no vapour phase. We are interested, however, in the embryo distribution in the initial stages of the transformation, when almost all the assembly is in the vapour phase, and the most probable states are thus excluded. Volmer's assumption is that nuclei of size greater than n_c grow rapidly, and may be regarded as removed from the assembly so far as the calculation of nucleation rates is concerned. For embryos of size smaller than n_c, the distribution function is assumed to be given by eqn. (47.1). We shall use the symbol Z_n for the number of embryos of size n present in a metastable assembly, making a distinction between this distribution function and N_n which applies to a stable assembly. The original Volmer theory thus requires

$$\left.\begin{aligned} n < n_c \quad & Z_n = N\exp(-\Delta G_n/kT), \\ n > n_n \quad & Z_n = 0. \end{aligned}\right\} \qquad (47.11)$$

This distribution can be obtained formally by erecting an infinite free energy barrier at $n = n_c$, but there would not then, of course, be any nucleation. To allow nucleation, we must suppose that each embryo of size n_c can penetrate the barrier when a vapour atom condenses on it, and it then becomes a nucleus which is removed from the assembly. The distribution of embryos will be maintained if n_c+1 vapour atoms are added to the assembly for each nucleus removed. This rather unreal situation is sometimes called a quasi-steady-state distribution. The rate of nucleation will be given by the product of the number of critical size embryos in the quasi-steady-state, Z_c, and the probability in unit time of a vapour atom condensing on one of these embryos.

The preceding paragraph gives the mathematical conditions required to ensure a quasi-steady-state distribution. Physically, there will clearly be a close approximation to such a distribution at the beginning of transformation, when the supply of vapour atoms is virtually infinite, provided the rate of nucleation is small. Steady-state nucleation is illustrated diagrammatically in Fig. 10.2, the upper part of which shows the curve of $\Delta G_n/kT$ against n. The horizontal lines are drawn with a weight proportional to $\exp(-\Delta G_n/kT)$, so that they illustrate the relative numbers of embryos which would be present if ΔG_n continued to increase beyond $n = n_c$. Nuclei of size n_c just spill over the free energy barrier and, in the steady state, the rate at which they do so will be equal to the rate at which new embryos of size n_c are formed. The lower part of the figure shows the distributions given by eqn. (47.11) (full line) and eqn. (47.1) (dotted line).

From (47.11) we can now write the number of stable nuclei formed in the assembly in unit time as

$$I = q_0 O_c Z_c = N q_0 O_c \exp(-\Delta G_c/kT). \qquad (47.12)$$

These nuclei are formed by a chain of the forward processes of (47.8). It should be noted that most embryos evaporate again before becoming nuclei; only occasionally will there be a series of favourable fluctuations. The quantity I, called the nucleation rate or nucleation current, is of great importance. It has dimensions s^{-1}, and is proportional to the total

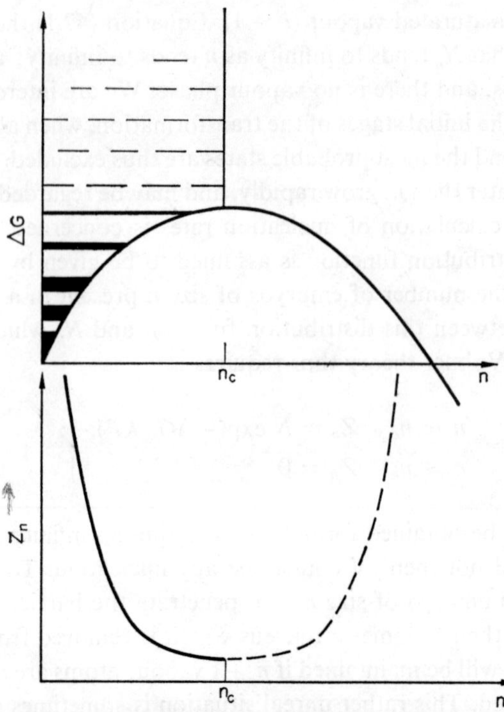

FIG. 10.2. To illustrate the quasi-steady distribution of embryos of size $n < n_c$ (after Fisher *et al.*, 1948). The upper figure shows the curve of free energy of formation vs. nucleus size, and the horizontal lines are drawn to indicate qualitatively the numbers of embryos of various sizes present in a quasi-steady distribution. The lower figure shows the distribution assumed in the Volmer theory (eq. (47.11)).

number of atoms in the assembly; this is true of all homogeneous nucleation phenomena. The number of nuclei formed in unit time in unit volume is thus constant, and is given by

$$^{v}I = \frac{I}{Nv^{l}} = \frac{q_0 O_c}{v^{l}} \exp(-\Delta G_c/kT). \tag{47.13}$$

We emphasize again that this constant nucleation rate only applies to conditions under which I is small and little or no transformation has occurred. For the vapour–liquid change, and in some other transformations, the growth rate at constant supersaturation is much greater than the steady nucleation rate. The kinetics of the transformation are then effectively governed by the nucleation phenomenon, since a small number of nuclei effect the whole change. The transformation rate will be greater for large assemblies, and under specified conditions will be roughly proportional to the number of systems in the assembly. In the extreme case, the whole assembly will transform when a single nucleus has formed, and the reciprocal of I is the time required to do this. Since nucleation is a statistical process, however, the time to form one nucleus will show small variations in similar assemblies, or in the same assembly in different experiments.

From eqns. (46.19) and (47.12), the nucleation rate varies as

$$\exp(-1/(\ln i)^2) = \exp(-1/j^2),$$

and is thus extremely sensitive to the degree of supersaturation. The curve of I against i is shown in Fig. 10.3; the nucleation rate is very small until a critical supersaturation is reached, and then it rises abruptly to a much larger value. The critical supersaturation corresponds to the breakdown of the metastable state of the assembly, and the quasi-steady-state distribution of embryos can no longer be maintained.

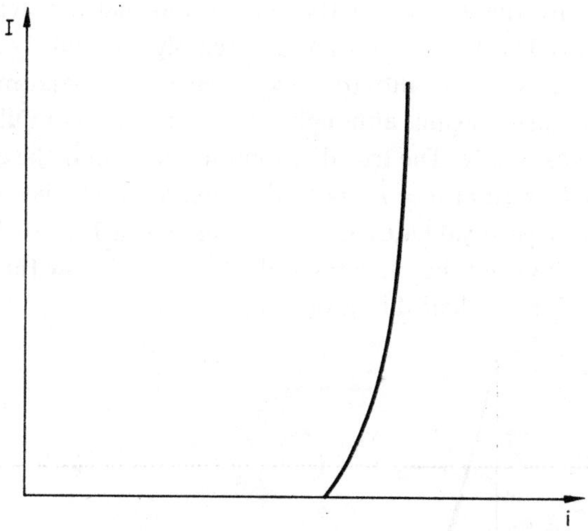

FIG. 10.3. Schematic variation of nucleation rate with degree of supersaturation.

As will be seen in the following sections, a value for a steady-state nucleation rate can be obtained for all kinds of transformations. The overall rate of transformation under conditions in which both the nucleation frequency and the growth rate have some influence was considered in Section 4, Chapter 1. The question then arises whether the empirical nucleation frequency used in the theory of Section 4 can be identified with the steady-state nucleation rate per unit volume ${}^vI$. The phenomenological theory uses the nucleation rate in the *untransformed* regions of the assembly, which we may treat as sub-assemblies where the conditions for quasi-steady-state embryo distribution are satisfied. Provided that the relaxation time for the establishment of the quasi-steady state is not too large, we may thus expect the nucleation rate of Section 4 to be the steady-state nucleation rate appropriate to the untransformed regions at any stage of the transformation. During the course of the transformation, however, the intensive parameters such as density, internal strains, composition, etc., which specify the condition of the untransformed part of the assembly, may be altered, producing corresponding changes in ${}^vI$. Thus although ${}^vI$ of Section 4 may usually be identified with a steady nucleation rate, it cannot be assumed to be time independent, as has already been emphasized. Furthermore, if the assembly is suddenly changed (quenched) from one condition to another at time $t = 0$, there may be a finite period during which the quasi-steady-state embryo distribution appropriate to the new condition is established. The initial nucleation rate will then vary rapidly with time.

48. THE BECKER–DÖRING THEORY OF NUCLEATION†

Volmer and Weber's expression for the rate of nucleation consists of two terms—an exponential factor involving the free energy increase due to the formation of a critical size nucleus, and a factor which is proportional to the frequency of collision of vapour atoms. The main defect of the theory lies in the assumption that the steady state distribution function is given by (47.11). In the metastable assembly, critical size embryos can either grow or shrink with equal probability ($q_n = q_0$ when $n = n_c$), and nuclei of greater than ciritical size may also shrink again, although they are rather more likely to grow and thus be removed from the assembly. The true distribution function in the quasi-steady state will thus not fall abruptly to zero at $n = n_c$, but will decrease slowly, becoming effectively zero when n is large. This quasi-steady distribution is shown in Fig. 10.4; it approximates to (47.11) for very small n values, but progressively decreases below this distribution as n increases, the value of Z_c being half that given by (47.11).

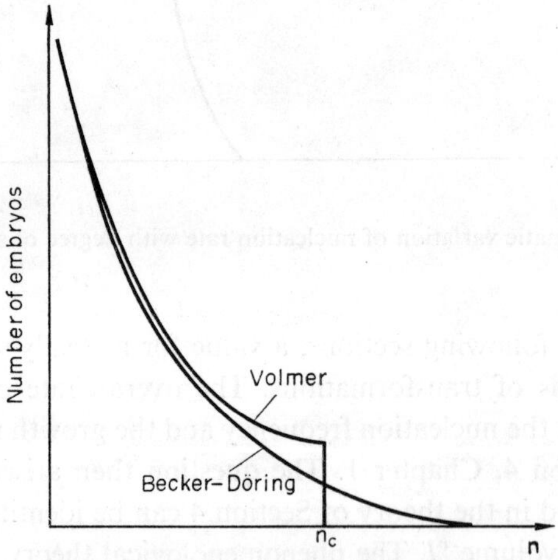

FIG. 10.4. Distribution functions for embryos of different sizes according to Volmer and Becker–Döring theories of nucleation.

Improvements of Volmer's equation were suggested by many workers, and notably by Becker and Döring (1935), who developed a kinetic theory of nucleation which is the basis of almost all subsequent treatments. The importance of this theory arises more from the correct formulation of the kinetic problem (within the limits discussed on pp. 418–9) than

† Much of the credit for the kinetic theory of nucleation properly belongs to Farkas (1927) who determined the so-called non-equilibrium or Zeldovich factor (see p. 435) well before the work of Becker and Döring. However, the theory is usually ascribed to the latter authors who developed it in a different way.

in the precise solution obtained; the result modifies only the non-exponential factor of eqn. (47.12), and the nucleation rate is much more sensitive to slight changes in ΔG_c than to large changes in this factor. There is, moreover, no general agreement on the best value of the pre-exponential factor, and slightly different results are obtained in different developments of the Becker–Döring theory. This situation arises because the quasi-steady state is found to be characterized by a set of difference equations. An analytical solution can only be obtained by using approximations in which summations are replaced by integrals, and the results differ somewhat according to the procedure adopted. The derivation given here uses an approach due originally to Zeldovich (1943) and followed by Frenkel (1946), although the final result differs from that which Frenkel obtained.

Consider again the interchange of embryos E_n, E_{n+1} represented in eqn. (47.9). In place of (47.10), the net transfer is now given quite generally by

$$I_{n,t} = Z_{n,t} q_0 O_n - Z_{n+1,t} q_{n+1} O_{n+1}, \tag{48.1}$$

where $Z_{n,t}$ is the number of embryos of size n present in the assembly at time t. Using the result of (47.10), we may write this kinetic equation in the useful form

$$I_{n,t} = N_n q_0 O_n \left\{ \frac{Z_{n,t}}{N_n} - \frac{Z_{n+1,t}}{N_{n+1}} \right\}, \tag{48.2}$$

and there will be a set of such equations for all allowable values of n. In general, the equations are very difficult to handle, but for the present we are interested only in the limiting conditions which lead to the quasi-steady-state distribution. As in the previous discussion, the required conditions are that $I_{n,t}$ is small, and all the $Z_{n,t}$ are small compared with the number of vapour atoms. The distribution function is then $Z_{n,t} = Z_n$, independent of time, and the steady state is maintained if $I_{n,t} = I$, a constant. The net rate, I, at which embryos E_{n-1} change to embryos E_n is equal to the rate at which E_n change to E_{n+1}, and so on. I is thus the rate at which nuclei are produced in the assembly.

We have no information about the details of the distribution function, Z_n, but we know that for very small embryos it must be effectively identical with the equilibrium distribution function which would be obtained in the absence of nucleation, whilst for large embryos, the distribution function must approach zero. Thus

$$n \to 0, \quad Z_n \to N_n: \; n \to \infty, \quad Z_n \to 0.$$

It is convenient to suppose that $Z_n = N_n$ for all values $n \leqslant p$, and $Z_n = 0$ for all values $n \geqslant s$, where s is greater than n_c. The latter condition replaces Volmer's assumption that $Z_n = 0$ for $n = (n_c+1)$, and we shall see that it is not necessary to specify the values of either p or s.

From (48.2), we may now write a series of equations

$$\frac{I}{N_n q_0 O_n} = \frac{Z_n}{N_n} - \frac{Z_{n+1}}{N_{n+1}}$$

for all values of n from p to s. Adding these equations together,

$$I\sum_{p}^{s}\frac{1}{N_n q_0 O_n} = \frac{Z_p}{N_p} - \frac{Z_{s+1}}{N_{s+1}} = 1$$

or

$$I = \frac{q_0}{\sum_{p}^{s}\left\{\frac{1}{N_n O_n}\right\}}. \tag{48.3}$$

In this expression, N_n is determined by eqns. (47.1) and (46.3), and it follows that the discontinuous function $1/N_n$ has a sharp maximum at $n = n_c$.† The only effective contributions to the summation will come from terms where $n-n_c$ is small, and the effect of terms with large or small n may be neglected. In all the important terms, O_n will thus differ only slightly from the surface area of the critical nucleus, O_c, and it is thus permissible to treat this as a constant factor and take it outside the summation giving

$$I \simeq \frac{q_0 O_c}{\sum_{p}^{s}\frac{1}{N_n}}. \tag{48.4}$$

In order to proceed further, we make the approximation of treating N_n as a continuous function of n, although strictly it is only defined for integral values. The summation can then be replaced by an integral. Furthermore, the function $1/N_n$ has appreciable values only near $n = n_c$ so that it is permissible to expand ΔG_n as

$$\Delta G_n \simeq \Delta G_c + \frac{\xi^2}{2}\left(\frac{\partial^2 \Delta G_c}{\partial \xi^2}\right)_{\xi=0}, \tag{48.5}$$

where $\xi = n-n_c$. The linear term in ξ is, of course, zero, since $(\partial\Delta G_n/\partial\xi)_{\xi=0} = 0$, ΔG_n having its maximum value at $n = n_c$. Using eqns. (48.5), (47.1), (46.3), and (46.20), we may thus write

$$\sum_{p}^{s}\frac{1}{N_n} \simeq \frac{1}{N}\exp\left(\frac{\Delta G_c}{kT}\right)\int_{p-n_c}^{s-n_c}\exp\left(\frac{-\Delta G_c \xi^2}{3n_c^2 kT}\right)d\xi. \tag{48.6}$$

Finally, since only values of $1/N_n$ near $\xi = 0$ have any appreciable influence on the value of the integral, the limits $\xi = -(n_c-p)$, $\xi = s-n_c$ may be changed to $\xi = \pm\infty$ without affecting the result. The integral is then transformed into the error integral and

$$\sum_{s}^{p}\frac{1}{N_n} \simeq \frac{1}{N}\exp\left(\frac{\Delta G_c}{kT}\right)\cdot\left(\frac{3\pi kT}{\Delta G_c}\right)^{1/2} n_c. \tag{48.7}$$

† All treatments of the Becker–Döring theory involve summations of sets of terms which have appreciable values only near $n = n_c$. It is interesting to note here that by taking the simplest and most drastic approximation — replacing the sum by its maximum term — the Volmer equation (47.12) for nucleation is obtained. This clearly cannot be correct, since the other terms near $n = n_c$ will make finite contributions, and the true nucleation rate must be smaller than that given by (47.12).

Substituting into (48.4), the nucleation current is finally obtained as

$$I = \frac{Nq_0O_c}{n_c}\left(\frac{\Delta G_c}{3\pi kT}\right)^{1/2}\exp\left(\frac{-\Delta G_c}{kT}\right). \tag{48.8}$$

The factor

$$\Gamma_Z = (1/n_c)(\Delta G_c/3\pi kT)^{1/2} \tag{48.9}$$

by which eqn. (48.8) differs from the Volmer eqn. (47.13) is sometimes called the Zeldovich factor, although as we have already noted it was first derived by Farkas (1927). A typical value is $\frac{1}{20}$ and it is unlikely to be smaller than 10^{-2} in conditions under which experimental observations may be made. Because of the very rapid variation of nucleation rate with supercooling or supersaturation ratio (see Fig. 10.3), changes in the pre-exponential factors of this magnitude have very little effect on observable features of the kinetics.

The derivation above is identical with that used by Reiss (1952), and differs from Becker and Döring's original method, which was to write (48.1) in the form

$$IR_n = \varphi_n - \varphi_{n+1}$$

where

$$R_n = \frac{1}{q_0}\prod_2^n\left(\frac{q_n}{q_0}\right), \quad \varphi_n = Z_nO_n\prod_2^n\left(\frac{q_n}{q_0}\right).$$

Summing for all values of n from 1 to s ($\varphi_1 = Z_1O_1$, $\varphi_s = 0$)

$$I\sum_1^s R_n = Z_1O_1. \tag{48.10}$$

The "growth resistances" R_n are evaluated by expressing the terms in the product as exponential functions of the reciprocal radii, using eqn. (46.14). This gives

$$R_n = \frac{1}{q_0}\exp\left\{\frac{2\sigma v_L}{kT}\left(\sum_2^n \frac{1}{r_n} - \frac{n-1}{r_c}\right)\right\} \tag{48.11}$$

and the sum is approximated by an integral. The resistances R_n have a sharp maximum at $n = n_c$, and the sum $\sum_1^s R_n$ may thus be transformed into the error integral by expanding about $n = n_c$ and changing the limits of integration, as above.

Volmer (1939) followed the Becker–Döring treatment, but he pointed out that the sum in (48.11) had previously been overestimated by an amount representing the heat of solution of one molecule in a large volume of liquid. His corrected expression for the nucleation rate then contained an additional factor $\exp(\Delta h^{lv}/kT)$, where Δh^{lv} is the molecular heat of condensation. This factor is not negligible (it may be 5–7 powers of ten!), but Volmer states it is accidentally cancelled by another previously neglected effect. This is the sudden decrease in the condensation coefficient, and hence of q_0, at very low n values, due to the heat released by condensation. Both of Volmer's "neglected" factors seem to concern the properties of embryos with very small n values, and it is a weakness of Becker and Döring's original procedure that the initial summation has to be made from $n = 1$ to $n = s$, instead of from $n = r$ to $n = s$. It is thus probable that the cancellation of the factors is not really accidental, since the treatment in terms of distribution functions shows how their introduction may be avoided. However, this argument only applies to the abnormally low values of the condensation coefficient for small embryos, and q_0 should undoubtedly contain a condensation coefficient appropriate to critical-sized embryos in addition to the collision factor. It has often been assumed that the condensation coefficient is essentially unity, but under some circumstances it may be appreciably smaller, as emphasized by Hirth and Pound (1963).

The effect of the latent heat of condensation was considered more carefully by Kantrowitz (1951) who pointed out that the mean temperature of embryos of size n is slightly greater than the temperature of the vapour itself because of the energy released by a condensing atom. His quantitative estimate of the temperature rise is very nearly zero for small embryos, but is a few degrees Celsius for values of n near n_c; the corresponding nucleation rate is changed by a factor which is a function of the appropriate heat content quantities. In a more detailed treatment of this problem, Feder *et al.* (1966) considered coupled currents of matter and thermal energy in a non-isothermal nucleation process. They found that the nucleation current of sub-critical embryos is largely carried by embryos which are colder than the vapour. In each size class of embryos there are more embryos which are at a temperature above that of the vapour than there are at a lower temperature, thus giving a mean temperature in excess of that of the vapour. However, the contribution to the nucleation rate of these more numerous "warm" embryos is more than counterbalanced by the higher probability of growth of the "cold" embryos. The net effect on the nucleation rate is to multiply by a factor

$$\Gamma_{\mathrm{Th}} = \frac{(c_v^l + \frac{1}{2}k)kT^2}{(c_v^l + \frac{1}{2}k)kT^2 + (\Delta h^{lv})^2} \tag{48.12}$$

in which c_v^l is the specific heat of the liquid at constant volume. (The expression is modified somewhat when nucleation is carried out in the presence of an inert carrier gas.) A typical value of Γ_{Th} is $\sim\frac{1}{5}$, so that it has a very small influence in the pre-exponential term.

The final expression for the nucleation rate may now be written in the equivalent forms

$$I = \Gamma_Z \Gamma_{\mathrm{Th}} N_c q_0 O_c \tag{48.13a}$$

$$= N\Gamma_Z \Gamma_{\mathrm{Th}} q_0 O_c \exp(-\Delta G_n/kT) \tag{48.13b}$$

$$= N\Gamma_Z \Gamma_{\mathrm{Th}} \Gamma_{\mathrm{LP}} q_0 O_c \exp(-\Delta G'_n/kT). \tag{48.13c}$$

Accurate experimental measurements of I are not possible because of its rapid variation with supersaturation (or supercooling), and it is usually possible to determine only the critical supersaturation at which the nucleation rate changes from a negligible to a very large value. As already emphasized, this means that the results are not sensitive to the value assumed for the pre-exponential term to within a few powers of ten. This means that the Zeldovich factor and the non-isothermal factor may be effectively ignored, and to within a sufficient approximation

$$\boxed{I = (\text{collision frequency of vapour atoms}) \exp(-\Delta G_n/kT).} \tag{48.14}$$

However, the very large values calculated for Γ_{LP} by some authors certainly cannot be ignored in a comparison of theory and experiment. Becker and Döring applied their theory to the condensation of water vapour and obtained good agreement with the experimental results of Volmer and Flood (1934); the critical nucleus size was estimated at $\sim$100 molecules. Volmer and Flood also determined the critical supersaturation ratio for various organic vapours, and their results were compared with the theory by Hollomon and Turnbull (1953). The most convenient test is to work backwards from the observed supersaturation to calculate σ and then compare this with the macroscopic value σ_∞. Very good agreement is found, and for many years this was taken to imply that the Volmer–Becker–Döring theory is an adequate description of the nucleation process, despite the qualifications on pp. 418–9. The results do not agree with the predictions of Buff's modified theory that $\sigma \simeq 0{\cdot}8\sigma_0$.

If the Lothe–Pound theory is correct, the additional factor of 10^{17} clearly cannot be neglected, and in fact this seems to destroy the above agreement between theory and experiment. More recent experimental work, however, has given results which seem to agree with the predictions of the Lothe–Pound theory. Jaeger *et al.* (1969) and Dawson *et al.* (1969) used a supersonic air nozzle technique, instead of a cloud chamber, to measure the critical supersaturation ratio for water, ammonia, benzene, ethanol, chloroform, and freon. Their results for water confirm the previous work, which agrees with the classical theory, and those for ethanol were inconclusive; in all other cases, much higher nucleation rates (i.e. lower critical supersaturations) than those predicted by the classical theory were found, and these were in reasonable agreement with the estimated Lothe–Pound factor. The authors point out that the classical theory seems to apply for polar or rod-shaped molecules, and Abraham (1969) has noted that liquids which are mainly hydrogen bonded agree with classical theory, and that such liquids have abnormally low surface entropies. This leads to an increase in the surface free energy of the less-ordered embryo over the flat surface of the bulk liquid, and the predicted change is of the required magnitude to restore agreement with (48.13c). In other words, Abraham suggests that the error in $\Delta G_c'$ resulting from incorrect use of the macroscopic σ is just sufficient to cancel the other terms in (46.4).

49. NUCLEATION OF THE SOLID FROM THE VAPOUR OR THE LIQUID

The theory developed above applies specifically only to the condensation of a supersaturated vapour. This is a transformation rarely studied in metallic assemblies, and the space devoted to it may therefore seem excessive. The results obtained, however, provide a basis for the treatment of other transformations, and we can now consider the relation of the theory to transformations involving solid phases.

If we consider first the growth of solid crystals from the vapour phase, we expect the nucleation rate to be given by an equation similar to (48.8) or (48.13). The embryo crystal of minimum surface energy, however, will not have a spherical form, but a shape conforming to Wulff's construction, described in Chapter 5. This modification to the form of the surface energy term is not the only alteration needed to apply nucleation theory to the formation of a solid crystal. In deriving the preceding expressions, we assumed that the shape of an embryo is not a function of n, the number of atoms which it contains. Except for very small n (where the approximation is unimportant), this assumption is justified for liquid droplets, but in a solid crystal, each atom has to fit into a fixed position relative to its neighbours. The configuration presented by the crystal embryo to the vapour thus varies periodically as successive rows and planes of atoms are completed.

Becker and Döring's original paper included a treatment of this problem, assuming the crystal nucleus to be cubical in shape. The expression they derived contained terms for the nucleation of rows and planes of atoms, but these have very little influence on the three-dimensional nucleation rate. We shall show in the next chapter that these terms, moreover, vanish for real crystals. The two-dimensional nucleation of new atomic planes is an important factor in the theory of growth of an *ideal* crystal.

When the extra terms are omitted, Becker and Döring's expression reduces to

$$I = O_1 q_c \frac{\Delta G_c}{kT} \exp\left\{\frac{-\Delta G_c}{kT}\right\}, \tag{49.1}$$

which differs slightly from (48.8) in the pre-exponential term. The change in Zeldovich factor is unimportant, and eqn. (48.14) gives the nucleation rate to sufficient accuracy for most purposes. According to the latest estimate of the replacement partition factor for small crystals (Nishioka *et al.* 1971), the Lothe–Pound correction factor for vapour-crystal nucleation is $\sim 10^{15}$, but depends rather strongly on the difference is surface free energy of macroscopic and small crystals.

We must now consider the problem of nucleation in liquid–solid or in solid–solid transformations, i.e. in completely condensed assemblies. The general form of the preceding equations will obviously be maintained, and the nucleation rate will again contain an exponential term involving the increase in free energy for a nucleus of critical size. The pre-exponential term will now be proportional to an encounter rate of atoms in the condensed assembly, but the kinetic theory of gaseous collisions is no longer applicable. As shown in Chapter 9, the encounter rate will be determined essentially by the product of an atomic vibrational frequency and an exponential term containing the activation energy for an elementary atomic movement. This kind of reasoning led Becker (1940) to suggest that the nucleation rate should be written

$$I = C_9 \exp(-\varepsilon/kT) \exp(-\Delta G_c/kT), \tag{49.2}$$

where ε is the activation energy for diffusion. The magnitude of the constant C_9 remains unspecified in the general case in Becker's theory, but an explicit value was obtained by Turnbull and Fisher (1949), who applied the reaction rate theory of Chapter 4 to the case where the product phase has the same composition as the parent phase. The simplest example of such a transformation in a condensed assembly is the freezing of a pure metal, and this is considered in the following development of the theory. The theory applies equally to solid–solid transformations, but has to be modified to allow for strain energies.

The excess free energy of an embryo solid crystal is now

$$\Delta G = n(g^s - g^l) + \eta\sigma n^{2/3}, \tag{49.3}$$

where η is a shape factor, as in Section 46. The maximum value of the free energy increase and the size of the critical nucleus are thus

$$\Delta G_c = \frac{4\eta^3\sigma^3}{27(g^l - g^s)^2}, \tag{49.4}$$

$$n_c = \left\{\frac{2\eta\sigma}{3(g^l - g^s)}\right\}^3. \tag{49.5}$$

These equations are the same as those used previously with $(g^v - g^l)$ replaced by $(g^l - g^s)$ and the additional assumption that (46.1) is valid. When both phases are condensed, it seems intuitively improbable that the factors of eqn. (47.6) or (47.7) can be large, so that additional terms in ΔG, analogous to those in (46.4), should not be important. However, one reservation must be made in the case of liquid→solid nucleation; Lothe and Pound (1962) have suggested that a 10^7 discrepancy between the experimental and theoretical pre-exponential factors might be attributable to nearly free rotation of the solid nuclei within the liquid. This does not seem entirely logical in view of their estimate of Q^l_{rep}, which depends on the assumption that nearly free rotation of liquid droplets within liquid does not take place.

The growth of an individual embryo may be assumed to involve a large number of small fluctuations, represented in eqns. (47.8), and only rarely will there be a sufficiently long chain of forward fluctuations for an embryo to reach critical size. In the kinetic equation (48.1), for the interchange of E_n and E_{n+1} embryos, we now have to find new expressions for the probabilities $O_n q_0$ and $O_{n+1} q_{n+1}$ that in unit time embryos E_n, E_{n+1} will change into each other. We make the reasonable assumption that during the addition

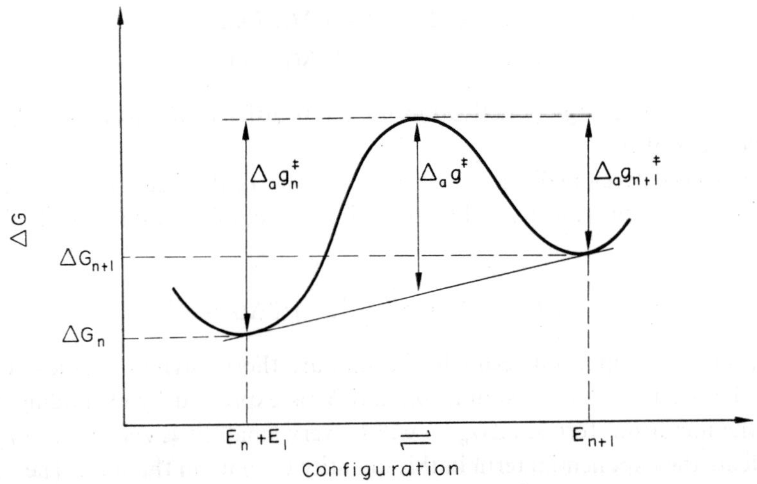

FIG. 10.5. To illustrate energy relations in the Turnbull–Fisher treatment of solid-state nucleation theory.

of an atom to an E_n embryo, the assembly passes through energy states which are higher than either the initial or final states of the process. There will thus be a maximum free energy at some intermediate stage, and the theory of Chapter 3 can be applied. The energy relations are shown in Fig. 10.5 as a function of the change in configuration from E_n to E_{n+1}. The energy ΔG_{n+1} is greater than ΔG_n since the figure is drawn for $n < n_c$. If $\Delta_a g_n^{\ddagger}$ is the height of the free energy maximum (the activated complex) above the energy $G+\Delta G_n$ of the assembly containing the E_n embryo, we may write the rate at which an atom in contact with the embryo will transfer into it as

$$(kT/h)\exp(-\Delta_a g_n^{\ddagger}/kT).$$

The energy has been written as $\Delta_a g_n^{\ddagger}$ rather than as $\Delta g_n^{\ddagger}$ (cf. eqn. (13.9)) in order to emphasize the straightforward physical interpretation of eqn. (13.10). As already emphasized, it is preferable to replace kT/h by a simple frequency ν, the magnitude of which is not specified exactly. We retain kT/h in the equations for nucleation rate in this section, however, to facilitate direct comparison with the equations given by Turnbull and Fisher.

Let the number of atoms in the surface of E_n be written o_n. The basic kinetic equation, in place of (48.1), is now

$$\begin{aligned} I_{n,t} = (kT/h)\,[&Z_{n,t}o_n \exp(-\Delta_a g_n^{\ddagger}/kT) \\ &-Z_{n+1,t}o_{n+1}\exp(-\Delta_a g_{n+1}^{\ddagger}/kT)]. \end{aligned} \tag{49.6}$$

Writing down the corresponding equation for the equilibrium distribution N_n, when the liquid phase is stable, we may again express (49.6) in the form

$$I_{n,t} = \frac{kT}{h}\exp\left(\frac{-\Delta_a g_n^{\ddagger}}{kT}\right)N_n o_n\left[\frac{Z_{n,t}}{N_n}-\frac{Z_{n+1,t}}{N_{n+1}}\right]. \tag{49.7}$$

It is to be noted that $\Delta_a g_n^{\ddagger}$ is a function of n only because of the difference in the energies ΔG_n and ΔG_{n+1}. In fact we may write approximately:

$$\begin{aligned} \Delta_a g_n^{\ddagger} &= \Delta_a g^{\ddagger}+\tfrac{1}{2}(\partial\Delta G_n/\partial n), \\ \Delta_a g_{n+1}^{\ddagger} &= \Delta_a g^{\ddagger}-\tfrac{1}{2}(\partial\Delta G_n/\partial n), \end{aligned}$$

where $\Delta_a g^{\ddagger}$ is the free energy of activation for the transfer of atoms across the interface, and is independent of n.

The theory is now developed exactly as in Section 48. In the quasi-steady state, we have the same conditions as previously, and by summing the set of equations we obtain

$$\frac{Ih}{kT}\exp\frac{\Delta_a g^{\ddagger}}{kT}\sum_{p}^{s}\frac{1}{N_n o_n \exp\{(\frac{1}{2}kT)(\partial\Delta G_n/\partial n)\}} = 1. \tag{49.8}$$

Once again, the only important terms in the sum are those having n values near $n = n_c$. Then o_n can be taken out of the sum as o_c, and N_n is expressed by expanding ΔG_n about ΔG_c. In all the important terms, $\partial\Delta G_n/\partial n$ will be very small (it is zero at $n = n_c$), and we may thus equate the exponential term in this quantity to unity in the sum. The summation is thus reduced to (48.7) again, and on substituting into (49.8), the nucleation rate is obtained as

$$I = N\frac{kT}{h}\frac{o_c}{n_c}\left(\frac{\Delta G_c}{3\pi kT}\right)^{1/2}\exp-\left(\frac{\Delta G_c+\Delta_a g^{\ddagger}}{kT}\right). \qquad (49.9)$$

We have obtained (39.9) in a slightly different and simpler manner from that used by Turnbull and Fisher, mainly in order to emphasize that their calculation is essentially a form of the Becker–Döring theory of the last section. According to Turnbull and Fisher, the quantity $(o_c/n_c)(\Delta G_c/3kT)^{1/2}$ is within one or two powers of ten for all nucleation problems of interest, so eqn. (49.9) may be written with sufficient accuracy as

$$I = N\frac{kT}{h}\exp\left\{\frac{-(\Delta G_c+\Delta_a g^{\ddagger})}{kT}\right\}. \qquad (49.10)$$

This has the same form as Becker's eqn. (49.2), except that the activation energy involved is that for transfer across an interface, rather than for diffusion. The activation energy for diffusion, however, will clearly be involved if the transformation produces a change of composition; this is dealt with in a later chapter.

The magnitude of the predicted nucleation rate is readily estimated. For the nucleation rate per unit volume, we have

$$^{v}I = {}^{v}N(kT/h)\exp\{-(\Delta G_c+\Delta_a g^{\ddagger})/kT\}, \qquad (49.11)$$

and reasonable values at ordinary temperatures are $^{v}N \simeq 10^{28}-10^{29}$ m^{-3} and $kT/h \simeq 10^{13}$ s^{-1}. The value of $\Delta_a g^{\ddagger}$ is less certain, but it is estimated by Turnbull that it is approximately equal to the activation energy for viscous flow, giving $\exp(-\Delta_a g^{\ddagger}/kT) \simeq 10^{-2}$ for liquid–solid changes in metals. Thus

$$^{v}I \approx 10^{39}\exp(-\Delta G_c/kT)\,\text{m}^{-3}\,\text{s}^{-1}. \qquad (49.12)$$

Some of the published papers give the uncertainty in the numerical factor of this equation as one power of ten, but it is probably two to four powers of ten when allowance is made both for the neglected factors of (49.9) and for the unknown value of $\Delta_a g^{\ddagger}$. Since the variations of the exponential term with change in ΔG_c is so rapid, the value of ΔG_c required to give a fixed nucleation rate is insensitive to the exact value of the pre-exponential term. A reasonable lower limit to the nucleation rate observable under normal conditions (within one or two powers of ten) is 10^6 m^{-3} s^{-1}, i.e. one nucleus per cm^3 per s, which corresponds to a free energy of activation $\Delta G_c \simeq 74kT \simeq 10^{-18}$ J at 1000 K. With $v^s = 5\times10^{-29}$ m^3 and assuming $\sigma = 0{\cdot}1$ J m^{-2} (100 ergs cm^{-2}), (46.19) gives $g^l - g^s \simeq 6\times10^{-21}$ J/atom, which is equivalent to 150 cal cm^{-3} or 1–3 kg cal/mole for most metals. This driving force varies as $\sigma^{3/2}$, but should not differ from this estimate by more than one order of magnitude, since σ for most metals probably lies in the range 0·02–0·25 J m^{-2} (20–250 ergs cm^{-2}).

In a condensed assembly, the degree of instability of a metastable phase with respect to a stable phase of the same composition is conveniently expressed by the supercooling (ΔT^-) or the superheating (ΔT^+) below or above the thermodynamic transformation temperature. For a liquid–solid change, we have $g^l = g^s$ at the freezing point (T^{ls}), and $\Delta s^{sl} = \Delta h^{sl}/T^{ls}$, where Δs^{sl} and Δh^{sl} are respectively the entropy and heat of fusion per atom. At a temperature $T < T^{ls}$,

$$g^l - g^s = (\Delta h^{sl})_T - T(\Delta s^{sl})_T,$$

where the heat and entropy of fusion now strictly refer to liquid formed from solid at temperature T. The heat of fusion, however, only varies with temperature through changes in the pV terms, which are negligible, and it is also quite a good approximation to regard the entropy of fusion as temperature independent. This gives

$$g^l - g^s = (T^{ls} - T)\,\Delta s^{sl} = \Delta h^{sl}\frac{\Delta T^-}{T^{ls}}, \tag{49.13}$$

where the supercooling $\Delta T^- = T^{ls} - T$.

Substituting into (49.4) and (49.12), the nucleation rate becomes

$${}^vI \simeq 10^{39}\exp\left\{\frac{-4\eta^3\sigma^3}{27(\Delta s^{sl})^2 T}\,\frac{(T^{ls})^2}{(\Delta T^-)^2}\right\}\,\mathrm{m^{-3}\,s^{-1}}. \tag{49.14}$$

We cannot easily obtain a general estimate of the value of $\Delta T^-/T^{ls}$ at which the nucleation rate becomes appreciable, since this is so dependent on σ. It will be seen, however, that if Δs^{sl} has the same value for all metals and if σ is proportional to Δh^{sl}, a given value of ${}^vI$ would correspond to a constant ratio of $\Delta T^-/T^{ls}$. Both these conditions are approximately true, at least for metals of simple crystal structure, and in Part II, Chapter 14, we shall find that experimental results show that the amounts of supercooling which can be obtained with different metals are proportional to the melting points of the metals on the absolute scale. The very rapid variation of nucleation rate with supercooling is illustrated by the fact that changing $g^l - g^s$ to one-half the value required to give ${}^vI \simeq 10^6\ \mathrm{m^{-3}\,s^{-1}}$ reduces the nucleation rate to ${}^vI \simeq 10^{-90}\ \mathrm{m^{-3}\,s^{-1}}$.

50. TIME-DEPENDENT NUCLEATION

The above solutions to eqns. (48.1) and and (49.6) apply only to the limiting conditions in which a quasi-steady distribution of embryos has been established. In many experiments, the assembly is suddenly changed from a stable to a metastable condition, and the nucleation rate is then a function of time until the quasi-steady state is attained. This effect will be unimportant if the transient is of short duration compared with the period of observation, and the steady state solutions (48.8) and (49.10) then provide a good description of the nucleation process. In condensed phases, however, the existence of an activation energy barrier to the addition and removal of atoms from embryos may mean that the quasi-steady distribution Z_n is only approached slowly, and it is then necessary to investigate the time-dependent nucleation rate. The transient has also been shown to be important in considering vapour condensation effects in supersonic wind tunnels, where the supersaturation is increased extremely rapidly.

Consideration of the embryo distribution and nucleation rate as a function of time requires some specification of the initial state of the assembly at time $t = 0$. If the assembly has been quenched from a stable condition, the initial concentration of embryos of any size will be very small (curve A in Fig. 10.6), and a sufficient approximation to this distribution is to assume that there are no embryos present at $t = 0$. We have then effectively to calculate how $Z_{n,t}$ changes from 0 to the quasi-steady Z_n (curve C) as t increases from 0 to ∞.

The establishment of the quasi-steady state is represented qualitatively in Fig. 10.6, which shows the number of embryos E_n as a function of time. With the initial condition of curve B in Fig. 10.6, the rate of production of embryos E_2 will obviously be a maximum

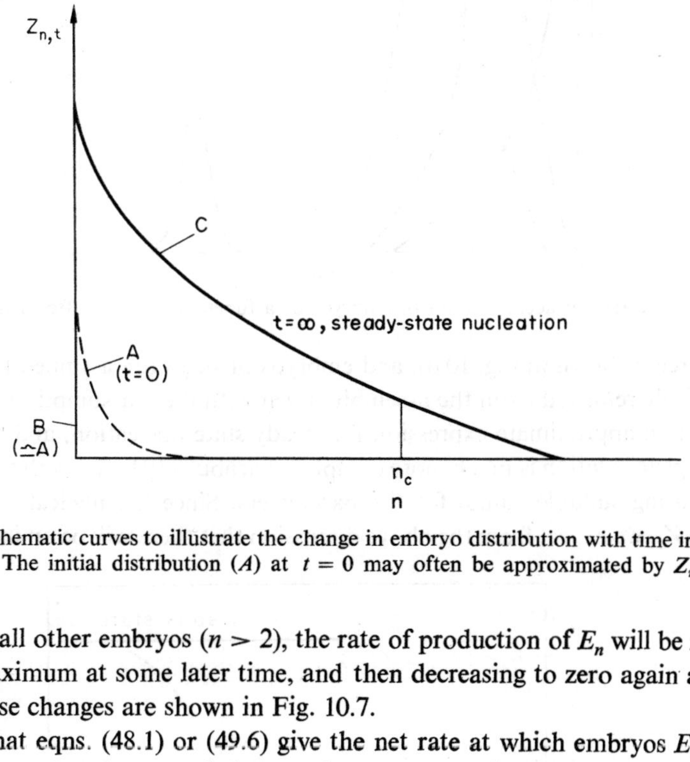

FIG. 10.6. Schematic curves to illustrate the change in embryo distribution with time in a quenched specimen. The initial distribution (A) at $t = 0$ may often be approximated by $Z_{n,t} = 0$ (B).

at $t = 0$. For all other embryos ($n > 2$), the rate of production of E_n will be zero at $t = 0$, rising to a maximum at some later time, and then decreasing to zero again as $Z_{n,t}$ approaches Z_n. These changes are shown in Fig. 10.7.

Recalling that eqns. (48.1) or (49.6) give the net rate at which embryos E_n are changed into E_{n+1}, we see that the rate of change of the number of E_n embryos present is given by

$$\frac{\partial Z_{n,t}}{\partial t} = I_{n-1,t} - I_{n,t}. \tag{50.1}$$

This set of equations has to be solved subject to the initial conditions

$$Z_{1,0} = N, \quad Z_{n,0} = 0 \quad (n \geqslant 2)$$

and the boundary conditions (50.2)

$$Z_{1,t} = N, \quad Z_{s,t} = 0, \quad \text{for all } t.$$

These boundary conditions are, of course, the same as those used in the steady-state solutions above; the number of atoms in the parent phase is treated as constant (neglecting

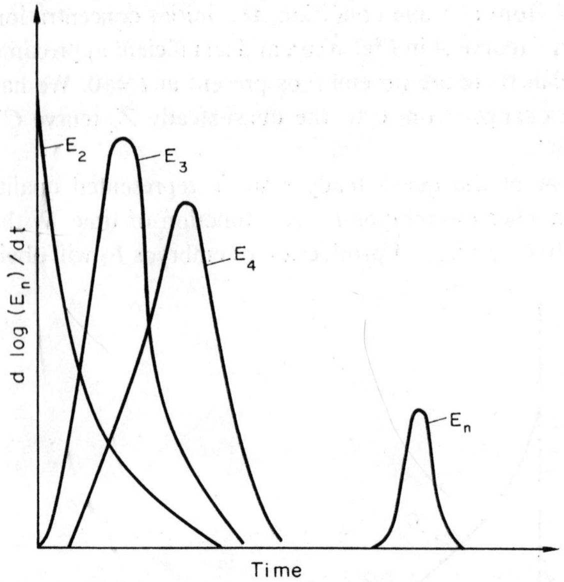

FIG. 10.7. The rate of accumulation of embryos as a function of time (after Turnbull, 1948).

the small decrease shown in Fig. 10.6), and embryos of size s are assumed to grow rapidly and be effectively removed from the assembly. Even with these assumptions, we were only able to obtain an approximate expression for steady state nucleation, and the problem of finding a complete solution is much more complex. Turnbull (1948) has solved the equations numerically, using suitable values for the parameters. Since the nucleation rate depends essentially on Z_c, we expect from the above discussion that there will be an induction period

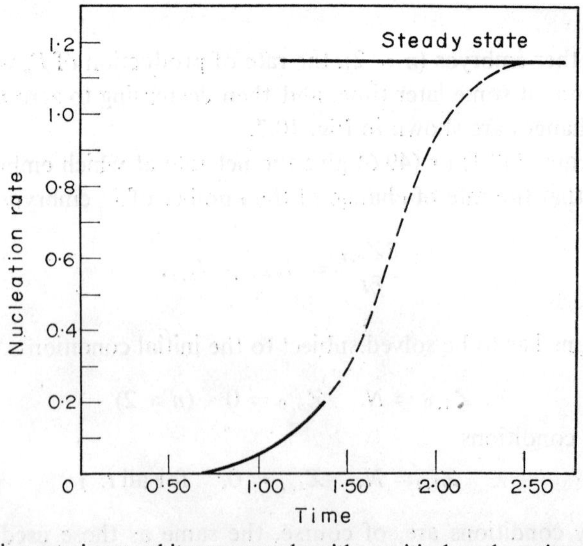

FIG. 10.8. Nucleation rate in an arbitrary example with a critical nucleus size of 25 atoms (after Turnbull, 1948).

during which $I_t = 0$, followed by a gradual rise to the steady state value of I. Turnbull's results, shown in Fig. 10.8, confirm the expected shape of the curve.

Approximate analytical expressions can be obtained only if the difference equations are reduced to differential equations; (48.2) and (49.7) may both be written in the form

$$I_{n,t} = N_n D_n \left[\frac{Z_{n,t}}{N_n} - \frac{Z_{n+1,t}}{N_{n+1}} \right], \tag{50.3}$$

where D_n is given the appropriate value, corresponding to nucleation in a vapour or in a condensed phase. The quantities in this equation are defined only at a set of integral values of n, but we now regard them as functions of a continuous variable n, this being a good approximation except for very small n. The equation can thus be written in differential form as

$$I_{n,t} = -N_n D_n \frac{\partial}{\partial n}\left(\frac{Z_{n,t}}{N_n}\right).$$

Substituting from (47.1), this becomes

$$I_{n,t} = -D_n \frac{\partial Z_{n,t}}{\partial n} - \frac{D_n Z_{n,t}}{kT} \frac{\partial \Delta G_n}{\partial n}. \tag{50.4}$$

Finally, from (50.1) and (50.4),

$$\frac{\partial Z}{\partial t} = \frac{\partial}{\partial n}\left(D \frac{\partial Z}{\partial n}\right) + \frac{1}{kT} \frac{\partial}{\partial n}\left(DZ \frac{\partial \Delta G}{\partial n}\right), \tag{50.5}$$

in which we have omitted the subscripts n and t.

As pointed out by Zeldovich (1943) and Frenkel (1946), eqn. (50.5) is formally equivalent to the diffusion equation for a set of particles distributed along an axis n and moving in the influence of a force field specified by a potential ΔG_n. It is sometimes called the Fokker–Planck equation, and has been studied in connection with the theory of Brownian motion. D_n is the formal analogue of a diffusion coefficient, but it is not independent of n.

The initial and boundary conditions to be satisfied by the solution to (50.5), replacing those of the more exact (50.2), are

$$Z_{n,0} = 0 \quad (n > 0), \qquad Z_{0,t} = N, \qquad Z_{s,t} = 0. \tag{50.6}$$

The earliest treatment of the problem is due to Zeldovich (1943). He noted that in the region $n < n_c$, the exact solution of eqn. (50.5) for the case of harmonically bound particles executing Brownian motion leads to the approximate result

$$Z_{n,t} = Z_n \exp(-n_c^2/4D_c t), \tag{50.7}$$

where Z_n is the quasi-equilibrium value. Since the nucleation rate will be approximately proportional to Z_c, this suggests the nucleation current may be written

$$I_t \approx I \exp(-\tau/t) \tag{50.8}$$

and Zeldovich's result is sometimes quoted in this form. It is not usually emphasized, however, that this equation is based on a solution which assumes an invalid relation between ΔG_n and n, and some further justification of (50.8) is thus needed.

Equation (50.5) has also been investigated by Kantrowitz (1951), who pointed out that when t is small, the second term on the right, representing the thermodynamic work barrier to the formation of embryos, will be relatively unimportant compared with the first term, which represents the kinetic obstacles to growth. An approximate solution can thus be obtained by neglecting this second term, and also treating D_n as a constant, since its variation with n is much smaller than that of $Z_{n,t}$. Equation (50.5) then reduces to the ordinary diffusion equation

$$\frac{\partial Z}{\partial t} = D\frac{\partial^2 Z}{\partial n^2}$$

and with the appropriate conditions (50.6), the solution is (Carslaw and Jaeger, 1947)

$$Z = N(1-n/s) - \frac{2N}{\pi}\sum_{m=1}^{\infty}\frac{1}{m}\sin\left(\frac{m\pi n}{s}\right)\exp\left(\frac{-m^2\pi^2 Dt}{s^2}\right). \tag{50.9}$$

The nucleation rate is $I_{s,t}$, and from (50.4) this is approximately $-D(\partial Z/\partial n)_s$. This gives

$$I_{s,t} = \frac{ND}{2}\left[1+2\sum_{m=1}^{\infty}(-1)^m \exp(-m^2\pi^2 Dt/s^2)\right].$$

The series converges slowly at first, but may be transformed by means of the Poisson summation formula (Courant and Hilbert, 1953), into the more rapidly converging form

$$I_{s,t} = 2N(D/\pi t)^{1/2}\sum_{m=1}^{\infty}\exp(-s^2/4Dt(2m-1)^2).$$

For small t, only the first term need be retained, and

$$I_{s,t} = 2N(D/\pi t)^{1/2}\exp(-s^2/4Dt), \tag{50.10}$$

which is Kantrowitz's expression for the nucleation rate. It shows that the nucleation rate does not become appreciable until a time $\tau \simeq s^2/4D_c \simeq n_c^2/D_c$ has elapsed. The result is only valid for small t, and it is not possible to obtain the relation between $I_{s,t}$ and the quasi-steady-state nucleation rate I by allowing t to tend to infinity in eqn. (50.10). Nevertheless, the general similarity of the expression to Zeldovich's result suggests again that (50.8) may be a reasonable approximation to the nucleation rate in the early part of the transient. It is evident that the duration of the transient is determined by a characteristic time, given to an order of magnitude by

$$\tau = n_c^2/D_c = n_c^2/q_0 A_c. \tag{50.11}$$

A physical argument which leads to a similar approximate value for τ has been developed by Russell (1968, 1969). In the range of sizes $n_1 < n < n_2$ within which $\Delta G_n + kT \geqslant \Delta G_c$, an embryo executes a nearly random walk, i.e. it gains and loses atoms at the same average rate (Feder *et al.*, 1966). Nuclei of sizes $n > n_2$ and embryos of sizes $n < n_1$, on the other hand, have very small probabilities of shrinkage or growth respectively. Consider a nucleus of size n_2 which ultimately disappears. The time taken will be the sum of the time τ_1 for the random walk plus the time τ_2 for the "drift flow", and it can be shown that $\tau_1 > \tau_2$. The principle of time reversal requires that these same average times would be those for a nucleus of size n_2 to form *ab initio*, and Russell therefore takes the larger time τ_1 as the relaxation time τ for the establishment of the steady state nucleation rate.

The size range $n_1 - n_2$ may be derived from the expansion (48.5) which gives

$$n_2 - n_1 = (8kT)^{1/2}(-\partial^2\Delta G_n/\partial\xi^2)_{\xi=0}^{-1/2}, \tag{50.12}$$

and from (46.3) and (46.20), this becomes†

$$n_2 - n_1 = (-12kT/\Delta G_c)^{1/2}\, n_c. \tag{50.13}$$

The relaxation time for the random walk is

$$\tau_1 = (n_2 - n_1)^2/2q_0A_c, \tag{50.14}$$

and with the typical values for observable nucleation rates, $\Delta G_c = 60kT$, this becomes

$$\tau_1 = n_c^2/10q_0A_c. \tag{50.15}$$

Russell developed this description for a precipitation reaction involving long-range diffusion, in which case the factor q_0 involves the activation energy for volume diffusion, as already emphasized. An alternative estimate by Hillig (1962) takes better account of the long-range diffusion aspects of the problem, but neglects the effects of ΔG_n; it therefore applies only to precipitation from dilute solutions which is considered in a later chapter. Equations (50.11) and (50.15) differ by only one order of magnitude, and greater accuracy cannot be expected; eqn. (50.15) is certainly an underestimate of the relaxation time.

A more complete solution to eqn. (50.5), valid in the important region where n is not very different from n_c, has been given by Probstein (1951). The equation is expanded, and the terms in $\partial\Delta G_n/\partial n$ are omitted, since this is close to zero when n is near to n_c. Treating D as constant, the equation in then reduced to

$$\frac{\partial Z}{\partial t} = D\frac{\partial^2 Z}{\partial n^2} + \left(\frac{D}{kT}\frac{\partial^2 \Delta G}{\partial n^2}\right)Z.$$

The solution to this equation with the conditions (50.5) and the additional assumption $s = 4n_c/3$ has two terms, the first giving the quasi-equilibrium distribution, and the second being a slowly converging series. As $\partial^2\Delta G/\partial n^2 \to 0$ the solution changes into (50.9). Probstein confirmed that the influence of the work function is relatively unimportant in the early part of the transient; in a typical example, the time to reach a nucleation rate of one-tenth of the quasi-steady rate is 50% greater than predicted by Kantrowitz's equation. Such a

† Russell states that $(\partial^2\Delta G_n/\partial\xi^2)_{\xi=0} = 2\,\Delta G_c/3n_c^2$ only for spherical nuclei, but the development of Section 46 shows it to be generally valid within the assumptions used there.

discrepancy is insignificant, since these calculations can only claim to give an order of magnitude to the delay time.

In an isothermal transformation, the significant factor will often be the ratio of the time taken to establish the steady-state nucleation rate to the effective time taken to complete the transformation. When this ratio is small, the transformation kinetics will be described satisfactorily by the assumption that ${}^{v}I$ is constant; if it approaches unity, the time dependence of ${}^{v}I$ will influence the overall kinetics. From eqn. (50.11) it follows that the ratio increases with $(\Delta G_c)^2$, so that transient effects are important when nucleation is difficult. The ratio may also approach unity quite independently of ΔG_c if the free energy of activation for growth of a macroscopic region is appreciably less than $\Delta_a g^{+}$. This means that macroscopic regions grow much more rapidly than do embryos, so that the whole transformation is complete before the embryo distribution reaches a steady state. This is further discussed in Section 68.

51. HETEROGENOUS NUCLEATION AND NUCLEATION ON GRAIN BOUNDARIES

In the preceding sections, the formation of a nucleus has been regarded as a homogeneous process occurring with equal probability in all parts of the assembly. In practice, this is unlikely to happen unless the assembly is extremely pure, and also contains (if in the solid state) very few structural defects. More usually, the presence of impurity particles or strained regions of lattice enable nuclei to be formed with a much smaller free energy of activation than that of the homogeneous nuclei. In this section, we describe the modifications to the theory required when nuclei form on foreign particles present in the assembly, on the walls of the container, or at grain boundaries.

When strain energy effects are not important, the catalysing of a nucleation process must depend on a reduction in the net surface energy needed to form a nucleus. This can happen if the formation of an embryo involves the destruction of part of an existing surface, the free energy of which helps to provide the free energy needed for the new surface. The calculations in this section all depend on this assumption.

We begin by making the rather general assumption that the α phase is in contact with a solid surface, S, and we calculate the energy required to produce a β embryo, also in contact with S. If the surface energy, $\sigma^{\alpha\beta}$, of the α–β interface is isotropic, the β embryo will be bounded by spherical surfaces of radius r, except where it is in contact with S. The volume of the embryo may be written as $\eta^{\beta} r^3$, and its surface area of contact with the α phase as $\eta^{\alpha\beta} r^2$, where η^{β}, $\eta^{\alpha\beta}$ are shape factors. The area of contact of the embryo with S is equal to $\eta^{\alpha S} r^2$, the area of α–S interface destroyed when the β region is formed. The free energy of formation may thus be written

$$\Delta G^S = (\eta^{\beta} r^3/v^{\beta})(g^{\beta}-g^{\alpha})+r^2\{\eta^{\alpha\beta}\sigma^{\alpha\beta}+\eta^{\alpha S}(\sigma^{\beta S}-\sigma^{\alpha S})\},$$

where $\sigma^{\alpha\beta}$, $\sigma^{\alpha S}$, $\sigma^{\beta S}$ are the free energies per unit area of the various interfaces. The free energy of formation of the critical size nucleus is found by equating $\partial \Delta G^S/\partial r = 0$ to find r_c.

This gives

$$\Delta G_c^S = \frac{4}{27} \frac{\{\eta^{\alpha\beta}\sigma^{\alpha\beta} + \eta^{\alpha S}(\sigma^{\beta S} - \sigma^{\alpha S})\}^3 (v^\beta)^2}{(\eta^\beta)^2 (g^\alpha - g^\beta)^2}. \tag{51.1}$$

The problem of heterogeneous nucleation thus reduces mainly to evaluation of η^β, $\eta^{\alpha\beta}$, $\eta^{\alpha S}$ for various particular cases. The problem is determinate, since surface energy conditions of equilibrium have also to be satisfied whenever three surfaces meet along a line.

The simplest possibility is a solid impurity with a flat surface S, as in Fig. 10.9. The formation of a stable nucleus of β in contact with S was first considered by Volmer (1929). A β embryo will obviously be the segment of a sphere, since $\sigma^{\alpha\beta}$ is assumed constant, and

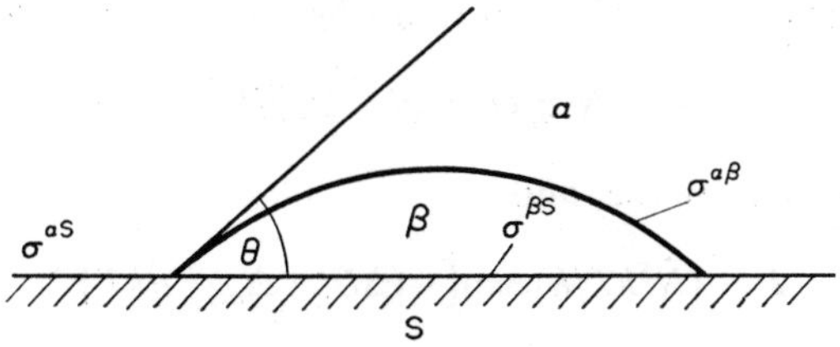

FIG. 10.9. The formation of a β embryo on a flat impurity surface S.

if the contact angle between the embryo and the surface is θ, the condition for static equilibrium is

$$\sigma^{\alpha S} = \sigma^{\beta S} + \sigma^{\alpha\beta} \cos\theta \qquad (0 \leqslant \theta \leqslant \pi). \tag{51.2}$$

When θ lies outside the stated limits, there can be no equilibrium of the surface tension forces, and either the α or the β phase will spread over the surface.

From the geometry of the figure, we see that $\eta^\beta = \pi(2 - 3\cos\theta + \cos^3\theta)/3$, $\eta^{\alpha\beta} = 2\pi(1 - \cos\theta)$ and $\eta^{\alpha S} = \pi \sin^2\theta$. Hence from (51.1) and (51.2)

$$\Delta G_c^S = \frac{4\pi}{3} \frac{(\sigma^{\alpha\beta})^3 (v^\beta)^2 (2 - 3\cos\theta + \cos^3\theta)}{(g^\alpha - g^\beta)^2}. \tag{51.3}$$

Comparing with (46.19), we see that the effect of S is to reduce the formation energy of the corresponding critical spherical nucleus by a factor $(2-3\cos\theta+\cos^3\theta)/4$. For $\theta > 0$, this term is always positive, so the presence of the impurity S cannot enable β embryos to remain stable in the region where $g^\alpha < g^\beta$. When $\theta = \frac{1}{2}\pi$, the free energy to form a critical nucleus in the interior of the α phase is twice that to form one on the surface S. As $\theta \to \pi$, the free energy of the heterogeneous nucleus increases to that required for homogeneous nucleation; as $\theta \to 0$, the free energy decreases to zero. When $\theta = 0$, the β phase "wets" the substrate S in the presence of the α phase, and the only energy required for the formation of a nucleus is that of its periphery, neglected above. It is believed that this condition is satisfied for the formation of liquid embryos on solid surfaces of the same composition, and that this explains why solids cannot be superheated above their melting points (see Part II, Chapter 14).

If $\sigma^{\alpha S} > \sigma^{\beta S} + \sigma^{\alpha\beta}$, θ does not satisfy eqn. (51.2), and there is a negative free energy change in forming the embryo (neglecting the peripheral energy). The embryos in contact with S may thus be stable above the thermodynamic transition temperature of the bulk phases,† where $g^{\alpha} < g^{\beta}$. This condition, however, will not often occur.

The nucleation rate for embryos formed on the surface S may now be found by methods analogous to those used above for homogeneous nucleation. The processes leading to the growth and decay of an embryo are a little uncertain; it seems most probable that the embryos are in quasi-steady equilibrium with atoms from the parent phase which are in contact with S, or effectively adsorbed on S. If the number of such atoms is N^S, then in the region where the α phase is stable, there will be an equilibrium distribution of embryos of size n

$$N_n = N^S \exp(-\Delta G_n^S/kT).$$

The corresponding steady-state nucleation rate when α is metastable may thus be written

$$I \simeq N^S q_0 O_c \exp(-\Delta G_c^S/kT), \tag{51.4}$$

for nucleation from a vapour phase, and

$$I \simeq (N^S kT/h) \exp\{-(\Delta G_c^S + \Delta_a g^*)/kT\} \tag{51.5}$$

for nucleation from condensed phases. Thus the only modifications required for changing from homogeneous nucleation equations to those for heterogeneous nucleation consists in substituting N^S for N and multiplying ΔG_c by a function of a single parameter θ. Equation (51.5) can be expressed in the generally more useful form

$$^S I = (N^S/O^S)(kT/h) \exp\{-(\Delta G_c^S + \Delta_a g^*)/kT\}, \tag{51.6}$$

where O^S is the total surface area of S in the assembly, and $^S I$ is the nucleation rate per unit area of S. The nucleation rate per unit volume is clearly

$$^v I^S = {}^v O^S\, {}^S I = {}^v N^S (kT/h) \exp\{-(\Delta G_c^S + \Delta_a g^*)/kT\}, \tag{51.7}$$

where $^v O^S$, $^v N^S$ are respectively the surface area and number of surface atoms of S present in unit volume. Note that whereas the homogeneous nucleation rate is proportional to the volume of the assembly, the heterogeneous rate is proportional to the surface area of the impurity which catalyses the transformation. Thus the state of dispersion of the impurity, as well as the total amount, is effective in determining the nucleation rate.

A uniform flat substrate is not a good approximation to practical conditions. Surface roughening will always be present to an appreciable extent, even on a supposedly smooth flat surface, and the possibility of having β embryos present in surface cavities was first pointed out by Volmer (1939). For $\theta < \frac{1}{2}\pi$, $\sigma^{\alpha S} > \sigma^{\beta S}$ (eqn. (51.2)), and the part of the surface energy change due to the substitution of the β–S surface for the previous α–S surface is thus negative. If the embryo can be so shaped that its surface is bounded mainly by S, rather than by α, the total surface energy change may become negative. The ratio of the α–β surface to the β–S surface is decreased when the embryo fills a cavity in S, and under suitable conditions, such an embryo may remain stable at temperatures considerably

† We are assuming for definiteness that α is stable at higher temperatures than β. This and subsequent statements are equally applicable below the transformation temperature if α is the low-temperature phase.

above the thermodynamic transformation temperature. The retention of embryos in cylindrical and conical cavities has been considered in some detail by Turnbull (1950). He has used his results to explain the variation of the rate of nucleation in liquid–solid transformations with the extent to which the liquid is previously heated above the melting point.

As an illustration of the effect of nucleation in a cavity, consider a cylindrical hole of radius r containing a height h of the β phase (Fig. 10.10). We then have $\eta^{\beta} = \pi h/r$, $\eta^{\alpha\beta} =$

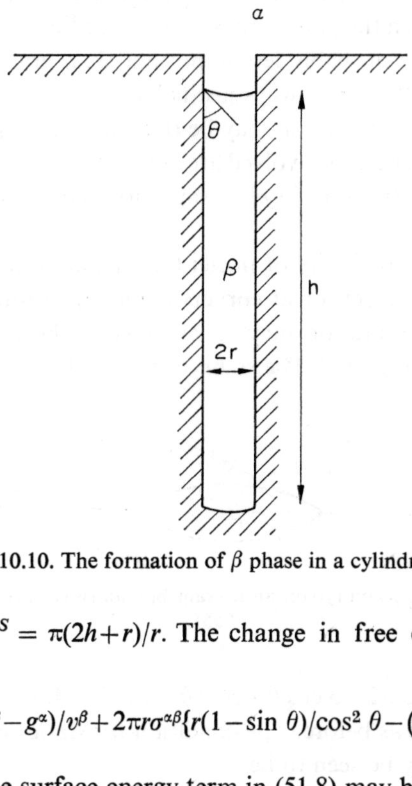

FIG. 10.10. The formation of β phase in a cylindrical cavity.

$2\pi(1-\sin\theta)/\cos^2\theta$ and $\eta^{\alpha S} = \pi(2h+r)/r$. The change in free energy after formation of the embryo is thus

$$\Delta G = \pi r^2 h(g^{\beta}-g^{\alpha})/v^{\beta}+2\pi r\sigma^{\alpha\beta}\{r(1-\sin\theta)/\cos^2\theta-(h+\tfrac{1}{2}r)\cos\theta\}. \tag{51.8}$$

If h is sufficiently large, the surface energy term in (51.8) may be negative. The stability of an embryo in a cavity above the transformation temperature, however, does not depend on the sign of ΔG, but on that of the coefficient of the term in h.† This is because the term in r will be unchanged when a small quantity of the embryo is absorbed into the α phase, and the change in free energy thus produced will depend only on Δh. Clearly, the embryo will be stable if the change in ΔG due to a decrease in h is positive, i.e. if

$$\left.\begin{aligned} &(\pi r^2/v^{\beta})(g^{\beta}-g^{\alpha})-2\pi r\sigma^{\alpha\beta}\cos\theta < 0 \\ \text{or}\quad &r < 2v^{\beta}\sigma^{\alpha\beta}\cos\theta/(g^{\beta}-g^{\alpha}). \end{aligned}\right\} \tag{51.9}$$

The smaller the cavity radius r, the greater is the chance of the embryo remaining stable.

† This statement is, of course, true only in a limited sense. In principle, an embryo will ultimately disappear if ΔG is positive, but the probability of this happening in any reasonable time is nearly negligible unless ΔG decreases when a small quantity of the embryo disappears.

As the temperature is increased above the transformation temperature, $(g^\beta - g^\alpha)$ increases, and more and more cavities become unstable, until eventually all are empty.

Once a cavity is empty of β, a β embryo can fill it again only by first forming on the nearly flat bottom or sides of the hole. The critical free energy for this process is thus given not by (51.8) but by (51.1), so that embryos in cavities will not form again until $(g^\beta - g^\alpha)$ is sufficiently negative for eqns. (51.4) or (51.5) to give an appreciable rate.

The above discussion of heterogeneous nucleation applies in principle to all transformations, but is most useful when the parent phase is a gas or liquid. In solid–solid transformations, impurity particles are probably less important than grain boundaries or structural defects as nucleation sites. It is not always justifiable to neglect strain energy effects, but if this is done, grain boundary nucleation may be treated as a simple extension of the above problems. The only surface energies involved are now $\sigma^{\alpha\beta}$ and $\sigma^{\alpha\alpha}$, the grain boundary energy of the α phase. In eqn. (51.1) the term in $\sigma^{\beta S}$ disappears, and $\eta^{\alpha S}$, $\sigma^{\alpha S}$ are replaced by $\eta^{\alpha\alpha}$, $\sigma^{\alpha\alpha}$ respectively.

We have to consider the possibility of nuclei forming at surfaces separating two grains, at edges where three grains meet, or at corners common to four grains. For a two-grain boundary, the surface of separation may be assumed to be planar, and the problem is essentially similar to that of Fig. 10.9. The embryo (Fig. 10.11) will be a symmetrical doubly-

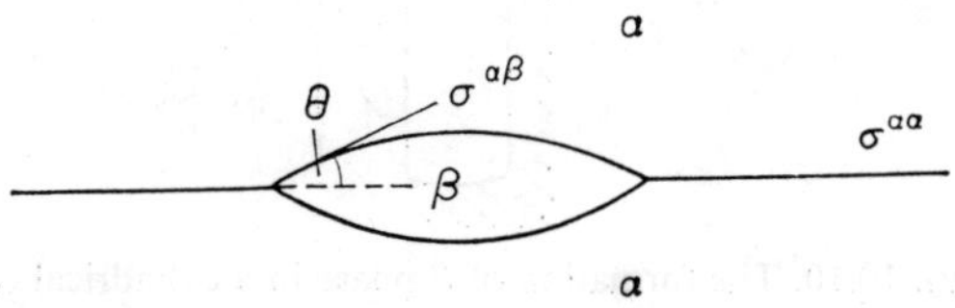

FIG. 10.11. The formation of a β embryo on an α grain boundary surface (after Clemm and Fisher, 1955).

spherical lens, so that $\eta^\beta = 2\pi(2-3\cos\theta+\cos^3\theta)/3$, $\eta^{\alpha\beta} = 4\pi(1-\cos\theta)$ and $\eta^{\alpha\alpha} = \pi\sin^2\theta$, where θ is the contact angle as before. The condition for static equilibrium is slightly different; from Fig. 10.11 it may be seen to be

$$\sigma^{\alpha\alpha} = 2\sigma^{\alpha\beta}\cos\theta. \tag{51.10}$$

The free energy of formation of the critical nucleus is thus found to be twice that given by eqn. (51.3), and the ratio of this energy to the energy for homogeneous nucleation is

$$\Delta G_c^B/\Delta G_c^H = \tfrac{1}{2}(2-3\cos\theta+\cos^3\theta), \tag{51.11}$$

where we have written ΔG_c^B for the critical free energy needed to form a nucleus on the boundary, and ΔG_c^H for the critical free energy for homogeneous nucleation (previously written just as ΔG_c).

At a three-grain junction, we assume the equilibrium configuration (p. 329) in which three planar boundaries meet in a line at angles of 120° to each other. The shape of an embryo will then be a figure bounded by three spherical surfaces; a section normal to the line is shown in Fig. 10.12. From this we see that the condition for static equilibrium is again given by eqn. (51.10), where θ is the dihedral angle at the edge between two α–β surfaces and an α–α surface, as before. The geometry is a little more tedious, but it is readily seen that if the radius of the spherical surfaces is r, the length of the edge along which the embryo forms is $2r(1-4\cos^2\theta/3)^{1/2}$. The three edges, in each of which two spherical

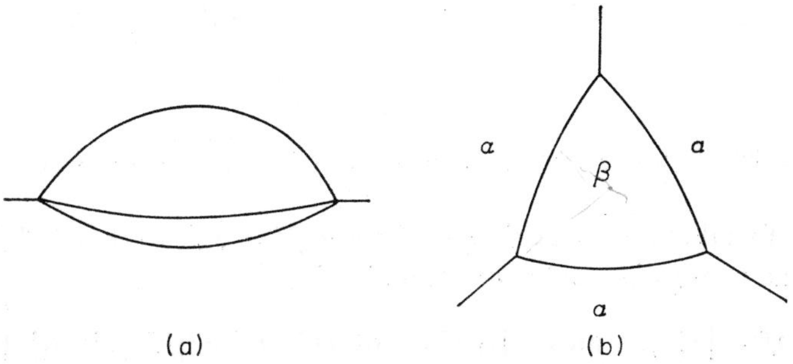

FIG. 10.12. The shape of a β embryo formed at a three-grain junction in α. (*a*) General view, (*b*) section normal to grain edge (after Clemm and Fisher, 1955).

boundaries of the embryo meet a planar grain boundary, are small circles of radius $r \sin\theta$. Calculation of the volume and of the planar and spherical surface areas then gives

$$\left.\begin{aligned} \eta^{\beta} &= 2[\pi-2 \text{ arc } \sin(\tfrac{1}{2}\text{cosec } \theta)+\tfrac{1}{3}\cos^2\theta(4\sin^2\theta-1)^{1/2} \\ &\quad -\text{arc } \cos(\cot\theta/\sqrt{3})\cos\theta(3-\cos^2\theta)], \\ \eta^{\alpha\beta} &= 6\pi-12 \text{ arc } \sin(\tfrac{1}{2}\text{cosec } \theta)-12\cos\theta \text{ arc } \cos(\cot\theta/\sqrt{3}), \\ \eta^{\alpha\alpha} &= 3\sin^2\theta \text{ arc } \cos(\cot\theta/\sqrt{3})-\cos\theta(4\sin^2\theta-1)^{1/2}. \end{aligned}\right\} \quad (51.12)$$

From this, $\eta^{\alpha\beta}-2\eta^{\alpha\alpha}\cos\theta = 3\eta^{\beta}$, and hence

$$\Delta G_c^E = 4\eta^{\beta}\sigma^{\alpha\beta}(v^{\beta})^2/(g^{\beta}-g^{\alpha})^2.$$

The ratio of the nucleation energy to that for homogeneous nucleation is now

$$\Delta G_c^E/\Delta G_c^H = (3/4\pi)\eta^{\beta}, \quad (51.13)$$

where η^{β} is given by (51.12) above. We should note that in deriving ΔG_c^E we have neglected any extra edge energy, and have assumed that the energy gain is given entirely by the free energies of the planar boundaries eliminated.

Finally, we have to consider a grain corner where four different grains meet. The four grain edges, each being the junction of three of the grains, may be assumed to radiate symmetrically from the corner, so that an embryo bounded by spherical surfaces will have the shape of a spherical tetrahedron (Fig. 10.13).† When the volume and area shape factors are

† It will be realized that this condition, and the condition assumed for edge nucleation, are not likely to be exactly realized in practice, for reasons discussed on pp. 330–2. However, the assumption is a good approximation to the average actual configuration.

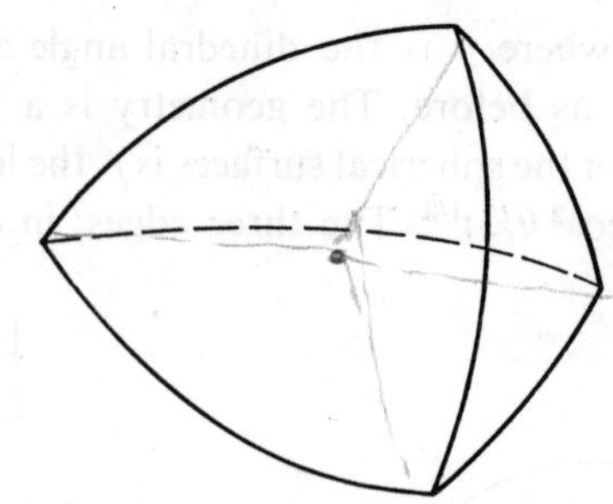

FIG. 10.13. Possible shape of a β embryo forming at a grain corner (after Clemm and Fisher, 1955).

evaluated, it is found that once again $\eta^{\alpha\beta} - 2\cos\theta\,\eta^{\alpha\alpha} = 3\eta^{\beta}$, so that ΔG_c^C is also given by (51.13), with the appropriate value of η^{β}, which is

$$\left.\begin{aligned}\eta^{\beta} = [8\{\pi/3 - \text{arc cos}[\{\sqrt{(2)} - \cos\theta\,(3 - C_{10}^2)^{1/2}\}/C_{10}\sin\theta]\}\\ + C_{10}\cos\theta\,\{(4\sin^2\theta - C_{10}^2)^{1/2} - C_{10}^2/\sqrt{2}\}\\ -4\cos\theta\,(3 - \cos^2\theta)\,\text{arc cos}\,C_{10}/(2\sin\theta)],\end{aligned}\right\} \tag{51.14}$$

where $C_{10} = 2\{\sqrt{(2)}\,(4\sin^2\theta - 1)^{1/2} - \cos\theta\}/3$.

The ratio of the free energy required to form a grain boundary nucleus to that needed to form a homogeneous nucleus obviously decreases as the ratio of the grain boundary energy to the interphase boundary energy increases. This ratio, given by eqn. (51.13) with appropriate values of η^{β}, is plotted in Fig. 10.14 as a function of $\cos\theta = \frac{1}{2}\sigma^{\alpha\alpha}/\sigma^{\alpha\beta}$. It will be noted that for all values of $\cos\theta$, $\Delta G_c^C < \Delta G_c^E < \Delta G_c^B < \Delta G_c^H$. The ratio of the energies

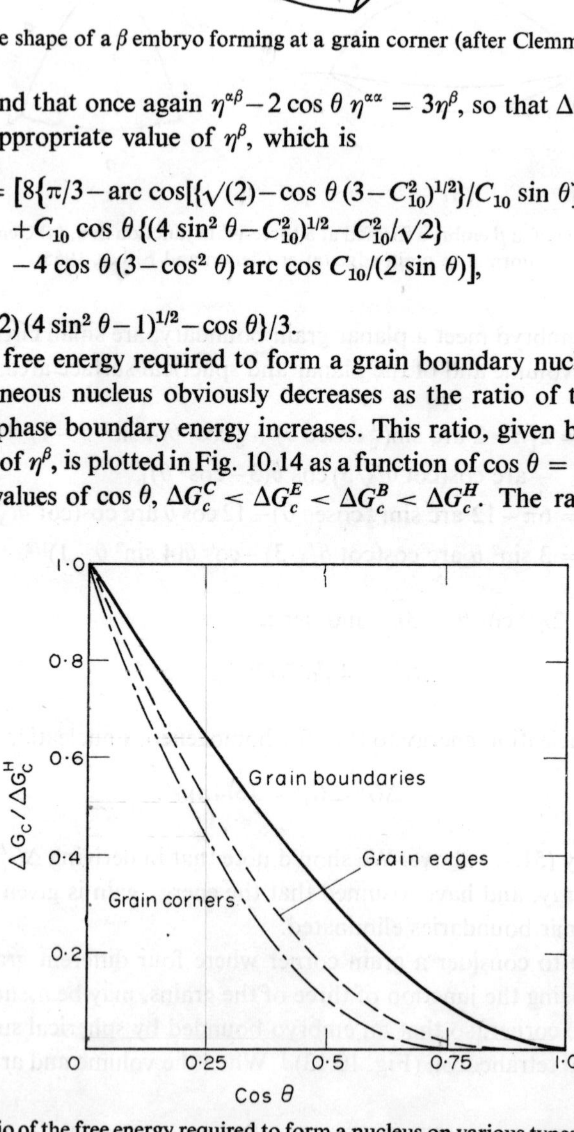

FIG. 10.14. The ratio of the free energy required to form a nucleus on various types of grain boundary site to that required to form a nucleus in the interior of a grain is plotted as a function of $\cos\theta = \frac{1}{2}\sigma^{\alpha\alpha}/\sigma^{\alpha\beta}$ (after Cahn, 1956).

$$\Delta G_c = \Delta G_c^{H} \cdot f(\theta)$$
$$f(\theta) = o \rightarrow$$

becomes zero at some finite value of $\sigma^{\alpha\alpha}/\sigma^{\alpha\beta}$, and for boundary energies which are relatively higher than this, no equilibrium is possible. There is then a continuous energy decrease as the new phase grows from zero size, and the amount of transformation depends only on a growth rate. From Figs. 10.11, 10.12, and 10.13 we see that the minimum values of θ which allow surface energy equilibrium correspond to infinite radii of the spherical surfaces, and are $0°$, $30°$, and $\sin^{-1}(1/\sqrt{3})$. The critical values of $\sigma^{\alpha\alpha}/\sigma^{\alpha\beta}$, above which ΔG_c is zero, are thus 2, $\sqrt{3}$ and $2\sqrt{2}/\sqrt{3}$ for boundaries, edges, and corners respectively.

The relative nucleation rates do not necessarily increase in the same order as the activation energies for nucleation decrease, since the density of sites also decreases as the mode of nucleation changes from homogeneous to grain corners. We have already seen that for boundary nucleation, N in eqn. (49.10) must be replaced by N^B, and we may similarly expect factors N^E, N^C for edge and corner nucleation. It must be admitted that these pre-exponential factors are rather ill-defined, since the way in which atoms within an embryo forming at a surface, edge, or corner interchange with other atoms is somewhat uncertain. However, if we suppose that the effective thickness of a grain boundary is δ^B, and the mean grain diameter is L^B, we can write the number of atoms per unit volume on the various sites as

$$\left. \begin{array}{l} {}^v N^B = {}^v N(\delta^B/L^B), \\ {}^v N^E = {}^v N(\delta^B/L^B)^2, \\ {}^v N^C = {}^v N(\delta^B/L^B)^3, \end{array} \right\} \tag{51.15}$$

to a fair approximation. The nucleation rate per unit volume due to grain boundary surfaces is then related to the corresponding homogeneous nucleation rate ${}^v I^H$ by

$$\frac{{}^v I^B}{{}^v I^H} = \frac{\delta^B}{L^B} \exp\left\{ \frac{\Delta G_c^H - \Delta G_c^B}{kT} \right\} \tag{51.16}$$

with similar equations for ${}^v I^E$ and ${}^v I^C$. From these equations we find the conditions under which homogeneous nuclei, grain boundary nuclei, grain edge nuclei and grain corner nuclei respectively make the greatest contribution to the overall volume nucleation rate ${}^v I$. These conditions may readily be expressed in terms of a quantity $R^B = kT \ln(L^B/\delta^B)$ and are:

Greatest nucleation rate	
${}^v I^H$	$R^B > \Delta G_c^H - \Delta G_c^B$
${}^v I^B$	$\Delta G_c^H - \Delta G_c^B > R^B > \Delta G_c^B - \Delta G_c^E$
${}^v I^E$	$\Delta G_c^B - \Delta G_c^E > R^B > \Delta G_c^E - \Delta G_c^C$
${}^v I^C$	$\Delta G_c^E - \Delta G_c^C > R^B$

The corresponding limits on $R^B/\Delta G_c^H$ may be expressed in terms of $\sigma^{\alpha\alpha}/\sigma^{\alpha\beta}$ by means of Fig. 10.14. Figure 10.15 shows these regions for different values of $\sigma^{\alpha\alpha}/\sigma^{\alpha\beta}$, as calculated in this way by Cahn.

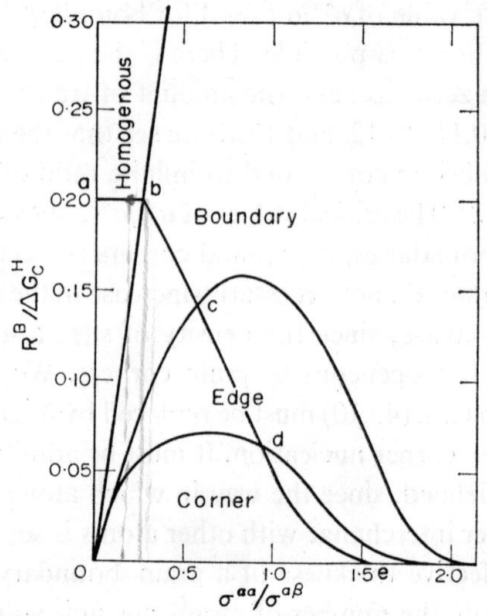

FIG. 10.15. To illustrate the conditions under which various types of site make the greatest contribution to the initial nucleation rate (after Cahn, 1956). The curve *abcde* represents the conditions for the minimum observable nucleation rate. The type of nucleation first observed will depend upon the region in which this curve lies for the given $\sigma^{\alpha\alpha}/\sigma^{\alpha\beta}$.

It remains to examine the magnitude of the predicted effects. For homogeneous nucleation in the solid, reasonable values of the parameters give eqn. (49.11) as

$$ {}^{v}I^{H} \simeq 10^{36} \exp(-\Delta G_c^H/kT) \text{ m}^{-3}\text{ s}^{-1} \tag{51.17}$$

and for ${}^{v}I^{H} \sim 10^{6}$ m^{-3} s^{-1}, we thus obtain $\Delta G_c^H \simeq 69\,kT$. If the grain diameter is $\sim 0{\cdot}1$ mm, $L^B/\delta^B \simeq 10^6$, and the boundary nucleation rate is

$$ {}^{v}I^{B} \simeq 10^{30} \exp(-\Delta G_c^B/kT) \text{ m}^{-3}\text{ s}^{-1}, \tag{51.18}$$

so that for ${}^{v}I^{B} \sim 10^{6}$ m^{-3} s^{-1}, $\Delta G_c^B \simeq 57kT$. For any particular ratio of $\sigma^{\alpha\alpha}/\sigma^{\alpha\beta}$, the ratio $\Delta G_c^B/\Delta G_c^H$ of the activation energies for boundary and homogeneous nucleation under the same driving force may be read from the upper curve of Fig. 10.15. From the value of ΔG_c^H when $\Delta G_c^B = 57kT$, we can plot a curve of $kT \ln(L^B/\delta^B)/\Delta G_c^H$ (assuming the above parameters) against $\sigma^{\alpha\alpha}/\sigma^{\alpha\beta}$ on Fig. 10.15. The curve gives the relation between ΔG_c^H and $\sigma^{\alpha\alpha}/\sigma^{\alpha\beta}$ needed to sustain the assumed nucleation rate; part of it is shown as the curve $b-c$ on the figure. When the value of ΔG_c^H determined in this way equals $69kT$, the homogeneous and boundary volume nucleation rates are both equal to the assumed unit nucleation rate, and this gives the point b on the figure. At values of $\sigma^{\alpha\alpha}/\sigma^{\alpha\beta}$ smaller than this, a smaller value of ΔG_c^H (i.e. a larger driving force $g^{\alpha}-g^{\beta}$) is needed to give ${}^{v}I^{B} \sim 10^{6}$ m^{-3} s^{-1}, but this rate can of course be obtained by homogeneous nucleation at a driving force corresponding to $\Delta G_c^H = 69kT$.

For grain-edge and grain-corner nucleation we have the corresponding equations

$$\left.\begin{aligned} {}^{v}I^{E} &\simeq 10^{24}\exp(-\Delta G_c^E/\boldsymbol{k}T)\ \mathrm{m^{-3}\,s^{-1}}, \\ {}^{v}I^{C} &\simeq 10^{18}\exp(-\Delta G_c^C/\boldsymbol{k}T)\ \mathrm{m^{-3}\,s^{-1}}, \end{aligned}\right\} \quad (51.19)$$

giving 10^6 nuclei $\mathrm{m^{-3}\,s^{-1}}$ when $\Delta G_c^H \simeq 41\boldsymbol{k}T$ and $\Delta G_c^C \simeq 28\boldsymbol{k}T$ respectively. The corresponding values of ΔG_c^H for various ratios of boundary energies may be derived from Fig. 10.14, and the corresponding values of $\boldsymbol{k}T\ln(L^B/\delta^B)/\Delta G_c^H$ are plotted in Fig. 10.15 over the regions *cd* (for edges) and *de* (for corners). The whole curve *abcde* thus shows the maximum value of ΔG_c^H, and hence (using eqn. (46.19)) the minimum value of $g^\alpha - g^\beta$, which will allow the just perceptible nucleation rate of 10^6 nuclei $\mathrm{m^{-3}\,s^{-1}}$ to be attained. It follows from the curve that the type of nucleation which occurs when $g_\alpha - g_\beta$ becomes large enough to give a measurable rate of transformation depends on the value of $\sigma^{\alpha\alpha}/\sigma^{\alpha\beta}$. In principle, if $g^\alpha - g^\beta$ is slowly increased from zero (e.g. by continuous cooling), nucleation will initially be greatest on corners, then on edges, then boundaries, and then homogeneously. But if $\sigma^{\alpha\alpha} < 0{\cdot}9\sigma^{\alpha\beta}$, the corner nucleation rate will be too small to be observed until the driving force has increased to such a value that edge nucleation is more rapid than corners; if $\sigma^{\alpha\alpha} < 0{\cdot}6\sigma^{\alpha\beta}$, the amount of edge nucleation will similarly be too small to be observed before boundary nucleation has become predominant; and if $\sigma^{\alpha\alpha} < 0{\cdot}25\,\sigma^{\alpha\beta}$, only effectively random homogeneous nucleation will ever be observed.

52. NUCLEATION IN THE SOLID STATE

A change of volume accompanies most phase transformations, and its effects have specifically to be considered when the transformation occurs entirely in the solid state. If the parent phase is a gas or a liquid, the volume change is achieved with negligible increase in energy by flow of the surrounding fluid. When a small region in a crystal transforms, however, the rigidity of the material may enable it to withstand very considerable stresses with negligible flow or creep rate. The volume change has then to be accommodated in the assembly, and the associated strain energy is an important factor in the transformation. In this section, we consider its importance in nucleation theory.

We consider the following sequences of imaginary operations, leading to the production of a small β crystal in the middle of a large amount of parent α material.

(1) Remove a small volume from the centre of the α and allow it to undergo an unconstrained transformation to β.

(2) Apply surface tractions to the β to return it to its original size and shape, and insert it into the hole in the α again.

(3) Weld together the α and (constrained) β regions over their surface of contact.

(4) Allow the assembly to relax; i.e. remove the built-in layer of surface force by applying an equal and opposite layer of surface force.

The final assembly of matrix α + small β crystal is clearly in a state of self-stress, the strain energy of which is a factor influencing the nucleation rate. The operations we have specified lead to a β nucleus which is coherent with the α, in the sense that the displacements and tractions are continuous across the common interface. The unconstrained transformation to β has been assumed to occur without mass transfer, so that there is a one-to-one correspondence between the atomic positions in the original α and final β crystals. When the final β crystal approximates to a flat plate, we may usefully distinguish two types of coherence. The first possibility is that the principal axes of the matrix which specifies the free transformation $\alpha \rightarrow \beta$ lie in the plane of the plate and perpendicular to it. In the final state, much of the strain will remain in the β particle, which will be extended or compressed so as to match the α phase along their common boundaries. This type of coherent nucleus is believed to be formed in many nucleation and growth reactions. Alternatively, the principal axes may be inclined to the plate in such a way that the displacements correspond to a shear on the plane of the plate, possibly combined with an expansion or contraction normal to the plate. The strain in the final state will then consist largely of shears in the surrounding matrix. This type of coherency is characteristic of martensitic reactions, but may also occur in the nucleation stage of other transformations.

The formation of a small β crystal with an incoherent interface is represented by the above steps if we assume the shear modulus of the transformed region to be zero. The appropriate model is thus to make a hole in the α and fill it with a compressible fluid, of natural volume equal to the volume of the freely transformed α. The only condition now being imposed is that the same total number of atoms are removed from and reinserted into the hole; there is no correspondence of atomic positions. Physically, this means that atoms will migrate so as to minimize the energy; in a flat plate of a β phase having larger specific volume than the α phase, for example, atoms migrate from the edges to the centres of the faces, which can readily bulge outwards. In a sense, the formation of an incoherent interface requires the continuous recrystallization of the β, and will occur at temperatures such that the atoms can migrate fairly rapidly under the (very large) transformation stresses.

If the β particle is a sphere, the distinction between the coherent and incoherent particles disappears. However, we shall find that the strain energy of an incoherent β particle is a function of its shape, and in suitable conditions can be made very small. The strain energy of a coherent particle also varies with its shape, but only a limited reduction can usually be achieved by changing the shape. Thus we might expect that all nuclei will form incoherently except when the volume change is very small. We have noted previously, however, that the surface energy of an incoherent boundary is much larger than that of a coherent boundary, and this opposes the strain energy factor. Since the strain energy is proportional to the volume of the β crystal, the surface energy term will predominate at sufficiently small sizes, and the first nuclei may be coherent. As the nucleus grows, the strain energy will increase until it becomes more favourable energetically for the particle to "break away" from the matrix. This breaking away will usually occur quite early in a nucleation and growth reaction, providing the temperature is high enough to ensure that equilibrium is eventually reached. It cannot happen in a martensitic reaction since the mechanism of continued growth in this case depends on the maintenance of coherence.

All the above virtual processes lead to some strain energy unless the volume change is zero. Zero strain energy is obtained only if we imagine that the original hole in the α is enlarged or reduced by the removal or addition of slices of α material until the freely transformed regionjust fits into it again. There is then no built-in surface force layer, and α and β are both unstressed, although there may be atomic readjustments at the interface to minimize the surface energy. This is the process used in defining the interfacial energy (Section 38), and the elastic energy discussed in this section is understood to be the energy of the constrained assembly additional to this quantity. The total number of atoms in the region which is finally β is now not conserved, and the possibility of producing a β region without strain energy in this way depends on whether diffusion or plastic flow can readily accommodate the changes in size and shape.

At sufficiently high temperatures, flow rates in a crystal should be sufficient to relieve all but the smallest transformation stresses, and strain energy factors will not be important. At low temperatures, where diffusion rates are slow and ineffective, some strain energy will be created, and the magnitude will be determined by the maximum stress which can be supported by the α phase before plastic flow begins. This is not an easy problem. The ordinary yield stress criteria are of little value, since a nucleus is so small that there is a negligible chance of its containing a Frank–Read source (this only applies, of course, if it is formed randomly within the assembly; heterogeneous nucleation on preferred sites, including dislocations, is considered later). The surrounding matrix will thus probably be perfect crystal, and should support a shear stress about 1000 times greater than that corresponding to the activation of a dislocation source. If a nucleus does form in a perfect crystal in this way, the transformation stresses will not usually be able to activate Frank–Read sources in the vicinity, since they are applied over such a small volume ($\sim 10^{-26}$ m^3).

The conclusion suggested by the above arguments is that the strain energy for nucleation is obtained by assuming the nucleus to be formed in a region of entirely good crystal. However, although single dislocations cannot readily reduce stresses developed in small volumes, certain larger defects such as grain boundaries may be able to do this. In cases where the transformation strain energy in a perfect crystal would be very large, heterogeneous nucleation at places where local flow can occur is to be anticipated, since at these sites the work of nucleation is so much reduced. This effect is additional to the reduction in the surface energy term, discussed in the last section. It shows, however, that when grain boundary nucleation is to be considered, the treatment of the last section is probably adequate, even with considerable transformation volume changes.

We shall now give the standard nucleation theory for a transformation where strain energy effects are not negligible. As in earlier sections, we find it convenient to develop the theory for homogeneous nucleation, since the modifications to allow for nucleation at preferred sites are easily made. The formation of a nucleus of given size will now require an increase in free energy given by (46.3) with the addition of the elastic energy term. The elastic energy is proportional to the number of atoms in the nucleus, so that we may write

$$\Delta G = n(g^{\beta}-g^{\alpha}+\Delta g_s)+\eta\sigma n^{2/3}, \tag{52.1}$$

where Δg_s is the elastic energy per atom. In contrast to the previous treatment, the shape of the nucleus is not determined by the condition that the surface energy term shall have its lowest value, since this shape also affects the value of Δg_s. If ΔG is plotted as a function of size and shape, we have to find the combination of size and shape which gives a "minimax" (saddle-point) value to ΔG. The value ΔG_c at this point in the energy field represents the activation energy for nucleation by the most favourable path from α to β.

From eqn. (52.1), we see that the elastic energy reduces the effective driving force $(g^\alpha - g^\beta)$ of the reaction. Clearly, nucleation will not occur at all unless

$$g^\alpha - g^\beta > \Delta g_s .$$

Consider first the formation of an incoherent β nucleus. If the β crystal is spherical, the misfitting sphere model of Section 25 may be applied immediately to give the strain energy. The continuum approximation obviously has much more justification in the present application, where the inclusion really is spherical, and has better defined elastic properties; the main limitation is now in the neglect of crystal anisotropy, made in the interests of simplicity. From eqn. (25.16) we see that the strain energy per atom of nucleus is given by

$$\Delta g_s = 2\mu^\alpha C_6 (v^\beta - v^\alpha)^2 / 3v^\beta , \tag{52.2}$$

where $C_6 = 3K^\beta/(3K^\beta + 4\mu^\alpha)$, and v^α, v^β are the specific volumes of atoms in the α and β phases.

When the nucleus has some other shape, we may proceed as follows. Let Δg_s^α, Δg_s^β be the strain energies per atom when the *whole* of the volume misfit 3ε is taken up respectively by the surrounding α matrix or by compression (expansion) of the β particle. Then if the actual volume of the β nucleus is $(1 + 3C_6\varepsilon)$ times that of the hole, the compression in it is $3(C_6 - 1)\varepsilon$. The energy in the particle is $n(C_6 - 1)^2 \Delta_s^\beta$; and that in the matrix is $nC_6^2 \Delta g_s^\alpha$. The total strain energy is thus given by

$$\Delta g_s = (C_6 - 1)^2 \Delta g_s^\beta + C_6^2 \Delta g_s^\alpha$$

and this has its minimum value when

$$C_6 = \Delta g_s^\beta / (\Delta g_s^\alpha + \Delta g_s^\beta)$$

giving

$$\Delta g_s = \Delta g_s^\alpha \Delta g_s^\beta / (\Delta g_s^\alpha + \Delta g_s^\beta) = C_6 \Delta g_s^\alpha = (1 - C_6) \Delta g_s^\beta . \tag{52.3}$$

In particular, if $\Delta g_s^\beta \gg \Delta g_s^\alpha$, $\Delta g_s \simeq \Delta g_s^\alpha$ and the inclusion is virtually unstrained. We note that C_6 is a measure of the partition of energy between the matrix and the β nucleus, which is

$$\frac{\text{Energy in } \alpha}{\text{Energy in } \beta} = \frac{C_6^2 \Delta g_s^\alpha}{(C_6 - 1)^2 \Delta g_s^\beta} = \frac{C_6}{1 - C_6} \tag{52.4}$$

in agreement with eqns. (25.14) and (25.15) for the case of a sphere. From (52.3), C_6 is also a measure of the extent to which the matrix is able to accommodate the volume change. Since the strain energy is $(1 - C_6)$ times its value is an incompressible medium, good accomodation is specified by values of C_6 approaching unity, and most of the remaining energy then resides in the matrix.

For a spherical particle, the energy when all the strain is taken by the matrix is

$$\Delta g_s^\alpha = 2\mu^\alpha (v^\beta - v^\alpha)^2 / 3v^\beta .$$

Nabarro proposed that for the more general shapes represented by the family of ellipsoids of revolution having semi-axes R, R, y, the corresponding strain energy should be written

$$\Delta g_s^\alpha = \{2\mu^\alpha(v^\beta - v^\alpha)^2/3v^\beta\}\, E(y/R), \tag{52.5}$$

and he obtained values for the function $E(y/R)$ for some limiting cases. Since, in all the calculations, Δg_s^α was shown to depend only on μ^α and not on the other elastic constant of the isotropic matrix, it seems reasonable to assume that this is true for all y/R.

When $y/R = \infty$, the ellipsoid becomes a cylinder (the mathematical model of a needle precipitate particle). This can be treated in the same way as the sphere; the displacements have the form $\mathbf{w} = A\mathbf{r}$ inside the inclusion and $\mathbf{w} = A(r_0^2/r^2)\mathbf{r}$ outside the inclusion, which is forced into a cylindrical hole of radius r_0. When the β inclusion is incompressible, $A = 3\varepsilon/2$, giving a fractional volume change of 3ε and an increase of energy of

$$\Delta g_s^\alpha = \mu^\alpha(v^\beta - v^\alpha)^2/2v^\beta, \tag{52.6}$$

so that $E(y/R) = \frac{3}{4}$.

For a sphere, $E(y/R) = 1$. Nabarro showed that for a spheroid with $y/R = 1+\xi$, $E(y/R)$ differs from unity only in terms of order ξ^2 when ξ is small, so that $E(y/R)$ is a slowly changing function near $y/R = 1$. Finally, when $y/R \ll 1$, the ellipsoid approximates to a thin plate or disc. Nabarro obtained an approximate solution for this case in which

$$E(y/R) \simeq 3\pi y/4R \quad (y/R \ll 1). \tag{52.7}$$

A smooth graph through these points leads to the variation of $E(y/R)$ shown in Fig. 10.16. It must be emphasized that $E(y/R)$ gives only the variation of the energy Δg_s^α, and this is equivalent to Δg_s only when $\Delta g_s^\beta \gg \Delta g_s^\alpha$. Clearly this condition is satisfied for a plate nucleus, where the energy resides almost entirely in the matrix, and tends to zero as y/R decreases. Minimum strain energy is obtained by making the plate as thin as possible.

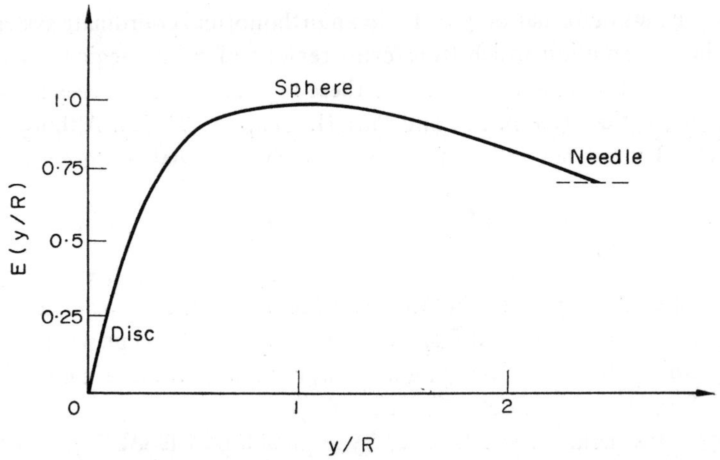

FIG. 10.16. To illustrate the variation of the strain energy of an incoherent nucleus with its shape (after Nabarro, 1940a).

As already stressed, the main limitation of the above results comes from the assumption of elastic isotropy. However, Kröner (1954) has made essentially equivalent calculations, making full allowance for crystal anisotropy. Although the details of the energy variation then depend on the individual elastic properties, the general shape of Fig. 10.16 is obtained for all curves of strain energy against y/R. Incoherent nuclei formed in a solid under constrained conditions will thus always have a shape approximating to a flat plate (oblate spheroid) if the volume change is appreciable. The smaller the y/R ratio, the smaller is the strain energy, and the larger is the surface energy for a given volume of nucleus. The most favourable nucleation path will utilize the shape (of finite y/R) which minimizes the total free energy. Substituting expressions for volume and surface area into eqn. (52.1), we find

$$\Delta G = n\left[g^{\beta}-g^{\alpha}+\frac{\pi}{4}\frac{y}{R}\mu^{\alpha}\frac{(v^{\beta}-v^{\alpha})^2}{v^{\beta}}\right]+\pi^{1/3}(3v^{\beta}y/4R)^{1/3}$$
$$\times\left[2+\frac{y^2}{R^2(1-y^2/R^2)^{1/2}}\ln\left(\frac{1+(1-y^2/R^2)^{1/2}}{1-(1-y^2/R^2)^{1/2}}\right)\right]\sigma n^{2/3}, \tag{52.8}$$

where the axial ratio y/R specifies the shape of the ellipsoid. The saddle point in the energy field is determined by the simultaneous equations

$$(\partial\Delta G/\partial n)_{y/R} = 0, \quad [\partial\Delta G/\partial(y/R)]_n = 0, \tag{52.9}$$

which give the values n_c, y_c, R_c characteristic of the critical size nucleus of most favourable shape. The free energy of formation of this nucleus is obtained by substituting these values back into eqn. (52.8) and the homogeneous nucleation rate is then given by an equation of the form of (49.10).

We now consider the energy change when the β nucleus forms coherently within the α phase. A general method of treating this problem in the approximation of isotropic elasticity has been given by Eshelby (1957). Suppose the free transformation $\alpha \rightarrow \beta$ is represented by the homogeneous deformation $\mathbf{y} = \mathbf{Ex}$ in an orthonormal coordinate system. The tensor $\mathbf{E}$ specifies the deformation which transforms regions of α into regions of β; it gives the change of shape and volume undergone by these regions, but not necessarily the relations between the two lattices (see p. 319 and Part II, Chapters 21–22). Although the displacements $\mathbf{w} = (\mathbf{E}-\mathbf{I})\mathbf{x}$ are not infinitesimal, we use them to define an elastic strain tensor

$$e_{ij}^{T} = \frac{1}{2}\left\{\frac{\partial w_i}{\partial x_j}+\frac{\partial w_j}{\partial x_i}\right\} = \frac{1}{2}(E_{ij}+E_{ji})-\delta_{ij}. \tag{52.10}$$

The suffix T is used to emphasize that the quantities e_{ij}^{T} are the strains in a free transformation which are to be distinguished from the strains e_{ij}^{C} which occur when the transformed region is constrained by the surrounding matrix. The main problem in considering the constrained transformation is the determination of e_{ij}^{C}.

Let X_{ij}^{T} be a stress field derived from e_{ij}^{T} by application of Hooke's law. Then from eqn. (11.26)

$$X_{j}^{T} = \lambda\Delta^{T}\delta_{ij}+2\mu e_{ij}^{T}. \tag{52.11}$$

The elastic constants used should be those appropriate to the β region, but for simplicity we shall assume that the α and β phases have the same elastic properties for the remainder of this section. Now let the unit vector **n** represent the outward normal at any point of the surface S separating the β particle and the α matrix, and consider the sequence of operations detailed on p. 457. Since both the β and α are unstressed at the end of stage 1, it follows that application of surface tractions $-X^T_{ij}n_j$ over the surface of the β region will restore it elastically to its original size and shape, as required for stage 2. Thus after welding the surfaces together again, we are left with a layer of body force $-X^T_{ij}n_j$ spread over the boundary surface, and this has to be annulled by a further distribution $+X^T_{ij}n_j$. We take the state of the assembly before adding this distribution (i.e. at the end of stage 3) as a state of zero displacement. The stress and strain in the matrix are then zero, and the β particle, although not stress-free, has the same external geometrical form as it had before transformation began. The displacements $\mathbf{w}^C$ produced in matrix and β particle by the distribution $X^T_{ij}n_j$ over S are thus the actual displacements involved in the whole process, and they define an elastic field given in all parts of the assembly by

$$e^C_{ij} = \tfrac{1}{2}\{(\partial w^C_i/\partial x_j)+(\partial w^C_j/\partial x_i)\}.$$

The final stress in the matrix is written X^C_{ij}, and is derived from the elastic field e^C_{ij} by application of Hooke's law. However, the β nucleus had a stress field $-X^T_{ij}$, and a corresponding strain field $-e^T_{ij}$ at the beginning of stage 4, so that its final stress field is

$$X^I_{ij} = X^C_{ij} - X^T_{ij} \tag{52.12}$$

and its final strain field is

$$e^I_{ij} = e^C_{ij} - e^T_{ij}.$$

If the β occupies a volume V, the elastic energy contained in it is

$$\tfrac{1}{2}\int_V (X^I_{ij}e^I_{ij})\,dv = \tfrac{1}{2}\int_V X^I_{ij}(e^C_{ij}-e^T_{ij})\,dv. \tag{52.13}$$

The elastic energy in the matrix is the integral $\frac{1}{2}\int (X^C_{ij}e^C_{ij})\,dv$ taken over the remaining volume of the assembly. However, this energy is also the work done on the surface S in setting up the elastic field, and may be written

$$-\tfrac{1}{2}\int_S (X^C_{ij}w^C_i n_j)\,dS = -\tfrac{1}{2}\int_S (X^I_{ij}w^C_i n_j)\,dS.$$

The negative sign corresponds to an outward direction of the positive normal to the element dS, and the two expressions are equal since the tractions and displacements are continuous across S. Use of Gauss's theorem, with the auxiliary conditions (eqns. (11.6)) $\partial X^I_{ij}/\partial x_j = 0$ and $X^I_{ij} = X^I_{ji}$, enables the second surface integral to be transformed into

$$-\tfrac{1}{2}\int_V (X^I_{ij}e^C_{ij})\,dv. \tag{52.14}$$

Adding together (52.13) and (52.14), we find the total strain energy in the β nucleus and α matrix is given by

$$-\tfrac{1}{2}\int_V X^I_{ij}e^T_{ij}\,dv. \tag{52.15}$$

X_{ij}^T is defined by eqn. (52.11), and X_{ij}^C is given by a similar expression, so that X_{ij}^I may be written

$$X_{ij}^I = \lambda(\Delta^C - \Delta^T)\delta_{ij} + 2\mu(e_{ij}^C - e_{ij}^T). \tag{52.16}$$

The quantities e_{ij}^T being specified for a given transformation, the strain energy can be calculated from eqns. (52.15) and (52.16) if w_i^C and hence e_{ij}^C can be found. Eshelby showed that a general expression for w_i^C may be written in terms of the derivatives of two functions which are respectively the Newtonian potential and the biharmonic potential of attracting matter filling the volume V bounded by S. We confine ourselves to an inclusion of ellipsoidal shape, where explicit solutions were possible. The strains e_{ij}^C are then uniform, and depend only on the shape (this result remains true in an anisotropic medium), so that the strain energy per atom may be obtained from (52.15) as

$$\Delta g_s = -\tfrac{1}{2} X_{ij}^I e_{ij}^T v^\beta. \tag{52.17}$$

It is convenient to use the axes of the ellipsoid as coordinate axes, and Eshelby proved that the relation between the constrained and stress-free strains can be written

$$e_{il}^C = s'_{ilmn} e_{mn}^T, \tag{52.18}$$

where the quantities s'_{ilmn} are determined by the values of certain elliptic integrals. Coefficients linking shears to extensions (s'_{1123}, s'_{2221}, etc.) are zero, as also are those connecting two shears (s'_{1223}, etc.), but $s'_{1122} \neq s'_{2211}$, etc. We consider only the simple model in which the ellipsoid is a flat oblate spheroid having axes R, R, y with $R \gg y$. This corresponds to the model of the flat plate precipitate used by Nabarro for incoherent precipitation. The elliptic integrals then reduce to simple forms, and the following results for s'_{ilmn} may be derived (the x_3 axis is directed along y, and ν is Poisson's ratio).

$$\left.\begin{aligned}
s'_{1111} &= s'_{2222} = \frac{(13-8\nu)}{32(1-\nu)}\pi\frac{y}{R},\\
s'_{1122} &= s'_{2211} = \frac{(8\nu-1)}{32(1-\nu)}\pi\frac{y}{R},\\
s'_{1133} &= s'_{2233} = -\frac{(1-2\nu)}{8(1-\nu)}\pi\frac{y}{R},\\
s'_{3333} &= 1-\frac{(1-2\nu)}{4(1-\nu)}\pi\frac{y}{R},\\
s'_{3311} &= s'_{3322} = \frac{\nu}{(1-\nu)}-\frac{(1+4\nu)}{8(1-\nu)}\pi\frac{y}{R},\\
s'_{1313} &= s'_{2332} = \frac{1}{2}-\frac{(2-\nu)}{8(1-\nu)}\pi\frac{y}{R},\\
s'_{1212} &= \frac{(7-8\nu)}{32(1-\nu)}\pi\frac{y}{R}.
\end{aligned}\right\} \tag{52.19}$$

We are now able to calculate Δg_s for various possibilities of physical interest. For example, if the transformation is a pure dilatation, we have from (52.18)

$$\Delta^C = (2\Delta^T/3)(s'_{1111}+s'_{1122}+s'_{1133}+s'_{3311}+\tfrac{1}{2}s'_{3333}) = \frac{(1+\nu)}{3(1-\nu)}\Delta^T.$$

On substituting into (52.16) and eliminating λ, we find

$$\Delta g_s = \frac{2}{9}\frac{(1+\nu)}{(1-\nu)}\mu(\Delta^T)^2 v^\beta. \tag{52.20}$$

This equation does not contain y/R, and is in fact identical with the energy of the misfitting sphere model (25.16) or (52.2) if the β and α regions have the same elastic constants. It may be proved quite generally that (52.20) always gives the strain energy for a uniform dilatation, whatever the shape of the particle, a result due originally to Crum and quoted by Nabarro (1940a). Since the strain energy in a rigid matrix would be $\Delta g_s^\beta = \frac{1}{2}K(\Delta^T)^2 v^\beta$, the accommodation factor is

$$1-C_6 = \frac{4}{9}\frac{(1+\nu)}{(1-\nu)}\frac{\mu}{K} = \frac{2}{3}\frac{(1-2\nu)}{(1-\nu)} \simeq \frac{1}{3}. \tag{52.21}$$

This accommodation factor is identical with that for the incoherent spherical precipitate except that we are now ignoring the differences in elastic properties. In contrast to the behaviour of the incoherent nucleus, however, the accommodation factor for a coherent nucleus stays constant for all shapes if the transformation is a uniform dilatation.

If we now consider other types of transformation, it is physically obvious that good accommodation (low energy) is possible for displacements perpendicular to the face of a flat plate, but not for displacements parallel to the plate. In the most general case, the transformation has principal axes parallel to those of the ellipsoid, but the changes in length are unequal. The transformation is specified by

$$e_{11}^T = \varepsilon_1, \quad e_{22}^T = \varepsilon_2, \quad e_{33}^T = \varepsilon_3,$$

and some rather lengthy algebra leads to the strain energy

$$\Delta g_s = \frac{\mu v^\beta}{(1-\nu)}\{\varepsilon_1^2+\varepsilon_2^2+2\nu\varepsilon_1\varepsilon_2\} - \frac{\mu v^\beta}{32(1-\nu)}\pi\frac{y}{R}\{13(\varepsilon_1^2+\varepsilon_2^2)+2(16\nu-1)\varepsilon_1\varepsilon_2 - 8(1+2\nu)(\varepsilon_1+\varepsilon_2)\varepsilon_3 - 8\varepsilon_3^2\}. \tag{52.22}$$

The limiting value of Δg_s as $c/a \to 0$ is thus finite, except when $\varepsilon_1 = \varepsilon_2 = 0$, and only in this case is complete accommodation possible. On the other hand, when $\varepsilon_3 = 0$, the energy is not greatly reduced below the value $\Delta g_s^\beta = [\mu/(1-2\nu)]\{(1-\nu)(\varepsilon_1^2+\varepsilon_2^2)+2\nu\varepsilon_1\varepsilon_2\}$ which would result from transformation in a rigid matrix. Two special cases are $\varepsilon_2 = 0$, giving a transformation in which the two planes are constrained into coherency by stretching along one atomic direction, and $\varepsilon_1 = \varepsilon_2$, giving a transformation in which the two phases have the same arrangement of atoms on some plane, but slightly different interatomic distances. In the first case, the accommodation factor is

$$1-C_6 = \frac{(1-2\nu)}{(1-\nu)^2} \simeq \frac{3}{4} \tag{52.23}$$

and in the second case it is

$$1-C_6 = \frac{(1+\nu)(1-2\nu)}{(1-\nu)} \simeq \frac{2}{3}. \tag{52.24}$$

Thus for these two types of coherency, the strain energy resides mainly in the nucleus, and change of shape is rather ineffective in reducing the energy. Nabarro (1940a) gave a rough calculation for an anisotropic (cubic) crystal to show that the strain energy of a coherent flat precipitate with $\varepsilon_1 = \varepsilon_2 = \varepsilon_3$ is one-sixth that of a spherical precipitate in which $\varepsilon_1 = \varepsilon_2 = \varepsilon_3 = \varepsilon$. Equation (52.22) shows that in the limit of very small y/R these two energies are identical in the isotropic approximation, as would also be expected from the result that the strain energy of a uniform dilatation is independent of shape.

It is important to note that the strain energy of a coherent nucleus can be very large in comparison with the energies which can usually be stored in a metal by deformation. If the transformation involves a 10% volume change, for example, the strain energy of a uniform dilatation with $\mu = 8{\cdot}10^{10}$ N m^{-2} and $\nu = \frac{1}{4}$ is given by (52.20) as $3{\cdot}10^8$ J m^{-3}, or (say) 400–500 cal mole^{-1}. Strain energies of the same order are obtained for any transformation in which there is a misfit of a few per cent in any direction in the plane of the flat ellipsoid representing the nucleus. These energies are comparable with the chemical free energies which provide the driving force for transformation, and it follows that in all such cases the nucleus must break away to relieve its strain energy at an early stage in transformation. As already emphasized, the existence of such high strain energies and corresponding high internal stresses is only possible because of the small volumes of the particles concerned. Very small coherent particles are important in precipitation reactions in the solid state (Part II, Chapter 16), and they can harden the material very considerably. Because of the magnitude of the strain energy, however, reasonably large precipitates can remain coherent only if macroscopic length changes in the interface are less than about 1%.

If in (52.22) we now put $\varepsilon_1 = \varepsilon_2 = \varepsilon_3 = \Delta^T/3$, we recover eqn. (52.20). A slightly different possibility is the combination of a uniform dilation Δ_1 with a uniaxial strain normal to the plate. Substituting $\varepsilon_1 = \varepsilon_2 = \Delta_1/3$, $\varepsilon_3 = \xi+\Delta_1/3$, we find the strain energy

$$\Delta g_s = \frac{\mu v^\beta}{(1-\nu)}\left\{\frac{2}{9}(1+\nu)\Delta_1^2 + \frac{\pi}{4}\frac{y}{R}\xi^2 + \frac{(1+\nu)}{3}\pi\frac{y}{R}\Delta_1\xi\right\}. \tag{52.25}$$

This expression is important in the theory of martensite nucleation (Christian, 1958).

In the above examples, the principal directions of e_{ij}^T coincided with the axes of the ellipsoid, and the two phases were constrained to coherency by expansions or contractions parallel to these directions. Now consider the strain energy when the β phase is formed from the α phase by a shear transformation in which $e_{13}^T = e_{31}^T = s/2$ are the only non-zero components of e_{ij}^T. This gives

$$\Delta g_s = \tfrac{1}{2}\mu v^\beta(1-2s'_{1313})s^2 \tag{52.26}$$

and the accommodation factor is

$$1-C_6 = 1-2s'_{1313} = \frac{(2-\nu)}{4(1-\nu)}\pi\frac{y}{R} \simeq 2\frac{y}{R}, \tag{52.27}$$

so that the strain energy is small when y/R is very small. For reasonably small y/R, there is good accommodation, and most of the strain energy is contained in the surrounding matrix. Since the coefficients s'_{ilmn} linking shear and non-shear components of e^C_{il} and e^T_{mn} are zero, the strain energy in a transformation in which the change is an invariant plane strain is simply the sum of (52.26) and (52.25) with $\Delta = 0$. This energy tends to zero as y/R decreases, so that the shape change of an invariant plane strain can be accommodated in the matrix with very small energy if y/R is small enough, a result which is, perhaps, physically obvious. Note that if the operative shear components are $e^T_{12} = e^T_{21}$, the corresponding accommodation factor $1-2s'_{1212}$ approaches unity as y/R decreases. There is thus no accommodation when the shear tends to deform the plate in its own plane.

The application of these results to the theory of nucleation may now be made in the same way as before. Suppose, for example, that a coherent nucleus forms in such a way that two crystallographic planes from the two phases are parallel and have similar atomic arrangements, but different atomic spacing. The simplest case is when the atoms are in disregistry along one atomic direction only by an amount ε_1, so that the strain energy when the phases remain completely coherent is

$$\Delta g_s = \mu v^\beta \varepsilon_1^2/(1-\nu). \tag{52.28}$$

The rate of nucleation is then given by an equation similar to (52.8), with the above expression for Δg_s, replacing the strain energy of the incoherent particle. Obviously, Δg_s is much larger for coherent precipitation, but the surface energy σ is correspondingly smaller. When the nucleus is small, it is probable that the surface term is predominant, and coherent nucleation is favoured. Nevertheless, the condition that the most favourable nucleation path be chosen also requires consideration of the intermediate case of semi-coherent nucleation. As a model for this, we may assume that the strain in the β lattice in the x_1 direction is ε'_1 instead of ε_1, the remaining difference in length of $\varepsilon_1-\varepsilon'_1$ being obtained by an array of dislocations of density proportional to $\varepsilon_1-\varepsilon'_1$. This introduces an extra surface energy term, which may be taken to be proportional to the dislocation density in a first approximation, so that for semi-coherent nucleation, the surface free energy per unit area becomes

$$\sigma = \sigma' + C_{11}(\varepsilon_1-\varepsilon'_1). \tag{52.29}$$

At the same time, the accommodation strain energy is reduced to

$$\Delta g_s = \mu v^\beta (\varepsilon'_1)^2/(1-\nu). \tag{52.30}$$

It is clear that a semi-coherent nucleus will form if it is cheaper in energy to accommodate some of the elastic misfit by the introduction of dislocations into the surface region. Equation (52.8) gives the energy of this semi-coherent nucleus if (52.30) is used in place of (52.5) for the strain energy and (52.29) is substituted for σ. The saddle point free energy change is now defined so that ΔG should be a maximum with respect to changes in size and a minimum with respect to changes in shape and degree of coherency. This gives the auxiliary conditions

$$\left.\begin{aligned} [\partial \Delta G/\partial n]_{y/R,\, \varepsilon_1'} &= 0, \\ [\partial \Delta G/\partial (y/R)]_{n,\, \varepsilon_1'} &= 0, \\ [\partial \Delta G/\partial \varepsilon_1']_{n,\, y/R} &= 0, \end{aligned}\right\} \tag{52.31}$$

which together define the combination of size, shape, and degree of coherency for the critical nucleus.

Essentially equivalent considerations are involved when the misfit is not confined to a single atomic direction. In principle, the free energy for the most general case of a semi-coherent precipitate will be a function of n, y/R, and three parameters ε_1', ε_2', ε_3' representing the degree of coherency. Very detailed data would be required to establish the favoured mode of nucleation in any particular case, since a knowledge of the anisotropy of both elastic properties and surface energy would be needed.

We have already emphasized that real nucleation in the solid state, as in other assemblies of atoms, is nearly always heterogeneous. It is, indeed, rather more difficult to obtain homogeneous solid nucleation, since the production of crystals nearly free from defects is harder than the production of assemblies nearly free from impurities; some results have, however, been achieved by the small particle method. The effectiveness of nucleation catalysts in reducing the surface energy term has already been fully discussed, but there are extra considerations in the solid state. As mentioned on p. 459, certain preferred sites, in particular grain boundaries, may be able to accommodate the shape and volume changes by flow at much lower temperatures than is possible in a region of nearly perfect lattice. The reduction of the Δg_s term to nearly zero then gives a lower critical free energy of nucleation, and a correspondingly more rapid nucleation rate.

On p. 459, we gave reasons for believing that single dislocations or Frank–Read sources would be ineffective in causing flow of this kind. Nevertheless, a large number of experimental observations show conclusively that dislocations do act as preferred sites for the formation of nuclei. Most results are for precipitation from solid solutions, where a composition change has to be produced, and dislocation sites are presumably favoured because the Cottrell atmosphere effect makes them suitable centres for segregation of solute atoms. However, there should be an analogous effect even in a pure component; solute atoms are attracted to a dislocation because they can lower its strain energy, and the same result can be obtained if the atoms around the dislocation line are rearranged into some new stable pattern.

A theory of nucleation on dislocations has been given by Cahn (1957). He assumes that the nucleus lies along the dislocation and has a circular section perpendicular to the dislocation line. The radius of the section is not constant, but varies with distance along the line, so that the longitudinal section if the nucleus is approximately as shown in Fig. 10.17. In addition to the usual volume and surface energy terms in the expression for the energy of formation of a nucleus of given size, there is a term representing the strain energy of the dislocation in the region now occupied by the new phase. In effect, the atomic rearrangement

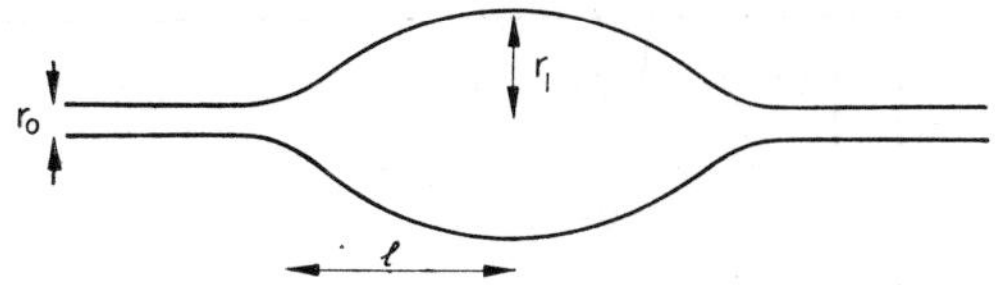

FIG. 10.17. The formation of a nucleus on a dislocation line (after Cahn, 1957).

within the nucleus is assumed to destroy that part of the elastic energy of the dislocation located in the volume of the nucleus, and the energy thus gained is available to help the nucleation process.

Suppose a small length of the nucleus is effectively a cylinder of radius r centred on the dislocation. The change in free energy per unit length when the nucleus is formed is given by

$$\Delta G = \pi r^2(g^\beta - g^\alpha)/v^\beta + C_{12} - \tfrac{1}{2}Bb \ln r + 2\pi\sigma r, \tag{52.32}$$

where the first term is the (negative) volume free energy change, the term $C_{12} - \frac{1}{2}Bb \ln r$ (see eqns. (30.16) and (30.18)) represents the dislocation energy within the radius r, and the last term is the positive surface energy. The constant B varies between values B_e for an edge dislocation, and B_s for a screw dislocation. If the transformation in a region of good α crystal involves an appreciable strain energy Δg_s, this should be added to the first term, as before.

The important difference between eqn. (52.32) and previous expressions for the change in energy on forming a nucleus is that there are two negative terms, the chemical free energy change, which becomes predominant at large values of r, and the dislocation strain energy released, which is most important at small values of r. At intermediate radii, the positive surface energy may force an increase in ΔG, but this does not necessarily happen. Differentiating the equation, and equating $\partial \Delta G/\partial r = 0$ gives

$$r = \frac{\sigma v^\beta}{2(g^\alpha - g^\beta)}\left[1 \pm \left\{1 - \frac{(g^\alpha - g^\beta)Bb}{\pi v^\beta \sigma^2}\right\}^{1/2}\right]. \tag{52.33}$$

The behaviour thus depends on whether the quantity $\alpha^D = (g^\alpha - g^\beta)Bb/\pi v^\beta \sigma^2$ is greater or less than one. If $\alpha^D > 1$, there are no turning points in the $\Delta G - r$ relation, and the energy of the whole nucleus (whatever its exact shape) decreases continually as r increases. If $\alpha^D < 1$, there is a minimum free energy at a value $r = r_0$ corresponding to the negative sign in eqn. (52.33), followed by a maximum at a value of r corresponding to the positive sign. These two possibilities are shown in Fig. 10.18. As α^D increases from 0 to 1, the value of r_0 changes only by a factor of two, from $Bb/4\pi\sigma$ to $Bb/2\pi\sigma$.

When the chemical free energy and dislocation energy factors are sufficiently large in comparison with the surface energy, there is thus no energy barrier to nucleation on dislocations, and transformation rates will be governed only by growth conditions. When the free energy curve has the alternative form A (Fig. 10.18), there will be a sub-critical metastable cylinder of the β phase surrounding the dislocation line to the radius of the minimum in the curve, and this is roughly analogous to the Cottrell atmosphere of solute atoms in a segregation problem. When the two phases are in equilibrium, $\alpha^D = 0$, and the maximum

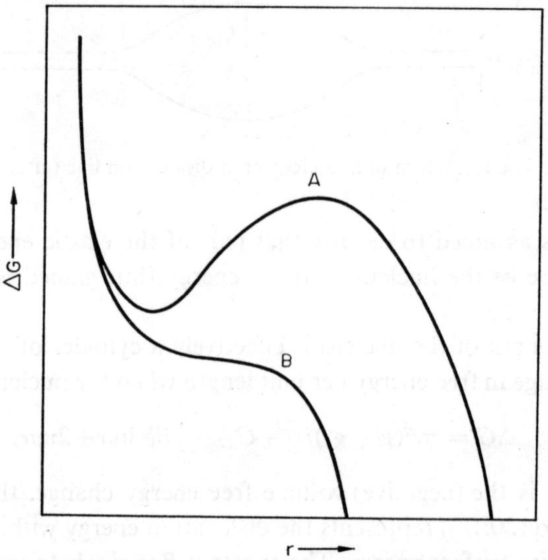

FIG. 10.18. Schematic forms for the variation of the free energy of formation of a dislocation nucleus with its radius (after Cahn, 1957). Curve A, $\alpha^D < 1$. Curve B, $\alpha^D > 1$.

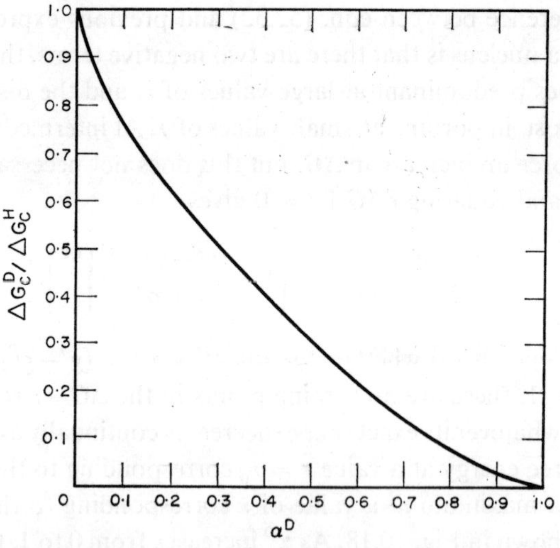

FIG. 10.19. The ratio of the free energy required to form a nucleus on a dislocation to that required to form a homogeneous spherical nucleus is plotted as a function of the parameter α^D (after Cahn, 1957).

in the free energy change of eqn. (52.32) is infinite, as for homogeneous nucleation. As the β phase becomes relatively more stable, α^D increases until the nucleation rate becomes appreciable; it is shown below that this usually requires α^D to be in the range 0·4–0·7. The situation in curve B (Fig. 10.18) will only be obtained if the conditions are altered so rapidly that α^D exceeds unity before appreciable transformation has occurred.

Equation (52.32) and the curves of Fig. 10.18 refer only to the energy per unit length of a small element of the nucleus. The energy of the whole nucleus will be obtained by integrating along the effective length of the nucleus, shown in Fig. 10.17. To find the nucleation rate, the combination of size and shape which give the saddle point energy change for the whole nucleus is required. Cahn solved this problem by applying the appropriate boundary conditions; the critical nucleus (Fig. 10.17) is defined in terms of a maximum radius r_1 and a measure of its effective length l. His results are shown in Fig. 10.19, in which the ratio of the critical free energy of nucleation to the corresponding free energy of a homogeneous spherical nucleus of critical size is plotted against the functions α^D.

Let ΔG_c^D be the critical free energy of nucleus formation on a dislocation. Then the total nucleation rate will be given by eqn. (49.10), replacing ΔG_c by ΔG_c^D and N by N^D, the total number of atoms on dislocation lines. More usefully, the nucleation rate per unit volume, ${}^vI^D$, will be given by

$$\begin{aligned} {}^vI^D &= {}^vN^D (kT/h) \exp\{-(\Delta G_c^D + \Delta_a g^{\ddagger})/kT\} \\ &\approx ({}^vN)^{1/3}\, {}^vL^D (kT/h) \exp\{-(\Delta G_c^D + \Delta_a g^{\ddagger})/kT\}, \end{aligned} \tag{52.34}$$

where ${}^vN^D$ is the number of atoms per unit volume which are on dislocation lines, and ${}^vL^D$ is the dislocation density, that is the length of dislocation line per unit volume. The second expression for ${}^vI^D$ is correct only if dislocations are regarded essentially as lines one atom diameter in cross-section; more realistically, it should be multiplied by a small numerical factor, say about five. As always, the pre-exponential term in the nucleation rate is ill-defined, but relatively unimportant to one or two orders of magnitude. We see that the volume nucleation rate is linearly proportional to the dislocation density, and the condition for the nucleation rate to become appreciable (${}^vI^D \sim 10^6\ \text{m}^{-3}\ \text{s}^{-1}$) is, for ${}^vL^D$ in m^{-2},

$$\Delta G_c^D + \Delta g^{\ddagger} = kT(39 + \ln {}^vL^D). \tag{52.35}$$

The function $f(\alpha^D)$ shown in Fig. 10.19 gives the ratio of ΔG_c^D to ΔG_c of eqn. (46.19), so that in terms of this function, condition (52.35) becomes

$$\frac{1}{(\alpha^D)^2} f(\alpha^D) = \frac{3\pi\sigma kT}{16(Bb)^2}\left[39 + \ln {}^vL^D - \frac{\Delta_a g^{\ddagger}}{kT}\right]. \tag{52.36}$$

With typical values of ${}^vL^D = 10^{12}\ \text{m}^{-2}$, $\Delta_a g^{\ddagger} \simeq 10kT$, this gives $(1/\alpha^D)^2 f(\alpha^D) \simeq 34\sigma kT/(Bb)^2$, and the numerical factor is rather insensitive to the actual value of ${}^vL^D$ and $\Delta_a g^{\ddagger}/kT$. A graph of $(1/\alpha^D)^2 f(\alpha^D)$ against α^D thus gives the relation between $34\sigma kT/(Bb)^2$ and α^D needed to ensure rapid nucleation. Reasonable values of B and σ give $34\sigma kT/(Bb)^2 \simeq 1$ at 1000 K, so that for most transformations of practical interest $(1/\alpha^D)^2 f(\alpha^D)$ probably falls in the range 0·5–5, corresponding to α^D values between 0·4 and 0·7. From the value of α^D, the required driving force $g^\alpha - g^\beta$ may be estimated.

It is interesting to compare the nucleation rate on dislocations, given by (52.34), with the corresponding homogeneous nucleation rate. Using $\sigma = 0{\cdot}2\ \text{J m}^{-2}$ (200 ergs cm^{-2}), and a value of $g^\alpha - g^\beta$ corresponding to $\alpha \simeq 0{\cdot}6$, the homogeneous nucleation rate given by eqn. (49.10) may be estimated at $10^{-64}\ \text{m}^{-3}\ \text{s}^{-1}$, whereas the nucleation rate on dislocations under these conditions is 10^{+14} m^{-3} s^{-1}. Nucleation would thus be entirely confined to dislocations if the assumptions of the model are correct. The assumptions are physically self-consistent, since r_0, r_1 have the very reasonable values 0·2 nm, 1 nm (2 Å, 10 Å) when $\alpha^D \simeq 0{\cdot}6$. Since the ratio $\Delta G_c^D/\Delta G_c^H$ decreases as α^D increases, nucleation is easier on dislocations with large Burgers vector. Also, since α^D is proportional to $g^\alpha - g^\beta$, the rate of change of dislocation nucleation rate with increasing $g^\alpha - g^\beta$ (i.e. with temperature or supersaturation) must be even more rapid than that of the homogeneous nucleation rate.

In Cahn's calculation, the assumption that the whole of the strain energy of the dislocation within the volume occupied by the nucleus can be relaxed to zero requires that the nucleus be incoherent. For a coherent nucleus forming on or near dislocations, this assumption cannot be made; instead it is necessary to calculate the elastic interaction energy between the nucleus and the matrix. A calculation of this type for nuclei in which e_{ij}^T corresponds to a pure dilatation and the dislocation is edge in character has been made by Dollins (1970) and improved by Barnett (1971). The problem for a spherical nucleus with its centre at r, θ, z with respect to the dislocation line is clearly identical with the treatment given in Section 30; Barnett used the correct (Eshelby) formulation and obtained for the interaction energy in the case when nucleus and matrix have the same elastic properties

$$W_i = \tfrac{2}{3}B_e(1+\nu)\,\Delta^T v \sin\theta/r, \tag{52.37}$$

which is equivalent to (30.41) with the misfit volume replaced by $\Delta^T v$, where Δ^T is the cubical dilatation and v is the unconstrained nucleus volume. This same equation also represents the interaction energy for a coherent nucleus in the form of an ellipsoid of revolution if the stress-free strain is a pure dilatation, and it may therefore also be applied to a disc-like precipitate.

When the nucleus has different elastic properties from the matrix, the interaction energy is more difficult to estimate. The energy may be divided into two parts

$$W_i - W_i^{(1)} + W_i^{(2)}, \tag{52.38}$$

where $W_i^{(1)}$ results from the interaction of the stress field of the dislocation with the strain field of the nucleus, as above, whereas $W_i^{(2)}$ is the so-called modulus effect in which, for example, a region of higher elastic modulus increases the self-energy of the dislocation.

In the case of a sphere, an exact calculation is possible and yields

$$W_i^{(1)} = (\mu^\alpha b/\pi)\{3K^\beta/(3K^\beta + 4\mu^\alpha)\}\,\Delta^T v \sin\theta/r \tag{52.39}$$

where μ^α, K^β are elastic stiffness of the matrix and the nucleus respectively. This equation is equivalent to eqns. (30.42) and (25.8) with ΔV_{AB} in the latter replaced by $\Delta^T v$. It differs from (52.37) only by the factor

$$\{\varkappa^\alpha + (K^\alpha/K^\beta)(1-\varkappa^\alpha)\}^{-1} \tag{52.40}$$

where

$$\varkappa^\alpha = (1+\nu^\alpha)/3(1-\nu^\alpha)$$

When $K^\alpha = K^\beta$, the second bracketed term in (52.39) reduces to $\varkappa^\alpha$, and (52.39) becomes equivalent to (52.37).

The modulus effect was estimated by Dollins and Barnett by making the assumption that the strain field of the dislocation is not changed by the nucleus. Barnett gives for the ellipsoid of revolution of semi-axes R, R, y.

$$W_i^{(2)} = \frac{(\mu^\beta - \mu^\alpha)b^2 y}{2\pi(1-\nu^\alpha)}\{1-(\beta^2-1)^{1/2}\tan^{-1}(\beta^2-1)^{-1/2}\} \tag{52.41}$$

where $\beta = r/R > 1$. However, the approximation used to calculate $W_i^{(2)}$ is generally not a good one, since the strain field of the dislocation inside the nucleus is changed appreciably from its matrix value when the nucleus has different elastic properties.

In general $W_i^{(1)}$ can always be made negative by suitable choice of sin θ; i.e. the nucleus can form on the compression or tension side of the edge dislocation, depending on the sign of the volume change. The sign of $W_i^{(2)}$ depends on whether μ^β is greater or smaller than μ^α. Nucleation will occur preferentially in the vicinity of the nucleus provided W_i is negative, and the net driving force is then increased. In contrast to Cahn's theory, however, the nucleation barrier ΔG_c may be reduced but not eliminated by this type of dislocation-catalysed nucleation.

Since a screw dislocation has no hydrostatic stress field in first order approximation, nucleation of coherent precipitates near screws will not be expected when, as assumed above, the shape change is a pure dilatation. In the case of martensitic transformations, however, the stress free strains correspond mainly to shears, and preferential regions for nucleation may thus be found near to screw dislocations. A treatment of this problem would be similar to that given for the interaction of an interstitial defect with a screw dislocation; see eqn. (30.43).

Gomez–Ramirez and Pound (1973) have developed a model for nucleation on dislocations which is in some respects intermediate between Cahn's model and the model just described. It is assumed that the nucleus will form along the dislocation because of the importance of the core energy, but an Eshelby-type calculation is used to estimate the reduction in strain energy. Instead of Cahn's assumption that the elastic energy within the nucleus is released, it is supposed that for an incoherent nucleus the dislocation along the centre of the nucleus is replaced by a distribution of infinitesimal dislocations over its surface. The energy change resulting from formation of the nucleus along the dislocation rather than in the defect-free lattice then contains three terms, namely:

(1) the core energy, assumed to be uniformly distributed over the volume of the core;
(2) the elastic interaction energy, calculated as above, except that since sin θ/r varies rapidly it is necessary to take the integral of $X_{ij}^d e_{ij}^T$ over the volume of the embryo;
(3) the change in elastic energy caused by "spreading" the Burgers vector into the interface.

For the particular case of a screw dislocation and a cylindrical embryo, the elastic energy of the smeared out dislocations is calculated to be fortuitiously equal to the elastic energy of the original dislocation outside the embryo, and this is assumed to be generally a good estimate. Thus the contribution (3) is taken to be the negative of the elastic energy of the dislocation, integrated over the volume of the embryo, but excluding the core region. Note that this corresponds fairly closely to Cahn's assumption.

With this model it was found impossible to minimize ΔG analytically, but numerical calculation showed the energy barrier to be only slightly shape dependent. Numerical calculations were therefore carried out for assumed smooth shapes and isotropic surface free energies. Embryo shapes along screw dislocations are predicted to be similar to prolate spheroids with eccentricity $\leqslant 0{\cdot}83$, whilst similar closed shapes but with heart-shaped cross-sections to allow for the asymmetry of the dislocation field are proposed for edge dislocations. Metastable embryos form along dislocations in both stable and metastable α regions if the volume misfit is small and the shear modulus is sufficiently high; but the automatic formation of metastable tubes of β phase, as in Cahn's theory, is not predicted. (This difference presumably arises from different estimates of the core energy density, or of the effective cut off radius in the elastic calculation.)

As would be expected for a model in which the strain field of the embryo is a pure dilatation, the critical free energy of formation on an edge dislocation ΔG_c^{De}, is found to be smaller than the corresponding energy on a screw dislocation ΔG_c^{Ds}. However, in all cases

$$\Delta G_c^H - \Delta G_c^{Ds} > \Delta G_c^{Ds} - \Delta G_c^{De},$$

and with reasonable values of dislocation density the nucleation rates per unit volume were greatest for edge and smallest for homogeneous nucleation. Thus for the transformation in pure iron at a driving force of $\sim 4{\cdot}8 \times 10^8$ J m^{-3} (115 cal cm^{-3}) with an assumed interfacial free energy of $0{\cdot}24$ J m^{-2}, the authors calculate ΔG_c^H, ΔG_c^{Ds}, and ΔG_c^{De} as 7·1, 1·6, and 0·9 eV respectively. If the dislocation density is $\sim 10^{10}$ m^{-2}, the corresponding nucleation rates are $10^{5\cdot5}$, $10^{24\cdot5}$, and $10^{26\cdot8}$ m^{-3} s^{-1} respectively. It follows that nuclei will form first on dislocations of predominantly edge character, and that these may effect the whole transformation at low driving forces; at high driving forces, however, nucleation should take place on all types of dislocations and possibly also homogeneously in the matrix.

REFERENCES

ABRAHAM, F. F. (1969) *J. Chem. Phys.* **51,** 1632.
BARNETT, D. M. (1971) *Scripta metall.* **5,** 261.
BECKER, R. (1940) *Proc. Phys. Soc.* **52,** 71.
BECKER, R. and DÖRING, W. (1935) *Ann. Phys.* **24,** 719.
BENSON, G. C. and SHUTTLEWORTH, R. (1951) *J. Chem. Phys.* **19,** 130.
BUFF, F. P. (1951) *J. Chem. Phys.* **19,** 1591.
BURTON, J. J. (1973) *Acta metall.* **21,** 1225.
CAHN, J. W. (1956) *Acta metall.* **4,** 449; (1957) *Ibid.* **5,** 168.
CARSLAW, H. S. and JAEGER, J. C. (1947) *Conduction of Heat in Solids*, p. 82, Clarendon Press, Oxford.
CHRISTIAN, J. W. (1958) *Acta metall* **6,** 377.
CLEMM, P. J. and FISHER, J. C. (1955) *Acta metall.* **3,** 70.
COURANT, R. and HILBERT, D. (1953) *Methods of Mathematical Physics*, p. 76, Interscience, New York.
DAWSON, D. B., WILSON, E. J., HILL, P. G., and RUSSELL, K. C. (1969) *J. Chem. Phys.* **51,** 5389.
DOLLINS, C. C. (1970) *Acta metall.* **18,** 1209.
DUNNING, W. J. (1965) *Symposium on Nucleation*, p. 1, Case Institute of Technology, Cleveland, Ohio; (1969) *Nucleation*, p. **1,** Dekker, New York.
ESHELBY, J. D. (1957) *Proc. R. Soc.* A, **241,** 376; (1961) *Prog. Solid Mech.* **2,** 89.

FARKAS, L. (1927) *Z. phys. Chem.* **125**, 239.
FEDER, J., RUSSELL, K. C., LOTHE, J., and POUND, G. M. (1966) *Adv. Phys.* **15**, 111.
FISHER, J. C., HOLLOMON, J. H., and TURNBULL, D. (1948) *J. Appl. Phys.* **19**, 775.
FRENKEL, J. (1939) *J. Chem. Phys.* **1**, 200, 538; (1946) *Kinetic Theory of Liquids*, Oxford University Press.
GOMEZ-RAMIREZ, R. and POUND, G. M. (1973) *Metall. Trans.* **4**, 1563.
HILLIG, W. B. (1962) *Symposium on Nucleation and Crystallisation in Glasses and Melts*, p. 77, American Ceramic Society.
HIRTH, J. P. and POUND, G. M. (1963) *Prog. Mater. Sci.* **11**, 1.
HOLLOMON, J. H. and TURNBULL, D. (1953) *Prog. Metal Phys.* **4**, 333.
JAEGER, H. L., WILSON, E. J., HILL, P. G., and RUSSELL, K. C. (1969) *J. Chem. Phys.* **51**, 5380.
KANTROWITZ, A. (1951) *J. Chem. Phys.* **19**, 1097.
KIRKWOOD, J. G. and BUFF, F. P. (1949) *J. Chem. Phys.* **17**, 338.
KRÖNER, E. (1954) *Acta metall.* **2**, 301.
KUHRT, F. (1952) *Z. Phys.* **131**, 185.
LIN, J. (1968) *J. Chem. Phys.* **48**, 4128.
LOTHE, J. and POUND, G. M. (1962) *J. Chem. Phys.* **36**, 2080; (1966) *Ibid.* **45**, 630; (1969) *Nucleation*, p. 109, Dekker, New York.
NABARRO, F. R. N. (1940a) *Proc. Phys. Soc.* **52**, 90; (1940b) *Proc. R. Soc.* A, **175**, 519.
NISHIOKA, K., SHAWYER, R., BIENENSTOCK, A., and POUND, G. M. (1971) *J. Chem. Phys.* **55**, 5082.
PROBSTEIN, R. F. (1951) *J. Chem. Phys.* **19**, 619.
REISS, H. (1952) *Ind. Eng. Chem.* **44**, 1284.
REISS, H. and KATZ, J. L. (1967) *J. Chem. Phys.* **46**, 2496.
REISS, H., KATZ, J. L., and COHEN, E. R. (1968) *J. Chem. Phys.* **48**, 5553.
RODEBUSH, W. H. (1952) *Ind. Eng. Chem.* **44**, 1289.
RUSSELL, K. C. (1968) *Acta metall.* **16**, 761; (1969) *Ibid.* **17**, 1123.
TOLMAN, R. C. (1949) *J. Chem. Phys.* **17**, 333.
TURNBULL, D. (1948) *Trans. Am. Inst. Min. (Metall.) Engr.* **175**, 774; (1950) *J. Chem. Phys.* **18**, 198.
TURNBULL, D., and FISHER, J. C. (1949) *J. Chem. Phys.* **17**, 71.
VOLMER, M. (1929) *Z. Elektrochem.* **35**, 555; (1939) *Kinetik der Phasenbildung*, Steinkopf, Dresden.
VOLMER, M. and FLOOD, H. (1934) *Z. phys. Chem.* **170A**, 273.
VOLMER, M. and WEBER A, (1926) *Z. phys. Chem.* **119**, 227.
ZELDOVICH, J. B. (1943) *Acta, Physicochim. URSS* **18**, 17.

CHAPTER 11

Theory of Thermally Activated Growth

53. GROWTH CONTROLLED BY PROCESSES AT THE INTERFACE

This chapter is concerned with the descriptive theory of growth in condensed phases. In Chapter 1, we classified transformations by considering the operative growth mechanism, and we shall now discuss the various types of thermally activated growth in more detail. Martensitic growth is treated separately in Part II, Chapter 21, and growth in liquid–solid transformations, which is usually controlled by heat flow, is discussed in Part II, Sections 66 and 67. The theory of the growth of a crystal from the vapour, which is more highly developed than are other examples of interface controlled growth, is also described separately in Part II, Chapter 13.

We begin by considering the type of growth which is controlled by processes in the immediate vicinity of the interface. For convenience we may suppose that there is a negligible change of volume and no change of composition as the boundary moves; this is nearly true for many transformations, and even in interface controlled precipitation processes, the boundary conditions of the diffusion equation may be such that the composition gradients can be virtually ignored.

There are two possible ways in which an incoherent crystal boundary might migrate normal to itself. In the first of these, atoms are able to cross the interface, adding themselves to the crystalline region on one side of the interface, simultaneously and independently at all points of the interface. This means that there is continuous growth at all points of the boundary. In the second mechanism, the interface is stepped on an atomic scale, and atoms are transferred from one phase to the other only at these steps. The interface then grows by the lateral motion of the steps, an element of surface undergoing no change until a step passes over it, when it moves forward through a distance equal to the step height.

In discussing the free surface of a crystal we have already pointed out that singular surfaces (surface orientations corresponding to sharp minima of surface free energy) will usually be atomically flat, whereas other surfaces will consist of atomic facets, or stepped sections of singular surfaces. Isolated atoms adding to a crystal on one side of a singular interface will be unstable and will tend to be removed again; the motion of such an interface thus requires a step mechanism, as we discuss in detail in Part II, Chapter 13. On the other hand, it is usually considered that continuous growth of a non-singular interface is possible, provided the atomic disordering (faceting) is sufficiently extensive for atoms to be able to find a stable position in all parts of the interface.

The distinction between stepped and continuous growth is sometimes made on the basis of the diffuseness of the interface, i.e. the degree of atomic disorder and the extent of the transition region, rather than on the grounds of the singularity of the surface free energy. Of course, as we have pointed out in Chapter 5, these two criteria often go together, inasmuch as a singular interface is not disordered and is likely to be atomically sharp, whilst a non-singular interface is much more likely to be diffuse. However, Cahn (1960) has pointed out that although similar, the two distinctions are not necessarily equivalent, and he has derived a more general condition for deciding whether growth will be continuous or stepped. This criterion depends on the driving force (difference in free energy of the bulk phases on the two sides of the interface).

In introducing the idea of a step in Section 18, we assumed the interface to be sharp. However, a more general definition of a step may be given as the transition between two adjacent areas of surface which are parallel to each other and have identical atomic configurations, and which are displaced from each other by an integral number of lattice planes. This includes steps in diffuse interfaces, as already briefly mentioned for a special case on p. 292, even though the step height may be less than the thickness of the interface.

Cahn's theory of growth is based on the idea that a lateral growth (step) mechanism will be required whenever the interface is able to attain a metastable equilibrium configuration in the presence of the driving force. It will then tend to remain in such a configuration, advancing only by the passage of a step which does not change the configuration. If no such metastable equilibrium is possible, the boundary will move forwards continuously. This is thus a rather general thermodynamic criterion for growth mechanism, in contrast to the more atomistic theory based on the nature of the interface. The application of the driving force distinction, however, brings out the importance of the interface structure, which was previously recognized in a semi-intuitive way.

We shall not reproduce the details of Cahn's calculations here, but only quote the main results. For all types of interface, continuous growth is possible at a sufficiently large driving force, but stepped growth is required for driving forces less than some critical value. The magnitude of the critical driving force is very dependent on the nature of the interface; for very diffuse interfaces, it is so low that almost any driving force will lead to continuous growth, whereas for sharp interfaces it may be so high that it is never achieved in practice. Although the intuitive feeling that non-singular interfaces should always be able to grow continuously is not quite justified, the driving force required for such growth may be less than that attributable to the deviation of the actual crystal shape from the equilibrium (Wulff) shape. In other cases, it is possible that non-singular interfaces may grow by a step mechanism.

Let the position of the interface be given relative to some fixed lattice plane by a parameter z, and let the surface free energy of the interface as a function of z be

$$\sigma(z) = \{1+\varphi(z)\}\sigma_0, \tag{53.1}$$

where σ_0 is the minimum value of σ. Now let the interface move forward through a distance δz, so that the change in free energy per unit area of boundary is

$$\delta G = \left[\left((\Delta g^{\alpha\beta}/v)+\sigma_0 \frac{d\varphi}{dz}\right)\right] \delta z, \tag{53.2}$$

where $-\Delta g^{\alpha\beta}$ is the driving force per atom and v is the atomic volume. The step growth mechanism is required only if for some value of z the change in free energy is zero. If the interface region contains a large number of lattice planes, Cahn shows that the maximum value of $\sigma_0(d\varphi/dz)$ is $\pi\sigma_0\varphi_{max}/d$, where φ_{max} is the maximum value of φ and d is the interplanar spacing of the lattice parallel to the interface. If the negative of $\Delta g^{\alpha\beta}/v$ is greater than this maximum variation in surface free energy, the interface should be able to advance continuously, since equilibrium configurations will not be attained. This gives a condition

$$\Delta g^{\beta\alpha}/v = -\Delta g^{\alpha\beta}/v > \pi\sigma_0\varphi_{max}/d \tag{53.3}$$

for continuous growth. For diffuse interfaces, n lattice planes in extent, Cahn shows that the function $\varphi(z)$ is given approximately by

$$\varphi(z) = (\pi^4 n^3/16)(1-\cos 2\pi z/d)\exp(-\tfrac{1}{2}\pi^2 n) \tag{53.4}$$

and is very small when n is large. For very sharp interfaces, the maximum value of φ is of order unity, and this corresponds to very high critical driving forces according to the inequality (53.3).

Step growth will be discussed in detail in Part II, Chapter 13; for the remainder of this section we consider continuous growth on an atomic scale. A plausible model is one in which atoms cross the boundary independently of each other, each atom having to surmount an energy barrier in order to do this. In effect, we have already treated this problem in discussing the theory of nucleation in condensed assemblies (Section 49).

Consider first a phase transformation, in which a β region is growing into a metastable α region. Figure 11.1 shows schematically the variation in free energy as an atom transfers from the α phase to the β phase; the question of whether the activation energy $\Delta_a g^*$ is to

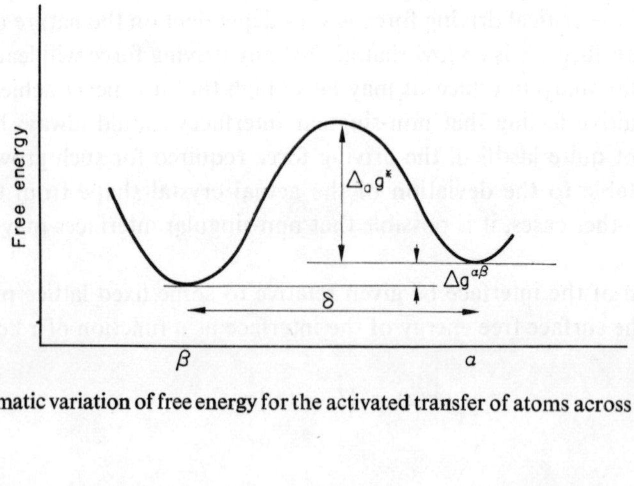

FIG. 11.1. Schematic variation of free energy for the activated transfer of atoms across the α–β interface.

be identified with the energy $\Delta_a g^{\ddagger}$ of Section 49 is discussed below. In accordance with the theory developed in Chapter 3, the frequency with which an individual atom will transfer from the α phase to the β phase is

$$\nu \exp(-\Delta_a g^*/\boldsymbol{k}T),$$

where ν is a characteristic frequency, and is given the value $\boldsymbol{k}T/\boldsymbol{h}$ in Eyring's theory. As already explained, it is preferable to regard ν as a frequency of known order of magnitude, the exact value of which is to be determined. This enables a straightforward physical interpretation to be given to both $\Delta_a g^*$ and to ν.

The frequency of the reverse transition of atoms from β to α is similarly

$$\nu \exp\{-(\Delta_a g^* + \Delta g^{\beta\alpha})/\boldsymbol{k}T\},$$

where, for simplicity, the frequency factors are assumed to be identical. The difference of these two expressions gives the net rate at which atoms transfer from the α to the β phase. Thus if the distance across the interface is δ^B, the velocity of the interface is

$$\Upsilon = \delta^B \nu \exp(-\Delta_a g^*/\boldsymbol{k}T)\,[1-\exp(-\Delta g^{\beta\alpha}/\boldsymbol{k}T)]. \tag{53.5}$$

The formal similarity of this equation to the net rate of a chemical reaction (eqn. (15.11)) should be noted. At small values of the driving energy, we may expand the equation to give

$$\Upsilon \simeq (\delta^B \nu/\boldsymbol{k})(\Delta g^{\beta\alpha}/T)\exp(-\Delta_a g^*/\boldsymbol{k}T) \tag{53.6}$$

and the growth velocity is directly proportional to the difference in free energy of the two phases. The validity of this equation is limited to conditions where $\Delta g^{\beta\alpha} \ll \boldsymbol{k}T$, a restriction which is usually satisfied in practice.

We have derived eqn. (53.6) by considering a specific model, but its form suggests that a much more general treatment in terms of the theory of Chapter 4 should be possible. Such a theory has been proposed by Machlin (1953), who pointed out that the motion of an interface separating two regions of different free energy may be regarded as a single irreversible process. The thermodynamical theory of irreversible processes can then be applied to the growth.

Consider a spherical element of the interface, having area O and radius of curvature r. If this moves normal to itself through a distance δr, the free energy change will be

$$\begin{aligned}\delta G &= (O\,\Delta g^{\alpha\beta}/v)\,\delta r + \sigma\,\delta O \\ &= -O\,\delta r\{\Delta g^{\beta\alpha}/v) - (2\sigma/r)\},\end{aligned}$$

where v is the volume per atom and σ the surface energy of unit area of the interface. In accordance with the theory developed in Section 15, we require an expression for the rate of production of irreversible entropy as the boundary advances. This is given by $-(1/T)(\partial G/\partial t)$, so that the irreversible entropy per unit area of the boundary increases at a rate

$$\frac{\mathrm{d}_i S}{\mathrm{d}t} = \frac{1}{T}\left\{\frac{\Delta g^{\beta\alpha}}{v} - \frac{2\sigma}{r}\right\}\Upsilon. \tag{53.7}$$

This equation has the expected form of a "force" and a conjugate "flux" Υ, and the theory assumes that two such quantities are linearly related, so that

$$\Upsilon = (C_{13}/T)\{\Delta g^{\beta\alpha}/v) - (2\sigma/r)\}. \tag{53.8}$$

For a plane interface, a suitable choice of C_{13} gives eqn. (53.6). For a curved interface we see that we should strictly include the change in surface energy in the overall free energy change used as the "driving force" for the motion. This is usually unimportant when r becomes large, but is always significant in grain growth, where it provides the only driving force.

The impression that the above theory gives a rather more general justification of the assumption that the boundary velocity is proportional to the free energy released when unit area of the boundary moves through unit distance, must be regarded as misleading. There is no necessity for the force and flux of eqn. (53.7) to be linearly related, and the validity of such a relation can only be settled experimentally. Eqn. (53.6) was derived on the basis of a particular atomic model, and will be correct if the real atomic processes approximate to those envisaged in the model. The corresponding equation (53.8) is justified only on the rather vague grounds that there are many circumstances in which the rate of entropy production is given by an equation like (53.7), and in which there is a proportionality between the "force" and the "flux". Quite evidently, there are other atomic models which do not give this linear relation between velocity and driving force. For example, Hillig and Turnbull (1956) considered as a very simple approximation that for step growth centred on dislocations (see Part II, Chapter 13) eqn. (53.6) should be multiplied by the fraction of boundary sites which are at growth steps. In a spiral step, the minimum radius of curvature cannot be smaller than the critical size of a two-dimensional nucleus, so that the total length of step in unit area is a function of the driving force. Thus in this model, the growth velocity is proportional to the square of the undercooling at the interface, whereas eqn. (53.8) predicts it is linearly proportional to the undercooling. A rather different example is provided by the diffusion controlled linear growth of dendrites (see Section 54). Some theories require a proportionality between growth velocity and driving force (or undercooling of the parent phase), but this assumption appears to be incorrect.

The quantity in curly brackets in eqn. (53.8) is the net change in free energy per unit volume produced by migration of the interface. If we write Δg^{Υ} as the corresponding free energy change per atom, we may write the interface velocity as

$$\Upsilon = M^{B} \Delta g^{\Upsilon} \tag{53.9}$$

as the formal equivalent of either (53.6) or (53.8). We call M^B the mobility of the boundary. This concept of a boundary moving at a speed given by the product of a mobility and a driving force may be compared to the similar formal representation of a diffusion flow as the product of the mobility of an atom and the thermodynamic force acting on it (p. 381).

Returning to eqn. (53.6) we now have to consider the magnitude of $\Delta_a g^*$. In general, this may be equal to the free energy of activation for the migration of atoms within the α phase, but it can scarcely be greater than this, since the interface is disordered relative to the α structure. If the boundary between the α and β phases is incoherent, as it will usually be for macroscopic crystals in a non-martensitic transformation, $\Delta_a g^*$ may reasonably be expected to approximate to the activation energy for grain boundary diffusion, rather than lattice diffusion. Under these circumstances, $\Delta_a g^*$ may be appreciably smaller than the activation free energy for growth at the nucleation stage $\Delta_a g^{+}$, since the latter will frequently relate to coherent or semi-coherent growth. In a dislocation-free region, $\Delta_a g^{+}$ will then correspond to the energy for migration through the lattice.

For an incoherent boundary, the activation energy for growth should be independent of the relative orientations of the two crystals on each side of the boundary, or of the boundary itself, and the same should be true for a high angle grain boundary. Experiments on recrystallization and grain growth, however, show that some high angle grain boundaries of special orientations may be much more mobile than the other random boundaries. Very mobile boundaries appear to be "coincidence site boundaries" and the relative mobilities of special and random boundaries may be sensitive to very small quantities of impurities. These effects are discussed more fully in Part II, Chapter 19, but they show that the boundary motion can often not be represented as a simple thermally activated process, as assumed above. For boundary motion in grain growth, the driving force is almost independent of temperature, and the above theory implies that ln Υ should be a linear function of $1/T$, the slope of which gives the heat of activation. Activation energies measured in this way are often much larger than is predicted by the above theory, and Mott (1948) suggested that atoms may be activated in groups rather than singly. The more probable explanation is that if impurities or inclusions are present, the temperature dependence of the growth rate does not give the activation enthalpy for boundary movement.

In recrystallization, the free energy $\Delta g^{\beta\alpha}$ will be replaced by the difference in free energies per atom in the strained and strain free regions. During grain growth there is no volume free energy term of this kind, and $\Upsilon = \partial r/\partial t$ is negative; the boundary migrates towards its centre of curvature. As the grain structure gradually coarsens, the effective driving force decreases, and hence the mean growth rate decreases as $1/r$. These ideal kinetics are usually observed only in zone-refined metals. It is to be noted that according to eqn. (53.8), interface controlled growth is strictly linear only when the interface is plane.

Returning to the application of eqn. (53.6) to phase transformations, we see that for a transition on heating Υ increases continually with increasing temperature. For a transition on cooling, the rate is zero at the transition temperature (eqn. (53.5)), increases to a maximum value, and then ultimately decreases again because of the overriding influence of the $\exp(-\Delta_a g^*/kT)$ factor. This behaviour is observed in many transformations, e.g. that in pure tin, where the growth rate can be measured reasonably accurately. However, it is often difficult to obtain numerical agreement with theoretical predictions, and this has led to various suggestions for modifying the theory. One possibility is that the boundary mobility is restricted by inclusions which act as obstacles, or by impurities segregated to the boundary region, and this effect is known to be important for grain boundaries (see Section 83). Alternatively, atoms or molecules may be unable to transfer from one phase to another at all parts of the boundary, but only at certain active sites. These sites could be places where the local activation energy for crossing the boundary is reduced, perhaps because of the presence of imperfections in the boundary, or places where the local driving force is increased. The latter possibility corresponds to the presence of steps in the interface, as discussed above in general terms, and this is the situation in growth from the vapour phase.

If the fraction of active sites is x_a, and the transfer of atoms is treated as before, the growth velocity of eqn. (53.6) will simply be multiplied by x_a. This has no effect on the functional form of the curves if x_a is independent of temperature, but it is more plausible to assume that the temperature dependence of x_a is important. Becker (1958) has given a theory of this kind for the growth of tin in which x_a decreases as the temperature increases.

When the variation of the driving force with temperature is known, it may be possible to deduce the heat of activation from experimental measurements. Writing

$$X = \Upsilon/\delta^B\nu[1-\exp(-\Delta g^{\beta\alpha}/kT)]$$

and assuming that Υ is equal to x_a times the right-hand side of (53.5), we have

$$\frac{d(\ln X)}{d(1/T)} = -\frac{\Delta_a h^*}{k} - \frac{T^2}{x_a}\frac{dx_a}{dT}. \tag{53.10}$$

If a graph of ln X against $1/T$ is a straight line, it follows that the second term on the right of (53.10) is either negligible in comparison with $\Delta_a h^*/k$, or is constant independent of temperature. The latter seems improbable, so that an experimental straight line plot may give the true heat of activation $\Delta_a h^*$. In view of the other possibilities mentioned above however, and of known examples from grain growth experiments in which the measured activation energy varies with purity, this assumption cannot be made with much confidence. The data required for this type of analysis are not available for many phase transformations.

54. DIFFUSION-CONTROLLED GROWTH

In the preliminary discussion of isothermal transformation kinetics, given in Section 4, we assumed a linear growth law (constant growth rate). This assumption is appropriate under all circumstances in which the interface advances into a region of matrix of constant mean composition, so that the rate of the process controlling the growth is independent of the interface position, and hence of the time.

This situation does not necessarily apply when an isolated precipitate particle (β) grows into a phase (α') of different composition. If β is richer in solute atoms than the equilibrium α, there may be a region depleted in solute formed around the β particle as growth proceeds; if the equilibrium α contains more solute, then the region around the β particle may be enriched. In either case, the continual growth of the particle requires chemical diffusion in the surrounding α', and as the particle increases in size, the effective distances over which diffusion takes place may also increase. When the particle is first formed, it is probable that processes near the interface will control the net rate of growth, but the volume diffusion will eventually become the dominant factor if the interface mobility is reasonably high. The particle will then grow just as fast as the diffusion rate allows.

In contrast to the linear growth law obtained when processes near the interface are slowest, the position of the interface may be proportional to the square root of the time when diffusion is decisive. This may be seen from dimensional arguments, since (with certain simplifying assumptions) the controlling diffusion equation and all the boundary conditions are homogeneous in the concentration. The concentrations of solute B atoms in the equilibrium α and β phases c^α and c^β, and the initial concentration in the metastable α phase c^m, thus enter into the growth equations only in dimensionless combination. The linear dimensions of the β region are then functions only of the diffusion coefficient and the time, and possibly of each other. If all linear dimensions increase in the same way, each such dimension must be proportional to $(Dt)^{1/2}$ and the growth is parabolic with a velocity varying as $(D/t)^{1/2}$. Alternatively, if some critical dimension r is constant, an increasing dimension may be proportional to Dt/r and the growth is linear with a constant velocity. In the first case, the effective diffusion distance is proportional to $(Dt)^{1/2}$, as in Section 40, whereas in the second case, this distance is constant and proportional to r.

In any transformation involving long-range transport, the diffusion equation

$$\frac{\partial c(\mathbf{x}, t)}{\partial t} = D\,\nabla^2 c(\mathbf{x}, t) \tag{54.1}$$

must be satisfied at each point in the metastable phase, the concentration $c(\mathbf{x}, t)$ being a function of both position $\mathbf{x}$ and time t (cf. eqn. (40.15)). The diffusion coefficient D will be assumed to be independent of concentration; the case of non-constant D is considered separately below. If there is no precipitate present at time $t = 0$, there is an initial condition

$$c(\mathbf{x}, 0) = c^m. \tag{54.2}$$

In order to find the distribution of solute from eqn. (54.1), we have to specify additional boundary conditions, and these are not immediately obvious. It seems reasonable to assume that there is a constant concentration of solute in the region of the α' phase immediately adjacent to a growing particle. If the interface is mobile, this concentration cannot deviate much from the equilibrium α concentration, and in the limit in which growth is entirely diffusion controlled, it may be put equal to this equilibrium concentration. However, if the interface has an appreciable curvature, the compositions of the α and β phases in equilibrium are displaced from the values given by the phase diagram because of the Gibbs–Thomson effect. It is often convenient to make the simplifying assumption that c^β is independent of r, and we introduce the notation c^α_∞ for the concentration of solute in equilibrium with c^β across a planar interface whilst c^α_r will denote the local equilibrium composition which changes with the curvature. The distinction between c^α_∞ and c^α_r is important only for particles not much larger than the critical nucleus size, and is neglected in many growth theories. The assumption of local equilibrium is one limiting form of this boundary condition; the other limiting form is for a very sluggish interface, in which case the concentration near the boundary is almost equal to c^m. This leads to interface controlled growth. In any real transformation, the boundary condition must lie between these extremes, but we are often justified in approximating it by one or other of them.

A second boundary condition may be needed to ensure that the solute concentration behaves properly in regions between growing β particles. Finally, the fact that we have a moving boundary must be incorporated into the diffusion equation. This is done by finding a relation between the velocity of the interface, and the flux of solute to or from the interface region.

When dealing with interface controlled linear growth, the theory of the growth rate is quite distinct from the theory of the mutual interference or impingement of growing regions. In any direction, the growth rate is constant until impingement occurs, after which it is zero. This is not so with diffusion limited growth. The interference of different growing crystals results from competition for the available excess solute, and becomes important when the diffusion fields of the two particles begin to overlap to an appreciable extent. This interference is thus part of the diffusion problem, and results in a growth rate which gradually becomes zero. The interference of different β regions is implicit in the second boundary condition, which deals with the concentration in α' regions remote from the β. The two ways in which growing particles can interfere with one another are sometimes called "hard" and "soft" impingement respectively.

We find it convenient to discuss the formal kinetics of isothermal transformations separately in Chapter 12, and the effects of hard impingement are discussed there in a similar manner to the elementary treatment already given in Section 4. Although soft impingement is not really to be distinguished from the general problem of diffusion limited growth, we shall make an equivalent separation of this problem. The theory of diffusional growth in this section will thus be effectively confined to the growth of an isolated particle in an infinite matrix. Exact treatments of the diffusion problem for a sphere have been given by Zener (1949) and Frank (1950), and for particles of more general shape by Ham (1958, 1959) and Horvay and Cahn (1961). We shall begin with the earlier and much simpler treatment due to Zener, since Ham has shown that his method gives the correct result in most cases. All of these theories assume that the interfaces are smooth so that the diffusion flux is constant over a planar interface: the problem of diffusion-controlled growth of a stepped interface is considered later.

Zener considered one-dimensional growth (the thickening of a plate), isotropic two-dimensional growth (the radial growth of a cylinder) and three-dimensional growth (the growth of a sphere). His treatment assumes that the phases at the interface have equilibrium compositions, any variation with the curvature of the interface being neglected. The variation of composition along a line normal to the interface is assumed to be represented by Fig. 11.2.

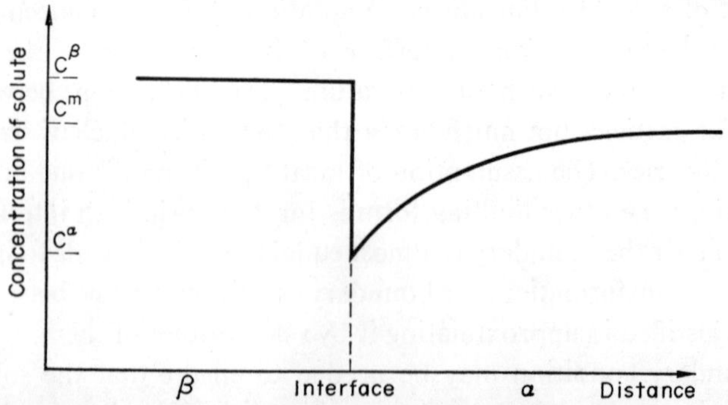

FIG. 11.2. Schematic variation of solute composition assumed in theory of diffusion controlled growth.

The precipitated β phase may be richer or poorer in B atoms than the α phase, and the corresponding net diffusion current of B may be towards or away from the β particle. In either case, the concentration c^m is intermediate between c^α and c^β. Changes in volume per atom as the β phase grows are ignored.

Let r be a coordinate normal to the interface, measuring the semi-thickness of the plate, or the radius of the cylinder or sphere. Then when the boundary is in a position $r = r^I$, the diffusion flux of B atoms across unit area in a time δt will be

$$D\left(\frac{\partial c}{\partial r}\right)_{r=r^I} \cdot \delta t,$$

where the concentration c of atoms in the α phase is a function of r and t. If the interface advances a distance δr during this time, the extra number of solute atoms in a volume δr will be $(c^\beta - c^\alpha)\,\delta r$. These atoms can only come from the diffusion flow, so the two expressions may be equated. The limit of $\delta r/\delta t$ gives the growth velocity, which is

$$\Upsilon = \frac{dr^I}{dt} = \frac{D}{c^\beta - c^\alpha}\left(\frac{\partial c}{\partial r}\right)_{r=r^I}. \tag{54.3}$$

It is sometimes convenient to represent the concentration gradient in this equation (rather loosely) as a ratio $\Delta c/y^D$, where Δc is the difference in solute concentration in matrix regions near to and remote from the interface, and y^D is an effective diffusion distance. The equation then becomes

$$\Upsilon = \frac{\Delta c}{c^\beta - c^\alpha}\,\frac{D}{y^D}. \tag{54.4}$$

We have already used the fact that eqns. (54.3) and (54.4) are homogeneous in the composition terms in our dimensional argument at the beginning of this section, and certain qualitative conclusions about the growth rate may be reached by consideration of (54.4). If the phase boundary is planar, y^D must continually increase as the boundary advances, leading to a diminishing growth rate, as mentioned above. For an interface which has a convex curvature towards the matrix, y^D will be approximately proportional to the radius of curvature. The increasing diffusion distance will thus also lead to a diminishing rate for the growth of a spherical particle or the radial growth of a cylinder. The growth conditions are not obvious, however, for the lengthening of a cylinder or the edgewise growth of a plate.

When a plate grows in its own plane, or a needle extends, it seems intuitively probable that the interface is always moving into fresh regions of constant composition, so that the growth rate is constant. This is certainly true if the rate is controlled by interface processes, composition gradients in the matrix being then virtually non-existent, but Zener (1949) and Wert (1949) implied that linear growth rates controlled by diffusion may also be obtained for particles of these shapes. A constant growth rate with diffusion control implies the existence of a steady-state solution to the diffusion equation, and from eqn. (54.4) we see that this should be possible if the growth conditions are such that the radius of curvature at the edge of a plate or the tip of a needle remains constant. Zener considered a flat plate to have such a constant radius equal to half the plate thickness, but he noted that the restriction of the growth theory to dimensions much larger than those of a critical nucleus prevents us from applying eqn. (54.4) indiscriminately to a very thin plate, even though at first sight the reduction in y^D gives a high diffusion controlled rate of growth to such a plate. When the radius of curvature is small, the change in chemical free energy has to be subtracted from the chemical driving force, and there is a minimum radius below which growth is not possible. Another way of expressing this correction is that the effective difference in pressure across the curved interface changes the concentration of solute in the matrix with which the particle is in equilibrium (Gibbs–Thomson effect), and for the critical radius of curvature the particle is in equilibrium with the average composition of the matrix.

Steady-state solutions of the diffusion equation are important in the theory of discontinuous precipitation and eutectoidal decompositions (see Section 55), and also in such problems as the inward growth of a Widmanstätten plate from a grain boundary. More recent work has made it clear that there are two types of solution to be considered in relation to variations in particle shape. Plates and needles may be represented by prolate and oblate spheroids respectively, so that a continuous change of shape is obtained mathematically by varying the axial ratio of a family of ellipsoids with two equal axes. These particles all have closed form, and may thus be enclosed by a matrix. An alternative representation of shape variations is given by a family of paraboloids of elliptical cross-section, but this can only be used for dendrites, or for crystals growing in from a grain boundary. The limiting forms of the flat disc and the circular cylinder might apply to either enclosed or dendritic particles, but Zener's assumptions are incorrect for enclosed particles and rather doubtful for dendritic forms.

According to Zener and Wert's assumption, the eccentricity of an ellipsoidal particle increases during diffusion controlled growth. An opposite possibility was suggested by Doremus (1957) who assumed that a spherically symmetrical diffusion field will be established around any particle, independent of its shape. This implies that $(\partial c/\partial r)$ in eqn. (54.3) is constant over the surface of a β particle, and growth in all directions is parabolic, the particles becoming more spherical as they grow. An exact treatment first given by Ham (1958) shows that both these predictions are incorrect; subject to certain assumptions, the particles have a fixed eccentricity and do not change shape.

We now have to obtain a more quantitative treatment of parabolic growth by evaluating eqn. (54.3). This means that $c(r, t)$ has to be found by solving the general diffusion eqn. (54.1) which takes the form

$$\partial c/\partial t = D(\partial^2 c/\partial r^2)+(j-1)(D/r)(\partial c/\partial r), \tag{54.5}$$

where $j = 1, 2$, or 3 for one-, two-, or three-dimensional growth respectively. The boundary conditions for the solution of this equation are (54.2) and

$$c(r^I, t) = c^\alpha, \tag{54.6}$$

which corresponds to the concentration at the interface always being c^α. Note that r^I, in this second condition, is itself a function of t.

Zener showed that if c^α has the constant value c^α_∞ the solution of eqns. (54.5) and (54.6) is

$$c = c^m + (c^\alpha_\infty - c^m)\frac{\varphi_j\{r/(Dt)^{1/2}\}}{\varphi_j\{r^I/(Dt)^{1/2}\}}, \tag{54.7}$$

where

$$\varphi_j(x) = \int_x^\infty \xi^{1-j}\exp(-\xi^2/4)\,d\xi.$$

When $j = 1$, this reduces to a standard solution of the diffusion equation similar to those already given in Section 40. Since $r^I \propto (Dt)^{1/2}$, the quantity $r^I/(Dt)^{1/2}$ is a dimensionless function of the concentrations which determines the growth velocity; we write it α_j and relate it to the dimensionless supersaturation $\bar{\alpha}$, where

$$\bar{\alpha} = \frac{c^m - c^\alpha_\infty}{c^\beta - c^\alpha_\infty}. \tag{54.8}$$

From eqn. (54.3) we now find

$$\frac{1}{\bar{\alpha}}\frac{dr^I}{dt} = \frac{(\alpha_j)^{1-j}}{\varphi_j(\alpha_j)}\exp(-\alpha_j^2/4)\,.\,D^{1/2}t^{1/2},$$

and hence

$$(\alpha_j)^j = \{2\bar{\alpha}\exp(-\alpha_j^2/4)\}/\varphi_j(\alpha_j). \tag{54.9}$$

This equation is an implicit expression for α_j in terms of the quantity $\bar{\alpha}$, and for any supersaturation, the growth law is given by

$$r^I = \alpha_j(Dt)^{1/2}. \tag{54.10}$$

The relation in eqn. (54.10) is exact within the limitations of the assumptions mentioned above, but α_j is not a simple function of $\bar{\alpha}$, varying from 0 to ∞ as the latter varies from 0 to 1. Zener showed that simpler equations for α_j could be obtained by using asymptotic expansions of $\varphi_j(\alpha_j)$, valid in the limits $\alpha_j \ll 1$ and $\alpha_j \gg 1$. For one-dimensional growth, this gives

$$\left.\begin{aligned} \alpha_1 &= (2/\pi^{1/2})\bar{\alpha} && (\alpha_1 \ll 1),\\ \alpha_1 &= \{2(c^\beta - c^\alpha_\infty)/(c^\beta - c^m)\}^{1/2} && (\alpha_1 \gg 1), \end{aligned}\right\} \tag{54.11}$$

and for three-dimensional growth

$$\left.\begin{aligned} \alpha_3 &= (2\bar{\alpha})^{1/2} && (\alpha_3 \ll 1),\\ \alpha_3 &= \{6(c^\beta - c^\alpha_\infty)/(c^\beta - c^m)\}^{1/2} && (\alpha_3 \gg 1). \end{aligned}\right\} \tag{54.12}$$

Zener also showed that it is convenient to express α_j in terms of different ("natural") concentration parameters, which may be obtained by considering approximate solutions. For one-dimensional growth, such a solution follows from the assumption that the concentration decreases linearly from $r = r^I$ to some higher value of r, after which it is constant at $c = c^m$. The condition that the total extra number of solute atoms in the β region has come from the surrounding depleted region then fixes the outer limit of the latter, and hence the assumed constant value of $\partial c/\partial r$. This gives

$$\partial c/\partial r = (c^m - c^\alpha_\infty)^2/2(c^\beta - c^m)r^I$$

and on substituting into (54.3), we find

$$r^I \simeq \frac{(c^m - c^\alpha_\infty)}{(c^\beta - c^\alpha_\infty)^{1/2}(c^\beta - c^m)^{1/2}}(Dt)^{1/2} = \alpha_1^*(Dt)^{1/2}. \quad (54.13)$$

The quantity α_1^* is thus a natural parameter in which to express the growth law for a plane interface. The true growth parameter, α_1, given by (54.9), is related to α_1^* by

$$\alpha_1 = C\alpha_1^*,$$

where C varies from $2/\pi^{1/2} \simeq 1{\cdot}13$ to $\sqrt{2}$ as c^m varies from c^α_∞ to c^β (compare asymptotic expressions (54.11) above). Equation (54.9) was solved graphically by Zener, and plotted in the form of a smooth α_1–α_1^* relation, but it is clear that eqn. (54.13) is a sufficient approximation for all practical purposes. For the growth of a spherical particle, an estimate of $(\partial c/\partial r)$ may be made in the same way. This is equivalent to supposing that the solute atoms are taken from a thin spherical shell surrounding the precipitate particle, and the concentration gradient is assumed constant across this shell. This will only be a good approximation when there is a high degree of supersaturation, so that

$$c^\beta - c^m \ll c^\beta - c^\alpha_\infty.$$

The result may be expressed in the form of a growth coefficient which is $\sqrt{(3)}\alpha_1^*$.

At low degrees of supersaturation, i.e. when

$$c^\beta - c^m \gg c^m - c^\alpha_\infty \quad (54.14)$$

the approximation of a linear change of concentration will not be valid. In this case, the limiting form of the steady-state solution for a stationary boundary is appropriate, so that the diffusion problem reduces to finding the solution of Laplace's equation

$$\nabla^2 c = 0 \quad (54.15)$$

obtained from (40.15) or (54.1) by setting $\partial c/\partial t = 0$. With the above boundary conditions, the solution for spherical growth is

$$c = c^m(1 - r^I/r) + c^\alpha_\infty r^I/r \quad (54.16)$$

so that

$$(\partial c/\partial r)_{r=r^I} = (c^m - c^\alpha_\infty)/r^I$$

and the effective diffusion distance (eqn. 54.4) is given by

$$y^D = r^I. \quad (54.17)$$

This gives a growth coefficient

$$\alpha_3^* = \sqrt{(2)}\,(c^m - c^\alpha_\infty)^{1/2}/(c^\beta - c^m)^{1/2} \quad (54.18)$$

so that there is an apparent difference between the "natural" growth parameters for small and large supersaturations. However at high supersaturations, the coefficient $\surd(3)\alpha_1^* \simeq (\frac{3}{2})^{1/2}\alpha_3^*$ so that for both high and low supersaturations, the quantity α_3^* forms a suitable parameter for the growth law. Zener's graphical computation shows that the true growth coefficient $\alpha_3 = C\alpha_3^*$, where C varies slowly from 1 to $\surd 3$ as c^m changes from c_∞^α to c^β (see eqns. (54.12)).

When the approximation of the steady-state solution is valid, it may also be used to deduce the growth rate for some other geometries. An example is the thickening of a circular cylinder, where, however, it is necessary to replace the boundary condition at infinity by

$$c = c^m \quad \text{for} \quad r \geqslant r_e,$$

where r_e is an outer cut-off radius, large in comparison with the radius of the cylinder r^I. The solution of the steady state diffusion equation is then

$$c = c_\infty^\alpha + (c^m - c_\infty^\alpha)\ln(r/r^I)/\ln(r_e/r^I), \tag{54.19}$$

and the effective diffusion distance in eqn. (54.4) becomes

$$y^D = r^I \ln(r_e/r^I). \tag{54.20}$$

Solutions are also available for more complex geometries, such as a parabolic cylinder or a paraboloid of revolution (Trivedi and Pound, 1967).

As noted in Chapter 9, the diffusion coefficient is often a function $D(c)$ of composition, and we consider next the effect of this variation on the computed growth rate. This problem has been considered by Trivedi and Pound (1967) and Atkinson (1967, 1968). The conservation equation (54.3) still applies with D having the value $D(c_\infty^\alpha)$ appropriate to the interface, but the diffusion equation has to be solved in the general form (40.15) rather than in the form (54.1).

Trivedi and Pound considered first the case in which the steady-state solution for a stationary boundary is an adequate approximation, i.e. when eqn. (54.14) is valid. The diffusion equation then reduces to the time-independent form

$$\nabla(D\,\nabla c) = 0 \tag{54.21}$$

and on making the substitution

$$\psi(c) = \{D(0)\}^{-1}\int_0^c D\,\mathrm{d}c. \tag{54.22}$$

this becomes

$$\nabla^2\psi = 0. \tag{54.23}$$

Also from (54.3)

$$\Upsilon = \frac{D(0)}{(c^\beta - c_\infty^\alpha)}\left(\frac{\partial\psi}{\partial r}\right)_{r=r^I}, \tag{54.24}$$

which can be written in a form similar to (54.4) as

$$\Upsilon = \frac{D(0)}{(c^\beta - c_\infty^\alpha)}\,\frac{\psi(c^m) - \psi(c_\infty^\alpha)}{y^D}. \tag{54.25}$$

Since (54.23) has the same form as (54.15), the value of y^D appropriate to (54.25) is identical with that in (54.4) for the case of D constant, e.g. $y^D = r^I$ for spherical growth. Moreover, it follows from (54.22) that (54.25) can be placed in the form (54.4) if D in the latter equation is given the value

$$\bar{D} = \int_{c^\alpha}^{c^m} D \, \mathrm{d}c / (c^m - c^\alpha_\infty). \tag{54.26}$$

Thus the important result of this theory is that for a concentration dependent diffusion coefficient, the growth rate corresponds to that calculated for a constant diffusion coefficient evaluated as a weighted average over the composition range found in the matrix. This corrects earlier suggestions that either the value of the diffusion coefficient at $c = c^\alpha_\infty$ or the maximum value in the range $c^\alpha_\infty - c^m$ should be used in place of the constant D of the simple solutions.

Trivedi and Pound also considered the effect of variations in D with composition on the edgewise growth of a plate-shaped particle. The growth theory is more complex (see below), but with certain assumptions they were able to obtain numerical solutions for the growth rate. The computed values differed only slightly from those calculated from an assumed constant D given by the weighted average of (54.26).

Atkinson considered one-dimensional growth when the steady-state approximation is invalid. He assumed that the growth rate continues to be given by an equation of form (54.10), and he was able to obtain analytical solutions for the growth coefficient α_1 for a class of exact solutions originally considered by Philip (1960). An approximation to these solutions is given by Zener's theory with D replaced by an average value which is determined mainly by $D(c^\alpha_\infty)$ and $D(c^m)$. In view of the difficulty of matching experimental results for $D(c)$ with a form giving an exact solution, Atkinson (1968) also developed an iterative procedure for the numerical solution of the diffusion equation for one-dimensional growth. The one-dimensional form of eqn. (54.5) is first transformed into a moving system of co-ordinates with the origin fixed in the interface, and the Boltzmann substitution $\lambda = x/t^{1/2}$ is then used to give a relation between c and λ which involves the experimental function $D(c)/D(c^\alpha_\infty)$. The differential equation is then solved numerically by a technique which involves beginning with an arbitrary value of the growth coefficient α_1 which defines the interface position through

$$r^I = \alpha_1 \{D(c^\alpha_\infty)t\}^{1/2} \tag{54.27}$$

(compare eqn. (54.10)) and iterating until a constant value is obtained for α_1. However, as already indicated, it is probably adequate in most cases to use the solution for a constant diffusion coefficient with an appropriate average value of D.

We now turn to a brief consideration of the treatment by Ham. He uses a more complex mathematical procedure, in which the solute density is expanded in terms of the eigenfunctions of an appropriate boundary value problem. This enables him to obtain solutions of the diffusion equation for particles of any shape provided they are distributed periodically in the matrix. The form of the solution shows that it may be applied also to a non-uniform distribution, so that the method includes a proper treatment of the problem of soft impingement. We shall not describe the calculations in detail, but we shall summarize some of the important results; others will be mentioned in Chapter 12.

Ham assumes that the particles grow from initially negligible dimensions, that the concentration of solute in the matrix adjacent to the particle is constant at all parts of the surface, and that accumulation of solute occurs at the point of contact with the particle. As already mentioned, this leads to the result that particles growing as spheroids of any shape maintain constant eccentricity under diffusion limiting conditions. Each dimension of a particle is proportional to the square root of the time, and the volume varies as $t^{3/2}$, just as for a sphere. This contrasts with the earlier incorrect suggestions that the volume varies as $t^{5/2}$ for plates and t^2 for rods. Ham also showed that these results for spheroids may be obtained by Zener's approximate method. As described above, this involves the calculation of the concentration gradient by the steady-state solution of the diffusion equation near a particle of fixed dimensions equal to the instantaneous dimensions of the growing particle.

Ham also considered the problem of small rods and plates of finite initial dimensions which remain highly eccentric and do not alter their long dimensions appreciably during the diffusion growth. (This is appropriate, for example, if the shape is established by rapid interface controlled growth, as suggested above.) In both cases the volume increases as t^1 during the subsequent diffusion growth. This contradicts the assumption of Zener and Wert that a plate thickens as $t^{1/2}$, as predicted from the above theory of one-dimensional growth. The theory is correct, but is not applicable to the thickening of a plate of finite initial dimensions since the actual diffusion field is then quite different from that of one-dimensional diffusion. There is an initial transient whilst the diffusion field is established, during which the growth changes from a $t^{1/2}$ to a t^1 law.

The existence of shape-preserving solutions does not, of course, imply that growth is necessarily of this form for all enclosed particles. If the interface processes are dependent on the orientation of the boundary, for example, or if thermal gradients exist in the matrix, the assumption of constant concentration at the boundary would not be valid. This would lead to a change of shape as the particle grew. Similar effects occur if the curvature of the boundary is large, since this displaces the equilibrium composition because of the Gibbs–Thomson effect. However the most important reservation, as Ham himself recognized, relates to the stability of a growth front under diffusion-limited conditions.

The problem of interface stability first came into prominence in work on the solidification of liquids, and in Part II, Chapter 14, we shall find that an initially planar solid–liquid interface may become non-planar under certain conditions of solute diffusion. Rather more extreme conditions then lead to "projections" on the solid becoming unstable and growing out rapidly into the liquid (dendritic growth). Comparable situations should arise in diffusion-limited growth in the solid state. The first systematic treatment of the problem of morphological stability was given by Mullins and Sekerka (1963, 1964), and this branch of growth theory has been important ever since their work.

We give here a necessarily over-simplified account of the theory of stability which is adapted primarily to the solid-state diffusion problem; the liquid–solid problem is similar but is complicated by heat flow and will be considered separately in Part II, Chapter 14.

Consider a planar interface $x_1 = 0$, moving with velocity Υ, and superimpose on this interface a shape perturbation

$$x_1 = \delta(t) \sin \omega x_2, \tag{54.28}$$

where x_2 defines a direction in the interface. The wavelength of this perturbation $2\pi/\omega$ is not necessarily large compared with the amplitude δ, so that the effect of curvature on the equilibrium composition can not be neglected. Figure 11.3 shows two extreme conditions; in (a) the composition in the α phase remains constant along the perturbed interface, so that the concentration contours in front of the interface are also sinusoidal with diminishing amplitude, whilst in (b) the concentration contours remain planar, so that the composition varies along the non-planar interface. In Fig. 11.3(a) the concentration gradient in the x_1 direc-

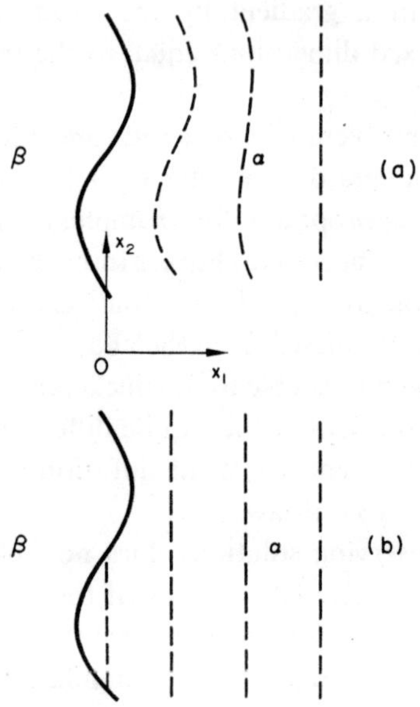

FIG. 11.3. To illustrate the instability of a planar inteface when growth is diffusion-controlled (after Shewmon, 1965).

tion is increased ahead of the "hills" (i.e. the iso-concentration surfaces are compressed together) and reduced in front of the valleys, whilst in Fig. 11.3(b) the concentration gradient normal to the overall interface is constant.

If the situation is as envisaged in Fig. 11.3(a) the hills will tend to grow faster and the valleys more slowly compared with the growth rate of a flat interface. Thus the amplitude of the sine wave will tend to increase with time, and the flat interface is inherently unstable. However, as the amplitude increases so also does the curvature and the equilibrium composition of the α phase begins to increase ahead of the hills and to decrease ahead of the valleys. Thus the Gibbs–Thomson effect exerts a stabilizing influence and tends to diminish the perturbation; the situation of Fig. 11.3(b) will be obtained, in principle, if the variation of composition along the interface is just that required for equilibrium of α and β planes over a curved interface.

For small values of δ the curvature is approximately d^2x_1/dx_2^2, so that the composition of the α phase in equilibrium with the β becomes (see eqn. (22.38))

$$c^\alpha = c_\infty^\alpha[1+\Gamma\delta(t)\omega^2 \sin \omega x_2]. \tag{54.29}$$

The composition variation in the α phase may be obtained in the steady-state approximation (i.e. when (54.14) is valid) by writing down the general solution to Laplace's equation and imposing the boundary conditions (54.29). This gives (Shewmon, 1965)

$$c(x_1, x_2) = c_\infty^\alpha+[c_\infty^\alpha\Gamma\omega^2-(\partial c/\partial x_1)_I]\,\delta \sin(\omega x_2) \exp(-\omega x_1)+x_1(\partial c/\partial x_1)_I, \tag{54.30}$$

where $(\partial c/\partial x_1)_I$ is the composition gradient at the interface. The velocity of the interface eqn. (54.3) is thus

$$\Upsilon(x_2) = \frac{D}{c^\beta-c^\alpha}\left[\left(\frac{\partial c}{\partial x_1}\right)_I-\left\{c_\infty^\alpha\Gamma\omega^2-\left(\frac{\partial c}{\partial x_1}\right)_I\right\}\omega\delta \sin \omega x_1\right]. \tag{54.31}$$

The growth rate of a fluctuation, i.e. the velocity of a peak relative to that of the mean interface, is

$$\dot{\delta} = \frac{D\omega}{c^\beta-c^\alpha}\left[\left(\frac{\partial c}{\partial x_1}\right)_I-c_\infty^\alpha\Gamma\omega^2\right]\delta. \tag{54.32}$$

According to this simple theory, the expression for the growth of a fluctuation contains a positive term, linear in ω and $(\partial c/\partial x_1)_I$, and a stabilizing negative term which depends on $\Gamma\omega^3$. The critical wave number ω_c above which $\dot{\delta} < 0$ and below which $\dot{\delta} > 0$ is given by

$$\omega_c = \{(\partial c/\partial x_1)_I/c_\infty^\alpha\Gamma\}^{1/2} \tag{54.33}$$

and the maximum growth rate occurs at

$$\omega_1 = \omega_c/3^{1/2}. \tag{54.34}$$

The growth rate is negative at large ω (small wavelengths) because of the Gibbs–Thomson effect and is small at small ω (large wavelengths) because of the large diffusion distance.

In their original paper, Mullins and Sekerka (1963) treated the growth of a sphere with a superimposed perturbation $\delta_l Y_{lm}$ in the form of a spherical harmonic $Y_{lm}(\theta, \varphi)$ of amplitude δ_l. For a particular harmonic, they found the growth rate of the fluctuation to be given by

$$\dot{\delta}_l = \frac{D(l-1)}{(c^\beta-c^\alpha)r^I}\left[\left(\frac{\partial c}{\partial r}\right)_{r^I}-\frac{c_\infty^\alpha\Gamma}{(r^I)^2}(l+1)(l+2)\right]\delta_l, \tag{54.35}$$

which is of the same form as (54.32). It follows from this equation that a harmonic of given l will grow if the radius of the sphere exceeds a value $r_1(l)$ given by

$$r_1(l) = (\tfrac{1}{2}(l+1)(l+2)+1)r_c, \tag{54.36}$$

where

$$r_c = 2c_\infty^\alpha\Gamma/(c^m-c_\infty^\alpha) \tag{54.37}$$

is the critical radius at which a β precipitate is in equilibrium with the matrix of composition c^m (see eqn. (22.33)) and thus represents the critical nucleus size for this degree of supersaturation (see Part II, Chapter 18). The first harmonic which distorts the spherical shape has $l = 2$ and hence $r_1 = 7r_c$. A sphere perturbed by δY_{20} is to first order an ellipsoid of eccentricity $2\delta/r$, and application of eqn. (54.35) to this shape when $r \gg r_c$ (Gibbs–Thomson effect negligible) shows that the solution is shape preserving, i.e. a sphere perturbed only by this harmonic does not change shape as it grows. This result is thus consistent with Ham's solutions, which were obtained by neglecting the effect of curvature. More severe perturbations of the sphere give non-shape preserving solutions and for $l = 3$, the condition is

$$r_1 > 21r_c. \tag{54.38}$$

These solutions indicate that for both spherical and non-spherical particles instability is favoured by high supersaturations and large particle sizes. An individual sphere growing in a supersaturated matrix tends to dissolve for $r < r_c$ and the spherical shape is unstable for $r > r_1$; similar considerations apply to an ellipsoidal particle. Thus there should be a comparatively small range of sizes for which Ham's shape preserving solutions are valid; near $r = r_c$, the neglect of Gibbs–Thomson effects is a serious limitation and for $r > r_1$ the shape preserving solution is unstable against fluctuations.

Shewmon (1965) pointed out that the experimental evidence suggests that stable polyhedral growth forms are found to persist over a much wider range of sizes and supersaturations than is implied by the above theory, and hence that additional stabilizing factors must be considered. Neglected effects include diffusion in the β phase if its composition is non-uniform, faster diffusion along the α–β interface than in the bulk α phase, the effects of transformation stresses, and the possible existence of a slow interfacial reaction (i.e. growth which is not completely diffusion controlled); impurities, however, will increase the tendency to growth instability. Surface diffusion adds a negative term in ω^4 to eqn. (54.32), and this has the effect of decreasing ω_c and thus extending the range of ω over which a planar growth front is stable against fluctuations, but it does not remove the instability at small ω. Shewmon concluded that transformation stresses have a rather weak stabilizing effect, and that observed non-dendritic growth forms in solid-state precipitation reactions probably indicate some degree of interface control of the growth rate.

A growth morphology which bears some resemblance to dendritic growth is the formation of long, thin needles or plates (Widmanstätten side plates) by inward growth from a grain boundary. Experimental observations show that the growth of such plates is linear rather than parabolic, so that it is necessary to look for steady-state solutions of the diffusion equation. One model of this type, due to Hillert (1957), is a development of Zener's theory for the growth of a plate of constant thickness or a needle of constant diameter; other more

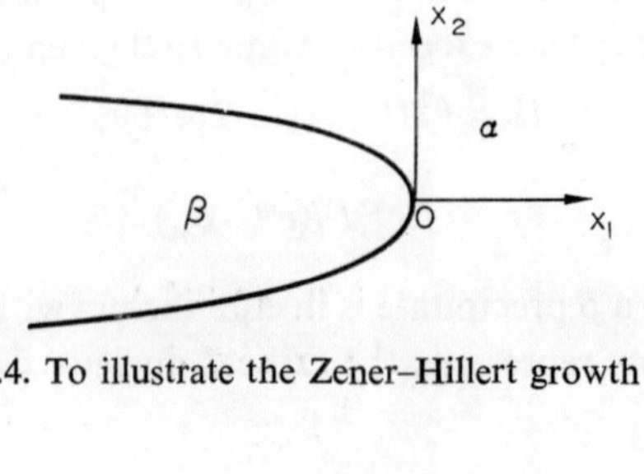

FIG. 11.4. To illustrate the Zener–Hillert growth model.

recent models consider plates of parabolic section. The Zener–Hillert model, which is shown in Fig. 11.4, is admittedly inexact, but it has assumed a prominent place in the literature and will therefore be discussed first.

Zener assumed that the tip of a needle is hemispherical and the edge of a plate is hemicylindrical so that the composition of the α phase at the interface, c_r^α is constant over the curved surface. In eqn. (54.4) Δc is now given by $(c^m - c_r^\alpha)$ and it follows from eqns. (22.33), (22.38), and (54.37) that this is equal to $(c^m - c_\infty^\alpha)(1 - r_c/r)$. The composition term in the denominator of (54.4) is $c_r^\beta - c_r^\alpha$ but it follows from the discussion of the Gibbs–Thomson effect in Section 22 that this is nearly equal to $c_\infty^\beta - c_\infty^\alpha$. The diffusion distance y^D is assumed to be proportional to r and hence to remain constant with time for a needle or plate of given thickness. The growth rate may thus be written

$$\Upsilon = (D\bar{\alpha}/C_{14}r)\{1-(r_c/r)\}, \tag{54.39}$$

where $y^D = C_{14}r$ and $\bar{\alpha}$ is given by (54.8).

The growth rate given by (54.39) is zero for $r = r_c$ and tends to zero as $r \to \infty$. The maximum growth rate occurs when

$$r = 2r_c \tag{54.40}$$

and is given by

$$\Upsilon = D\bar{\alpha}/4C_{14}r_c. \tag{54.41}$$

Note that eqns. (54.39)–(54.41) are valid whatever approximations are made to fix Γ in (22.33) or (22.38).

Equation (54.39) gives a relation between velocity and tip radius but does not predict either uniquely. Zener assumed that the observed velocity will be the maximum possible velocity, given by eqn. (54.41), and correspondingly that the diameter of a needle or the thickness of a plate will be given by $2r = 4r_c$ from (54.40). This assumption is one of the more contentious points of the theory. In view of the approximate nature of the treatment, it is not possible to give a reliable estimate for C_{14}, but if eqn. (54.17) remains valid, $C_{14} \sim 1$.

In the theory of heat flow it is often convenient to use a dimensionless velocity parameter known as the Péclet number. The corresponding Péclet number $\bar{p}$ for diffusional growth is obtained by replacing the thermal diffusivity by the diffusion coefficient and is defined as

$$\bar{p} = \Upsilon r/2D \tag{54.42}$$

The Zener growth theory is given by the condition

$$\bar{\alpha}_{\text{eff}} = \bar{\alpha}\{1-(r_c/r)\} = 2C_{14}\bar{p} \tag{54.43}$$

and the maximum velocity condition is

$$\bar{\alpha} = 4C_{14}\bar{p}, \quad r = 2r_c. \tag{54.44}$$

Hillert (1957) attempted to give a more rigorous treatment of the diffusion problem. In Fig. 11.4 composition gradients along x_3 may be neglected in first approximation. Consider an origin in the moving boundary and let x_1, x_2 be coordinates in the direction of growth and normal to the plane of the plate respectively. Then at any point, the rate of change of concentration caused by diffusion processes is $D\{\partial^2 c/\partial x_1^2)+(\partial^2 c/\partial x_2^2)\}$ (see eqn. (54.1)). The corresponding change of concentration due to the boundary velocity, Υ, is $\Upsilon(\partial c/\partial x_1)$. If the growth of the plate is a steady-state process with constant Υ, the concentration of any point with fixed coordinates relative to the moving edge of the plate remains unchanged. Thus we have the diffusion equation

$$D\{(\partial^2 c/\partial x_1^2)+(\partial^2 c/\partial x_2^2)\}+\Upsilon(\partial c/\partial x_1) = 0. \tag{54.45}$$

This may be contrasted with eqn. (54.5) which describes a growth process in which a steady state is not attained.

A solution of this equation is obtained by separating the variables, that is by making the assumption that

$$c(x_1, x_2) = g(x_1)\, h(x_2). \tag{54.46}$$

Note that c is not a function of the time in a steady-state solution, since the time has been eliminated by using a moving coordinate system. Substituting (54.46) into (54.45) gives

$$\frac{1}{g}\left\{\frac{\mathrm{d}^2 g}{\mathrm{d}x_1^2}+\frac{\Upsilon}{D}\frac{\mathrm{d}g}{\mathrm{d}x_1}\right\} = -\frac{1}{h}\frac{\mathrm{d}^2 h}{\mathrm{d}x_2^2} = b^2$$

and the particular solutions are

$$g = \exp(-\lambda x_1), \quad h = \cos(bx_2), \tag{54.47}$$

where

$$\lambda = (\Upsilon/2D)\,(1+\sqrt{[1+4b^2D^2/Y^2]}) \tag{54.48}$$

and b can take any constant value. The complete solution is thus

$$c-c^\alpha = \int_0^\infty A(b)\exp(-\lambda x_1)\cos(bx_2)\,\mathrm{d}b. \tag{54.49}$$

The problem now reduces to the determination of the coefficients $A(b)$ from the boundary conditions of the model, the assumption of local equilibrium at the interface, etc. Unfortunately this cannot be done because the boundary conditions are in fact not compatible with the existence of the steady-state solution. For this reason, the value of the whole calculation is rather doubtful, and it seems unlikely that the model is a good approximation to the real physical situation. Hillert avoids the difficulty by specifying an internal condition which satisfies the boundary conditions only in the x_2 direction. The concentration in the plane $x_1 = 0$ is assumed to fit the equation

$$c-c^\alpha = C_{15}\exp(-\pi x_2^2/16m^2). \tag{54.50}$$

where C_{15} and m are arbitrary constants. Although this is a plausible form for the concentration change, no real justification can be given for this assumption. If it is used to replace the real boundary conditions, however, a unique solution to the diffusion problem is obtained, the steady-state growth rate being given by

$$\Upsilon = \frac{D\bar{\alpha}}{2}\frac{(c^\beta-c_r^\alpha)}{(c^\beta-c^m)}\frac{1}{m}\left(1-\frac{m_c}{m}\right), \quad (m \simeq r). \tag{54.51}$$

This corresponds to Zener's equation (54.39) with

$$C_{14} = 2(c^\beta - c^m)/(c^\beta - c_r^\alpha).$$

After substituting into (54.43) and rearranging,

$$\bar{\alpha}_{\text{eff}} = 4\bar{p}(1+4\bar{p})^{-1}. \tag{54.52}$$

Trivedi and Pound (1969) refer to this as the Zener–Hillert equation and point out that at low supersaturations (small $\bar{p}$)

$$\bar{\alpha}_{\text{eff}} = 4\bar{p}, \qquad (\bar{\alpha}_{\text{eff}} \to 0), \tag{54.53}$$

whereas at high supersaturations (large $\bar{p}$)

$$\bar{\alpha}_{\text{eff}} = 1-(1/4\bar{p}) \qquad (\bar{\alpha}_{\text{eff}} \to 1). \tag{54.54}$$

The Zener theory gives a linear relation between velocity and supersaturation, but this is true for Hillert's result only in the limit of (54.53). At very high supersaturations the Péclet number tends to infinity as $\bar{\alpha} \to 1$; this is expected physically because as $c^m \to c^\beta$, the diffusion required becomes negligible. Previous treatments of the Hillert theory have generally approximated $(c^\beta - c_r^\alpha)/(c^\beta - c^m)$ by unity and have thus concluded that it is equivalent to the Zener theory with $C_{14} = 2$. The maximum growth rate then occurs at $r = 2r_c$ and $\bar{\alpha}_{\text{eff}} = \frac{1}{2}\bar{\alpha}$, but these results are valid only in the low supersaturation limit.

Hillert also gave a similar treatment for the growth of a needle-shaped particle and for the growth of a lamellar aggregate (pearlite). The pearlite problem, to which the steady-state solution is appropriate, is discussed in Section 55. The boundary conditions do not permit a steady-state solution for the growth of a cylindrical needle of constant thickness, and a similar approximate method has to be used. The maximum growth rate obtained by Hillert for a needle is 1·5 times that for a plate.

The minimum radius of curvature is given by the condition that the net free energy change is zero when a plate of this thickness extends. Let $-\Delta g^t$ be the chemical driving force per atom for the growth of the plate. The chemical free energy released when unit length of the plate moves through a distance $\mathrm{d}x$ is thus $-(y\,\Delta g^t/v)\,\mathrm{d}x$ where y is the thickness of the plate and v is the volume per atom. The change in surface energy is $2\sigma\,\mathrm{d}x$, and in the limiting case when this equals the chemical free energy released $y = 2r_c$. Hence

$$r_c = -2\sigma v/\Delta g^t. \tag{54.55}$$

An exact solution of the diffusion problem for a particle growing with constant velocity in one direction may be obtained by neglecting the Gibbs–Thomson effect of the curvature, which is emphasized in the Zener–Hillert model, and returning to the boundary conditions used by Ham. For the problem of dendritic growth controlled by heat flow, Papapetrou (1935) first pointed out the existence of such a linear solution when the dendrite has the shape of a paraboloid of revolution. The most complete treatment of this problem (Horvay and Cahn, 1961) is based on an analysis due to Ivantsov (1947) and shows that exact solutions can be obtained for a family of elliptical paraboloids, including as special cases needles, or dendrites with aspect ratio of the ellipse near unity, and dendritic plates. For the case in which the shape is a parabolic cylinder, i.e. an infinite plate, the Ivantsov theory gives

$$\bar{\alpha} = (\pi\bar{p})^{1/2} \exp(\bar{p}) \operatorname{erfc}(\bar{p})^{1/2}, \tag{54.56}$$

where $\operatorname{erfc}(\bar{p})$ is the complementary error function defined in eqn. (40.21). The corresponding solution for a paraboloid of revolution, i.e. a needle with a circular cross-section, was obtained by Ivantsov as

$$\bar{\alpha} = \bar{p} \exp(\bar{p}) \int_{\bar{p}}^{\infty} (1/\eta) \exp(-\eta)\, d\eta. \tag{54.57}$$

Horvay and Cahn give the general solution for the elliptical paraboloid which we can write in the form

$$\bar{\alpha} = f(\bar{p}), \tag{54.58}$$

where $f(\bar{p})$ has limiting values (54.56) and (54.57) when the aspect ratio of the elliptical cross-section is ∞ and 1 respectively.

In contrast to eqn. (54.52), the growth rate given by (54.56) for small supersaturations is now proportional to the square of the supersaturation, or equivalently to the square of the supercooling ΔT^- from the equilibrium condition. Horvay and Cahn express the general result for the elliptical paraboloids in the form

$$\Upsilon r \propto \bar{\alpha}^z \propto (\Delta T^-)^z, \tag{54.59}$$

where z is a slowly varying function of ΔT^-. The Zener theory gives $z = 1$, whilst the Ivantsov theory gives $z \sim 2$ for a flat platelet and $\sim 1{\cdot}2$ for a needle with a circular cross-section.

Neglect of the Gibbs–Thomson effect would not be significant if the radius of curvature at the tip of a needle or plate were large in comparison with the critical radius of eqn. (54.37) or (54.55). However, measurements show that this is frequently not so, and there must therefore be a variation in c^α over the surface of the dendrite. This variation will in turn modify the steady-state shape of the particle, an effect first treated by Temkin (1960). Bolling and Tiller (1961) proposed that a reasonable approximation would be to retain the parabolic shape with an assumed constant concentration at the interface of c_r^α appropriate to the tip of radius r. This simply means that eqn. (54.58) becomes

$$\bar{\alpha}_{\text{eff}} = f(\bar{p})$$

or

$$\bar{\alpha} = f(\bar{p})\{1 + (r_c/r)\bar{\alpha} f(\bar{p})\}. \tag{54.60}$$

Trivedi (1970a) has given an exact solution for the diffusion problem which allows the composition to vary along the interface of a plate but assumes that the shape remains that of a parabolic cylinder. His growth equation may be written in the form

$$\bar{\alpha} = f(\bar{p})\{1 + (r_c/r)\bar{\alpha} S_2(\bar{p})\}, \tag{54.61}$$

where $S_2(\bar{p})$ is a function of the Péclet number $\bar{p}$ which Trivedi and Pound (1969) estimated as $3[2\bar{p}(1+2\bar{p})]^{-1}$. Trivedi's exact solution leads to a complex expression for $S_2(\bar{p})$ with numerical values varying from ~ 10 at $\bar{p} = 0{\cdot}07$ to $\sim 0{\cdot}45$ at $\bar{p} = 10$.

Bolling and Tiller also considered that interface processes should not be neglected in dendritic growth, and the theory of morphological stability described above emphasized the possible significant of interface kinetics. If we assume that a finite free energy change is required to move the interface and that the velocity is linearly proportional to this driving force (eqn. 53.9), we can suppose that this results in a deviation Δc^B of the interface composition from the equilibrium value which is additional to that required by the Gibbs–Thomson effect. The composition at the interface is thus

$$c^{\alpha, B} = c^{\alpha}_{\infty} + \Delta c^{\alpha}_{r} + \Delta c^{B}, \tag{54.62}$$

and since the free energy charge on crossing the boundary is in first approximation linearly proportional to Δc^B, we can also write this equation (see eqn. (22.38))

$$c^{\alpha, B} = c^{\alpha}_{\infty}(1 + \Gamma/r) + \Upsilon/M^c, \tag{54.63}$$

where M^c is the boundary mobility per unit concentration difference. If this expression is used in the Zener theory, eqn. (54.43) is modified to

$$\bar{\alpha}_{\text{eff}} = \bar{\alpha}\{1 - (r_c/r) - (\Upsilon/\Upsilon_c)\} = 2C_{14}\bar{p}, \tag{54.64}$$

where Υ_c is defined as the velocity of a flat interface in the limit that the growth is entirely interface-controlled, i.e.

$$\Upsilon_c = M^c(c^m - c^{\alpha}_{\infty}). \tag{54.65}$$

However, it is the normal velocity of the interface which is proportional to Δc^B and so for constant velocity of a parabolic dendrite the interface driving force Δc^B must also vary along the surface. This complication is included in Trivedi's treatment and the final form which replaces eqn. (54.61) is

$$\bar{\alpha} = f(\bar{p})\{1 + (r_c/r)\bar{\alpha}S_2(\bar{p}) + (\Upsilon/\Upsilon_c)\bar{\alpha}S_1(\bar{p})\}, \tag{54.66}$$

where $S_1(\bar{p})$ has a range of values similar to that of $S_2(\bar{p})$. In effect, the function $S_2(\bar{p})\cdot f(\bar{p})$ corrects for the variation in composition due to changing curvature along the interface and $S_1(\bar{p})\cdot f(\bar{p})$ similarly corrects for the required variation of Δc^B due to the changing orientation of the interface.

Trivedi (1970b) also obtained an exact solution for the paraboloid of revolution. The growth equation has the same form as eqn. (54.66), but $f(\bar{p})$ is given by (54.57) and $S_1(\bar{p})$ and $S_2(\bar{p})$ are replaced by different functions with roughly the same range of values. Presumably this form of the growth equation also applies to the general elliptical paraboloid with $f(\bar{p})$ given by the Horvay and Cahn isoconcentrate solution. We must remember that the solutions are exact only within the framework of the imposed boundary conditions (just as the Ham and Ivantsov solutions are "exact" for simpler boundary conditions) and the actual variation of concentration over the interface will lead to a deviation from a parabolic shape. According to Trivedi, this effect is not serious provided the tip radius is greater than 2 or 3 times r_c.

For a given shape and value of $\bar{\alpha}$ there is an infinity of possible growth rates, corresponding to different tip radii. The observation that plates and needles grow at a constant rate presumably indicates that there is a close approximation to a steady-state diffusion field with respect to the advancing interface, and that the tip radius of curvature remains constant and is stable against fluctuations. Zener's assumption that the plate or needle evolves to the form corresponding to maximum velocity has no detailed justification, but attempts to remove the degree of freedom in the system by the use of other optimization principles, such as those discussed in Chapter 4, are not generally accepted. The maximum velocity condition thus has the status of a working hypothesis, to be tested against experiment.

The maximum velocity at given $\bar{\alpha}$ is obtained by differentiating eqn. (54.66) with respect to r and putting $\partial \Upsilon/\partial r = 0$. This gives a further relation between Υ and r which in combination with (54.66) enables Υ and r to be fixed uniquely. The value of r which corresponds to maximum velocity now depends on both $\bar{\alpha}$ and on the degree of interface control; Trivedi expresses this latter by a dimensionless parameter

$$\left.\begin{aligned} q &= M^c(c^m - c^\alpha_\infty)r^c/2D \quad \text{(plates)},\\ q^1 &= 2q \quad \text{(needles)},\end{aligned}\right\} \tag{54.67}$$

where a large value of q means the interface kinetics terms is negligible and a small value means it predominates. If the growth is entirely diffusion-controlled $(q = \infty)$ the value of r/r_c approaches the Zener value of 2 at high supersaturations, but increases to a value of ~ 20 as $\bar{\alpha}$ is reduced to 0·4. At any value of $\bar{\alpha}$, r/r_c increases as the kinetics term becomes more important, i.e. as q decreases; also for sufficiently small q $(\lesssim 0{\cdot}1)$ the dependence of r/r_c on $\bar{\alpha}$ is reversed so that the ratio increases with increasing $\bar{\alpha}$. For intermediate values (e.g. $q = 1{\cdot}0$) there is a minimum in the curve of r/r_c against $\bar{\alpha}$. However the most striking result is that at moderate supersaturations $r \geqslant 10r_c$, and this in turn justifies the assumption made about the shape. Similar results, though with somewhat smaller values of r/r_c, are obtained for needle growth.

The appreciably larger radius of curvature predicted by the Trivedi theory in comparison with the Zener–Hillert theory is associated with corresponding predictions of appreciably smaller maximum velocities. Tests of the theory require simultaneous measurements of tip geometry and velocity and are now beginning to be undertaken (e.g. Purdy, 1971). The theory predicts that needles will always grow faster than plates, and hence should be the preferred growth form, but Trivedi points out that experimental results indicate needles form only at small supersaturations and that plates form at higher supersaturations. He suggests this may be caused by the variation of the diffusion coefficient with concentration, which will lead to a different weighted average D (see p. 490) in the two cases, or alternatively to the smaller transformation strain energy associated with a plate. Clearly anisotropy of surface free energy or of interfacial kinetics might also be significant factors.

The various theories of diffusional growth described above are based on the assumption that growth is continuous at all parts of the interface. As discussed at the beginning of this chapter, this is expected if the interface is disordered on an atomic scale, but for semi-coherent interfaces there is the possibility that growth takes place, even under diffusion control, only at favourable sites on the interface. Several electron microscopic studies of the structures of the planar interfaces of precipitate plates, for example, have indicated that such interfaces contain steps of atomic height and have led to the concept of a growth process controlled by the diffusion flux to the steps (for summary, see Aaronson et al., 1970). The interpretation of these observations is rather controversial and they will be discussed in Part II, Chapter 16, but it is useful here to note how the theory of diffusional growth has been modified to allow for a step mechanism.

For a series of linear steps (ledges) of height d and step density k^S per unit length, the velocity of the interface in the direction normal to the broad face of the plate is simply

$$\Upsilon = dk^S u, \tag{54.68}$$

where u is the mean velocity of a step (Cahn *et al.*, 1964). A first approximation to the velocity of step growth may be to treat the step as the edge of a half-plate, and to use the Zener–Hillert equation (54.52) with $r \simeq d$. A better theory was developed by Jones and Trivedi (1971) who assumed that an isolated edge is square-ended and that the growth is steady state, i.e. that the solute distribution in a coordinate system moving with the step is independent of the time. In order to simplify the complex boundary conditions, they considered only the limit of low supersaturation (54.14) so that Laplace's equation (54.15) could be used to find the concentration along the step interface. This concentration variation was then used to solve the complete diffusion equation.

Jones and Trivedi define a Peclet number for the step as

$$\bar{p}^S = ud/2d \tag{54.69}$$

[see (54.42)] and also use a parameter q^S to express the degree of interface control which is identical with (54.67) except that r_c is replaced by the step height d. Their solution takes the form

$$u = M^c(c^m - c^\alpha)\,[1 + 2q^S\alpha(\bar{p}^S) - 2\bar{p}^S\alpha(\bar{p}^S)]^{-1}, \tag{54.70}$$

where $\alpha(\bar{p}^S)$ is a function which was evaluated numerically and varies from ~ 6 at $\bar{p}^S = 0{\cdot}002$ to $\sim 0{\cdot}5$ at $\bar{p}^S = 0{\cdot}5$. This equation may be expressed in the same form as (54.66), i.e.

$$\bar{\alpha} = 2\bar{p}^S\alpha(\bar{p}^S) + (\bar{p}^S/q^S)\,\{1 - 2\bar{p}^S\alpha(\bar{p}^S)\} \tag{54.71}$$

and in the limit of pure diffusion control ($q^S \to \infty$)

$$u = D\bar{\alpha}/d\alpha(\bar{p}^S), \tag{54.72}$$

where $\alpha(\bar{p}^S)$ is ~ 3 in the range in which the theory is applicable. In the opposite limit of pure interface control, the velocity is given by

$$u = M^c(c^m - c^\alpha). \tag{54.73}$$

Combination of eqns. (54.68) and (54.72) suggests that the diffusion controlled growth rate of a stepped interface will be constant if k^S is constant, i.e. if there is some mechanism for the formation of new steps at a constant rate. Alternatively, with a spiral growth mechanism as discussed in Part II, Chapter 13, the total step length is kept constant by a topological property of the interface.

The small proportion of the sites at which atoms can be added to the growing β-plate seem physically to imply that the step mechanism under diffusion control must always give a growth rate smaller than that for a disordered interface, but it is noteworthy that after a sufficiently long time, the above theory gives the opposite prediction since the growth rate given by (54.71) decreases with time. This has been discussed by Atkinson *et al.* (1973) who believe this prediction may represent a real result of the greater flux achieved by the point effect in diffusion, and these authors cite some experimental evidence that the growth velocity of a stepped interface sometimes exceeds the calculated velocity for continuous growth. On the other hand, it is not clear that the changes' in boundary conditions produced when a large number of steps sweep successively across an interface have been properly considered; this may set an upper limit given by (54.71), to the velocity of the interface.

The last topic which we shall discuss in this section is that of diffusion in a stress field, which is of considerable interest in view of effects arising from the interaction between solute atoms (or point defects) and dislocations. Consider a dislocated crystal which initially contains a uniform distribution of solute atoms, so that there is a net flow of atoms of one type towards the dislocation lines. There are two distinct types of boundary condition which may be imposed at the dislocation cores. In the first of these, the equilibrium state of the dislocated crystal is a non-uniform solute distribution, with solute atoms all remaining mobile, but building up "atmospheres" around the dislocation lines (see p. 267). (The dislocation lines must be regarded as frozen into the structure.) The second possibility is that the dislocation lines are effectively sinks for solute atoms which are removed from solution. This boundary condition corresponds to the nucleation and growth of precipitate particles on dislocations, but there is no reason why this process should not be preceded by a stage of segregation to dislocations within the parent matrix.

The migration of solute atoms to dislocations is controlled by a drift flow resulting from the interaction energy W_i of an individual solute atom with a dislocation and by an opposing diffusion flow. The appropriate flux equation is

$$\mathbf{I} = -D \nabla c - (Dc/kT) \nabla W_j, \tag{54.74}$$

where Einstein's relation (41.15) between the diffusion coefficient and the mobility has been used to give the mean drift velocity of an atom. Equation (54.74) is thus valid only under conditions such that ∇W_i is sufficiently small for the drift velocity to have the same activation energy as that needed for diffusion, and it may not apply close to the dislocation core. The simple elastic model gives $W_i = -(C_{16} \sin \theta)/r$ for an edge dislocation (eqn. (30.24)), where C_{16} is a constant and r, θ are the polar coordinates of the atom relative to the dislocation. The equipotential lines are thus circles passing through the centres of the dislocations, and the lines of force normal to these are also circles defined by $(\cos \theta)/r =$ const. Cottrell and Bilby assumed that in the early stages of ageing only the drift flow need be considered, so that on the average the atoms move along these equipotential lines until they reach the position of maximum binding near the centre of the dislocation line.

Consider a particular flow line of radius R. The mean gradient of W_i on this line is approximately $dW_i/dR = -C_{16}/R^2$, and hence the atoms may be considered to move along this circular line of force with a constant velocity $(C_{16}D/kT)/R^2$. An atom on average will move a distance of order R in travelling round the flow line, and this will take a time of about $R^3kT/C_{16}D$. Correspondingly, after a time t has elapsed all flow lines with radii less than $(C_{16}Dt/kT)^{1/3}$ have been completely drained of solute atoms.

It is now possible to write an expression for the number of solute atoms which arrive at the dislocation in a time interval dt by integrating over all active flow lines, that is for all R greater than the critical radius which has just ceased to operate at time t. Qualitatively, it is obvious that the volume of crystal supplying solute atoms to unit length of the dislocation increases as R^2 and hence as $(C_{16}Dt/kT)^{2/3}$.

In a more accurate treatment, the differential equation

$$\frac{\partial c}{\partial t} = \frac{C_{16}D}{kTr^2}\left[\sin\theta\frac{\partial c}{\partial r} - \frac{\cos\theta}{r}\frac{\partial c}{\partial\theta}\right] \tag{54.75}$$

obtained from (54.74) by neglecting the $D\nabla c$ terms has to be solved subject to the boundary condition $c = 0$ at $r = 0$. The solution is discontinuous across the curve

$$kTr^3\{\tfrac{1}{2}\pi-\theta-\tfrac{1}{2}\sin 2\theta\}/2\cos^3\theta = C_{16}Dt \tag{54.76}$$

with $c = 0$ inside this curve and $c = c^m$ outside the curve. The total number of impurity atoms removed from solution to the dislocation is thus equal to c^m times the area enclosed by (54.76) and this gives the original Cottrell–Bilby result

$$N(t) = 3(\pi/2)^{1/2}c^m(C_{16}Dt/kT)^{2/3}. \tag{54.78}$$

Bullough and Newman (1962) pointed out that it is not possible to control the core boundary condition in this treatment, i.e. it is necessary to postulate an ideal sink at $r = 0$. They refer to this situation as a "Cottrell atmosphere" and distinguish it from a true Maxwellian atmosphere which is a distribution of solute atoms

$$c = c'\exp(-W_i/kT), \tag{54.79}$$

where $c(x_1, x_2, x_3)$ depends only on $W_i(x_1, x_2, x_3)$ and $c' \simeq c^m$. A Maxwellian atmosphere is the equilibrium configuration if the solubility limit is nowhere exceeded, but it may only be obtained kinetically when the diffusion flow term is included in the flux equation. Bullough and Newman's usage of Cottrell atmosphere for the particular situation when a precipitation type of boundary condition is applied at the dislocation core (or rather an undefined perfect sink is postulated at the core) is not universal; as noted on p. 270, Cottrell originally introduced the concepts of a Maxwellian atmosphere and a condensed atmosphere, and any preferred distribution of solute atoms around a dislocation is often referred to as a Cottrell atmosphere.

An alternative to the above treatment of the drift flow is to drop the angular terms in the interaction energy and to write $W_i = -C_{16}/r$; this should be equally valid when applied to measurements which are averaged over many dislocations. The governing equation now becomes

$$\partial c/\partial t = (C_{16}D/kTr^2)\{(\partial c/\partial r)-(c/r)\} \tag{54.80}$$

which has the solution

$$c(r, t) = c^m r / \{r^3 + (3C_{16}Dt/kT)\}^{1/3}. \qquad (54.81)$$

This solution is physically more acceptable than the solution of (54.75) since the concentration field (54.81) contains no discontinuities. The number of solute atoms removed to the dislocation in time t is now given by

$$N(t) = 3^{2/3}\pi c^m (C_{16}Dt/kT)^{2/3} \qquad (54.82)$$

which differs from (54.78) only by a small numerical factor. Thus the apparently large change from circular flow lines to radial flow lines makes very little difference to the overall kinetics of the initial segregation.

A related problem which has assumed some importance in the annealing of slightly impure materials after fast neutron irradiation is the segregation of solute atoms to an assembly of small dislocation loops, rather than to a quasi-linear set of dislocations. Bullough, Stanley and Williams (1968) have shown that the assumption of pure drift flow in this case gives an initial dependence of the fraction of available solute which segregates to the dislocation loops as

$$\zeta(t) = 0{\cdot}38\pi\varrho_1(6\pi r^2 C_{16}Dt)^{3/5}, \qquad (54.83)$$

where ϱ_1 is the density of loops of mean radius r. Thus this modification of the dislocation geometry changes the time dependence of the drift flow from $t^{2/3}$ to $t^{3/5}$.

As the dislocation begins to draw solute from the surrounding region, concentration gradients will be established in the matrix, and the net flux of solute atoms at any point will then consists of contributions from both the drift flow and the normal diffusion flow. This was clearly recognized in the original paper of Cottrell and Bilby, but it is by no means obvious at what stage in the segregation the diffusion flow becomes significant. Further consideration of this problem is postponed to Section 58, but we may mention here that later work has shown that the proportionality of c to $t^{2/3}$ should be valid only for the very early stages of any segregation process, and should not include the range of times over which observable precipitation takes place.

55. LINEAR GROWTH OF DUPLEX REGIONS

In the two previous sections we have discussed linear growth in transitions which do not involve a composition change, and parabolic growth in transitions which are diffusion controlled. A transformation involving long-range transport of atoms, however, can take place in such a way that the mean composition of the transformed regions is equal to that of the untransformed matrix. The conditions for linear growth are then satisfied, the composition of the untransformed matrix remaining uniform except near the boundary of a growing product region. This concept covers eutectoidal decompositions and precipitation reactions of the discontinuous type, in both of which the transformed regions are duplex and usually consist of one or more colonies of parallel lamellar crystals of the product phases. The transformed regions generally originate on grain boundaries and grow hemispherically into one of the grains; they are often called nodules. We shall frequently adopt a terminology due to Turnbull and Treaftis (1955) and refer to "cells" when we wish to specify transformed regions of duplex structure originating from a single nucleus. All of the crystals of one type in a cell are approximately parallel and have the same lattice orientation; a growing hemispherical nodule may consist of one or more sets of cells. It is believed that often each cell consists of two interwoven single crystals, one of which continually develops new branches to give the lamellar microstructure, and evidence for this is available in some transformations (Hillert, 1962). However, Tu and Turnbull (1969) found the tin lamellae from a lead–tin alloy not to be interconnected after dissolution of the lead matrix, and they suggest that independent nucleation may make a large contribution to the multiplication of lamellae.

The cells have an incoherent boundary with the untransformed matrix, even when one of the equilibrium phases has the same structure as the parent phase. The two types of reaction which are important in practice are shown schematically in Fig. 11.5. In the first

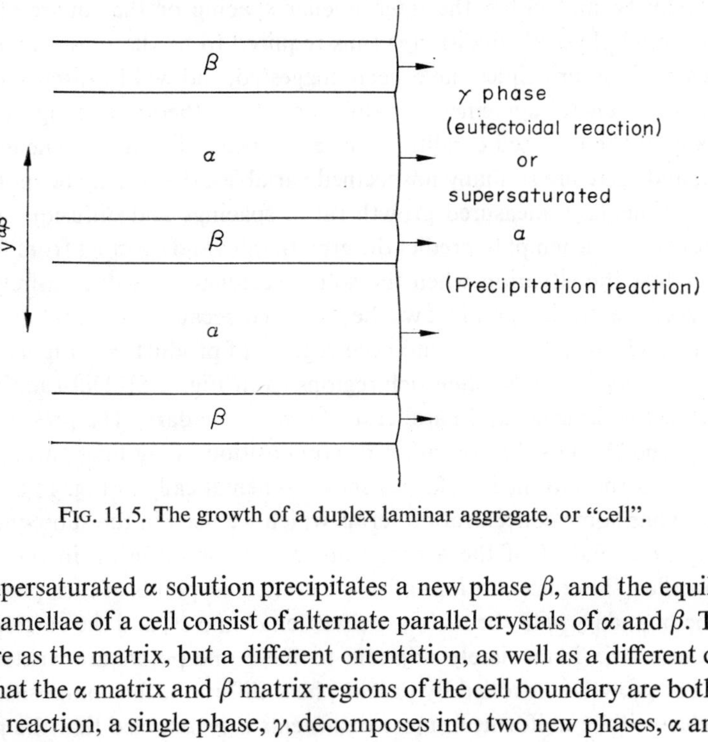

FIG. 11.5. The growth of a duplex laminar aggregate, or "cell".

of these, a supersaturated α solution precipitates a new phase β, and the equilibrium state is $\alpha+\beta$. The lamellae of a cell consist of alternate parallel crystals of α and β. The α has the same structure as the matrix, but a different orientation, as well as a different composition. This means that the α matrix and β matrix regions of the cell boundary are both incoherent. In the second reaction, a single phase, γ, decomposes into two new phases, α and β, and the duplex $\alpha+\beta$ cell has an incoherent boundary with the γ matrix.

In this section we discuss theories of growth for these duplex product regions. It should be noted that concepts such as nucleation rate and growth rate may be applied to the cell as a whole, as well as to individual crystals within the cell. As the cell grows, the individual plates maintain an approximately constant spacing; this must require the branching of existing plates, or the nucleation of new plates, or both processes. A distinction is sometimes made between the edgewise extension of a set of plates, and the sideways growth which requires "lamellation"—i.e. eithernucleation or branching. The growth theories we discuss here are concerned entirely with edgewise growth. In some cases it has been shown directly that sideways growth takes place at a similar rate to edgewise growth, and other experiments show that either increasing or decreasing the temperature of reaction produces an immediate corresponding change in the spacing of the lamellae, so that lamellation is not a limiting process in the growth (Cahn, 1959).

Mehl (1938) noted that the growth of a lamellar aggregate is a steady-state process, and he suggested that the growth rate should be proportional to both the diffusion coefficient and to an effective concentration gradient in the parent phase. This assumption that solute segregation is obtained by volume diffusion in the matrix has been made in the majority of papers on this kind of growth process, but the possibility of diffusion in the product phases has also been discussed, and the importance of the diffusion short circuit provided by the incoherent boundary of the aggregate has been increasingly recognized. In all theoretical treatments, the solution of the diffusion equation leads to an expression for the product of the interface velocity and either the interlamellar spacing or the square of this spacing, and some additional physical condition is thus required to fix these two quantities separately. Although various principles have been suggested, and will be discussed later in this section, there is no general agreement on this part of the theory, and rigorous arguments in favour of any of the imposed conditions have not been discovered. The growth process is so complex, and there are so many unspecified variables, that it may be more informative to test the experimentally measured growth rates, spacings and diffusion coefficients for self-consistency than to attempt to predict the growth rates and spacings from first principles.

We consider first the situation when the solute segregation is due entirely to diffusion in the matrix. A concentration gradient will be produced because the parent phase is enriched in solute immediately ahead of the solute-poor regions of product (α in Fig. 11.5) and depleted in solute just ahead of the solute-rich regions (β in Fig. 11.5). Diffusion then transfers atoms normal to the lamellae and parallel to the cell boundary. The process is illustrated in Fig. 11.6(a), and the possible variation of composition along lines through the centres of the plates and normal to the interface is shown schematically in Fig. 11.6(b). This figure also serves to define some concentration terms which we need in the subsequent discussion and which are independent of the assumption of volume diffusion in the parent phase. Within the lamellae, the concentrations of solute at the mid-points of each plate immediately adjacent to the boundary with the matrix are written c^{α} and c^{β}, and the corresponding compositions in the matrix immediately adjacent to these mid-points are written $c^{m\alpha}$ and $c^{m\beta}$ respectively. For growth controlled by parent phase diffusion, the compositions within the product lamellae will be uniform; in most treatments it is assumed that the products have their equilibrium compositions at the reaction temperature considered, but we emphasize that there is no *a priori* reason for such an assumption. When the growth is controlled by the diffusion rate, the assumption of local equilibrium at the interface should be valid. For the particular case of a eutectoidal reaction, Hultgren (1938) and others have suggested that the compositions of the γ phase adjacent to the mid-points of the α and β plates should be those obtained by extrapolation of the $\alpha+\gamma/\gamma$ and $\beta+\gamma/\gamma$ boundaries to the reaction temperature. In effect, this assumption gives a maximum value to the composition difference $c^{m\alpha}-c^{m\beta}$ for a binary eutectoidal system.

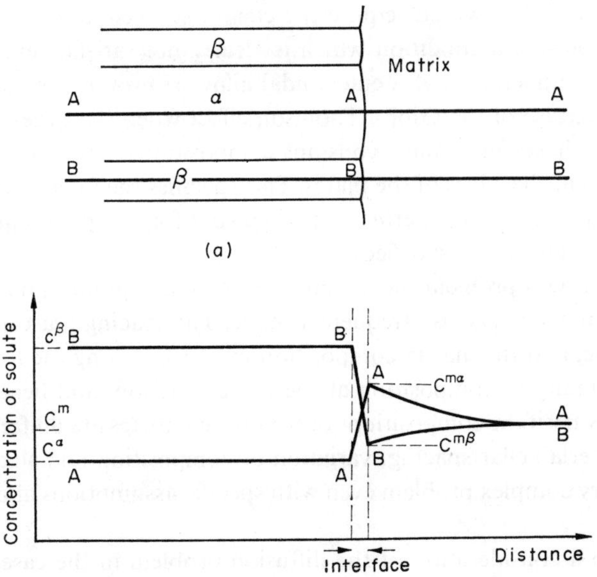

FIG. 11.6. Schematic variation of solute composition for cellular growth controlled by diffusion in the parent phase.

The diffusion gradient and the growth rate may be assumed to be determined by the interlamellar spacing, $y^{\alpha\beta}$, since the diffusion effects the segregation of the solute between the α and β phases; this result is common to all theories in which growth is attributed to volume diffusion. The growth rate for given solute separation does not continue to increase as $y^{\alpha\beta}$ decreases, however, since the net driving force for the reaction also decreases; this effect is identical with that already considered on p. 497. When the cell boundary advances through a distance dx, the chemical free energy released in the volume swept out by unit length of each pair of plates cannot exceed $-(y^{\alpha\beta}\,\Delta g^{t}/v)\,dx$, where v is the volume per atom and Δg^{t} is the change in free energy per atom produced by transformation of the whole structure to the equilibrium $\alpha+\beta$ state, neglecting the energy of the interfaces. To avoid the complications of volume changes, we assume that the volume per atom is approximately equal in all three phases. The increase of surface energy in the volume element is $2\sigma^{\alpha\beta}\,dx$, and in the limiting case when all the available driving force is needed to supply this interfacial energy

$$(y^{\alpha\beta})_{\min} = -2\sigma^{\alpha\beta}v/\Delta g^{t}. \qquad (55.1)$$

A smaller interlamellar spacing is not possible since there would then be an increase in free energy as the cell grows. If the growth process leads to α and β phases of non-equilibrium compositions, the chemical driving force may be written $-P^{t}\,\Delta g^{t}$, where P^{t} specifies the fractional amount of chemical free energy released. If the growth velocity increases at fixed spacing, the degree of segregation achieved, and hence P^{t}, will decrease.

The minimum spacing for a fixed degree of segregation is $(y^{\alpha\beta})_{\min}/P^{t}$, and corresponds to zero growth velocity, that is to an equilibrium between the product phases and the matrix.

For a planar interface, this would require the chemical potential of any component to be equal in all three phases, a condition which is clearly not satisfied except at the thermodynamic transition temperature of a eutectoidal alloy. However, a metastable equilibrium may be possible because of the Gibbs–Thomson effect when the edges of the lamellae are curved. The three phases must have constant compositions, and there must therefore be constant curvatures at the edges of the plates. This specifies the interface shape, and ensures three-phase equilibrium at α–β interface junctions, but for complete equilibrium the static condition (35.1) has also to be satisfied.

In treating the growth problem the assumption of local equilibrium at the interface and at the three phase junctions is also frequently made. The spacing must now be greater than the minimum value, and the matrix composition must vary along the interface, since there can be no discontinuity in composition at the triple junction, and hence the curvature of the interface must vary if the compositions of the product plates are uniform. The calculation of growth rate, interlamellar spacing, variation of composition and shape of the interface thus becomes a very complex problem even with specific assumptions about rate controlling mechanisms.

We now turn to a consideration of the diffusion problem in the case when diffusion in the parent phase produces the segregation. The earliest theory to give independent expressions for both growth rate and spacing was that of Zener (1946), and we find it convenient to give an outline of his argument here, although it will be evident that it contains many approximations or inconsistencies. Zener made no attempt to calculate the growth rate by a rigorous solution of the diffusion equation, but he assumed that eqn. (54.4) may be applied to each plate and hence to the whole interface. This is clearly unsatisfactory, since the equation was derived for conditions in which the concentration gradients are normal to the interface, whereas in cellular growth the gradients parallel to the interface are more important. The assumption is equivalent to one implicit in an earlier discussion by Mehl, namely that the complex diffusion field may be replaced by a simple effective concentration gradient, and we may tentatively accept it if we recognize that the meanings of the composition terms are rather uncertain. Presumably $(c^{\beta}-c^{\alpha})$ in eqn. (54.4) has to be replaced by either $(c^{\beta}-c^{m\beta})$ or by $(c^{m\alpha}-c^{\alpha})$, depending on whether we choose to apply the equation to the centre parts of the β or α plates. The equation requires that Υ is proportional to a characteristic concentration difference in the matrix, Δc, and Zener supposed that Δc is linearly dependent on the curvature of the α-matrix or β-matrix boundary respectively. This is effectively equivalent to the assumption that the diffusion flux, and hence Δc, is proportional to the net change in free energy produced by growth. Both chemical and surface terms must be included in this net free energy change, which may thus be regarded as the effective driving force for growth; in the case that the equilibrium segregation is achieved, it is given by

$$-\Delta g^{\Upsilon} = -\Delta g^{t} - 2(\sigma^{\alpha\beta} v / y^{\alpha\beta}). \tag{55.2}$$

If we now make Δc proportional to $-\Delta g^{\Upsilon}$ we obtain

$$\Delta c = [1 - \{(y^{\alpha\beta})_{\min} / y^{\alpha\beta}\}] \Delta c_{\infty}, \tag{55.3}$$

where Δc_∞ is the concentration difference for a plane boundary. The diffusion distance of eqn. (54.4) is taken to be proportional to $y^{\alpha\beta}$ and is written $C_{17}y^{\alpha\beta}$, where C_{17} is a constant. The equation for the growth rate of a β plate then becomes

$$\frac{\Upsilon}{D} = \frac{\Delta c_\infty}{(c^\beta - c^{m\beta})} \frac{(y^{\alpha\beta} - (y^{\alpha\beta})_{\min})}{C_{17}(y^{\alpha\beta})^2}, \tag{55.4}$$

and the maximum value of Υ corresponds to $y^{\alpha\beta} = 2(y^{\alpha\beta})_{\min}$. This maximum growth velocity is given by

$$\Upsilon = \frac{D}{2y^D} \frac{\Delta c_\infty}{(c^\beta - c^{m\beta})} = \frac{D}{4C_{17}(y^{\alpha\beta})_{\min}} \frac{\Delta c_\infty}{(c^\beta - c^{m\beta})}. \tag{55.5}$$

Zener's condition for maximum growth rate thus requires that one-half of the driving force is transformed reversibly into the surface energy of the α–β interfaces. This treatment is, of course, virtually identical with that on p. 495.

From eqn. (55.1) we may estimate the variation of $(y_{\alpha\beta})_{\min}$ with temperature, and hence also the variation of $y^{\alpha\beta}$. Using a common approximation, though perhaps with less justification than on p. 442, we write $\Delta g^t = \Delta h^t\,(T^E - T)/T^E$, where T^E is the thermodynamic transition temperature. This gives the temperature variation

$$y^{\alpha\beta} \propto (\Delta T^-)^{-1}, \tag{55.6}$$

where $\Delta T^- = T^E - T$ is the supercooling. In the eutectoidal reaction in steels, the spacing of the plates in pearlite apparently fits this law, but the slope of the experimental line cannot be reconciled with the theory since it leads to an unreasonably large value for $\sigma^{\alpha\beta}$. Data for both the kinds of reaction illustrated in Fig. 11.5 indicates that $y^{\alpha\beta}$ is appreciably larger than $2(y^{\alpha\beta})_{\min}$.

If a numerical value for the growth rate is to be deduced, the composition terms and the diffusion distance have to be specified, and here there is considerable confusion in the literature. As already noted, Zener applied eqn. (54.4) to the growth of individual plates of the aggregate, and he considered specifically the case of cementite plates in pearlite. If we call these plates the β plates, the composition term in the denominator becomes $c^\beta - c^{m\beta}$, as in eqn. (55.4), and y^D was rather arbitrarily taken to be equal to the thickness of the β plate, so that C_{17} is the fractional width of these plates. Zener further assumed that the concentration difference Δc is equal to $c^{m\alpha} - c^{m\beta}$, and the used the Hultgren extrapolation to obtain these quantities. He recognized that this treatment would have to give equal growth velocities to both α and β plates, and gave this as his reason for making y^D equal to the thickness of only one of these plates when the composition terms in the denominator are expressed in the form of $c^\beta - c^\alpha$. In a later review of the theory, Fisher (1950) wrote the growth rate as

$$\Upsilon = \frac{2D}{y^{\alpha\beta}} \frac{(c^{m\alpha} - c^{m\beta})}{c^{m\alpha}}, \tag{55.7}$$

which is nearly the equivalent of Zener's equation applied to the growth of ferrite (α) plates if it is assumed that c^α is neglected in comparison with $c^{m\alpha}$, and that $y^D = \frac{1}{2}y^{\alpha\beta}$. The difficulties of reconciling the diffusion distances and composition terms illustrate the limitations of this semiqualitative approach, but if eqn. (54.4) is to be used, it would seem sensible for reasons of symmetry to take the diffusion distance as proportional to $y^{\alpha\beta}$ and the composition term in the denominator as $(c^{\beta}-c^{\alpha})$ for both phases. We shall now consider the attempts which have been made to obtain a rigorous solution to the diffusion equation, and we shall find that these compositions do indeed define a convenient growth parameter.

Consider the edgewise growth of a set of parallel lamellae when all the diffusion takes place in the parent phase. Using an origin in the moving boundary, we have to solve eqn. (54.45) in order to obtain a steady-state growth process. The complete solution may be obtained by separating variables as before, and is now periodic in the x_2 direction; as first shown by Brandt (1945), it may be written as an infinite series

$$c-c^m = \sum_{n=0}^{\infty} A_n \exp-(\lambda_n x_1)\cos(2\pi n x_2/y^{\alpha\beta}), \tag{55.8}$$

wher the continuous variable b of eqn. (54.49) has been replaced by $b_n = 2\pi n/y^{\alpha\beta}$. The corresponding values of the coefficients λ are

$$\lambda_n = (\Upsilon/2D)\,[1+\{1+(4\pi n D/\Upsilon y^{\alpha\beta})^2\}^{1/2}]. \tag{55.9}$$

In a multi-component system, a solution of form (55.8) exists for each component, c^m being the initial concentration of that component in the metastable matrix, and the values of A_n for that component being determined by the boundary conditions. In the present application, the boundary conditions are completely compatible with the existence of a steady-state solution, but unfortunately our knowledge of these conditions is insufficient to determine the coefficients A_n. If we suppose that the composition is uniform within the α and β plates, the steady-state solution imposes a condition that the mean concentration over any plane normal to the growth direction in the product shall be the original concentration of that element c^m, so that

$$y^{\alpha}c^{\alpha}+y^{\beta}c^{\beta} = y^{\alpha\beta}c^m, \tag{55.10}$$

where y^{α}, y^{β} are the mean widths of the individual α and β plates respectively. Since $y_{\alpha\beta} = y^{\alpha}+y^{\beta}$, this condition can be written

$$y^{\alpha}(c^m-c^{\alpha}) = y^{\beta}(c^{\beta}-c^m) = (y^{\alpha}y^{\beta}/y^{\alpha\beta})\,(c^{\beta}-c^{\alpha}). \tag{55.11}$$

In the more general case where the composition in the product phases is not taken to be constant in the x_2 direction, eqn. (55.10) would have to be replaced by

$$\int_{y^{\alpha}} c(\alpha)\,dx_2+\int_{y^{\beta}} c(\beta)\,dx_2 = y^{\alpha\beta}c^m. \tag{55.12}$$

Although the mean composition in the product must equal that in the original matrix, this is not necessarily true of the diffusion zone ahead of the growing interface. The average concentration over any plane of constant x_1 in the matrix phase is given by

$$\bar{c}(x_1) = (1/y^{\alpha\beta})\int_0^{y^{\alpha\beta}} c\,dx_2 = c^m+A_0\exp(-\Upsilon x_1/D). \tag{55.13}$$

There will thus be a zone ahead of the interface which is enriched or depleted in the component considered, except when A_0 for this component is zero. If there is local equilibrium at the interface, A_0 will be zero for some particular matrix composition at each reaction temperature; in the case of a eutectoidal reaction, this matrix composition will be represented by a line in the equilibrium diagram beginning at the eutectoid point. When a depleted or enriched region exists, its extent in the x_1 direction for a particular component will be increased by a high diffusion coefficient, and decreased by a high interface velocity; it is clear from eqn. (55.13) that we may use D/Υ as a measure of the thickness of the region.

By retaining only the first two terms of the expansion in eqn. (55.8), Brandt obtained an expression for the growth rate in the form

$$\Upsilon = (2\pi D/y^{\alpha\beta})\alpha^m \tag{55.14}$$

in which D refers to the rate-controlling diffusion process and α^m is a rather complex homogeneous function of the various concentrations shown in Fig. 11.6. The factor 2π is introduced purely for convenience. The parameter α^m may be compared to the growth parameters used in eqns. (54.8) – (54.13) to characterize growth under parabolic conditions. When α^m is small in comparison with unity, it is possible to show (Cahn and Hagel, 1962) that

$$\alpha^m = \frac{c^{m\alpha} - c^{m\beta}}{2(c^\beta - c^\alpha)}, \tag{55.15}$$

which are just the composition terms which we should expect to find in a "natural" parameter for this type of growth. Combination of eqns. (55.14) and (55.15) now gives an expression for the growth rate which is very similar to that of Zener, and which is independent of the original composition c^m. The more accurate expression given by Brandt does include a small dependence on c^m, but the error in neglecting this is less significant than is that resulting from the retention of only two terms in (55.8).

We can now use eqns. (55.11) and (55.15) to see how the growth rate, spacing, diffusion coefficient, and degree of segregation achieved in the product are interrelated. When D is small or the product $\Upsilon y^{\alpha\beta}$ is large, α^m is relatively large. There is an upper limit to $(c^{m\alpha} - c^{m\beta})$ which corresponds to the assumption of local equilibrium at the interface, so that $(c^\beta - c^\alpha)$ must become small in these circumstances. In the other extreme, α^m is small when D is large or the product $\Upsilon y^{\alpha\beta}$ is small; since $(c^\beta - c^\alpha)$ cannot exceed the difference given by the equilibrium phases, it follows that $(c^{m\alpha} - c^{m\beta})$ must then become small, so that the composition gradients along the interface are reduced. In most treatments is has simply been assumed that the equilibrium phases are produced by the growth process; the realization that this need not be so is largely due to Cahn (1959) and greatly complicates the theory of the growth. Thermodynamic arguments tell us only that the structures and compositions of the product phases must be such as to produce a net decrease in free energy; within this limitation, the boundary could move slowly, producing a high degree of solute separation, or rapidly, producing less solute segregation. This degree of freedom is applicable to a fixed value of interlamellar spacing, and implies that the growth rate may be dependent on some additional kinetic parameter, such as the mobility of the interface. Some authors dispute this dilemma, regarding it as self-evident that equilibrium conditions obtain at the interface, including the static condition for the equilibrium of the interphase boundaries at the triple junctions, but there does not seem to be any satisfactory way of proving the necessity of such an equilibrium in a kinetic situations.

Even if the segregation achieved is known or may be derived, the theory of the growth still contains a further degree of freedom, which is needed to fix either the spacing or the growth velocity. We have already described Zener's assumption that the spacing is that which corresponds to maximum growth rate. This yields absolute expressions for growth rate and spacing when combined with the Gibbs–Thomson condition, which reduces the effective composition difference available for the diffusion path in the matrix. However, although the assumption of maximum growth rate is intuitively attractive, it is difficult to give any rigorous justification for it, and Cahn comments that it is not obvious why some other physical quantity such as the rate of production of entropy should not be maximized. In this theory of cellular growth by interface diffusion, Cahn suggested that the spacing should be fixed by minimizing the rate of decrease of free energy; this is discussed below. An alternative suggestion (Kirkaldy, 1958, 1962) is that the rate of entropy production should be minimized; this is based on arguments derived from the theory of the thermodynamics of the steady, state, as described in Section 15. Theories based on any of these variational principles are rather unconvincing. The important physical condition is presumably that the spacing, interface shape, compositions, etc. are stable against fluctuations of all types (see below).

We have already noted that the Zener spacing is not observed under experimental conditions, which frequently correspond to $y^{\alpha\beta} \sim 5(y^{\alpha\beta})_{\text{min}}$. There is an equally large discrepancy in computed growth rates, although this should not perhaps be taken too seriously because of the approximations made in Brandt's treatment of the diffusion problem. A more accurate solution of the diffusion equation was obtained by Hillert (1957), using the results of eqns. (55.8), (55.9), and (55.10). In order to obtain values for the coefficients A_n, he assumed that the interface is so plane that $\exp-(\lambda_n x_1)$ is approximately unity along the whole front. Assuming equilibrium conditions as defined by the Gibbs–Thomson relations and the surface free energies, he deduced a very reasonable approximate shape for the interface in the case of pearlite, and Fourier coefficients given by

$$A_n = \frac{\lambda_n (y^{\alpha\beta})^2}{2\pi^3 n^3 D}(c^{\beta}-c^{\alpha})\sin(n\pi y^{\alpha}/y^{\alpha\beta}) \qquad (n>0), \tag{55.16}$$

and this yields growth velocity

$$\frac{\Upsilon}{D} = \pi^3 \frac{y^{\alpha}y^{\beta}}{(y^{\alpha\beta})^2 \sum_{1}^{\infty}(1/n^3)\sin^2(n\pi y^{\alpha}/y^{\alpha\beta})}\frac{2\alpha^m}{y^{\alpha\beta}}, \tag{55.17}$$

where α^m is the parameter of eqn. (55.15). This equation differs from Zener's equation, or from (55.14) only by a factor dependent on the ratio of the thicknesses of the α and β plates, i.e. on the original composition c^m.

Hillert makes Zener's assumption that the parameter α^m is equal to the maximum value of this parameter α^{∞}, obtained from the equilibrium diagram, multiplied by $1-[(y^{\alpha\beta})_{\text{min}}/y^{\alpha\beta}]$

to take care of the effects of curvature. He thus obtains an expression for maximum growth rate which is of the same form as eqn. (55.5). When applied to experimental results for many eutectoidal and precipitation reactions, all of the equations we have derived give growth rates which are very much slower than those observed, that is, they would require coefficients of diffusion orders of magnitude greater than the experimental coefficients of diffusion in the parent phase. The evidence thus shows that the Zener spacing is often not observed and that volume diffusion in the parent frequently does not effect the segregation.

We shall now consider briefly the possibility that the main diffusion paths are in the product rather than in the parent. This seems an attractive possibility for pearlite formation, as pointed out by Fisher (1950), since the diffusion rate of carbon is about 100 times larger in ferrite than it is in austenite. Fisher wrote the solute flux through the ferrite–matrix boundary as Υc^m (apparently neglecting c^α in comparison with c^m) and he gave the growth rate as

$$\Upsilon = (2D/y^{\alpha\beta})(\Delta c/c^m), \tag{55.18}$$

where D is now the diffusion coefficient in the α, Δc is the difference in concentration in the α phase adjacent to the matrix and to the β phase respectively, and the diffusion distance has been takes as $\frac{1}{2}y^{\alpha\beta}$. A more rigorous treatment gives similar results. The diffusion equation (54.45) yields the solution

$$c - c^\alpha = \sum_{n=-\infty}^{0} A_n \exp(-\lambda_n x_1)(\cos 2\pi n x_2/y^{\alpha\beta}) \tag{55.19}$$

with

$$\lambda_n = (\Upsilon/2D)\,[1-\{1+(4\pi n D/Y y^{\alpha\beta})^2\}^{1/2}] \qquad (n < 0),$$

which may be interpreted in a similar way to that already discussed for the case of parent phase diffusion. Diffusion in the product must involve composition gradients in both phases, which is not too probable, or else the thickness of the β plate must decrease towards zero as the interface is approached. Although such tapering or ending of the β plates is occasionally observed (Darken and Fisher, 1962), it does not seem to be generally found in sufficiently rapidly quenched specimens. Moreover, although the diffusion coefficient in one of the product phases may be appreciably larger than in the matrix, as in the example just quoted, the available composition difference for diffusion is also smaller. Thus it does not appear likely that any appreciable fraction of the solute segregation is achieved by volume diffusion in the region of product immediately behind the interface. High growth rates can then only be explained by some form of diffusion short circuit, and it now seems certain that in many transformations the dominant diffusion process is along the cell boundary and not through the parent or product lattices.

The use of the cell boundary as a diffusion short circuit was suggested by several workers, and an attempt at a quantitative treatment was first made by Turnbull (1955). The incoherent boundary is particularly effective in a reaction of this nature, since it sweeps through the matrix as the cell grows, and is favourably orientated for achieving segregation in the required direction. Since the activation energy for cell boundary diffusion will be appreciably less than that for lattice diffusion, the relative importance of the processes may change with temperature, boundary diffusion becoming dominant as the temperature is lowered.

In Turnbull's theory of cell growth, eqn. (55.18) is modified to give

$$\Upsilon = \{(D^B\delta^B/(y^{\alpha\beta})^2\}\{(c^m - c^\alpha)/c^m\}, \qquad (55.20)$$

where D^B is the diffusion coefficient in the boundary and δ^B is the thickness of the boundary. Equation (55.20) contains Fisher's approximate treatment of the concentration terms and a re-evaluation by Aaronson and Liu (1968) leads to the replacements of the last brackets by $4(c^\beta - c^\alpha)/(c^\beta - c^m) \simeq 4$ when (54.14) applies. Cahn and Hagel (1962) use a slightly different formulation, writing instead of eqn. (55.14)

$$\alpha^B = \Upsilon(y^{\alpha\beta})^2/4\pi^2 D^B\delta^B \qquad (55.21)$$

since the thickness of the band of sideways diffusion has changed by approximately $2\pi\delta^B/y^{\alpha\beta}$. This follows since the diffusion in the parent phase takes place over a band of thickness $1/\lambda_1$. When the value of α^m given by eqn. (55.14) is much smaller than unity, $1/\lambda_1 = y^{\alpha\beta}/2\pi$ and the thickness of the diffusion zone is much smaller than the extent of the depleted or enriched zone D/Υ. If α^B is much greater than unity, the diffusion band is equal to D/Υ and smaller than $y^{\alpha\beta}/2\pi$; little segregation then occurs by volume diffusion. If segregation is achieved by a combination of volume and interface diffusion, the growth rate spacing and composition terms will be related by equating the sum of α^m and α^B to the value given by eqn. (55.15). The dominant diffusion mechanism depends on the ratio of the diffusion coefficients, the condition for boundary diffusion being the more important, being

$$D^B/D > y^{\alpha\beta}/2\pi\delta^B. \qquad (55.22)$$

A theory of growth when the boundary mechanism is operative is due to Cahn (1959). He pointed out that with boundary diffusion it is impossible to achieve the equilibrium segregation at any non-zero value of the growth rate, since there must be a composition gradient in the boundary itself, and hence in the α and β plates, unless the growth rate tends to zero. This means that P^t (p. 507) must always be less than unity, and the net free energy decrease of the reaction becomes

$$\Delta g^\Upsilon = P^t\,\Delta g^t + 2\sigma^{\alpha\beta}v/y^{\alpha\beta} \qquad (55.23)$$

in place of (55.2). For fixed $y^{\alpha\beta}$, there is a minimum value of P^t, and a corresponding maximum growth rate, Υ_{max}, which gives $\Delta g^\Upsilon = 0$. Equally for fixed P^t, there is a minimum spacing of $1/P^t$ times the minimum spacing in eqn. (55.1).

Suppose we fix the spacing and consider the variation in the amount of segregation with boundary velocity. We then encounter the difficulty discussed on p. 511 that any value of Υ smaller than Υ_{max} is thermodynamically possible. A full specification of the growth process now requires a second kinetic parameter in addition to the diffusion coefficient, and the choice of this is not obvious. Cahn assumed that the boundary may be treated in the same way as an ordinary boundary (Section 53) and by analogy with eqn. (53.9), the growth rate is regarded as the product of a mobility and a free energy driving force. Using the net driving force Δg^{Υ}, the growth rate becomes

$$\Upsilon = -M^B \Delta g^{\Upsilon}, \tag{55.24}$$

where M^B is the boundary mobility and Δg^{Υ} is given by eqn. (55.23). Combination of this equation with the diffusion equations (55.21) and (55.15) then enables the growth rate and degree of segregation to be determined if the spacing is known.

Some physical principle is still needed to fix the spacing itself. In Zener's theory the spacing was chosen to maximize the growth rate. Cahn adopted the hypothesis that the spacing maximizes the decrease in free energy Δg^{Υ}, and points out that the spacing which achieves this also maximizes the growth rate Υ and the rate of decrease of free energy per unit area of cell boundary. This is thus a very reasonable assumption although justification in a more fundamental manner seems desirable. In Zener's treatment the maxima in the above three quantities occur at quite different spacings, and it is not obvious which should be chosen.

The permissible variation in segregation achieved by cellular growth has quite different forms for precipitation and eutectoidal reactions. This is evident from Fig. 11.7. which

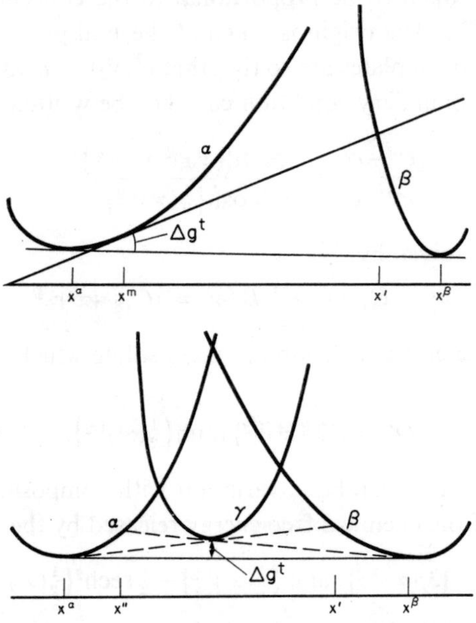

FIG. 11.7. Schematic free energy vs. composition curves for eutectoidal and precipitation reactions (after Cahn, 1959).

shows hypothetical free energy curves for the two kinds of reaction. For the precipitation reaction, a lowering of free energy results from the formation of β phase of any composition richer in solute than x', even if the amount of β phase formed is so small that the α phase has a composition almost identical with that of the original matrix composition x^m. Thus a small amount of segregation is quite feasible. For the eutectoidal reaction, the composition of the α phase must be less than x'', and that of the β phase greater than x' if the free energy is to be reduced. This means that neither phase can approximate to the matrix composition, and a large fraction of the equilibrium segregation must be accomplished as the cells grow.

To put the above considerations into quantitative form it is necessary to solve the diffusion equation. Cahn has done this, using an idealized model for each type of reaction. The diffusion is assumed to be confined to a plane boundary of thickness δ^B and diffusion coefficient D^B, independent of concentration. The concentration of solute in the boundary is $c^B(x_2)$, where x_2 is a coordinate along the boundary normal to the lamellae. The corresponding concentration in the product regions of α and β plate is written $c^P(x_2)$. Equating the diffusion flow along the boundary to the flow due to the motion of the boundary, in order to obtain a steady state, we then find

$$D^B\delta^B(\mathrm{d}^2c^B/\mathrm{d}x_2^2)+\Upsilon(c^m-c^P) = 0, \tag{55.25}$$

which is the equivalent in this model of the volume diffusion equation (54.45).

In solving this equation, Cahn assumes that the α and β phases both have their equilibrium concentrations at the α/β boundaries. For a precipitation reaction, the concentration of the solute in the α phase is taken to be proportional to the concentration in the boundary, so that $c^P(x_2)/c^B(x_2) = C_{18}$. The origin is taken at the mid-point of an α plate, and if the supersaturation is small, the β plates are so thin that $c^P(x_2) = c^\alpha$ at $x_2 = \frac{1}{2}y^{\alpha\beta}$. The solution to eqn. (55.25) with this boundary condition can then be written

$$\frac{c^m-c^P}{c^m-c^\alpha} = \frac{\cosh\{(\alpha^t)^{1/2}\,x_2/y^{\alpha\beta}\}}{\cosh\{\frac{1}{2}(\alpha^t)^{1/2}\}}, \tag{55.26}$$

where the parameter α^t is given by

$$\alpha^t = C_{18}\Upsilon(y^{\alpha\beta})^2/D^B\delta^B = (C_{18}/4\pi^2)\alpha^B. \tag{55.27}$$

In terms of this parameter, the fraction of excess solute which is precipitated as the cell grows is given by

$$Q^t = \{(2/(\alpha^t)^{1/2}\}\tanh\{\tfrac{1}{2}(\alpha^t)^{1/2}\}. \tag{55.28}$$

If the free energy curve has a parabolic variation with composition, as is true for dilute solutions, the fraction of the chemical free energy released by the growth is given by

$$P^t = \{3/(\alpha^t)^{1/2}\}\tanh\{\tfrac{1}{2}(\alpha^t)^{1/2}\}-\tfrac{1}{2}\operatorname{sech}^2\{\tfrac{1}{2}(\alpha^t)^{1/2}\}. \tag{55.29}$$

These equations both contain the growth rate and the spacing. An expression for α^t in terms of independent quantities is obtained by maximizing $-\Delta g^\Upsilon$ in eqn. (55.23), and using eqn. (55.24). This gives

$$\Delta g^\Upsilon/\Delta g^t = P^t+2\alpha^t(\mathrm{d}P^t/\mathrm{d}\alpha^t)$$

assuming P^t to be a function of α^t only, and leads to a relation

$$\frac{(\alpha^t)^3(\mathrm{d}P^t/\mathrm{d}\alpha^t)^2}{\alpha^t+2\alpha^t(\mathrm{d}P^t/\mathrm{d}\alpha^t)} = -\frac{C_{18}M^B(\sigma^{\alpha\beta})^2\,v^2}{D^B\delta^B\,\Delta g^t} = \beta^t, \tag{55.30}$$

which implicitly gives α^t in terms of a parameter β^t which may be evaluated independently. Figure 11.8 shows the way in which log α^t, P^t and Q^t vary with β^t, increasing β^t representing conditions in which the boundary mobility becomes larger in relation to the diffusion coefficient.

Figure 11.8. also includes plots of the quantities R^t and $(P^t - R^t)$. R^t is the fraction of the chemical driving force which is converted to surface energy, and is given by

$$R^t = -(2\sigma^{\alpha\beta}v/y^{\alpha\beta}\,\Delta g^t) = (y^{\alpha\beta})_{\min}/y^{\alpha\beta}$$
$$= -2\alpha^t(\mathrm{d}P^t/\mathrm{d}\alpha^t). \tag{55.31}$$

In Zener's theory, $R^t = \frac{1}{2}$, but as may be seen from the figure, Cahn's model allows it to

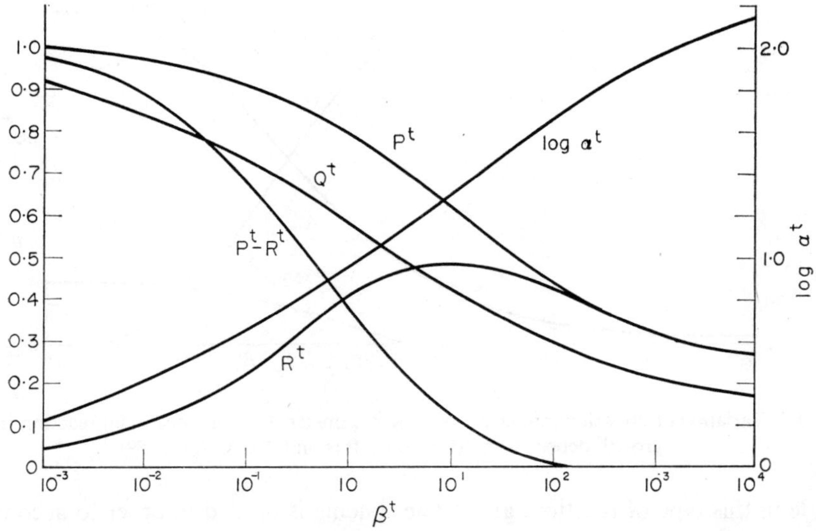

FIG. 11.8. Variation of growth parameters with the parameter β^t of (55.30) for a discontinuous precipitation reaction in which growth depends on boundary diffusion (after Cahn, 1959).

have any value from 0 to a maximum of 0·49. The quantity $(P^t - R^t)$ is the driving force available to exert pressure on the boundary.

Examination of Fig. 11.8 shows that for a highly mobile boundary with a low diffusion coefficient, $P^t = R^t$, and all the free energy released is converted to surface energy. The growth is almost reversible, and the boundary velocity is governed only by the thermodynamic requirement that $P^t - R^t$ shall be positive. The boundary velocity is thus $\Upsilon = \Upsilon_{max}$. Conversely, if the diffusion coefficient is relatively high, and the mobility low, the proportion of the chemical energy released is much higher. Since the boundary can now achieve segregation efficiently, the spacing can also be large, and R^t becomes very small. A large spacing permits more of the chemical driving force to be utilized in moving the rather immobile boundary.

The other idealized case considered by Cahn is that of a symmetrical eutectoid, with the eutectoid composition at $x_B = \frac{1}{2}$ and the two new phases having compositions x^α and x^β =

$1-x^{\alpha}$. We shall not discuss this in detail; the results are shown in Fig. 11.9 in terms of a parameter $\beta' = -\beta^t \Delta g^t / C_{19}$, where C_{19} is a constant specifying the (assumed parabolic) free energy curves. The parameter β' has some advantage over β^t in being nearly independent of temperature. Figure 11.9 is plotted for a particular assumption, $-\Delta g^t / C_{19} = 10^{-2}$, so that it may readily be compared with Fig. 11.8 for the precipitation reaction. It is seen that for high diffusion rate and low mobility, the sets of curves are very similar, but this is not true of high mobilities and low diffusion rates. As mentioned earlier, Q^t is able to

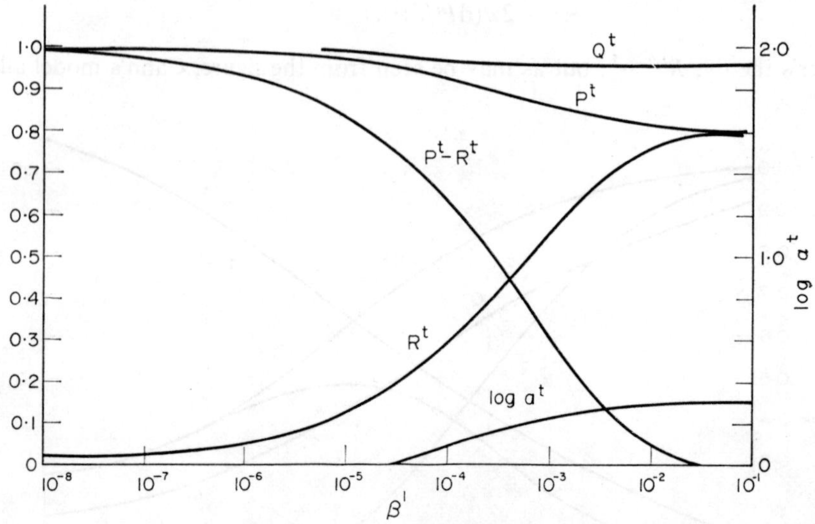

FIG. 11.9. Variation of growth parameters with the parameter β' for a eutectoidal reaction in which growth depends on boundary diffusion (after Cahn, 1959).

vary little in this type of reaction, and a fine spacing is needed in order to accomplish the required segregation when the mobility is high. The requirement that $(P^t - R^t)$ be positive can then only be met if about 80% of the available free energy is converted into surface energy. Thus P^t and R^t both tend to $\sim 0{\cdot}8$, instead of to the much lower value in Fig. 11.8. This high value of R^t means that the spacing in a eutectoidal reaction can be much finer even than Zener's prediction, approaching a limit of $y^{\alpha\beta} \simeq 1{\cdot}25(y^{\alpha\beta})_{\min}$.

Cahn's theory contains some unverified assumptions but is clearly more satisfactory than earlier treatments. Although the detailed results in Figs. 11.8 and 11.9 depend on the properties of the idealized models, the general trends will be common to all reactions of the two types provided the assumptions are correct. Modern techniques allow the theory to be tested, since Υ, $y^{\alpha\beta}$ and Q^t may all be measured. Liu and Aaronson (1968) conclude that both the modified Turnbull theory and the Cahn theory give reasonable values for D^B when used in conjunction with experimental results on lead–tin alloys, but that predicted values of $\sigma^{\alpha\beta}$ are much too high and temperature sensitive. Speich (1968) found his experimental results on iron–zinc alloys implied that the interface velocity depends on the cube of the driving force, so that eqn. (55.24) is replaced by

$$\Upsilon = -M^1(\Delta g^{\Upsilon})^3. \tag{55.32}$$

The modified Cahn theory which resulted from this growth has predicted spacings $y^{\alpha\beta}$ within a factor of two of those observed.

Some authors (e.g. Hillert, 1968) have criticized the Turnbull and Cahn theories of discontinuous precipitation because they do not consider explicitly the forces acting on the α'–α boundary and thus give no clear indication why the boundary should move. The difficulty is not encountered in a rather special model (Shapiro and Kirkaldy, 1968) which is based on the assumption that there is a metastable monoeutectoid in systems exhibiting discontinuous precipitation, but this seems unsatisfactory as a general condition. A more general way of avoiding the difficulty (Hillert, 1968; Sundquist, 1973) is to use a model in which the grain boundary is treated as a separate phase.

The most recent theories of lamellar growth (Sundquist 1968, 1973) are based on the hypothesis that the solute segregation is produced by boundary diffusion, but local equilibrium at the interface is assumed for eutectoidal decomposition, whereas a deviation from equilibrium is required in discontinuous precipitation. The interface spacing in these theories is determined by stability criteria rather than by optimization principles.

In his theory of eutectoidal growth, which is applied specifically to pearlite in steels, Sundquist (1968) uses the Gibbs–Thomson effect to relate the interface shape with the composition variation along the interface, and he does not introduce a mobility parameter. For a given spacing, solution of the diffusion equation enables the shape and velocity of the interface to be approximately computed. This theory is thus similar to the growth theory developed much earlier by Hillert (1957), but is modified to take account of interface diffusion rather than volume diffusion.

Since the interface is no longer considered to be planar and diffusion is allowed only along the interface, it is convenient to use the interface to define a curvilinear coordinate l in place of x_2 in the diffusion equation (55.25). The angle between the local direction of l and x_2 is θ, and it follows also that the velocity in the equation is $\Upsilon_\theta = \Upsilon \cos\theta$ where Υ is the velocity in the overall growth direction (see p. 499). In this theory the interface diffusion coefficient D^B is allowed to have separate values $D^{B\alpha}$ and $D^{B\beta}$ in the γ–α and γ–β interfaces. The interface concentration is assumed to be proportional to the concentration in the austenite adjacent to the interface, and this in turn is governed by the Gibbs–Thomson effect. This gives (cf. eqn. (22.38))

$$c^B = K^B c_r^{\gamma P} = K^B c_\infty^{\gamma P}(1+\Gamma^P/r), \qquad (55.33)$$

where P means either α or β depending on whether the α–γ or β–γ interface is considered, $c_\infty^{\gamma P}$ represents the composition of γ in equilibrium with either α or β at a planar interface, r is the local radius of curvature, and Γ^P is given by eqn. (22.34) or one of its approximate forms. The diffusion equation (55.25) becomes

$$\Upsilon \cos\theta (c^m - c^P) = -D^B K^B \delta^B \Gamma^P \{\partial^2(1/r)/\partial l^2\}. \qquad (55.34)$$

Since $1/r = \mathrm{d}\theta/\mathrm{d}l$, this equation may be put in the form

$$B^P \cos\theta = \partial^3\theta/\partial l^3. \qquad (55.35)$$

Equations (55.33)–(55.35) may each be applied separately to the γ–α and γ–β boundaries. However, at the three phase junction, $c^{\gamma\alpha}$ must equal $c^{\gamma\beta}$, so that if the origin of l is chosen at this point

$$c_{\infty}^{\gamma\alpha}\{1+\Gamma^{\alpha}(\partial\theta^{\alpha}/\partial l)_{l=0}\} = c_{\infty}^{\gamma\beta}\{1+\Gamma^{\beta}(\partial\theta^{\beta}/\partial l)_{l=0}\}. \tag{55.36}$$

There are also two other boundary conditions, namely that $(\theta^{P})_{l=0}$ has values θ_0^P which satisfy the condition for local equilibrium of surface forces (eqn. 35.1), and that

$$(\partial^n\theta/\partial l^n)_{l=L^P} = 0 \qquad (n = 2, 4, \ldots), \tag{55.37}$$

where the superscript P, as before, identifies α or β, and L^P is the value of l at the mid-points of either α or β plates respectively. Clearly, the approximation $\cos\theta \simeq 1$ requires

$$L^{\alpha} \simeq \tfrac{1}{2}y^{\alpha}, \quad L^{\beta} \simeq \tfrac{1}{2}y^{\beta},$$

but the exact relation is

$$y^P = 2\int_0^{L^P} \cos\theta^P \,\mathrm{d}l. \tag{55.38}$$

Condition (55.37) ensures that the interface profiles are symmetrical about the mid-points of the plates.

Sundquist obtained a numerical solution for the interface velocity and shape given by eqn. (55.35) and an approximate analytical solution by setting $\cos\theta = 1$. In terms of the following parameters

$$R_D = D^{B\alpha}/D^{B\beta}, \quad R^P = c_{\infty}^{\gamma P}\Gamma^P\theta_0^P, \tag{55.39}$$
$$U^B = \gamma/D^{B\alpha}K^B\delta^B,$$

the analytical solution is

$$U^B = \frac{3\{c_{\infty}^{\gamma\beta}-c_{\infty}^{\gamma\alpha}+(R^{\alpha}/L^{\alpha})-(R^{\beta}/L^{\beta})\}}{\{(c^m-c^{\beta})(L^{\beta})^2/R_D\}-(c^m-c^{\alpha})(L^{\alpha})^2}. \tag{55.40}$$

The quantity K^BU^B thus replaces $4\pi^2\alpha^B/(y^{\alpha\beta})^2$ of eqn. (55.21).

A plot of U^B against $y^{\alpha\beta}$ is negative for small $y^{\alpha\beta}$, rises steeply to a positive maximum, and then decreases slowly towards zero when $y^{\alpha\beta}$ becomes large. Sundquist's numerical calculations showed that this curve is insensitive to R_D and to the values of θ_0. With one set of assumptions ($R_D = 3{\cdot}0$, $\theta_0^{\alpha} = \theta_0^{\beta} = 30°$), he plotted a family of curves for different temperatures and found that the experimental values of $y^{\alpha\beta}$ fitted the result

$$U^B(y^{\alpha\beta})^3 = \text{const.} \tag{55.41}$$

Since $y^{\alpha\beta} \propto (\Delta T^-)^{-1}$ it follows that the temperature dependence of the growth velocity is given by

$$\Upsilon \propto (\Delta T^-)^3\, D^BK^B. \tag{55.42}$$

Assuming that D^B is given by an equation of type (44.13) and that $K^B = K_0^B\exp(\Delta h_K/kT)$, where Δh_K is the heat evolved on depositing a solute atom on the interface, experimental results on the variation of velocity with temperature may now be used to derive an apparent activation energy which represents the difference between the activation energy for diffusion and Δh_K. The results for pearlite give a value for this energy of the order of the activation energy for lattice diffusion of carbon in austenite (or twice that for carbon in ferrite), and this is clearly not consistent with the assumption of growth controlled by interface diffusion. Sundquist attributes the discrepancy to the effects of impurities in reducing the growth rate.

As in the case of the theories of growth controlled by volume diffusion, the observed interlamellar spacings are of the order of five times the spacing which would give maximum growth rate. Following Cahn (1959) and Jackson and Hunt (1966), Sundquist shows that spacing smaller than that of maximum growth rate will be unstable and will rapidly converge to the spacing of maximum growth, unwanted lamellae being simply left behind or "grown over". He refers to this as the lower catastrophic limit. Spacings which are very large are also unstable because there is then no steady-state shape to the interface. As shown schematically in Fig. 11.10, the interface doubles back on itself at large spacings, and at a

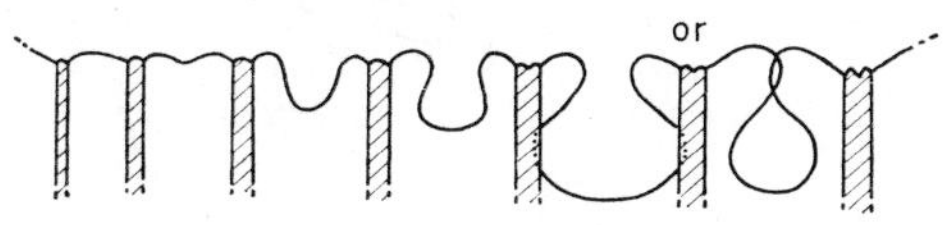

FIG. 11.10. Schematic representation of pearlite–austenite, interface shapes as a function of interlamellar spacing (all other conditions held constant) (Sundquist, 1968).

certain spacing the maximum negative angle increases rapidly with further increase of spacing. This leads to the formation of deep recesses in the centre of the α plates, either exposing new β–γ interfaces which leads to branching, or enclosing volumes of γ phase which then nucleate new lamellae. In either case, the result is to produce a rapid decrease in spacing down to a limit (the upper catastrophic limit) at which a steady-state solution (velocity constant along the interface) is possible.

Between these two catastropic limits, spacing changes are slower and continuous and are produced by the motion of lamellar faults (internally terminating lamellae). Thus the theory makes no definite prediction about observed spacings in this region, but the experimental results seem to correspond to spacings near the upper catastrophic limit. The reason for this is suggested to be that pearlite grows as colonies which have overall interfaces which are not planar but are convex towards the austenite. This requires a constant supply of new lamellae to maintain a given spacing and the lamellar fault mechanism may be inadequate for this; hence, growth takes place near the upper limit of stability.

The mechanism for maintaining the spacing is also a prominent feature of the theory of discontinuous precipitation developed by Sundquist (1973). He adapts a theory due to Tu and Turnbull (1967) based on experimental results which indicate that the β particles in an $\alpha+\beta$ cell growing from α' are not formed by branching from a singular nucleus but by successive nucleation of favourably orientated regions of α–α' boundary. In Sundquist's description, this nucleation occurs wherever the interlamellar spacing becomes sufficiently large (because of the diverging growth of a cell with a curved interface) to produce recesses in the α'–α boundary. The theory is thus essentially similar to the earlier explanation for the maintenance of the pearlite spacing near the upper limit of stability. The suggested process is shown schematically in Fig. 11.11.

The quantitative theory developed by Sundquist depends on rather detailed asumptions and will not be described here. Four "forces" are considered to act on the α'–α boundary, namely a curvature (Gibbs–Thomson) force, a solute drag from adsorbed solute which moves with the boundary, the force resulting from departure from local equilibrium (described as a "negative solute drag"), and the intrinsic drag or grain boundary kinetics effect. This last mobility force is included in Cahn's theory but is here considered negligible. The

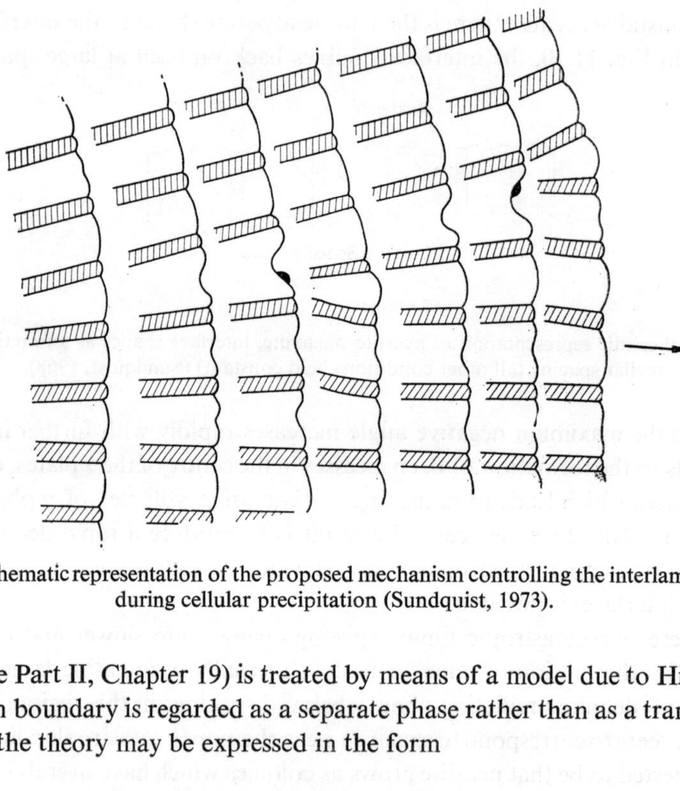

FIG. 11.11. Schematic representation of the proposed mechanism controlling the interlamellar spacing during cellular precipitation (Sundquist, 1973).

solute drag (see Part II, Chapter 19) is treated by means of a model due to Hillert (1968) in which the grain boundary is regarded as a separate phase rather than as a transition region. The results of the theory may be expressed in the form

$$D^B \delta^B K_0^B = \Upsilon (y^\alpha)^2 \bar{Q}^2 / 4(1-\bar{Q}^4)^2 \cos \theta, \tag{55.43}$$

and this equation should be compared with the corresponding Turnbull and Cahn expressions eqns. ((55.20) and (55.21)) and with Sundquist's theory of eutectoidal growth (eqn. 55.40). In eqn. (55.43) K_0 is a coefficient which depends on the interaction between the solute and the boundary, $\bar{Q}$ is related to the fractional amount of excess solute precipitated Q^t but is modified to take account of the deviation from the equilibrium value c^α of the composition of the α phase adjacent to the β lamellae immediately behind the interface, and $\bar{\theta}$ is the average value of the interface inclination θ.

Sundquist has compared his theory with the experimental results of Liu and Aaronson (1968) and Speich (1958) referred to above and concludes that it is consistent with reasonable values for grain boundary diffusivities and other data. The available experimental results on discontinuous precipitation and eutectoidal reactions are considered more fully in Sections 74 and 77. However, we should note here that as for eutectoidal reactions, the results strongly suggest that the interface velocity is much more than linearly dependent on driving force. A detailed relation between spacing $y^{\alpha\beta}$ and supercooling ΔT^{-} is not derived, but the experimental results are claimed to be consistent with the criterion shown in Fig. 11.11.

REFERENCES

AARONSON, H. I. and LIU, Y. C. (1968) *Scripta metall.* **2,** 1.

AARONSON, H. I., LAIRD, C., and KINSMAN, K. R. (1970) *Phase Transformations*, p. 313. A.S.M. Metals Park, Ohio.

ATKINSON, C. (1967) *Acta metall.* **15,** 1207; (1968) *Ibid.* **16,** 1019.

ATKINSON, C., KINSMAN, K. R. and AARONSON, H. I. (1973) *Scripta metall.* **7,** 1105.

BECKER, J. H. (1958) *J. Appl. Phys.* **29,** 110.

BOLLING, G. F. and TILLER, W. A. (1961) *J. Appl. Phys.* **32,** 2587.

BRANDT, W. H. (1945) *J. Appl. Phys.* **16,** 139.

BULLOUGH, R. and NEWMAN, R. C. (1962) *Acta metall.* **10,** 971.

BULLOUGH, R. STANLEY, J. T. and WILLIAMS, J. M. (1968) *Metal Sci. J.* **2,** 93.

CAHN, J. W. (1959) *Acta metall.* **7,** 20; (1960) *Ibid.* **8,** 556.

CAHN, J. W. and HAGEL, W. C. (1962) *Decomposition of Austenite by Diffusional Processes*, p. 131, Interscience, New York.

CAHN, J. W., HILLIG, W. B. and SEARS G. W. (1964) *Acta metall.* **12,** 1421.

COTTRELL, A. H. and BILBY, B. A. (1949) *Proc. Phys. Soc.* **62,** 49.

DARKEN, L. S. and FISHER, R. M. (1962) *Decomposition of Austenite by Diffusional Processes*, p. 249, Interscience, New York.

DOREMUS, R. H. (1957) *Acta metall.* **5,** 393.

FISHER, J. C. (1950) *Thermodynamics in Physical Metallurgy*, p. 201, American Society for Metals, Cleveland, Ohio.

FRANK, F. C. (1950) *Proc. R. Soc.* A, **201,** 586.

HAM, F. S. (1958) *J. Phys. Chem. Solids* **6,** 335; (1959a) *Q. J. Appl. Math.* **17,** 137; (1959b) *J. Appl. Phys.* **30,** 915; (1959c) *Ibid.* **30,** 1518.

HILLERT, M. (1957) *Jernkontorets Ann.* **141,** 757; (1962) *Decomposition of Austenite by Diffusional Processes*, p. 197. Interscience, New York; (1968) *The Mechanism of Phase Transformations in Crystalline Solids*, p. 231, Institute of Metals, London.

HILLIG, W. B. and TURNBULL, D. (1956) *J. Chem. Phys.* **24,** 914.

HORVAY, G. and CAHN, J. W. (1961) *Acta metall.* **9,** 695.

HULTGREN, A. (1938) *Hardenability of Alloy Steels*, p. 55, American Society for Metals, Cleveland, Ohio.

IVANTSOV, G. P. (1947) *Dokl. Akad. Nauk. SSSR* **58,** 567.

JACKSON, K. A. and HUNT, J. D. (1966) *Trans. Am. Inst. Min. (Metall.) Engrs.* **236,** 1129.

JONES, G. J. and TRIVEDI, R. (1971) *J. Appl. Phys.* **42,** 4299.

KIRKALDY, J. S. (1958) *Can. J. Phys.* **36,** 907; (1959) *Ibid.* **37,** 739; (1962) *Decomposition of Austenite by Diffusional Processes*, p. 39, Wiley, New York.

LIU, Y. C. and AARONSON, H. I. (1968) *Acta metall.* **16,** 1343.

MACHLIN, E. S. (1953) *Trans. Am. Inst. Min. (Metall.) Engrs.* **197,** 437.

MEHL, R. F. (1938) *Hardenability of Alloy Steels*, p. 1, American Society for Metals, Cleveland, Ohio.

MOTT, N. F. (1948) *Proc. Phys. Soc.* **60,** 391.

MULLINS, W. W. and SEKERKA, R. F. (1963) *J. Appl. Phys.* **34,** 323; (1964) *Ibid.*, **35,** 444.

PAPAPETROU, A. (1935) *Z. Kristallogr.* **92,** 108.

PHILIP, J. R. (1960) *Aust. J. Phys.* **13,** 1.

PURDY, G. R. (1971) *Met. Sci. J.*, **5,** 81.

Shapiro, J. M. and Kirkaldy, J. S. (1968) *Acta Met.* **16** 1239.
Shewmon, P. G. (1965) *Trans. Met. Soc. AIME* **233,** 736.
Speich, G. R. (1968) *Trans. Met. Soc. AIME* **242,** 1359.
Sundquist, B. E. (1968) *Acta metall.* **16,** 1413; (1973) *Metall. Trans.* **4,** 1919.
Temkin, D. E. (1960) *Soviet Physics-Doklady* **5,** 609.
Trivedi, R. (1970a) *Metall. Trans.* **1,** 921; (1970b) *Acta metall.* **18,** 287.
Trivedi, R. and Pound, G. M., (1967) *J. Appl. Phys.* **38,** 3569; (1969) *Ibid.* **40,** 4293.
Tu, K. N. and Turnbull, D. (1967) *Acta metall.* **15,** 369, 1317; (1969) *Ibid.* **17,** 1263.
Turnbull, D. (1955) *Acta metall.* **3,** 55.
Turnbull, D. and Treaftis, H. C. (1955) *Acta metall.* **3,** 43.
Wert, C. (1949) *J. Appl. Phys.* **20,** 943.
Zener, C. (1946) *Trans. Amer. Inst. Min.* (*Metall.*) *Engrs.* **167,** 550; (1949) *J. Appl. Phys.* **20,** 950.

CHAPTER 12

Formal Theory of Transformation Kinetics

56. TRANSFORMATIONS NUCLEATED ON GRAIN BOUNDARIES

A brief introduction to the theory of isothermal transformation curves was given in Section 4, and we now attempt to remove some of the restrictive assumptions. We begin by considering transformations which do not involve a change in mean composition, including polymorphic phase changes, single phase processes such as recrystallization, and reactions of the type considered in Section 55. In all these changes, concentration gradients either do not exist or are present only in the immediate vicinity of a boundary, and their extent does not depend on the position of the boundary. Steady-state conditions should then be quickly established, so that growth rates are constant. The conclusion that any dimension of a growing region is a linear function of the time has been verified experimentally for a number of reactions. The theory of Chapter 10 also indicates that under some circumstances a steady-state, time-independent nucleation rate per unit volume of untransformed material should be attained, but there are many reactions for which this will not be true.

An isotropic growth rate Υ was assumed in Section 4 and we shall continue to use this assumption in most of the subsequent development of this chapter. This is in agreement with experimental evidence that in many transformations the reaction product grows approximately as spherical nodules. The extension to the general case of anisotropic growth is readily made if the shape of the growing region stays constant. We can then represent the growth rate in any direction in terms of the principal growth velocities Υ_1, Υ_2, Υ_3 in three mutually perpendicular directions and the volume of a region originating at time τ is

$$v_\tau = \eta\Upsilon_1\Upsilon_2\Upsilon_3(t-\tau)^3 \qquad (t > \tau),$$
$$v_\tau = 0 \qquad (t < \tau),$$

where η is a shape factor. For homogeneous nucleation throughout the body of the assembly, the treatment of Section 4 still applies, and the volume fraction transformed at time t is

$$\zeta = 1-\exp\left[\eta\Upsilon_1\Upsilon_2\Upsilon_3\int_0^t {}^vI(t-\tau)^3\,d\tau\right] \qquad (56.1)$$

and in particular, for constant ${}^vI$

$$\zeta = 1-\exp(-\eta\Upsilon_1\Upsilon_2\Upsilon_3\,{}^vIt^4/4). \qquad (56.2)$$

A more general assumption about nucleation is that there are ${}^vN_0$ nuclei pre-existing at time $t = 0$, and in addition there is a subsequent nucleation rate ${}^vI = C_{20}t^n$. Substituting into eqn. (56.1), the form of the kinetic law becomes

$$\zeta = 1-\exp-[\eta\Upsilon_1\Upsilon_2\Upsilon_3\{{}^vN_0 t^3+C_{20}t^{4+n}\}] \tag{56.3}$$

(see also (4.10)).

In view of the conclusions of Section 51, we must next examine the possibility that nucleation occurs only at grain boundary surfaces, grain edges, or grain corners. Johnson and Mehl (1939) gave a treatment of grain boundary nucleation in which they assumed that nodules of reaction product grow only in the grains in which they nucleate, and cannot cross grain boundaries. This assumption was not made in deriving eqn. (56.2), and if it is valid, the previous results for random volume nucleation will have to be modified for transformations in fine grained material. However, the available experimental evidence seems more in favour of the alternative assumption that grain boundaries offer no resistance to growing nodules. If a boundary does stop a transformed region, nucleation on the other side of the boundary at this place is much more probable than at random points along the boundary, and the assumption of continued growth is a fair approximation. We shall thus not give Johnson and Mehl's analysis here, and we shall neglect grain boundary hindrance to growth.

The calculation of the isothermal reaction curves for nuclei forming preferentially at either grain boundaries (where two grains meet) or at edges or corners (three and four grains meeting respectively) is due to Cahn (1956a). In Section 4 we introduced the extended volume, which is the volume of all transformed regions, assuming that each one never stops growing, and that nuclei continue to form in transformed as well as untransformed regions. Now consider any plane surface, of total area O. The extended area O_e^β is defined as the sum of the areas of intersection of the extended nodules with this plane. Let the area of real transformed regions intersected by this plane be O^β. and consider a small period of time, during which these two quantities change by $\mathrm{d}O_e^\beta$ and $\mathrm{d}O^\beta$. Then if the intersections which together comprise O_e^β are randomly distributed on the plane, a fraction $(1-O^\beta/O)$ of the elements which make up $\mathrm{d}O_e^\beta$ will also contribute to $\mathrm{d}O^\beta$, so that

$$\left.\begin{aligned} \mathrm{d}O^\beta &= (1-O^\beta/O)\,\mathrm{d}O_e^\beta \\ \text{or}\qquad O_e^\beta/O &= -\ln(1-O^\beta/O). \end{aligned}\right\} \tag{56.4}$$

This equation, and its derivation, are identical with that given previously for V_e^β (eqn. (4.5)). Clearly, we may also define an extended line intercept L_e^β, which is the sum of the lengths of the intercepts cut off on length L by the extended volumes. It is related to the sum of the lengths cut off by real transformed regions by

$$L_e^\beta/L = -\ln(1-L^\beta/L).$$

For simplicity, we assume an isotropic growth rate Υ in deriving an expression for grain boundary nucleated transformation. A nodule nucleated on a plane boundary at time τ will then intersect an arbitrary plane parallel to the boundary and distant y away from it in a circle. At time t, the radius of this circle is $[\Upsilon^2(t-\tau)^2-y^2]^{1/2}$ when $\Upsilon(t-\tau) > y$, and zero when $\Upsilon(t-\tau) < y$. Suppose that ${}^BI$ is the specific grain boundary nucleation rate per unit area of the boundary (defined operationally in the same way as ${}^vI$ in Section 4), and let the boundary have area O^b. Then, at time t, the extended transformed area intersected by our reference plane is O_e^β, and the contribution to this extended area from regions nucleated between times $t = \tau, \tau+d\tau$ is

$$dO_e^\beta = \pi O^b\,{}^BI[\Upsilon^2(t-\tau)^2 - y^2] \quad [\Upsilon(t-\tau) > y],$$
$$dO_e^\beta = 0 \quad [\Upsilon(t-\tau) < y],$$

and the whole extended area is

$$O_e^\beta = \int_{\tau=0}^{t} dO_e^\beta = \pi O^b \int_0^{(t-y/\Upsilon)} \{\Upsilon^2(t-\tau)^2 - y^2\}\,{}^BI\,d\tau.$$

This expression can be integrated if ${}^BI$ is assumed to be constant. Introducing the new variable $\xi = y/\Upsilon t$ for convenience, we find

$$O_e^\beta = \pi O^b\,{}^BI\Upsilon^2 t^3(1-3\xi^2-2\xi^3)/3 \quad (\xi < 1),$$
$$O_e^\beta = 0 \quad (\xi > 1).$$

Since the intersections making up O_e^β are randomly distributed, the true area of transformed material intersected by this plane is related to O_e^β by eqn. (56.4).

We next calculate the total volume of all transformed material originating from this grain boundary, assuming there is no interference with growth by regions originating in other boundaries. By treating y as a variable, and allowing it to take all values from $-\infty$ to $+\infty$, we find that this volume is given by

$$2\int_0^\infty O^\beta\,dy = 2\Upsilon t \int_0^1 \{1-\exp(-O_e^\beta)\}\,d\xi$$
$$= 2O^b(\Upsilon/{}^BI)^{1/3} f^B(a^B), \tag{56.5}$$

where

$$a^B = ({}^BI\Upsilon^2)^{1/3} t$$

and

$$f^B(a^B) = a^B \int_0^1 [1-\exp\{(-\pi/3)(a^B)^3(1-3\xi^2-2\xi^3)\}]\,d\xi.$$

Now consider the whole assembly to contain a large number of planar grain boundaries of total area $O^B = \Sigma O^b$. Replacing O^b in (56.5) by O^B gives the total volume of transformed material nucleated at the boundaries, on the assumption that regions from different boundaries do not interfere. This volume is thus an extended volume in which allowance has been made for mutual impingement of nodules starting from one planar boundary, but not for impingement of regions starting from different boundaries. This extended volume can be related to the true transformed volume if it is assumed that the planar boundaries are themselves randomly distributed in space. Equation (4.5) then applies, and the transformed volume fraction is

$$\zeta = 1-\exp\{-(b^B)^{-1/3} f^B(a)^B\}, \tag{56.6}$$

where the further abbreviation

$$b^B = {}^B I/\{8({}^v O^B)^3\,\Upsilon\} \tag{56.7}$$

has been introduced. The quantity ${}^v O^B$ is the grain boundary area per unit volume.

The case of edge nucleation is treated similarly. The concept of extended line intercept is used to calculate the total volume of nodules originating from one straight edge, assuming impingement only with other nodules nucleated on the same edge. A random distribution of edges is then considered, in order to allow for impingement of regions nucleated on different edges. Since the calculation is virtually identical with that above, we shall only quote Cahn's final result, which is

$$\zeta = 1-\exp\{-(b^E)^{-1} f^E(a^E)\}, \tag{56.8}$$

where

$$a^E = ({}^E I\,\Upsilon)^{1/2} t, \quad b^E = {}^E I/\{2\pi\,{}^v L^E\,\Upsilon\}, \tag{56.9}$$

and

$$f^E(a^E) = (a^E)^2 \int_0^1 \xi[1-\exp\{-(a^E)^2\,[(1-\xi^2)^{1/2}-\xi^2 \ln\{(1+(1-\xi^2)^{1/2})/\xi\}]\}]\,d\xi,$$

and ${}^E I$ is the specific edge nucleation rate per unit length of edge, and ${}^v L^E$ the boundary edge length per unit volume.

Finally, there is the possibility that nucleation occurs only at grain corners. This is equivalent to Avrami's assumption of a limited number of randomly distributed nucleation sites, and the transformation eqn. (4.9) is immediately applicable. To facilitate comparison with the results in this section, we write it in the form

$$\zeta = 1-\exp\{-(b^C)^{-3} f^C(a^C)\}, \tag{56.10}$$

where

$$a^C = {}^C I t, \quad b^C = (3/4\pi\,{}^v N^C)^{1/3}\,({}^C I/\Upsilon) \tag{56.11}$$

and

$$f^C(a^C) = (a^C)^3-3(a^C)^2+6a^C-6\{1-\exp(-a^C)\}.$$

${}^C I$ is the specific corner nucleation rate (i.e. the nucleation rate per corner site) and corresponds to the quantity ν_1 used in Section 4. ${}^v N^C$ is the density of corner sites, and corresponds to the quantity ${}^v N_0$ of Section 4.

57. ANALYSIS OF ISOTHERMAL TRANSFORMATION CURVES

Experimental determinations of some physical quantity such as electrical resistivity, specimen length, or relative intensities of X-ray diffraction lines, enable the ζ–t relations to be found with fair accuracy in wany transformations. If the theory of Sections 4 and 53 is applicable, the most useful may of analysing the data is usually to plot curves of log log $[1/(1-\zeta)]$ against log t. When the general equation (4.11) applies, such curves should be straight lines of slope n. Figure 12.1 is an example, taken from unpublished work on the transformation from β to α manganese.

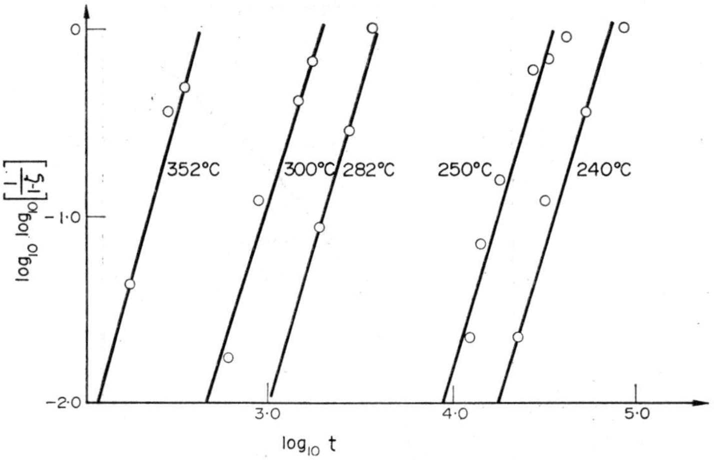

FIG. 12.1. Kinetics of the transformation from β to α manganese (Husband *et al.*, 1959).

The existence of a straight line relation might be thought to imply random volume nucleation, since the functional dependence in eqns. (56.6), (56.8), and (56.10) cannot be expressed in the simple form (4.11). In fact we shall find this is not so. Consider first that nucleation is on grain boundaries. When a^B is very small, eqn. (56.6) approaches the limiting form

$$\zeta = 1-\exp(-\pi\, {}^vI^B\, \Upsilon^3 t^4/3), \tag{57.1}$$

where ${}^vI^B = {}^vO^B\, {}^BI$ is the grain boundary nucleation rate per unit volume of the assembly. This expression is identical with the equation (4.7) or (56.2)) for random volume nucleation, so that ζ depends only on the nucleation rate per unit volume, irrespective of where the nuclei are formed. When a^B is very large, (56.6) has another limiting form

$$\zeta = 1-\exp(-2\, {}^vO^B\, \Upsilon t). \tag{57.2}$$

The log log $[1/(1-\zeta)]$–log t plot thus consists of two straight lines, of slopes four and one, with an intermediate region over which the slope decreases. From eqn. (56.6) we see that a $\{\log f^B(a^B)\}$–$\{\log a^B\}$ plot is equivalent to a $\{\log \ln[1/(1-\zeta)] + \frac{1}{3}\log b^B\}$–$\{\log t + \frac{2}{3}\log({}^BI\Upsilon^2)\}$ plot. This curve, part of which is shown in Fig. 12.2, is thus a master curve for all grain boundary nucleated reactions, and it is related to an actual $\{\log\log[1/(1-\zeta)]\}$–$\{\log t\}$ plot only by two additive constants. The experimental curve for any reaction to which the theory applies should thus fit this curve merely by moving the origin.

The physical explanation for the change in slope in Fig. 12.2. was termed by Cahn "site saturation". It occurs because the nucleation sites on the boundaries are not randomly distributed in the volume, but are concentrated near other nucleation sites. This means that at any stage the fraction of the boundary area transformed is greater than the volume fraction transformed, and because of this, the overall nucleation rate, which depends on the untransformed boundary area, decreases more rapidly than does the untransformed volume. Before saturation occurs, the fact that nuclei are confined to grain boundaries scarcely affects the overall volume fraction transformed, and eqn. (57.1) applies. After satu-

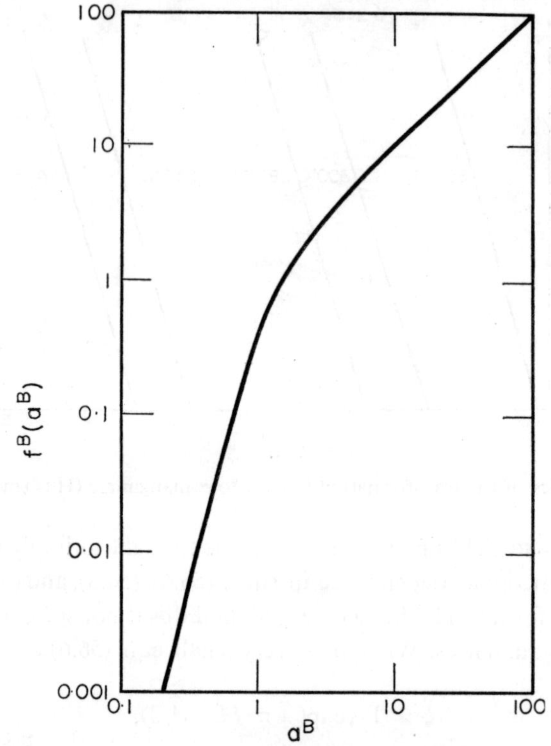

FIG. 12.2. Master curve for transformations which nucleate on grain boundary surfaces (after Cahn, 1956a).

ration, the later stages of the reaction correspond to effectively zero nucleation rate. Equation (57.2) can, in fact, be obtained simply by considering the growth laws of slabs of transformation product spreading into the grains from the grain boundaries.

The above discussion might lead us to expect that the experimental {log log [1/(1−ζ)]}–{log t} curves will show the bend in the middle of Fig. 12.2. However, variation of ζ from 0·01 to 0·99, which represents the maximum observable experimental range, only covers a range of 2·7 in the ordinates of Fig. 12.2. It is thus probable that the whole observable range of reaction will correspond to one or other of the straight line regions even when nucleation is confined to the grain boundaries. Site saturation occurs when $a^B \simeq 1$, i.e. at times $t \simeq 1/({}^B I \Upsilon^2)^{1/3}$. This is independent of ${}^v O^B$, and hence of the grain size, but the grain size is important in determining whether or not this time corresponds to an observable stage of the reaction.

Site saturation will be observed if it occurs when $\zeta \simeq 0.5$, i.e. when

$$b^B = \{f^B(a^B)/\ln 2\}^3 \tag{57.3}$$

and $f^B(a^B)$ is given a value $\sim 1{\cdot}25$ corresponding to the bend in Fig. 12.2. From (56.7) we see that site saturation occurs at half reaction if

$${}^B I \simeq \{125({}^v O^B)^3 \Upsilon\}/\{8(\ln 2)^3\}.$$

Substituting the approximate value for ${}^vO^B$ given on p. 332, and writing ${}^vI^B = {}^vO^{B\,B}I$, this equation becomes

$$ {}^vI^B \simeq 6\times 10^3\,\Upsilon/(L^B)^4. \tag{57.4}$$

For values of ${}^vI^B$ smaller than some value near that given by this equation, saturation of nucleation sites will not occur until a late stage of the reaction, and the kinetics are equivalent to those of random volume nucleation. This was realized by Avrami, who pointed out that this treatment also applied to nucleation which is only "locally random". For larger values of ${}^vI^B$, saturation occurs early in the reaction. Only for a small critical range, where the condition (57.4) holds almost exactly, should the change of slope be discernible on a $\{\log\log[1/(1-\zeta)]\}$–$\{\log t\}$ curve. In the previous chapter we have noted that the nucleation rate changes very rapidly with the degree of supercooling (or superheating) from the thermodynamical transition temperature. It follows that the transition between the two linear functions representing the extremes of Fig. 12.2 will occur in a very small temperature interval. Effectively we may say that there is a critical temperature, defined by the condition (57.4). At temperatures nearer to the thermodynamical transition temperature, the kinetics of isothermal transformation will be indistinguishable from those resulting from random volume nucleation. At temperatures more remote than the critical temperature, the kinetics will suggest that the grain boundaries all constitute nuclei pre-existing at the beginning of the transformation.

Reactions which are nucleated at grain edges or grain corners may be treated in the same way. The master curves, which are effectively plots of $\{\log\ln[1/(1-\zeta)]+\log b^E\}$ against $\{\log t+\frac{1}{2}\log {}^EI\Upsilon\}$ and of $\{\log\ln[1/(1-\zeta)]+3\log b^C\}$ against $\{\log t+\log {}^CI\}$, are shown in Figs. 12.3 and 12.4. Site saturation occurs when the edges or corners all lie in transformed

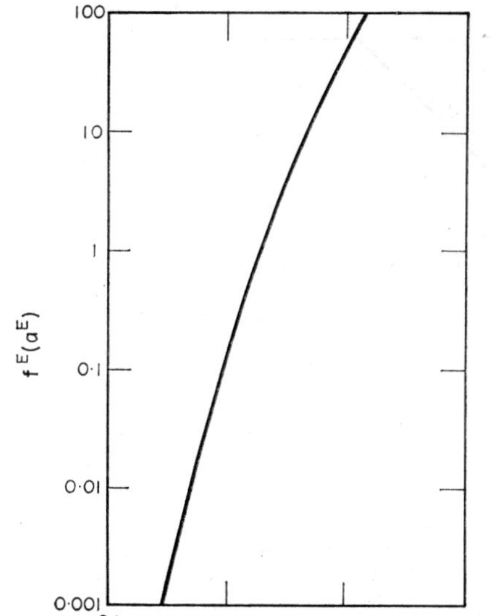

FIG. 12.3. Master curve for transformations which nucleate on grain edges (after Cahn, 1956a).

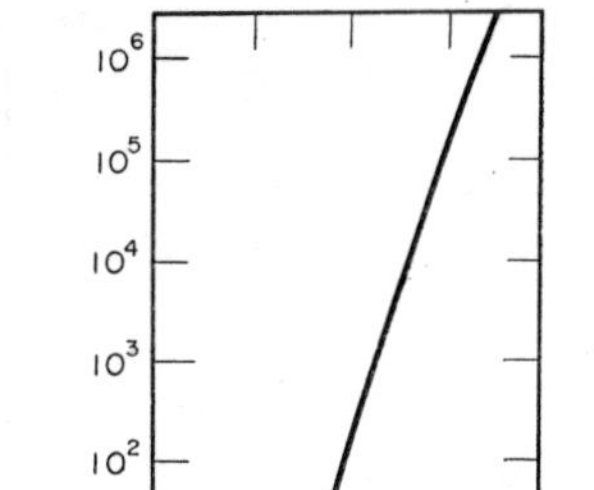

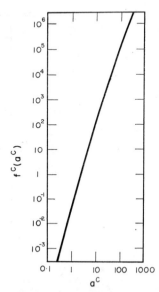

FIG. 12.4. Master curve for transformations which nucleate on grain corners (after Cahn, 1956a).

material. When a^E or a^C is small, the equation for ζ in both cases again approaches (57.1), so that before saturation is reached, the kinetics are identical with those for random volume nucleation. When a^E is large, eqn. (56.8) becomes

$$\zeta = 1-\exp(-\pi\, {}^vL^E\, \Upsilon^2 t^2) \tag{57.5}$$

and when a^C is large, the corresponding limit of (56.10) is

$$\zeta = 1-\exp(-4\pi\, {}^vN^C\, \Upsilon^3 t^3/3) \tag{57.6}$$

(see also eqn. (4.10)), Once again, there is a bend in the transformation curves in the region of a^E or $a^C \simeq 1$, but this is less pronounced for edges and still less for corners. Even if the constants are such that site saturation occurs at about half transformation, it is unlikely that the bend in the curve will be detected experimentally; instead a straight line of intermediate slope may be found.

Equation (57.3), and the equivalent expressions for edge and corner nucleation, may be used to determine the dependence of the time to half transformation $t_{1/2}$, on the grain size L^B. Using the appropriate expressions for ${}^vO^B$ and substituting for b^B

$$({}^BI/\Upsilon)^{1/3}\, L^B \simeq 9{\cdot}7 f^B(a^B_{1/2}), \tag{57.7}$$

where $a^B_{1/2} = ({}^BI/\Upsilon)^{1/3}\,\Upsilon t_{1/2}$. A plot of $\log\{9{\cdot}7 f^B(a^B)\}$ against $\log a^B$ is thus equivalent to plotting $\{\log L^B + \frac{1}{3}\log({}^BI/\Upsilon)\}$ against $\{\log \Upsilon t_{1/2} + \frac{1}{3}\log({}^BI/\Upsilon)\}$. This curve is shown in Fig. 12.5; it differs from Fig. 12.2 only in a change of origin. Provided the quantities L^B and $\Upsilon t_{1/2}$ are measured in the same units, any line of unit slope in this figure represents a

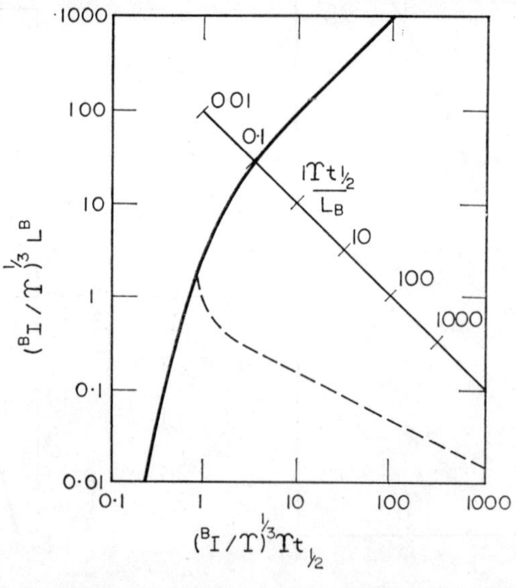

FIG. 12.5. Theoretical curve for half-times of transformations which nucleate on grain boundary surfaces (after Cahn, 1956a). The broken line gives the results of Johnson and Mehl (1939).

set of constant values of $(\Upsilon t_{1/2}/L^B)$ with varying $({}^BI/\Upsilon)$. A scale of $(\Upsilon t_{1/2}/L^B)$ can thus also be drawn. From this, we see that when site saturation has occurred, $\{({}^BI/\Upsilon)^{1/2} L^B >\sim 12\}$:

$$t_{1/2} \simeq 0{\cdot}1(L^B/\Upsilon) \tag{57.8}$$

and that if saturation does not occur, the half-time of the reaction must be greater than $0{\cdot}1(L^B/\Upsilon)$. If ${}^BI$ and Υ are fixed (at constant temperature), $t_{1/2}$ decreases as L^B decreases, but once the grain size is sufficiently small to prevent saturation of nucleation sites by half-transformation time, the decrease in $t_{1/2}$ is less rapid than the decrease in L^B. At sufficiently low values of $({}^BI/\Upsilon)^{1/2} L^B$, the curve of Fig. 12.5 has a slope of four as we have seen previously. In this region

$$t_{1/2} \simeq (6/9{\cdot}7\pi)(L^B/{}^BI\Upsilon^3)^{1/4}.$$

This result, expressed in the form $t_{1/2} \propto ({}^vI^B\Upsilon^3)^{-1/4}$, is the same as that for random volume nucleation, but the nucleation rate per unit volume, ${}^vI$, is of course independent of grain size for volume nucleation.

The continual increase of $t_{1/2}$ with L^B in Fig. 12.5 is simply a result of the decreasing ratio of grain boundary area to volume, since we have assumed that the transformed volume starting from a single nucleus is not restricted by the grain boundaries. The alternative assumption of Johnson and Mehl that transformed regions cannot cross grain boundaries gives an almost identical curve in the limit when the grain size is large. For small grain sizes, however, the amount of transformation per nucleus is limited not by impingement but by the grain boundaries, and is thus proportional to $(L^B)^3$. Since the volume nucleation rate only increases as $(L^B)^{-1}$, the overall transformation rate decreases as $(L^B)^2$. Figure 12.5 also shows the curve obtained from Johnson and Mehl's calculation; it will be seen that there is an optimum grain size of $L^B \simeq (\Upsilon/{}^BI)^{1/3}$ for rapid transformation, and this corresponds to a transformation time of $t_{1/2} \simeq (1/{}^BI\Upsilon^2)^{1/2}$.

Figures equivalent to 12.4 can be obtained for edge and corner nucleation in the same way. For edge nucleation, a curve of $\log\{8{\cdot}8[f^E(a^E)]^{1/2}\}$ against $\log a^E$ is equivalent to plotting $\{\log L^B + \frac{1}{2}\log({}^EI/\Upsilon)\}$ against $\{\log \Upsilon t_{1/2} + \frac{1}{2}\log({}^EI/\Upsilon)\}$. For corner nucleation, a curve of $\log\{4{\cdot}2[f^C(a^C)]^{1/3}\}$ against $\log a^C$ is equivalent to plotting $\{\log L^B + \log({}^CI/\Upsilon)\}$ against $\{\log \Upsilon t_{1/2} + \log({}^CI/\Upsilon)\}$. In all cases, site saturation leads to half-times given by (57.8), and very small grain sizes lead to $t_{1/2} \propto ({}^vI\Upsilon^3)^{-1/4}$. Since ${}^vI \propto (L^B)^{-2}$ for edges and $(L^B)^{-3}$ for corners, we can now write down the functional dependence of $t_{1/2}$ on L^B when other parameters are fixed. The results are gathered together in Table VIII.

We may also generalize this result by considering a nucleation rate which is a power function of the time, of the form

$${}^vI = C_{20}t^n. \tag{57.9}$$

It is readily seen that $t_{1/2} \propto (L^B)^m$, where

$$m = (3-i)/(4+n) \tag{57.10}$$

and i is the dimensionality of the site. The values of i for homogeneous nucleation and for nucleation on grain boundaries, edges and corners are 3, 2, 1, and 0 respectively.

TABLE VIII. DEPENDENCE OF TIME TO HALF TRANSFORMATION ON GRAIN SIZE, ASSUMING CONSTANT NUCLEATION RATE $t_{1/2} \propto (L^B)^m$

Type of nucleation	m
Random volume	0
Grain boundary	$\frac{1}{4}$
Grain edge	$\frac{1}{2}$
Grain corner	$\frac{3}{4}$
Site saturation (all types)	1

It follows from this discussion that experimental investigation of the variation of $t_{1/2}$ with grain size may give useful information about the type of nucleation which is operative. Few investigations of this type appear to have been made.

In a real transformation, it is possible that several kinds of site are active at the same time, all with different characteristic nucleation rates. When all sites are unsaturated, the rate law is approximately that for random nucleation

$$\zeta = 1 - \exp(-\pi\, {}^vI/\Upsilon^3 t^4/3)$$

with ${}^vI$ the sum of the separate contributions ${}^vI^H$, ${}^vI^B$, ${}^vI^E$, etc. The grain size dependence of the half-time will be approximately that characteristic of the type of nucleation making the largest contribution to the total ${}^vI$. When saturation occurs, the sites which saturate first must either be those with largest ${}^vI$ or those of lower "dimensionality"; e.g. if edges have the highest ${}^vI$, corners may saturate first, but not boundary surfaces. After one type of site has saturated, only those of higher dimensionality remain; these will eventually control the reaction if their volume nucleation rates are comparable to that of the site which initially had the largest ${}^vI$. If one type of site saturates early in the reaction, the sites of higher dimensionality which remain will either contribute little to the overall transformation rate (if they have small ${}^vI$), or else they will also saturate early in the reaction. The half-time will thus be given by $0{\cdot}1(L^B/\Upsilon)$ if any site saturates, and by $({}^vI\Upsilon^3)^{-1/4}$ if none of the sites saturate.

58. TRANSFORMATIONS WITH PARABOLIC GROWTH LAWS

The kinetics of a reaction in which the growth rate is controlled by long-range diffusion processes, as in Section 54, will now be considered. Equation (54.9) shows that for a spherical particle, nucleated at time τ, the volume at time t will be given by

$$v_\tau = (4\pi/3)(\alpha_3)^3 D^{3/2}(t-\tau)^{3/2} \quad (t > \tau). \tag{58.1}$$

In most applications, it is assumed that the fraction transformed may be obtained by the method used previously, which gave

$$\zeta = 1 - \exp\left[-\int_0^t v_\tau\, {}^vI\, d\tau\right]$$

for a linear growth process in which the nucleation is effectively randomly distributed throughout the assembly. For a diffusion controlled reaction, such as continuous precipitation, however, we can no longer define ζ as the fraction of the whole assembly which has transformed. If V^β is the equilibrium volume of β in the whole assembly of volume V, conservation of solute atoms requires

$$V^\beta/V = (c^m - c^\alpha)/(c^\beta - c^\alpha).$$

During reaction, the total volume of β may be written $V^\beta(t)$, and the volume fraction of transformation is now

$$\zeta = V^\beta(t)/V^\beta = \{V^\beta(t)/V\}\{(c^\beta - c^\alpha)/(c^m - c^\alpha)\}. \tag{58.2}$$

In the particular case of a constant number of pre-existing nuclei, ${}^vN$ per unit volume, we have

$$\zeta = {}^vNv_\tau(c^\beta - c^\alpha)/(c^m - c^\alpha) \tag{58.3}$$

whereas, more generally, the method previously used to treat impingement would give

$$\zeta = 1 - \exp\left[-\{(c^\beta - c^\alpha)/(c^m - c^\alpha)\}\int_0^t v_\tau\, {}^vI\, d\tau\right]. \tag{58.4}$$

This leads to the usual form of transformation equation, with $n = \frac{5}{2}$ for ${}^vI$ constant, and $n = \frac{3}{2}$ for early site saturation of randomly placed heterogeneous nuclei.

Clearly, however, this treatment of the isothermal reaction rate cannot be justified without much more detailed discussion. As pointed out in Section 54, two growing regions interfere with each other when the volumes from which they are drawing solute atoms begin to overlap. This "soft impingement" is a diffusion problem, and must be treated by solving the diffusion equation with an appropriate boundary condition. Until this has been done, we cannot predict with certainty whether or not eqn. (58.4) will prove to be a reasonable approximation in the early stages of precipitation.

The first approximate treatment of the soft impingement of spherical particles growing parabolically was given by Wert and Zener (1950). The calculation applies only to solutions with a low degree of supersaturation, as defined by (54.14), and assumes moreover that all the nuclei are present at the beginning of transformation. This latter assumption, corresponding to Avrami's model when ν_1 is very large (p. 19), is not so restrictive as first appears. Since the volume nucleation rate ${}^vI$ is so sensitive to the degree of supersaturation (Chapter 10), a very small amount of continuous precipitation will change the mean concentration of solute in the untransformed matrix by an amount sufficient to decrease ${}^vI$ by one or more orders of magnitude. This conclusion is also supported by experimental evidence, since results obtained for continuous precipitation of the type discussed here can generally be interpreted only by assuming that all nuclei were present at the beginning of transformation.

For a solution which is only slightly supersaturated, we use the steady-state approximation (p. 488) for the concentration gradient in the α phase at the β interface. In order to treat the impingement problem, we must assume that a β particle is growing, not in an infinite matrix, but in an α phase where the concentration tends to a steady mean value in regions remote from an interface. For slightly supersaturated solutions, the interfaces will all move sufficiently slowly for the steady-state solution to be appropriate, and the distances between particles of precipitate will all be large, so the assumption seems reasonable. Writing $c^\infty(t)$ for the mean concentration of B atoms at large distances from any precipitate at time t, we have

$$c^\infty(0) = c^m \quad \text{and} \quad c^\infty(\infty) = c^\alpha$$

and eqn. (54.3) becomes

$$(c^\beta - c^\alpha)(\mathrm{d}r^I/\mathrm{d}t) = D\{c^\infty(t) - c^\alpha\}/r^I. \tag{58.5}$$

Now if all nuclei are present at time $t = 0$, and are well separated, the final size of all regions will be the same and may be written r^f. Then the volume fraction is $\zeta = (r^I/r^f)^3$, and we also have the relation $c^\infty(t) - c^\alpha = (c^m - c^\alpha)(1-\zeta)$. Substituting these relations into (58.5) gives

$$\frac{\mathrm{d}\zeta}{\mathrm{d}t} = \frac{3D}{(r^f)^2}\frac{(c^m - c^\alpha)}{(c^\beta - c^\alpha)}\zeta^{1/3}(1-\zeta). \tag{58.6}$$

The relation of this equation to our former treatment of impingement may readily be seen. For spherical growth in the absence of impingement, we have

$$\zeta = kt^{3/2} \tag{58.7}$$

if all the nuclei are present at time $t = 0$. The corresponding growth rate is

$$\mathrm{d}\zeta/\mathrm{d}t = (\tfrac{3}{2})kt^{1/2}. \tag{58.8}$$

The usual treatment of the impingement problem is equivalent to multiplying this equation by $(1-\zeta)$

$$\mathrm{d}\zeta/\mathrm{d}t = (\tfrac{3}{2})kt^{1/2}(1-\zeta) \tag{58.9}$$

and this leads to $\zeta = 1 - \exp(-kt^{3/2})$.

Equation (58.8) can also be written in the completely equivalent form

$$\mathrm{d}\zeta/\mathrm{d}t = (\tfrac{3}{2})k^{2/3}\zeta^{1/3} \tag{58.10}$$

and we might thus, by analogy, expect that impingement can be included by writing

$$\mathrm{d}\zeta/\mathrm{d}t = (\tfrac{3}{2})k^{2/3}\zeta^{1/3}(1-\zeta). \tag{58.11}$$

Equation (58.6) is of this form, with

$$k^{2/3} = 2(c^m - c^\alpha)D/(c^\beta - c^\alpha)(r^f)^2. \tag{58.12}$$

Note that whilst (58.8) is equivalent to (58.10), (58.9) is *not* equivalent to (58.11) although they approach each other as $t \to 0$. Wert and Zener plotted ζ against t from (58.6) by numerical integration† and the resulting curve is compared with that given by eqn. (58.9) in Fig. 12.6. There is appreciable deviation towards the end of transformation.

Although, as we have seen, the assumption of zero nucleation rate is probably justified, it is to be expected that nuclei will not be distributed completely at random. This does not

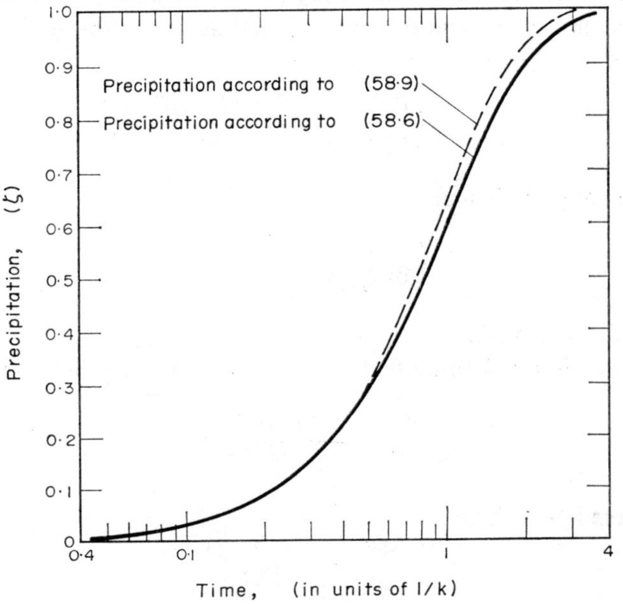

FIG. 12.6. Reaction curve for diffusion controlled growth, eqn. (58.7) compared with the Avrami type curve eqn. (58.9) (after Wert and Zener, 1950).

affect the analysis, since it may be shown that $c^{\infty}(t)$ is almost constant despite fluctuations in the density of nuclei. If there are ${}^{v}N^{C}$ nuclei per unit volume, the relation

$${}^{v}N^{C}\,(4\pi/3)\,(r^{f})^{3}\,(c^{\beta}-c^{\alpha}) = c^{m}-c^{\alpha},$$

may be used to give the alternative expression

$$k^{2/3} = (8\pi\,{}^{v}N^{C}\,r^{f}D)/3 \qquad (58.13)$$

and eliminating r^{f} between (58.12) and (58.13)

$$\frac{D^{3/2}\,{}^{v}N^{C}}{k} = \frac{3}{8\sqrt{(2)}\pi}\left[\frac{c^{\beta}-c^{\alpha}}{c^{m}-c^{\alpha}}\right]^{1/2}, \qquad (58.14)$$

so that ${}^{v}N^{C}$ may be calculated from the experimental quantities D and k.

† Wert and Zener failed to obtain an analytical solution of (58.6), but one was given by Markovitz (1950). His result is

$$\left(\frac{3}{2}\right)k^{2/3}t = \frac{1}{2}\ln(1+x+x^{2})/(1-x)^{2}+\sqrt{(3)}\arctan\frac{-\sqrt{(3)}x}{2+x}, \text{ where } \zeta = x^{3}.$$

Doremus (1957) has attempted to generalize the Zener–Wert treatment of the impingement problem to non-spherical particles, making use of the asymptotic solutions (54.11), (54.12) instead of the assumption of linear concentration change in the α phase. When impingement begins, the usual assumption of randomly distributed particles is equivalent to a reduction of the concentration in remote regions of the matrix by a factor $(1-\zeta)$. Using the asymptotic values of α_j valid for small α_j, and assuming the equilibrium solubility c^α is so small that it may be neglected, we then have

$$\left(\frac{\partial c}{\partial r}\right)_{r=r^I} = \frac{2\{c^m(1-\zeta)\}^2}{\pi r^I c^\beta}$$

for one-dimensional growth, and

$$\left(\frac{\partial c}{\partial r}\right)_{r=r^I} = \frac{c^m(1-\zeta)}{r^I}$$

for three-dimensional growth.

Substituting back into (54.3), we find the velocity of the interface is

$$\frac{\mathrm{d}r^I}{\mathrm{d}t} = \frac{2D(c^m)^2(1-\zeta)^2}{\pi(c^\beta)^2 r^I} \tag{58.15}$$

for one-dimensional growth and

$$\frac{\mathrm{d}r^I}{\mathrm{d}t} = \frac{Dc^m(1-\zeta)}{c^\beta r^I} \tag{58.16}$$

for three-dimensional growth.

If there are ${}^vN^C$ nuclei per unit volume, all present at the beginning of transformation, eqn. (58.3) gives

$$\zeta = {}^vN^C(c^\beta/c^m)v, \tag{58.17}$$

where v is the volume of the particle concerned. For the one-dimensional growth of a plate, growing only on its faces, Doremus puts

$$v = 2\pi R^2 r^I \tag{58.18}$$

at a stage in growth when the semi-thickness is r^I. From (58.15), (58.17) and (58.18),

$$[\zeta/(1-\zeta^2)](\mathrm{d}\zeta/\mathrm{d}t) = 8\pi DR^4({}^vN^C)^2,$$

and integrating from 0 to t, the growth equation is

$$\ln(1-\zeta)+[\zeta/(1-\zeta)] = 8\pi DR^4({}^vN^C)^2 t. \tag{58.19}$$

Unfortunately, as pointed out on p. 491, the work of Ham (1958) shows that eqn. (58.15) is incorrect if applied to a plate of finite dimensions growing under diffusion controlled conditions, at least if R is much smaller than the distance between particles. However, we include eqn. (58.19) since it may be appropriate to transformations in which complete edge impingement has occurred at an early stage. The growth conditions would then approximate to one-dimensional growth.

Similarly, for the three-dimensional growth of spheres, eqns. (58.16) and (58.17) give

$$d\zeta/dt = D(4\pi\, {}^vN^D)^{2/3}(3c^m/c^\beta)^{1/3}\zeta^{1/3}(1-\zeta).$$

This equation is identical with Zener and Wert's expression for spherical growth (58.6) when the substitution $4\pi\, {}^vN^C (r^f)^3 c^\beta = 3c^m$ is made, remembering that c^α is zero in the present approximation.

Doremus also derived growth laws for the three-dimensional diffusion growth of plates and needles of finite initial dimensions. The volume of a disc growing in three dimensions with negligible initial thickness is given by

$$v = 2\pi R^2 r^I + \pi^2 R(r^I)^2 + \tfrac{4}{3}\pi(r^I)^3, \qquad (58.20)$$

where R is the initial radius and r^I the semi-thickness, provided the assumption of constant growth velocity in all directions is made (see p. 485). Combination of eqns. (58.16), (58.17), and (58.20) then gives

$$\int_0^z \frac{z\,dz}{a(1-\zeta)} = \pi^2 D\, {}^vN^C\, Rt, \qquad (58.21)$$

where $a = c^m/(c^\beta\, {}^vN^C R^3)$ and $z = \pi r^I/R$. The relation between ζ and z is

$$a\zeta = 2z + z^2 + (4z^3/3\pi^2). \qquad (58.22)$$

Although eqns. (58.21) and (58.22) have a complex solution, a numerical integration is possible. Ham's work shows that the assumption of a spherically symmetrical diffusion field used in deriving eqn. (58.20) is quite wrong, and the only reason for including eqn. (58.21) here is that as an empirical equation it seems to give a rather good fit with the experimental data for the precipitation of carbon from iron. We shall not reproduce the corresponding equation which Doremus derived for the growth of a cylindrical particle of finite initial length but negligible initial thickness.

We have referred to the results obtained by Ham several times in this section and in Section 54. He has given a more rigorous treatment of the diffusion impingement problem, and he found the time dependence of the precipitation rate for an array of spherical β particles arranged on a regular cubic lattice. The method of solution utilizes the symmetry properties of this array, and introduces the boundary condition that the normal component of the solute flux vanishes on the surface of the cubic "cell" surrounding each particle.† Apart from an initial transient, of short duration, the result is identical with that given by the Zener–Wert method. Moreover, the growth law is not appreciably affected by a non-uniform distribution of particles.

Ham's work shows, as already emphasized, that a spheroid grows from an initial infinitesimal size with constant eccentricity provided each precipitating atom remains at the

† The procedure used has certain formal similarities with the Wigner–Seitz method of calculating electron energy bands in solids. As in the simpler applications of that method, the boundary condition is simplified by replacing the cubic "cell" by a spherical "cell' of the same volume.

point on the surface at which it settles. This result arises from the marked non-uniformity of the gradient $\partial c/\partial r$ over the interface, the gradient being greatest at regions of high surface curvature. It follows, correspondingly, that the time dependence of precipitation for diffusion controlled growth of plates or needles of constant eccentricity is qualitatively similar to the Zener–Wert result for spheres, the only differences being in the parameters in the formula.

From Fig. 12.6 we see that the usual expression

$$\zeta = 1 - \exp(-kt^n)$$

still provides a reasonable approximation for the growth law in the early stages of the reaction, although Ham emphasizes that a law of this kind has no fundamental significance in diffusion-limited reactions, except as an approximation for small t. Analysis of experimental results in the form of log log$[1/(1-\zeta)]$ vs. log t curves, as described in Section 57, will still give straight lines initially, even for general precipitation reactions. Prior to Ham's work, it was thought that the value of n gives useful information about the shape of the precipitating particles. Thus the assumption made by Wert and Zener about the growth laws for plates and needles (p. 485) led to $n = \frac{5}{2}$ and 2 respectively for particles of these shapes. It now appears that $n = \frac{3}{2}$ for the diffusion-limited growth of spheroids of any shape (plates, needles, spheres), provided all particles were present at $t = 0$ and had negligible initial dimensions.

It is also of interest to consider the growth laws for precipitate particles in which one or more of the initial dimensions is finite, since this situation may be produced by special nucleation conditions, or by a stage of interface controlled growth. For long cylinders, thickening radially, Ham finds $n = 1$; this result is also given by the Zener–Wert and Doremus treatments. For rods of nearly constant length, or discs of nearly constant radius, n is again equal to unity, provided that the long dimensions are small compared with the particle separation. If this condition is not satisfied, the initial transient during the establishment of the diffusion field is not negligible in comparison with the remainder of the transformation. Finally, if diffusion-limited growth begins from particles of initial volume more than about one-tenth of the final volume, n has a value intermediate between 1 and 1·5.

We conclude this section by considering briefly the very complex problem of the kinetics expected when precipitate particles form on dislocation lines. A semi-empirical equation of the Avrami type was proposed by Harper (1951) as an extension of the Cottrell–Bilby formula (eqn. (54.68)). As already noted, the variation of $c(t)$ with $t^{2/3}$ is valid only for the early stages of segregation of solute atoms to dislocations, and Harper suggested that at later times the equation

$$c(t)/c^m = 1 - \exp(-kt^{2/3}), \tag{58.23}$$

where k is specified from (54.48), should be used. Although this equation seems to give good agreement with experiment in some alloys, more detailed analyses show that the reasoning on which it is based is untenable (Ham, 1959; Bullough and Newman, 1959, 1961, 1962, 1972).

The first complete solution of (54.74) was given by Ham (1959) whose treatment assumed the dislocations to be ideal sinks, as in the Cottrell–Bilby model. The boundary conditions he applied were thus $c = 0$ at $r = 0$ and $\mathbf{I} = 0$ at $r = r_e$ where r_e is the effective radius from which each dislocation draws solute. As expected, the initial characteristics of his solution correspond to the pure drift flow treatment, but at large times he obtained a simple exponential variation with time. Bullough and Newman (1972) point out that this result applies to various problems for impurity segregation from a finite volume of crystal provided the accumulation of the impurities at the dislocation is expressed by a simple boundary condition; the experimental observation that first-order kinetics are rarely observed in such processes indicates that the boundary conditions are complex. Ham also showed that the angular terms in W_i and the assumption of a random distribution of defects rather than a regular array have little effect on the predicted kinetics. An important result of his work was that the $t^{2/3}$ dependence of (54.78) or (54.82) is valid only at a very early stage of the transformation, and his results also show that precipitation should be complete at a much earlier time than is predicted by the Harper equation. Thus the agreement of (58.23) with experiment must be in some sense fortuitous, since the equation has no firm physical basis.

Bullough and Newman (1959, 1962a) and Meisel (1967) investigated a different problem, namely the kinetics of the formation of a Maxwellian atmosphere at the dislocation core using a gas-like boundary condition of constant concentration in the core radius, or alternatively an interaction potential which is bounded in the core region. For weak interactions W_i, Bullough and Newman (1959) obtained an analytical solution by treating the drift flow as a perturbation of the diffusion flow; strong interactions were treated later by numerical methods. In all cases, the Cottrell and Bilby $t^{2/3}$ kinetics were found initially, which illustrates that this result depends only on the dominance of the drift flow and is independent of the boundary conditions. However, even with very strong interactions, the number of atoms removed to the core region of the dislocation does not follow the Harper equation, and eventually first-order kinetics are predicted, i.e. (4.11) with $n = 1$. It must therefore be concluded that Maxwellian atmosphere formation does not correspond to the physical situation.

The failure of the simple precipitation model of Ham and the atmosphere model led Bullough and Newman (1962b) to consider more complex precipitation models in which there is a finite rate of transfer of solute atoms from the dislocation core into the precipitate particles. For a model in which discrete particles are nucleated along dislocations, they again found that the kinetics did not follow the Harper equation except in the early stages of the process, and at long times the simple exponential behaviour was again found. However in another model of continuous rod like precipitates forming along the dislocation line, an approximation to the Harper type kinetics was found by using a transfer velocity from core to precipitate which decreases during the precipitation process. Physically, this effect was considered to result from the internal pressure generated by the change of volume when the precipitate is formed; this in turn affects the diffusion coefficient near to the core. Other effects of the same type may also give deviations from first order kinetics; for example the field of the precipitate may partially cancel that of the dislocation. These more complex boundary conditions and their effect on the kinetics cannot be considered further here and reference should be made to review papers by Bullough (1968) and Bullough and Newman (1970).

The exponential growth law summarized in Avrami's equation (4.11) is valid for linear growth under most circumstances, and approximately valid for the early stages of diffusion controlled growth. Table IX summarizes the values of n which may be obtained in

TABLE IX. VALUES OF n IN KINETIC LAW $\zeta = 1-\exp(-kt^n)$

(a) Polymorphic changes, discontinuous precipitation, eutectoid reactions, interface controlled growth, etc.

Conditions	n
Increasing nucleation rate	>4
Constant nucleation rate	4
Decreasing nucleation rate	3–4
Zero nucleation rate (saturation of point sites)	3
Grain edge nucleation after saturation	2
Grain boundary nucleation after saturation	1

(b) Diffusion controlled growth

Conditions	n
All shapes growing from small dimensions, increasing nucleation rate	$>2\frac{1}{2}$
All shapes growing from small dimensions, constant nucleation rate	$2\frac{1}{2}$
All shapes growing from small dimensions, decreasing nucleation rate	$1\frac{1}{2}$–$2\frac{1}{2}$
All shapes growing from small dimensions, zero nucleation rate	$1\frac{1}{2}$
Growth of particles of appreciable initial volume	1–$1\frac{1}{2}$
Needles and plates of finite long dimensions, small in comparison with their separation	1
Thickening of long cylinders (needles) (e.g. after complete end impingement)	1
Thickening of very large plates (e.g. after complete edge impingement)	$\frac{1}{2}$
Precipitation on dislocations (very early stages)	$\sim\frac{2}{3}$

various experimental situations. The conditions listed are not meant to be exhaustive; for example, the effects of an external surface which may reduce n by up to 1 for a foil and 2 for a wire have not been included. It is evident from the table that a kinetic investigation which is limited to the establishment of the value of n most appropriate to the assumed growth law does not, as once assumed, give sufficient information for the growth habit to be deduced.

59. EFFECTS OF TEMPERATURE: NON-ISOTHERMAL TRANSFORMATIOMS

If the kinetics of a transformation are found experimentally at a number of different constant temperatures, a complete isothermal transformation diagram may be drawn. This figure, also known as a time–temperature–transformation (*TTT*) diagram, gives the relation between the temperature (plotted linearly) and the time (plotted logarithmically) for fixed fractional amounts of transformation to be attained. Thus the complete diagram consists of a number of curves of T against log t_ζ, where t_ζ is the time required for the transformation to reach stage ζ. Frequently, only two or three such curves are given in a *TTT* diagram, measuring the times (t_0, t_1) for the beginning and end of transformation, and sometimes for 50% transformation ($t_{1/2}$). However, it must be emphasized that t_0 and t_1 cannot really be measured experimentally, and the quantities peotted are something like $t_{0\cdot05}$ and $t_{0\cdot95}$.

When both nucleation and growth rates are temperature dependent, the isothermal transformation rate will not be a simple function of the temperature. In very many reactions on cooling, the nucleation rate is determined mainly by a Boltzmann type equation with an activation energy which decreases more than linearly with temperature (Chapter 10). This gives a rapidly increasing nucleation rate as the undercooling (or supersaturation) increases. The growth rate, in contrast, is controlled by an activation energy which is nearly independent of temperature, and hence the rate decreases as the temperature decreases. These opposing factors give an overall transformation rate which first increases and then decreases again as the temperature falls, leading to the *C* curves characteristic of so many *TTT* diagrams. At sufficiently low temperatures, the nucleation rate may be so large that the nucleation sites saturate early in the reaction. The growth rate alone then controls the overall reaction rate.

The isothermal reaction curves are usually interpreted in terms of eqn. (4.11), which we have seen is a good approximation in almost all modes of transformation. For most transformations, the value of n is independent of temperature over appreciable temperature ranges, as is expected from the previous analysis. Since n depends only on the growth geometry, it should only change when this geometry alters; this happens, for example, when a heterogeneously nucleated reaction reaches the degree of supercooling at which the nucleation sites are saturated at an early stage of transformation.

The value of k, in contrast, is found to vary markedly with temperature. Differentiation of eqn (4.11) leads to an expression for the rate of transformation

$$d\zeta/dt = nk(1-\zeta)t^{n-1} \tag{59.1}$$

in which $(1-\zeta)$ is the impingement factor, and $n\,k\,t^{n-1}$ gives the rate law in the absence of impingement. Many writers have used the (rather misleading) analogy of a chemical rate equation, and interpreted the factor nk of this equation in the same way as a chemical rate constant. A plot of log $(n\,k)$ against $1/T$ is then made (see Chapter 3) with the object of deriving an activation energy for the reaction.

In the early stages of the transformation, the reaction rate is often assumed to be controlled by two different activation energies, one for formation of critical nuclei (ΔG_c) and the other for their subsequent growth $\Delta_a g^*$. The time to some fixed amount of transformation would then be given by

$$\ln t_\zeta = C_{21}+(\Delta G_c+\Delta_a g^*)/kT. \tag{59.2}$$

Note that this is really a form of Becker's equation (49.2), since the nucleation rate and the initial transformation rate are interdependent. The reciprocal rate plot is then

$$\frac{d \ln t_\zeta}{d(1/T)} = \frac{\Delta_a g^*}{k} + \frac{\Delta G_c}{k} + \frac{1}{kT}\frac{\partial \Delta G_c}{\partial(1/T)}. \tag{59.3}$$

A plot of $\ln t_\zeta$ against $1/T$ will have a *C*-shape, as noted above, but will approximate to a straight line at low temperatures when $\Delta G_c \ll \Delta_a g^*$. The slope of the straight line may be less than $\Delta_a g^*$ if ΔG_c decreases to zero more rapidly than the (negative) third quantity on the right-hand side of this equation.

Interpreting the initial transformation rate in terms of our equation (4.11), we see that before saturation occurs k is proportional to the nucleation rate to the power one and to either a growth rate or a diffusion coefficient to the power $(n-1)$. Thus we have

$$\frac{d \ln k}{d(1/T)} = \frac{-(n-1)\varepsilon}{k} + \frac{\Delta G_c}{k} + \frac{1}{kT}\frac{\partial \Delta G_c}{\partial(1/T)}, \tag{59.4}$$

where ε is the growth or diffusion coefficient activation energy. Note that this equation gives a more accurate indication of the roles played by the two kinds of activation energy in determining the overall transformation rate than does (59.3). In the region where the curve becomes a straight line, nucleation saturation has occurred, and the value of n has decreased to a lower value n'. The relation is then

$$\frac{d \ln k}{d(1/T)} = \frac{-n'\varepsilon}{k} \tag{59.5}$$

from which ε may be determined.

Equation (59.5) implicitly assumes that the number of nuclei formed is independent of temperature, so that the other factors determining k (for example eqns. (56.3) or (56.6) are constant. In the same way, for continuous precipitation processes k depends on the mean diffusion distance between precipitate particles, so eqn. (59.5) should strictly be written

$$\left[\frac{d \ln k}{d(1/T)}\right]_{\zeta, L^B} = \frac{-n'\varepsilon}{k}$$

to emphasize that the grain size L^B or the mean diffusion distance should be kept constant. Since these quantities usually have strong temperature variation (because of nucleation rate variation, or grain size variation), results for activation energies frequently have no significance unless special precautions are taken. One way of doing this is to start transforming at a fixed low temperature, to produce nucleation site saturation, and then to measure k for subsequent transformation at various higher temperatures. This effectively measures the kinetics of the growth part of the transformation only.

Since the pioneer work of Davenport and Bain (1930), *TTT* diagrams have been widely used in industry as a guide to heat treatment procedures. Nevertheless, in industrial practice, the kinetic behaviour of an assembly at constant temperature is frequently of less importance than its behaviour during constant heating or cooling through a transformation range. The general theory of transformation kinetics is largely confined to isothermal reaction, but there have been several attempts to predict the course of a non-isothermal reaction from an experimentally determined set of isothermal transformation curves. The idealized problem is thus to calculate the ζ–t curves from the isothermal ζ–t curves and some given t–T relation.

The difficulties in treating non-isothermal reactions are mainly due to the independent variations of growth and nucleation rate with temperature, mentioned above. The problem is tractable only when the instantaneous transformation rate can be shown to be a function solely of the amount of transformation and the temperature. This leads to the concept of additivity which we shall now describe.

Consider the simplest type of non-isothermal reaction, obtained by combining two isothermal treatments. The assembly is transformed at temperature T_1, where the kinetic law is $\zeta = f_1(t)$ for a time t_1, and is then suddenly transferred to a second temperature T_2. If the reaction is additive, the course of the transformation at T_2 is exactly the same as if the transformed fraction $f_1(t_1)$ had all been formed at T_2. Thus if t_2 is the time taken at T_2 to produce the same amount of transformation as is produced at T_1 in a time t_1, we have $f_1(t_1) = f_2(t_2)$, and the course of the whole reaction is

$$\left.\begin{aligned} \zeta &= f_1(t) \qquad && (t < t_1), \\ \zeta &= f_2(t+t_2-t_1) \qquad && (t > t_1). \end{aligned}\right\} \tag{59.6}$$

Suppose that t_{a1} is the time taken to produce a fixed amount of transformation ζ_a at T_1, and t_{a2} is the corresponding time to produce the same amount of transformation at T_2. Then in the composite process above, an amount ζ_a of transformation will be produced in a time

$$t = t_{a2} - t_2 + t_1 \tag{59.7}$$

if the reaction is additive. The time can also be found from the rule

$$\frac{t_1}{t_{a1}} + \frac{(t-t_1)}{t_{a2}} = 1 \tag{59.8}$$

and we can see that (59.7) and (59.8) are equivalent provided $t_1/t_2 = t_{a1}/t_{a2}$. We shall see that when (59.6) is true, this condition is satisfied.

An additive reaction thus implies that the total time to reach a specified stage of transformation is obtained by adding the fractions of the time to reach this stage isothermally until the sum reaches unity. The generalization of eqn. (59.8) to any time–temperature path is clearly

$$\int_0^t \frac{dt}{t_a(T)} = 1, \tag{59.9}$$

where $t_a(T)$ is the isothermal time to stage ζ_a (as plotted on a TTT diagram), and t is the time to ζ_a for the non-isothermal reaction. Note than an additive reaction does *not* imply (as sometimes stated) that for the two-temperature transformation path above $\zeta = f_1((t_1)+f_2(t-t_1)$.

It is obvious that (59.6) will be true if the reaction rate depends only on ζ and T, that is, only on the state of the assembly, and not on the thermal path by which it reached that state. An analytical proof of eqn. (59.9) also follows from this statement. Consider a transformation for which the instantaneous reaction rate may be written

$$d\zeta/dt = h(T)/g(\zeta), \tag{59.10}$$

where $h(T)$, $g(\zeta)$ are respectively functions only of temperature and volume fraction transformed. Then we may write

$$\int h(T)\,dt = \int g(\zeta)\,d\zeta = G(\zeta) \tag{59.11}$$

for any transformation path. This equation is equivalent to

$$\zeta = F\left\{\int h(T)\,dt\right\} \tag{59.12}$$

and in particular for an isothermal transformation

$$\zeta = F\{h(T)t\}. \tag{59.13}$$

According to eqn. (59.13), the fraction transformed at a fixed temperature is dependent only on the time and on a single function of the temperature. This function $h(T)$ might specify the growth rate or the diffusion coefficient, for example. Transformations at different temperatures then differ only in the time scale, as we assumed above in asserting the identity of (59.7) and (59.8). Reactions of this type are called isokinetic, after Avrami (1939, 1940). Avrami defined an isokinetic reaction by the condition that the nucleation and growth rates are proportional to each other (i.e. they have same temperature variation). Following Cahn (1956b), however, we shall take (59.12) as a more general definition of an isokinetic reaction. We now show that an isokinetic reaction is additive in the sense defined by (59.9).

From (59.11) we see that

$$h(T) = G(\zeta_a)/t_a(T)$$

and on substituting into (59.10)

$$t_a(T)\,(d\zeta/dt) = G(\zeta_a)/g(\zeta). \tag{59.14}$$

Now consider the integral

$$\int_0^t \frac{dt}{t_a(T)} = \int_0^\zeta \frac{d\zeta}{t_a(T)\,(d\zeta/dt)} = \frac{G(\zeta)}{G(\zeta_a)}. \tag{59.15}$$

This gives the relation between the time t and the fraction transformed for the whole non-isothermal reaction, and is the general expression of the concept of additivity. In particular, if $\zeta = \zeta_a$, (59.15) reduces to (59.9).

Avrami's condition for an isokinetic reaction will rarely be satisfied, since a fortuitous coincidence is required. For many reactions, however, we have seen that the nucleation rate may saturate early in the transformation. Provided that the growth rate at any instant is then only dependent on the temperature, the reaction will be isokinetic in the general sense defined by Cahn. In the case of discontinuous reactions with duplex cell formation, a direct test is possible if the assembly is allowed partially to transform at one temperature and then quenched suddenly to another temperature and held there for a further time. An abrupt change in the interlamellar spacing indicates that the growth rate $\Upsilon = h(T)$, but a gradual change, as has also been observed, implies that Υ depends also on dT/dt. If this is so, the reaction cannot be additive.

In continuous reactions, the nucleation often saturates at a very early stage. The size of a transformed region is then proportional to a diffusion coefficient, raised to an appropriate power, and to a growth factor α, which is some function of the concentrations, as described in Section 55. A condition for additivity is thus that this factor is not dependent on temperature. If the compositions are not constant, a change from one temperature to another will not only change ζ but disturb the equilibrium at the interface of the already precipitated regions. Thus additivity can only be expected at large degrees of undercooling, where approximate constancy of equilibrium concentrations may be found. This factor is not so important in discontinuous reactions, since the average compositions of transformed and untransformed regions are the same. Strictly, the transformed regions in such reactions will not all be in equilibrium without further diffusional adjustments if the compositions of the final phases vary with temperature.

A more important restriction on the notion of additivity must now be mentioned. A reaction in which the nucleation sites saturate can only be written in the form (59.12) if the number of nuclei is not a function of temperature. Thus for grain boundary nucleated reactions, eqn. (59.9) is valid only if the grain size is constant for all the isothermal transformation data (which give $t_a(T)$) and also for the non-isothermal process considered. The dependence of transformation rate on grain size after saturation for various types of nucleation has been given previously (eqns. (57.2), (57.5), and (57.6)). From these equations we can modify (59.9) for the case where the non-isothermal specimen has a grain size different from that used in the isothermal experiments. For nucleation on grain boundaries, edges or corners, the integral in eqn. (59.9) has to equal the ratio of the non-isothermal to the isothermal mean grain diameter when the transformation reaches stage ζ_a. For nucleation on preferred sites randomly distributed through the volume, this integral equals the cube root of the ratio of the number of such sites per unit volume of the isothermal specimens to the number per unit volume of the non-isothermal specimen.

From eqns. (57.2), (57.5), and (57.6) we may also write down expressions for $G(\zeta)$ by direct comparison with (59.11). These equations are

$$\left.\begin{aligned} G(\zeta) &= -\ln(1-\zeta)/(2\,{}^{v}O^{B}), \\ G(\zeta) &= [-\ln(1-\zeta)/(\pi\,{}^{v}L^{E})]^{1/2}, \\ G(\zeta) &= [-3\ln(1-\zeta)/(4\pi\,{}^{v}N^{C})]^{1/3}, \end{aligned}\right\} \tag{59.16}$$

for nucleation on surfaces, edges and corners of grains respectively. Combining these equations with (59.15),

$$\left.\begin{aligned} \ln(1-\zeta) &= \ln(1-\zeta_a)\int_0^t [\mathrm{d}t/t_a(T)], \\ \ln(1-\zeta) &= \ln(1-\zeta_a)\left\{\int_0^t [\mathrm{d}t/t_a(T)]\right\}^2, \\ \ln(1-\zeta) &= \ln(1-\zeta_a)\left\{\int_0^t [\mathrm{d}t/t_a(T)]\right\}^3, \end{aligned}\right\} \tag{59.17}$$

for the three types of site. This last set of equations gives directly the amount of transformation ζ in terms of the experimentally determined quantities $t_a(T)$ and the path $(t-T)$ of the reaction. They are readily modified to allow for a difference of grain size between the non-isothermal and the isothermal specimens.

REFERENCES

AVRAMI, M. (1939) *J. Chem. Phys.* **7,** 1103; (1940) *Ibid.* **8,** 212; (1941) *Ibid.* **9,** 177.

BULLOUGH, R. (1968) *The Interactions Between Point Defects and Dislocations, p.* 22, HMSO, London.

BULLOUGH, R. and NEWMAN, R. C. (1959) *Proc. R. Soc.* A **249,** 427; (1961) *Phil. Mag.* **6,** 403; (1962a) *Proc. R. Soc.* A, **266,** 198; (1962b) *Ibid.* **266,** 209; (1970) *Rep. Prog. Phys.*, **33,** 101.

CAHN, J. W. (1956a) *Acta metall.* **4,** 449; (1956b) *Ibid.* **4,** 572.

DAVENPORT, E. S. and BAIN, E. C. (1930) *Trans. Am. Inst. Min. (Metall.) Engrs.* **90,** 117.

DOREMUS, R. H. (1957) *Acta metall.* **5,** 393.

HAM, F. S. (1958) *J. Phys. Chem. Solids* **6,** 335; (1959) *J. Appl. Phys.* **30,** 915.

HARPER, S. (1951) *Phys. Rev.* **83,** 709.

HUSBAND, J. N., MASON, K. L., and CHRISTIAN, J. W. (1959) (Unpublished work).

JOHNSON, W. A. and MEHL, R. F. (1939) *Trans. Am. Inst. Min. (Metall.) Engrs.* **135,** 416.

MARKOVITZ, H. (1950) *J. Appl. Phys.* **21,** 1198.

MEISEL, L. V. (1967) *J. Appl. Phys.* **38,** 4780.

WERT, C. and ZENER, C. (1950) *J. Appl. Phys.* **21,** 5.

Problems

THE problems are arranged in approximately the same order as the relevant text, but vary considerably in difficulty.

1. Find the metric tensors **G** and $\mathbf{G}^{-1}$ for a basis defined by the primitive vectors $\frac{1}{2}a$ [101], $\frac{1}{2}a$ [110] and $\frac{1}{2}a$ [011] of the f.c.c. lattice and verify for typical directions and planes the formulae in Table I. Show that the f.c.c. and the b.c.c. lattices are reciprocal to each other.

2. Express the vectors $[10\bar{1}1]$, $[1\bar{1}01]$ and $[22\bar{4}3]$ of a hexagonal lattice in three-component form using the vectors of (a) the primitive unit cell and (b) the orthorhombic unit cell as bases. Write down the metrics for these two bases and find the lengths of the above vectors using both sets of components. What are the angles between $[1\bar{1}01]$ and $[22\bar{4}3]$ and between $[10\bar{1}1]$ and $[22\bar{4}3]$?

3. Using an orthonormal coordinate system, find the rotation matrices which represent respectively a right-handed rotation through an angle θ about [001] and a left-handed rotation through the same angle about [010].

4. Show that in an orthonormal coordinate system the matrix

$$\frac{1}{4}\begin{pmatrix} 3 & 1 & \sqrt{6} \\ 1 & 3 & -\sqrt{6} \\ -\sqrt{6} & \sqrt{6} & 2 \end{pmatrix}$$

represents a rotation of 60° about [110].

5. Show that the eigenvalues of the rotation matrices in questions 3 and 4 are 1, $e^{i\theta}$ and $e^{-i\theta}$ in all cases. Find the corresponding eigenvectors, and show that each set of three eigenvectors may be regarded mathematically as orthonormal, even though two of the vectors are imaginary. (The scalar product of two imaginary vectors is defined by eqn. (5.15) when u_i^*, v_i^* represent the complex conjugates of u_i, v_i.)

6. Using the result of the previous question, or otherwise, show that if a rotation is represented in an orthonormal axis system by an orthogonal matrix which is symmetrical, the angle of rotation must be $n\pi$ where n is any integer.

7. Derive the form of the matrix S representing a simple shear in an orthonormal system in which the plane of shear is perpendicular to the x_2 axis and the direction of shear makes an angle θ with the x_3 axis.

8. Prove that two successive simple shears, $\mathbf{S}_1$ and $\mathbf{S}_2$, combine to give a resultant deformation which is also a simple shear if *either* (a) the invariant planes of the two shears, *or* (b) the two shear directions, coincide. Express the shear direction **e**, the normal **n** to the invariant plane and the magnitude s of the resultant shear in terms of $\mathbf{e}_1$, $\mathbf{n}_1$, s_1 and $\mathbf{e}_2$, $\mathbf{n}_2$, s_2 for the component shears. Is the condition necessary as well as sufficient?

9. Show that if two successive shears have the same plane of shear, the resultant deformation is a simple shear combined with a rotation. Use the result of question 7 to find expressions for the angle φ between the invariant plane of the first shear and the undistorted plane of the resultant deformation, the magnitude s of the shear in the composite deformation, and the magnitude ψ of the rotation in the composite deformation.

10. Prove the statement on p. 42 that two matrices S and JSJ^{-1} which are related by a similarity transform have the same eigenvalues. Use the result of question 5 to show that the trace of a rotation matrix is a scalar invariant of magnitude $1+2\cos\theta$.

11. A lattice transformation is represented in an orthonormal coordinate system by

$$\mathsf{S} = \frac{1}{4}\begin{pmatrix} 3\sqrt{2} & -\sqrt{2} & 0 \\ -1 & 3 & -4-2\sqrt{2} \\ -1 & 3 & 4+2\sqrt{2} \end{pmatrix}$$

Find the principal deformations and the directions of the principal axes and factorise S into a pure deformation **P** and a pure rotation **R**.

12. In the lattice transformation from f.c.c. to h.c.p. structures, encountered for example in cobalt alloys or in certain alloy steels, the basal plane of the hexagonal structure is derived from a {111} plane of the cubic structure, and close-packed directions within these two planes are corresponding directions. Write down *two* variants of the correspondence matrix (a) using a primitive unit cell to define the hexagonal lattice and (b) using an orthorhombic unit cell for this lattice.

13. The lattice transformation described in question 12 may be accomplished by a shear of magnitude $1/2\sqrt{2}$ on the cubic system $\{111\}\langle 11\bar{2}\rangle$, whilst a shear of twice this magnitude produces a twin of the f.c.c. lattice. Using cubic axes, write down expressions for the matrix representing the lattice deformation **S** in the two cases. Show also that the transformation to the hexagonal lattice may be accomplished by a shear of the same magnitude on the system $\{55\bar{7}\}\langle 7, 7, 10\rangle$. [Assume the hexagonal lattice has the 'ideal' axial ratio of $(8/3)^{1/2}$.]

14. Tensile stresses of 5, −2 and 3 MN m^{-2} are applied across the (100), (010) and (001) faces of a cubic crystal. Calculate the magnitude of the normal stress and the magnitude and direction of the shear stress acting across the planes (110), (211) and (123) and derive the general formula for the stresses on the plane of unit normal **n** when tensile stresses X_i are applied in the directions of the cube axes.

15. Prove that the stiffnesses c_{pq} and compliances s_{pq} may be derived from the tensor quantities c_{ijkl} and s_{ijkl} by contracting ij to p and kl to q in accordance with the scheme of equation (11.14) and imposing the conditions $c_{pq} = c_{ijkl}$ and $s_{pq} = \alpha\, s_{ijkl}$ where $\alpha = 1$ if both p and $q \leq 3$, $\alpha = 2$ if either p or $q > 3$ and $\alpha = 4$ if both p and $q > 3$. Find the form of the matrices c_{pq} and s_{pq} for a hexagonal crystal, and show that the elastic properties of such a crystal (for example, Young's modulus) are isotropic for directions within the basal plane.

16. Show that if a tensile stress T is applied in the direction of the unit plane normal **n**, the components of the stress tensor are $X_{ij} = n_i n_j T$. Hence show that Young's modulus in the direction **n** is equal to $(s_{ijkl} n_i n_j n_k n_l)^{-1}$ where s_{ijkl} are the tensor compliances.

17. Apply the result of question 16 to obtain an equation for the variation of Young's modulus with direction in a cubic crystal and deduce the condition for elastic isotropy in terms of the compliances s_{pq}. The values of s_{pq} in $T\,Pa^{-1}$ (10^{-11} m^2 N^{-1}) for copper and molybdenum are

	s_{11}	s_{12}	s_{44}
Cu	15	−6	13
Mo	2	−0.7	9

Calculate Young's modulus for the ⟨100⟩ ⟨110⟩ and ⟨111⟩ directions of both metals.

18. Find expressions in terms of c_{pq} and s_{pq} for the shear moduli of cubic crystals subjected to shear stresses in the following planes and directions: (a) (100) [001], (b) (110) [001], (c) (110) [$1\bar{1}0$], (d) (110) [$1\bar{1}2$], (e) (111) [$1\bar{1}0$], (f) (112) [$11\bar{1}$]. Show also that the shear modulus is isotropic for shear stresses applied in any direction of the {111} plane or any plane of the ⟨111⟩ zone.

19. Express Poisson's ratio for an isotropic material in terms of the ratio of the bulk modulus to the shear modulus, and hence prove that $-1 \leq \nu \leq \frac{1}{2}$. What ratio of bulk modulus to shear modulus gives a Poisson's ratio of zero?

20. The atoms of a crystalline solid may exist in either of two energy states. If the difference in energy of these two states is ε, show that there is a contribution to the molar specific heat which at low temperatures

($\varepsilon \gg kT$) is given by $(N\varepsilon^2/kT^2) \exp(-\varepsilon/kT)$, where N is Avogadro's number. (Assume that Boltzman statistics apply.)

21. The enthalpies, in J mol^{-1}, of the α and γ forms of manganese at 300K are given in terms of the energies of these phases at 0K by $H^{\alpha}_{300} - H^{\alpha}_{0} = 5040$. $H^{\gamma}_{300} - H^{\gamma}_{0} = 5170$, and the corresponding entropies, in J K^{-1} mol^{-1}, are $S^{\alpha}_{300} = 32$, $S^{\gamma}_{300} = 32.5$. The specific heats of these phases at constant pressure for $T > 300$K are $C^{\alpha} = 24 + 0.014T - 160{,}000/T^2$ and $C^{\gamma} = 25.3 + 0.015T - 186{,}000/T^2$, and a calorimetric experiment gives $H^{\gamma}_{600} - H^{\alpha}_{300} = 10{,}900$. Find the latent heat of the α–γ transformation at 0K and 1000K. The β phase has $S^{\beta}_{300} = 28$ and its specific heat for $T > 300$K is $C^{\beta} = 35 + 0.0028T$. Given that the β–γ transformation temperature is 1374K, find G^{α}, G^{β} and G^{γ} in terms of H^{α}_{0} for $T = 1000$K and 1374K.

22. A sphere of radius r_0 is removed from the interior of an isotropic elastic solid of shear modulus μ; Poisson's ratio ν and surface energy per unit area σ. Show that the equilibrium radius of the spherical hole is approximately $r_0 - \sigma/2\mu$ provided $\mu r_0 \gg \sigma$. If this model is used to represent a vacancy, estimate the surface and strain contributions to the formation energy in a solid for which $r_0 = 0.1$ nm, $\mu = 60$ GN m^{-2}, $\sigma = 1.2$ J m^{-2}. If the vacancy is created by removing a spherical atom from the interior and replacing it on the surface of a finite solid, find the change in volume as a fraction of the atomic volume, given that $\nu = \frac{1}{3}$.

23. Estimate the ratio of vacancy-silver complexes to free vacancies in an alloy of copper containing 0.1 at.% silver held at a temperature of 600°C, given that the binding energy of a vacancy to a silver atom is 0.21 eV. (Assume $kT \simeq 0.025$ eV at room temperature.)

24. Use eqn. (30.34) to find the ratio of the line tensions of edge and screw dislocations in an elastically isotropic medium.

25. Show that the macroscopic shear strain produced by the displacement of N dislocation segments per unit volume through an average area A of a single set of parallel slip planes is NAb where b is the Burgers vector, and that the tensile strain in the x_1 direction produced by small shear strains of this type on i independent slip systems is $\sum_i N_i A_i b_{i1} n_{i1}$ where b_{i1} and n_{i1} are the x_1 components of the Burgers vector $\mathbf{b}_i$ and the unit normal to the slip plane $\mathbf{n}_i$ of the ith system.
Show that for dislocations on a single set of slip planes and pinned at average distances l along the lines, the apparent shear modulus for small strains is $\mu(1 + Nl^3/6)^{-1}$ where μ is the true modulus.

26. A crystal contains a mobile dislocation density of 10^{12} m^{-2}. If the dislocations are all on a single set of parallel glide planes but are pinned at points along their lengths of average separation 1 μm calculate (1) the maximum shear strain which can be recovered on removal of stress and (2) the resolved shear stress at which irreversible strain occurs. (Shear modulus $= 2.8 \times 10^{10}$ nm^{-2}, Burgers vector $= 0.25$ nm, assume line tension $= \frac{1}{2}\mu b^2$.)

27. Two opposite screw dislocations with Burgers vectors $\pm\mathbf{b}$ are moving on parallel slip planes under the action of an applied shear stress, τ, in the slip direction which is just sufficient to overcome the frictional resistance to dislocation motion. Show that if this frictional resistance is identical on a cross-slip plane, the dislocations will be able to annihilate each other provided the separation d of the primary slip planes satisfies the relation

$$d \leq (\mu b \cot \tfrac{1}{2}\theta)/2\pi\tau$$

where μ is the shear modulus and θ the angle between the primary and cross-slip planes.
In a b.c.c. structure, a dislocation moving on a $(\bar{1}01)$ plane might cross-slip on either $(\bar{2}11)$ or $(\bar{1}10)$. If the critical value of d is equal for these two planes, calculate the ratio of the resistances to glide on the two cross-slip planes.

28. Show that the angular term neglected in eqn. (33.5) is $(2 - \nu - 2\nu \cos 2\varphi/2(1-\nu)$ where ν is Poisson's ratio and φ is the angle between the Burgers vector of the undissociated dislocation and the direction of the dislocation line. Derive the corresponding expression for the width of a strip of antiphase boundary in an extended $a\langle 111\rangle$ dislocation of the ordered β-brass (B2) structure.

29. Obtain a formula for the elastic energy per unit length of a Lomer–Cottrell dislocation and hence show that the separation r of each Shockley partial dislocation from the stair-rod dislocation is given by

$$r = \mu a^2/\{72\pi\sigma^f(1-\nu)\}$$

where σ^f is the stacking fault energy and the other symbols have their usual meanings.

30. Two parallel straight dislocations with identical Burgers vectors each inclined at an angle φ to the line direction are situated in parallel slip planes. Find the condition that there is a position of stable equilibrium with respect to glide when the mutual perpendicular to both lines is normal to the slip planes and show that when this condition is not satisfied the dislocations will separate to infinity if they can glide freely. If the angle between the mutual perpendicular and the slip plane is θ, find the values of θ for which the glide force is zero and those for which this force has a maximum value.

31. Two parallel edge dislocations on different slip planes which intersect at an angle θ are situated respectively along the line of intersection of the two slip planes and at a distance r from this line. Find the total force and the glide component of the force acting on each dislocation.

32. The displacement field of a wedge disclination of strength ω is given by

$$w_r = (-\omega/2\pi)\{\mu/(\lambda+2\mu)\}r \ln r$$
$$w_\theta = (-\omega/2\pi)r\theta$$

Verify that these displacements satisfy the elastic equations of equilibrium (11.29) and find the corresponding stress field.

Two wedge disclinations of strengths $+\omega$ and $-\omega$ are both parallel to the x_3 axis and have x_1, x_2 coordinates $(0, \frac{1}{2}L)$, and $(0, -\frac{1}{2}L)$ respectively. By superposing the displacements, prove the result assumed in Fig. 8.10, namely that for small L the disclination dipole represents an edge dislocation of Burgers vector $L\omega$.

33. Calculate the values of θ and Σ for coincidence site lattices produced by rotations of cubic crystals about [210] and [211].

34. In the first edition of this book, it was stated that a $\Sigma = 7$ coincidence site lattice, represented in Fig. 8.11, may be obtained by a rotation of either $\sim 38°$ or $\sim 22°$ about $\langle 111 \rangle$. Explain why this is incorrect.

35. Prove the relations stated in equations (36.14) and (36.15).

36. Two parallel small-angle tilt boundaries with tilt angles θ_1 and θ_2 about a common axis unite to form a single boundary. Show that the energy is reduced irrespective of whether θ_1 and θ_2 have the same or opposite signs.

37. A silver wire of radius 50 μm possesses a 'bamboo' structure with a mean grain length of 0.1 mm. Thermal grooving develops where the grain boundaries intersect the surface of the wire and the included angle is 162°. The wire is in equilibrium when supporting a load of 20 mg. Calculate the energies of the free surface and of the grain boundaries.

38. A [001] twist boundary in a f.c.c. metal of cell edge 0.42 nm consists of a crossed grid of lattice screw dislocations with a mean dislocation–dislocation spacing of 10 nm. The boundary intersects the free surface of the crystal and after annealing at a high temperature, a groove develops with an included angle at its base of 179.5°. Estimate the surface energy of the crystal, given that the energy of the twist boundary is represented by equation (37.3) with $\sigma_0 = 0.15$ J m^{-2} and $A = 1.0$.

39. Consider M solute atoms to be initially concentrated in a very small volume of an infinite medium. Show that for subsequent diffusion along a line, the number of atoms per unit length, c'', is given by eqn. (40.17) with an additional factor of $\frac{1}{2}$, whilst for diffusion over a plane, the number of atoms per unit area, c', at a distance r from the source is

$$c'(r, t) = (M/4\pi Dt) \exp(-r^2/4Dt),$$

and for three-dimensional diffusion, the concentration, c, is

$$c(r, t) = \{M/(8\pi Dt)^{3/2}\} \exp(-r^2/4Dt)$$

40. Use the results of question 39 to find the distribution of solute around a line source of M solute atoms per unit length (a) after diffusion over an infinite surface and (b) after diffusion in an infinite volume.

41. The following data were obtained in a Kirkendall type experiment:
atomic fraction x_A of A at markers 0.4,
slope of penetration curve $x_A = 0.4$ mm^{-1},
diffusion time = 100 hr
marker displacement = 0.36 mm

If the interdiffusion coefficient D_{chem}, is 2.0×10^{-11} m^{2}s^{-1}, find the intrinsic diffusion coefficients D_A and D_B.

42. A b.c.c. metal with a lattice parameter of 0.3175 nm yields at a stress of 0.1 GN m^{-2} and work-hardens very little during subsequent plastic deformation. If 5% of the plastic work is stored in the metal, estimate the increase in energy in eV per atom for a specimen deformed to a true strain of 0.4. The energy of a high angle grain boundary in this metal is 0.5 J m^{-2}. Estimate the size of a critical nucleus for the homogeneous formation of a strain-free grain and the corresponding increase in free energy associated with the formation of such a nucleus.

43. The atomic fraction of carbon in austenite in pure iron–carbon alloys is 0.036 at the eutectoid temperature of 723°C and is 0.089 at 1147°C; the $\alpha \rightarrow \gamma$ transition in pure iron occurs at 910°C. The growth rate of pearlite at 690°C is 20 μm s^{-1} and the interlamellar spacing is 525 nm. If growth is controlled by the volume diffusion of carbon in austenite, estimate the apparent diffusion coefficient.

Solutions to Problems

1. $\mathsf{G} = \frac{a^2}{4}\begin{pmatrix} 2 & 1 & 1 \\ 1 & 2 & 1 \\ 1 & 1 & 2 \end{pmatrix} \qquad \mathsf{G}^{-1} = a^{-2}\begin{pmatrix} 3 & -1 & -1 \\ -1 & 3 & -1 \\ -1 & -1 & 3 \end{pmatrix}.$

2. (a) [211] $[1\bar{1}1]$ [663]. (b) $\frac{1}{2}[312]$ $\frac{1}{2}[3\bar{1}2]$ [333].
$\cos^{-1}\{c^4/(3a^2+c^2)(4a^2+c^2)\}^{1/2}$ $\cos^{-1}\{(3a^2+c^2)/(4a^2+c^2)\}^{1/2}$.

5. [001] axis: [001], $2^{-1/2}[i10]$, $2^{-1/2}[-i10]$
[110] axis: $2^{-1/2}[110]$, $4^{-1}[-3^{1/2}-i, 3^{1/2}+i, -2^{1/2}+6^{1/2}i]$,
$4^{-1}[-3^{1/2}+i, 3^{1/2}-i, -2^{1/2}-6^{1/2}i]$.

7. $\begin{pmatrix} 1+s\sin\theta\cos\theta & 0 & s\cos^2\theta \\ 0 & 1 & 0 \\ -s\sin^2\theta & 0 & 1-s\sin\theta\cos\theta \end{pmatrix}.$

9. $\cot\varphi = \{[s_2(\cos\theta - s_1\sin\theta) - D] \pm [s_1s_2 + s_1s_2D\sin\theta + D^2]^{1/2}\}/s_2\sin\theta$
where $D = (s_1-s_2)/(s_2\sin\theta - 2\cos\theta)$
$s = \{s_1s_2(s_1\sin\theta - 2\cos\theta)(s_2\sin\theta - 2\cos\theta) + (s_1-s_2)^2\}^{1/2}$
$\cos\psi = 1 - s_1s_2\sin\varphi\sin\theta\cos(\varphi-\theta) + \frac{1}{2}[s_1\sin 2\varphi + s_2\sin 2(\varphi-\theta)]$.

11. $2^{1/2}, 2^{-1/2}, 1+2^{1/2}$
$2^{-1/2}[\bar{1}10]$, $2^{-1/2}[110]$, [001]

$$\mathsf{P} = 8^{-1/2}\begin{pmatrix} 3 & -1 & 0 \\ -1 & 3 & 0 \\ 0 & 0 & 4+8^{1/2} \end{pmatrix}, \quad \mathsf{R} = \begin{pmatrix} 1 & 0 & 0 \\ 0 & 2^{-1/2} & -2^{-1/2} \\ 0 & 2^{-1/2} & 2^{-1/2} \end{pmatrix}.$$

12. (a) One variant is $(\mathrm{cCh}) = \frac{1}{2}\begin{pmatrix} 0 & 1 & 1 \\ -1 & 0 & 1 \\ 1 & -1 & 2 \end{pmatrix} \qquad (\mathrm{hCc}) = \frac{1}{2}\begin{pmatrix} 1 & -3 & 1 \\ 3 & -1 & -1 \\ 1 & 1 & 1 \end{pmatrix}.$

(b) $(\mathrm{cCo}) = \frac{1}{2}\begin{pmatrix} 0 & 2 & 1 \\ -1 & -1 & 1 \\ 1 & -1 & 2 \end{pmatrix} \qquad (\mathrm{oCc}) = \frac{1}{4}\begin{pmatrix} -1 & -5 & 3 \\ 3 & -1 & -1 \\ 2 & 2 & 2 \end{pmatrix}.$

13. $\frac{1}{12}\begin{pmatrix} 13 & 1 & 1 \\ 1 & 13 & 1 \\ -2 & -2 & 10 \end{pmatrix} \quad \frac{1}{6}\begin{pmatrix} 7 & 1 & 1 \\ 1 & 7 & 1 \\ -2 & -2 & 5 \end{pmatrix}.$

14. (110): 1.5 MN m^{-2}; 3.5 MN m^{-2} along $[1\bar{1}0]$.
(211): 3.5 MN m^{-2}; $(79/12)^{1/2}$ MN m^{-2} along $[6\ \overline{11}\ \bar{1}]$.

(123): (12/7) MN m^{-2}; $(283)^{1/2}/7$ MN m^{-2} along [23 $\overline{52}$ 27].
Stress in direction **n** has magnitude $X_n = X_i n^2_i$.
Shear stress components are $(X_1 - X_n)n_1$, $(X_2 - X_n)n_2$, $(X_3 - X_n)n_3$.

17. $E^{-1} = s_{11} + \{s_{44} - 2(s_{11} - s_{12})\}(n_1^2 n_2^2 + n_2^2 n_3^2 + n_3^2 n_1^2)$.
$s_{44} = 2(s_{11} - s_{12})$
Cu: E = 67, 127, 188 G Pa.
Mo: E = 500, 345, 314 G Pa.

18. (a) s_{44}^{-1}, c_{44} (b) s_{44}^{-1}, c_{44} (c) $2(s_{11} - s_{12})^{-1}$, $\frac{1}{2}(c_{11} - c_{12})$
(d) $3/2(s_{11} - s_{12} + s_{44})$, $3c_{44}(c_{11} - c_{12})/2(c_{11} - c_{12} + c_{44})$
(e) $3/\{s_{44} + 4(s_{11} - s_{12})\}$, $3c_{44}(c_{11} - c_{12})/(c_{11} - c_{12} + 4c_{44})$
(f) see (e).

19. 2/3.

21. 1465, 2900 J mol^{-1}.
At 1000K −42,050, −42,175, −41,785 J mol^{-1}.
At 1374K −70,660, −71,530, −71,530 J mol^{-1}.

22. surface: $3.84\pi \times 10^{-20}$ J $\approx$ 0.75 eV.
strain: $0.48\pi \times 10^{-20}$ J $\approx$ 0.094 eV.
31/40.

23. 0.2.

24. $(1-2v)/(1+v)$, where v is Poisson's ratio.

26. 10^{-4}; 7 MN m^{-2}.

27. $2/3^{1/2}$.

28. $(3\mu a^2/8\pi\sigma')(1 - v\cos\phi)/(1-v)$, where σ' = APB energy.

30. $\alpha = \tan^2\varphi/(1-v) \geq 1$.
Zero force: $\theta = 90°$ or $\cos^{-1}(-1/\alpha)$
Maximum force: θ satisfies $(\cos 2\theta/\cos 4\theta) = -\alpha$

31. Total force has components $(B_e b/r)\cos 2\theta$ and $(B_e b_e/r)\sin 2\theta$ normal and parallel to glide plane of first dislocation. Glide force = $(B_e b/r)\{\cos 2\theta \cos\theta + \sin 2\theta \sin\theta\}$.

33. [210]: 48.2°−15, 73.6°−7, 96.4°−9, 131.8°−3, 180°−5.
[211]: 63.0°−11, 78.5°−15, 101.5°−5, 135.6°−7, 180°−3.

37. 1.5 J m^{-2}; 0.46 J m^{-2}.

38. 0.62 J m^{-2}.

40. $\{M/2\pi Dt)^{1/2}\}\exp(-x^2/4Dt)$, $(M/4\pi Dt)\exp(-r^2/4Dt)]$.

41. 1.85×10^{-11} m^2 s^{-1}, 2.1×10^{-11} m^2 s^{-1}.

42. 2×10^{-4} eV; 500 nm; 0.5×10^{-12} J $\approx 3 \times 10^6$ eV.

43. 6.4×10^{-11} m^2 s^{-1}.

Index of Symbols

The index gives the page on which the symbol is introduced or defined. Page references are not given for universal constants and general symbols such as G.

Symbol	*Meaning*	*Page*
Δh^{sv}	Heat of vaporization per atom	149
Δh^{sl}	Latent heat of melting per atom	153
$\Delta_m h$	Heat of mixing per atom	178
$\Delta_m h_A$	Partial heat of mixing per A atom	178
$\Delta h_{\square}$, $\Delta h_{\square\square}$	Enthalpy of formation of vacancy, divacancy	126
Δh_n	Formation energy of vacancy cluster	129
Δh_1, Δh_2	Contributions to enthalpy of diffusion from vacancies and divacancies	407
$\Delta h_{\bullet}$	Enthalpy of formation of interstitial	127
$\Delta_a h$	Enthalpy of activation	135
$\Delta_a h^*$	Enthalpy of activation, especially for incoherent growth	482
$\mathbf{h}$	Normal to a set of lattice planes	27
h_i, ${}^A h_i$	Components of $\mathbf{h}$ referred to reciprocal base A^*	27
h', $(\mathbf{h}; \mathrm{A}^*)$	Representation of $\mathbf{h}$ as a row matrix	28
H	Enthalpy function	—
$\Delta_m H$	Heat of mixing	186
$\Delta H^{\ddagger}$	Molar enthalpy of activation	87
$\Delta_a H$	Molar enthalpy of activation	88
H	Function in theory of non-isothermal kinetics	546
i	Molecularity of unit step	86
i	Supersaturation ratio	423
i	Dimensionality of nucleation site	533
$\mathbf{i}_i$	Set of orthonormal vectors	22
$\bar{\mathbf{i}}$	Direction of dislocation line element	253
I, $\mathbf{I}$	Generalized current or flux	94
$\mathbf{I}_A$	Flux of A atoms	381
$I_{\square}$	Flux of vacancies	389
I	Steady state nucleation rate	429
$I(t)$	Time-dependent nucleation rate	445
${}^v I$, ${}^o I^H$	Homogeneous nucleation rate per unit volume	16, 455
${}^v I^B$, ${}^v I^E$, ${}^v I^C$	Grain boundary, edge and corner nucleation rate per unit volume of assembly	455
${}^S I$, ${}^B I$	Nucleation rate per unit area of substrate, grain boundary	450, 527
${}^E I$	Nucleation rate per unit length of grain edge	528
${}^C I$	Nucleation rate per grain corner	528
${}^v I^D$	Dislocation nucleation rate per unit volume of assembly	471
I	Moment of inertia	421
I, I'	Cell factors for centred unit cells	36
I	Orthonormal basis $\mathbf{i}_i$ of co-ordinate system	23
$\mathbf{I}$, δ_{ij}	Unit tensor	42
$\mathbf{I}_i$	Flux of ith component	389
j	Degree of supersaturation	423
J, $(\mathrm{A}\ \mathbf{J}\ \mathrm{B})$	Transformation matrix relating bases A and B	29
J_{ij}	Components of J	29
$\boldsymbol{k}$	Boltzmann's constant	—
k, k_f, k_b	Reaction rate constants	15, 79
k_A	Frequency of motion of A atom into adjacent vacant site	382
k_{A1}, k_{A2}	Jump frequencies of specified atoms into vacant site	401
k_e	Jump frequency for dissociation of vacancy B atom complex	403
k_r, k_w	Rate constants in theory of ordering	218
k	Constant in Avrami equation	19
k^S	Interface step density	501
$\mathbf{k}^F$	Wave vector at Fermi surface	116
K, $K_c^{\ddagger}$, K_c	Equilibrium constants	86
K	Bulk modulus of elasticity	75
K_B	Bulk modulus of component B	200
K_1	Twinning (invariant) plane of a mechanical twin	50

Symbol	*Meaning*	*Page*
K_2	Conjugate twinning plane (second undistorted plane) of a mechanical twin	51
K	Gradient energy coefficient	183
K	Factor in anisotropic elastic energy of dislocation	260
K^B	Interface concentration ratio in cellular growth	519
K_0	Solute-boundary interaction coefficient	522
ΔK	Parameter in diffusion theory	91
l	Edge length of nucleus	471
e	Curvilinear coordinate along interface	519
$\mathbf{l}$	Local Burgers vector	317
$\mathbf{l}_+, \mathbf{l}_-$	Axes of rotation	341
L	Separation of elements of disclination dipole	343
L_1, L_2	Macroscopic lengths	252
L	Long range order parameter	215
L^B	Mean grain diameter	332
${}^vL^E$	Grain edge length per unit volume	332
${}^vL^D$	Length of dislocation line per unit volume (dislocation density)	471
L^β	Length of a line intersected by β regions	526
L^β_e	Extended line intercept	526
L^P	Value of l at mid-points of plates	519
L, (ɪ **L** ᴋ)	Orthogonal matrix relating bases which are connected by a pure rotation	44
L_{ij}	Direction cosines (elements of L)	44
L_{klm}	Coefficients of connection	319
m_i	Atomic mass of ith atom	89
m	Effective mass of complex	84
m	Attractive exponent in force law	113
m	Arbitrary constant	496
$\mathbf{m}_i$	Vector reciprocal to unit normal $\mathbf{n}_i$	155
$M^{\ddagger}$	Activated complex	81
M	Mobility or conductance	94
$\mathbf{M}$, M_{ij}	Generalized mobility or conductance	94
M'_{ij}	Atomic mobilities	389
M_{ij}	Diffusional mobilities per unit volume	389, 395
M^B	Boundary mobility	480
M_s	Martensite start temperature	13
M_d	Temperature below which martensite forms under stress	14
M	Number of atoms per unit area of planar diffusion source	385
M^c	Boundary mobility per unit concentration difference	499
n	Any undetermined small number	—
n	Number of vacancies	126
n	Number of atoms in embryo or nucleus	420
n_c	Number of atoms in nucleus of critical size	422
n	Repulsive exponent in force law	113
n	Time exponent in Avrami equation	19
n_{ik}	Number of paths with $(i-1)$ jumps to site k	393
$\mathbf{n}$	Unit vector normal to surface or lattice plane	66
N_0	Avogadro's number	—
N	Any undetermined large number	—
N	Total number of atoms (molecules) in assembly	126
N_A	Total number of A atoms in assembly	99, 171
N^v, N^α	Total number of atoms in vapour phase, α phase, etc.	425
N^S	Number of atoms adsorbed on substrate	450
${}^vN^S$	Number of atoms on substrate per unit volume of assembly	450
${}^vN^D$	Number of atoms at dislocation cores per unit volume of assembly	471
${}^vN^C$	Number of grain corners per unit volume	332
${}^vN_0$	Number of pre-existing nuclei or nucleation sites	18

Symbol	*Meaning*	*Page*
N_n	Number of embryos of size n in stable assembly	425
N_{AB}	Number of unlike nearest neighbour contacts per unit coordination number	186
N_{AA}, N_{BB}	Number of like nearest neighbour contacts per unit coordination number	186
N_{AB}	Equilibrium value of N_{AB}	189
N^0_{AB}	Maximum value of N_{AB}	190
o_n	Number of surface atoms in embryo E_n	440
O	Area of a face	155
O^b	Area of a boundary	527
O^B	Total area of grain boundaries	527
${}^vO^B$	Grain boundary area per unit volume of assembly	332
O^S	Area of substrate	450
${}^vO^S$	Substrate area per unit volume of assembly	450
O^β	Area of plane intersected by β regions	526
O^β_e	Extended area of plane	526
O_n	Surface area of an embryo E_n	420
O_c	Surface area of nucleus of critical size	429
p	Pressure	181
p_r, p_∞	Vapour pressure in equilibrium with surfaces of radii $r, \ldots \infty$	423
p	Effective pressure	113
$\bar{p}$	Péclet number for diffusional growth	495
$\bar{p}^S$	Péclet number for step growth	501
p	Minimum number of atoms in heterophase fluctuation	420
$\mathbf{p}$	Stress vector	67
$\mathbf{p}$	A large vector in a boundary	340
P	Probability of a fluctuation	97
$P_{AA}(r), P_{AB}(r)$	Pair density functions	170
P	Probability of a site being vacant	382
P^t	Fraction of free energy released in a transformation	507
P	Measured value of some physical property, especially electrical resistance	137
P_k	Jump probability for direction $\theta_{1(k)}$	393
$\mathbf{P}$	Pure deformation tensor, especially pure lattice deformation	44
$\bar{\mathsf{P}}$	Representation of $\mathbf{P}$ in principal axis system	44
$\mathrm{d}q$	Heat flow	98
q_0	Probability per unit area per unit time of an atom adding itself to an embryo	428
q_n	Probability per unit area per unit time of an atom leaving an embryo of n atoms	428
q	Dimensionless parameter relating interface and diffusion control	500
q^S	Corresponding dimensionless parameter for step growth	501
q_i	Normal coordinates	89
Q'_A, Q'_B	Partition functions referred to a common energy zero	84
Q_A, Q_B	Partition functions referred to separate zeros	84
$Q^{\ddagger}, Q^{\ddagger\prime}, Q^{\ddagger\prime\prime}$	Partition functions of activated complex	84–5
Q_x	Translational partition function	85
Q_v	Vibrational partition function	87
Q_{tr}	Translational partition function for embryos	421
Q_{rot}	Rotational partition function for embryos	421
Q_{rep}	Replacement partition function for embryos	422
Q_n	Partition function for embryo in vapour phase	426
Q^t	Fraction of excess solute precipitated	516
$\bar{Q}$	Modified value of Q^t	522
r	Polar co-ordinate	257
r_0	Equilibrium separation of two atoms	111
r_1	Nearest neighbour separation in crystal	111
r_2, r_3	Second and third nearest neighbour separations	114
r^0_A	Atomic radius in pure A	200
r_B	Radius of B atom in infinite crystal of A	200

Symbol	*Meaning*	*Page*
r^*_{AA}, r^*_{AB}	Characteristic atomic lengths	197
r_0	Limiting radius for Hookean displacements around a centre of dilatation	198
r_i	Radius of dislocation core	257
r_e	Radius of dislocation strain field	257
r	Radius of droplet or spherical nucleus	420
r_c	Radius of critical nucleus	422
r	Co-ordinate normal to moving interface	485
r^I	Value of r giving position of interface	485
r	Final radius of transformed region	536
r_A	Probability of an A atom being in a "right" site of a superlattice	215
r	Critical radius for given supersaturation	493
r	Critical semi-thickness of plate or needle	497
$\mathbf{r}$	Position vector	170
$\boldsymbol{R}$	Gas constant	—
R	Radius of sphere	201
R	Semi-axis of ellipsoid representing particles of varying shape	461
R_n	Growth resistance	435
R^B	Parameter depending on temperature and grain size	455
R	Radius of flow line	502
R^ι	Fraction of chemical driving force converted to surface energy	517
$R(\beta)$	Amplification factor	396
R_D, R^P	Parameters in theory of cellular growth	520
$\mathbf{R}$	Pure rotation tensor	45
$\mathsf{R}_+\mathsf{R}_-$	Orthogonal matrices	340
s	Amount of shear, especially twinning shear	51
s	Unpaired spin per atom	107
s	Limiting nucleus size above which $Z_n = 0$	433
$\Delta_m s$	Entropy of mixing per atom	176
$\Delta_e s$	Excess entropy of mixing per atom	178
$\Delta_e s_A$	Excess partial entropy of mixing per atom of A	178
$\Delta s_{\square}$, $\Delta s_{\square\square}$	Entropy of formation of a vacancy, divancy	126
Δs_1, Δs_2	Contributions to entropy of diffusion from vacancies and divacancies	408
$\Delta s_{\bullet}$	Entropy of formation of an interstitial	127
$\Delta_a s$	Entropy of activation per atom	135
s, s_{ijkl}	Elastic compliances (or elastic moduli) tensor	70
s, s_{ij}	Elastic compliances ($i, j = 1 \ldots 6$)	71
s'_{ilmn}	Coefficients in elastic theory of constrained transformation	464
$\Delta S^{\ddagger}$	Molar entropy of activation	87
$\Delta_a S$	Molar entropy of activation	88
$d_e S$	Change in entropy of environment	98
$\mathrm{d}_i S$	Irreversible entropy created during a natural process	99
$S_1(p)$, $S_2(p)$	Functions used in Trivedi theory of growth	498–9
$\mathbf{S}$	Lattice deformation tensor	40
$\mathsf{AS} = (\mathrm{A}\ \mathbf{S}\ \mathrm{A})$	Matrix representation of **S** in basis A	40
$\overline{\mathsf{S}}$	Diagonal matrix representation of **S** in principal axis system	43
t	Time	15
t_ζ	Time to fixed volume fraction ζ of transformation	543
$t_a(T)$	Time to volume fraction ζ_a of transformation at constant temperature T	545
$\mathbf{t}$	Translation vector in lattice relations	55, 370
T	Temperature	—
T	Line tension of dislocation	265
T_c	Curie temperature	107
T_c	Critical temperature for surface disordering	151
T_λ	Disordering temperature of a superlattice	216

Symbol	*Meaning*	*Page*
T^{sl}	Melting point (solidus) temperature	152
T^{ls}	Freezing point (liquidus) temperature	441
ΔT^-, ΔT^+	Supercooling or superheating from equilibrium temperature	441
T_{ijk}	Torsion tensor	318
$\Delta_m u$	Internal energy of mixing per atom	186
u	Velocity of a dislocation	280
u_0	Parameter in empirical equation for dislocation velocity	283
u	Mean velocity of step	501
$\mathbf{u}$	Lattice vector	21
u_i, u_i^*	Contravariant and covariant components of $\mathbf{u}$	21, 26
$\mathbf{u}^{(0)}$	Vector of DSC lattice	371
U	Internal energy	—
U_0, U_∞	Internal energy at absolute zero, and at very high temperatures	219
U_T, U_λ	Internal energy at temperature T, T_λ	220
$\Delta_m U$	Energy of mixing	186
u, [A; $\mathbf{u}$]	Column matrix of components of $\mathbf{u}$ in base A	22
u′, ($\mathbf{u}$; A)	Row matrix of components of $\mathbf{u}$ in base A	23
$\Delta U^{\ddagger}$	Molar energy of activation	88
U^B	Parameter in theory of cellular growth	520
U_D	Deformation energy in pairwise model	112
U_i	Orthogonal matrix representing symmetry operation	345
U	Unimodular matrix	348
v_a, v_a^*	Volume of unit cell and reciprocal cell	25
v	Atomic volume	113
v_A^0	Atomic volume of pure A	200
v_B	Volume of B dissolved in A	200
v^l, v^α, etc.	Volume per atom in liquid, alpha phase, etc.	420
Δv^∞	Change in volume within closed surface around solute B atom	201
Δv_{AB}	Difference in atomic volumes of pure B and pure A	201
Δv^i	Increase of volume due to image forces in elastic model	202
Δv	Total difference in volume of assembly in elastic model	202
v_{AB}	Mean volume per atom in solid solution	203
v_τ	Volume at time t of transformed region which nucleated at time τ	16
v_d	A characteristic volume in bulk liquid	426
$\mathbf{v}$	Lattice vector, especially vector formed from $\mathbf{u}$ by a lattice deformation	40
v_i, v_i^*	Contravariant and covariant components of $\mathbf{v}$	40
V	Volume of whole assembly	15
$V^\beta(t)$	Volume of β phase at time t	15
w_A, w	Probability of an A atom being in a wrong superlattice site	215
$\mathbf{w}$	Displacement vector	61
W	Potential energy per atom in solid	112
W_s	Total strain energy in misfitting sphere model	202
W_B	Strain energy in compressed B sphere	202
W_A	Strain energy in surrounding A matrix	202
${}^LW_d$	Dislocation energy per unit length	260
${}^LW_s$	Dislocation elastic energy per unit length	259
${}^LW_c$	Dislocation core energy per unit length	260
${}^LW_i$	Interaction energy per unit length of dislocation	262
W_i	Interaction potential between dislocation and solute	267
W_d	Energy of dislocation loop	264
W_c	Critical activation energy for formation of a loop under stress	279
${}^LW_s'$	Strain energy per unit length of moving dislocation	281
${}^LW_k'$	Kinetic energy per unit length of moving dislocation	281
${}^LW_d'$	Total energy per unit length of moving dislocation	281
${}^LW_P$	Peierls energy	274

Symbol	*Meaning*	*Page*
λ	Boltzmann variable in diffusion equation	386
λ_A	Absolute activity	177
λ	Elastic constant of isotropic continuum	73
λ, λ_n	Coefficients in solution of diffusion equation	496, 510
λ_c	Critical wave-length	397
λ_i	Eigenvalues of a matrix	42
λ_i	Principal shape deformations	45
μ	Shear modulus of isotropic elastic continuum	73
ν_1	Rate at which a site becomes a nucleus	18
ν	Poisson's ratio	74
ν	Characteristic vibration frequency	88
ν_i	Normal atomic vibration frequencies	89
ν_i^0	Normal frequencies for constrained assembly	90
ν_D	Debye frequency	422
ν	Normal to an invariant plane	58
ξ	Degree of advancement	100
ξ	Strain normal to martensite plate	466
$\boldsymbol{\xi}$	Atomic position vector	24
$2\Xi_1$, $2\Xi_2$	Binding energies (negatives of interatomic potentials) for nearest neighbour and second nearest neighbour atoms	109
$2\Xi_{AA}$	Binding energy of a nearest neighbour *A–A* atom pair	184
$2\Xi_{AB}$	Binding energy of a nearest neighbour *A–B* pair	184
Ξ	Characteristic potential of nearest neighbour model	184
$\Xi_0(r)$	Function in theory of conformal solutions	196
$\Xi_{AA}{}^*$	Characteristic energies in average potential model	197
$\pi_{A, 12}$	Probability of an *A* atom transferring from plane 1 to plane 2 in unit time	381
$\pi_{\Box}(\xi\eta)$	Probability of a vacancy moving from a plane of concentration ξ to a plane of concentration η in unit time	398
$\pi_{\Box A}(\xi\eta)$	Probability of a vacancy on a plane of concentration ξ exchanging places with an *A* atom on a plane of concentration η in unit time	158
$\varrho(r)$	Radial distribution function	158
ϱ	Parameter of average potential model	197
ϱ	Density	280
ϱ	Mobile dislocation density	284
ϱ_1	Density of dislocation loops	514
$\sigma(\mathbf{n})$	Surface free energy per unit area of surface with normal **n**	153
$\sigma^{\alpha\alpha}$	Free energy per unit area of an α—α grain boundary	164
σ^{sl}, $\sigma^{\alpha\beta}$	Free energy per unit area of solid–liquid, alpha–beta interface	164, 448
σ_∞	Surface free energy per unit area of a planar interface	419
σ_0	Minimum value of σ as an interface advances	478
σ_{kl}	Effective stress in pairwise model	113
$\sigma^f(x, y)$	Energy of generalised stacking fault	124
σ^f	Stacking fault (free) energy	296
σ	Short range order parameter	221
Σ	Reciprocal density of coincidence sites	344
τ_s	Effective shear stress on slip plane	260
τ_0	Parameter in empirical dislocation velocity equation	283
τ	Time at which a region nucleates	15
τ	Time to cross an activation barrier	84
τ^D	Mean time of stay at a dislocation	414
τ	Relaxation time for establishment of a steady state nucleation rate	445
τ_0, τ_1	Coefficients in expansion of shear stress	356
φ	Angle giving deviation of tilt boundary from symmetrical position	336
$\varphi(z)$	Function giving periodic variation in energy with interface position	477
$\varphi_j(\alpha_j)$	Function used in theory of growth	487

Symbol	*Meaning*	*Page*
2φ	Angle between K_1 and K_2 planes after twinning shear	51
φ	Angle between interface normal and rotation axis	347
φ_D, φ'_D	"Driving force" for diffusion	395–6
Φ	Disregistry in Peierls–Nabarro theory	272
ζ	Fractional amount of transformation; usually volume fraction transformed	17
ζ	Width of dislocation	272
ψ	Angle between curve of constant composition and grain boundary	412
ω	Strength of disclination dipoles	343
ω	Wavenumber of shape perturbation	492
ω_c	Critical wavenumber for stability	493
$\boldsymbol{\omega}$	Rigid body rotation	62
$\Omega(N_A N_B N_{AB})$	Number of arrangements for given N_{AB}	16
Ω_a, Ω_b	Plane waves radiated from moving dislocation	282
Υ	Velocity of an interface	16
Υ_c	Velocity of planar interface under interface control	499
Υ_θ	Local velocity of interface	519

Author Index

Subject Index

Alloy systems are indexed alphabetically, irrespective of major and minor components